Risk Analysis in Engineering and Economics

SECOND EDITION

Risk Analysis in Engineering and Economics

SECOND EDITION

Bilal M. Ayyub

University of Maryland
College Park, USA

CRC Press is an imprint of the
Taylor & Francis Group, an **informa** business
A CHAPMAN & HALL BOOK

First published in paperback 2024

First published 2014 by CRC Press
2385 NW Executive Center Drive, Suite 320, Boca Raton FL 33431

and by CRC Press
4 Park Square, Milton Park, Abingdon, Oxon, OX14 4RN

First issued in hardback 2019

CRC Press is an imprint of Taylor & Francis Group, LLC

Publisher's Note
The publisher has gone to great lengths to ensure the quality of this reprint but points out that some imperfections in the original copies may be apparent.

ISBN 13: 978-1-46-651825-4 (hbk)
ISBN 13: 978-1-03-291800-6 (pbk)
ISBN 13: 978-0-42-909847-5 (ebk)

DOI: 10.1201/b16663

Visit the Taylor & Francis Web site at
http://www.taylorandfrancis.com

and the CRC Press Web site at
http://www.crcpress.com

Dedicated to the vibrant and resilient people of Palestine

Contents

Preface

Societies increasingly rely on complex human-made systems and new technologies, and decisions are commonly made under conditions of uncertainty. Although people have some control over the levels of technology-caused risk to which they are exposed, reduction of risk also generally entails reduction of benefit, thus posing a serious dilemma. The public and its policy makers are required, with increasing frequency, to weigh the benefits objectively against risks and to assess the associated uncertainties when making decisions. When decision makers and the general public lack a systems engineering approach to risk, they are apt to overpay to reduce one set of risks and in doing so offset the benefit gained by introducing larger risks of another kind.

Life is definitely a risky business in all its aspects, from start to end. Newspapers are filled with accounts of mishaps—some are significant, while others are minor. Some of the more dramatic incidents that stick in our memory include the loss of the space shuttle Columbia on February 1, 2003, during reentry into the Earth's atmosphere, killing its seven crew members; the attack by hijackers who slammed passenger jets into the World Trade Center and the Pentagon on September 11, 2001, killing thousands and causing billions of dollars of damage to the world economy; the destruction of the space shuttle Challenger on January 28, 1986, and the death of its seven crew members resulting from the failure of the solid rocket boosters at launch; and the explosion at the Union Carbide plant in Bhopal, India, on December 3, 1984, that released a toxic cloud of methyl isocyanate gas, enveloping hundreds of shanties and huts surrounding the pesticide plant, exposing more than 500,000 people to the poisons; this resulted in the death of 1,430 people according to the Indian government that was updated in 1991 to more than 3,800 dead and ~11,000 with disabilities and many others suffering.

Most risk situations are more mundane. Each day we encounter risk-filled circumstances, for example, delays caused by an electric power outage, files lost and appointments missed due to the breakdown of a personal computer, loss of investments in high-technology Internet stocks, and jeopardizing one's health by trying to maintain a stressful schedule to meet sales targets and due dates in a competitive market. The urgent need to help society deal intelligently with problems of risk has led to the development of the discipline known as *risk analysis*. The complexity of most problems of risk requires a cooperative effort by specialists from diverse fields to model the uncertainties underlying various components of risk. For example, the resolution of technical aspects of risk demands the efforts of specialists such as physicists, biologists, chemists, and engineers. Resolving social aspects of risk may require efforts from public policy experts, lawyers, political scientists, geographers, economists, and psychologists. In addition, the introduction of new technologies can involve making decisions about issues with which technical and social concerns are intertwined. To practice risk assessment, decision-making specialists must coordinate this diverse expertise and organize it so that optimal decisions can be reached and risk can be managed by properly treating uncertainty. Furthermore, risk assessors must use formal risk management and communication tools in a clear, open manner to encourage public support and understanding.

Ideally, risk analysis should invoke methods that offer a systematic and consistent performance for evaluating and managing uncertainty in a risk-focused technology. Risk assessment should measure risk and all its associated uncertainties. Answers to questions about the acceptability of risk or when a risk is sufficiently significant to require public regulation clearly involve social values. However, the information in quantitative risk assessments should be relatively objective. In deciding on acceptable levels of risk, the question of credible or justifiable evidence becomes more scientific than political.

As regulatory activity proliferates, those in the regulated communities complain that risk analyses are neither rigorous nor balanced, noting that risk analysis can be an inexact science. Where data are lacking on some parameters of interest—for example, the direct impact of a substance on human health or the environment—these data gaps may be filled with tests of laboratory animals, computer simulations, expert opinions, and other extrapolations. Despite these limitations, risk assessment will certainly play a major role in prioritizing future expenditures of scarce public and private resources on issues related to health, safety, security, and the environment.

In preparing the second edition of this textbook, I strove to achieve the following additional or enhanced objectives: (1) make it an easy-to-read study book and practical guide; (2) target the audience of analysts, researchers, and/or anyone who deals with risk analysis to solve problems and for decision making; (3) introduce the fundamentals of risk analysis and management in a meaningful manner to facilitate quantification and rational decision making; (4) emphasize the practical use of these methods; and (5) establish the limitations, advantages, and disadvantages of various methods. The book was developed with an emphasis on solving real-world technological problems.

The book can also be used in education in the fields of risk analysis, reliability assessment, decision analysis, economics, finance, engineering, forecasting, probability and statistics, and social sciences. For example, engineers could use the book for assessing failure probabilities and consequences, life expectancy assessment, condition assessment, failure and accident rate analysis, and technology forecasting. Economists and financial analysts can use concepts covered in the book in comparing alternatives and decision making. Other uses could include environmental and ecological risk studies, biosecurity, food safety, threat reduction, terrorism, financial planning, land use, and wildlife habitat in environmental engineering and sciences, consequence assessment, market dynamics, competition assessment in economics, decision analysis for investment, conflict management, litigation issues such as strategies and tactics, and diagnoses and treatment selection in health services. The readers will especially appreciate the real-world case studies that are used to illustrate the basic principles. The book was purposely designed to emphasize the applications of the concepts presented. The fundamental concepts are properly balanced in eight well-organized chapters. The ninth chapter provides a case study.

Structure, Format, and Main Features

The second edition was written with a dual use in mind, as both a self-learning guidebook and a textbook for a course. In either case, the text will be designed to achieve the important educational objectives of introducing theoretical bases and guidance on and applications of risk methods.

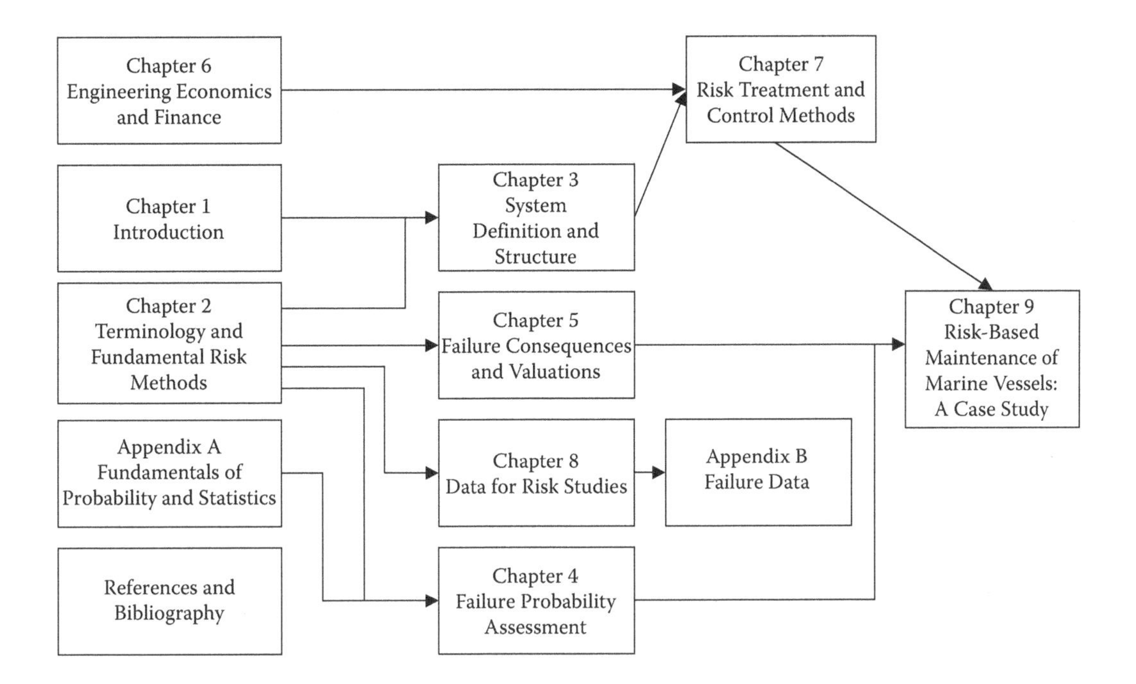

The nine chapters of this book lead the readers from the definition of needs, to the foundations of the concepts covered in the book, to theory and applications, and finally to data needs and sources with a case study.

The first chapter provides historical perspectives on the origins of risk analysis and management, defines the risk domain, and discusses knowledge, and its sources and acquisition, as well as ignorance and its categories, in order to offer a rational justification and bases for risk analysis and management in the context of systems. Risk and system concepts are briefly introduced in this chapter and are covered in detail in Chapters 2 and 3, respectively. Chapters 4 and 5 are devoted to failure probability assessment, and severity and consequence assessment, respectively. Chapter 4 includes both analytical and empirical methods. Chapter 5 offers a broad coverage of valuation and many consequence types including property damage, human loss and injury, and environmental, ecological, and health effects. Chapter 6 provides a practical coverage of engineering economics and finance. Chapter 7 describes decision analysis methods for risk treatment and control by presenting the fundamental concepts of utility, risk attitude, benefit–cost analysis, and applications. Chapter 8 covers data sources, and needs and collection for risk analysis including elicitation of expert opinions. Chapter 9, as the last chapter, provides a case study on risk-based maintenance of marine vessels. The book also includes two appendices: Appendix A summarizes the fundamentals of probability and statistics, and Appendix B summarizes the failure data. Examples and applications are included in all the chapters covering all key subjects and concepts. Also, each chapter includes a set of exercise problems that cover the materials of the chapter. The problems were carefully designed to meet the needs of instructors in assigning homework and the readers in practicing the fundamental concepts.

For the purposes of teaching, the book can be covered in one semester. The chapter sequence can be followed as a recommended sequence. However, if needed, instructors can choose a subset of the chapters for courses that do not permit a complete coverage of all chapters, or a coverage that cannot follow the order presented. In addition, selected chapters can be used to supplement courses that do not deal directly with risk analysis, such as reliability assessment, economic analysis, systems analysis, health and environmental risks, and social research courses. Chapters 1, 2, and 6 can be covered concurrently. Chapters 3 through 5 build on some of the materials covered in Chapter 2. Chapter 7 builds on Chapters 3 and 6. Chapter 8 provides information on data sources and failure that can be covered independently of other chapters. Chapter 9 offers a case study to illustrate how to bring different concepts together to plan maintenance activities using risk information. The book also contains references and bibliography at its end. The accompanying schematic diagram illustrates the possible sequences of these chapters in terms of their interdependencies.

I invite the users of the book to send any comments on the book to the e-mail address ba@umd.edu. These comments will be used to develop future editions of the book. Also, I invite the users of the book to visit the Web site of the Center for Technology and Systems Management at the University of Maryland at College Park to find the information posted on various projects and publications that can be related to risk analysis. The URL address is http://ctsm.umd.edu

Bilal M. Ayyub
University of Maryland

Acknowledgments

This book was developed over many years and draws on my experiences in teaching courses related to risk analysis, uncertainty modeling and analysis, probability and statistics, numerical methods and mathematics, reliability assessment, and decision analysis. Drafts of most sections of the book were tested in several courses at the University of Maryland at College Park before its publication. This testing period has proved to be a very valuable tool in establishing its contents and the final format and structure.

I was very fortunate to receive direct and indirect help from many individuals over many years that has greatly affected this book. Students who took my courses and used portions of this book provided me with great insight into how to effectively communicate various theoretical concepts. Also, advising and interacting with students on their research projects stimulated the generation of some examples used in the book. The students who took courses on structural reliability, risk analysis, and mathematical methods in civil engineering from 1995 to 2002 and 2010 to 2013 contributed to this endeavor. Their feedback was very helpful and contributed significantly to the final product. I greatly appreciate the input and comments provided by Professor R. H. McCuen and the permission to use materials from our book *Probability, Statistics, and Reliability for Engineers and Scientists, Third Edition*, published in 2011 by CRC Press for the development of Appendix A. The assistance of Dr. M. Kaminskiy in developing Chapters 4 and 7 and Dr. M. Morcos in developing project management examples and end-of-chapter exercise problems is most appreciated. In addition, comments provided by Dr. A. Blair, Dr. I. Assakkaf, Mr N. Rihani, and Dr. R. Wilcox on selected chapters are greatly appreciated. I acknowledge the assistance of the former students—H. M. Al-Humaidi, R. Chan, C.-Y. Chang, C.-A. Chavakis, H. Kamal, E. Lillie, Y. Fukuda, G. Lawrence, W. L. McGill, R. B. Narayan, K. S. Nejaim, D. Olawuni, S. M. Robbins, S. Tiku, and D. Webb—and the class of the Spring 2013 in preparing solutions to selected problems consisting of H. Allison, S. Dahle, G. Khirallah, A. Laripour, R. Li, N. Qureshi, J. Tembunkiart, and H.-H. Wei.

The reviewers' comments provided by the publisher were used to improve the book to meet the needs of readers and enhance the educational process. The input from the publisher and the book reviewers enhanced this second edition.

The financial support that I received from the US Navy, the Coast Guard, the Army Corps of Engineers, the Air Force Office of Scientific Research, the Office of Naval Research, and the American Society of Mechanical Engineers over more than 20 years has contributed immensely to this book by providing me with a wealth of information and ideas for formalizing the theory and developing applications. In particular, I acknowledge the opportunity and support provided by L. Almodovar, A. Ang, R. Art, K. Balkey, J. Beach, M. Black, P. Bowen, P. Capple, J. Crisp, S. Davis, D. Dressler, G. Feigel, M. Firebaugh, J. Foster, P. Hess III, M. Houston, Z. Karaszewski, T. S. Koko, D. Moser, N. Nappi, Jr., W. Melton, L. Parker, G. Remmers, T. Shugar, J. R. Sims, R. Taylor, S. Wehr, M. Wade, and G. White.

The University of Maryland at College Park provided me with the platform, support, and freedom that made such a project possible.

Author

Bilal M. Ayyub is a professor of civil and environmental engineering at the University of Maryland, College Park, and the director of the Center for Technology and Systems Management at the A. James Clark School of Engineering. Dr. Ayyub has been at the University of Maryland since 1983. He has been serving on the State of Maryland Governor's Emergency Management Advisory Council since 2011. He is specialized in risk analysis, uncertainty modeling, decision analysis, and systems engineering. He received degrees from Kuwait University and the Georgia Institute of Technology. He is a fellow of the American Society of Civil Engineers (ASCE), the American Society of Mechanical Engineers, Society of Risk Analysis, and the Society of Naval Architects and Marine Engineers, and is a senior member of the Institute of Electrical and Electronics Engineers (IEEE). He completed many research and development projects for many governmental and private entities—including the National Science Foundation; the US Air Force, the Coast Guard, the Army Corps of Engineers, the US Navy, and the Department of Homeland Security—and insurance and engineering firms. Dr. Ayyub is a multiple recipient of the American Society of Naval Engineers (ASNE) Jimmie Hamilton Award for the best papers in the *Naval Engineers Journal* in 1985, 1992, 2000, and 2003. Also, he was the recipient of the ASCE Outstanding Research-Oriented Paper in the *Journal of Water Resources Planning and Management* in 1987, the ASCE Edmund Friedman Award in 1989, the ASCE Walter L. Huber Research Prize in 1997, and the K. S. Fu Award of North American Fuzzy Information Processing Society (NAFIPS) in 1995. He received the Department of the Army Public Service Award in 2007. He was appointed to many national committees and investigation boards. Dr. Ayyub is the author and coauthor of more than 550 publications in journals, conference proceedings, and reports. Among his publications are more than 20 books, including the following selected textbooks: *Uncertainty Modeling and Analysis for Engineers and Scientists* (Chapman & Hall/CRC, 2006, with G. Klir); *Risk Analysis in Engineering and Economics* (Chapman & Hall/CRC, 2003); *Elicitation of Expert Opinions for Uncertainty and Risks* (CRC Press, 2002); *Probability, Statistics, and Reliability for Engineers and Scientists, Second Edition* (Chapman & Hall/CRC, 2011, with R. H. McCuen); and *Numerical Methods for Engineers* (Prentice Hall, 1996, with R. H. McCuen).

1

Introduction

This chapter provides historical perspectives on the origins of risk analysis and management, defines the risk domain, and discusses knowledge and its sources and acquisition, as well as ignorance and its categories, in order to offer a rational justification and bases for risk analysis and management in the context of systems. The practical use of concepts and tools presented in the book requires a framework and frame of thinking that deal holistically with problems and issues as systems. Risk and system concepts are briefly introduced in this chapter and are covered in detail in Chapters 2 and 3, respectively.

CONTENTS

1.1 Societal Needs and Demands

Citizens are becoming increasingly aware of and sensitive to the harsh and discomforting reality that information abundance does not necessarily give us certainty. This holds not only in knowledge-based economies but also in traditional industrial economies. In fact, this abundance of information can sometimes lead to errors in decision making and undesirable outcomes due to either overwhelmingly confusing situations or a sense of overconfidence that leads to improper use of information. The former situation can be an outcome of both the limited capacity of the human mind to deal with complexity in some situations and information abundance in other cases, whereas the latter can be attributed to a higher order of ignorance, referred to as the *ignorance of self-ignorance*.

Our society advances in many scientific dimensions and invents new technologies by expanding our knowledge through observation, discovery, information gathering, and logic. Access to newly generated information is becoming easier than ever as a result of computers and the Internet. We have entered an exciting era where electronic libraries, online databases, and information on every aspect of our civilization—patents, engineering products, literature, mathematics, economics, physics, medicine, philosophy, and public opinions, to name a few—are only a mouse click or screen touch away. In this era, computers can generate even more information from the abundance of online information. Society can act or react based on this information at the speed of its generation, sometimes creating undesirable situations, for example, price or political volatilities.

It is important to assess uncertainties associated with information and to quantitatively qualify our state of knowledge by measuring its deficiency, termed herein ignorance. The accuracy, quality, and incorrectness of such information, as well as knowledge deficiencies or inconsistency or incoherence, are being closely examined by our philosophers, scientists, engineers, economists, technologists, decision and policy makers, regulators and lawmakers, and our society as a whole. As a result, uncertainty and ignorance analyses are receiving increased attention (Ayyub and Klir 2006). We are moving from a state of emphasizing knowledge expansion and information creation to a state that includes knowledge and information assessment by critically evaluating them in terms of relevance, completeness, nondistortion, consistency, and other key measures.

Our society is becoming less forgiving and more demanding in regard to our knowledge base than ever. The processing of available information or acting on its results, even if its results might be inconclusive, is ultimately regarded as less excusable than a simple lack of knowledge and ignorance. In 2000, the US Congress and the Justice Department investigated Firestone and Ford companies for allegedly knowing that perhaps defective tires were suspected in causing accidents, claiming more than 88 lives worldwide without taking appropriate actions. The investigation and media coverage elevated the problem to a full-blown scandal as a result of inaction in light of the available information. Both Firestone and Ford argued that the test results conducted after they knew about the potential problem were inconclusive. Such an approach can often be regarded by our demanding society as a deliberate cover-up.

People have some control over the levels of technology-caused risks to which they are exposed and willing to undertake. Attempts to reduce risk by governments and corporations in response to the increasing demands by our society generally can, however, entail a reduction in benefits, thus posing a serious dilemma. The public and policy makers are required with increasing frequency to weigh benefits against risks and assess associated uncertainties when making decisions. Further, the lack of a systems or holistic approach leads to vulnerabilities when the reduction of one set of risks introduces offsetting or larger risks of other kinds.

1.2 Historical Perspectives on the Origins of Risk Analysis and Management

Within the context of projects, risk is commonly associated with an uncertain event or condition that—if it occurs or materializes, as possibly a result of an interaction with a hazard—has a negative effect on a project's objective. Risks resulting from such events can, therefore, be characterized by *the occurrence likelihood and the occurrence consequences* of such events. Assessing risk requires its quantification by developing models that represent a system of interest, events or scenarios that are of concern or interest, assessing likelihoods, and assessing consequences. The results of these assessments can be graphically represented using an x–y plot of consequences (x: ranges of dollars, injuries, or lives lost) versus likelihoods (y: probability, exceedence probability, rate, or exceedence rate). The outcomes of risk assessment can then be used in economic models or decision structures to perform trade-offs among risk control and management options available to keep risk within acceptable levels.

Risk analysis should be performed using a systems framework that accounts for uncertainties in modeling, behavior, prediction models, interaction among components of a system, and impacts on the system and its surrounding environment. Risk assessment constitutes a necessary prerequisite to risk management and communication. Risk acceptance is a complex socioeconomic, technological decision that can partly be based on previous human and societal behavior and actions. Risk communication can follow the risk assessment and management stages for informing other analysts, decision makers, engineers, scientists, system users, and the public about the risks associated with the system; therefore, subsequent decisions relating to the system are made with risk awareness, not blindly. Primary risk methods are described in detail in Chapter 2 that can form the bases for developing risk quantification and management methodologies.

Risk analysis and management, as a field of study, has many origins at different times for different motives in active evolutionary tracks. The emergence of this field obviously cannot be separated from the human experience and its evolution behaviorally, culturally, economically, financially, technologically, and intellectually. The history of risk is, therefore, quite complex and entails uncertainty, and any attempt to summarize it would only produce an incomplete and biased account of it. The best source available is the 1998 Bernstein's book *Against the Gods*. An informative summary is provided by Vesper (2006).

The word *risk* is of obscure origin, perhaps dating back to 1655–1665 based on an old French word *risqué* or an Italian word *risc*, whereas the word *hazard* comes from the Arabic word *al-zahr*, which means the dice (from http://dictionary.reference.com). The study of risk emerged from the games of chance and gambling that were depicted in Egyptian tomb paintings from 3500 BCE and powered by the changes in mathematical numbering based on the Hindu–Arabic numbering system, an understanding of the statistical basis of probability, and the rise in popularity of gambling in Europe between 1000 and 1200 leading to works during the Renaissance period on the games of chance and playing dice. Other mathematicians, with the availability of large bodies of data such as birth and death records, produced properties and rules concerning sampling, actuarial tables, and prediction methods of events, behavior, and populations.

A key origin of risk analysis is money and financial interests as early as Aristotle, who in his treatise *Politics* discusses the concept of options as a financial instrument that allows individuals to buy and sell goods from one another at prearranged prices. Options were traded in Holland as early as the 1630s, and in the United States in the 1790s in what would later become the New York Stock Exchange. As for trading in futures, their use in Europe dates to the medieval times to help reduce risk for farmers and commodity buyers. Futures on products such as grain, copper, and pork bellies were sold on the Chicago Board of Trade starting in 1865.

Insurance, as a financial tool to manage risk for a person or party by sharing potential financial losses with others who are compensated for taking on the added risk by insurance premiums, was started to help finance voyages of ships and was used for life insurance by trade and craft guilds in Greece and Rome, and against natural hazards to protect farmers and traders from droughts and floods. Lloyd's of London was born in a coffee shop near the

Tower of London in 1687. The shop was a gathering place for ship captains who shared information about past and upcoming voyages, routes, weather, and hazards, and eventually the sharing extended to the risks by signing their names on a board under the terms of a contract for all parties to see. This practice of signing under the risk-sharing contract gave rise to the term *underwriters*. Insurance coverage has continued to expand in scope, parties covered, layering, and complexity for a variety of hazards in the form of new products.

Between the 1970s and the 1990s, derivatives were created as financial contracts, so named because they derive their value from one or more assets, used to hedge or protect against a financial loss, and are particularly useful in conditions in which there is significant volatility, that is, financial risk. Examples are futures and options.

A technological origin of risk analysis and management can be associated with the industrial revolution out of concern over risks stemming from new technologies, in particular the invention of steam-powered engines of the late 1700s and 1800s. Steam engines, which used onboard ships, had the potential to cause a greater number of casualties compared to other technologies invented until the late 1700s. High-pressure steam engines resulted in thousands of fatalities in a couple of hundred steamboat accidents that occurred between the years 1816 and 1848 (adapted from Vesper 2006). In 1838, the US Congress passed the first law regulating an industry, by establishing the Steamboat Inspection Service, and subsequently strengthened the law until 1852 when Congress moved the Inspection Service from the oversight of the Department of Justice to that of the Department of Treasury.

Building and building codes have a long history that can be used to establish an origin of risk methods. The Code of Hammurabi, a well-preserved Babylonian law code of many articles, dating back to about 1772 BCE (King 2011), includes several articles intended to manage risks associated with building practices as follows: Article 229—if a builder builds a house for someone and does not construct it properly, and the house that he built falls in and kills its owner, then that builder shall be put to death; Article 230—if it kills the son of the owner, then the son of that builder shall be put to death; and so on. Building codes nowadays are based on probabilistic methods and probabilistic risk analysis, such as the standards by the American Concrete Institute (ACI), the American Society of Civil Engineers (ASCE), and the American Society of Mechanical Engineers (ASME).

The increase in hazards and risks to individuals, society, and the environment has been a driving force behind vigilance by regulators, industry, and other entities involved in managing and controlling risks by creating and evolving risk methods and a variety of approaches to manage and control risks. An interesting, but extreme, approach that emerged in the United States was the *Delaney Clause* that was added to the Federal Food, Drug, and Cosmetic Acts, in 1954, 1958, and 1960, respectively, to prohibit the addition of any pesticide, additive, or coloring agents, which are shown to be carcinogenic in humans or animals, to foods based on the belief that thresholds do not exist below which respective chemicals do not provoke carcinogenic responses. It was initially based on an *absolute safety requirement* defined by cancer occurrence due to a single contact between a carcinogenic substance of any amount and a cell, without accounting for the fact that test animals are subjected to toxicity levels exceeding a person's consumption over a lifetime or that the animal species had a metabolic pathway very different from that of humans. This approach was changed to threshold limit values established, interpreted, and applied by professionals to prevent an unreasonable risk of disease or injury, for example, an 8-hour time-weighted average limit would be set to mean that a normal, otherwise healthy, worker exposed to this level over a 40-hour work week would not be expected to develop any type of health injury (Vesper 2006).

The scope of hazards and risks has greatly increased reflecting increases in the technology applications and emergence of new technologies covering diverse areas including nuclear energy, bridges, airplanes, marine systems, buildings, energy-related advancements, power and chemical plants, pesticides, biologically active agents, subatomic particles, nanotechnology, foods and drugs, pharmaceutical products and devices, and so on. As a result, many agencies have adopted risk methods in their regulations, such as the US Nuclear Regulatory Commission, Department of Energy, Environmental Protection Agency, Coast Guard, Department of the Interior, Transportation Security Agency, and Department of Defense, to name a few. The adoption and use by the private sector has increasingly kept up with this trend and proliferated in many directions including risk analysis and management methods for enhancing technologies, project management practices, project execution, strategic planning, and enterprise growth and management.

1.3 Risk Types with Varying Impacts: From Nuisances to Existential

All living systems have a simple, but challenging, genetically encoded mission in life to survive and reproduce to preserve their kind in hostile environments that are full of hazards, adversaries, and predators, and most importantly dynamic in their nature. Humans are not different from other species, however, with an added aspiration of enhancing the quality of life, longevity, and happiness. Luckily, these risks faced by humans differ in occurrence rates (or odds), impacts, and in terms of our ability to recover from any impacts if they materialize. Figure 1.1 shows the scope and impacts of these risks on ordinal scales from the perspective of humans. The figure, inspired by the work of Bostrom (2002), does not show the important dimension of rates (or odds) for the various cases identified but shows only the impact and scope with transgenerational and time dimensions, and is intended to offer a basis for exhaustively scoping out risks for the purposes of this book. The impact categories displayed are human centric, that is, anthropocentric, starting from nuisance to human health with the components of bodily injuries, mental health, and death, to property, and finally to the environment. The second scale of scope is also human centric, which shows four categories from an individual to a group to a locality to worldwide, that is, global, for each the time and transgenerational dimensions are implicitly included. The examples provided in the figure cover many illustrative cases identified, from the mundane, tolerable risk of car damage due to a falling tree to the existential risk from an asteroid impacting the Earth. Some of these risks are manageable, others are tolerable, some are inevitable, and a few could make us feel helpless.

A differentiator among living systems is their intelligence level to achieve their respective objectives through appropriate decision making that entails coping with these risks through an effective and efficient use of resources. The objective of this book is to develop, present, and demonstrate analytical methods for structuring decision situations to quantify and manage risks using rational, and perhaps can be described as biologically inspired, ways of thinking to achieve superior intelligence for achieving prescribed objectives or a mission. The methods included in the book can handle all the cases presented in Figure 1.1; however, the limited size of the book does not permit coverage of all the cases. Readers can draw parallelism among different cases to construct methodologies in addressing any case of interest including the ones that are not covered.

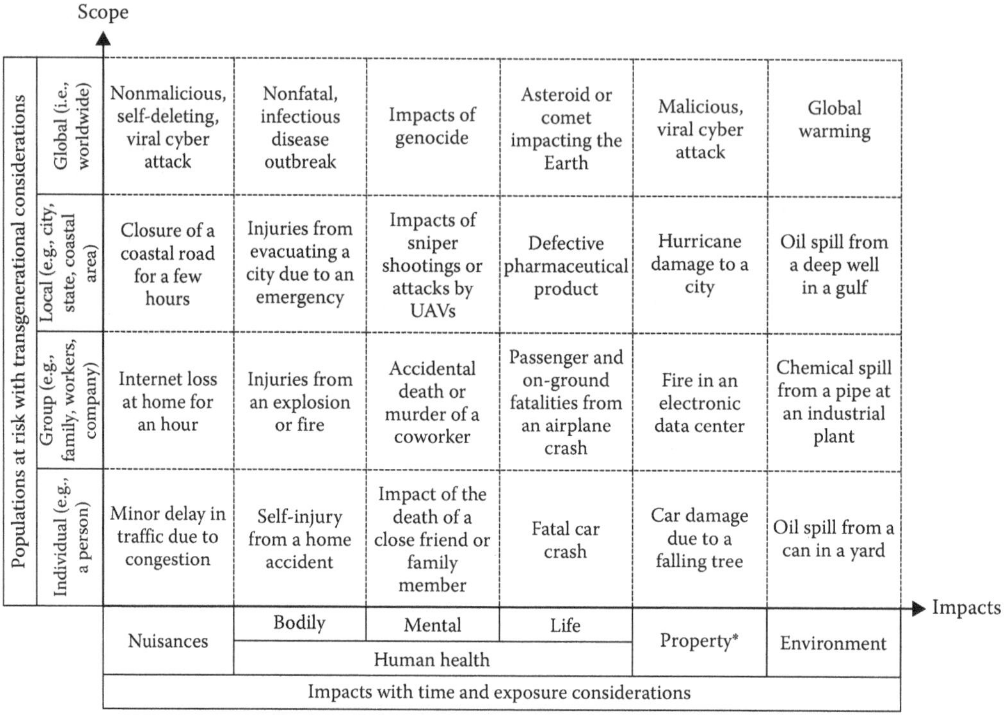

Scope

Populations at risk with transgenerational considerations	Nuisances	Bodily (Human health)	Mental (Human health)	Life (Human health)	Property*	Environment
Global (i.e., worldwide)	Nonmalicious, self-deleting, viral cyber attack	Nonfatal, infectious disease outbreak	Impacts of genocide	Asteroid or comet impacting the Earth	Malicious, viral cyber attack	Global warming
Local (e.g., city, state, coastal area)	Closure of a coastal road for a few hours	Injuries from evacuating a city due to an emergency	Impacts of sniper shootings or attacks by UAVs	Defective pharmaceutical product	Hurricane damage to a city	Oil spill from a deep well in a gulf
Group (e.g., family, workers, company)	Internet loss at home for an hour	Injuries from an explosion or fire	Accidental death or murder of a coworker	Passenger and on-ground fatalities from an airplane crash	Fire in an electronic data center	Chemical spill from a pipe at an industrial plant
Individual (e.g., a person)	Minor delay in traffic due to congestion	Self-injury from a home accident	Impact of the death of a close friend or family member	Fatal car crash	Car damage due to a falling tree	Oil spill from a can in a yard

Impacts

Impacts with time and exposure considerations

FIGURE 1.1
Anthropocentric risks. The symbol * indicates that financial markets and economics are included under property.

Humans, in their aspiration for enhancing their quality of life, longevity, and happiness, take premeditated actions, undertake projects, alter environments, consume resources, and so on, which entail the risks for the potential achievement of these aspirations as benefits. Trade-offs among resource consumption, environmental changes, benefits, and potential adverse impacts associated with risks are complex, and their analysis and quantification require the concepts and methods presented in this book.

Events, when they occur, or systems of interest to humans, when they interact with a hazard, could lead to impacts that could span multiple cases shown in Figure 1.1. For example, when a hurricane passes over a city, the impacts could cover the cases of individual, group, and local populations under nuisances, human health, property, and environment, and if it is of an extreme intensity, the impact might extend to a global level through interconnected world economies. The figure, therefore, is intended to illustrate various cases by examples with each listed under a particular case for convenience, while the impacts could knowingly extend to other cases listed in the figure. Moreover, when events occur, or systems of interest interact with a hazard, adverse financial, monetary, and economic effects may result (Figure 1.1). These effects are included under property, although they exhibit some intangible characteristics that could make their analysis challenging, and quantification and valuation illusive.

According to the United Nations Office for Disaster Risk Reduction (UNISDR 2012), half of the world's inhabitants, expected to increase to roughly two-thirds by 2015, and the vast majority of property and wealth are concentrated in urban centers situated in locations already prone to major disasters, such as earthquakes and severe droughts, and along flood-prone coastlines. It also reported that the 2011 natural disasters, including the earthquake and tsunami that struck Japan, resulted in US$366 billion in direct damages and 29,782 fatalities worldwide. Storms and floods accounted for up to 70% of the 302 natural disasters worldwide in 2011, with earthquakes producing the greatest number of fatalities. It is anticipated that such disasters would occur in increasing trends of storm rates and disaster impacts due to a combined effect of a changing climate and an increased coastal inventory of assets (Ayyub et al. 2012). Although no population center or a geographic area can ever be risk free from natural or human-caused hazards, communities should strive to manage risk and enhance resilience to the destructive forces or the impacts of resulting events that may claim lives and damage property. Risk perceptions of the risk landscape as assessed from the 2011 World Economic Forum places storms and climate change at high levels as summarized in Figure 1.2. Gilbert (2010) provided population-and-wealth-adjusted loss and fatality count trends from 1960 to 2009 to demonstrate that both are about flat without significant slopes; however, he noted that the United States is becoming more vulnerable due to increased population concentration in areas prone to natural hazards (Berke et al. 2008; Burby 1998) and persisting inadequate condition of infrastructure (ASCE 2009).

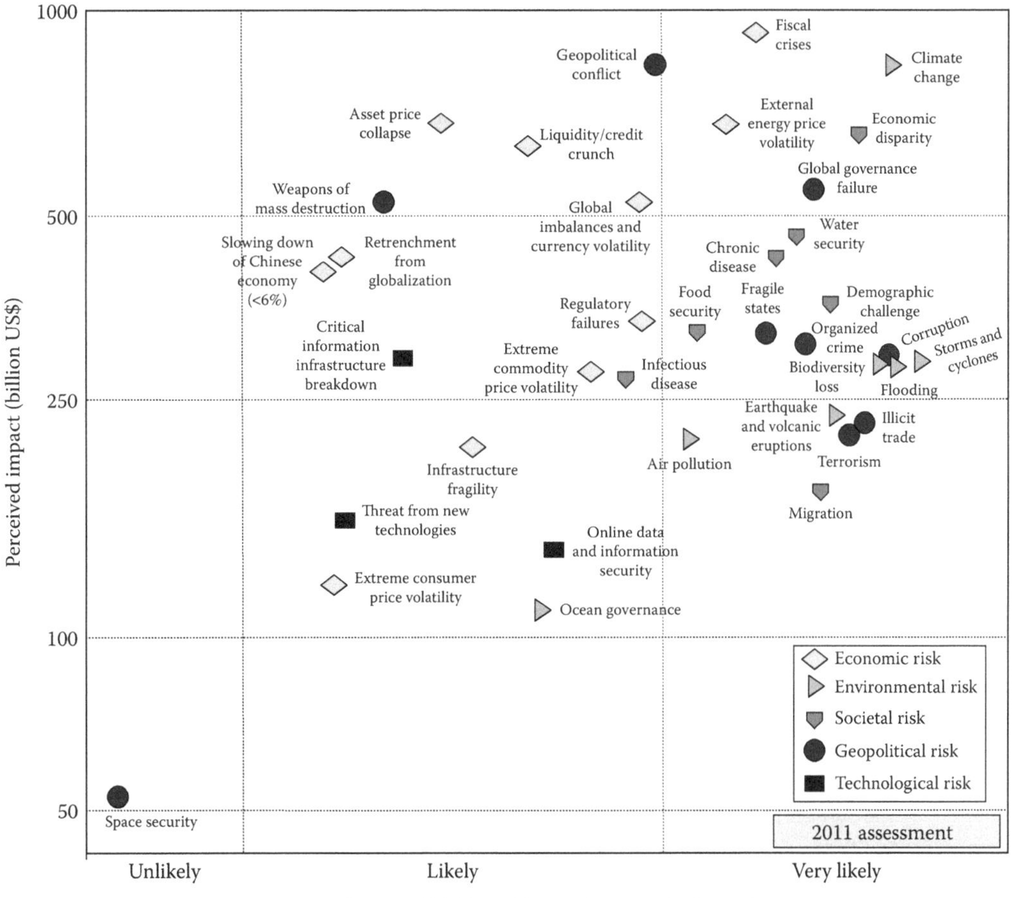

FIGURE 1.2
Perceived risks in 2011.

Example 1.1: Nuisances

Figure 1.1 shows risks that fall under the nuisance impact level. These risks could affect individuals, groups, populations of cities, or humans worldwide. The figure shows selected examples from the following list:

- Individual: minor delay in traffic due to congestion, leading to missing an appointment or starting a task; minor allergic reaction to a medication; minor loss in the stock market due to daily fluctuation; and so on
- Group: brief Internet loss at home for an hour, momentary power loss at a meeting room during a meeting, brief delayed arrival of a speaker at a meeting, delayed arrival of a coworker on a crew, and so on
- Local population: closure of the only coastal road for a few hours, brief Internet loss at town for an hour, brief power loss at town for an hour, limited supplies of snow removal equipment during a snow storm, and so on
- Worldwide population: nonmalicious, self-deleting viral cyber attack; small oil price increase; and so on

Example 1.2: Human Health

Human health impacts are grouped in Figure 1.1 as physical injuries, metal health, and life loss. Figure 1.1 provides examples for the four cases of individual, group, local population, and worldwide population. This category has temporal and transgenerational dimensions, for example, death due to aging or birth defects due to exposure to radiations, respectively. The following additional examples are provided for several impact types:

- Individual: broken hip due to slipping and falling at home, substantial financial loss in the stock market, drowning during a recreational water activity, and so on
- Group: injuries from exposure to gas released from riot-control grenades at a civil disobedience demonstration, experiences of war veterans or torture victims, fatalities from a bus crash or a train crash, and so on
- Local population: injuries from evacuating a city due to a dam breach, psychological impacts of random sniper shootings or psychological impacts on children as a result of attacks by unmanned aerial vehicles (UAVs), defective pharmaceutical products or defective toys, and so on
- Worldwide population: nonfatal, infectious disease outbreaks; impacts of genocide or large-producing fatality event; an asteroid or comet impacting the Earth, leading to human extinction or fatal, infectious disease outbreaks or global nuclear war; and so on

Example 1.3: Property

Impacts on property are illustrated in Figure 1.1 based on the following list:

- Individual: damage to a vehicle due to a falling tree, damage to a cellular phone by accidently dropping it on a hard floor from the hand of a user while standing up, and so on
- Group: fire in an electronic data center of a corporation, fire at a residence while the owners are on travel, flooding of an office due to a water pipe breakage, and so on
- Local population: hurricane damage to a city, flooding of a town due to extreme precipitation, tornado damage to a town, and so on
- Worldwide population: malicious, viral cyber attack; rising sea level; and so on

Example 1.4: Environment

Impacts on the environment are illustrated in Figures 1.1 and 1.2 based on the following list:

- Individual: oil spill from a can in the backyard of a house; minor gasoline spill from a pump in a gas station; contamination of water in a private, small tank by sewage water; and so on
- Group: chemical spill from a pipe at an industrial plant, gasoline release from an underground tank of a gas station, contamination of water in a storage tank by sewage water, and so on
- Local population: oil spill from an deep well in a gulf, oil spill from a oil tanker close to the coast, release of gaseous chemical to the air, radioactive material release, and so on
- Worldwide population: global warming, major release of radioactive material, nuclear war, and so on

Example 1.5: A Truss Structural System

A truss as shown in Figure 1.3 can be viewed as a structural system that must be designed to an acceptable safety level to support traffic loads on a bridge. The system in this case can be thought of as a system in series, meaning that if 1 out of 29 members fails, then the entire system would fail to function properly and may collapse. The failure potential is a serious matter that designers consider carefully in the design stage of the truss. Designers have to identify potential modes of failure and assess the associated consequences and risks. The design stage includes studying the possible scenarios of failure of the members in order to enhance the design and manage the risks. For example, a design could be enhanced to allow for partial failures instead of catastrophic failures and to introduce redundancy through the addition of other members to work as standby or load-sharing members to critical members in the structure. The benefits of such enhancements, which are intended to reduce the likelihood of failure, could include increasing design and construction costs to such an extent that the structure becomes economically unfeasible. Bridge failure consequences may be included in this analysis. Trade-off analyses can be performed to make the structure economically feasible and achieve the acceptable safety levels. This example demonstrates the potential of risk analyses during the design process to provide the acceptable risk levels.

Example 1.6: A Water Pipeline System

The primary water supply system of a city is shown in Figure 1.4. The water delivery system of city C has two sources, A and B, from which water passes through a pumping station. Pipelines (branches 1–3) are used to transport water as shown in Figure 1.4. Assuming that either source alone is sufficient to supply the city with water, failure can happen in branch 1, 2, or 3. Designers and planners of the pipeline system, therefore, have to identify possible water loss scenarios to assess the associated risks. The simplistic failure scenarios given in

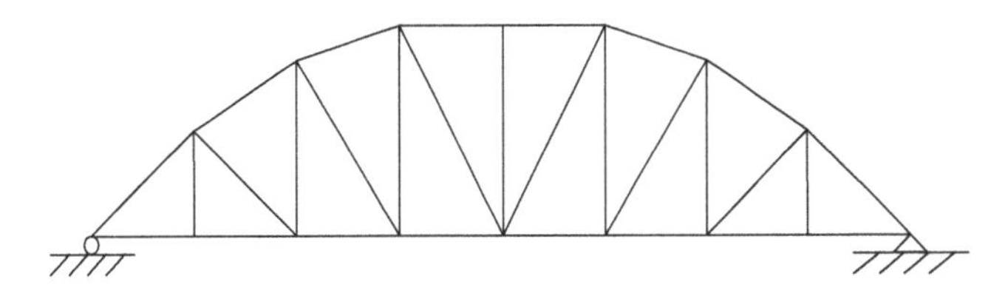

FIGURE 1.3
Truss structural system.

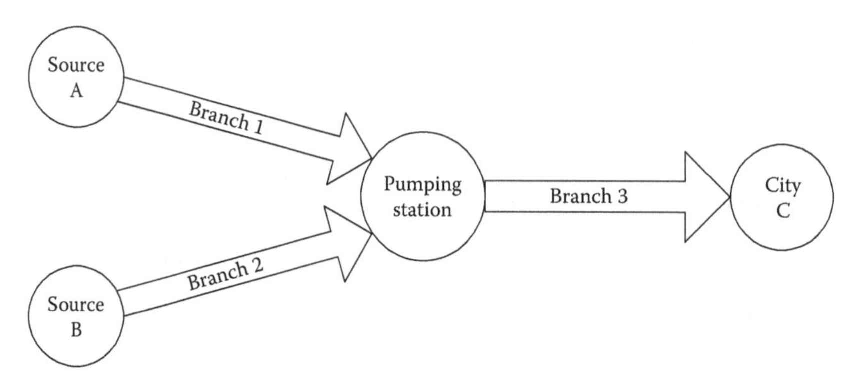

FIGURE 1.4
City water pipeline system.

TABLE 1.1
Failure Possibilities and Their Impact on a Water Pipeline System

Source of Failure	Failure	Impact on System or Consequences
Failure of branch 1 only	T	P
Failure of branch 2 only	T	P
Failure of branch 3 only	T	T
Failure of branches 1 and 2 only	T	T
Failure of branches 1 and 3 only	T	T
Failure of branches 2 and 3 only	T	T
Failure of branches 1–3	T	T

P, partial; T, total.

Table 1.1 can be used for risk analysis studies of the supply pipelines. Table 1.1 is limited only to cases where total failure happens in each of the three branches. The table can be expanded to include partial failures of a branch including water loss from leaks.

Example 1.7: A Fire Escape System

In the event of a fire in an apartment that is equipped with a smoke detector, the potential consequences of the fire to occupants may be analyzed using qualitative risk analysis methods. The consequences of the fire depend on whether the smoke detector operates successfully during the fire and whether the occupants are able to escape. Table 1.2 shows possible qualitative scenarios that can be thought of as results of a fire. The table can be extended further to perform quantitative risk analyses by assigning probability

TABLE 1.2
Possible Escape Scenarios and Their Risk Consequences

Source of Risk as an Adverse Event	Escape Scenario	Smoke Detector Working Successfully?	Occupants Managed to Escape?	Consequences in Terms of Loss of Life
Fire initiated in an apartment	1	Yes	Yes	No injury
	2	Yes	No	Death
	3	No	Yes	Severe injury
	4	No	No	Death

values to the various events in paths (i.e., rows of the table). An additional column before the last column can be inserted to calculate the total path probability of each scenario. Such an analysis can assist planners and designers in computing the overall probability of each consequence for the purpose of planning, designing, and constructing escape routes more efficiently. Such analysis can reduce risks and increase safety to occupants of the apartments, leading to reduction in insurance premiums and enhancement of market values of the apartments. A formal approach for such analysis can involve fault tree analysis as will be discussed in detail in Chapter 2.

Example 1.8: Project Management

Risk analysis can be a very useful technique when applied in the field of project management. In construction projects, managers and clients commonly pursue areas and sources of risks in all the five phases of a project from feasibility to disposal or termination as listed in Table 1.3. The methods can be applied by developing risk scenarios associated with failure states for all project phases by using methods that examine causes and effects as shown in Table 1.3. The failure states in this example were selected to illustrate the causes and effects for respective stages.

Example 1.9: Organizational Hierarchy of a Corporation

Risk methods can be used to analyze potential failures in managing an organization due to errors, inappropriate decision, or incorrect decisions. Organizational failures can lead to significant adverse consequences. Executives and managers of organizations are

TABLE 1.3

Cause-and-Effect Risk Scenarios for Project Phases

Source of Risk in Project Stages	Failure State	Cause of Failure	Effect on Project
Feasibility study	Delay	Feasibility stage is delayed due to complexities and uncertainties associated with the system.	The four subsequent stages of the project will be delayed, thus causing problems in regard to the client's financial and investment obligations.
Preliminary design	Approval not granted	The preliminary design is not approved for various reasons; failure can be attributed to the architect, engineer, project planner, or project manager.	The detailed design will not be ready for zoning and planning approval or for the selection process of contractors, thus causing accumulated delays in finishing the project, leading to additional financial burdens on the client.
Design details	Delay	Detailed design performed by the architect/engineer is delayed.	The project management activities cannot be performed efficiently, and the contractor cannot start work properly, thus causing delays in the execution of the project.
Execution and implementation	Delay or disruption	Execution and implementation stage is delayed or disrupted as a result of accidents.	The project will definitely not be finished on time and will be completed over budget, thus creating serious financial difficulties for the client.
Disposal or termination	Delay	The termination stage is delayed or not scheduled.	The system will become unreliable and hazardous, thus causing customer complaints and exacerbating the client's contractual obligation problems.

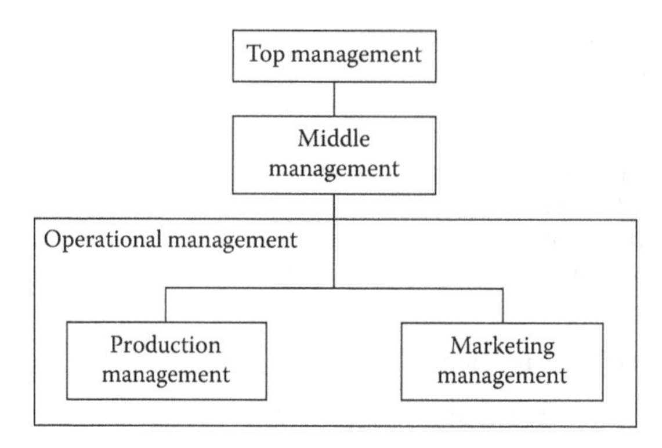

FIGURE 1.5
A series–parallel model of organizational structural hierarchy.

responsible for designing the hierarchical structure of their organization. They should rigorously study the implications for designing their organizational structure as a system with in-series, in-parallel, or mixed series–parallel lines of authority and communications among departments and management levels. These lines represent the flow of instructions and the feedback channels that could fail and potentially lead to damage to the entire organization. For the series–parallel structure shown in Figure 1.5, managers need to analyze the risks associated with the structure and perform failure analysis, such as that provided in Table 1.4. Performing qualitative failure analysis in organizational management systems poses a unique challenge, as managed departments are not mechanical devices that show a crisp transition from a functioning state to a failure state but rather exhibit partial failures and blurred transitions from one state to another. Therefore, in analyzing such structures, the percentage of failure at every management level has to be assessed through brainstorming and interviewing sessions. The qualitative analyses are usually a prelude to quantitative analyses to quantify these partial to disruptive failures (Table 1.4).

Example 1.10: Project Cost and Profitability

An important management consideration is to execute a project within a prescribed budget, thereby realizing profits or benefits as anticipated from the project. Cost overrun is an undesirable outcome and can be treated as failure in a risk

TABLE 1.4

Possible Failure Scenarios for a Multilevel Organizational Structure

Source of Risk as an Adverse Event	Failure Scenarios	Failure of Top Management?	Failure of Middle Management?	Failure of Operational Management?	Performance of the Organizational Structure
Failure of existing structural hierarchy to achieve organizational goals	1	Yes	Yes	Yes	D
	2	Yes	Yes	No	P
	3	Yes	No	Yes	P
	4	Yes	No	No	P
	5	No	Yes	Yes	P
	6	No	Yes	No	P
	7	No	No	Yes	P
	8	No	No	No	S

D, disruptive failure; P, partial; S, success.

analysis framework. The underlying top events that can contribute to this failure type are as follows:

- Uncertainties associated with cost estimates
- Underlying assumptions and factors that introduce randomness and correlations in the actual costs compared to estimated costs
- Various hazards that can be classified broadly into natural hazards, human-caused events, and external events
- Project management decisions relating to insurance and contract terms, contingency allocations, and skill of the management team
- Enterprise management decisions relating to portfolio considerations and correlations, and the responsiveness and nimbleness of the senior management team

These top events require separate analytical models to facilitate risk quantification in order to inform and enhance decision making.

1.4 Systems Framework

The definition and articulation of problems in engineering and sciences are critical tasks in the processes of analysis and design, and can be systematically performed within a systems framework. Albert Einstein said, "The mere formulation of a problem is often far more essential than its solution"; however, according to Werner Karl Heisenberg, this definition entails uncertainty by stating thus: "What we observe is not nature itself, but nature exposed to our method of questioning." Generally, a system as abstracted by an analyst or envisioned by an engineer, such as a power plant, can be modeled to include a segment of its environment that interacts significantly with it to define an underlying system. The boundaries and the extent of details, that is, resolution of the system as defined, are drawn based on the objectives of the analysis and the class of performances (including failures) under consideration.

A generalized system formulation allows scientists and engineers to develop a comprehensive understanding of the nature of a problem and the underlying physics, processes, and activities. In a system formulation, an image or a model of an object that emphasizes some important and critical properties is defined. System definition is usually the first step in an overall methodology formulated for achieving a set of objectives. This definition can be based on observations at different system abstraction levels that are established based on these objectives. The observations can be about the different elements (or components) of the system, interactions among these elements, and the expected behavior of the system. Each level of abstraction is treated as a knowledge layer. These knowledge layers are defined for a problem or a project of interest, and define a system to represent the problem or the project. As additional layers of knowledge are added to previous ones, higher epistemological levels of system definition and description are attained which, taken together, form a hierarchy of system descriptions. Two views have emerged in systems science for the purpose of system definition: (1) *realism* and (2) *constructivism*, as described by Ayyub and Klir (2006) and summarized in this section.

According to realism, a system that is obtained by applying correctly the principles and methods of science represents some aspects of the real world. This representation is only approximate, due to limited capability or resolution of our sensors and measuring instruments, and is viewed as a homomorphic image of its counterpart in the real world. Using more enhanced capability or refined instruments, the homomorphic mapping between the entities of the system of concern and those of its real-world counterpart (the corresponding *real system*) also becomes more refined, and the system becomes a better representation of its real-world counterpart. Realism thus assumes the existence of systems in the real world, which are usually referred to as real systems. It claims that any system obtained by sound scientific inquiry is an approximate (simplified) representation of a real system via an appropriate mapping.

According to constructivism, all systems are artificial abstractions. They are not made by nature and presented to us to be discovered, but we construct them by our perceptual and mental capabilities within the domain of our experiences. The concept of a system that requires correspondence to the real world is illusory because there is no way of checking such correspondence. We have no access to the real world except through experience. It seems that the constructivist view has become predominant, at least in systems science, particularly in the way formulated by von Glasersfeld (1995). According to this formulation, constructivism does not deal with ontological questions regarding the real world. It is intended as a theory of knowing, not as a theory of being. It does not require analysts to deny ontological reality. Moreover, the constructed systems are not arbitrary, that is, they must not collide with the constraints of the experiential domain. The aim of constructing systems is to organize our experiences in useful ways. A system is useful if it helps us to achieve some aims, for example, to predict, retrodict, control, and make proper decisions.

We perceive reality as a continuum in its composition of objects, concepts, and propositions. We construct knowledge in quanta to meet constraints related to our cognitive abilities and limitations, producing what can be termed as quantum knowledge defined by fields, practices, and so on. This quantum knowledge leads to and contains deficiencies, that is, *ignorance* as termed in this book—manifested in two forms: (1) *ignorance of self-ignorance* and (2) knowledge deficiencies in the form of *incompleteness* and/or *inconsistency*, as discussed in detail in subsequent sections. The ignorance of self-ignorance is called *blind ignorance*. The incompleteness form of ignorance stems from quantum knowledge that does not cover the entire domain of inquiry. The inconsistency form of ignorance rises from specialization and focuses on a particular specialty discipline, or science, or a phenomenon without, for example, accounting for interactions with or from other sciences or disciplines or phenomena.

Methods for system definition and analysis are described in Chapter 3. The examples presented in this section include systems of varying complexity levels from various fields. They are intended to demonstrate domain diversity and variations in representation styles.

Example 1.11: Carbon Inventory and Cycle for Assessing Global Warming Risks

Carbon, as a primary element for all living systems, is present in pools (or reservoirs) in the Earth's atmosphere, soils, oceans, and crust, and is in flux as it moves from one pool to another at different rates. The overall movement of carbon can be described as a cycle. Starting with the carbon in the atmosphere (Figure 1.6), it is used in a photosynthesis process with other elements to create new plant material. As a result, this process transfers large amounts of carbon from the atmosphere's pool to the plants' pool. These plants, similar to other living systems, eventually die and decay, or are consumed by fire, or are harvested by humans or other living systems for consumption, placing carbon in fluxes to other pools, and eventually released back to the atmosphere. This cycle is linked to other cycles of the oceans' microbes, fossil rocks, volcanoes, and so on. Earth has many individual cycles linked to each other on spatial and temporal scales to form an integrated global carbon cycle as shown schematically in Figure 1.6 according to the National Oceanic and Atmospheric Administration (NOAA)'s data (Intergovernmental Panel on Climate Change or IPCC 2007). A similar figure is provided by the University of New Hampshire (2011) with arrows indicating fluxes. Pan et al. (2011) estimated a global carbon budget for two time periods of 1990–1999 and

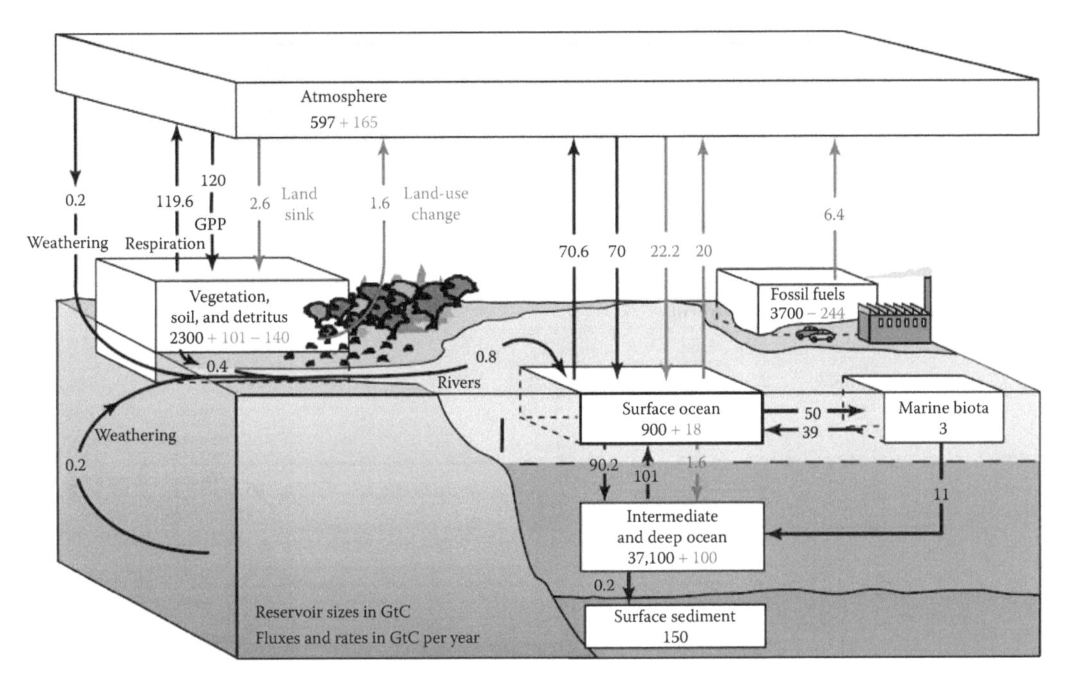

FIGURE 1.6
The anthropogenic carbon cycle carbon pools and fluxes according to the NOAA. GPP, gross primary production; GtC, gigatonnes of carbon; 1 GtC is equal to 10^9 tonnes of carbon or 10^{12} kg; 3.7 Gt CO_2 will give 1 GtC. (Adapted from University of New Hampshire, Global carbon cycle, http://globecarboncycle.unh.edu/CarbonCycleBackground.pdf based on http://globe.gov/science/topics/carbon-cycle#Overview, 2011; IPCC 2007.)

2000–2007 as shown in Figure 1.7 that clearly shows the increase of carbon emission under fossil fuel and cement over time. This increase goes unmatched in the carbon uptakes in efficiency with a potential for creating a prolonged time lag from emissions to uptakes. The result is an increase in the atmospheric carbon dioxide (CO_2) concentration as shown in Figure 1.8a based on the data of ancient air trapped in Antarctic ice gathered from ice cores from glacier averaged using 75-year smoothing (Etheridge et al. 2012) (Figure 1.8b). Such an increase due to this time lag drives other processes leading to global temperature increases. Carbon dioxide is predicted to double by 2100, leading to global temperature increase of 1.5°C–3.5°C that in turn leads to sea-level rise, storm extremes, and so on. This example illustrates the identification of the boundaries of a complex system as provided by Ayyub (2012).

Example 1.12: Linked Ecological and Socioeconomic Systems for Assessing Associated Risks

A link between terrestrial and aquatic systems can be defined by wetlands. The critical functions of wetlands can be defined by their landscape position, and their ecological and socioeconomic services as schematically represented in Figure 1.9. Researchers have characterized wetlands as keystone ecosystems in linked ecological-socioeconomic systems based on their importance for improving water quality, flood storage, storm protection, and biological diversity support. Factors affecting wetlands include qualities of the ecosystem, the surrounding environment defined by the underlying environmental processes, and human activities competing with these processes. Modeling this interaction and the factors requires the development of a system model, as illustrated in Figure 1.9, in order to sustain multiple desirable outcomes and assess any risks resulting from imbalances created by engineering projects, changes in human

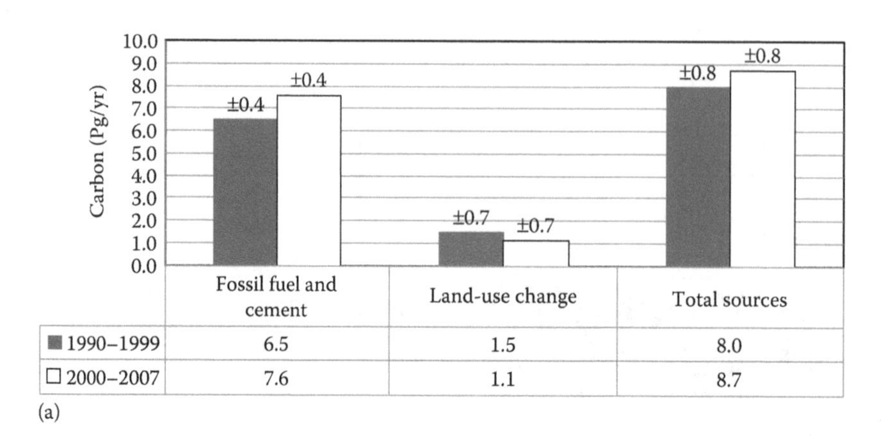

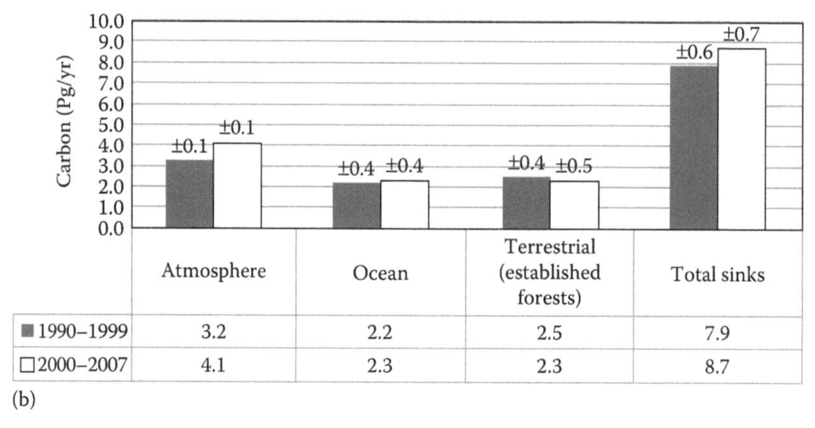

FIGURE 1.7
Carbon budget for two time periods: (a) sources (carbon emissions) and (b) sinks (carbon uptakes). (Data from Pan et al., *Science*, 133, 988–993, 2011.)

behavior or activities, and changes in policy. The cumulative effect over some time period could lead to adverse effects, such as causing the decline of all or some species, and in other instances acting in concert with other stressors impairing the biogeochemical cycles of a wetland such that it is no longer a viable ecosystem. Other considerations based on Example 1.11 are (1) to link the treatment to climate changes leading to warmer temperatures, changes in regional precipitation, and sea-level rise or changes in lake levels and (2) to assess the long-term impacts and our ability to develop response scenarios that are becoming challenging and entail great uncertainties. Figure 1.9 shows examples of positive (+) ecosystem functions and services, such as flood storage, clean water provision, carbon sequestration, nutrient processing, and biodiversity, and negative (−) impacts, such as toxins, physical disturbance, nonnative invasive species, and climate change. These are only examples. Such a functional and impact structure could help define interactions and the system boundaries and resolution level to meet modeling and decision-making objectives.

Example 1.13: System Boundaries of a Structural Truss

For the truss system used in Example 1.5, as shown in Figure 1.3, the system boundaries must be defined in order to establish the limits on the scope of a study and the extent of coverage by risk analysis methods. For this truss system, some analysts or designers may consider the system boundaries to include only the 29 members under

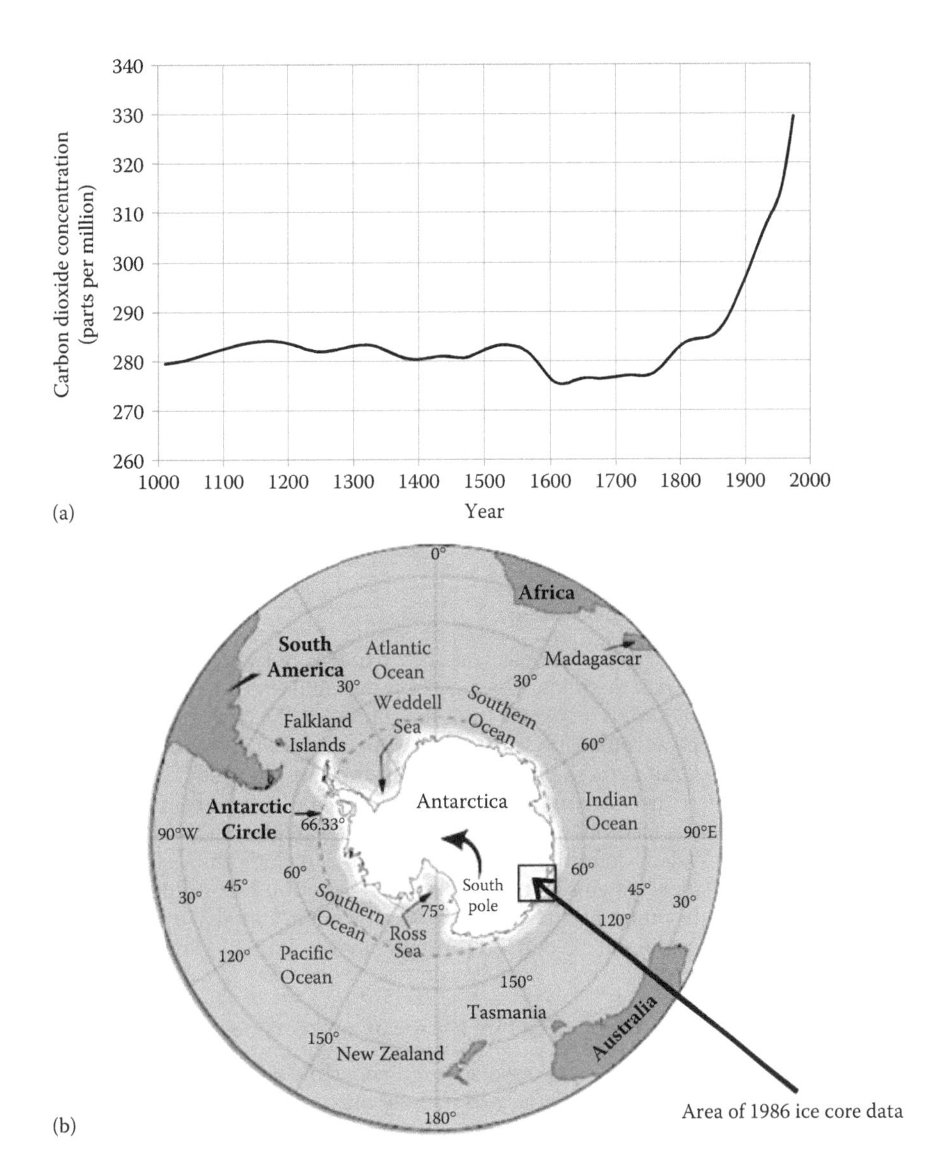

FIGURE 1.8
Atmospheric CO_2 concentration trends and data source. (a) Atmospheric carbon dioxide concentration; (b) area of 1986 ice data. (http://cdiac.ornl.gov; Adapted from Hamley et al. 1986; Etheridge et al. 2012.)

study. Others, however, may include the two supporting rollers and pins, and the system boundaries can be extended further to include supporting walls or piers. Other analyses might require extending the boundaries even further to include the foundations and their types, such as shallow, concrete, or piles. Moreover, other risk analysts may include the landscaping around the walls or columns and their effects on the type of concrete. Another extension of boundaries might require inclusion of a group of similar trusses to create a hanger, a roofing system for a factory, or a multiple-lane bridge. In the case of multiple trusses, bracing members or roofing structures connected to the trusses must be included. Hence, the responsibility of analysts or designers includes proper identification of the boundaries and limits for such a system that are necessary to perform relevant risk studies.

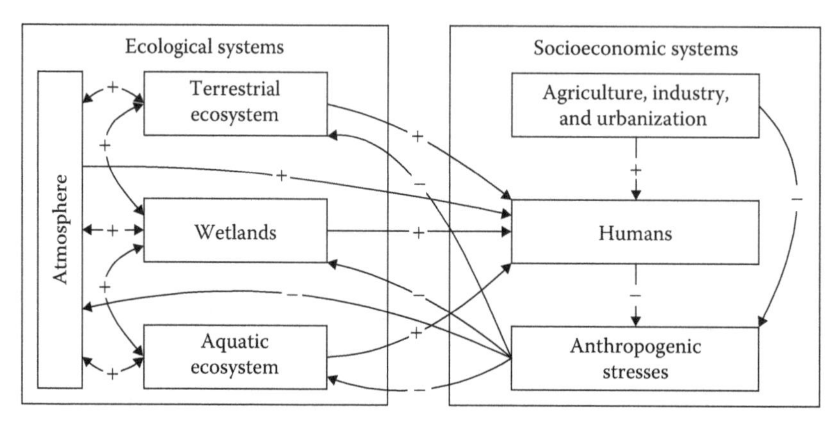

FIGURE 1.9
Linking ecological and socioeconomic systems.

Example 1.14: A Water Pipeline System

Example 1.6 utilized a water delivery system, as shown in Figure 1.4, to illustrate the need for and potential uses of risk methods. The goal of water delivery system is to meet the water needs of the city by supplying water from the sources to the city. For this situation, the system can be defined as consisting of three long pipes. Some analyses might consider the shape and various sizes of these pipes, or whether they are connected by intermediate valves. This city water network might require planners and designers to consider the effect of other obstructing facilities or infrastructures, for example, other crossing pipes, cables, or roads. The system definition can be extended to include the supports of the pipes and their foundation systems. Some studies might require expanding the system boundaries to include the pumping station, operators, and environmental conditions. The system boundaries, therefore, can be defined through having clear study objectives.

Example 1.15: System Boundaries of a Fire Escape of an Apartment Building

Referring to Example 1.7, the fire escape system for an apartment building, planners and designers may view the system boundary to include only the fire escape system from the inside to the outside of the apartments. Another perspective might be to consider other escape routes inside the building that are not designated as fire escape routes, especially for those apartments in higher levels of the building. The system boundaries can be extended to include external escape routes. High-rise building apartments with internal constraints may need to be designed to include egress to the roof of the building with an appropriate rescue plan that could include direct alarm links with fire and rescue departments. In this case, the system boundaries extend beyond the location of the building to include communication links and the response of fire and rescue units and personnel.

Example 1.16: System Boundary Identification in Project Management

Example 1.8 dealt with a risk analysis in project management. The system boundary in this case can include all people involved in the five stages of a project. They can be limited, according to traditional project management, to a client, an engineer, and a contractor. If a project management team is introduced, the definition of the system would have to be extended to include the management team. The system can be extended further to include suppliers, vendors, subcontractors, shareholders, and/or all stakeholders having an interest in the project. In this case, the client and the project management team need to clearly identify the parties that should be considered in such an analysis.

Example 1.17: System Boundary Identification in Organizational Structural Hierarchy

Using the information provided in Example 1.9, the system under consideration can include a subset of the management levels for performing a failure analysis, for example, operational management having two departments (production and marketing). Other analyses might require including only the middle management level, a critical level through which all instructions and information pass within an organization. Some studies might require including only the top management level for analysis, as its failure would lead to disruption of functions of the entire organization. In general, any combination of the two management levels can be included within a system to meet various analytical objectives. The system definition can be extended to include all levels, even the board of directors. The objective of any analysis must be defined in order to delineate the boundaries of a system and should be used as the basis for including and excluding the management level in examining the risks associated with failures of the organization.

1.5 Knowledge

Risk studies in many disciplines of engineering and the sciences rely on the development and use of predictive models that in turn require knowledge and information and sometimes subjective opinions of experts. Working definitions for *knowledge*, *information*, and *opinions* are required for this purpose. The reliability of the results from such predictive models greatly depends on the level of deficiency in the underlying knowledge, information, and opinions. In this section and subsequent sections, these concepts are discussed and examples are provided.

According to evolutionary epistemology, knowledge is a product and a construct of the human experience. Knowledge can be viewed as two types: *nonpropositional* and *propositional*. Nonpropositional knowledge can be further broken down into *know-how and concept knowledge* and *familiarity knowledge* (commonly called *object knowledge*). Know-how and concept knowledge requires someone to know how to do a specific activity, function, procedure, and so on, such as riding a bicycle. The concept knowledge can be empirical in nature. In evolutionary epistemology, know-how knowledge is viewed as a historical antecedent to propositional knowledge. Object knowledge is based on a direct acquaintance with a person, place, or thing; for example, Mr Smith knows the President of the United States. Propositional knowledge is based on propositions that can be either true or false; for example, Mr Smith knows that the Rockies are in North America (di Carlo 1998; Sober 1991). This proposition can be expressed as, where *knows* is intended to mean *claims that*:

$$\text{Mr Smith knows that the Rockies are in North America} \tag{1.1a}$$

$$S \text{ knows } P \tag{1.1b}$$

where:
S is the subject (Mr Smith)
P is the proposition or claim that "the Rockies are in North America"

Epistemologists require the following three conditions for making this claim and having a true proposition:

1. *S* must believe *P*.
2. *P* must be true.
3. *S* must have a reason to believe *P*; that is, *S* must be justified in believing *P*.

The justification in the third condition can take various forms; however, to simplify, it can be taken as justification through rational reasoning or empirical evidence. Therefore, propositional knowledge is defined as a body of propositions that meet the conditions of *justified true belief* (JTB) based on these three items. This general definition does not satisfy a class of examples called the *Gettier problem*, initially revealed in 1963 by Edmund Gettier (Austin 1998). Gettier showed that we can have highly reliable evidence and still cannot have knowledge. In addition, someone can skeptically argue that, as long as it is possible for *S* to be mistaken in believing *P* (i.e., the third condition is not met), the proposition can be false. This argument, sometimes called a *Cartesian argument*, undermines empirical knowledge. In evolutionary epistemology, this high level of scrutiny is not required. According to evolutionary epistemology, true beliefs can be justified by cause–effect examination and from reliably attained, law-governed procedures, where law refers to a natural law. Sober (1991) noted that there are very few instances, if ever, where we have perfectly infallible evidence. Almost all of our common sense beliefs are based on the evidence that is not infallible even though some may have overwhelming reliability. The presence of a small doubt in meeting the justification condition does not make our evidence infallible but only reliable. Evidence reliability and infallibility arguments form the bases of the *reliability theory of knowledge*. Figure 1.10 shows a breakdown of knowledge by types, sources, and objects that were based on a summary provided by Honderich (1995).

As stated earlier, *knowledge* is defined as a body of JTBs, such as physical laws, models, objects, concepts, know-hows, processes, and principles, acquired by humans about a system of interest, where the justification condition can be met based on the reliability theory of knowledge. The most basic and reliable knowledge (RK) category is cognitive knowledge (*episteme*), which can be acquired by human senses. The second category is based on the correct reasoning from hypotheses such as mathematics and logic (*dianoi*). The third category, which moves us from intellectual categories to categories that are based on the realm of appearances and deception, is based on the propositions and is known as belief (*pistis*). *Pistis*, the Greek word for faith, denotes intellectual and/or emotional acceptance of a proposition. The fourth category is conjecture (*eikasia*), in which knowledge is based on inference, theorization, or prediction using incomplete or unreliable evidences. These four categories are shown in Figure 1.11. They also define the knowledge box in Figure 1.12. These categories constitute the human cognition for constructing knowledge, which might be different from a future state of knowledge achieved by an evolutionary process, as shown in Figure 1.12. The *pistis* and *eikasia* categories are based on expert judgment and opinions regarding the system issues of interest. Although the *pistis* and *eikasia* knowledge categories might be marred with uncertainty, they are certainly sought after in many engineering disciplines and the sciences, especially by decision and policy makers.

Information can be defined as sensed objects, things, places, processes, and information and knowledge communicated by language and multimedia. Information can be viewed

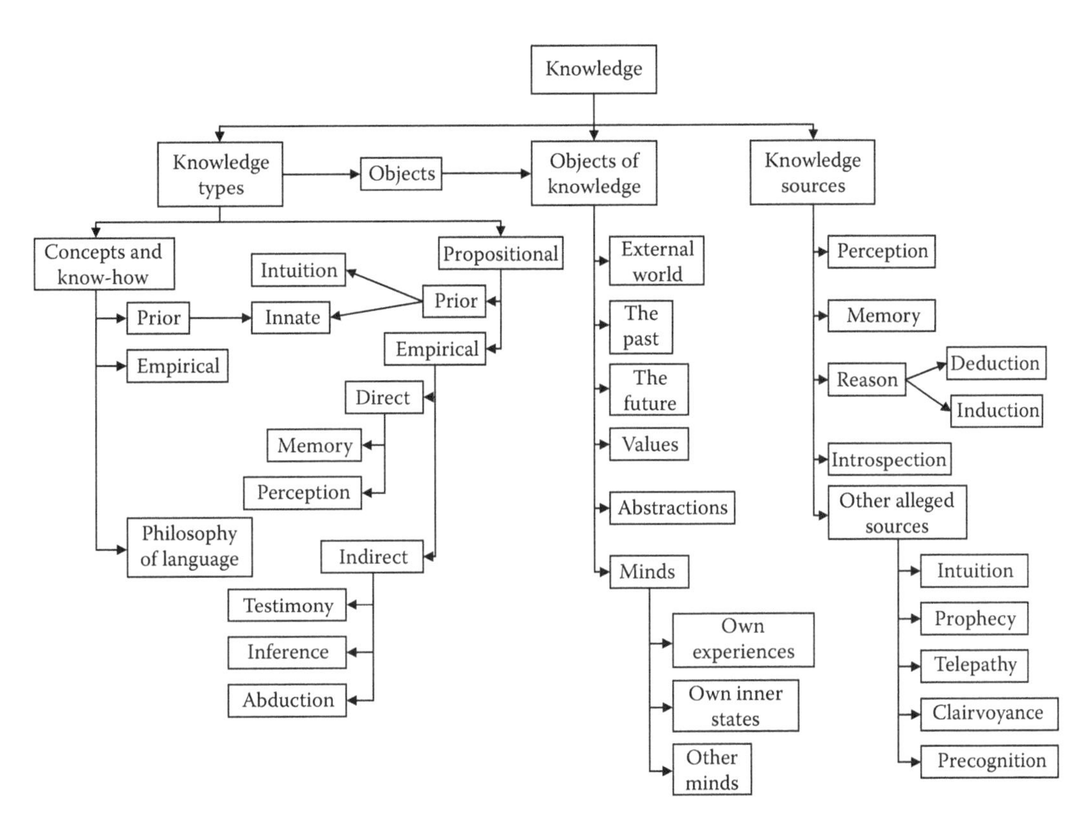

FIGURE 1.10
Knowledge types, sources, and objects.

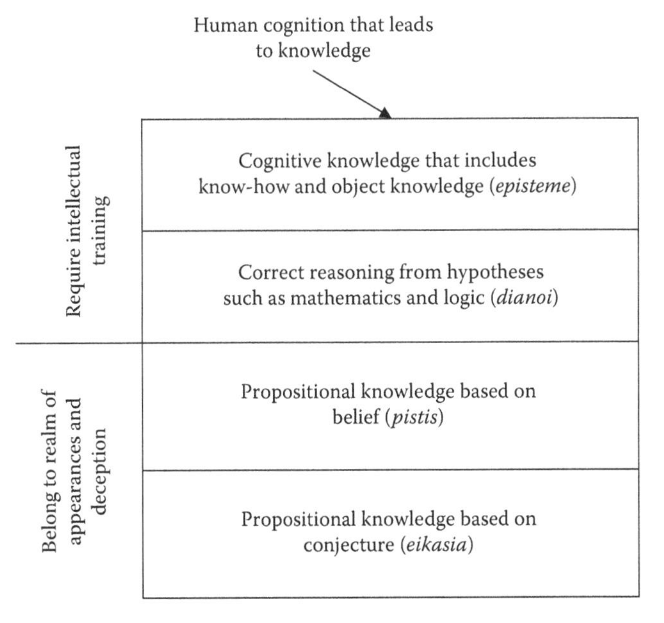

FIGURE 1.11
Knowledge categories and sources.

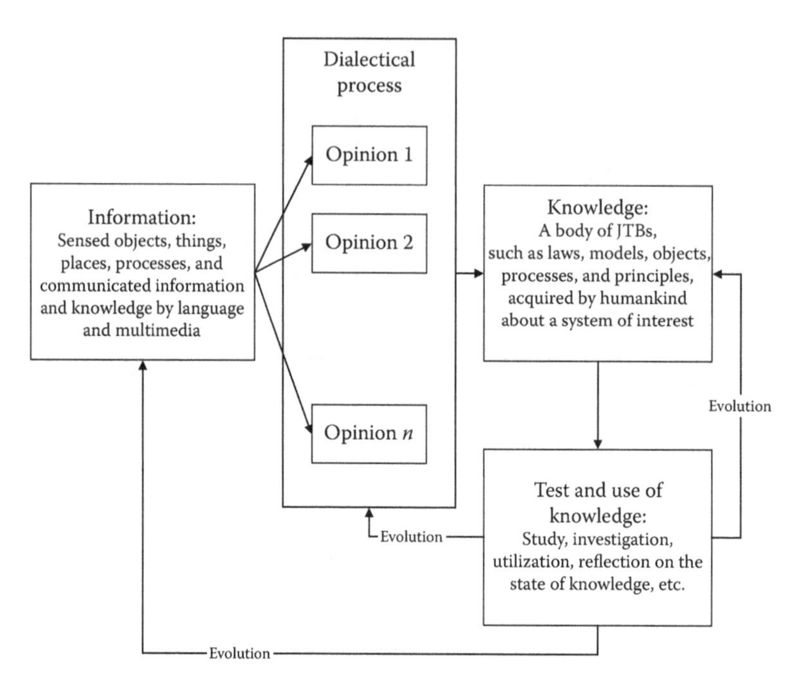

FIGURE 1.12
Knowledge, information, opinions, and evolutionary epistemology.

as a preprocessed input to our intellect system of cognition, and knowledge acquisition and creation. Information can lead to knowledge through investigation, study, and reflection. However, knowledge and information about the system might not lead to discovery nor to an evolutionary, knowledge-state advancement about the system as a result of not meeting the justification condition in the JTB or the ongoing evolutionary process, or both. Knowledge is defined in the context of humankind, evolution, language, and communication methods, as well as social and economic dialectical processes; knowledge cannot be removed from them. As a result, knowledge always reflects the imperfect and evolutionary nature of humans, which can be attributed to their reliance on their senses for information acquisition; their dialectical processes; and their mind for extrapolation, creativity, reflection, and imagination, with the associated biases as a result of preconceived notions due to time asymmetry, specialization, and other factors. An important dimension in defining the state of knowledge and truth about a system is nonknowledge or knowledge deficiency or ignorance due to many factors and reasons including information deficiency, that is, uncertainty.

Opinions rendered by experts that are based on the information and existing knowledge can be defined as preliminary propositions with claims that are not fully justified or are justified with adequate reliability. Expert opinions are seeds of propositional knowledge that do not meet one or more conditions required for the JTB with the reliability theory of knowledge. They provide a valuable, and sometimes necessary, input to a decision situation, as they might lead to knowledge expansion; but decisions that are made based on them might be risky sometimes, and due to their preliminary nature, they might be proven false by others in the future.

The relationships among knowledge, information, opinions, and evolutionary epistemology are shown in Figure 1.12. The dialectical processes include communication methods such as languages, visual and audio means, and other forms. They also include economic class, schools of thought, and political and social dialectical processes within peers, groups, colonies, societies, and the world.

1.6 Cognition and Cognitive Science

Cognition can be defined as the mental processes of receiving and processing information for knowledge creation and behavioral actions. Cognitive science is the interdisciplinary study of mind and intelligence (Stillings 1995). Cognitive science deals with many disciplines including philosophy, psychology, artificial intelligence, neuroscience, linguistics, and anthropology. The intellectual origins of cognitive science started in the mid-1950s, when researchers in several fields began to develop theories on how the mind works based on complex representations and computational procedures.

The origin of cognitive science can be taken as the theories of knowledge and reality proposed by the ancient Greeks, when philosophers such as Plato and Aristotle tried to explain the nature of human knowledge. The study of the mind remained the province of philosophy until the nineteenth century, when experimental psychology was developed by Wilhelm Wundt and his students, who initiated laboratory methods for the systematic study of mental operations. A few decades later, experimental psychology was dominated by behaviorism, which virtually limited the examination of the mind as it relates to psychology. Behaviorists, such as J. B. Watson, argued that psychology should restrict itself to examining the relationship among observable stimuli and observable behavioral responses and should not deal with consciousness and mental representations. The intellectual landscape began to change dramatically in 1956, when George Miller summarized numerous studies showing that the capacity of human thinking is limited, with short-term memory, for example, being limited to around seven items. He proposed that memory limitations are compensated for by humans through their ability to recode information into chunks and through mental representations that require mental procedures for encoding and decoding the information. Although at this time primitive computers had been around for only a few years, pioneers such as John McCarthy, Marvin Minsky, Allen Newell, and Herbert Simon were founding the field of artificial intelligence. Moreover, Noam Chomsky rejected behaviorist assumptions about language as a learned habit and proposed instead to explain language comprehension in terms of mental grammars consisting of rules.

Cognitive science is based on a central hypothesis that thinking can best be understood in terms of representational structures in the mind and computational procedures that operate on those structures (Johnson-Laird 1988). The nature of the representations and computations that constitute thinking is not fully understood. The central hypothesis is general enough to encompass the current range of thinking in cognitive science, including connectionist theories, which model thinking using artificial neural networks. This hypothesis assumes that the mind has mental representations analogous to computer data structures and computational procedures similar to computational algorithms. The mind is considered to contain such mental representations as logical propositions, rules, concepts, images, and analogies. It uses mental procedures such as deduction, search, matching, rotating, and retrieval for interpretation, generation of knowledge, and decision making. The dominant mind/computer analogy in cognitive science has taken on a novel twist from the use of another analog, that is, of the brain. Cognitive science then works with a complex three-way analogy among the mind, the brain, and computers. Connectionists have proposed a brain-like structure that uses neurons and their connections as inspiration for data structures and neuron firing and spreading activation as inspirations for algorithms. No single computational model for the mind exists, as the various programming approaches suggest different ways in which the mind might work, ranging from serial processors such as the commonly used computers that perform one instruction at a time to parallel processors such as some recently developed computers that are capable of doing many operations at the same time.

Cognitive science claims that the human mind works by representation and computation using empirical conjecture. Although the computational–representational approach to cognitive science has been successful in explaining many aspects of human problem solving, learning, and language use, some philosophical critics argue that it is fundamentally flawed based on the following limitations (Thagard 1996; von Eckardt 1993):

- Emotions: Cognitive science neglects the important role of emotions in human thinking.
- Consciousness: Cognitive science ignores the importance of consciousness in human thinking.
- Physical environments: Cognitive science disregards the significant role of physical environments in human thinking.
- Social factors: Human thought is inherently social and has to deal with various dialectical processes in ways that cognitive science ignores.
- Dynamic nature: The mind is a dynamic system, not a computational system.
- Quantum nature: Researchers argue that human thinking cannot be computational in the standard sense, so the brain must operate differently, perhaps as a quantum computer that is yet to be defined.

These open issues need to be considered by philosophers and scientists in developing new cognitive theories and for a better understanding of how the human mind works to process information and construct knowledge.

1.7 Ignorance as Knowledge Deficiency

1.7.1 Evolutionary Nature of Knowledge

Knowledge deficiency, in its broadest term, is called *ignorance* for the purpose of constructing an analytical structure for its examination (Ayyub 2010; Ayyub and Klir 2006). Ignorance, therefore, can be defined as knowledge deficiency; is considered to encompass notions covered by many other terms relating to knowledge deficiency, such as uncertainty, randomness, contradiction, inconsistency, and incompleteness; and is inclusive of all deficiency forms, types, and extents.

Humans tend to focus on what is known and not on the unknowns. The English language, similar to many other languages, evolved creating this emphasis. For example, we can easily state that "John *informed* Mark," whereas we cannot directly state the contrary. We can only state it by using the negation of the earlier statement: "John *did not inform* Mark." Statements such as "John *misinformed* Mark" or "John *ignored* Mark" do not convey the same (intended) meaning. Another example is "I *know* Professor Ayyub," for which a meaningful direct contrary statement does not exist in many languages as illustrated in Figure 1.13 based on a survey conducted by the author, Ayyub, in 2010 at a class of 18 mostly doctoral students

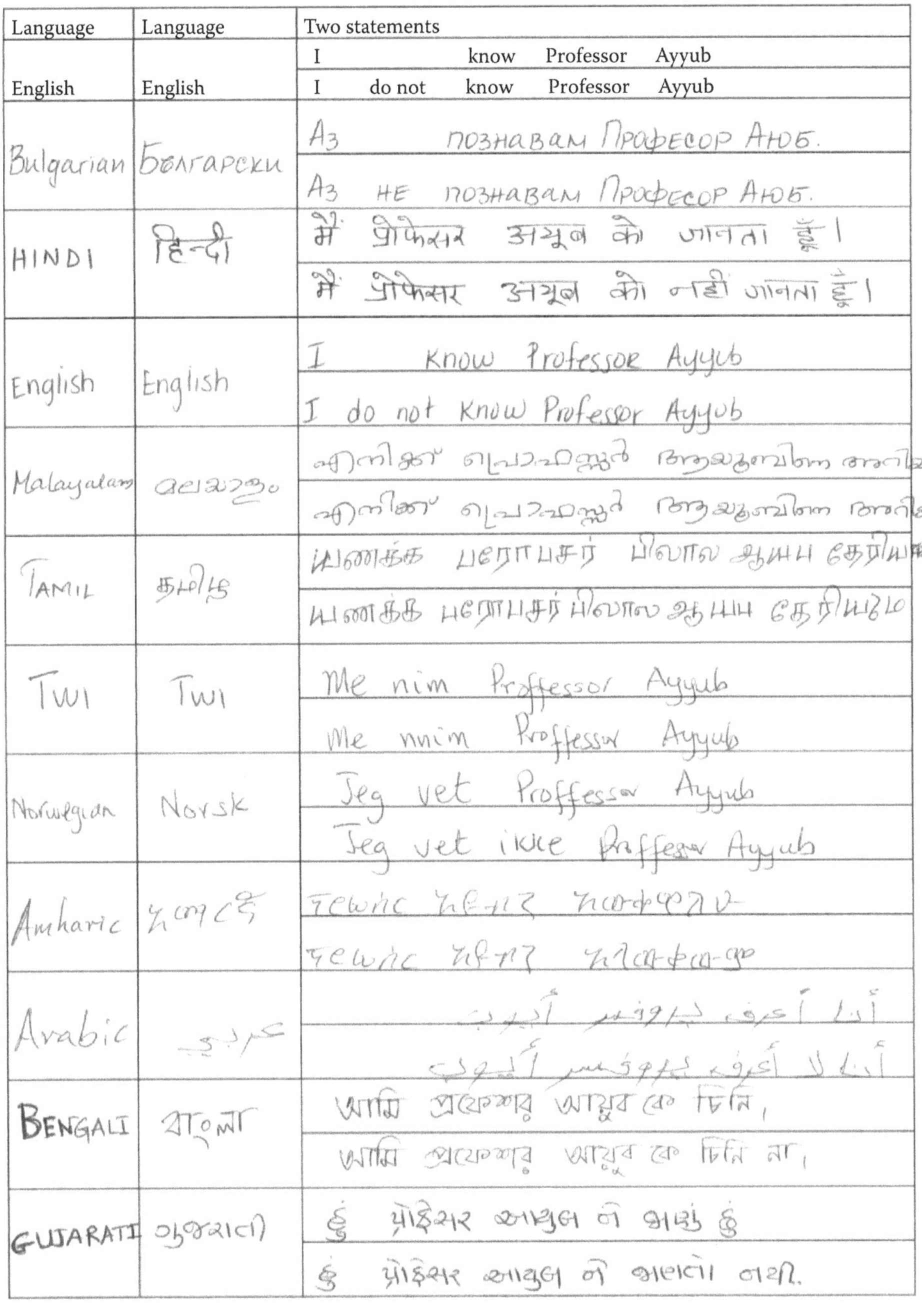

Language	Language	Two statements
		I know Professor Ayyub
English	English	I do not know Professor Ayyub
Bulgarian	Български	Аз познавам Професор Аюб.
		Аз не познавам Професор Аюб.
HINDI	हिन्दी	मैं प्रोफेसर अयूब को जानता हूँ।
		मैं प्रोफेसर अयूब को नहीं जानता हूँ।
English	English	I know Professor Ayyub
		I do not know Professor Ayyub
Malayalam	മലയാളം	എനിക്ക് പ്രൊഫസറ് അയ്യൂബിനെ അറിയാം
		എനിക്ക് പ്രൊഫസറ് അയ്യൂബിനെ അറിയില്ല
TAMIL	தமிழ்	எனக்கு பேராசிரியர் [illegible] ஆயுப் தெரியும்
		எனக்கு பேராசிரியர் [illegible] ஆயுப் தெரியாது
Twi	Twi	Me nim Proffessor Ayyub
		Me nnim Proffessor Ayyub
Norwegian	Norsk	Jeg vet Proffessor Ayyub
		Jeg vet ikke Proffessor Ayyub
Amharic	አማርኛ	ፕሮፌሰር አዩብን አውቀዋለሁ
		ፕሮፌሰር አዩብን አላውቀውም
Arabic	عربي	أنا أعرف بروفسر أيوب
		أنا لا أعرف بروفسر أيوب
BENGALI	বাংলা	আমি প্রফেসর আয়ুব কে চিনি।
		আমি প্রফেসর আয়ুব কে চিনি না।
GUJARATI	ગુજરાતી	હું પ્રોફેસર આયુબ ને ઓળખું છું
		હું પ્રોફેસર આયુબ ને ઓળખતો નથી.

FIGURE 1.13
Emphasis of languages on the known things, and not the unknowns.

<table>
<tr><td rowspan="2">Chinese</td><td rowspan="2">中文</td><td>我 认识 Ayyub</td></tr>
<tr><td>我 不 认识 Ayyub</td></tr>
<tr><td rowspan="2">Tigringa</td><td rowspan="2">ትግርኛ</td><td>[illegible]</td></tr>
<tr><td>[illegible]</td></tr>
<tr><td rowspan="2">GHANAIN LANGUAGE</td><td rowspan="2">AKAN</td><td>ME NIM PROFESSOR AYYUB</td></tr>
<tr><td>ME INIM PROFESSOR AYTUB</td></tr>
<tr><td rowspan="2">Ancient Chinese</td><td rowspan="2">古文</td><td>吾 识、阿育甫 教授</td></tr>
<tr><td>吾 阿育甫 教授 不识</td></tr>
<tr><td rowspan="2">German</td><td rowspan="2">Deutsh</td><td>Ich weiss Professor Ayyub</td></tr>
<tr><td>Ich weiss Professor Ayyub nicht.</td></tr>
<tr><td rowspan="2">Indonesian</td><td rowspan="2">Indonesian</td><td>Saya tahu Profesor Ayyub</td></tr>
<tr><td>Saya tidak tahu Profesor Ayyub</td></tr>
<tr><td rowspan="2">Korean</td><td rowspan="2">한국어</td><td>나는 있다 알고 교수 아유브</td></tr>
<tr><td>나는 못한다 알지 교수 아유브</td></tr>
</table>

FIGURE 1.13
(Continued) Emphasis of languages on the known things, and not the unknowns.

taking a graduate-level course on uncertainty modeling and analysis. From this class, students voluntarily completed a form that was presented to them to translate the two typed statements from English to their respective native languages and indicate if a direct negation can be stated. None of these languages have a verb for the direct negation of "know." The languages covered by the class in the order listed in Figure 1.13 are Bulgarian, Hindi, English, Malayalam, Tamil, Twi, Norwegian, Amharic, Arabic, Bengali, Gujarati, Chinese, Tigrigna, Ghanaian, Ancient Chinese, German, Indonesian, and Korean.

The emphasis on knowledge and not on ignorance can also be noted in sociology, which has a field of study called the *sociology of knowledge* but not the *sociology of ignorance*, although Weinstein and Weinstein (1978) introduced the *sociology of nonknowledge*, and Smithson (1985) introduced the *theory of ignorance*.

Engineers and scientists tend to emphasize knowledge and information, and sometimes intentionally or unintentionally brush aside ignorance. In addition, information (or knowledge) can be misleading in some situations because it does not have the truth content that was assigned to it leading potentially to overconfidence. In general, knowledge and ignorance can be classified as shown in Figure 1.14 using squares with crisp boundaries for the purpose of illustration. The shapes and boundaries can be made multidimensional, irregular, and/or fuzzy. The *evolutionary infallible knowledge* (EIK) about a system is shown as the top-right square in the figure and can be intrinsically unattainable due to the fallacy of humans and the evolutionary nature of knowledge. The state of RK is shown using another square (the bottom-left square) for illustration purposes. RK represents the present state of

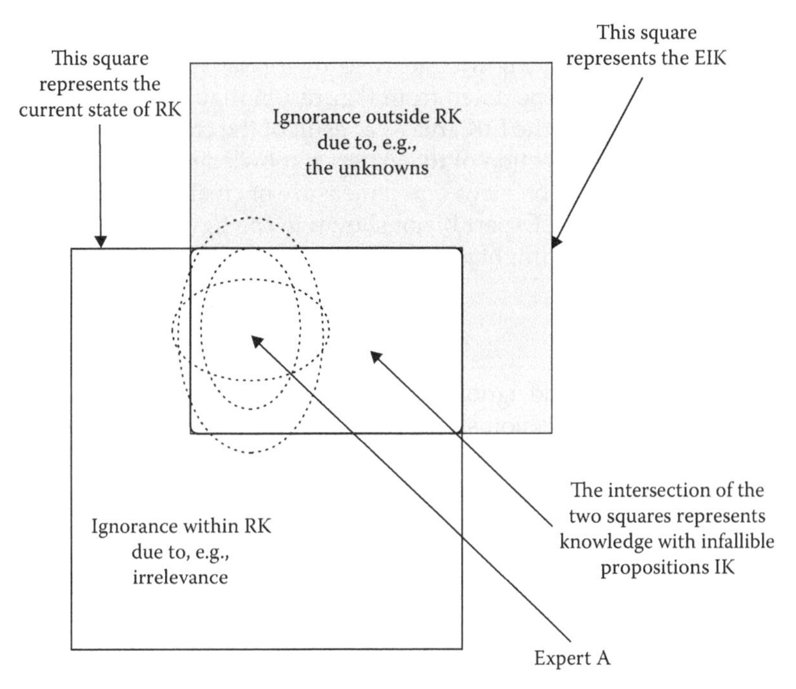

FIGURE 1.14
Knowledge and ignorance.

knowledge in an evolutionary process; that is, it is a snapshot of knowledge, a set of know-hows, objects, and propositions that meet justifiable true beliefs within reasonable reliability levels. At any stage of human knowledge development, this knowledge base about the system is a mixture of truth and fallacy. The intersection of EIK and RK represents a knowledge base with infallible knowledge (IK) components (i.e., know-hows, objects, and propositions). Therefore, the following relationship can be stated using the notations of set theory:

$$\text{IK} = \text{EIK} \cap \text{RK} \tag{1.2}$$

where:
$\cap$ indicates intersection

IK is defined as knowledge that can survive the dialectical processes of humans and societies and passes the test of time and use. This IK can be schematically defined by the intersection of the two squares representing EIK and RK. Based on this representation, two primary types of ignorance can be identified: (1) ignorance within the knowledge base RK due to factors such as irrelevance and (2) ignorance outside the knowledge base due to unknown objects, interactions, laws, dynamics, and know-hows.

Expert A, who has some knowledge about the system, can be represented elliptically as shown in Figure 1.14 for illustrative purposes. Three types of ellipticals can be identified: (1) a subset of the EIK that the expert has learned, captured, and/or created; (2) self-perceived knowledge by the expert; and (3) perception by others of the expert's knowledge. The EIK of the expert might be smaller than the self-perceived knowledge by the expert, and the difference between the two types is a measure of overconfidence that can be partially related to the expert's ego. Ideally, the three ellipticals should be the same, but commonly they are not. They are greatly affected by the communication skills of experts and their successes in dialectical processes that with time might lead to marginal advances or quantum leaps in evolutionary knowledge. Also, their relative sizes and positions within the IK base are unknown. It can be noted from Figure 1.14 that the expert's knowledge can extend beyond the RK base into the EIK area as a result of the creativity and imagination of the expert. Therefore, the intersection of the expert's knowledge with the ignorance space outside the knowledge base can be viewed as a measure of creativity and imagination. The ellipticals of another expert (say Expert B, not shown in the figure) might overlap with the ellipticals of Expert A, and they might overlap with other regions by varying magnitudes.

1.7.2 Classifying Ignorance

In the context of knowledge and ignorance, risks are based on the comprehension and abstractions by humans of a decision situation; hence, they are all based on their perceptions. Risks, if quantified, have varying levels of uncertainty that may be tied back to the state of knowledge and associated deficiencies not always with success except in cases of retrospective examinations. Ignorance contributes to the motivation for assessing and managing risk.

The state of *ignorance* for a person or society can be (1) of the unintentional type due to an erroneous cognition state and not knowing relevant information or (2) of the deliberate type by ignoring either information or deliberate inattention to something for various reasons such as limited resources and cultural opposition, respectively. The latter type is a state of *conscious ignorance*, which is intentional; and once recognized, evolutionary species try to correct for that state for survival reasons with varying levels of success. The former ignorance type belongs to the *blind ignorance* category; therefore, ignoring means that someone can either unconsciously or deliberately refuse to acknowledge or regard or leave out an account or consideration for relevant information (di Carlo 1998). These two states should be addressed when developing a hierarchical breakdown of ignorance.

Using the concepts and definitions from evolutionary knowledge and epistemology, ignorance can be classified based on the three knowledge sources as follows:

1. *Know-how ignorance* can be related to the lack of, or having erroneous, know-how knowledge. Know-how knowledge requires someone to know how to do a specific activity, function, procedure, and so on, such as riding a bicycle.
2. *Object ignorance* can be related to the lack of, or having erroneous, object knowledge. Object knowledge is based on a direct acquaintance with a person, place, or thing; for example, Mr. Smith knows the President of the United States.
3. *Propositional ignorance* can be related to the lack of, or having erroneous, propositional knowledge. Propositional knowledge is based on propositions that can be either true or false; for example, Mr. Smith knows that the Rockies are in North America.

The above three ignorance types can be cross-classified against two possible states for knowledge agents (such as a person) in regard to knowing their state of ignorance. These two states are as follows:

1. *Nonreflective* (or *blind*) *state*, where the person does not know of self-ignorance—a case of ignorance of ignorance, also called *metaignorance*
2. *Reflective state*, where the person knows and recognizes self-ignorance, also called *conscience ignorance*

Smithson (1985) termed the latter type of ignorance as conscious ignorance, and the blind ignorance was referred to as metaignorance. As a result, in some cases the person might formulate a proposition but still be ignorant of the existence of a proof or disproof (i.e., *ignoratio elenchi*). A knowledge agent's response to reflective ignorance can be either passive acceptance or a guided attempt to remedy one's ignorance, which can lead to four possible outcomes: (1) a successful remedy that is recognized by the knowledge agent as a success, leading to fulfillment; (2) a successful remedy that is not recognized by the knowledge agent as a success, leading to search for a new remedy; (3) a failed remedy that is recognized by the knowledge agent as a failure, leading to search for a new remedy; and (4) a failed remedy that is recognized by the knowledge agent as a success, leading to blind ignorance, such as *ignoratio elenchi* or drawing an irrelevant conclusion.

The cross-classification of ignorance is shown in Figure 1.15 in two possible forms that can be used interchangeably. Although the blind state does not feed directly into the evolutionary process for knowledge, it does represent a becoming knowledge reserve. The reflective state has survival value to evolutionary species; otherwise, it can be argued that it would never have flourished (Campbell 1974). Ignorance emerges as a lack of knowledge relative to a particular perspective from which such gaps emerge. Accordingly, the accumulation of beliefs and the emergence of ignorance constitute a dynamic process resulting in old ideas perishing and new ones flourishing (Bouissac 1992). According to Bouissac (1992), the process of scientific discovery can be metaphorically described not only as a cumulative sum (positivism) of beliefs but also as an activity geared toward relentless construction of ignorance (negativism), producing as architecture of holes, gaps, and lacunae, so to speak.

Hallden (1986) examined the concept of evolutionary ignorance in decision theoretic terms. He introduced the notion of gambling to deal with blind ignorance or lack of

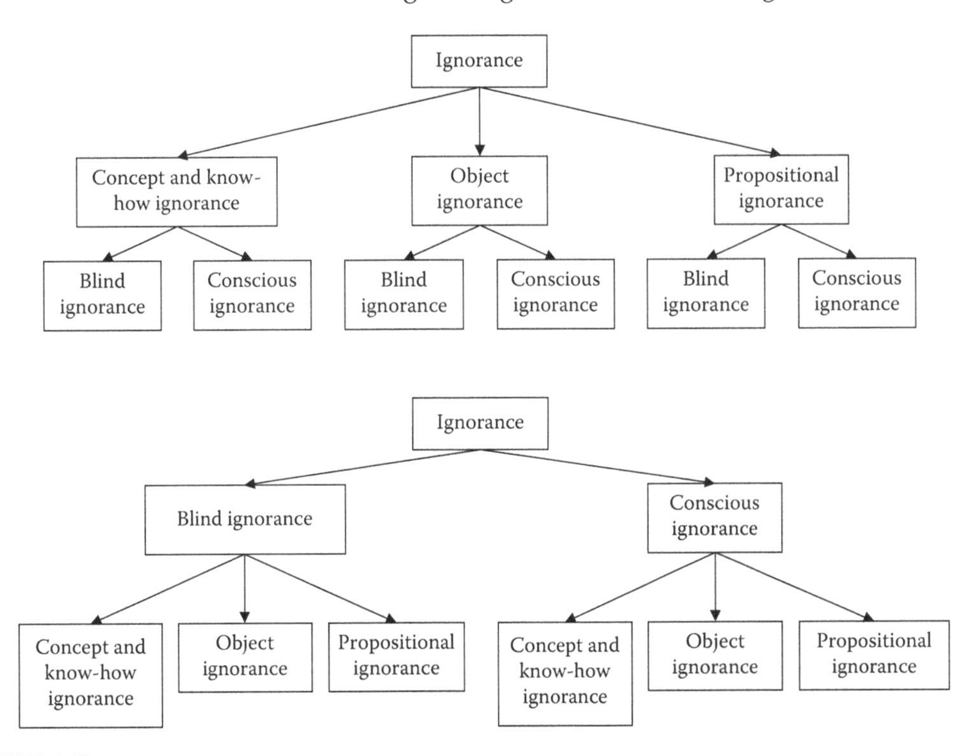

FIGURE 1.15
Classifying ignorance.

knowledge by proposing that there are times when, in lacking knowledge, gambles must be taken. Sometimes gambles pay off with success (i.e., continued survival) and sometimes they do not, leading to sickness or death.

According to evolutionary epistemology, ignorance is factitious; that is, it has human-made perspectives. Smithson (1988) provided a working definition of ignorance based on the following: "Expert A is ignorant from B's viewpoint if A fails to agree with or show awareness of ideas that B defines as actually or potentially valid." This definition allows for self-attributed ignorance, and either Expert A or Expert B can be the attributer or perpetrator of ignorance. Our ignorance and claimed knowledge depend on our current historical setting, which is relative to various natural and cultural factors such as language, logical systems, technologies, and standards that have developed and evolved over time. Therefore, humans evolved from blind ignorance through gambles to a state of incomplete knowledge with reflective ignorance recognized through factitious perspectives. In many scientific fields, the level of reflective ignorance becomes larger as the level of knowledge increases. Duncan and Weston-Smith (1997) stated in the *Encyclopedia of Ignorance* that "compared to our bond of knowledge, our ignorance remains Atlantic." They invited scientists to state what they would like to know in their respective fields and noted that the more eminent they were, the more readily and generously they described their ignorance. Clearly, before solving a problem, it must be articulated.

1.7.3 Ignorance Hierarchy

Figures 1.12 and 1.14 express knowledge and ignorance in evolutionary terms as they are socially or factually constructed and negotiated. Ignorance can be viewed as having a hierarchical classification based on its sources and nature, as shown in Figure 1.16; brief definitions are provided in Table 1.5. Figure 1.16 is intended to be exhaustive of all possibilities within the knowledge and cognitive limitations of the author. Ignorance can be classified into two types: (1) blind ignorance (or metaignorance) and conscious ignorance (or reflective ignorance).

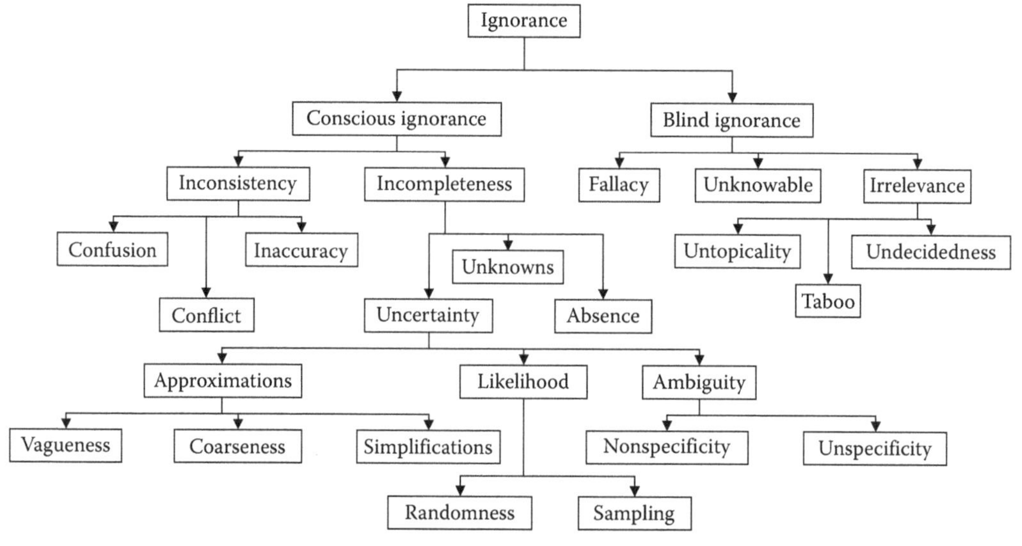

FIGURE 1.16
Ignorance hierarchy.

TABLE 1.5

Taxonomy of Ignorance

Term	Meaning
Blind ignorance	Ignorance of self-ignorance or metaignorance
Unknowable	Knowledge that cannot be attained by humans based on current evolutionary progressions, or cannot be attained at all due to human limitations, or can only be attained through quantum leaps by humans
Irrelevance	Ignoring something
Untopicality	Intuitions of experts that could not be negotiated with others in terms of cognitive relevance
Taboo	Socially reinforced irrelevance; issues that people must not know, deal with, inquire about, or investigate
Undecidedness	Issues that cannot be designated true or false because they are considered unsolvable, or solutions that are not verifiable, or *ignoratio elenchi*
Fallacy	Erroneous belief due to misleading notions
Conscious ignorance	A recognized self-ignorance through reflection
Inconsistency	Inconsistency in knowledge that can be attributed to distorted information and propositions as a result of inaccuracy, conflict, contradiction, and/or confusion
Confusion	Wrongful substitutions
Conflict	Conflicting or contradictory assignments or substitutions
Inaccuracy	Bias and distortion in degree
Incompleteness	Lacking or nonwhole knowledge in its extent due to absence or uncertainty
Absence	Incompleteness in kind
Unknowns	The difference between the *becoming* knowledge state and *current* knowledge state
Uncertainty	Inherent deficiencies in information used in acquiring knowledge
Ambiguity	The possibility of having multiple outcomes for processes or systems
Unspecificity	Outcomes or assignments that are incompletely defined
Nonspecificity	Outcomes or assignments that are improperly or incorrectly defined
Approximations	A process that involves the use of vague semantics in language, approximate reasoning, and dealing with complexity by emphasizing relevance
Vagueness	Noncrispness of belonging and nonbelonging of elements to a set or a notion of interest
Coarseness	Approximating a crisp set by subsets of an underlying partition of the set's universe that would bind the set of interest
Simplifications	Assumptions needed to make problems and solutions tractable
Likelihood	Defined by its components of randomness and sampling
Randomness	Nonpredictability of outcomes
Sampling	Samples versus populations

Blind ignorance includes not knowing relevant know-how, objects-related information, and relevant propositions that can be justified. The unknowable knowledge can be defined as knowledge that cannot be attained by humans based on current evolutionary progressions, or cannot be attained at all due to human limitations or can only be attained through quantum leaps by humans. Blind ignorance also includes irrelevant knowledge that can be of two types: (1) relevant knowledge that is dismissed as irrelevant or ignored and (2) irrelevant knowledge that is believed to be relevant through unreliable or weak justification or as a result of *ignoratio elenchi*. The irrelevance type can be due to a lack of topicality, taboo, or undecidedness. A lack of topicality can be attributed to intuitions of experts that could not be negotiated with others in terms of cognitive relevance. Taboo is due to socially reinforced irrelevance; issues that people must not know, deal with, inquire about, or investigate define the domain of taboo. The undecidedness type deals with issues that cannot be designated either true or false because they are considered unsolvable, solutions that are not verifiable, or a result of *ignoratio elenchi*. A third component of blind ignorance is fallacy, which can be defined as erroneous beliefs due to misleading notions.

To illustrate the blind ignorance type, consider the 9/11 terrorist attacks that resulted in nearly 3000 deaths in Lower Manhattan, on a field in Pennsylvania, and at the Pentagon along the banks of the Potomac River, and historically recorded as the single largest loss of life from an attack on the US soil. The 9/11 National Commission on Terrorist Attacks on the United States (2002), as an independent, bipartisan panel, examined the facts and circumstances and provided recommendations to safeguard against future acts. In its report, a chapter was devoted to "foresight and hindsight" with a key statement that captured the essence of root causes as "We believe the 9/11 attacks revealed four kinds of failures: in imagination, policy, capabilities, and management." The failure in imagination component speaks to the blind ignorance. Another example can be made based on the statement made by Roberta Wohlstetter (1962), a historian of military intelligence, in her most influential work *Pearl Harbor: Warning and Decision*: it is "much easier after the event to sort the relevant from the irrelevant signals. After the event, of course, a signal is always crystal clear; we can now see what disaster it was signaling since the disaster has occurred. But before the event, it is obscure and pregnant with conflicting meanings." This second example speaks to the irrelevance as a component of blind ignorance.

Kurt Gödel (1906–1978) proved the incompleteness of axioms for arithmetic, as well as the relative consistency of the axiom of choice and continuum hypothesis with the other axioms of set theory (Hofstadter 1999; Nagel and Newman 2001). According to Gödel, mathematicians hoped that their axioms could be proven consistent, that is, free from contradictions, and complete, that is, strong enough to provide proofs of all true statements. Gödel, however, showed that these hopes were overly naive by proving that any consistent formal system strong enough to axiomatize arithmetic must be incomplete; that is, there are statements that are true but not provable. Also, one cannot hope to prove the consistency of such a system using the axioms themselves (Hofstadter 1999; Nagel and Newman 2001). Moreover, many systems defined in engineering and the sciences are not based on a closed universal space defined by sets and accompanying axioms, but potentially an open universal space defined by sets, including vague sets, and axioms as constructed and deemed appropriate by analysts.

Within the context of the collective propositional knowledge of humans (noting that all knowledge is attributable to humans), we could state that humans cannot be both consistent and complete, and could not prove completeness without proving inconsistency and vice versa. This view can be used as a basis for classifying the conscious ignorance into inconsistency and incompleteness. This classification is also consistent with the concept of quantum knowledge discussed earlier.

Inconsistency in knowledge can be attributed to distorted information and propositions as a result of inaccuracy, conflict, contradiction, and/or confusion as shown in Figure 1.16. Inconsistency can result from assignments and substitutions that are wrong, conflicting, or biased causing confusion, conflict, or inaccuracy, respectively. The confusion and conflict result from in-kind inconsistent assignments and substitutions; whereas inaccuracy results from a level bias or error in these assignments and substitutions.

Incompleteness is defined as lacking or nonwhole knowledge in its extent. Knowledge incompleteness consists of (1) absence and unknowns as incompleteness in kind and

(2) uncertainty. The unknowns or unknown knowledge can be viewed in evolutionary epistemology as the difference between the becoming knowledge state and the current knowledge state. The knowledge absence component can lead to one of the following scenarios: (1) no action and working without the knowledge, (2) unintentionally acquiring irrelevant knowledge leading to blind ignorance, or (3) acquiring relevant knowledge that can be with various uncertainties and levels. The fourth possible scenario of deliberately acquiring irrelevant knowledge is not listed since it is not realistic.

Uncertainty can be defined as incompleteness due to inherent deficiencies in information used to acquire knowledge. Klir (2006) formally defines uncertainty as information deficiency including deficiency types of incompleteness, imprecision, fragmentation, unreliability, vagueness, or contradiction. Uncertainty can be classified into three types based on its sources: (1) ambiguity, (2) approximations, and (3) likelihood. The ambiguity comes from the possibility of having multiple outcomes for processes or systems. Recognizing only some of the possible outcomes creates uncertainty. The recognized outcomes might constitute only a partial list of all possible outcomes leading to unspecificity. In this context, unspecificity results from outcomes or assignments that are incompletely defined. The improper or incorrect definition of outcomes, that is, error in defining outcomes, can be called nonspecificity. In this context, nonspecificity results from outcomes or assignments that are improperly defined. The unspecificity is a form of knowledge absence and can be treated similar to the absence category under incompleteness. The nonspecificity can be viewed as a state of blind ignorance.

The human mind has the ability to perform approximations through reduction and generalizations, that is, induction and deduction, respectively, in developing knowledge. The process of approximation can involve the use of vague semantics in language, approximate reasoning, and dealing with complexity by emphasizing relevance. Approximations can be viewed to include vagueness, coarseness, and simplification. Vagueness results from the imprecise nature of belonging and nonbelonging of elements to a set or a notion of interest, whereas coarseness results from approximating a set by subsets of an underlying partition of the set's universe that would bind the crisp set of interest. Simplifications are assumptions introduced to make problems and solutions tractable.

Likelihood can be defined in the context of chance, odds, and gambling. Likelihood has primary components of randomness and sampling. Randomness stems from the nonpredictability of outcomes. Engineers and scientists commonly use samples to characterize populations—hence, the last type.

Example 1.18: Examples corresponding to Ignorance Types

Tables 1.6 and 1.7 build on the ignorance hierarchy of Figure 1.16 and the definitions provided in Table 1.5 give examples from engineering and the sciences. The challenge in coming up with illustrative examples is that each ignorance type does not occur in isolation from other types. Generally, these examples intrinsically include multiple types at the same time, and someone could advance a credible argument to place an example under a different ignorance type. The intent herein is to place each example under a particular type that dominates an example cited.

Example 1.19: Surprise as an Indicator of Ignorance

Expectation is imagination constrained by bounded uncertainty (Shackle 1970). Rational decisions are based on expectations, and expectations are largely influenced by the bounds a decision maker's imagination places on possible outcomes. When these

TABLE 1.6

Blind Ignorance: Examples

Term	Example
Blind ignorance	Approving a design of a system based on design standards without considering those failure modes that may not be included in the development of the standards
Fallacy	The erroneous belief, pre-Galilean concepts, that understood the Earth to be the center of the universe The steel columnar support structure of the World Trade Center towers can withstand a progressive failure induced by a plane striking the towers. The belief in the early twentieth century that there exists ubiquitous ether as a transmitting medium for light
Unknowable	The behavior of an n-dimensional creature and its way of life that cannot be ascertained by humans The use of atomic differentiation techniques for growing molecular electronic circuits at the nanoscale
Irrelevance	Some scientists once ignored the behavior and events of planets far distant from the Earth, citing these events as irrelevant to the conditions in our solar system; now, these effects are recognized.
Untopicality	Since the existence of fairy magic is deemed highly unlikely, researchers will ignore its existence when trying to explain natural phenomena.
Undecidedness	An approach to solving a problem may not follow the logic of current problem-solving methodologies leading to undecidedness on the appropriateness of the logic used. The quandary whether life exists on other planets in the universe
Taboo	The use of cloning techniques for humans to produce identical human beings using bioengineering Fetal research for finding genetic reasons for diseases occurring in humans

TABLE 1.7

Conscious Ignorance: Examples

Term	Example
Conscious ignorance	DNA provides the genetic makeup of living things, yet we are unable to completely decode its meaning.
Inconsistency	Analysis using a linear model, yet true behavior is nonlinear.
Confusion	In their attempt to develop explanations of previously unexplained phenomena, fresh graduate students often lack the understanding to clearly identify and define the problem. Whether light has a particle nature or it behaves as a wave was a matter of confusion until a theory about the dual nature of light was accepted.
Inaccuracy	Design equations often lend much insight into a physical problem, yet most often their results, though may be on the same order of magnitude, are inaccurate.
Conflict	Light is either particle or energy, but can never be both. Conflicting theories regarding how high temperatures affect the solder joint failure characteristics in electronic products that are subjected to vibrations The heliocentric theory forwarded by Copernicus was in direct conflict with the geocentric theory of solar system commonly accepted at that time.
Incompleteness	The theory of everything, also called the unified field theory rather, attempts to link all forces together into a single theory—though much progress has been made in this field (so we believe), our ideas are still incomplete.

(*Continued*)

TABLE 1.7

(Continued) Conscious Ignorance: Examples

Term	Example
Unknowns	The idea in 1980s that computers throughout the world can be connected for information flow. The notion in the early twentieth century that atom is indivisible was disproved when it was found to consist of a positive core surrounded by negatively charged particles.
Absence	Defining bulk characteristics of composite materials for high stress applications The maps of the world were without the American continents before Europe discovered it in 1492.
Approximations	Most techniques in structural analysis are approximate in nature since they are known not to be able to characterize all aspects of structural behavior.
Vagueness	Meaning of the phrase "premium paint quality" Characterizing a person as a highly experienced executive
Coarseness	In structural analysis, crude calculations are often made to validate highly complex numerical models—if the order of magnitude is achieved, the model is assumed valid. In thermodynamics, macroscopic properties of ensembles are used to predict behavior instead of microscopic analysis at the molecular level.
Simplifications	Taking a nonlinear problem and simplifying it enough to justify using linear analysis
Likelihood	If a probability of an event occurring is <1, the event is inherently uncertain. For example, scientists quote a probability that a given asteroid will hit the Earth. Unless such a probability is 0 or 1, it is uncertain whether or not the event will occur. Since such an event has been unlikely, scientists often ignore these asteroids if their apparent path does not tend toward the Earth.
Randomness	Wave heights in open seas Wind speed or direction
Sampling	Assessing the structural strength of concrete by testing three specimens In pharmaceutical development, a guinea pig or rat is randomly selected for evaluating the effect of a drug on it, and the effect is taken as a representative for the whole population.
Ambiguity	In an experiment testing the fidelity of an electrical connection, specifying failure as an open circuit instead of quantified increase in resistance typically precedes it. Defining failure of a component as a single event, whereas it can fail in two different modes Assignment of all hereditarily transmitted traits to the genes that pass them over

bounds on expectation are prescribed incorrectly, such as through the illusion of knowledge manifesting from overconfidence, blind sightedness, or faulty reasoning, a failure of imagination could result, which in some contexts could prove harmful to the decision situation. For any decision problem, it is thus important to clearly articulate what is known, what is thought to be known, and what is not known, and acknowledge the possibility of unknown unknowns. The following two quotes are relevant here, noting that the illusion of knowledge falls within the realm of blind ignorance:

> The greatest enemy of knowledge is not *ignorance*, it is the Illusion of knowledge.
>
> **Stephen Hawking**
> *Theoretical physicist and cosmologist*

> The most difficult subjects can be explained to the most slow-witted man if has not formed any idea of them already; but the simplest thing cannot be made clear to the most intelligent man if he is firmly persuaded that he knows already, without a shadow of doubt, what is laid before him.
>
> **Leo Tolstoy**
> *Novelist*

McGill and Ayyub (2009) discuss the menacing problem of surprise in the context of critical infrastructure protection and propose several strategies for defeating the potential for and effects of surprise. It is suggested that adversaries achieve surprise by exploiting ignorance and lack of preparedness. They advocate two complementary approaches for defeating the surprise of (1) threat anticipation aimed at increasing awareness of plausible threat scenarios and (2) possibility management aimed at improving preparedness to deal with loss irrespective of its cause.

Surprise manifests itself in the unknown, unrecognized, and unrealized, and is a direct by-product of a failure of imagination. In the context of homeland security and warfare, surprise occurs when a defender is either unaware of potential hazards or unprepared to defend or respond to unexpected consequences from known, but ignored hazards. Adversaries seek to leverage a defender's ignorance about their intent, capabilities, and operations to achieve an asymmetric advantage over their targets. For instance, the use of airplanes to attack the World Trade Center and Pentagon on September 9, 2001 was arguably a surprise given that defenders were unaware that such vehicles would be used as ballistic missiles to attack buildings. For natural hazards, Woo (1999) notes that "there are many arcane geological hazard phenomena which are beyond the testimony of the living, which would be met with incredulity and awe were they to recur in our own time." Such "black swans" are highly consequential scenarios that are either unknown or have a perceived probability so low as to be considered negligible, yet would result in significant surprise were they to occur (Taleb 2004). In highly complex technical systems, surprise occurs due to unexpected emergent behaviors stemming from the interaction between system components and their environment. Critical infrastructure is among such highly complex technical systems, where unknown interdependencies between infrastructure services may lead to unpredictable and potentially cascading consequences (Rinaldi et al. 2001).

1.7.4 Mathematical Models for Ignorance Types

Any identified ignorance types according to Figure 1.16 and Table 1.5 would require the use of a mix of mathematical theories appropriately selected to effectively model this ignorance content. Table 1.8 shows a matrix of applications and mathematical theories and methodologies for illustrative purposes as given by Ayyub and Klir (2006). For example, classical sets theory can effectively deal with ambiguity by modeling nonspecificity, whereas fuzzy and rough sets can be used to model vagueness, coarseness, and simplifications. The theories of probability and statistics are regularly used to model randomness and sampling uncertainty applied to quality control and reliability analysis. Bayesian methods can be used to combine randomness or sampling uncertainty with subjective information that can be viewed as a form of simplification, and their results can be applied to reliability analysis. Ambiguity, as an ignorance type, forms a basis for randomness and sampling, as shown in the table, in conjunction with classical sets, probability, statistics, Bayesian, evidence, and interval analysis methods. Inaccuracy, as an ignorance type, can be present in many problems, such as forecasting, risk analysis, and validation. The theories of evidence, possibility, and monotone measures can be used to model confusion and conflict in diagnostics, and vagueness in control. Interval probabilities and interval analysis can be used to model inaccuracy in risk analysis and validation, and vagueness and simplification in risk analysis.

TABLE 1.8

Theories and Example Applications to Model and Analyze Ignorance Types

	Ignorance Type						
Selected Theories and Methodologies	**Confusion and Conflict**	**Inaccuracy**	**Ambiguity**	**Randomness and Sampling**	**Vagueness**	**Coarseness**	**Simplification**
Classical sets			Modeling				
Probability		Forecasting	Modeling	Quality control			Modeling
Statistics			Analysis	Sampling			
Bayesian			Modeling	Reliability analysis			Modeling
Fuzzy sets					Control	Modeling	Modeling
Rough sets						Classification	Modeling
Evidence	Diagnostics		Modeling				
Possibility	Target tracking	Forecasting			Control		
Monotone measure							
Interval probabilities	Risk analysis	Risk analysis	Modeling		Risk analysis		Risk analysis
Interval analysis	Risk analysis	Validation	Analysis				Risk analysis

1.8 Aleatory and Epistemic Uncertainties

Traditional uncertainty analysis and modeling in engineering are commonly defined as knowledge incompleteness due to inherent deficiencies in acquired knowledge. It can also be used to characterize the state of a system as being unsettled or in doubt, such as the uncertainty of the outcome. Uncertainty as an added dimension in risk analysis accounts for deficiencies in the definitions or quantifications of the hazards, threats and threat scenarios, vulnerabilities, failure consequences, prediction models, underlying assumptions, effectiveness of countermeasures and consequence mitigation strategies, decision metrics, and appropriateness of the decision criteria. Traditionally, uncertainty in risk analysis processes is treated as being of two types: (1) inherent, called aleatory, uncertainty and (2) subjective, called epistemic, uncertainty as will be described in the sections that follow. These two uncertainty types are not in agreement with the ignorance classification of Figure 1.16; however, readers should be aware of their meaning and use in the literature.

1.8.1 Inherent Uncertainty

Some events and modeling variables are perceived to be inherently random and are treated to be nondeterministic in nature. The uncertainty in this case is attributed to the physical world because humans failed to reduce or eliminate it by enhancing the underlying knowledge base. This type of uncertainty is sometimes referred to as *aleatory uncertainty*. An example of this uncertainty type is strength properties of materials such as steel and concrete, and structural load characteristics such as wave loads on an offshore platform. For a probability or consequence parameter of interest, the aleatory uncertainty is commonly represented probabilistically by a random variable U_a.

1.8.2 Subjective Uncertainty

In many situations, uncertainty is also present as a result of a lack of or deficiency in knowledge, that is, epistemic in nature. In this case, the uncertainty magnitude could be reduced as a result of enhancing the state of knowledge by expending resources. Sometimes, this uncertainty cannot be reduced due to resource limitations, technological infeasibility, or sociopolitical constraints. This type of uncertainty, sometimes referred to as epistemic uncertainty, is the most dominant type in risk analysis. For example, the probability of an event can be computed based on many assumptions. A subjective estimate of this probability can be used in risk analysis; however, the uncertainty in this value should be recognized. With some additional modeling effort, this value can be treated as a random variable bounded using probability intervals or percentile ranges. By enhancing our knowledge base in this potential event, these ranges can be updated. For a probability or consequence parameter of interest, the epistemic uncertainty is commonly represented probabilistically by a random variable U_e.

Where uncertainty is recognizable and quantifiable, the framework of probability can be used to represent it. Objective or frequency-based probability measures can describe uncertainties of the aleatory type, and subjective probability measures can describe uncertainties of the epistemic type. Sometimes, however, uncertainty is recognized, but cannot be quantified in statistical terms. Examples include risks far into the future, such as those for radioactive waste repositories where risks are computed over design periods of 1,000 or 10,000 years, or risks aggregated across industries, sectors, population groups, or over the world, such as the cascading effects of a successful terrorist attack on a critical asset including consequent government changes and wars (National Research Council 1995).

The two primary uncertainty types of aleatory and epistemic for a parameter of interest can be combined as follows (Ang and Tang 1984, 2007):

$$U = U_a U_e \tag{1.3}$$

where:

U is a random variable representing both uncertainty types, that is, the combined uncertainty

U_a is a random variable to represent the aleatory uncertainty

U_e is a random variable to represent the epistemic uncertainty

For example, the following lognormal distribution (LN) can be used for this purpose:

$$U_e = \text{LN}(1.0, \text{COV}_e) \tag{1.4}$$

where:

COV_e is the coefficient of variation (COV) of U_e

LN is the lognormal probability distribution

In Equation 1.4, the random variable is assumed to be an unbiased estimate of the true value, that is, with a mean value of 1. The aleatory uncertainty can be represented in a similar manner using a COV_a. Using the first-order approximation, the total COV can be computed as follows:

$$\text{COV} = \sqrt{\text{COV}_a^2 + \text{COV}_e^2} \tag{1.5a}$$

The mean value (μ) can be computed using the first-order approximation based on Equation 1.3 as follows:

$$\mu = \mu_a \mu_e \tag{1.5b}$$

where:

μ is the mean value of the parameter that accounts for both uncertainty types

μ_a is the mean value of the parameter based on the aleatory uncertainty

μ_e is the mean value of the epistemic uncertainty, for example, 1 according to Equation 1.4

It is often important to treat the aleatory uncertainty separately from the epistemic uncertainty. For example, in light of the epistemic uncertainty, the pertinent result of the parameter will be a random variable. By combining the two uncertainty types as in Equations 1.5a and 1.15b, the expected value of the parameter is the best estimate that can be used as a basis for specifying values reflecting the risk attitude of a decision maker. In the case where the parameter of interest is the risk R, the decision maker may select or specify a risk-aversive value, such as the 90% value of the risk based on the total uncertainty computed. Such an approach is possible only if the two types of uncertainty are treated separately.

Example 1.20: Aleatory and Epistemic Uncertainty in Risk Analysis

A city examined flooding risk and determined that it could result in $100 million in property loss with a standard deviation of $50 million. The analysis was based on the aleatory uncertainty associated with storms, flooding, and damages. An expert panel convened by the city estimated that the model parameters entail epistemic uncertainties characterized in the form of a bias of 1.1 with a COV of 0.2. The city would like to account for the epistemic uncertainties in its risk estimates.

The uncertainty analysis in this case should examine both types as follows:

1. *Aleatory uncertainty.* The risk (R) can be modeled as a random variable with a mean value of $100 million and a standard deviation of $50 million. Therefore, the COV is 50/100 = 0.5. Assuming a lognormal probability distribution, the parameters, μ_Y and σ_Y, computed according to Equation A.66 of Appendix A are 4.493598 and 0.472381, respectively.
2. *Epistemic uncertainty.* The epistemic uncertainty can be represented as a factor (E) multiplied by the risk estimate (R) with a mean value of 1.1 and a COV of 0.2. Therefore, the standard deviation is 1.1(0.2) = 0.22. Assuming a lognormal probability distribution, the parameters, μ_Y and σ_Y, in this case computed according to Equation A.66 of Appendix A are 0.075699823 and 0.1980422, respectively.
3. *Combining the two uncertainty types.* Two solutions are offered herein: an approximate solution and an exact solution. Using the first-order approximations of Equations 1.15a and 1.15b, the risk with both uncertainty types combined has the following moments:

$$\text{Mean} = \$100 \text{ million } (1.1) = \$110 \text{ million} \tag{1.6a}$$

$$\text{COV} = \sqrt{0.5^2 + 0.2^2} = 0.5385 \tag{1.6b}$$

$$\text{Standard deviation} = 0.5385\ (\$110 \text{ million}) = \$59.24 \text{ million} \tag{1.6c}$$

The exact solution requires computing the lognormal parameters of the risk with both uncertainty types as follows:

$$\mu_Y = 4.494 + 0.0757 = 4.5697 \tag{1.7a}$$

$$\sigma_Y = \sqrt{0.4724^2 + 0.1980^2} = 0.5122 \tag{1.7b}$$

Then using Equation A.67 of Appendix A, the moments of the risk with both uncertainty types are as follows:

$$\text{Mean} = \$110 \text{ million} \tag{1.8a}$$

$$\text{Standard deviation} = \$60.25 \text{ million} \tag{1.8b}$$

$$\text{COV} = \frac{60.25}{110} = 0.5477 \tag{1.8c}$$

The approximate solution always produces the same mean as the exact solution; however, the standard deviations and COVs are not the same.

1.9 Knowledge and Ignorance in System Abstraction

Engineers use information for the purpose of system definition, analysis, and design. Information in this case are classified, sorted, analyzed, and used to predict system behavior and performances; however, classifying, sorting, and analyzing uncertainty in the information and using it to assess uncertainties in our predictions is far a more difficult task. Uncertainty in engineering was traditionally classified into objective and subjective types, that is, aleatory and epistemic uncertainties. This classification is deficient in completely capturing the nature of uncertainty and covering all its aspects. This difficulty stems from its complex nature and invasion of almost all epistemological levels of a system by varying degrees.

Analysis of an engineering system commonly starts with a definition of a system that can be viewed according to realism (or constructivism) as an abstract representation of an object of interest (or a construction from the experimental domain). The abstraction is performed at different epistemological levels (Ayyub 1992a, 1994; Ayyub and Klir 2006). A resulting model from this abstraction depends largely on the engineer (or analyst) who performed the abstraction, hence on the subjective nature of this process. During the process of abstraction, the engineer needs to make decisions regarding what aspects should or should not be included in the model. Aspects that are abstracted and not abstracted include the previously identified uncertainty types. In addition to the abstracted and nonabstracted aspects, unknown aspects of the system can exist due to blind ignorance, and they are more difficult to deal with because of their unknown nature, sources, extents, and impact on the system.

In engineering, uncertainty modeling and analysis is performed on the abstracted aspects of the system with some consideration of the nonabstracted aspects of a system. The division between abstracted and nonabstracted aspects can be a division of convenience that is driven by the objectives of the system modeling, or simplification of the model; however, the unknown aspects of the systems are due to blind ignorance that depends on the knowledge of the analyst and the state of knowledge about the system in general. The effects of the unknown aspects on the ability of the system model to predict the behavior of the object of interest can range from none to significant. These abstraction cases are described in subsequent sections.

1.9.1 Abstracted System Aspects

Engineers and researchers dealt with the ambiguity and likelihood types of uncertainty in predicting the behavior and designing engineering systems using the theories of probability and statistics, and Bayesian methods. Probability distributions were used to model system parameters that are uncertain. Probabilistic methods that include reliability methods, probabilistic engineering mechanics, stochastic finite element methods, reliability-based design formats, and other methods were developed and used for this purpose. In this treatment, however, a realization was established about the presence of the approximation type of uncertainty. Subjective probabilities were used to deal with this uncertainty type. Subjective probabilities are based on mathematics used for the frequency type of probability. Uniform and triangular probability distributions were used to model this type of uncertainty for some parameters. The Bayesian techniques were also used, for example, to deal with combining empirical and subjective information about these parameters. The underlying distributions and probabilities were, therefore, updated.

Regardless of the nature of uncertainty in the gained information, similar mathematical assumptions and tools that are based on probability theory were used. Approximations arise from human cognition and intelligence. They result in uncertainty in these abstractions. These abstractions are, therefore, subjective and can lack crispness, or they can be coarse in nature, or they might be based on simplifications. The lack of crispness, called vagueness, is distinct from ambiguity and likelihood in source and natural properties. The axioms of probability and statistics are limiting for the proper modeling and analysis of this uncertainty type and are not completely relevant, nor completely applicable. The vagueness type of uncertainty in engineering systems can be dealt with using appropriately fuzzy set theory. In engineering, this theory was proven to be a useful tool in solving problems involving vagueness.

To date, many applications of fuzzy set theory in engineering were developed, such as (1) strength assessment of existing structures and other structural engineering applications; (2) risk analysis and assessment in engineering; (3) analysis of construction failures, scheduling of construction activities, safety assessment of construction activities, decisions during construction, and tender evaluation; (4) the impact assessment of engineering projects on the quality of wildlife habitat; (5) planning of river basins; (6) control of engineering systems; (7) computer vision; and (8) optimization based on soft constraints (Ayyub 1991; Brown 1979, 1980; Brown and Yao 1983; Blockley 1975; Blockley et al. 1983; Furuta et al. 1985, 1986; Ishizuka et al. 1981, 1983; Itoh and Itagaki 1989; Kaneyoshi et al. 1990; Shiraishi and Furuta 1983; Shiraishi et al. 1985; Yao 1979, 1980; Yao and Furuta 1986). Coarseness in information can arise from approximating an unknown relationship or set by partitioning the universal space with associated belief levels for the partitioning subsets in representing the unknown relationship or set (Pawlak 1991). Such an approximation is based on *rough sets*. Pal and Skowron (1999) provide background and detailed information on rough set theory, its applications, and hybrid fuzzy–rough set modeling. Simplifying assumptions are common in developing engineering models. Errors resulting from these simplifications are commonly dealt with in engineering using bias random variables that are assessed empirically. A system can also be simplified by using knowledge-based if–then rules to represent its behavior based on fuzzy logic and approximate reasoning.

1.9.2 Nonabstracted System Aspects

In developing a model, an analyst needs to decide upon the aspects of the system that need to be abstracted at the different levels of modeling a system and the aspects that need not be abstracted. This division is for convenience and to simplify the model, and is subjective depending on the analysts, as a result of their background, and the general state of knowledge about the system. The abstracted aspects of a system and their uncertainty models can be developed to account for the nonabstracted aspects of the system to some extent. Generally, this accounting process is incomplete. Therefore, a source of uncertainty exists due to the nonabstracted aspects of the system. The ignorance categories and uncertainty types in this case are similar to the previous case of abstracted aspects of the system. The ignorance categories and uncertainty types due to the nonabstracted aspects of a system are more difficult to deal with than the uncertainty types due to the abstracted aspects of the system. The difficulty stems from a lack of knowledge or understanding of the effects of the nonabstracted aspects on the resulting model in terms of its ability to represent an object or its behavior. Poor judgment or human errors about the importance of the nonabstracted aspects of the system can partly contribute to these uncertainty types, in addition to contributing to the next category, uncertainty due to the unknown aspects of a system.

1.9.3 Unknown System Aspects

Some engineering failures have occurred because of failure modes that were not accounted for in the design stage of these systems. Failure modes were not accounted for due to various reasons that include (1) blind ignorance, negligence, using irrelevant information or knowledge, human errors, or organizational errors or (2) a general state of knowledge about a system that is incomplete. These unknown system aspects depend on the nature of the system under consideration, the knowledge of the analyst, and the state of knowledge about the system in general. Not accounting for these aspects in the models could result in varying levels of impact on the ability of these models in representing the behavior of the systems. The effects of the unknown aspects on these models can range from none to significant. In this case, the ignorance categories include wrong information and fallacy, irrelevant information, and unknowns. Engineers dealt with nonabstracted and unknown aspects of a system by assessing what is commonly called the modeling uncertainty, defined as the ratio of a predicted system's variables or parameter (based on the model) to the actual or measured value of the parameter. This empirical ratio, which is called the bias, is commonly treated as a random variable that can consist of objective and subjective components. Engineers use factors of safety that are intended to safeguard against uncertainty and associated potential failures. This approach of bias assessment is based on two implicit assumptions: (1) the value of the variable or parameter for the object of interest is known or can be accurately assessed from historical information or expert judgment and (2) the state of knowledge is complete and reliable. For some systems, the first assumption can be approximately examined through verification and validation, whereas the second assumption generally cannot be validated.

Example 1.21: Human Knowledge and Ignorance in Fire Escape Systems

Example 1.7 examines a fire escape system for an apartment building for risk analysis studies. The system definition can be extended to include the occupants of the building. The behavior of the occupants in the case of fire is uncertain. If the occupants of an apartment are not aware of the presence of smoke detectors or do not know the locations of the escape routes in the building, then catastrophic consequences might result due, in part, to their ignorance. The egress situation would also be serious if the occupants know the routes and are aware of the detectors, but the routes are blocked for various reasons. The results of the risk analysis in this case are greatly affected by assumptions made about the occupants. The group behavior of occupants under conditions of stress might be unpredictable and difficult to model. Some analysts might decide to simplify the situation through assumptions that are not realistic, thus leading to a fire escape system that might not work in case of a fire.

Example 1.22: Human Knowledge and Ignorance in Project Management Systems

In Example 1.8, risk analysis in project management, human knowledge, and ignorance can be the primary causes for delays in the completion of a project or budget overruns. Incompetent project managers or unqualified contractors can severely hamper a project and affect the investment of a client. Lack of knowledge or experience in managing a project, in regard to either technical or economical aspects, can cause delays and budget overruns. Sometimes engineers are assigned to manage a project who might concentrate only on the technical aspects of the project, without giving appropriate regard to the economical and managerial aspects of the project. Although the project might succeed in meeting its technical requirements, it might fail in meeting delivery and cost objectives. In this case, risk analysis requires constructing models that account for any lack of knowledge and properly represent uncertainties associated with the model structures and their inputs. These models should include in their assessments of risks the experience of personnel assigned to execute the project.

1.10 Exercise Problems

Problem 1.1 Provide engineering-related examples to demonstrate various risks defined in Figures 1.1 and 1.2 using a similar tabulation format. You may use multiple tables if needed.

Problem 1.2 Develop an example of linked systems for assessing associated risks similar to Example 1.12.

Problem 1.3 Define an engineering system and its breakdown. Provide an example of an engineering system with its breakdown.

Problem 1.4 Provide a definition of risk. What is risk assessment? What is risk management? Provide examples.

Problem 1.5 What are the differences between knowledge, information, and opinions?

Problem 1.6 What is ignorance?

Problem 1.7 Provide examples of primary knowledge types and sources.

Problem 1.8 Provide engineering examples of the various ignorance types using the hierarchy of Figure 1.16 in a table format.

Problem 1.9 Provide examples from the sciences of the various ignorance types using the hierarchy of Figure 1.16 in a table format.

Problem 1.10 What are the differences between an unknown and an unknowable? Provide examples.

Problem 1.11 Develop an ignorance identification and classification approach following the ignorance hierarchy of Figure 1.16 for a nuclear-powered, deep-space mission to search for an environment suitable for human life.

Problem 1.12 Develop an ignorance identification and classification approach following the ignorance hierarchy of Figure 1.16 to achieve sustainable human consumption of the Earth's resources.

Problem 1.13 Identify primary components from the ignorance hierarchy of Figure 1.16 necessary to quantify surprise according to Example 1.19. Use examples, as needed, to illustrate these components.

Problem 1.14 Provide examples of aleatory and epistemic uncertainty types.

Problem 1.15 A risk model results in an expected loss of $1000K and a standard deviation of $500K. The model accounts for only the aleatory uncertainty. An analyst believes that the parameters of the risk model are uncertain, and characterizes this uncertainty to be of the epistemic type. The analyst models the epistemic uncertainty as a multiplier to the risk estimate, and assigns a mean value of 1.05 and a COV of 0.25 to this multiplier. Estimate the risk that accounts for both the aleatory and epistemic uncertainties assuming normal probability distributions. Briefly discuss your results.

Problem 1.16 An unbiased prediction model accounts for only the aleatory uncertainty and has a COV of 0.1 in its prediction. An analyst believes that the parameters of this model are uncertain, and characterizes this uncertainty to be of the epistemic type as a multiplier with a mean value of 1.05 and a COV of 0.2. Estimate the total uncertainty in prediction by accounting for both the aleatory and epistemic uncertainties assuming normal probability distributions. Briefly discuss your results.

Problem 1.17 A biased prediction model accounts for only the aleatory uncertainty and has a bias ratio of 0.95, that is, underprediction, and a COV of 0.1 in its prediction. An analyst believes that the parameters of this model are uncertain, and characterizes this uncertainty to be of the epistemic type as a multiplier with a mean value of 1.05 and a COV of 0.2. Estimate the total uncertainty in prediction by accounting for both the aleatory and epistemic uncertainties assuming normal probability distributions. Briefly discuss your results.

Problem 1.18 A financial model predicts gain of an investment portfolio of $1000K and a standard deviation of $500K. The model accounts for only the aleatory uncertainty. An analyst believes that the parameters of the model are uncertain, and characterizes this uncertainty to be of the epistemic type. The analyst models the epistemic uncertainty as a multiplier to the gain estimate, and assigns a mean value of 1.1 and a COV of 0.15 to this multiplier. Estimate the gain that accounts for both the aleatory and epistemic uncertainties assuming normal probability distributions. Briefly discuss your results.

Problem 1.19 An unbiased prediction model accounts for only the aleatory uncertainty and has a COV of 0.25 in its prediction. An analyst believes that the parameters of this model are uncertain, and characterizes this uncertainty to be of the epistemic type as a multiplier with a mean value of 1.1 and a COV of 0.25. Estimate the total uncertainty in prediction by accounting for both the aleatory and epistemic uncertainties assuming normal probability distributions. Briefly discuss your results.

Problem 1.20 A biased prediction model accounts for only the aleatory uncertainty and has a bias ratio of 0.90, that is, underprediction and a COV of 0.2 in its prediction. An analyst believes that the parameters of this model are uncertain, and characterizes this uncertainty to be of the epistemic type as a multiplier with a mean value of 1.1 and a COV of 0.2. Estimate the total uncertainty in prediction by accounting for both the aleatory and epistemic uncertainties assuming normal probability distributions. Briefly discuss your results.

2

Terminology and Fundamental Risk Methods

This chapter has the objective of preparing readers for the rest of the book by developing working knowledge of risk terminology, analysis, quantification and management, and is followed by a series of chapters that cover the subject from system definition to probability estimation to consequence analysis and valuation, followed by the engineering economics and risk control, and end with data needs and sources. This chapter introduces the terminology and methods for performing risk analysis, management, and communication. It starts by defining risk and its dimensions, risk assessment processes, and fundamental analytical tools necessary for this purpose. The practical uses of these methods are illustrated to enhance the understanding of potential applications and limitations. The chapter is designed as a stand-alone, introductory primer to risk analysis and management.

CONTENTS

2.1 Introduction

Risk is associated with all projects and business ventures undertaken by individuals and organizations regardless of size, nature, or time and place of execution and utilization. Risk is present in various forms and levels from small domestic projects, such as adding a deck to a residential house, to large multibillion-dollar projects, such as developing and producing a space shuttle. These risks could result in significant budget overruns, delivery delays, failures, financial losses, environmental damages, and even injury and loss of life. Examples include budget overruns during construction of custom residential homes; budget overruns and delivery delays experienced in the development and implementation of a federally funded missile defense system; failures of space systems when launching military or satellite delivery rockets, such as the National Aeronautics and Space Administration (NASA) space shuttle challenger; and rollovers of sport utility vehicles (SUVs). In these examples, failures can lead to several consequence types simultaneously and could occur at any stage during the life cycle of a project induced by diverse hazards, errors, and other risk sources, whereas the success of a project can lead to benefits and rewards.

Risks are taken even though they could lead to devastating consequences because of potential benefits, rewards, survival, or future return on investment. Risk taking is a characteristic of intelligent living species, as it involves decision making, which is viewed as an expression of higher levels of intelligence. The fields of psychology and biology define intelligence as a behavioral strategy that gives each individual a means for maximizing the likelihood of success in achieving its goals in an uncertain and often hostile environment. These viewpoints consider intelligence as the integration of perception, reason, emotion, and behavior in a sensing, perceiving, knowing, feeling, caring, planning, and acting that can formulate and achieve goals. This process builds on risk-informed decision making at every step and stage toward achieving the goals.

2.2 Risk-Related Terminology and Definitions

Although we provide risk-related terminology throughout the book at appropriate locations or on first encounters, some foundational terminology and definitions are introduced at the outset in this section out of necessity to present risk analysis and management methods in subsequent sections.

2.2.1 Systems, Events, Scenarios, and Failure Modes

System is a group of interacting, interrelated, or interdependent elements, such as people, property, materials, environment, and processes. The elements together form a complex whole that can be a complex physical structure, process, or procedure of some attributes of interest. The interacting collection of discrete elements is commonly defined using deterministic models. The word *deterministic* implies that the system is identifiable and not uncertain in its architecture. The definition of the system is based on analyzing its functional and/or performance requirements. A description of a system may be a combination of functional and physical elements. Usually, functional descriptions are used to identify high levels of information or knowledge on a system. A system may be divided into subsystems that interact. Additional detail leads to a description of the physical elements, components, and various aspects of the system. Systems and their definitions are discussed in detail in Chapter 3.

Event is occurrence or outcome or change of a particular set of circumstances. An event can be used to define a notion of interest, such as failure and performance. According to the International Organization for Standardization (ISO) Guide 73 (2009b), an event can be one or more occurrences, can have several causes, can consist of something not happening, can sometimes be referred to as an "incident" or "accident," can be without consequences, and in this case can also be referred to as a "near miss," "near hit," or "close call."

Scenario is defined as joint events and system state that lead to an outcome of interest. A scenario defines a suite of circumstances of interest in a risk assessment. Thus, there may be loading scenarios, failure scenarios, or downstream flooding scenarios. A scenario can also be defined as the joint occurrence of events following a particular order or sequence in occurrence.

Initiating event is an event that appears at the beginning of a chain of events or a sequence of events, such as in an event tree.

Event tree analysis (ETA) is an inductive analysis method that utilizes an event tree graphical construct to show the logical sequence of the occurrence of events in, or states of, a system following an initiating event. Event trees can be used to define scenarios using a tree-like logic structure, and FTA limits the branching to the two cases of success and failure at each branching occurrence. A generalization of this method is *probability tree analysis* (PTA) that removes this limitation by allowing any number of branching.

Fault tree analysis (FTA) is a deductive analysis method for representing the logical combinations of various system states and possible causes that can contribute to a specified event, called the top event.

Failure mode is a way that failure can occur, described by the means or underlying physics by which element or component failures must occur to cause loss of the subsystem or system function. It can be treated as an event in an analysis.

Vulnerability is defined as the intrinsic properties of a system making it susceptible to a hazard or a threat or a risk source that can lead to an event with a consequence, or is an inherent state of the system, for example, physical, technical, organizational, or cultural, that can be exploited by an adversary to cause harm or damage.

2.2.2 Hazards and Threats

Hazard is a source of potential harm or a condition, which may result from an external cause (e.g., earthquake, flood, or human agency) or an internal vulnerability, with the potential to initiate a failure mode. It is a situation with a potential to cause loss, that is, a risk source. Depending on the nature of a project and its geographical location, some of the following natural hazards should be included within the scope of risk studies: flooding due to rivers, extreme precipitation and monsoons, coastal waves, dam/levee failure, sea-level rise, tidal, cyclones, drought including bushfires or forest fires, extreme wind, tornado, landslide, mudslide, subsidence and sinkholes, volcano, earthquakes and potential tsunamis, and coastal/shoreline erosion.

For example, the hazard can be uncontrolled fire, water, radioactive material, and strong wind. In order for the hazard to cause harm, it must interact with persons or things in a harmful manner. The magnitude of the hazard is its amount or intensity that could cause harm, such as wind speed, flooding depth, ground acceleration in the case of an earthquake, and quantity of radioactive release. Potential hazards must be identified and considered perhaps using life cycle analyses or some other approach necessary for an orderly and structured enumeration.

The interaction between a person (or a system) and a hazard can be voluntary or involuntary. For example, exposing a marine vessel to a sea environment might lead to its interaction with extreme waves in an uncontrollable manner (i.e., an involuntary manner). The decision of a navigator of the vessel to go through a developing storm system can be viewed as a voluntary act and might be necessary to meet schedule constraints or other constraints, and the potential rewards of delivery of shipment or avoidance of delay charges offer an incentive that warrants such an interaction. Other examples would include individuals who interact with hazards for potential financial rewards, fame, self-fulfillment, and satisfaction, ranging from investments to climbing cliffs.

Threat is the potential intent to cause harm or damage on, with, or through a system by exploiting its vulnerabilities. Threats can be associated with intentional human actions as provided in Table 2.1 that lists examples under several threat types including chemical, biological, radiological, nuclear, explosive, sabotage, and cyber.

TABLE 2.1

Threat Types and Examples

Selected Threat Type	Example Delivery Mode	Example Weapon/Agent	Example Quantity/Quality
Chemical	Outdoor dispersal	Ricin	Potent
		Mustard gas	Potent
	Crop duster	VX nerve agent	Potent
		Chlorine gas	Potent
	Missile	Any of the above	Potent
	Postal mail	Ricin	Potent
Biological	Outdoor dispersal	Anthrax	Potent
		Severe acute respiratory syndrome (SARS)	Potent
	Postal mail	Anthrax	Potent
	Food buffets	Hepatitis	Potent
		Salmonella	Potent
	Missile	Any of the above	Potent
Radiological	Standard deployment	Dirty bomb	Strong
		Radiological release	Strong
Nuclear	Standard deployment	Improvised nuclear device	In kilotons
		Strategic nuclear weapon	In kilotons

(*Continued*)

TABLE 2.1

(Continued) Threat Types and Examples

Selected Threat Type	Example Delivery Mode	Example Weapon/Agent	Example Quantity/Quality
Explosive	Standard deployment	Backpack bomb	In pounds of Trinitrotoluene (TNT)
		Missile	In tons
	Truck	Fertilizer bomb	In pounds
	Boat	Composition C4 explosives	In pounds
	Airplane	Jet fuel	In gallons
Sabotage	Physical	Cut power cable	Not applicable
		Cut bolts	Not applicable
		Improper operation or maintenance	Not applicable
	Cyber	Providing unauthorized access	Disruption of services
Cyber	Physical	Cut control cable	Not applicable
		Magnetic weapons	Power units
	Cyber	Worm virus	Disruption of services

2.2.3 Uncertainty and Ignorance

Uncertainty is the state of deficiency in information as discussed in detail in Chapter 1. It is a component of ignorance that broadly covers complete or partial deficiency in understanding or knowledge of an event, its consequence, or likelihood. In quantitative risk analysis, uncertainty is treated as a representation of the confidence in the state of knowledge about the models and parameter values used in risk quantification. In quantitative risk assessment, two uncertainty types are identified: aleatory uncertainty and epistemic uncertainty per respective definitions below:

Aleatory uncertainty is the inherent, random, or nonreducible uncertainty, such as material strength randomness.

Epistemic uncertainty is the knowledge-based, subjective uncertainty that can be reduced with the collection of data or attainment of additional knowledge.

Ignorance is deficiency in knowledge as discussed in detail in Chapter 1. Within the realm of conscience ignorance, incompleteness and inconsistency were described as the primary categories defining it.

2.2.4 Performance, Probability, and Reliability

Performance of a system or component is its ability to meet functional requirements. Performance should be specific to a particular domain and application and measurable, for example, speed measured in miles per hour, power, reliability, capability, efficiency, or maintainability. The design and operation of system could affect performance. Performance can be quantified by diverse measurement units, such as time, length, and force. The interest in performance and its measurement stems from fulfilling a need. These measurements, in diverse units, can be harmonized by assessing the probability of a particular performance type fulfilling a corresponding need.

Likelihood is the chance of something happening, whether defined, measured, or determined objectively or subjectively, qualitatively, or quantitatively, and described using general terms or mathematically, such as a probability or a frequency over a given time period.

Probability is a measure of chance of occurrence, likelihood, odds, or degree of belief that a particular outcome or event will occur, expressed as a number between 0 and 1, where 0 is impossibility and 1 is absolute certainty. This measure meets the axioms of probability theory as introduced in Appendix A. Probability has at least two primary interpretations: (1) a *frequency* representing the occurrence fraction of an outcome in repeated trials or an experiment as sometimes termed an *objective probability* and (2) *subjective probability* that is based on the state of knowledge.

Conditional probability is the occurrence probability of an event based on the assumption that another event (or multiple events) has already occurred, for example, the failure probability of a building under the condition that an earthquake of a particular intensity has occurred. Another example is the probability of an outcome at the end of a probability tree branch determined based on the fact, that is, under condition, that a particular event, or several events, has occurred.

Reliability for a system or a component is a performance of particular interest to engineers and can be defined as the ability of the system or component to fulfill its design functions under designated operating and/or environmental conditions for a specified time period. This ability is commonly measured using probabilities. Reliability is commonly interpreted as the occurrence probability of the complementary event to failure, as provided in the following expression:

$$\text{Reliability} = 1 - \text{failure probability} \tag{2.1}$$

2.2.5 Exposure, Consequences, and Contingency

Exposure is the extent to which an organization's and/or stakeholder's concerns are subject to an event and defined by *things at risk* that might include population at risk, property at risk, and ecological and environmental concerns at risk.

Consequence is the immediate, short-term, and long-term effects of an event affecting objectives, for example, an explosion of a chlorine storage tank. These effects may include human and property losses, environmental damages, and loss of lifelines. Property damage and losses, and operation interruption costs are generally directly expressed in ranges of monetary units, for example, dollars. Consequences involving loss of life, injury, loss of lifelines, and environmental damage might be measured in different units requiring seperate tracking. Consequences must be quantified using relative or absolute measures for various consequence types to facilitate risk analysis. Broadly stated, events may include successes, and the favorable consequences in this case can be defined as the degree of reward, return, or benefits from a success. Such an event could have, for example, beneficial economic outcomes or environmental effects.

Consequence mitigation is the preplanned and coordinated actions or system features that are designed to reduce or minimize the damage caused by an event; support and complement emergency forces, that is, first responders; facilitate field investigation and crisis management response; and facilitate recovery and reconstitution. Consequence mitigation may also include steps taken to reduce short- and long-term impacts, such as providing alternative sources of supply for critical goods and services.

Contingency is the set of organized and coordinated steps or actions to be taken to counter an increased threat level or if an emergency or disaster (such as fire, hurricane, injury, and robbery) strikes, and the associated costs.

2.2.6 Risk

2.2.6.1 Definition

The concept of risk can be linked to uncertainties associated with events. Within the context of projects, risk is commonly associated with an uncertain event or condition that, if it occurs, has a positive or a negative effect on the objectives of a project. Risk originates from the Latin term *risicum*, which means the challenge presented by a barrier reef to a sailor. The *Oxford Dictionary* defines risk as the chance of hazard, bad consequence, loss, and so on, or risk can be defined as the chance of a negative outcome.

Risk should be associated with a system and commonly defined as the potential loss resulting from an uncertain exposure to a hazard or resulting from an uncertain event that exploits the system's vulnerability. Risk should be based on identified risk events or event scenarios.

In 2009, the ISO provided a broadly applicable definition of risk in its standard (ISO 2009a) as the "effect of uncertainty on objectives" in order to cover the following considerations as noted in the standard:

- An effect is a deviation from the expected that can be positive and/or negative effect.
- Objectives can have different aspects, such as financial, health and safety, and environmental goals, and can apply at different levels, such as strategic, organization-wide, project, product, and process.
- Risk is often expressed in terms of a combination of the consequences of an event, including changes in circumstances, and the associated likelihood of occurrence as provided in the commonly used definition.

Providing two definitions of risk should not cause any confusion since most of the coverage in this book focuses on the adverse domain of effects, that is, using the former definition; however, readers must become familiar and comfortable with the latter, because of its broad applicability.

Risk context is the external and internal parameters or considerations to be taken into account when managing risk and setting the scope and risk criteria for the risk management policy as follows (ISO 2009a):

- External: the cultural, social, political, legal, regulatory, financial, technological, economic, natural, and competitive environment, whether international, national, regional, or local; key drivers and trends having impact on the objectives of the organization; and relationships with, and perceptions and values of external stakeholders
- Internal: governance, organizational structure, roles, and accountabilities; policies, objectives, and the strategies that are in place to achieve them; the capabilities, understood in terms of resources and knowledge (e.g., capital, time, people, processes, systems, and technologies); information systems, information flows, and decision-making processes (both formal and informal); relationships with, and perceptions and values of internal stakeholders; the organization's culture; standards, guidelines, and models adopted by the organization; and the form and extent of its contractual relationships

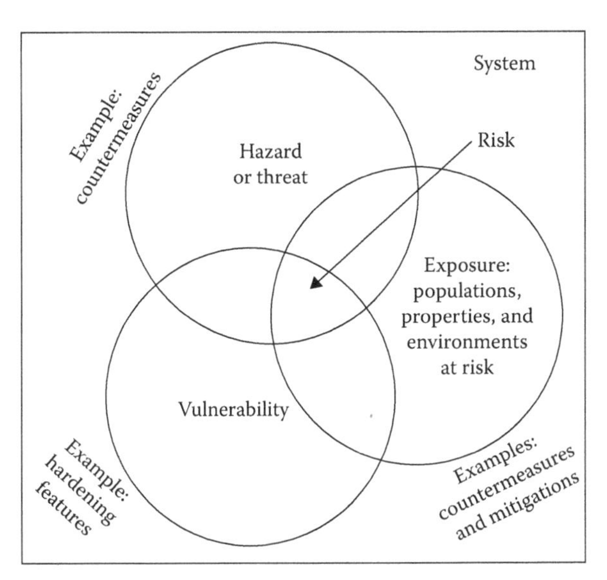

FIGURE 2.1
A schematic definition of risk.

In the context of the former risk definition, risk can be viewed as a multidimensional quantity that includes event occurrence probability, event occurrence consequences, consequence significance, and things at risk including populations, properties, and environmental concerns; however, it is commonly measured as a pair of the probability of occurrence of an event and the outcomes or consequences associated with the event's occurrence. Another common representation of risk is in the form of an exceedance probability (EP) function of consequences. Figure 2.1 schematically shows the definition of risk as the intersection of a hazard (or a threat) defined by scenarios with a system of interest that exploit its vulnerabilities and could impact things at risk including populations, properties, and environmental concerns. The figure also shows how each aspect of risk can be controlled or managed by countermeasures, system hardening, and mitigations as discussed in detail in later sections of this chapter.

Risk results from an event or sequence of events referred to as a *scenario*. The event or scenario can be viewed as a cause and, if it occurs, results in consequences with various severities. Sometimes these events or scenarios are called risk factors. For example, an event or cause may be a shortage of personnel necessary to perform a task required to complete a project. The event, in this case, of a personnel shortage for the task will have consequences in regard to the project cost, schedule, and/or quality. The events can reside in the project environment, which may contribute to project success or failure through project management (PM) practices, or in external partners or subcontractors.

Risk has defining characteristics that should be recognized in a risk assessment process. Risk is a characteristic of an uncertain future and is a characteristic of neither the present nor the past. Once uncertainties are resolved and/or the future is attained, the risk becomes nonexistent; therefore, we cannot describe risks for historical events or risks for events that are currently being realized. Moreover, risks cannot be directly associated with success. Although risk management through risk mitigation of selected events could result in project success, leading to rewards and benefits, these rewards and benefits cannot be considered as outcomes of only the nonoccurrence of events associated with the risks. The occurrence of particular events leads to adverse consequences that are clearly associated with their occurrence; however, their nonoccurrences are partial contributors to the project success that lead to rewards and benefits. The credit in the form of rewards and benefits cannot be given solely to the nonoccurrence of these events. Some risk assessment literature defines risk to include both potential losses and rewards, which should be treated separately as (1) risks leading to adverse consequences and (2) risks, if appropriately and successfully managed, contributing to benefits or rewards. Such a treatment utilizes the latter risk definition where a threat (or opportunity or a factor) could affect adversely (or favorably) the achievement of the objectives of a project and associated outcomes.

Developing an economic, analytical framework for a decision situation involving risks requires examining the economic and financial environments of a project. These environments can have significant impacts on the occurrence probabilities of events associated with risks. This added complexity might be necessary for particular projects in order to obtain justifiable and realistic results. The role of such environments in risk analysis is discussed in subsequent sections and chapters.

Risk, as the potential of losses for a system resulting from an uncertain exposure to a hazard or resulting from an uncertain event, can be viewed as a multidimensional quantity that includes event occurrence probability, event occurrence consequences, consequence significance, and commonly the population at risk; however, common practices minimally define it in terms of event probability and event outcomes or consequences. This pairing can be represented by the following equation:

$$\text{Risk} \equiv \{(p_1,c_1), (p_2,c_2),\ldots, (p_i,c_i),\ldots, (p_n,c_n)\} \tag{2.2}$$

where:

p_i is the occurrence probability of an outcome or event i
c_i is the occurrence consequences or outcomes of the event i

A generalized definition of risk is sometime expressed as follows:

$$\text{Risk} \equiv \{(l_1,o_1,u_1,\text{cs}_1,\text{po}_1), (l_2,o_2,u_2,\text{cs}_2,\text{po}_2),\ldots, (l_n,o_n,u_n,\text{cs}_n,\text{po}_n)\} \tag{2.3}$$

where:

l is the likelihood
o is the outcome
u is the utility (or significance)
cs is the causal scenario
po is the population affected by the outcome
n is the number of outcomes

The definition provided by Equation 2.3 covers key attributes measured in risk assessment that are described in this chapter and offers a practical description of risk, starting with the causing event to the affected population and consequences. The population size effect should be considered in risk studies as the society responds differently for risks associated with a large population in comparison to a small population. For example, a fatality rate of 1 in 100,000 per event for an affected population of 10 results in an expected fatality of 10^{-4} per event, whereas the same fatality rate per event for an affected population of 10,000,000 results in an expected fatality of 100 per event. Although the impact of the two scenarios might be the same on the society (same expected risk value), the total number of fatalities per event or accident is a factor in risk acceptance. Air travel may be safer than, for example, recreational boating, but 200–300 injuries per accident in the case of air travel are less acceptable to society in this case. Therefore, the size of the population at risk and the number of fatalities per event should be considered as factors in setting acceptable risk.

The dimension of likelihood that is not shown in Figure 1.1 can be illusive in nature due to two of its aspects: (1) the means of quantification and (2) the effect of time. The most common means of quantification are as follows:

- *Frequency.* It is defined as the count of an outcome of interest from a number of repeated observations of identical experiments or systems. If expressed as a fraction or percentage, it is called *relative frequency.*
- *Rate.* It is commonly defined as the count of an outcome of interest for a system occurring within a time period. The rate itself can be time dependent due to changes in the system's state, for example, due to aging. The term *frequency* is sometimes incorrectly used to mean the rate.
- *Probability.* It is defined as a measure of chance or likelihood.

The effects of time on these three quantification means are discussed, respectively, as follows:

- As for the frequency, by increasing the observation time, an estimate of the frequency tends toward a value, and for cases involving unbiased, consistent estimators, the estimate tends to the true value.
- As for the rate, by increasing the observation time, an estimate of the rate tends toward a value, and for cases involving unbiased, consistent estimators, the estimate tends to the true value.
- As for the probability, we are interested in a probability of an event in a time period. By increasing the length of this time period, this probability tends to one. As long as the event is possible, it has a sure eventual occurrence; otherwise it goes against the premise of being possible. All the example events shown in Figure 1.1 have probabilities tending to one as time extends indefinitely, even for global events including ones leading to human extinction. The length of a time period to reach a probability of one for practical purposes may vary from one event type to another, for example, nuisance events might require a few days at the most, whereas global events might require thousands to millions of years.

2.2.6.2 Risk Matrices or Heat Maps

Risk matrices, also called heat maps, are basically tools for representing and displaying risks by defining ranges for consequence and likelihood as a two-dimensional presentation of likelihood and consequences. According to this method, risk is characterized by categorizing probabilities and consequences on the two axes of a matrix. Risk matrices have been used extensively for screening of various risks. They may be used alone or as a first step in a quantitative analysis. Regardless of the approach used, risk analysis should be a dynamic process, that is, a living process where risk assessments are reexamined and adjusted. Actions or inactions in one area might affect the risk in another; therefore, continuous updating is necessary.

The likelihood metric can be constructed using the categories shown in Table 2.2, and the consequences metric can be constructed using the categories shown in Table 2.3; an example is provided in Table 2.4. The consequence categories of Table 2.2 focus on the health and environmental aspects of the consequences. The consequence categories of Table 2.4 focus on the economic impact and should be adjusted to meet the specific needs of an industry or application. An example risk matrix is shown in Figure 2.2. In the figure, each boxed area is shaded depending on a subjectively assessed risk level. Three risk levels are used here for illustration purposes: low (L), medium (M), and high (H). Other risk levels may be added using a scale of five instead of three, if necessary. These risk levels are known as *severity factors*. The high level can be considered unacceptable risk, the M level can be treated as either undesirable or acceptable with review, and the L level can be treated as acceptable without review.

TABLE 2.2

Likelihood Categories for a Risk Matrix

Category	Description	Annual Probability Range
A	Likely	≥0.1 (1 in 10)
B	Unlikely	≥0.01 (1 in 100) but <0.1
C	Very unlikely	≥0.001 (1 in 1,000) but <0.01
D	Doubtful	≥0.0001 (1 in 10,000) but <0.001
E	Highly unlikely	≥0.00001 (1 in 100,000) but <0.0001
F	Extremely unlikely	<0.00001 (1 in 100,000)

TABLE 2.3

Consequence Categories for a Risk Matrix

Category	Description	Examples
I	Catastrophic	Large number of fatalities and/or major long-term environmental impact
II	Major	Fatalities and/or major short-term environmental impact
III	Serious	Serious injuries and/or significant environmental impact
IV	Significant	Minor injuries and/or short-term environmental impact
V	Minor	First aid injuries only and/or minimal environmental impact
VI	None	No significant consequence

TABLE 2.4

Example Consequence Categories for a Risk Matrix in Monetary Amounts (US$)

Category	Description	Cost (US$)
I	Catastrophic loss	≥10,000,000,000
II	Major loss	≥1,000,000,000 but <10,000,000,000
III	Serious loss	≥100,000,000 but <1,000,000,000
IV	Significant loss	≥10,000,000 but <100,000,000
V	Minor loss	≥1,000,000 but <10,000,000
VI	Insignificant loss	<1,000,000

Probability category	A	L	M	M	H	H	H
	B	L	L	M	M	H	H
	C	L	L	L	M	M	H
	D	L	L	L	L	M	M
	E	L	L	L	L	L	M
	F	L	L	L	L	L	L
		VI	V	IV	III	II	I
	Consequence category						

FIGURE 2.2
Example risk matrix or heat map.

TABLE 2.5
Expanded Likelihood Categories for a Risk Matrix

Category	Description	Annual Probability Range
AA	Very likely	≥0.8
A	Likely	≥0.1 (1 in 10) but <0.8
B	Unlikely	≥0.01 (1 in 100) but <0.1
C	Very unlikely	≥0.001 (1 in 1,000) but <0.01
D	Highly unlikely	≥0.0001 (1 in 10,000) but <0.001
E	Very highly unlikely	≥0.00001 (1 in 100,000) but <0.0001
F	Extremely unlikely	<0.00001 (<1 in 100,000)

TABLE 2.6
Example Consequence Categories for a Risk Matrix in Monetary Amounts (US$)

Category	Description	Cost (US$)
I	Catastrophic loss	≥10,000,000,000
II	Major loss	≥1,000,000,000 but <10,000,000,000
III	Serious loss	≥100,000,000 but <1,000,000,000
IV	Significant loss	≥10,000,000 but <100,000,000
V	Minor loss	≥1,000,000 but <10,000,000
VI	Insignificant loss	<1,000,000
I+	Insignificant gain	<1,000,000
II+	Significant gain	≥1,000,000 but <10,000,000
III+	Major gain	≥10,000,000

The risk matrix presented so far does not account for potential gains due to nonoccurrence of an adverse event or the occurrence of a favorable event. As an example, the likelihood and monetary categories can be expanded, as shown in Tables 2.5 and 2.6, respectively, to permit the presentation of potential gain. The risk matrix can then be expanded as shown in Figure 2.2. Various events and scenarios can be assessed and allocated to various categories in the figure depending on their impact on the system as far as producing adverse consequences or favorable gains. The potential gains as provided in Figure 2.3 are grouped into illustrative three levels: low expected gain (L+), medium expected gain (M+),

H+	H+	M+	AA						
H+	M+	L+	A	L	M	M	H	H	H
M+	L+	L+	B	L	L	M	M	H	H
			C	L	L	L	M	M	H
			D	L	L	L	L	M	M
			E	L	L	L	L	L	M
			F	L	L	L	L	L	L
III+	II+	I+	Probability categories	VI	V	IV	III	II	I
Gain categories				Loss categories					

FIGURE 2.3
Example risk matrix with potential gains.

and high expected gain (H+). Scenarios that could lead to high expected gain should be targeted by project managers for facilitation and enhancement.

2.2.6.3 Risk Quantified Using Loss or Impact Probability Functions

To quantify risk, we must accordingly assess its defining components and measure the chance, its negativity, and potential rewards or benefits. Risk is commonly approximated by a point estimate as the expected value resulting from the multiplication of the conditional probability of the event occurring by the consequence of the event given that it has occurred as follows:

$$\text{Risk} = \text{likelihood} \times \text{impact} \tag{2.4}$$

The use of the expected value leads to a loss in information in terms of associated dispersion or variability. In Equation 2.4, the measurement scales, as bases for quantifying likelihood, impact, and risk, are as follows: likelihood is measured on an event rate scale in units of count of events per time period of interest, for example, events per year; impact is measured on a loss scale, such as monetary units or fatalities or any other units suitable for analysis or multiple units per event, for example, dollars per event; and risk is the product of (event per unit time) × (loss units per event) producing loss units per unit time. The likelihood in Equation 2.4 can also be expressed as a probability. Equation 2.4 presents risk as an expected value of loss per unit time or an average loss.

The product in Equation 2.4 is sometimes interpreted as the Cartesian product for scoping the space defined by the two dimensions of likelihood and impact for all underlying events and scenarios, which is a preferred interpretation. This interpretation preserves the complete nature of risk. Ideally, the entire probability distribution of consequences should be estimated.

A plot of occurrence probabilities and consequences is a *risk profile* or a *Farmer curve.* An example Farmer curve is given in Figure 2.4 based on a nuclear case study provided herein for illustration purposes (Kumamoto and Henley 1996). It should be noted that the abscissa provides the number of fatalities and the ordinate provides the annual frequency of exceedance for the corresponding number of fatalities. These curves are sometimes constructed using probabilities instead of rates. The curves represent the average or

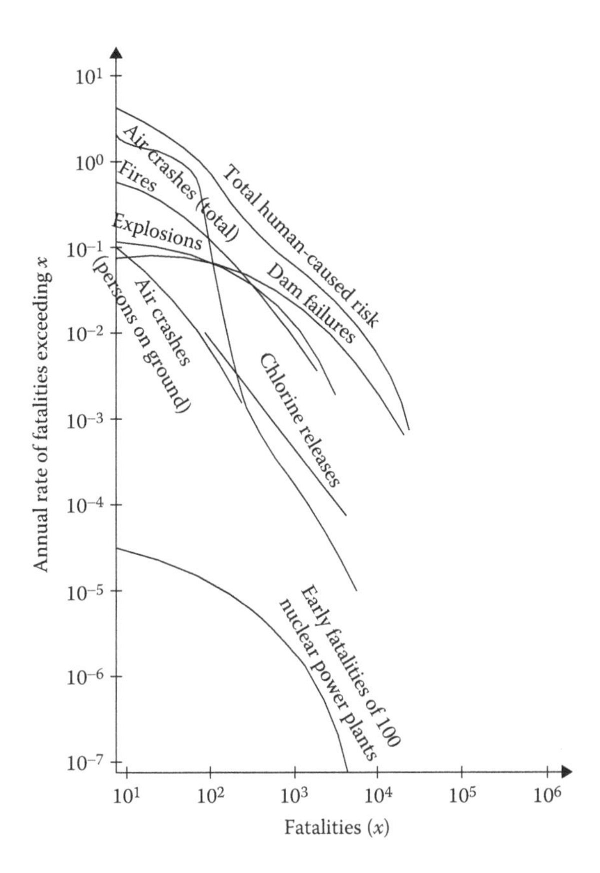

FIGURE 2.4
Example risk profile.

best estimate values. Figure 2.5 shows another example representing the gross margin of an investment project covering both potential loss and profits. This figure was generated using Monte Carlo simulation and presents the results in the form of a relative frequency histogram (i.e., the bar chart) and smoothed cumulative probability distribution (i.e., the solid curve). The figure shows that the loss probability is 0.047 and the probability of profits exceeding $200 million is 0.353.

Sometimes, bands or ranges are provided to represent uncertainty in these curves, and they represent confidence intervals for the average curve or the risk curve. Figure 2.6 shows examples of curves with bands (Kumamoto and Henley 1996). This uncertainty is sometimes called *epistemic uncertainty* or *meta-uncertainty*.

In cases involving deliberate threats, the occurrence probability (p) of an outcome (o) can be decomposed into an occurrence probability of an event or threat (t), a probability of success (s) given a threat ($s|t$), and an outcome probability given the occurrence of a successful event ($o|t,s$). The occurrence probability of an outcome can be expressed as follows using conditional probability concepts discussed in Appendix A on fundamentals of probability theory and statistics:

$$p(o) = p(t)p(s|t)p(o|t,s) \tag{2.5}$$

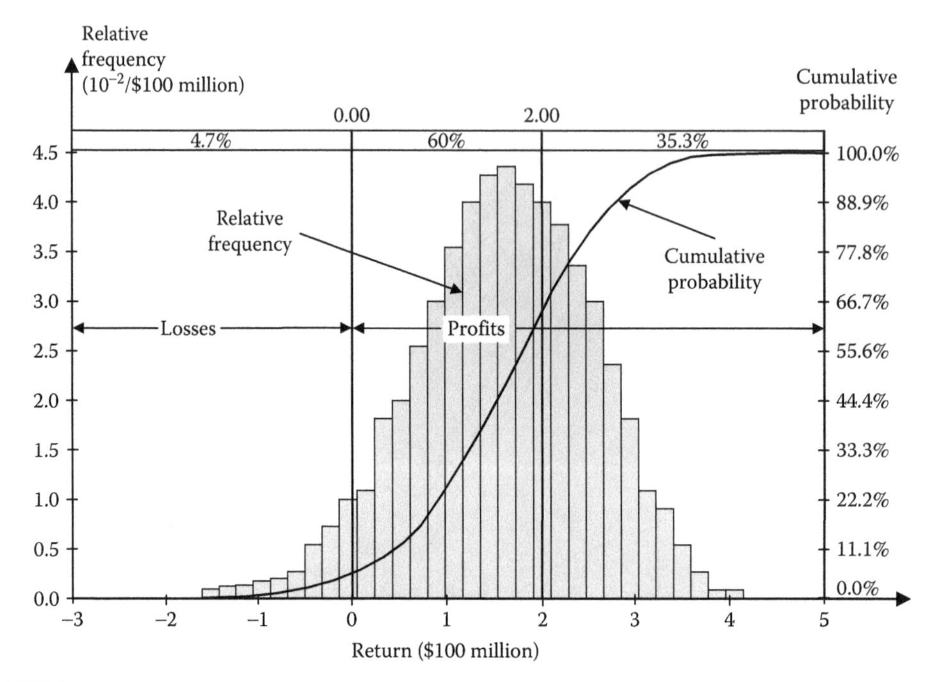

FIGURE 2.5
Example project risk profile.

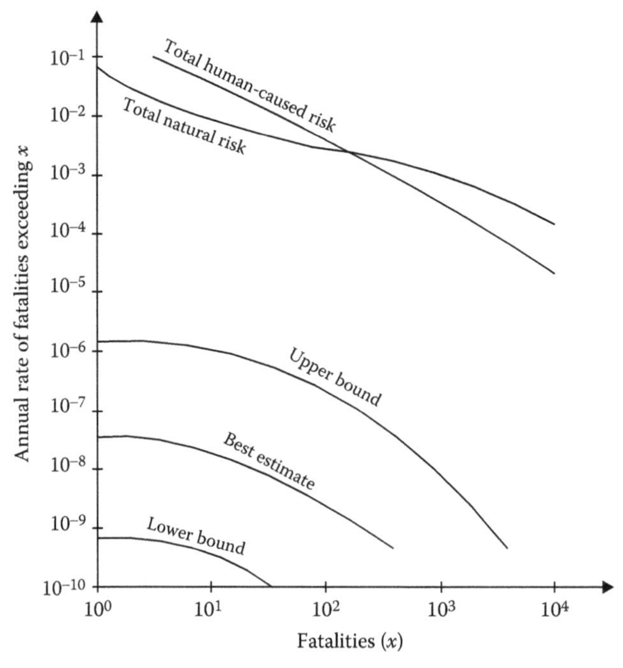

FIGURE 2.6
Uncertain risk profile. (Adapted from Imperial Chemical Industries Ltd., 1971.)

In this context, threat is defined as a hazard or the capability and intention of an adversary to undertake actions that are detrimental to a system or an organization's interest. In this case, threat is a function of only the adversary or competitor, and usually cannot be controlled by the owner of the system. The adversary's intention to exploit a situation may, however, be encouraged by vulnerability of the system or discouraged by an owner's countermeasures. The probability $p(o|t)$ can be decomposed further into two components: success probability of the adversary and conditional probability on this success in terms of consequences. This probability $p(o|t)$ can then be computed as the success probability of the adversary times the conditional probability of consequences given this success.

Risk register is a record of information about identified risks, sometimes called *risk log*.

Risk profile is a description of any set of risks that may relate to an entire organization, part of the organization, or a group of stakeholders or a region or a project.

Risk aggregation is the combination of a number of risk profiles into one risk profile to develop a more complete understanding of an overall risk; whereas *risk segregation* is the decomposition of an overall risk profile into a number of underlying risk profiles.

2.2.7 Asset Security and Protection

Asset is an item of value or importance. In the context of critical infrastructure and key resource (CI/KR) protection, a CI/KR asset is something of importance or value that if targeted, exploited, destroyed, or incapacitated could result in large-scale injury, death, economic damage, and destruction of property, or could profoundly damage a nation's prestige and confidence. Assets include physical elements (i.e., tangible property), cyber elements (i.e., information and communication systems), and human or living elements, (i.e., critical knowledge and functions of people).

Identifying critical assets requires defining the features that define criticality. Categories of critical assets are relatively broad and inclusive as shown in Table 2.7. The criticality of an asset should be based on features such as the impact of total destruction of or significant damage to an asset on the following:

- Public service and the operation of government
- Local, regional, and national economy
- Surrounding population
- National security
- Environment

It is noted that critical assets are identified primarily based on the consequences of a successful attack by an adversary rather than the probability that the attack will be successful. However, other asset features that should be considered include:

- Asset softness, that is, accessibility and inability to limit it
- Softness of targets within an asset
- Other specific features of these targets

TABLE 2.7

Asset Taxonomy

Agriculture and Food
Supply
Processing
Production
Packaging
Storage
Distribution
Transportation

Water
Dams, wells, reservoirs, and aqueducts
Transmission pipelines
Pumping stations
Sewer systems
Treatment facilities
Storage facilities

Public Health
National strategic stockpile
National Institutes of Health
State and local health departments
Hospitals
Health clinics
Mental health facilities
Nursing homes
Blood supply facilities
Laboratories
Mortuaries
Pharmaceutical stockpiles

Emergency Services
Fire houses and rescue
Federal Emergency Management Agency
Emergency medical services
Law enforcement
Mobile response
Communications systems

Defense Industry Base
Supply systems
Critical R&D facilities

Information and Telecommunications
Public Switched Telecommunications Network (PSTN)
Internet
Switch/router areas
Access tandems
Fiber/copper cable
Cellular, microwave, and satellite systems
Operations, administration, maintenance, and provisioning systems
Network operations centers
Underwater cables
Cable landing points
Collocation sites, peering points, and telecom hotels
Satellite control stations and radio cell towers

Energy
Electricity (Nonnuclear)
Hydroelectric dams
Fossil fuel electric power generation plants
Distribution systems
Key substations
Communications

Oil and Natural Gas
Off shore platforms
Refineries and pipelines
Storage facilities
Gas processing plants
Product terminals
Strategic petroleum reserve

Transportation
Aviation
Railways
Highways
Trucking
Busing
Bridges
Tunnels
Borders
Seaports
Pipelines
Maritime
Mass transit

Banking and Finance
Physical facilities (buildings)
Operations centers
Regulatory institutions
Physical repositories
Telecommunications networks
Emergency redundancy service areas

Chemical/Hazardous Materials Industry
Manufacturing plants
Transport systems
Distribution systems
Storage/stockpile/supply areas
Emergency response and communications systems

Postal and Shipping
Processing facilities
Distribution networks
Air, truck, rail, and boat transport systems
Security

National Monuments and Icons
National parks
Monuments
Historic buildings

Nuclear Power Plants
Commercial owned/operated
Government owned/operated
Physical facilities
Spent fuel storage facilities
Safety/security systems

Dams
Large
Small
Government owned
Private/corporate owned

Government Facilities
National Security Special Events
Commercial Assets
Prominent commercial centers
Office buildings
Sports centers/arenas
Theme parks
Processing/service centers

Asset owner is the primary person responsible for the safety, protection, and security of an asset.

Attack profile is the path and means by which a threat scenario is carried out. An attack is defined by a combination of intrusion path and hazard delivery system.

Attractiveness of an asset is an assessment of an adversary's perceived probability of success, gain from success, loss from failure, and cost to execute the attack. Attractiveness also considers the probability that the adversary is aware of the asset.

Countermeasure is an action taken or a physical capability provided with the principal purpose of reducing or eliminating vulnerabilities or reducing the occurrence of attacks.

Critical infrastructure consists of systems and assets, whether physical or virtual, which are vital to a nation that the incapacity or destruction of such systems and assets would have a debilitating impact on security, national economic security, national public health or safety, or any combination of those matters.

Key element is a hardware, software, organizational, or process element of a system that contributes directly to its mission.

Key resources are publicly or privately controlled resources essential to the operations of the economy or government. A nation possesses numerous key resources, whose exploitation or destruction could cause catastrophic health effects or mass casualties comparable to those from the use of a weapon of mass destruction or could profoundly affect national prestige and morale.

Security vulnerability is the inherent state of a security system that can be exploited by an adversary to undermine its effectiveness (see also *vulnerability*).

Security threat is a deliberate act, condition, or phenomenon that may result in the compromise of information, loss of life, damage, loss, or destruction of property, or disruption of vital services.

Susceptibility is an act, condition, or phenomenon capable of interacting with a specified target to cause disruption.

Threat scenario is the pairing of security threat type with a susceptible target. A threat scenario may consist of multiple possible attack profiles, each describing a way, that is, delivery system and intrusion path, in which the threat scenario can be executed.

Several definitions are available for the term *terrorism* without a globally accepted one as follows:

- US Code of Federal Regulations: "... the unlawful use of force and violence against persons or property to intimidate or coerce a government, the civilian population, or any segment thereof, in furtherance of political or social objectives" (28 C.F.R. Section 0.85).
- A US national security strategy: "premeditated, politically motivated violence against innocents."
- US Department of Defense: the "calculated use of unlawful violence to inculcate fear; intended to coerce or intimidate governments or societies in pursuit of goals that are generally political, religious, or ideological."

- British Terrorism Act 2000: It defines terrorism so as to include not only attacks on military personnel, but also acts not usually considered violent, such as shutting down a Website whose views one dislikes.
- 1984 US Army training manual states as follows: "Terrorism is the calculated use of violence, or the threat of violence, to produce goals that are political or ideological in nature."
- 1986 Vice-President's Task Force: "Terrorism is the unlawful use or threat of violence against persons or property to further political or social objectives. It is usually intended to intimidate or coerce a government, individuals or groups, or to modify their behavior or politics."
- Insurance documents: These documents define terrorism as "any act including, but not limited to, the use of force or violence and/or threat thereof of any person or group(s) of persons whether acting alone or on behalf of, or in connection with, any organization(s) or government(s) committed for political, religions, ideological or similar purposes, including the intention to influence any government and/or to put the public or any section of the public in fear."

2.2.8 Risk Management and Communication

Risk management is defined as the coordinated activities to direct and control an organization with regard to risk following a framework consisting of designing, implementing, monitoring, reviewing, and continually improving risk management throughout the organization. Risk management should be founded in strategic and operational policy, objectives, mandate, practices, and commitment through organizational arrangements including plans, relationships, accountabilities, resources, processes, and activities.

Stakeholder is a person, such as a decision maker and owner, or organization that can affect, be affected by, or perceive themselves to be affected by a decision or activity.

Risk owner is a person or entity with the accountability and authority to manage a risk.

Risk criteria are the terms of reference against which the significance of a risk is evaluated reflecting the organizational objectives expressed in external and internal contexts and in keeping with standards, laws, policies, and other requirements.

Resilience is defined by the Presidential Policy Directive (PPD)-21 (2013) as the ability to prepare for and adapt to changing conditions and withstand and recover rapidly from disruptions. Resilience includes the ability to withstand and recover from disturbances of the deliberate attack types, accidents, or naturally occurring threats or incidents. The resilience of a system's function can be measured based on the persistence of a corresponding functional performance under uncertainty in the face of disturbances (Ayyub 2013). This definition is consistent with the ISO (2009a) risk definition of the "effect of uncertainty on objectives."

Residual risk is the amount of risk remaining after realizing the net effect of risk reducing actions.

Risk tolerance is the degree of risk associated with normal daily activities that people tolerate, usually without making a conscious decision. As for organization or stakeholders, it is the readiness to bear the risk after risk treatment in order to achieve its objectives. Risk tolerance can be influenced by legal or regulatory requirements.

Risk acceptance is the degree of risk associated with a system or endeavor that a decision maker perceives and accepts the associated actions under a given set of circumstances and with the associated costs. A decision maker's risk tolerance and resources are the foundation of risk acceptance.

Risk attitude is an organization's approach to assess and eventually pursue, retain, take, or turn away from risk.

Risk appetite is the amount and type of risk that an organization is willing to pursue or retain.

Risk aversion is the attitude to turn away from risk.

Risk seeking is the attitude to pursue, retain, or undertake the risk for potential return.

Risk neutrality is having the same attitude regardless of the potential loss.

Safety is the judgment of risk tolerance, or acceptability in the case of decision making, for the system. Safety is a relative term since the decision of risk acceptance may vary depending on the individual or the group of people making the judgment. Different people are willing to accept various risks differently as demonstrated by factors such as location, method or system type, occupation, and life style. Examining various activities demonstrates an individual's safety preference despite a wide range of their risk values. Table 2.8 identifies varying annual risks for different activities based on typical exposure times for the respective activities. Also, Figure 2.7 illustrates risk exposure during a typical day that starts by waking up in the morning and getting ready to going to work, then commuting and working during the morning hours, followed by a lunch break, then additional work hours, followed by commuting back home to have dinner, and then a round trip on motorcycle to a local pub. The ordinate in this figure is the fatal accident frequency rate (FAFR) with an FAFR of 1.0 corresponding to 1 fatality in 11,415 years or 87.6 fatalities per 1 million years. The figure is based on an average number of deaths in 108 hours of exposure to a particular activity.

Risk retention is the acceptance of the potential benefit of gain, or burden of loss, from a particular risk that includes the acceptance of residual risks and depends on risk criteria.

TABLE 2.8

Relative Risk of Different Activities

Risk of Death	Occupation	Lifestyle	Accidents/Recreation	Environmental Risk
1 in 100	Stunt-person	–	–	–
1 in 1,000	Racecar driver	Smoking (one pack/day)	Skydiving Rock climbing Snowmobile	–
1 in 10,000	Firefighter Miner Farmer Police officer	Heavy drinking	Canoeing Automobiles All home accidents Frequent air travel	–
1 in 100,000	Truck driver Engineer Banker Insurance agent	Using contraceptive pills Light drinking	Skiing Home fire	Substance in drinking water Living downstream of a dam
1 in 1,000,000	–	Diagnostic X-rays Smallpox vaccination (per occasion)	Fishing Poisoning Occasional air travel (one flight per year)	Natural background radiation Living at the boundary of an NPP
1 in 10,000,000	–	Eating charcoal-broiled steak (once a week)	–	Hurricane Tornado Lightning Animal bite or insect sting

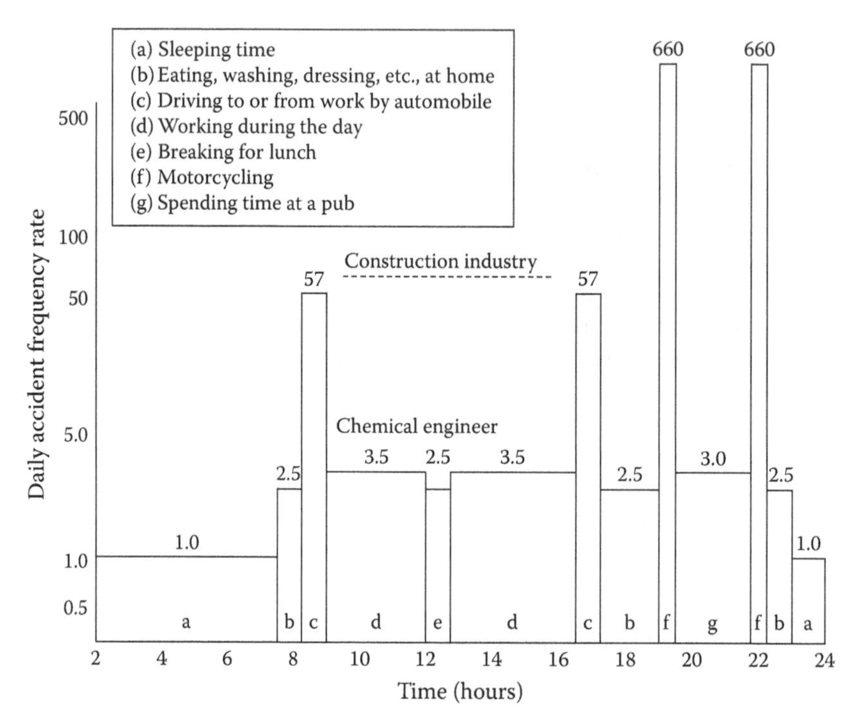

FIGURE 2.7
Daily death risk exposure for a working healthy adult. (Adapted from Imperial Chemical Industries Ltd., 1971.)

Risk perception is the stakeholders' view on a risk reflecting their needs, issues, knowledge, beliefs, and values. It is the manner and extent to which a decision maker or a person comprehends risk. The risk perception for a particular consequence and associated probability is a function of attributes such as a person's tolerance of (i.e., acceptance of or aversion to) the consequence and his or her ability to comprehend the assigned probability relative to other likely or unlikely events. Risk perception of safety may not reflect the actual level of risk in some activities. Table 2.9 shows the differences in risk perceptions by the League of Women Voters, college students, and experts for 29 risk items. Only the top items are listed in the table. Risk associated with nuclear power was ranked the highest by the League of Women Voters and college students, whereas it was placed 20th by the experts. Experts place motor vehicles as the highest risk. Public perception of risk and safety varies by age, gender, education, attitudes, and culture, among other factors. Individuals sometimes do not recognize uncertainties associated with a risk event or activity, which leads to unwarranted confidence in the individual's perception of risk or safety. Rare causes of death are often overestimated, and common causes of death are often underestimated. Perceived risk is often biased by the familiarity of the hazard. The significance or impact of safety perceptions stems from making decisions based on subjective judgments. If such judgments hold misconceptions about reality, this bias affects the decision. For example, choosing a transportation mode (train, automobile, motorcycle, bus, bicycle, etc.) is a decision based on many criteria, including items such as cost, speed, convenience, and safety. The weight and evaluation of the decision criteria in selecting a mode of transportation rely on the individual's perception of safety, which may deviate sometimes significantly from the actual values of the corresponding risks. Understanding these differences in risk and safety perceptions is vital to perform

TABLE 2.9

Risk Perception

Activity or Technology	League of Women Voters	College Students	Experts
Nuclear power	1	1	20
Motor vehicles	2	5	1
Hand guns	3	2	4
Smoking	4	3	2
Motorcycles	5	6	6
Alcoholic beverages	6	7	3
General aviation	7	15	12
Police work	8	8	17
Pesticides	9	4	8
Surgery	10	11	5
Firefighting	11	10	18
Heavy construction	12	14	13
Hunting	13	18	23
Spray cans	14	13	25
Mountain climbing	15	22	28
Bicycles	16	24	15
Commercial aviation	17	16	16
Electric (nonnuclear) power	18	19	9
Swimming	19	29	10
Contraceptives	20	9	11
Skiing	21	25	29
X-rays	22	17	7
High-school or college sports	23	26	26
Railroads	24	23	19
Food preservatives	25	12	14
Food coloring	26	20	21
Power mowers	27	28	27
Prescription antibiotics	28	21	24
Home applications	29	27	22

risk management decisions and risk communications, as discussed in Section 2.8 on risk treatment and control.

Risk treatment is the process of modifying risk by avoidance, removal of the risk source, countermeasures to change the likelihood, changing the consequences by mitigations, sharing the risk with another party or parties including contracts and risk financing, and retaining the risk by informed decisions. Risk treatments can include elimination, prevention, and reduction. Risk treatments can lead to the creation of new risks or the modification of other existing risks.

Risk financing is a form of risk treatment involving contingent arrangements for the provision of funds to meet or modify the financial consequences should they occur, such as insurance and bonds.

Risk avoidance is an informed decision not to be involved in, or to withdraw from, an activity in order not to be exposed to a particular risk.

Risk control is a measure in place that is risk modifying.

Risk sharing is a form of risk treatment involving the agreed distribution of risk with other parties, such as insurance and contracts. Sometimes, legal or regulatory requirements can limit, prohibit, or mandate risk sharing.

Risk transfer is a form of risk sharing.

A countermeasure is an action taken or a physical capability provided whose principal purpose is to reduce or eliminate one or more vulnerabilities or to reduce the frequency of attacks. *Consequence mitigation* is the preplanned and coordinated actions or system features that are designed to reduce or minimize the damage caused by hazards or attacks (consequences of a hazard or an attack); support and complement emergency forces (first responders); facilitate field investigation and crisis management response; and facilitate recovery and reconstitution. Consequence mitigation may also include steps taken to reduce short- and long-term impacts, such as providing alternative sources of supply for critical goods and services. Mitigation actions and strategies are intended to reduce the consequences (impacts) of an attack, whereas countermeasures are intended to reduce the probability that an attack will succeed in causing a failure or significant damage.

Risk monitoring is a process of continual checking, supervising, critically observing, or determining the status in order to identify the change from the performance level required or expected.

Risk communication is the continual and iterative processes that an organization conducts to provide, share or obtain information, and engage in dialog with stakeholders regarding the management of risk to achieve an interactive process of exchange of information and opinions among stakeholders such as individuals, groups, and institutions. It often involves multiple messages about the nature of risk or expressing concerns, opinions, or reactions to risk managers or to legal and institutional arrangements for risk management. Risk communication greatly affects risk acceptance and could determine the acceptance criteria for safety.

2.3 Risk Assessment

Risk assessment is an overall process of (1) risk identification, (2) risk analysis, and (3) risk evaluation. It is a systematic process for identifying risk sources and quantifying and describing the nature, likelihood, and magnitude of risks associated with some situation, action, or event that includes consideration of relevant uncertainties. Risk assessment can require and/or provide both qualitative and quantitative data to decision makers for use in risk management. This section provides additional risk-related terminology and provides a typical risk-informed methodology for analyzing a system. Subsequent sections offer details on the different steps involved in a typical methodology.

2.3.1 Risk Studies

Risk studies require the use of analytical methods at the system level that take into consideration subsystems and components when assessing their event probabilities and consequences. Systematic, quantitative, qualitative, or semiquantitative approaches for assessing event probabilities and consequences of engineering systems are used for this purpose. A systematic approach allows an analyst to evaluate expediently and easily complex systems for safety and risk under different operational and extreme conditions.

The ability to quantitatively evaluate these systems helps cut the cost of unnecessary and often expensive redesign, repair, strengthening, replacement, or corrective actions directed at components, subsystems, and systems. The results of risk studies can also be utilized in decision analysis methods that are based on benefit–cost trade-offs.

2.3.2 Risk Identification, Analysis, and Evaluation

Risk identification is the process of finding, recognizing, and describing risks including sources, events, scenarios, and their causes and potential consequences involving historical data, theoretical analysis, informed and expert opinions, and stakeholders' needs.

Risk analysis is the technical and scientific process to comprehend the nature of risk and to determine the level of risk by examining the underlying components or elements of risk. Risk analysis provides the basis for risk evaluation and decisions about risk treatment, and the processes for identifying hazards, event probability assessment, and consequence assessment. The risk analysis process traditionally focuses on answering three basic questions: (1) What can go wrong and how it can happen? (2) What is the likelihood that it will go wrong? and (3) What are the consequences if it does go wrong? Also, risk analysis can include examining the impact of making any changes to a system to control risks. Risk analysis generally contains the following steps: scope definition, hazard identification, and risk estimation.

Risk evaluation is the process of comparing the results of risk analysis with risk criteria to determine whether the risk and/or its magnitude is acceptable or tolerable in order to assist in decisions about risk treatments.

2.3.3 Qualitative versus Quantitative Risk Assessment

Risk assessment methods can be categorized into quantitative or qualitative analysis according to how the risk is determined. Qualitative risk analysis uses judgment and sometimes expert opinion to evaluate the probability and consequence values. This subjective approach may be sufficient to assess the risk of a system, depending on the available resources. Quantitative analysis relies on probabilistic and statistical methods, as well as databases that identify numerical probability values and consequence values for risk assessment. This objective approach examines the system in greater detail to assess risks.

The selection of a quantitative or qualitative method depends on the availability of data for evaluating the hazard and the level of analysis required to make an appropriate decision. Qualitative methods offer analyses without detailed information, but the intuitive and subjective processes may result in differences in outcomes by those who use them. Quantitative analysis generally provides a more uniform understanding among different individuals but requires quality data for useful and repeatable results. A combination of both qualitative and quantitative analyses can be used depending on the situation.

Risk assessment requires estimates of the event likelihood at some identified levels of decision making. The event likelihood can be estimated in the form of lifetime failure likelihood, annual failure likelihood, mean time between failures, or failure rate. The estimates can be in numeric or nonnumeric form. An example numeric form for an annual failure probability is 0.00015 and for a mean time between failures is 10 years. An example nonnumeric form for an annual failure likelihood is *large* and for a mean time between failures is *medium*. In the latter nonnumeric form, guidance should be provided regarding the meaning of such terms as large, medium, small, very large, and very small. The selection of the form should be based on the availability of information, the ability of the persons providing the needed information to express it in one form or another, and the importance of having numeric versus nonnumeric information when formulating the final decisions.

The types of event consequences that should be considered in a study include production loss, property damage, environmental damage, and safety loss in the form of human injury and death among other consequence types as discussed in Chapter 5. Event consequence estimates can be in numeric or nonnumeric form. An example numeric form for production loss is 1000 units; an example nonnumeric form for production loss is *large*. Again, guidance should be provided regarding the meaning of such terms as large, medium, small, very large, and very small. The selection of the form should be based on the availability of information, the ability of the persons providing the needed information to express it in one form or another, and the importance of having numeric versus nonnumeric information when formulating the final decisions.

Risk estimates can be determined by pairing likelihoods and consequences and computed simply and approximately as the arithmetic multiplication of the respective failure likelihoods and consequences for equipment, components, and other details. Alternatively, for all cases, plots of failure likelihood versus consequences can be developed, which enables to approximately rank them as groups according to risk estimates, failure likelihood, and/or failure consequences.

2.3.4 Risk-Based Technology

Risk-based technology (RBT) is used to describe methods or tools and processes for assessing and managing the risks of a component or system. RBT methods can be classified into risk management, which includes risk assessment and risk treatment using failure prevention and consequence mitigation, and risk communication, as shown in Figure 2.8. *Risk assessment* consists of risk identification, event probability assessment, consequence assessment, and risk evaluation. *Risk treatment* requires the definition of acceptable risk and comparative evaluation of options and/or alternatives through monitoring and

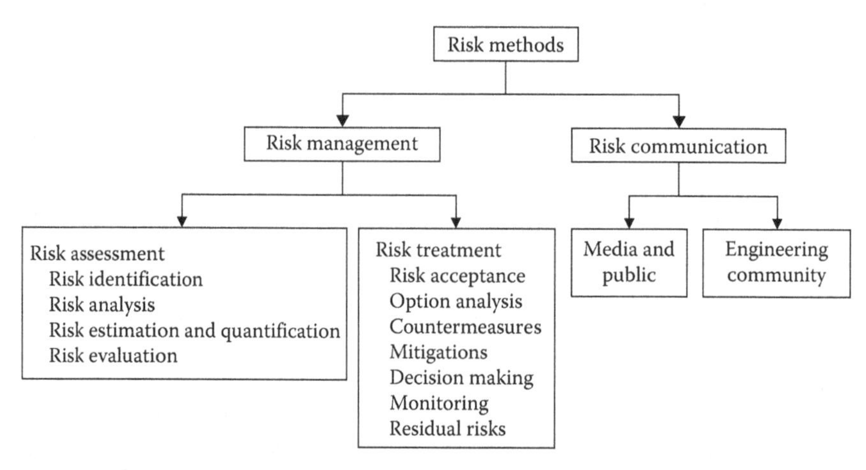

FIGURE 2.8
Risk-based technology methods.

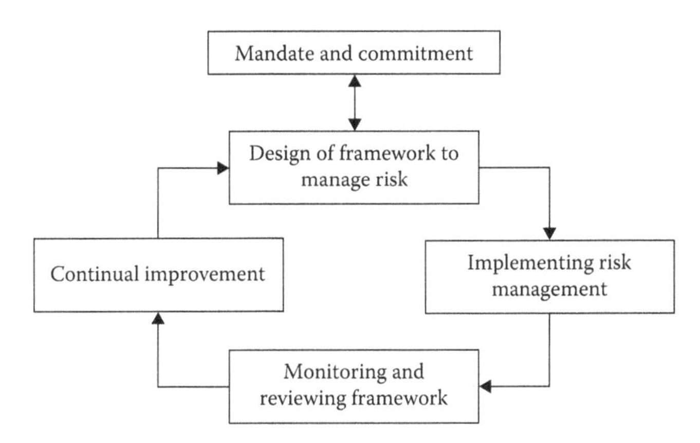

FIGURE 2.9
Risk management process.

decision analysis; risk treatment also includes failure prevention and consequence mitigation. *Risk communication* involves perceptions of risk and depends on the audience targeted; hence, it is classified into risk communication to the media, the public, and the engineering community.

The various components of managing risk can be structured in the form of a process as shown in Figure 2.9 (ISO 2009a). The process emphasizes the importance of continual improvement in risk management in order to ensure success.

2.3.5 Risk Assessment Methodologies

A risk assessment process should utilize experiences gathered from project personnel, other similar projects and data sources, previous risk assessment models, and other industries and experts, in conjunction with analysis and damage evaluation using various prediction tools. A risk assessment process is commonly part of a risk-based or risk-informed methodology that should be constructed as a synergistic combination of decision models, advanced probabilistic reliability analysis algorithms, failure consequence assessment methods, and conventional performance assessment methodologies that have been employed in a related industry for performance evaluation and management. The methodology should realistically account for the various sources and types of uncertainty involved in the decision-making process.

In developing a risk analysis methodology, the following requirements should be treated as key guiding principles:

- Analytic: The methodology must provide a system-based framework for assessing risk by decomposing it into its basic elements. It must enable a system-based framework for decomposition into its basic elements through a segregation process, and these elements are logically connected to enable aggregation. This requirement is called the *principle of consistent segregation and aggregation*.
- Quantitative: Risk is expressed in meaningful and consistent units (e.g., dollars and fatalities) so as to provide a basis for performing trade-offs and benefit–cost analysis.

- Probabilistic: The mathematics of probability theory is used for expressing uncertainty in all model parameters and assessing the likelihood of alternative scenarios.
- Consistent: It is consistent with established and accepted practices of probabilistic risk assessment (PRA) used in many other fields.
- Transparent: All assumptions and analytical steps are clearly defined.
- Defensible: Values for each parameter are supported by all available data, including knowledge from previous studies and expert opinion.

The ISO 31000 (2009a) defines the essential qualities for an effective risk management process, called principles to (1) create and protect value; (2) be an integral part of organizational processes; (3) be part of decision making; (4) explicitly addresses uncertainty; (5) be systematic, structured, and timely; (6) be based on the best available information; (7) be tailored; (8) take human and cultural factors into account; (9) be transparent and inclusive; (10) be dynamic, iterative, and responsive to change; and (11) facilitate continual improvement and enhancement of the organization. Also, the ISO 31000 (2009a) lists the following attributes of enhanced risk management: (1) continual improvement, (2) full accountability for risks, (3) application of risk management in all decision making, (4) continual communications, and (5) full integration in the organization's governance structure.

In this section, a typical overall methodology is provided in the form of a workflow or block diagram. The various components of the methodology are described in subsequent sections. Figure 2.10 provides an overall description of a methodology for risk-based life management of structural systems for the purpose of demonstration. The methodology consists of the following primary steps:

1. Definition of analysis objectives and systems
2. Hazard analysis, definition of failure scenarios, and hazardous sources and their terms
3. Collection of data in a life cycle framework
4. Qualitative risk assessment
5. Quantitative risk assessment
6. Management of system integrity through countermeasures, failure prevention, and consequence mitigation using risk-based decision making

These steps are described briefly below with additional background materials provided in subsequent sections and chapters.

The first step of the methodology is to define the system. Chapter 3 provides additional information on defining systems for the purpose of risk assessment. This definition should be based on a goal that is broken down into a set of analysis objectives. A system can be defined as an assemblage or combination of elements of various levels and/or details that act together for a specific purpose. Defining the system provides the risk-based methodology with the information required to achieve the analysis objectives. The system definition phase of the proposed methodology has five main activities:

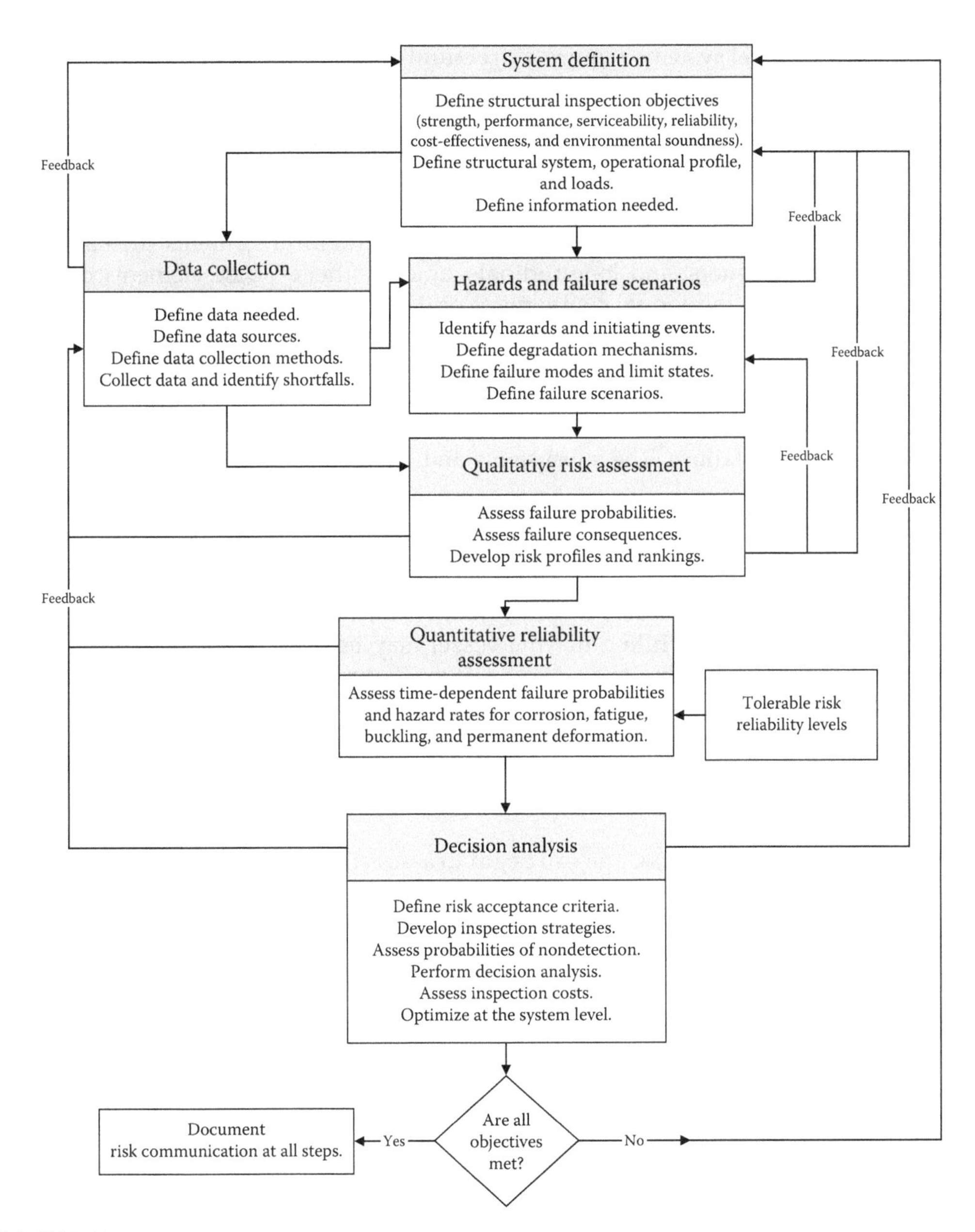

FIGURE 2.10
Methodology for risk-based life cycle management of structural systems.

1. Define the goal and objectives of the analysis
2. Define the system boundaries
3. Define the success criteria in terms of measurable performances
4. Collect information for assessing failure likelihood
5. Collect information for assessing failure consequences

For example, structural systems require a structural integrity goal that can include objectives stated in terms of strength, performance, serviceability, reliability, cost-effectiveness, and environmental soundness. The objectives can be broken down further to include other structural integrity attributes, such as alignment and watertightness for marine vessels. A system can be defined based on a stated set of objectives. The same system can be defined in various ways depending on these stated objectives. A marine vessel structural system can be considered to contain individual structural elements such as plates, stiffened panels, stiffeners, and longitudinals, among others. These elements could be further separated into individual components or details. Identifying all of the elements, components, and details allows an analysis team to collect the necessary operational, maintenance, and repair information throughout the life cycle of each item so that failure rates, repair frequencies, and failure consequences can be estimated. The system definition might need to include nonstructural subsystems and components that would be affected in case of failure. The subsystems and components are needed to assess the consequences.

In order to understand failure and the consequences of failure, the states of success must be defined. For the system to be successful, it must be able to perform its design functions by meeting its measurable performance requirements. The system, however, may be capable of various levels of performance, all of which might not be considered complete successes. While a marine vessel may be able to get from points A to B even at a reduced speed due to a fatigue failure that results in excessive vibration in the engine room, its performance would probably not be considered entirely successful. The same concept can be applied to individual elements, components, and details. It is clear from this example that the success and failure impacts of the vessel should be based on the overall vessel performance, which can easily extend beyond the structural systems.

With the definition of success, one can begin to assess the likelihood of occurrence and causes of failures. Most of the information required to develop an estimate of the likelihood of failure may exist in maintenance and operating histories available on the systems and equipment, and may be based on judgment and expert opinion. This information may not be readily accessible, and its extraction from its current source may be difficult. Also, assembling it in a manner that is suitable for the risk-based methodology can be a challenge.

Operation, maintenance, engineering, and corporate information on failure history should be collected and analyzed for the purpose of assessing the consequences of failures. The consequence information may not be available from the same sources as the information on the failure itself. Typically, there are documentations of repair costs, reinspection or recertification costs, lost person-hours of labor, and possibly even lost opportunity costs due to system failure. Much more difficult to find and assess are costs associated with effects on other systems, the cost of shifting resources to cover lost production, and other costs such as environmental, safety loss, or public relations costs. These may be determined through carefully organized discussions and interviews with cognizant personnel, including the use of expert opinion elicitation.

The ISO 31000 (2009a) provides risk management methodology, called a process, as shown in Figure 2.11. The arrows show the process flows, and the other lines are information and data flow. The entries at the various steps define these steps using the terminology in Section 2.2.

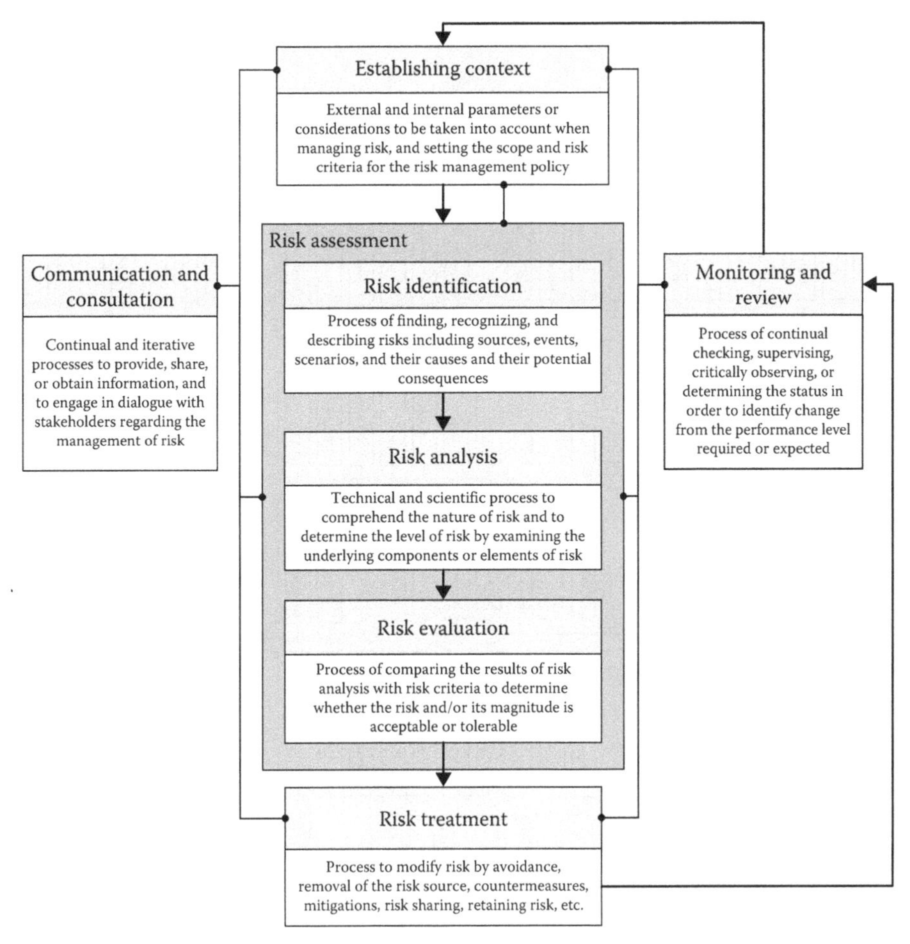

FIGURE 2.11
ISO 31000: 2009 risk management methodology.

2.4 Risk Events and Factors

2.4.1 Risk Categories and Breakdown

Starting a risk study based on an exhaustive set of possible events and factors that might adversely or favorably affect a system's objectives gives headway toward ensuring success of the study. Figure 2.12 provides categories for risk event and factor identification. These categories do not apply to all systems and are not necessarily exhaustive, and the structure is not necessarily unique. Simplified categories are shown in Figure 2.13 for illustration purposes. An analyst might use these figures to customize the categories for a system of interest and/or use the lists to identify events and factors that are applicable to a system of interest. Items within a category require further refinement in terms of definitions to make them relevant and applicable to the system.

Risk sources for a project can be organized and structured to provide a standard presentation that would facilitate understanding, communication, and management. The

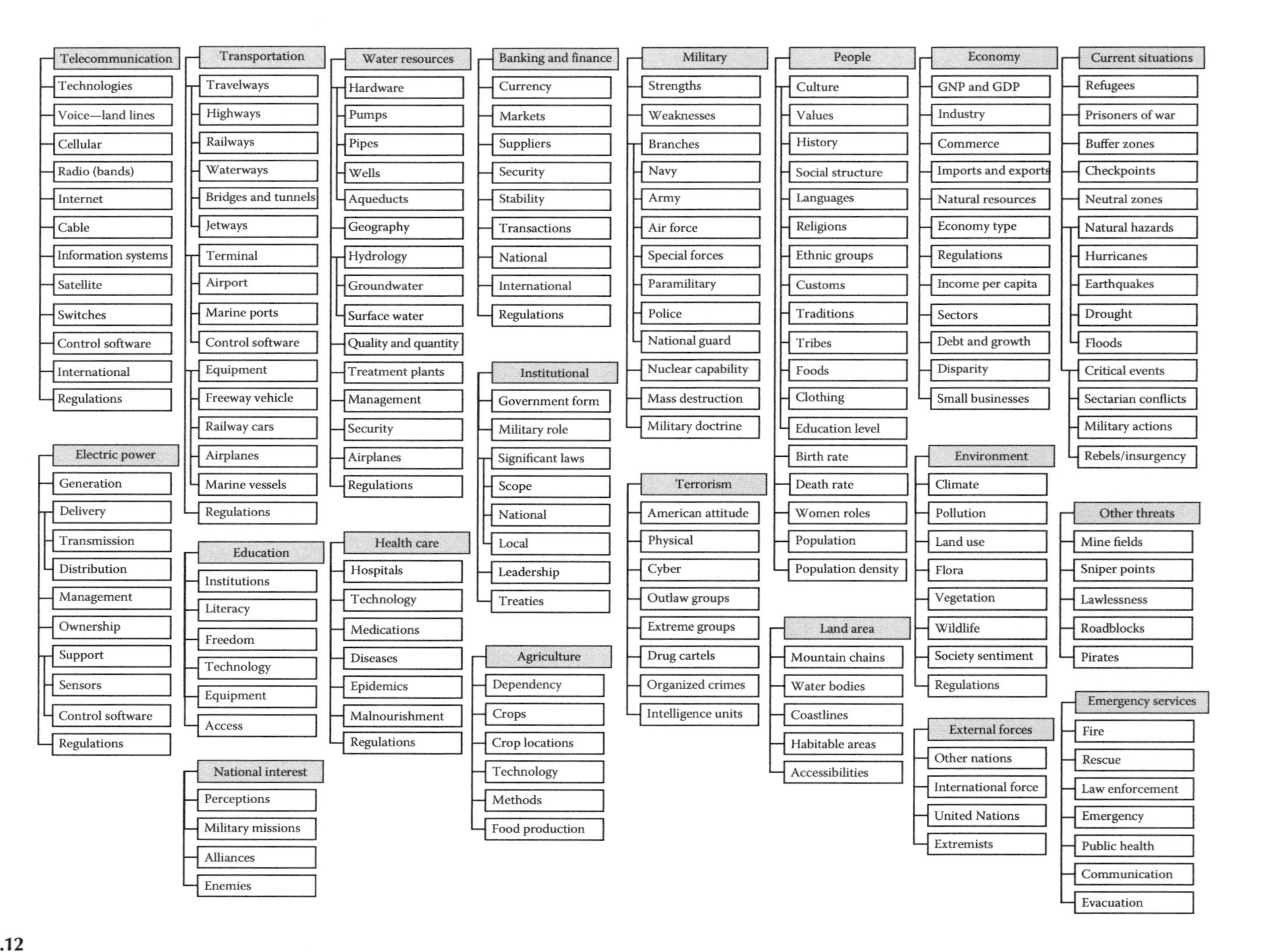

FIGURE 2.12
Categories for risk event and factor identification. Potential interdependencies among all categories should be examined.

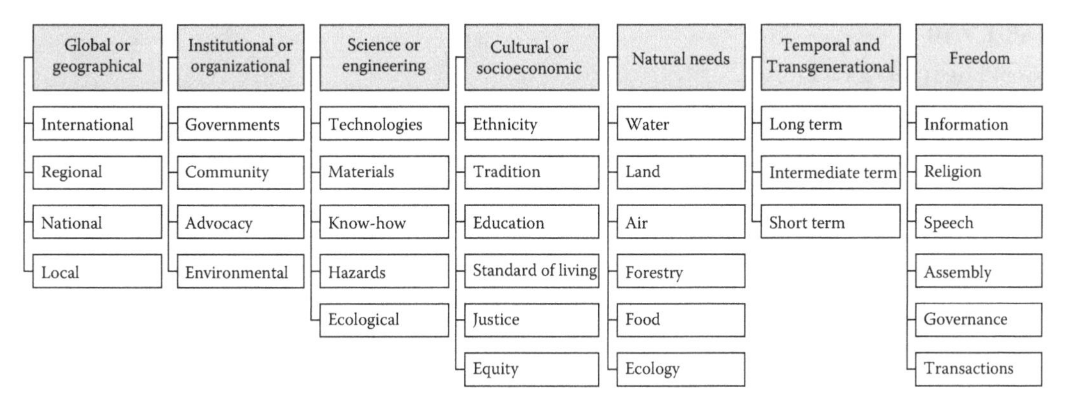

FIGURE 2.13
Simplified categories for risk event and factor identification. Potential interdependencies among all categories should be examined.

previously presented items and categories can be viewed as simple lists of potential sources of risk and provide a set of headings under which risks can be arranged. These lists are sometimes referred to as *risk taxonomy*. A simple list of risk sources might not provide the richness necessary for some decision situations, as it presents only a single level of organization. Some applications might require a full hierarchical approach to define the risk sources, with as many levels as required to provide the necessary understanding of risk exposure. Defining risk sources in such a hierarchical structure is referred to as a *risk breakdown structure* (RBS). The RBS is defined as a source-oriented grouping of project risks organized to define the total risk exposure of a project of interest. Each descending level represents an increasingly detailed definition of risk sources for the project. The RBS can help analysts understand the risks faced by the project or organization.

An example RBS is provided in Table 2.10. In this example, four risk levels are defined. The project risks are viewed as level 0. Three types of level 1 risks are provided in the table for the purpose of demonstration. The number of risk sources in each level varies

TABLE 2.10
RBS for a Project

Level 0	Level 1	Level 2	Level 3
Project risks	Management	Corporate	History, experiences, culture, personnel Organizational structure, stability, communication Finance conditions Other projects etc.
		Customers and stakeholders	History, experiences, culture, personnel Contracts and agreements Requirement definition Finances and credit etc.
	External	Natural environment	Physical environment Facilities, site, equipment, materials Local services etc.

(*Continued*)

TABLE 2.10
(Continued) RBS for a Project

Level 0	Level 1	Level 2	Level 3
		Cultural	Political Legal, regulatory Interest groups Society and communities etc.
		Economic	Labor market, conditions, competition Financial markets etc.
	Technology	Requirements	Scope and objectives Conditions of use, users Complexity etc.
		Performance	Technology maturity Technology limitations New technologies New hazards or threats etc.
		Applications	Organizational experience Personnel skill sets and experience Physical resources etc.

and depends on the application at hand. The subsequent level 2 risks are grouped and then detailed further in level 3. The RBS provides a means to identify systematically and exhaustively all relevant risk sources for a project.

The RBS should not be treated as a list of independent risk sources, as they are commonly interrelated and have common risk drivers. Identifying causes behind the risk sources is a key step toward an effective risk management plan, including mitigation actions. A process of risk interrelation assessment and root cause identification can be utilized to identify credible root cause factors and could lead to effective risk management.

2.4.2 Identification of Risk Events and Factors

The risk assessment process starts with the question, "What can go wrong?" The identification of what can go wrong entails defining hazards, risk events, and risk scenarios; the previous section provided the categories of risk events and scenarios. Risk identification involves determining which risks might affect a project or system and documenting their characteristics, and generally requires participation from a project team, risk management team, subject matter experts from other parts of an organization, customers, end users, other project managers, stakeholders, and/or outside experts on an as-needed basis. Risk identification can be an iterative process. The first iteration may be performed by selected members of the project team or by the risk management team. The entire project team and primary stakeholders may take a second iteration if needed. To achieve an unbiased analysis, persons who are not involved in the project may perform the final iteration. Risk identification can be a difficult task because it is often highly subjective, and no unerring procedures are available which may be used to identify risk events and scenarios other than relying heavily on the experience and insights by key project personnel.

For example, construction projects typically entail risk events and factors that can be grouped as follows:

- *Technical, technological, quality, or performance risks*, such as unproven or complex technology, unrealistic performance goals, and changes to the technology used or the industry standards during the project
- *Project-management risks*, such as poor allocation of time and resources, inadequate quality of the project plan, and poor use of PM disciplines
- *Organizational risks*, such as cost, time, and scope objectives that are internally inconsistent; lack of prioritization of projects; inadequacy or interruption of funding; resource conflicts with other projects in the organization; errors by individuals or by an organization; and inadequate expertise and experience by project personnel
- *External risks*, such as shifting legal or regulatory environment, labor issues, changing owner priorities, country risk, and weather
- *Natural hazards*, such as earthquakes, floods, strong wind, and waves that generally require disaster recovery actions in addition to risk management

Within these groups, several risk events and factors can be identified as listed in Table 2.11 at various stages of the life cycle of a project.

TABLE 2.11

Risk Events and Scenarios

Risk Event Category or Scenario	Description
Unmanaged assumptions	Unmanaged assumptions are neither visible nor apparent as recognizable risks. They are commonly introduced by organizational culture; when they are unrecognized in the project environment, they can bring about incorrect perceptions and unrealistic optimism.
Technological risk	A technological risk can arise from using unfamiliar or new technologies. At one end is application of the state-of-the-art and familiar technology, where the technological risk can be quite low. At the other end, a new technology is used which generates the greatest uncertainty and risk.
Economic climate	For example, uncertain inflation rates, changing currency rates, etc. affect the implementation of a project in terms of cash flow. A forecast of the relative valuations of currencies can be relevant for industries with multinational competitors and project partners.
Domestic climate	Risk events in this category include tendencies among political parties and local governments, attitudes and policies toward trade and investment, and any recurring governmental crises.
Social risks	Risks in this category are related to social values such as preservation of environment. Some projects have been aborted due to resistance from the local population.
Political risks	Political risks are associated with political stability both at home and abroad. A large investment may require looking ahead several years from the time the investment is made.
Conflicts among individuals	Conflicts can affect the success of a project. These conflicts could arise from cognitive differences or biases, including self-motivated bias.

(*Continued*)

TABLE 2.11

(Continued) Risk Events and Scenarios

Risk Event Category or Scenario	Description
Large and complex project risks	Large and complex projects usually call for multiple contracts, contractors, suppliers, outside agencies, and complex coordination systems and procedures. Complex coordination among the subprojects is itself a potential risk, as a delay in one area can cause a ripple effect in other areas.
Conceptual difficulty	A project may fail if the basic premise on which it was conceived is faulty. For example, if an investment is planned to remove some of the operational or maintenance bottlenecks that ignore market requirements and forces, the risk of such a project not yielding the desired financial benefits is extremely high.
Use of external agencies	Appointing an external agency as project manager without creating a large project organization may not ensure the kind of ownership required for successful implementation or elimination of defects that the client has observed.
Contract and legal risks	A contract is an instrument to transfer the risk from the owner to the contractor. The contractor risks only his fees, whereas the owner runs the risks, for example, of ending up with no plant at all. Although there are many contractual modes available (e.g., multiple split contracting, turnkey, engineering procurement/construction commissioning), none of these comes without risks.
Contractors	Contractor failure risk may originate from the lowest cost syndrome, lack of ownership, financial soundness, inadequate experience, etc. In the face of intense competition, contractors squeeze their profit margins to the maximum just to stay in business. Contractors sometimes siphon mobilization advances to other projects in which they have a greater business interest. If a contractor has difficulty with cash flow, then the project suffers.

Development of scenarios based on these events and factors for risk evaluation can be created deductively [e.g., fault tree (FT)] or inductively [e.g., failure mode and effect analysis (FMEA)]. These methods are discussed in Section 2.5. These methods assess the likelihood or frequency for scenarios expressed either deterministically or probabilistically. Also, they can be used to assess varying consequence categories, including items such as property loss and life loss or injuries.

Example 2.1: Project Risks for Warehouse Automation

ABC Grocery and Supermarket Outlets desires to automate its warehouse by installing a computer-controlled order-packing system, along with a conveyor system for moving goods from storage to the warehouse shipping area. Four parties are involved in this project: (1) client, (2) project manager, (3) engineer, and (4) contractor, as shown in Figure 2.14. The figure also shows the relationships among the parties, either by contract or as an exchange of technical information. The risk events and scenarios associated with this project can be constructed based on the perspectives of the four parties as provided in Tables 2.12 through 2.15.

The risk sources related to the automated warehouse project can be structured in a multilevel hierarchy. The project risks are divided into three risk levels for the purposes of managing risks in this case as follows: internal risks, external risks, and technology risks. The description of each risk level is provided in Table 2.16. The table shows the RBS for the entire project based on the total vulnerability of the project. This structure provides insights into how the parties involved in any project should take into consideration the three main levels of risk mentioned.

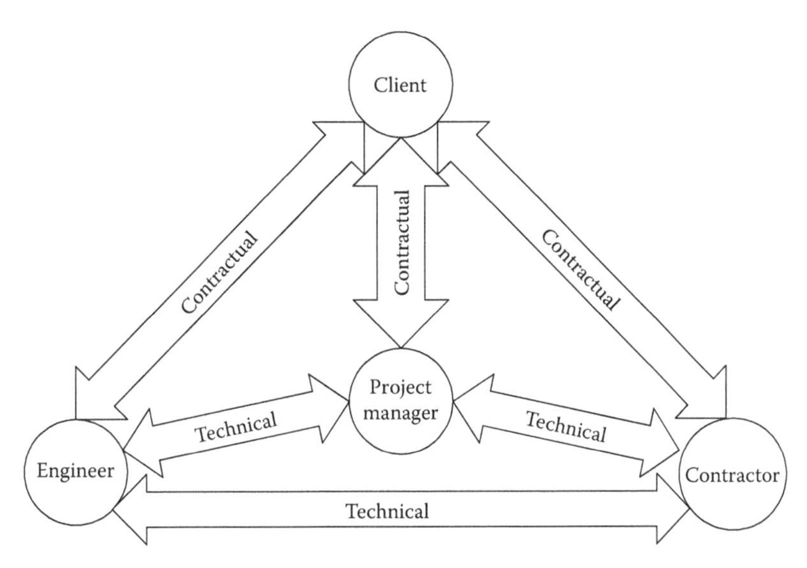

FIGURE 2.14
Relationships among the four parties involved in a project.

TABLE 2.12
Client Perspective of Risks Associated with the Project

Risk Category or Scenario	Description
Technological, quality, or performance risks	Client concerns include poor-quality products of various components of the project. The poor quality might result from using unfamiliar types of technology in construction. Additionally, the performance of other parties involved can be of great concern to the client during project execution, e.g., an incompetent project manager or engineer or a troublesome contractor.
Project management (PM) risks	The manager of the project can be a source of risk. Commonly, a PM company works on behalf of the client to handle all project aspects for a percentage of the total project cost. The client in this case is exposed to the risk of having an incompetent PM company.
Economic risks	This category includes uncertainty in inflation rates and/or changes in currency rates posing sources of risk to the client. The cash flow of the project would be affected, creating risks of delays or even bankruptcy, especially if the project is executed in another country.
Conflict among individuals	Conflicts among project parties pose a risk to the client. For example, a potential conflict in scheduling and work execution could materialize among the subcontractors or vendors of belt conveyor systems.
Contractor risks	An incompetent contractor with weak cash flow or inadequate personnel experience can be a source of risk for the client.
Contract and legal risks	This category covers the possibilities of having contractual and legal disputes among the parties. These disputes might lead to difficulties in executing or operating the project, including abandoning the project.
External risks	The client needs to be aware of any changes in regulations and laws related to the project, as licenses and permits for the project can be affected by changes in governmental regulations. If the project is constructed in a foreign country, this source of risk could be a significant one.

TABLE 2.13

Project Manager Perspective of Risks Associated with the Project

Selected Risk Category or Scenario	Description
Project management (PM) risks	PM companies should be concerned with proper allocation of budget, time, and personnel for completing a project on time and within budget. Risks in this category include improper allocation of resources.
Technological, quality, or performance risks	PM companies are concerned with the final outcome of the project. Although the project has to be finished on time and within budget, the best quality and performance must also be achieved in order to ensure a continuous workload for such companies from the same client or others. Establishing a reputation of quality work and a successful performance record are keys to success. Risks in this category include inadequate performance and improper use of technologies.
Contractor risks	Incompetent contractors or subcontractors with weak planning procedures and inefficiency in finishing tasks on time are risks to PM companies. For example, not completing some items related to the conveyor system could delay other tasks and completion of the entire project. The primary objective of the project manager is to fulfill relevant contracts with the client. Any events that lead to not fulfilling these contracts should be identified and risks and scenarios mitigated.
Contract and legal risks	This category covers the possibilities of having contractual and legal disputes with the other three parties. These disputes might lead to project delays and affect the performance of the project manager.
External risks	Political and governmental matters might affect the work of the project managers, especially when working internationally.

TABLE 2.14

Engineer Perspective of Risks Associated with the Project

Selected Risk Category or Scenario	Description
Technological, quality, or performance risks	Engineering companies working on-site for supervising contracted work have the objectives of completing tasks on time and within budget and complying with design and quality standards. Use of equipment and technological innovations, such as automation, might provide risk sources. Also, the performance of the contractor in these cases can be a critical issue that could affect engineering companies to a great extent. Risks associated with technology use, quality, and performance must be identified and mitigated.
Contractor risks	Engineers are responsible for accepting and signing off finished work as fully executed; therefore, the risk of accepting finished products of poor quality from a contractor exists, especially if the contractor is assigned or selected by the project manager and not within the contractual control of the engineer.
Contract and legal risks	This category covers the possibilities of having contractual and legal disputes with the other three parties, e.g., with the client.
External risks	Risks in this category might arise from working in a foreign environment or within complex governmental regulations.

TABLE 2.15

Contractor Perspective of Risks Associated with the Project

Selected Risk Category or Scenario	Description
Technological, quality, or performance risks	Risks might arise from using new technologies during construction as requested by the engineer or project manager. Additional risks include either producing poor-quality products or nonperformance related, e.g., to the use of new automation systems.
Conflict among individuals	Personnel of contractors can be a source of risk, especially in cases involving multinational or labor forces of diverse backgrounds at the same site. Another source of conflict is dealing with suppliers or vendors of different work attitudes or languages.
Contractor risks	This category includes an inadequate cash flow over the period of the project performance, improper scheduling of activities, or inadequate control of the contractor or subcontractors, leading to project delays and potentially defective products.
Contract and legal risks	Problems with the client can accumulate if the project manager reports to the client are not representative of actual performance. The risk of losing the contract or contractual disputes can arise from a lack of performance reports.
Use of external agencies	Working with subcontractors, suppliers, and vendors can produce risks to the contractor. Diligence is required when selecting subcontractors, and control and monitoring procedures must be placed over external agencies.
External risks	Political and governmental risks can also affect the contractor. International contractors could be exposed to additional risks associated with work in foreign countries. Additionally, the four parties share some common risk sources, such as earthquakes, flood, strong winds, or even uncertain political and economical events beyond their expectations or business models.

TABLE 2.16

RBS for the Warehouse Automation Project

Level 0	Level 1	Level 2	Level 3
Automated warehouse project risks	Management	Corporate	Risks related to retaining parties and personnel of all parties involved in the project within organizational structure flexibility
			Risks related to maintaining a structural flexibility
			Risks related to deciding on new projects
			Risks associated with continued financing of the project
			Risks associated with management interests and related conflicts
		Customers and stakeholders	Risks associated with lack of understanding of the intention of the project to serve customers and client requirements
			Failure to satisfy customers with regard to final packing of products including their satisfaction of on-time and adequate delivery of products
			Risks associated with conflicts in objectives of stakeholders and parties
			Risks associated with continued progression of the project
	External	Natural environment	Risks associated with the environment of execution of the project
			Risks associated with the site of the project, such as maneuvering and mobilizing equipment to and from the site
			Risks associated with local services, and planning procedures and permissions

(*Continued*)

TABLE 2.16

(Continued) RBS for the Warehouse Automation Project

Level 0	Level 1	Level 2	Level 3
		Cultural	Risks associated with cultural diversity among parties or even among personnel within a company
			Risks associated with political and governmental regulations, especially if executed in a foreign country
		Economical	Risks associated with working in an uncertain or risky market without good marketing study
			Risks associated with facing undesired financial situations because of competition
			Risks associated with changes in the currency rates
	Technology	Requirements	Risks associated with technology requirements and availability of resources needed for technology, such as personnel and funds
			Risks associated with complexity
			Risks associated with changes in project scope due to technology changes
			Risks associated with unfamiliarity with new technology
		Performance	Risks associated with changes in technology related to project leading to new demands on staff, equipment, and financial resources
			Risks associated with new technologies requiring staff training leading to high cost of operation beyond budgeted resources
			Risks associated with new hazards as a result of new technologies
		Application	Risks associated with applying newly introduced types of technologies
			Risks associated with maintaining key persons with experience needed for technologies
			Risks associated with staff resistance to a change to new technological applications
			Risks associated with increased demand on resources as a result of new technologies

Example 2.2: Enterprise RBS

The ISO definition of risk as "the effect of uncertainty on objectives" discussed earlier can be used as the basis for addressing enterprise risks, and the three keywords can be used in reverse order of their mention in the definition to develop a methodology as follows: (1) defining the objectives, (2) examining the uncertainty, and (3) assessing the effect. It should be noted that the effect can be a favorable or an adverse deviation from the objectives as defined. A four-level hierarchical definition of potential uncertainty and risk factors that could affect the objectives including portfolio-level and enterprise-level factors to account for interdependence is used herein. The effects of uncertainty on objectives, including adverse and favorable, should be quantified in the third step. The quantification of these effects combined with the identification of potential actions is a necessary step for decision making in order to enable benefit–cost analysis.

A multilevel structure, starting with programs to business units to subsidiaries to the entire enterprise as a system with self-similarity, ensures consistency and permits risk aggregation and segregation for examining risk profiles by program, unit, subsidiary, and the entire enterprise using several formats necessary to inform decision makers at various levels. Designing for self-similarity as illustrated in Figure 2.15 would drive consistency, offer simplicity in representing a complex framework, and enhance acceptance and embracement of a change of an organizational cultural. Such an organization-wide undertaking might require structural changes to the enterprise's organization, such as the formation of a risk management executive committee, appointing a chief risk officer, and creating risk functions at appropriate organizational levels.

FIGURE 2.15
Self-similarity in structuring enterprise risk.

Table 2.17 provides an outline of an RBS for an Engineering, Procurement, and Construction (EPC) enterprise. The enterprise-wide impacting factors are listed in level 0, whereas the project- and portfolio-level factors are listed in levels 1–3 in the form of a hierarchy. The entire hierarchy is not shown in the table, but only the first level branching, that is, level 1 includes site, technology and technical, labor, and so on (see Table 2.17); level 2 provides details on the first entry in level 1 of site by listing availability, suitability, transportation and logistics, and so on; and level 3 in turn details the first entry in level 2 of availability under site by showing delay to ownership, delay to permit, adequacy of permit, and so on. Other branches can be constructed in a similar manner. Only level 1 might be needed to examine prospects for bidding, whereas levels 1 and 2 are needed during bidding. As for project execution, levels 1–3 should be used to define risk factors for the activities in a project's work breakdown structure.

The statement on objective in the risk definition defines the scope of a risk assessment. The risk assessment scopes may include strategic risk assessment, operational risk

TABLE 2.17

Risk Breakdown for an EPC Enterprise

Level 0 Enterprise	Level 1 Prospects	Level 2 Bidding	Level 3 Execution
1. Strategic:	Site	Site (as an example):	Site/availability (as an example):
Reputational damage	Technology and technical	Availability	Delay to ownership
Competition	Labor	Suitability	Delay to permits
Customer wants	Materials and equipment	Transportation and logistics	Adequacy of permits
Demographic	Equipment for construction	Utilities	Regulatory agency requirements
Social/cultural trends	Procurement sources	Communications	Restrictions and easements
Technological innovation	Subcontractors	Conceptual difficulty	Residual war risks (mines, unexploded ordinance)
Capital availability	Commercial	First of a kind	Seashore use rights
Regulatory and political trends	Hazards	Unmanaged assumptions	Air use rights
2. Financial:	External	Local codes and standards	
Price		Scope definition	
Liquidity		Technical interfaces	
Credit		Fabrication and construction	
Inflation/purchasing power		Mining processes	
Hedging/basis risk		Environmental restoration	
3. Operational:		Services	
Business operations		Hazardous aspects	
Empowerment			
Information			
Information/business reporting			
4. Hazards:			
Fire			
Property damage			
Natural perils			
Theft and other crime			
Personal injury			
Business interruption			
Disease and disability			
Liability claims			
5. Assets:			
Physical and intellectual			
Financial			
Customer related			
Hires			
Organizational			
6. Environments:			
Markets			
Sovereign or political			
Legal or regulatory			
Attitudes or sentiments			
Acceptance or sensitivity			
Technological innovation			
Competition			
Catastrophic events			

assessment, compliance risk assessment, internal audit risk assessment, financial statement risk assessment, fraud risk assessment, market risk assessment, credit risk assessment, customer risk assessment, supply chain risk assessment, product risk assessment, security risk assessment, information technology risk assessment, project risk assessment, first-of-a-kind technology risk assessment, portfolio risk assessment, sector risk assessment, and logistics risk assessment.

2.4.3 Root Cause Analysis

Root cause analysis can be used to identify and correct the primary reasons for functional and operational problems that could lead to accidents or have led to accidents in case of accident investigation including forensic analysis. It effectively uncovers the fundamental issues (i.e., root causes) that generate a problem, as opposed to troubleshooting and problem solving that seek immediate solutions to resolve visible symptoms.

Causes, also called *causal factors*, are defined as conditions or events that result in or participate in the occurrence of an effect, such as failure, loss, and damage. They can be classified as follows:

- Direct cause: a cause that results in the occurrence
- Contributing cause: a cause that contributed to the occurrence, but would not have caused it by itself
- Root cause: a cause that, if corrected, would prevent recurrence of an event and similar occurrences; generally, the root cause usually has generic implications to a broad group of possible occurrences, and it is the most fundamental cause that can logically be identified and corrected.

Once these causes are identified, a cause-and-effect diagram representing the sequence of events and causal factors can be constructed. The diagram shows the specific actions that could have created a condition and contributed to an event. This event creates new conditions that, in turn, result in another event. Earlier events or conditions are placed in a sequence in the diagram and are called upstream factors.

Once the root causes with the contributing factors are identified, corrective and preventative actions can be identified for decision making based on cost-effectiveness and impacts on ongoing and future operations and systems. The root causes, contributing factors, and corrective and preventative actions can be summarized in a table to facilitate comparison and decision making. *Barrier analysis* can be used to identify actions. A barrier can be defined as an object or action to arrest the accident evolution so that the next event in the chain is never realized. Barrier systems are those maintaining barrier functions, such as an operator, an instruction, a physical separation, an emergency control system, and other safety-related systems, components, and human factors and organizational units. Examples of control barriers include conductors, approved work methods, job training, disconnection switches, pressure vessels, and so on. Examples of safety barriers are protective equipment, guard rails, safety training, work protection code, emergency contingency plans, and so on (Svenson 1991; Trost and Nertney 1985).

Several methods are available to help with the root cause analysis, such as cause-and-effect analysis using Ishikawa diagrams, events and causal factor analysis, event and FT diagrams that are discussed in Section 2.5.6, and events and causal factor analysis. These methods are discussed and illustrated in subsequent sections.

2.4.3.1 Events and Causal Factor Analysis with Barrier and Change Analyses

Causal factor charting provides a graphical or tabular structure for investigators or forensic analysts to organize and analyze the information gathered during the investigation as a sequence of boxes with logic tests that describe the events leading up to an incident occurrence. Time or timeline is not shown in these charts. Typical analytical steps include the following:

- Define the sequence of events before, during, and after the accident.
- Define any conditions that affected the events.
- As they become known, define any causes for each event.

For example, consider the case of a worker falling from a ladder installed at a manufacturing plant. In this case, the following initial event and cause table can be constructed:

Initial Event and Cause Table

Event	Condition	Causes of the Event
Fixed ladder installed in 2001	Ladder was not compliant with Occupational Safety and Health Administration (OSHA) requirements	(To be entered later once known)
Worker climbed ladder	Ladder steps were wet	
Worker fell from ladder		

In barrier analysis, barriers generally fall into the following categories: equipment, design, administrative (procedures and work processes), supervisory/management, warning devices, knowledge and skills, and physical. For example, consider the same case of a worker falling from a ladder. In this case, the following table can be constructed for barriers:

Barrier Analysis Table

Barriers to Prevent Accident	Barrier Performance	Reasons for Barrier Failure	Evaluation of Effect
Barrier 1: slip-resistant steps	The steps were smooth	The barrier did not exist	The steps did not provide traction
Barrier 2: proper climbing technique	Not used	Employee was carrying tools with one hand	Second hand was not available to stop fall

Change analysis is sometimes necessary to identify additional causes and factors for an accident. Considering the same case of a worker falling from a ladder, the following table can be constructed for changes:

Change Analysis Table

Situation at the Time of the Accident	Prior, Ideal, or Accident-Free Situation	Difference	Evaluation of Effect of the Change in Situation
Worker came early to avoid the heat	Worker started the day the same time as coworkers	No coworkers were available to help with the job	Change 1: Worker came to work early, so was working alone, carrying tools by one hand
Worker did not meet with supervisor the morning of the accident	Worker met with supervisor to discuss the day's work activities	Work activities were not discussed	Change 2: Since worker came to work early, job hazards were not discussed

These barrier and change analyses help to identify the causes that are classified into direct cause, contributing cause, and root cause. The table initial event and cause table can be revised as follows:

Updated Event and Cause Table

Event	Condition	Causes of the Event
Fixed ladder installed on building in 2001	Ladder was not compliant with Occupational Safety and Health Administration (OSHA) requirements	Ladder produced on-site by the workers
Worker climbed ladder	Ladder steps were wet	Weather conditions not considered before starting job
Worker fell from ladder		Barrier 1: Steps did not provide traction Barrier 2: Second hand was not available to stop fall Change 1: Employee came to work early, so was working alone, carrying tools by one hand Change 2: Since employee came to work early, job hazards were not discussed with supervisor

Based on these analyses, the level of responsibility for each causal factor can be determined as follows:

Responsibility Level Table

Management Tier	Causal Factor	Responsibility
President of company	Barrier 1: ladder rungs not Occupational Safety and Health Administration (OSHA) compliant	Failed to hold managers accountable for deficient items
Supervisor	Change 2: hazards not discussed	Allowed schedule change without determining impact
Worker	Change 1: carrying tools up ladder without assistance	Did not follow safety rules

2.4.3.2 Cause-and-Effect Analysis Using Ishikawa Diagrams

Cause-and-effect analysis examines the causes and effects leading to an occurrence, such as an accident or a particular event, and uses fishbone, herringbone, cause-and-effect, Fishikawa, or Ishikawa diagrams to show the causes leading to a particular event, attributed to Kaoru Ishikawa (in the 1960s) who pioneered quality management processes in the Kawasaki shipyards. Causes are usually grouped into major categories:

- People involved with the process
- Methods used in the process and the specific requirements including policies, procedures, rules, regulations, and laws
- Machines including any equipment, computers, tools, and so on required to accomplish the job
- Materials including raw ingredients parts, pens, paper, and so on used to produce the final product

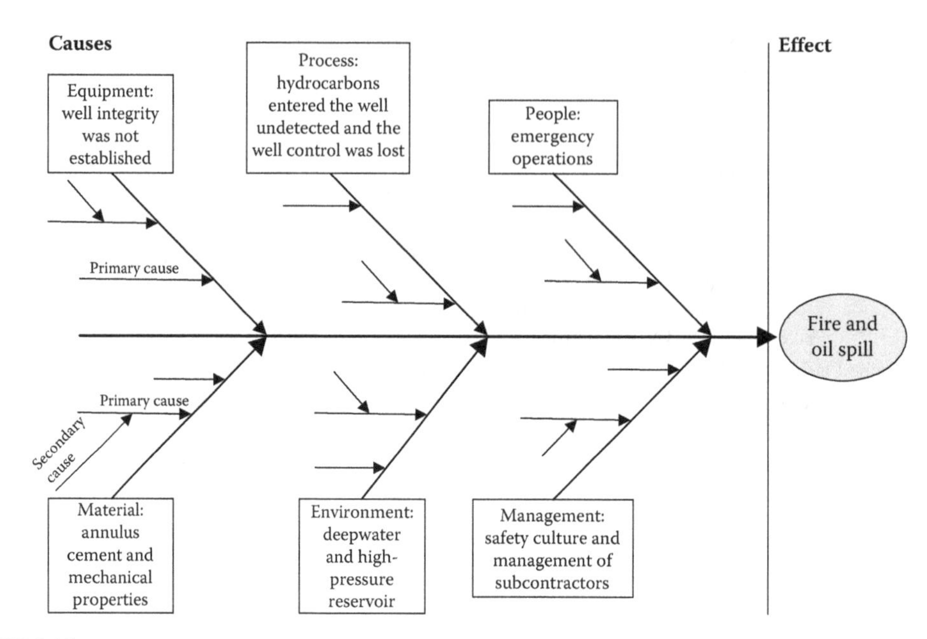

FIGURE 2.16
Ishikawa diagram for an oil spill.

- Measurements including data generated from the process that are used to evaluate its quality
- Environment including the conditions, such as location, time, temperature, and culture in which the process operates

The 2010 Deepwater Horizon oil spill (also referred to as the BP oil spill or the Macondo blowout) led to oil flowing unabated for 3 months until its control. This was the largest accidental marine oil spill in the history of the petroleum industry, stemming from a sea floor, and resulted in an explosion that burnt the Deepwater Horizon which drilled on the BP-operated Macondo Prospect, killed 11 men working on the platform, injured 17 others, and releasing about 4.9 million barrels of crude oil (see Whitehouse.gov 2010). The oil spill event causes can be analyzed as outlined in Figure 2.16 for the purpose of illustration.

2.4.3.3 Events and Causal Factors Analysis, and Pareto Analysis

Events and causal factors analysis consists of the identification of a series of events and factors using a time sequence leading to an incident occurrence or a particular outcome of interest. The factors include tasks and/or actions as well as their environmental conditions. The events and causal factors including tasks and/or actions are represented graphically on the timeline along with the relationships.

For example, consider a hypothetical drowning case of a two-year child wearing a flotation outfit in a 2.5-m-diameter inflatable pool in the backyard of a residential house with 0.25 m water depth. Figure 2.17 provides an events and causal factors diagram for this hypothetical case. The rectangles are the events and the ellipses are the factors (or tasks.) The event boxes show the date and the time where available. The numbers placed next to the ellipses classify the factors in terms of their criticality or effect on the respective events as (1) no impact on event, (2) contributor to event, and (3) directly impacting event.

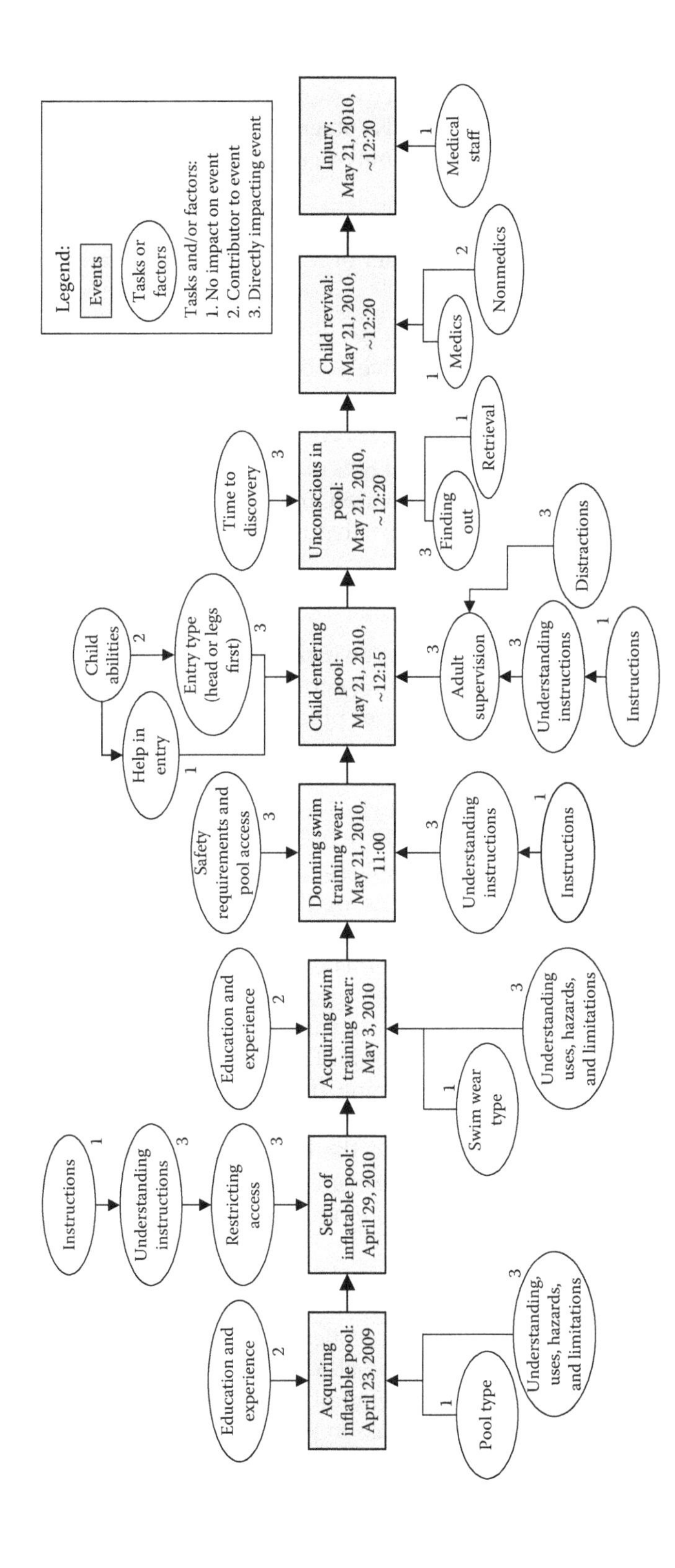

FIGURE 2.17
Events and causal factors diagram of child drowning.

Pareto analysis (also known as the 80-20 rule, the law of the vital few, named after an Italian economist Vilfredo Pareto as observed in 1906) is used in decision making for selection of a limited number of tasks that produce a significant overall effect. The idea is that 20% of work or factors provide 80% of the advantage or doing the entire job. The steps to identify the important causes using Pareto analysis are as follows:

- Develop a table listing the causes and their frequency as a percentage.
- Arrange the rows in the decreasing order of importance of the causes, that is, the most important cause first.
- Compute the cumulative percentage.
- Plot with causes on x-axis and cumulative percentage on y-axis.
- Sequentially select the causes that lead to a total cumulative frequency of 80%.

The subset of the causes selected accounts for 80% of the frequency and constitutes the list of causes that should receive careful examination and further analysis. The other causes excluded should be examined, but not necessarily to the same level of detail.

2.4.4 Precursor Event Analysis

A *precursor event* (PE) is an event that precedes an incident and substantially reduces safety margins, for example, a PE of materials cracking and pipe leaking that could have resulted in a loss-of-coolant accident in a nuclear power plant (NPP) or a PE of a nongovernment adversary successfully acquiring radioactive materials that could have resulted in the detonation of a dirty bomb. *Precursor event analysis* (PEA) is the process of breaking up the whole into its component parts to relate PEs to incidents and to assess related probabilities and uncertainties using available information. In PEA relating to accidents, PEs are operational events that may cause accidents in complex systems, such as core damage in an NPP. In the homeland security (HS) context, the PEs are the events, such as arrests of terrorists and disabling terrorist cells and network members, which prevent progression of potential scenarios that could lead to successful attacks. The PEA includes probabilistic modeling, statistical data analysis of PEs, and respective probabilistic projections. The PEA is a convenient tool for complex system safety monitoring and analysis, and is briefly discussed in this section based on the work for the safety of NPPs (e.g., Modarres et al. 1999).

PEA includes three major steps: (1) screening using *event trees*, that is, identification of events with anticipated high conditional probabilities of severe incident p_i given PE i; (2) quantification, that is, estimation of p_i and the observed rate of occurrence of severe incidents λ; and (3) trend analysis to assess the overall system performance and prediction. The three steps with appropriate estimation methods are described as follows:

- In the screening step, event trees can be used to identify events having high anticipated p_i values. The conditional probabilities of severe incident events given PE i ($i = 1, 2, \ldots$), p_i, are estimated based on data collected on observed operational events in order to identify those events that are above a conditional probability threshold level. Further, it is assumed that the number of precursors observed in exposure time t follows the homogeneous Poisson process (HPP) with a rate λ, and p_i is assumed to be based on an independently identically distributed (IID) continuous random variable having a truncated probability density function $h(p)$ defined by the said threshold.

- In the quantification step, p_i and λ are estimated based on HPPs. If n is the observed incident count during a time period of length t (e.g., real time, reactor-years, flight hours), then an estimate for the incident occurrence rate λ is n/t. This estimator unfortunately is not always useful since severe incidents are rare, and therefore, one must use postulated PEs. For NPPs, Apostolakis and Mosleh (1979) used observed data to compute the conditional probability p_i that the incident occurs given that PE i happens. It should be noted that PEs are usually distinct events of *different* types. When the p_i are added together—using only those that exceed 10^{-6} (a rule of thumb used in NPP safety analyses) to obtain the following estimator: $\hat{\lambda} = \sum_i p_i/t$.
- In the trend analysis and prediction step, the overall safety trend is assessed, and predictions are made. At this stage, the raw data include dates, PEs, and their p_i only. The data are used to estimate an annual rate of incidents based on these PEs. For example, Modarres et al. (1996) used PE data relating to NPPs to develop Table 2.18 that are plotted in Figure 2.18 to illustrate the trend. The final results of PE analyses applied to NPPs are expressed in the form of the annual rates of core damage. The figure displays increasing safety of NPPs after 1984.

TABLE 2.18

Analysis of Precursor Data for Nuclear Power Plants (NPPs)

Year	Estimate of Annual Core Damage Rate of Occurrence (per Year)
1984	1.1×10^{-4}
1985	2.0×10^{-4}
1986	1.6×10^{-4}
1987	1.3×10^{-4}
1988	1.1×10^{-4}
1989	9.3×10^{-5}
1990	8.6×10^{-5}
1991	9.0×10^{-5}
1992	8.1×10^{-5}
1993	7.2×10^{-5}

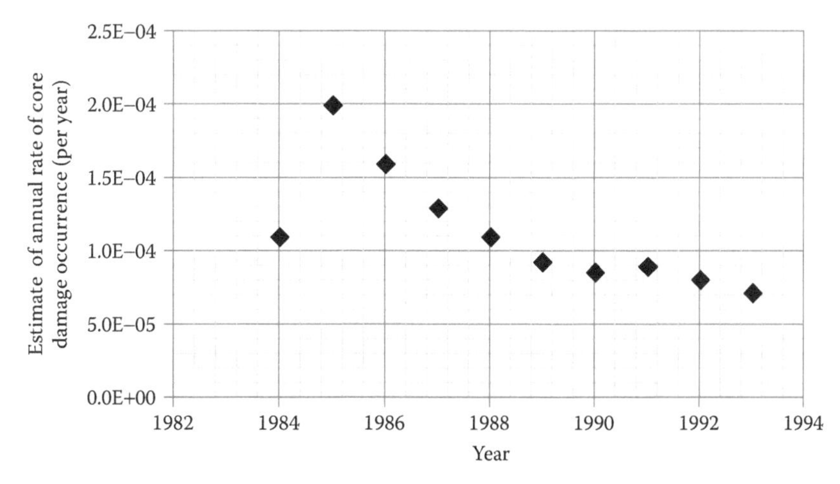

FIGURE 2.18
Precursor analysis: Trend of annual rate of core damage occurrence.

In the case of HS, the trend analysis requires data relating to arrests of terrorists and disabling terrorist cells and network members. Such data might be available from law enforcement officers and intelligence agencies. The data can then be used to estimate the annual rate of occurrence of relevant terrorist attacks similar to the core damage annual rates provided in the table.

In this type of analysis, the prediction of losses (i.e., consequences) using extreme value theory might be needed. Estimating the probability that a future loss S exceeds a given threshold s based on historical data can be treated as PEs. For each past attack, its associated loss S is treated as a continuous, IID random variable. If its cumulative distribution function is $F_S(s)$, then the maximum loss $F_{S\max}(s)$ among the nonrandom number of such PEs n is distributed according to the cumulative distribution function, $F_{S\max}(s) = (F_S(s))^n$.

As an example, an airport receives hundreds of security hits per year that can be grouped into three types in terms of increasing criticality (type 1, type 2, and type 3). These hits have not resulted recently in any incidents; however, they can be treated as PEs. The conditional probabilities of an incident given a hit of a particular type were estimated using event tress to be 0.001, 0.005, and 0.01, respectively. The annual rate of an incident based on the numbers of hits in two years for the three types of 200, 10, and 1, respectively, can be estimated approximately as $(200 \times 0.001 + 10 \times 0.005 + 1 \times 0.01)/2 = (0.2 + 0.05 + 0.01)/2 = 0.13$ per year. This approximate model is appropriate when dealing with small probabilities.

2.5 Risk Assessment Methods

2.5.1 Introduction

Defining an underlying system is an important first step in performing a risk assessment, as detailed in Chapter 3. The examination of a system must be made in a well-organized and repeatable fashion so that risk analysis can be performed consistently, thus ensuring that the important elements of a system are defined. Also, it ensures including relevant information and omitting extraneous information. The formation of system boundaries should be based on the objectives of the risk analysis.

Defining a system as a group of interacting, interrelated, or interdependent elements, such as people, property, materials, environment, and processes, requires addressing three of its aspects: (1) boundaries, (2) resolution or level of details, and (3) interactions among its components. These three aspects are introduced in this section.

Delineating *system boundaries* can assist in developing the system definition. Establishing the system boundary is partially based on what aspects of the system's performance are of concern. The selection of items to include within the external boundary region also depends on the objectives of the analysis. This is an important step of system modeling, as the comprehensiveness of the analysis depends on the system boundaries defined. Beyond the established system boundary is the environment external to the system.

Boundaries beyond the physical and/or functional system can also be established. For example, time may also be a boundary because an overall system model may change as a product progresses further along in its life cycle. The life cycle of a system is important because some potential hazards can change throughout the life cycle. For example, corrosion or fatigue may not be a problem early in the life of a system; however, they may become important concerns later in the life cycle of the system.

Along with identifying the boundaries, it is also important to establish a *resolution* limit for the system. The selected resolution is important as it limits the detail of the analysis. Providing too little detail might not provide enough information for the decision making. Too much information may make the analysis more difficult and costly due to the added complexity. The detail depth of the system model should be sufficient for the specific problem. Resolution is also limited by the feasibility of obtaining the required information for the specific problem and objectives. For failure analysis, the resolution should be set at the component level where failure data are available. Further resolution is not necessary and would only complicate the analysis unless demanded by the objectives or a decision-making process.

The system breakdown structure is the top-down division of a system into subsystems and components. This architecture provides internal boundaries within the system. Often the systems and subsystems are identified as functional requirements that eventually lead to defining the level of detail for components. The functional level of a system identifies the functions that must be performed for an appropriate operation of the system. Further decomposition of the system into discrete elements leads to the physical level of a system, which identifies the hardware within the system. By organizing a system hierarchy using a top-down approach rather than the fragmentation of specific systems, a rational, repeatable, and systematic approach to risk analysis can be achieved.

While the system model provides boundaries for the system, subsystems, and components, it might not provide for an integrated view. Systems integration is an important part in evaluating the ability of a system to perform in meeting the objectives. The problem with segregating a system is that, when the subsystems are assembled to form the overall system, failures may occur that are not obvious while viewing the individual subsystems or components. Therefore, the interfaces should be examined. This is especially important for consideration of human factors on the performance of a system. The potential for human error must be considered when performing a system analysis. Also, the potential for taking corrective actions from faults should be considered. Different people have varying views on how to operate and maintain systems that can affect the performance of a system.

Further system analysis using risk assessment methods is described in Table 2.19. These techniques develop processes that can assist in decision making about the system. The logic of modeling based on the interaction of the components of a system can be divided into induction and deduction. This distinction in the techniques of modeling and decision making is significant. Induction logic provides the reasoning of a general conclusion from individual cases. This logic is used when analyzing the effect of a fault or condition on the performance of a system. Inductive analysis answers the question, "What system states would result from a particular event?" In reliability and risk studies, this event is some fault in the system. Approaches using the inductive approach include preliminary hazard analysis (PrHA), FMEA, and ETA. Deductive approaches provide reasoning for a specific conclusion from general conditions. For system analysis, this technique attempts to identify what modes of a system, subsystem, or component failure can be used to contribute to the failure of the system. This technique answers the question, "How can a particular system state occur?" Deductive reasoning provides the basis for FTA or its complement, success tree analysis (STA). Table 2.19 lists the selected risk methods and describes the scope of each method. Most of these methods are discussed and illustrated in this section.

TABLE 2.19

Selected Risk Assessment Methods

Method	Scope
Safety/review audit	It identifies equipment conditions or operating procedures that could lead to a casualty or result in property damage or environmental impacts.
Checklist	It ensures that organizations are complying with standard practices.
What-if/then	It identifies hazards, hazardous situations, or specific accident events that could result in undesirable consequences.
Hazard and operability study (HAZOP)	It identifies system deviations and their causes that can lead to undesirable consequences and determine recommended actions to reduce the frequency and/or consequences of the deviations.
Preliminary hazard analysis (PrHA)	It identifies and prioritizes hazards leading to undesirable consequences early in the life of a system. It determines the recommended actions to reduce the frequency and/or the consequences of the prioritized hazards. This is an inductive modeling approach.
Failure mode and effect analysis (FMEA)	It identifies the component (equipment) failure modes and impacts on the surrounding components and the system. This is an inductive modeling approach.
Fault tree analysis (FTA)	It identifies the combinations of equipment failures and human errors that can result in an accident. This is a deductive modeling approach.
Event tree analysis (ETA)	It identifies various sequences of events, both failures and successes that can lead to an accident. This is an inductive modeling approach.
Events and causal factors charting	It describes graphically or textually the time sequence of contributing events associated with an accident. Ishikawa diagrams can be used to display the results, particularly in quality control.
Swiss cheese model	It organizes causes, analyzes, and represents the causes of systematic failures or accidents, and describes a scenario (or scenarios) leading to an accident as a series of events that must occur in a specific order and manner for an accident to occur.
Pareto analysis	It identifies and prioritizes, the most significant items among many. This technique employs the 80-20 rule, which states that ~80% of the problems or effects are produced by ~20% of the causes.
Probabilistic risk assessment (PRA)	It is a methodology for quantitative risk assessment developed by the nuclear engineering community for risk analysis. This comprehensive process may use a combination of risk assessment methods.
Risk register (or risk log)	It manages risk by acting as a central repository for risks identified by the project staff and, for each risk, tracks information such as risk factor, event, probability, impact, countermeasures, and risk owner.
Barrier analysis	It identifies objects and functions, within the categories of equipment, design, administrative (procedures and work processes), supervisory/management, warning devices, knowledge and skills, and physical, to prevent an event.
Change analysis	It looks systematically for possible risk impacts and appropriate risk management strategies in situations where change is occurring.
Delphi technique	It assists in reaching the consensus of experts on a subject such as project risk while maintaining anonymity by soliciting ideas about the important project risks that are collected and circulated to the experts for further comment. Consensus on the main project risks may be reached in a few rounds of this process.
Interviewing	It identifies risk events by interviews of experienced project managers or subject-matter experts. The interviewees identify risk events based on experience and project information.
Experience-based identification	It identifies risk events based on experience, including implicit assumptions.
Expert opinion elicitation	It is a structured process to collect data and obtain answers to specific questions on issues important to risk quantification as described in Chapter 8.
Brainstorming	It identifies risk events using facilitated sessions with stakeholders, project team members, and infrastructure support staff.

Example 2.3: Risk Assessment Methods for Warehouse Automation Project

This example discusses the selected risk assessment methods for various aspects of the warehouse automation project. Risk assessment methods include checklists, what-if/then analysis, FMEA, FTA, and ETA, as well as qualitative and quantitative risk assessments. Risk assessment also requires interviewing, brainstorming, and expert opinion elicitation to gather information required by these methods. The client risks identified in Example 2.1 are used here to illustrate the use of checklists and what-if/then analysis.

The representatives of the client can use checklists for listing all possible risks associated with the decision to automate the order packing process and to install a conveyer system for the warehouse. This checklist can be constructed to include all activities related to the five stages of a project as follows: feasibility study phase, preliminary design, detailed design, execution and implementation, and termination. The stage of termination that includes closure, decommissioning, and removal can entail unique or unusual risks. The what-if/then analysis can be performed to enhance the understanding of what could happen to this new system as a result of adverse events during the five stages of the project. The what-if/then analysis shown in Figure 2.19 can be constructed using brainstorming sessions among the client team. The figure shows an example what-if/then tabulation. These results help the team to realize the impact of various adverse events on the project. Also, these results can be used to perform further modeling using FMEA, ETA, or FTA, and the subsequent sections illustrate their uses. The results can also be used to ensure proper understanding, analysis, communication, and risk management. Figure 2.19 shows a schematic representation and a summary of the results. A similar approach can be applied to risks from the perspectives of the engineer, contractor, and project manager.

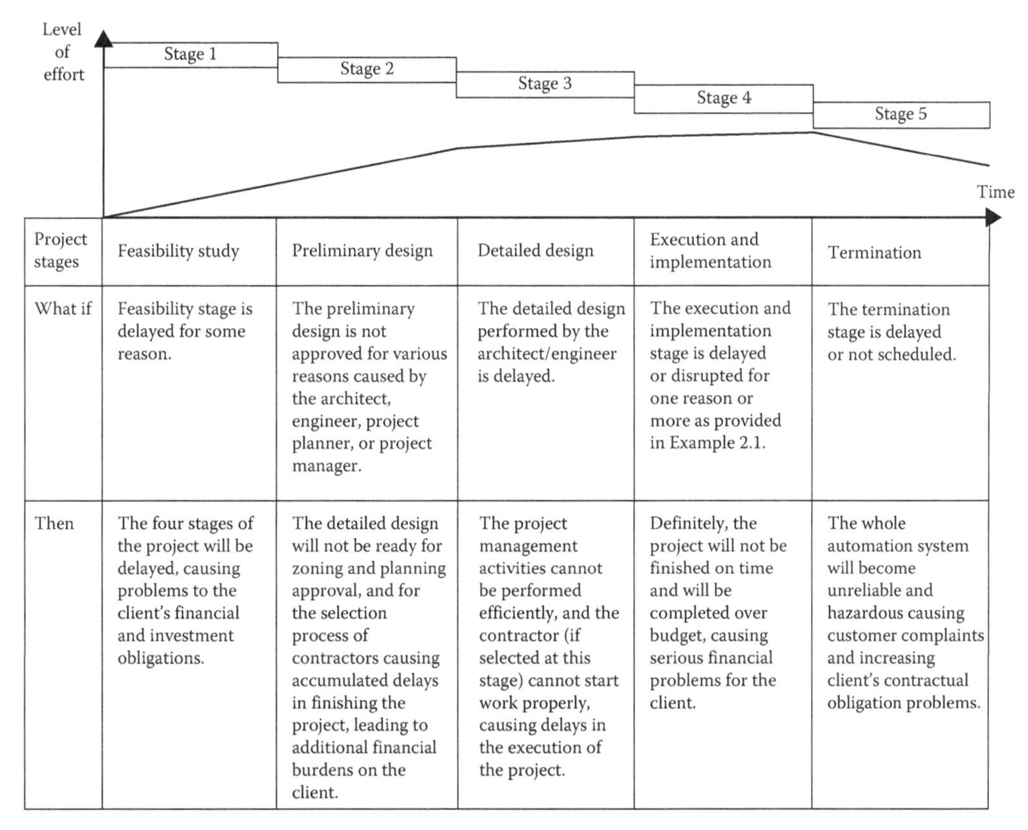

Project stages	Feasibility study	Preliminary design	Detailed design	Execution and implementation	Termination
What if	Feasibility stage is delayed for some reason.	The preliminary design is not approved for various reasons caused by the architect, engineer, project planner, or project manager.	The detailed design performed by the architect/engineer is delayed.	The execution and implementation stage is delayed or disrupted for one reason or more as provided in Example 2.1.	The termination stage is delayed or not scheduled.
Then	The four stages of the project will be delayed, causing problems to the client's financial and investment obligations.	The detailed design will not be ready for zoning and planning approval, and for the selection process of contractors causing accumulated delays in finishing the project, leading to additional financial burdens on the client.	The project management activities cannot be performed efficiently, and the contractor (if selected at this stage) cannot start work properly, causing delays in the execution of the project.	Definitely, the project will not be finished on time and will be completed over budget, causing serious financial problems for the client.	The whole automation system will become unreliable and hazardous causing customer complaints and increasing client's contractual obligation problems.

FIGURE 2.19
What-if/then analysis and results for various project stages.

2.5.2 Preliminary Hazard Analysis

PrHA is a commonly used method with many applications in manufacturing and industrial processes. The general process is shown in Figure 2.20. This technique requires experts as listed in Figure 2.20 to identify and rank the possible accident scenarios that could occur. It is frequently used as a preliminary method to identify and reduce the risks associated with major hazards of a system.

2.5.3 Risk Register

The risk register, sometimes called the register log, is commonly used in PM and can be defined as a framework for managing risk as a central repository for risks identified by the project staff. For each risk, the register tracks information such as risk factors, events, probabilities, impacts, countermeasures, and risk owners. The risk register does not have a standard format, and Table 2.20 shows an example risk register. The tabulated entries can be managed in a database on a computer server to facilitate the access by project staff. The deliberately designed risk register computations are simple in nature, as illustrated under various headings in Table 2.20, and are based on a scoring scheme. The probability and impact scores range from 1 (low) to 3 (high) and risk is computed as the product. Although such computations are of common use not only in this case, but also in the case of the FMEA, risk scoring could produce misleading results due to inherit limitations and shortcomings as discussed in Section 2.5.5.

2.5.4 Swiss Cheese Model

The Swiss cheese model is an organizational model used to analyze and represent the causes of systematic failures or accidents (Reason 2000). It is commonly used in the fields of aviation, engineering, and health care, and describes a scenario (or scenarios) leading to an accident as a series of events that must occur in a specific order and manner for an accident to occur. The scenario is represented by a lineup of the holes of several unique pieces of Swiss cheese slices stacked parallel to each other. The holes represent the opportunities for

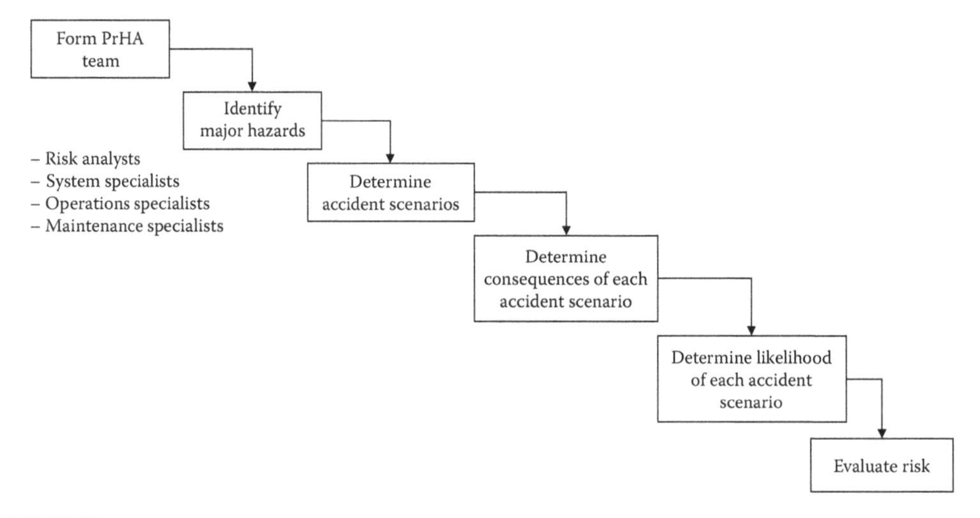

FIGURE 2.20
PrHA process.

TABLE 2.20
Risk Register

Risk Category	Risk Factor or Event	Identification Number	Probability (1–3)	Impact (1–3)	Risk Score	Mitigation or Countermeasure	Contingency	Risk Owner	Action Timing
Natural hazard	Strong wind	1.1	2 (medium)	2 (medium)	4	Avail hardware to secure equipment, supplies, and structure	Secure equipment, supplies, and structure	Jim	Within 2 hours
Natural hazard	High temperature	1.2	1 (low)	2 (high)	2	Access and ice to water suppliers	Offer frequent breaks, provide water, etc.	John	Within 2 hours
Materials	Delay in arrival	2.1	2 (medium)	2 (medium)	4	Identify points of contacts of suppliers	Check with suppliers	Janet	Within 2 hours
Labor	Strike	3.1	1 (low)	3 (high)	3	Monitor labor concerns and address early	Alternate labor providers	Everyone	According to plan
Labor	Low productivity	3.2	1 (low)	2 (medium)	2	Track and provide incentives	Increase or replace labor force	Susan Michael	Within a day
Start-up check	Low-power output	4.1	1 (low)	3 (high)	3	Perform component checks	Engage technical support	Mathew	Within 6 hours

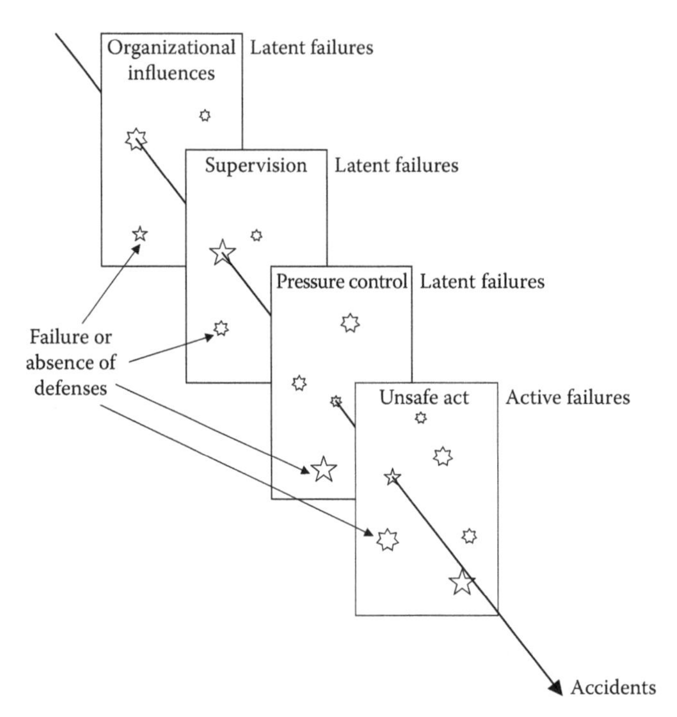

FIGURE 2.21
Swiss cheese model.

a failure or accident to occur, and the slices are the layers of the system defenses. Each layer represents a defense against the potential accident. Since the holes have random locations in each layer, an accident would occur only under the condition that a set of holes lines up so that a straight line representing the scenario can pass through the set. Under normal conditions, this lineup has a small probability of occurrence. Two types of holes are considered: latent failures and active failures. The former are the existing problems and the latter are the acute failures that lead to an accident. This method can be appropriately used where supported by other more rigorous methods. Figure 2.21 shows an outline of such a model.

The 2010 Deepwater Horizon oil spill was used to illustrate an Ishikawa diagram in Figure 2.16. The accident leading to the oil spill is approximately represented using a Swiss cheese model as shown in Figure 2.22.

2.5.5 Failure Mode and Effects Analysis

FMEA is another popular risk analysis method (Figure 2.23). This technique has been introduced in national and international regulations for aerospace (e.g., US MIL-STD-1629A), processing plants, and marine industries. In its recommended practices, the Society of Automotive Engineers introduces two types of FMEA: design and process FMEA. This analysis tool assumes a failure mode to occur in a system or component through some failure mechanism; the effect of this failure on other systems is then evaluated. A risk ranking can be developed for each failure mode according to its projected effect on the overall performance of the system. The various terms used in FMEA [with examples based

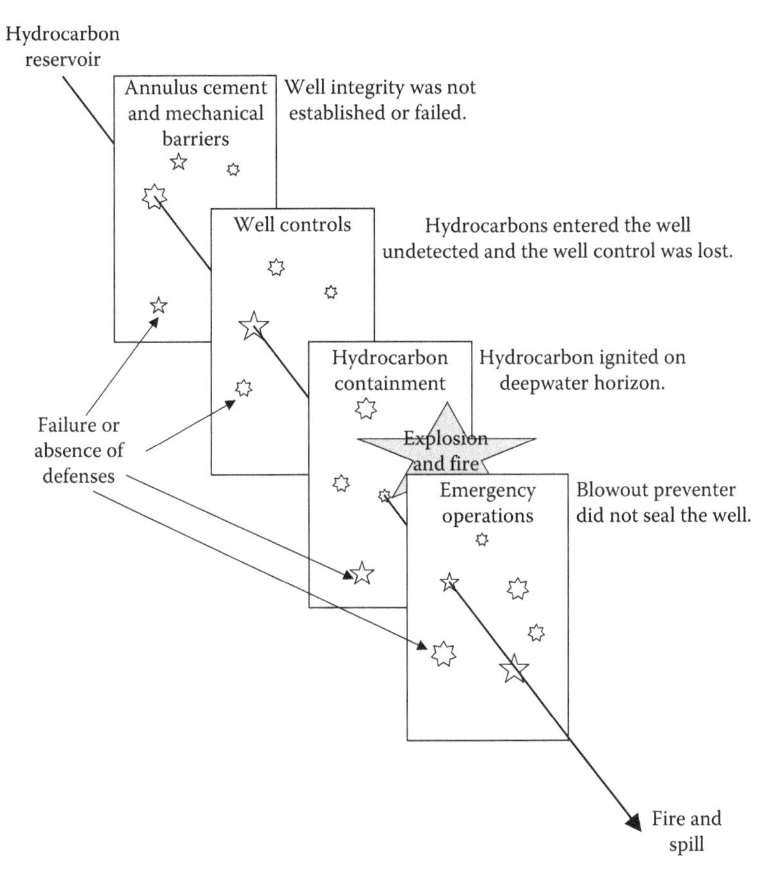

FIGURE 2.22
Oil spill represented using the Swiss cheese model.

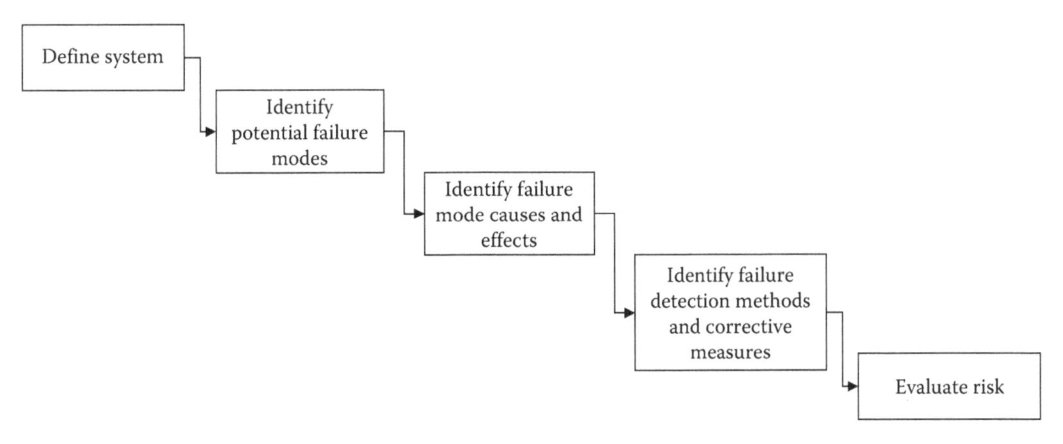

FIGURE 2.23
FMEA process.

on the manufacturing of personal flotation devices (PFDs)] are provided under subsequent headings and include failure mode, failure effect, severity rating, causes, occurrence rating, controls, detection rating, and risk priority number (RPN).

FMEA-related computations as provided under various headings are based on a scoring scheme. Although such computations are of common use not only in the case of the FMEA but also in risk scoring, they can produce misleading results due to inherit limitations and shortcomings. Generally, such approaches are not suitable for quantifying risk except for very simple cases. The primary reason for this limitation is the lack of an appropriate body of mathematics that is suitable for the respective score schemes. Moreover, they are commonly constructed in an *ad hoc* manner without the backing of a mathematical framework that should include axiomatic conditions for measurement and quantification. This section presents the FMEA computations in their original scoring scheme, although someone could easily suggest a probabilistic framework.

1. *Failure modes.* A failure mode is a way in which a specific process or product fails. It is a description of features that can be negatively affected by a process step or component. A failure mode may also be the cause of a potential failure mode in an upstream operation or the effect of one in a downstream operation. The assumption herein is made that the failure may occur but does not necessarily have to occur.
2. *Failure effects.* Failure effects are the impact on the end user or regulatory requirements. They are what the end user might experience or notice as a result of the failure mode. The effect is the outcome of the occurrence of the failure mode on the system.
3. *Severity ratings.* The severity rating is the importance of the effect on end-user requirements. It is concerned with safety and other risks if failure occurs. It is driven by failure effects and criticality and applies only to the effect. It should be the same each time the same failure effect occurs. A relative rating scale of 1–10 is commonly used (where 1 = not noticeable and 10 = extremely severe), as given in Table 2.21.

TABLE 2.21

Severity Rating Evaluation Criteria

Rating	Description
Minor	
1	Not noticeable; no effect on the product and end user
Low	
2	Not noticeable; no effect
3	Slightly noticeable; slight end-user annoyance
Moderate	
4–6	End user will notice immediately upon receipt. Noticeable effects on subsystem or product performance. Some end-user dissatisfaction; end user is uncomfortable with or annoyed by the failure
High	
7–8	Effects on major system, but not on safety or government-regulated compliance items; high degree of end-user dissatisfaction due to nature of failure
Extreme	
9–10	Effects on safety or involving noncompliance with government regulations (9, with warning; 10, without warning)

4. *Failure causes.* The failure causes are sources of process variations that cause the failure mode to occur. Potential causes describe how the failure could occur in terms of something that can be corrected or controlled. Potential causes should be thought of as potential root causes of a problem and point the way toward preventive/corrective action. Identification of causes should start with failure modes associated with the highest severity ratings.
5. *Occurrence rating.* The occurrence rating of a cause is the frequency with which a given cause occurs and creates the failure mode. It refers to the industry-wide average likelihood or probability that the failure cause will occur. A rating scale of 1–10 is used (Table 2.22).
6. *Definition of controls.* Current controls are those controls that either prevent the failure mode from occurring or detect the failure mode when it occurs. Prevention controls consist of mistake proofing and automated control. Controls also include inspections and tests to detect failures that may occur at a given process step or subsequently.
7. *Detection ratings.* The detection rating is a measure of the capability of current controls. It indicates the ability of the current control scheme to detect the causes before creating the failure mode and/or the failure modes before causing the effect. Detection ratings quantify the likelihood that current controls will prevent a defect from reaching the end user, given that a failure has occurred (Table 2.23).
8. *Risk priority number.* The RPN is a weighted assessment number used for prioritizing the risk items; the larger the RPN, the greater the risk. The RPN focuses the efforts on factors that provide opportunities to make the greatest improvement. The RPNs are sorted and the actions are recommended for the top issues. Risk assessment should be performed to determine when a corrective action is required.

TABLE 2.22

Occurrence Rating Criteria

Rating	Failure Consequence Description	Failure Rate
Minor		
1	Failure is unlikely; no failures ever associated with almost identical processes	<1 in 1,000,000
Low		
2	Only isolated failures associated with almost identical processes	1 in 20,000
3	Isolated failures associated with similar processes	1 in 4,000
Moderate		
4	Generally associated with similar	1 in 1,000
5	Processes that have experienced	1 in 400
6	Occasional failures, but not in major proportions	1 in 80
High		
7	Generally associated with similar	1 in 40
8	Processes that have often failed; process is not in control	1 in 20
Extreme		
9	Failure is almost inevitable	1 in 8
10		1 in 2

TABLE 2.23

Detection Rating Criteria for Likelihood That Defect Is Caught by Current Controls

Rating	Description
Certainty of nondetection	
10	Controls will not or cannot detect the existence of a defect.
Very low	
9	Controls probably will not detect the existence of a defect.
Low	
7–8	Controls have a poor chance of detecting the existence of a defect.
Moderate	
5–6	Controls may detect the existence of a defect.
High	
3–4	Controls have a good chance of detecting the existence of a defect; the process automatically detects failure.
Very high	
1–2	Controls will almost certainly detect the existence of a defect; the process automatically prevents further processing.

The RPN is calculated by multiplying the occurrence rating by the severity rating by the detection rating:

$$\text{RPN} = \text{occurrence rating} \times \text{severity rating} \times \text{detection rating} \tag{2.6}$$

Corrective actions should first be directed at the highest ranking concerns and critical items where causes are not well understood. The purpose of corrective actions is to reduce the ratings of severity, occurrence, and detection. Actions should be aimed at preventing the failure by controlling or eliminating the cause. A rule of thumb is to take a serious look at RPNs >125.

Example 2.4: Failure Mode and Effect Analysis of Manufacturing Personal Flotation Devices

Risk methods can be used to minimize the cost of follow-up tests during manufacturing of PFDs. The manufacturing of inherently buoyant PFDs requires the handling of several material types, such as foam, fabric, and hardware, and progression through several manufacturing steps. A prototype manufacturing process is presented in Figure 2.24. The process consists of the following six primary steps: (1) receiving incoming recognized components, (2) cutting operations, (3) preclosure assembly, (4) quality assurance checks and testing, (5) closure assembly, and (6) final tests. These steps are performed on three parallel tracks (Figure 2.24): fabric materials, foam materials, and hardware materials.

An FMEA of the PFD manufacturing process was performed. The various ratings were subjectively assessed as shown in Table 2.24. Of highest criticality are failure modes with RPNs of ≥125. The FMEA of the PFD manufacturing process and that of an inherently buoyant PFD product are combined in Table 2.24, where the failure modes are sorted in descending order of RPNs. The table ranks the selected product and process failure modes from the highest to the lowest criticality based on RPNs computed from the opinions of experts who participated in a workshop that was held for this purpose. Various tests can serve as controls for identified failure modes with ranks as provided in Table 2.24.

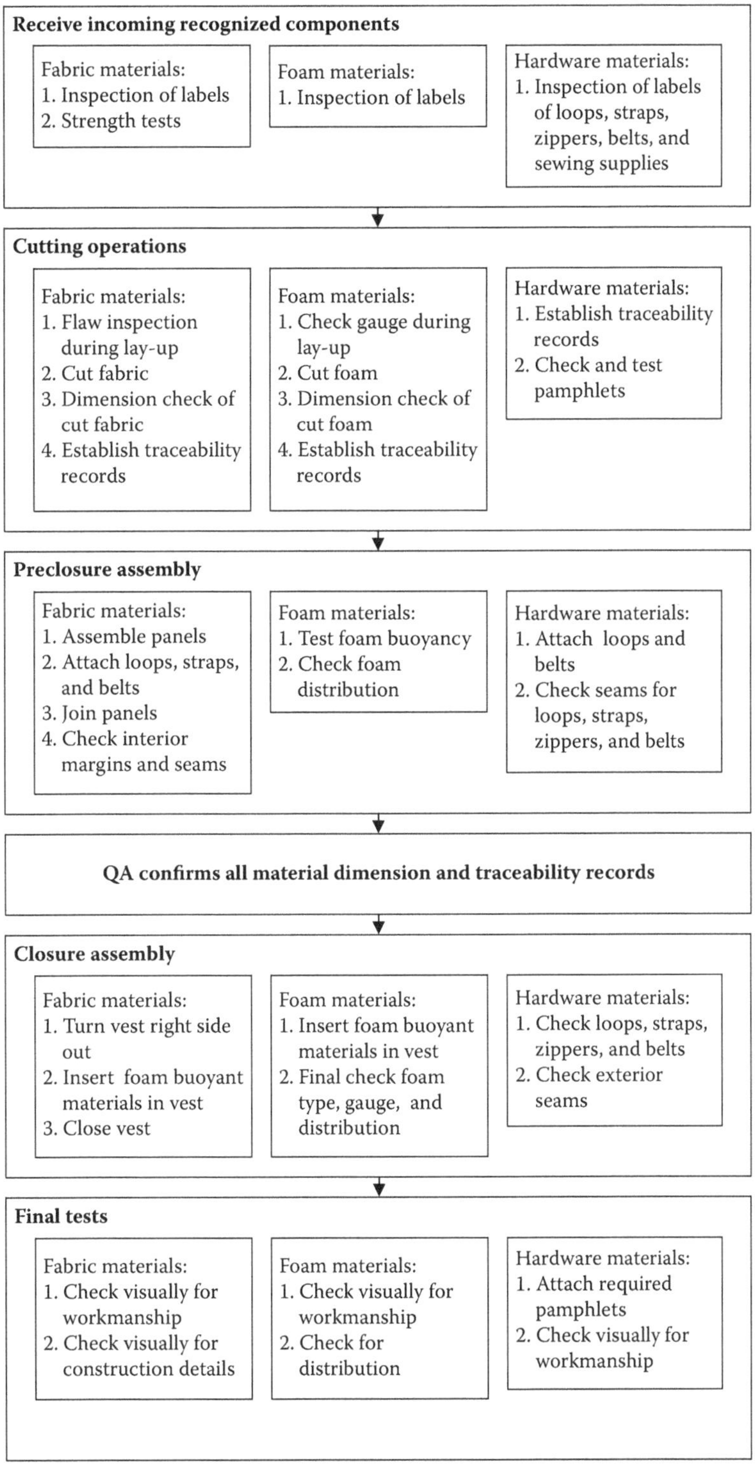

Receive incoming recognized components

Fabric materials:
1. Inspection of labels
2. Strength tests

Foam materials:
1. Inspection of labels

Hardware materials:
1. Inspection of labels of loops, straps, zippers, belts, and sewing supplies

Cutting operations

Fabric materials:
1. Flaw inspection during lay-up
2. Cut fabric
3. Dimension check of cut fabric
4. Establish traceability records

Foam materials:
1. Check gauge during lay-up
2. Cut foam
3. Dimension check of cut foam
4. Establish traceability records

Hardware materials:
1. Establish traceability records
2. Check and test pamphlets

Preclosure assembly

Fabric materials:
1. Assemble panels
2. Attach loops, straps, and belts
3. Join panels
4. Check interior margins and seams

Foam materials:
1. Test foam buoyancy
2. Check foam distribution

Hardware materials:
1. Attach loops and belts
2. Check seams for loops, straps, zippers, and belts

QA confirms all material dimension and traceability records

Closure assembly

Fabric materials:
1. Turn vest right side out
2. Insert foam buoyant materials in vest
3. Close vest

Foam materials:
1. Insert foam buoyant materials in vest
2. Final check foam type, gauge, and distribution

Hardware materials:
1. Check loops, straps, zippers, and belts
2. Check exterior seams

Final tests

Fabric materials:
1. Check visually for workmanship
2. Check visually for construction details

Foam materials:
1. Check visually for workmanship
2. Check for distribution

Hardware materials:
1. Attach required pamphlets
2. Check visually for workmanship

FIGURE 2.24
Typical manufacturing process of personal flotation devices.

TABLE 2.24

Failure Mode and Effect Analysis of the Personal Flotation Devices Manufacturing Process

Process Step or Product Component	Failure Mode	Failure Effects (Primary Performance Requirement Impacted)	Causes	Controls with Ranks	SEV	OCC	DET	RPN
Receiving incoming recognized components	Personal flotation device (PFD) does not turn unconscious wearer face-up in water.	Flotation	Excessive foam gauge variation, wrong material received, insufficient foam buoyancy	(1) Foam thickness test (2) Buoyancy distribution test (3) Component recognition	7.8	5	6	234
Other operations	Manufacturing process is out of control.	Flotation, security (or fit), comfort, longevity, identification for tracking	Ineffective organizational management style	Documentations to define process, training, internal audit, measurement system	8	3	8	192
Receiving incoming recognized components	Material breaks or deforms.	Security (or fit)	Inadequate strength tests, wrong material received	(1) Component recognition (2) Ultimate breaking strength test	8.5	3	6	153
Final tests	Buoyancy is distributed unevenly.	Flotation	Nonuniform foam not detected	Training trim and lock inspectors	7	3	7	147
Other operations	Manufacturing process is out of control.	Flotation, security (or fit), comfort, longevity, identification for tracking	Culture and attitude of workers (e.g., not process focused)	Providing operational definition	9	3.5	4.5	142
Receiving incoming recognized components	PFD components cannot be tracked to material lot.	Identification for tracking	Inadequate labeling of incoming components	(1) Regrouping of process lots (2) Sampling program and tabulation	5	4	7	140
Preclosure assembly	Buoyancy is distributed unevenly, with over 10% variation from design.	Flotation	Nonuniform foam distribution	(1) Gauge examination (2) Distribution test (3) Expanding tolerance through testing of two specimens	6	3	7	126
Component interfaces	Strap rips from fabric.	Security (or fit)	Weak fabric-to-strap connections	In-process examinations and inspections	7	3	6	126

Hardware materials	Closure adjuster gives way.	Security (or fit)	Weak closure adjuster	(1) Ultimate breaking strength test (2) Supplier testing (3) Component recognition	7	3	6	126
Foam materials	PFD buoyancy is distributed unevenly.	Flotation	Nonuniform foam thickness	(1) Thickness test (2) Distribution test	7	3	6	126
Preclosure assembly	PFD is unstable in water.	Flotation	Lack of pocket flotation stability	Gauge examination	6.5	3	6.3	123
Foam materials	PFD buoyancy is less than advertised.	Flotation	Foam not thick enough	(1) Thickness test (2) Gauge examination (3) Buoyancy test	8	3	5	120
Cutting operations	Cuttings cannot be tracked to material lot.	Identification for tracking	Incomplete tracking records	(1) Tracking ability (2) Sampling program and tabulation	6	3.8	5	114
Component interfaces	Seams rip or tear easily due to low seam strength.	Security (or fit)	Insufficient tensile strength of sewing threads	(1) Ultimate breaking strength test (2) Strength test (3) Component recognition (4) In-process inspections	7.3	3	5	110
Quality assurance checks and testing	PFD components are not traceable to lot.	Identification for tracking	Insufficient quality assurance tracking tests	–	6	3	6	108
Hardware materials	Hardware deforms or corrodes.	Longevity	Wrong hardware material, improper consumer use or storage	(1) Production examination (2) Qualitative infrared analysis (3) Differential scanning calorimetry and chemical analysis	6	3	6	108
Quality assurance checks and testing	PFD buoyancy is less than advertised.	Flotation	Insufficient quality assurance buoyancy tests	–	7.8	2	6.8	106
Fabric materials	Side adjustment breaks.	Security (or fit)	Low side adjustment tensile strength	(1) Strength test (2) Component recognition	7	3	5	105
Fabric materials	Fabric belt breaks.	Security (or fit)	Low fabric belt tensile strength	(1) Strength test (2) Component recognition	7	3	5	105

Notes: SEV, severity of the effects of the failure (1 = low, 10 = high).
OCC, probability of failure occurring (1 = low, 10 = high).
DET, likelihood failure is detected (10 = low, 1 = high).
RPN = SEV × OCC × DET.

Based on the PFD FMEA, controls for the highest criticality product failure modes include the following:

- Foam thickness test
- Buoyancy distribution test
- Component recognition
- Documentation to define process
- Training
- Internal audit and measurement system
- Ultimate breaking strength test
- Training trim and lock inspectors
- Regrouping of process lots
- Sampling program and tabulation
- Gauge examination
- Expand tolerance through testing of two specimens
- In-process examinations and inspections
- Supplier testing
- Traceability
- Strength test

Example 2.5: Failure Mode and Effect Analysis of the Warehouse Automation Project

The information provided and the results produced in Example 2.1 are used to develop a tabulated risk assessment using FMEA for key project risks. This example examines their severity degree and their effect on the performance of the entire project. Table 2.25 shows, from the project manager perspective (see Table 2.13), the failure modes, their effects on performance of the project, severity, causes, occurrence probability, detection likelihood, and RPN. The RPN can be used for risk control to eliminate or reduce the effects of these risks. The FMEA results can be used to prepare an FT model as demonstrated in Section 2.5.6.2.

2.5.6 Scenario Modeling and Logic Trees

Scenario modeling is a systematic and often a comprehensive way to identify how an accident or an event of interest could happen, and offers a basis to quantify risk. The methods under this class of methods provide frameworks for identifying scenarios to evaluate the performance of a system through system modeling. These methods are also called *logic trees.* The combination of event trees, success trees (STs), FTs, probability trees, and decision trees can provide a basis for the structured analysis of system safety.

2.5.6.1 Event Trees

ETA is often used if the successful operation of a component/system depends on a discrete (chronological) set of events. The initiating event is followed by other events, leading to an overall result (consequence). It offers the ability to identify a complete set of scenarios, as all combinations of both the success and the failure of the main events are included in the analysis. The probability of occurrence of the main events of the event tree can be determined using FTA or its complement, the STA. The scope of the analysis for event trees and FTs depends on the objective of the overall study.

ETA is appropriate if the operation of some systems or components depends on a successive group of events. Event trees identify the various combinations of event successes and

TABLE 2.25

Failure Mode and Effect Analysis of the Warehouse Automation Project from Project Manager Perspective

Source of Risk and Type	Failure Mode	Effect on Total Performance	Causes	Controls	SEV	OCC	DET	RPN
PM risks at corporate management risk level (internal type)	Budget overrun	Failure to finish the project within budget	Control of financial matters is lost, in addition to other technical problems.	Increase levels of financial and technical monitoring and auditing of project activities.	9	6	5	270
	Time overrun	Failure to start operation on time	Technical monitoring by project manager is reduced due to construction problems, design problems, or incompetent contractor.	Increase periodic technical control and tracking progress of activities.	9	5	8	360
	Party disputes	Arbitration, delay in finishing the project, and loss to client	Problems arise among parties for various reasons.	Resolve problems as they appear.	7	4	5	140
	Personnel problems on-site	Problems among personnel that can lead to total chaos	Organization on-site is lacking as a result of bad planning.	Organize periodic meetings to resolve organizational problems.	5	4	4	80
Technological, quality, or performance risks (technology level)	Changes in project technology	Failure to cope with changes	PM staff is not prepared to accept changes.	Organize meetings to make project manager staff aware of new changes.	6	6	6	216
	Quality problems	Failure to meet project requirements	Good-quality standards are not established at the beginning of the project.	Prepare quality manual and distribute to all parties involved.	8	5	6	240
Contractors risks (external type)	Contractor failure to finish project on time	Failure to deliver project to the client's expectations	Project manager lacks control over contractor.	Engage in the selection of the contractor at the beginning of the project.	7	4	6	168
	Incompetent contractor	Failure to meet project requirements	Project manager has no control over the chosen contractor.	Enforce adherence to PM procedures.	6	3	8	144
	Inefficient subcontractors	Problems in delivery and with subcontracted work	Contractor chosen is improper or problems appear between the contractor and his subcontractors.	Ask for a list of all selected subcontractors.	5	6	4	120

(Continued)

TABLE 2.25

(Continued) Failure Mode and Effect Analysis of the Warehouse Automation Project from Project Manager Perspective

Source of Risk and Type	Failure Mode	Effect on Total Performance	Causes	Controls	SEV	OCC	DET	RPN
Contractual and legal risks (management type)	Contractual problem with client	Disputes with the client	Project manager misunderstood the requirements.	Explain to client in detail the scope of services throughout the project.	4	4	5	80
		Failure to complete PM services	Project manager failed to fulfill his responsibilities.	Negotiate new terms or make provisional precautions before signing contract with client.	3	4	5	60
External risks (external type)	Political problems	Difficulty in providing PM services efficiently	Project manager did not anticipate political changes.	Perform uncertainty and risk analysis studies before accepting offer to work on the project.	4	7	3	84
	Economic problems	Failure to make anticipated profit	Project manager did not account for changes in currency rates or similar economical issues.	Perform effective marketing study before engaging in the project.	6	5	6	180

Notes: SEV, severity of the effects of the failure (1 = low, 10 = high).
OCC, probability of failure occurring (1 = low, 10 = high).
DET, likelihood failure is detected (10 = low, 1 = high).
RPN = SEV × OCC × DET.

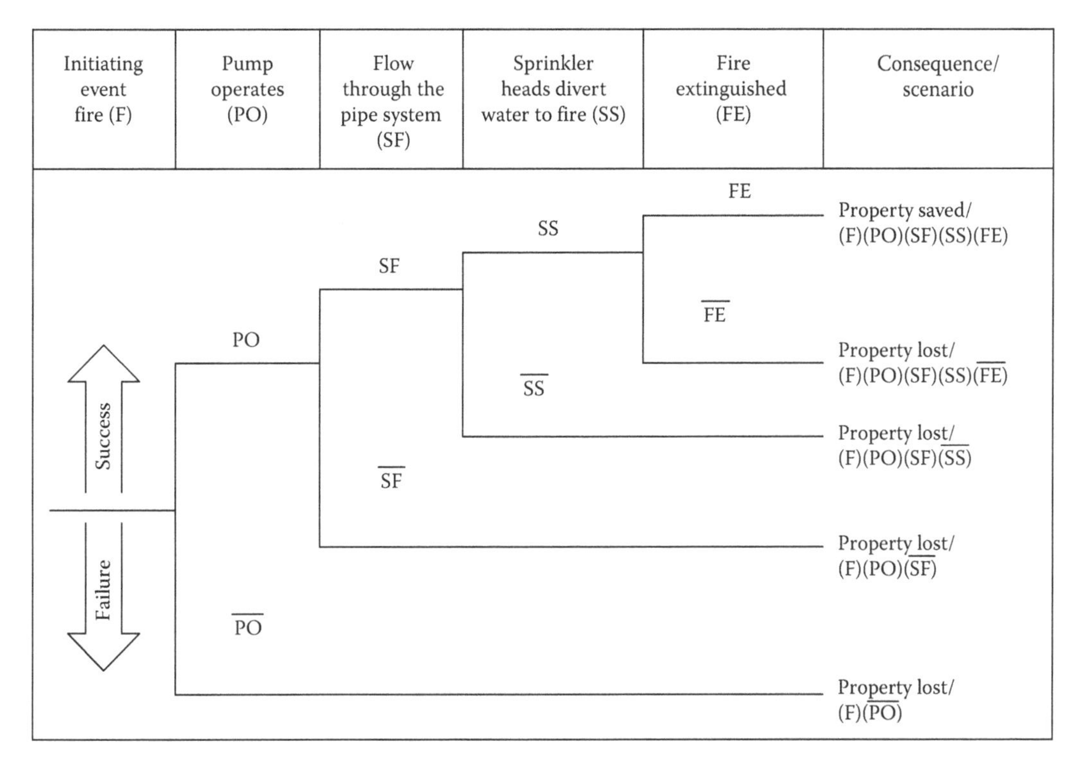

FIGURE 2.25
Event tree example for sprinkler system.

failures that could result from an initiating event. The event tree starts with an initiating event followed by some reactionary events. This reaction can be either a success or failure. If the event succeeds, the most commonly used indication is the upward movement of the path branch. A downward branch of the event tree marks a failure event, that is, its complement, as shown in Figure 2.25. The remaining events are evaluated to determine the different possible scenarios. The scope of the events can be functions or systems that can reduce the possible hazards resulting from the initiating event. The final outcome of a sequence of events identifies the overall state resulting from the scenario of events. Each path represents a failure scenario with varying levels of probability and risk. Event trees can be created for different event initiators. Figure 2.25 shows an example event tree for the basic elements of a sprinkler system that might be critical for maintaining the integrity of a marine vessel.

Based on the occurrence of an initiating event, the ETA examines possible system outcomes or consequences. This analysis tool is particularly effective in showing the interdependence of system components; such interdependence might at first appear to be insignificant but could result in devastating results if not recognized. ETA is similar to FTA because both methods use probabilistic data of the individual components and events along each path to compute the probability of each outcome.

Event tree probabilities can be evaluated in order to assess the probability of the overall system state. Probability values for the success or failure of the events can be used to identify the probability for a specific event tree sequence. The probabilities of the events in a sequence can be provided as inputs to the model or can be evaluated using FTs. These probabilities for various events in a sequence can be viewed as conditional probabilities and can therefore be multiplied to obtain the occurrence probability of the sequence. The probabilities of various sequences can be summed up to determine the overall probability of a particular outcome. The addition of consequence evaluation of all scenarios allows for generation of a risk value for the system. For example, the occurrence probability of the top branch (scenario) in Figure 2.25 is computed as the product of the probabilities of the underlying events to this scenario: $F \cap PO \cap SF \cap SS \cap FE$, or (F)(PO)(SF)(SS)(FE) for short.

2.5.6.2 Fault and Success Trees

Complex systems are often difficult to visualize, and the effect of individual components on the system as a whole is difficult to evaluate without an analytical tool. Two methods of modeling, which have greatly improved the ease of assessing system reliability or risk, are FTs and STs. An FT is a graphical model created by deductive reasoning that leads to various combinations of events that, in turn, lead to the occurrence of some top event failures. An ST shows the combinations of successful events leading to the success of the top event. An ST can be produced as the complement (opposite) of the FT as illustrated in this section. FTs and STs are used to further analyze the event tree headings (the main events in an event tree) to provide further detail to understand system complexities. In constructing the FT or ST, only those failure or success events, respectively, that are considered significant are modeled. This determination is assisted by defining system boundaries. For example, the pump operates (PO) event in Figure 2.25 can be analyzed by developing a top-down logical breakdown of failure or success using FTs or STs, respectively.

FTA starts by defining a top event, which is commonly selected as an adverse event. An engineering system can have more than one top event; for example, a ship might have the following top events for the purpose of reliability assessment: power failure, stability failure, mobility failure, or structural failure. Then, each top event needs to be examined using the following logic: in order for the top event to occur, what other events must occur. As a result, a set of lower level events is defined. Also, the form in which these lower level events are logically connected (i.e., in parallel or in series) should be defined. The connectivity of these events is expressed using AND or OR gates. Lower level events are classified into the following types:

- *Basic events* cannot be decomposed further into lower level events. They are the lowest events that can be obtained. For these events, failure probabilities must be obtained.
- *Events that can be decomposed further* are events that can be decomposed further into lower levels; therefore, they should be decomposed until the basic events are obtained.
- *Undeveloped events* are not basic and can be decomposed further; however, because they are not important, they are not developed further. Usually, the probabilities of these events are very small or the effect of their occurrence on the system is negligible or can be controlled or mediated.
- *Switch* (or *house*) *events* are not random and can be turned on or off with full control.

The symbols corresponding to these events are shown in Figure 2.26. Also, a continuation symbol is shown which is used to break up an FT into several parts for the purpose of fitting it on several pages.

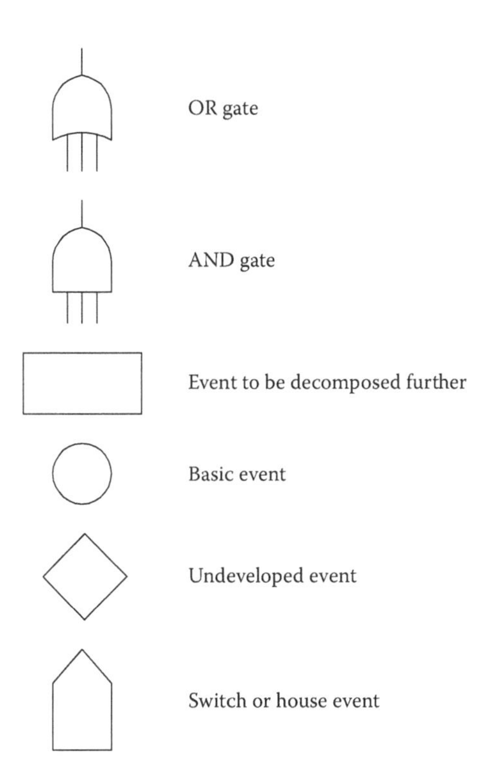

FIGURE 2.26
Symbols used in fault tree analysis.

FTA requires the development of a tree-looking diagram for the system that shows failure paths and scenarios that can lead to the occurrence of a top event. The construction of the tree should be based on the building blocks and Boolean logic gates.

The outcome of interest from the FTA is the occurrence probability of the top event. Because the top event was decomposed into basic events, its occurrence can be stated in the form of AND and OR logic operated on the basic events. The resulting statement can be restated by replacing the AND with the intersection of the corresponding basic events and the OR with the union of the corresponding basic events. Then, the occurrence probability of the top event can be computed by evaluating the probabilities of the unions and intersections of the basic events. The dependence between these events also affects the resulting probability of the system represented by the top event.

The computation of the occurrence probability of the top event in large FTs can be difficult because of the size of such trees. In this case, an efficient approach is required for the computation of probabilities and the identification of the minimal cut sets. For this purpose, a *cut set* is defined as a set of basic events where the joint occurrence of these basic events results in the occurrence of the top event. A minimal cut set is defined as a cut set with the condition that the nonoccurrence of any one basic event from this set results in the nonoccurrence of the top event. Therefore, a minimal cut set can be viewed as a subsystem in parallel. In general, systems have more than one minimal cut set. The occurrence of the top event of the system can, therefore, be due to any of these minimal cut sets. As a result, the system can be viewed as the union of all the minimal cut sets for the system. If the probability values are computed (or assigned) for the minimal cut sets, a probability for the top event can be determined by knowing (or assuming) the nature of dependence

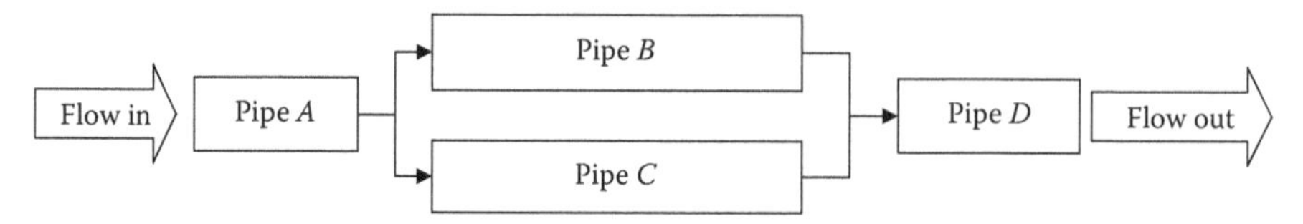

FIGURE 2.27
Reliability block diagram for a piping system.

among these sets and using Boolean algebra of probabilities of union and intersection of events. Two cases for this determination are provided herein for illustration purposes: (1) mutually exclusive sets permitting the addition of the probabilities of these sets and (2) independent sets permitting the use of the product rule for the event intersection probabilities as discussed in Appendix A.

A simple example of this type of modeling is shown in Figure 2.27 for a piping system using a reliability block diagram. If the goal of the system is to maintain water flow from one end of the system to the other, then the individual pipes can be related with a Boolean logic. Both pipes A and D, and pipe B or C must function for the system to meet their goal, as shown in the ST in Figure 2.28. The complement of the ST is the FT. The goal of the FT model is to construct the logic for system failure, as shown in Figure 2.29. Once these tree elements have been defined, possible failure scenarios of a system can be defined. Using the FT model, the top event (T) can be attained as follows:

$$T = A \text{ or } (B \text{ and } C) \text{or } D \tag{2.7}$$

Using the mathematics of probability as provided in Appendix A, the probability (P) of the top event can be computed as a function of pipes' failure probabilities:

$$P(T) = 1 - \left[1 - P(A)\right]\left[1 - P(B)P(C)\right]\left[1 - P(D)\right] \tag{2.8a}$$

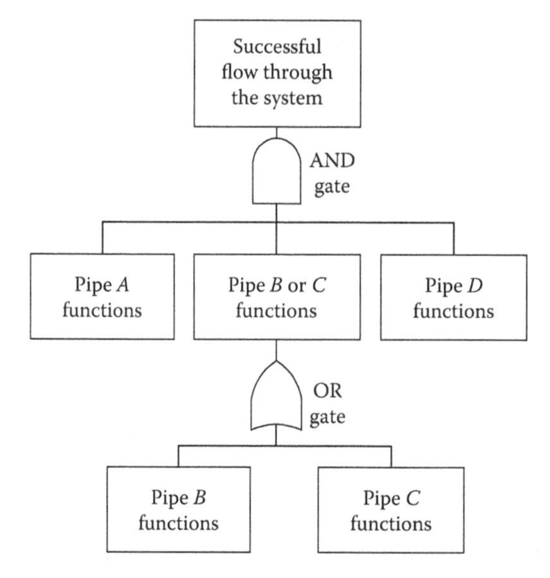

FIGURE 2.28
Success trees for the pipe system example.

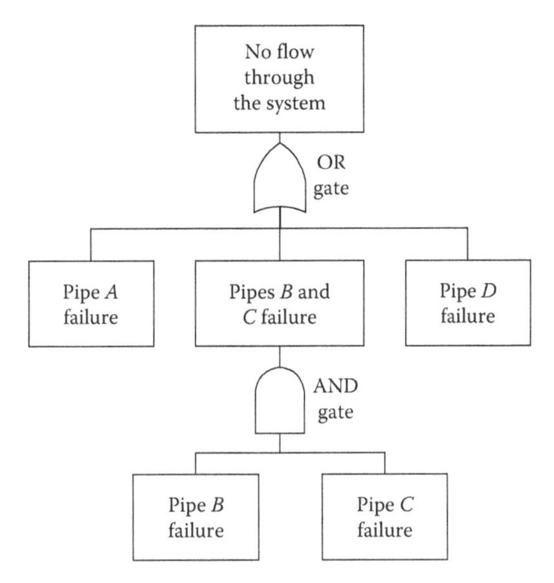

FIGURE 2.29
FT for the pipe system example.

The probability according to Equation 2.8a is based on the assumption of independent underlying events *A–D* and can be written for mutually exclusive failure scenarios, that is, mutually exclusive sets, as

$$P(T) = P(A) + P(B)P(C) + P(D) \tag{2.8b}$$

In some problems, mutually exclusive failure scenarios are justifiable in order to simplify computations. For example, assuming $P(A) = 0.1$, $P(B) = 0.15$, $P(C) = 0.1$, and $P(D) = 0.2$, the $P(T)$ can be computed to be $1 - (1 - 0.1)[1 - (0.15)(0.1)](1 - 0.2) = 0.2908$ according to Equation 2.8a and $0.1 + (0.15)(0.1) + 0.2 = 0.315$ according to Equation 2.8b. For complicated systems, the number of failure paths can be quite large. The number of possible failure scenarios (assuming only two possible outcomes for each basic event) is bounded by the following equation:

$$\text{Failure paths} = 2^n \tag{2.9}$$

where:
n is the number of basic events or components in the system

For a complex system, the number of failure paths can be very high.

As noted previously, a failure path is often referred to as a *cut set*. One objective of the analysis is to determine all minimal cut sets, where a minimal cut set is defined as a failure combination of all essential events that can result in the top event. A minimal cut set includes in its combination all essential events, that is, the nonoccurrence of any of these essential events in the combination of a minimal cut set results in the nonoccurrence of the minimal cut set. These failure combinations are used to compute the failure probability of the top event. The concept of the minimal cut sets applies only to the FTs. A similar concept can be developed in the complementary space of the STs and is called the *minimal pass set*. In this case, a minimal pass set is defined as a survival (or success) combination of all essential success events that can result in success as defined by the top event of the ST.

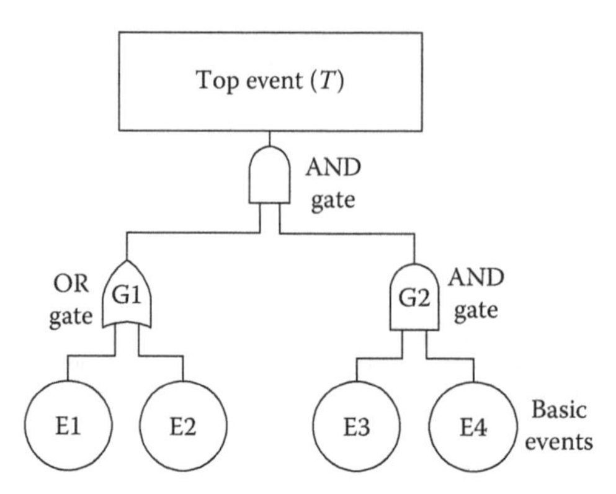

FIGURE 2.30
FT example for illustrating algorithm to generate the cut sets.

Several methods for generating the minimal cut sets are available. One of the methods is based on a top-down search of Boolean logic. Another algorithm for generating the cut sets is based on a bottom-up approach that substitutes the minimal cut sets from lower level gates into upper level gates. According to Equation 2.7, the minimal cut sets are as follows:

$$A \tag{2.10a}$$

$$D \tag{2.10b}$$

$$B \text{ and } C \tag{2.10c}$$

A minimal cut set includes events that are all necessary for the occurrence of the top event. For example, the following cut set is not a minimal cut set:

$$A \text{ and } B \tag{2.11}$$

The minimal cut sets can be systematically generated using the following algorithm with reference to the FT provided in Figure 2.30 as an example:

1. Provide a unique label for each gate as shown in Figure 2.30.
2. Label each basic event as noted in the figure.
3. Set up a two-cell array:

4. Place the top event gate label in the first row, first column:

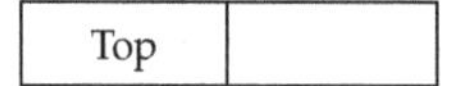

5. Scan each row from left to right replacing
   - Each OR gate by a vertical arrangement defining the input events to the gate.
   - Each AND gate by a horizontal arrangement defining the input events to the gate.

For example, the following table sequence can be generated for an AND top gate with two gates below it (gate 1 of the OR type and gate 2 of the AND type):

Top (AND)	

Leading to the following structure:

Gate 1 (OR)	Gate 2 (AND)

Gate 1 has two events (1 and 2), leading to the following structure:

Event 1	Gate 2
Event 2	Gate 2

Gate 2 has two events (3 and 4), leading to the following updated structure:

Event 1	Event 3	Event 4
Event 2	Event 3	Event 4

6. Once no gate and events remain, each row is a cut set.
7. Remove all nonminimal combinations of events such that only the minimal cut sets remain.
8. Compute the occurrence probability for each minimal cut set as the products of the probabilities of its underlying events.
9. Compute the system (top event) occurrence probabilities (see Equation 2.8a and 2.8b as examples).

   As an example of Figure 2.28, this algorithm can be used as follows:

Top event (T)

The top event has an OR gate with three branches. The top event should be replaced by the following three rows:

A
B and C
D

The middle row has an AND gate and should be replaced by the two events in one row as follows:

A	
B	C
D	

This table shows three minimal cut sets. Therefore, the probability of T is one minus the product of the nonoccurrence of three minimal cut sets as determined in Equation 2.8a.

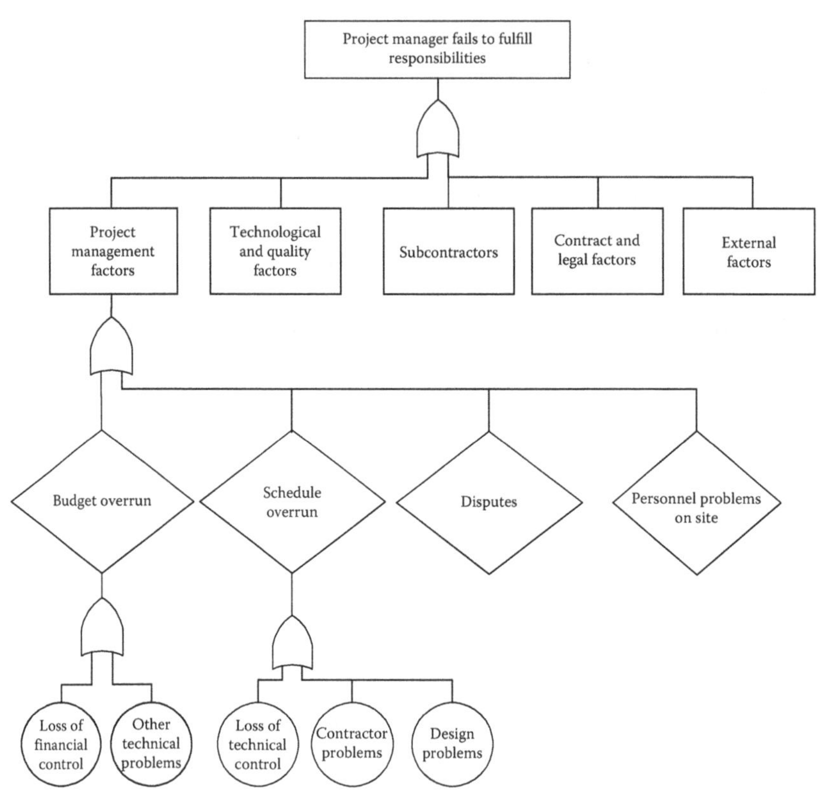

FIGURE 2.31
FT model of PM failure for the warehouse automation project.

Example 2.6: FT Model for the Warehouse Automation Project

The results of the FMEA for the warehouse automation project as provided in Example 2.5 can be used to develop an FT model for a selected top event. The top event is selected as the failure of the PM company to fulfill its responsibilities. This top event can be decomposed further to show the details of each intermediate event causing the top event. Figure 2.31 shows the decomposition into intermediate events to basic and undeveloped events.

Example 2.7: Several FT Models and Their Minimal Cut Sets

This example demonstrates how the cut sets can be identified and constructed for different arrangements of OR and AND gates to logically define a top event occurrence. Generally, the number of cut sets increases by increasing the number of OR gates in the tree. For example, Figure 2.32 demonstrates this increase by comparing cases *A*, *B*, and *D*. But increasing the number of AND gates results in increasing the number of events included in the cut sets, as shown in case *C* of Figure 2.32.

2.5.6.3 Common-Cause Scenarios

Common-cause scenarios are events or conditions that result in the failure of seemingly separate systems or components. Common-cause failures complicate the process of conducting risk analysis because a seemingly redundant system can be rendered ineffective by a common-cause failure. For example, an emergency diesel generator fed by the same fuel supply as the

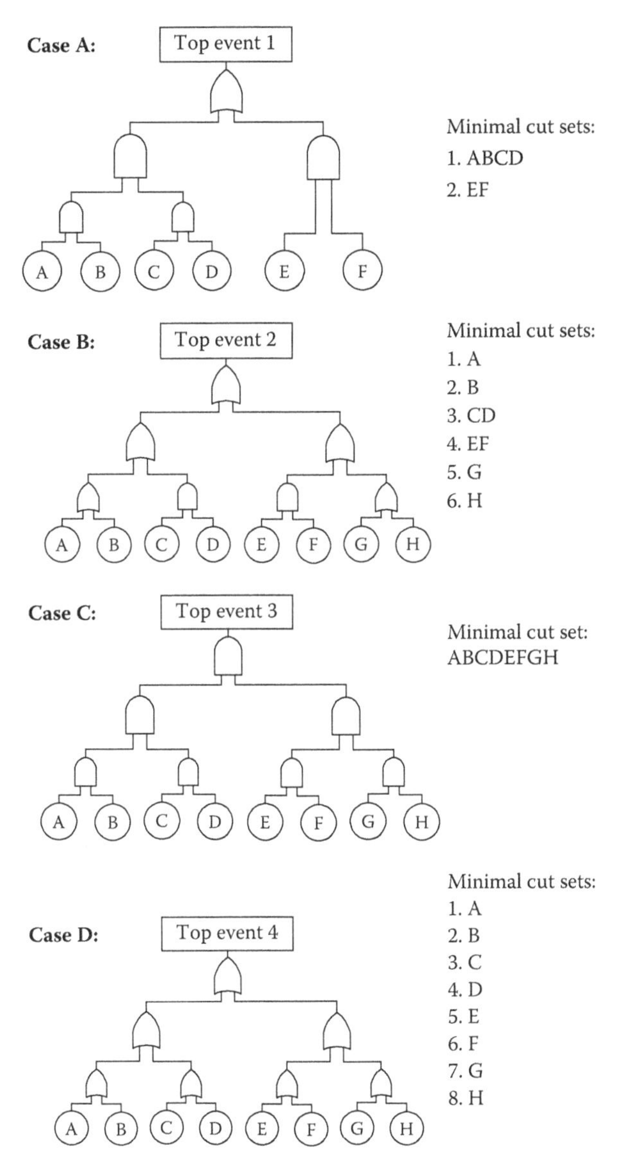

FIGURE 2.32
FT models and their minimal cut sets.

main diesel engine will fail with the main diesel generator if the fuel supply is the root source of failure. The redundant emergency diesel generator is not truly redundant in this case due to sharing of a cause failure with the primary diesel engine. Another example of common-cause events is the failure of two separate but similar pieces of machinery due to a common maintenance problem, two identical pieces of equipment failing due to a common manufacturing defect, or two pieces of equipment failing due to a common environmental condition such as the flooding of a compartment or a fire in the vicinity of both pieces of machinery. A method for calculating the reliability of a system while taking into account common-cause effects is the beta-factor model. Other methods include the multiple Greek letter model, alpha factor model, and beta binomial failure rate model as discussed by Kumamoto and Henley (1996).

2.5.6.4 Sensitivity Factors

Part of risk-based decision analysis is pinpointing the system components that can lead to high-risk scenarios. Commercial system reliability software provides this type of analysis in the form of sensitivity of system reliability to changes in the underlying component reliability values. In performing risk analysis, it is desirable to assess the importance of events in the model or the sensitivity of final results to changes in the input failure probabilities for the events. Several sensitivity or importance factors are available that can be used. The most commonly used factors include the Fussell–Vesely factor (FVF) and Birnbaum factor (BF). Also, a weighted combination of these factors can be used as an overall measure (Kumamoto and Henley 1996).

For any event (basic or undeveloped) in an FT, the FVF for the event is given by the following equation:

$$\text{FVF} = \frac{\displaystyle\sum_{\text{all sets containing the event}} \text{Occurrence probability of minimal cut set}}{\displaystyle\sum_{\text{all sets}} \text{Occurrence probability of minimal cut set}} \tag{2.12}$$

The FVF measures the contribution significance of the event in regard to the failure probability of the system. Events with large FVFs should be used to reduce the failure probability of the system by reducing their occurrence probabilities.

For any event (basic or undeveloped) in an FT, the BF for the event is given by the following equation:

$$\text{BF} = \frac{\displaystyle\sum_{\text{all sets containing the event}} \text{Occurrence probability of minimal cut set}}{\text{Occurrence probability of the event}} \tag{2.13}$$

The BF measures the sensitivity of the failure probability of the system to changes in the occurrence probability of the event. Events with large BFs should be targeted to reduce the failure probability of the system by reducing their occurrence probabilities.

2.5.6.5 Probability and Decision Trees

Probability trees can be used to develop scenarios and associated branch probabilities (p_i) and impacts (i.e., losses) L_i. In this section, a simple case is used to illustrate the development of scenarios. Figure 2.33 shows a generic probability tree for HS applications. A sequence of events constitutes a scenario in this tree. A probability can be assigned to a threat type; however, due to the difficulty in assigning such a probability to a threat type, someone may simply use a probability of one and consider the results as conditional scenario probabilities, that is, the scenario probability under the condition that the threat would occur. The threat probabilities are not required to add up to one. The asset attractiveness, as shown in Figure 2.33, can be quantified by the probability of an asset being selected by an adversary. Again these probabilities do not need to add up to one, and the corresponding events can be treated conditionally, if desired, as was illustrated in the case of the threats. The vulnerabilities can be quantified as conditional probabilities of an adversary success and system failure. By following a series of branches under each of the headings shown in the figure, a scenario can be identified or developed. The scenario probability as indicated in the figure

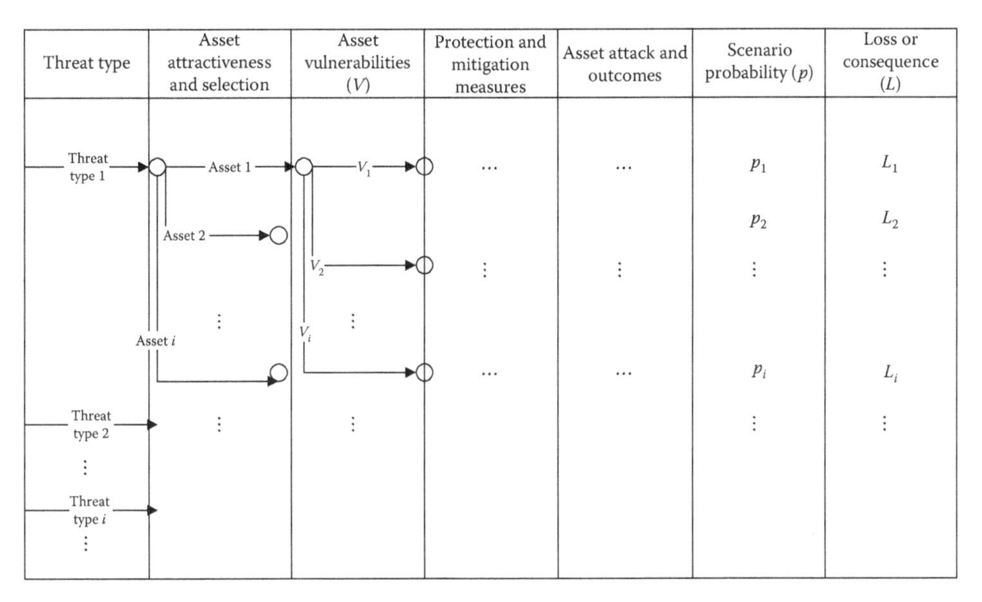

FIGURE 2.33
Construction of a generic probability tree.

can be evaluated as the product of the conditional probabilities of the branches appearing in the scenario. The conditional probability for a branch is the probability of the occurrence of the branch under the assumed conditions that all the branches leading to this particular branch in the scenario have occurred. This product can be viewed as the best (or point) estimate of the scenario probability. The loss or impact associated with a scenario is also conditional on the occurrence of the scenario and can be provided either as a best estimate or by a probability distribution. A percentile interval can instead be used.

Decision trees have a structure similar to probability trees with an added feature of branching through decisions or selections among available options at selected points along branching tracks, called decision nodes as illustrated in Figure 2.34. The chance branching points are called chance nodes. The losses or impacts can be assigned probability distributions. Each branch can be characterized probabilistically and various decisions can be evaluated in terms of expected values, standard deviations, dominance, or other criteria. Decision trees are covered in detail in Chapter 3.

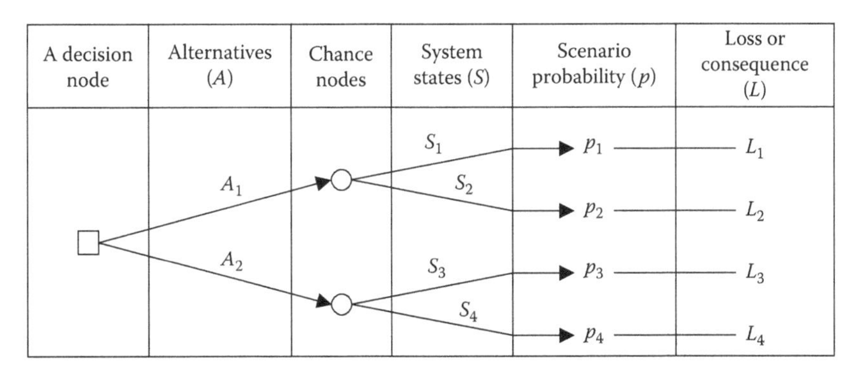

FIGURE 2.34
A decision tree.

TABLE 2.26

Logic Trees Compared

Logic Tree	Analysis Outcomes	Mathematical Foundation	Data Required	Advantages	Limitations
Fault tree	Probability of failure Determine the cut sets	Boolean logic Probability theory including reliability theory	System knowledge, failure modes, and probabilities	Focusing on components and failure modes	Complex systems requiring the use of specialized software
Success tree	Probability of success Determine the cut sets	Boolean logic Probability theory including reliability theory	System knowledge Success modes and probabilities	Focusing on success modes	Complex systems requiring the use of specialized software
Event tree	Probability of scenarios and consequences	Probability theory	Events and sequencing Outcome spaces	Multiple outcomes Conceptually simple to develop and solve	Binary outcomes
Probability tree	Probability of any uncertain event in a joint probability distribution	Probability theory Bayes theorem	Events and sequencing Outcome spaces Probabilities Consequences	Multiple outcomes Conceptually simple to develop and solve	Difficult to understand, display, and solve for large trees
Decision tree	Outcome values in order to determine the best decision strategy under uncertainty	Bayes theorem Utility theory	Events and sequencing Outcome spaces Probabilities Alternatives Consequences	Conceptually simple to develop and solve	Difficult to understand, display, and solve for large trees

Source: Dillon-Merrill, R.L., et al., 2009.

2.5.6.6 Logic Trees Compared

The logic trees presented in Section 2.5.6 have similarities, differences, appropriate uses, and limitations as summarized in Table 2.26 that was based on the work reported by Dillon-Merrill et al. (2009).

2.6 Human-Related Risks

Risk assessment requires the performance analysis of an entire system composed of a diverse set of components. The system definition readily includes the physical components of the system; however, humans are also part of most systems and provide significant contributions to risk. It has been estimated, for example, that human error contributes to ~90% of accidents at sea. The human contribution to risk can be estimated from examining the behavioral aspects of humans as individuals and groups by capitalizing on respective sciences. Hardware failure and human error should be addressed in the risk assessment, as they both contribute to risks associated with a system. Once the human error probabilities are determined, human error failures are treated with special attention to any differences compared to hardware failures in performing risk assessment quantification.

The human error contribution to risk is determined by using human reliability analysis (HRA) methods. HRA is the discipline that enables analysis of the impact of humans on the reliability and safety of systems. The important outcomes of HRA are determining the likelihood of a human error as well as ways in which human errors can be reduced. When combined with system risk analysis, HRA methods provide an assessment of the detrimental effects humans may have on the performance of a system. HRA is generally considered to be composed of three basic steps: error identification, modeling, and quantification.

Other sources of human-related risks are in the form of deliberate sabotage of a facility or a plant, for example, from within or from outside, such as the threat posed by a computer hacker or a terrorist actor. The hazard in this case is not simply random but also intelligent. The methods introduced in earlier sections might not be fully applicable for this risk type. The threat scenarios in this case have a dynamic nature that is impacted by the defense or risk mitigation and management scenarios that would be implemented by an owner of a facility. The use of game-theoretic methods might be necessary in this case, in combination with other risk analysis and management methods. Game theory is introduced in this section.

2.6.1 Human Errors

Human errors are unwanted circumstances caused by humans that result in deviations from expected norms that place systems at risk. It is important to identify the relevant errors to make a complete and accurate risk assessment. Human error identification techniques should provide a comprehensive structure for determining significant human errors within a system. Quality HRA allows for accuracy in both the HRA assessment and the overall system risk assessment.

Identification of human errors requires knowledge about the interactions of humans with other humans or machines (the physical world). It is the study of these interfaces that allows for the understanding of human errors. Potential sources of information for identifying human errors may be determined from task analysis, expert judgment, laboratory studies, simulation, and reports. Human errors may be considered active or latent, depending on the time delay between when the error occurs and when the system fails.

It is important to note the distinction between human errors and human factors. Human errors are generally considered separately from the analysis of human factors, which involves applying information about human behavior, abilities, limitations, and other characteristics to the design of tools, machines, systems tasks, jobs, and environments for productive, safe, comfortable, and effective human use. Human factors are determined by performing descriptive studies for characterizing populations and experimental research. However, human factors analysis may contribute to the HRA.

2.6.1.1 Human Error Modeling

Once human errors have been identified, they must be represented in a logical and quantifiable framework along with other components that contribute to the risk of the system. This framework can be determined from development of a risk model. Currently, no consensus has been reached on how to model humans reliably. Many models utilize human event trees and FTs to predict human reliability values. The identifications of human failure events can also be identified using FMEA. Estimates of human error rates are often based on simulation tests, models, and expert opinions.

2.6.1.2 Human Error Quantification

Quantification of human error probability (HEP) promotes inclusion of the human element in risk analysis. This is still a developing field of study that requires understanding of human performance, cognitive processing, and human perceptions as individuals and groups.

A reliable model for human cognition is unavailable. Therefore, much of the current human reliability data relies on accident databases, simulation, and other empirical approaches. Many of the existing data sources have been developed from specific industry data, such as from the nuclear and aviation industries. Applicability of these data sources to a specific problem should be thoroughly examined. The result of the quantification of human reliability in terms of probability of occurrence is typically referred to as a HEP. Many techniques have been developed to help predict HEP values. The technique for human error rate prediction (THERP) is one of the most widely used methods for HEP. This technique is based on the data gathered from the nuclear and chemical processing industries. THERP relies on HRA event tree modeling to identify the events of concern. Quantification is performed from data tables of basic HEPs for specific tasks that may be modified based on the circumstances affecting performance.

The degree of human reliability is influenced by many factors often referred to as performance shaping factors (PSFs). PSFs are those factors that affect the ability of people to carry out the required tasks. For example, the knowledge that someone has in regard to how to put on and activate a PFD will affect the performance of this task. Training (another PSF) in donning PFDs can assist in the ability to perform this task. Another example is the training that is given to passengers on airplanes before takeoff on using seat belts, emergency breathing devices, and flotation devices. Often the quantitative estimates of reliability are generated from a base error rate that is then altered based on the PSFs of the particular circumstances. Internal PSFs include an individual's own attributes (experience, training, skills, abilities, attitudes, etc.) that affect the ability of the person to perform certain tasks. External PSFs are the dynamic aspects of situation, tasks, and system that affect the ability to perform certain tasks. Typical external factors include environmental stress factors (such as heat, cold, noise, situational stress, time of day), management, procedures, time limitations, and quality of a human–machine interface. With these PSFs, it is easy to see the dynamic nature of HEP evaluation based on the circumstances of the analysis.

Human errors attributable to group behavior, dynamics, politics, and culture pose a great challenge to quantification and inclusion in risk studies.

2.6.1.3 Reducing Human Errors

Error reduction is concerned with lowering the likelihood for error in an attempt to reduce risk. The reduction of human errors may be achieved by human factor interventions or engineering means. Human factor interventions include improving training or the human–machine interface (such as alarms and codes) based on the understanding of the causes of error. Engineering means of error reduction may include automated safety systems or interlocks. Selection of the corrective actions to take can be done through decision analysis considering cost–benefit criteria.

2.6.1.4 Human Reliability Assessment

Figure 2.35 provides an illustrative procedure for assessing human reliability as outlined in the ISO 31010 (2009c). The procedure utilizes the concepts covered in Section 2.6.1 on human reliability, and error identification and reduction.

2.6.2 Deliberate Human Threats

Risk analysis against deliberate human threats, such as terror acts and sabotage, includes some features of typical probabilistic risk analysis methods, in which probability and consequence ranges are determined numerically; however, it relies heavily on many of the features of traditional qualitative approaches to balance the time and resources required to perform the analysis with the need for numerical risk measures that can be used to inform resource allocation decisions for security and protection. Some of the unique features of risk analysis related to assets, threats, countermeasures, and consequence mitigation are summarized in Table 2.27.

2.6.3 Game Theory for Intelligent Threats

Game theory is a study of multiagent decision problems based on the work of von Neumann and Morgenstern (1944) and Nash (1951), noting that it was first considered in the nineteenth century by Cournot (1838). Game theory was developed in economics to model the dynamic nature of two or more agents interacting to negotiate and execute the most favorable strategies based on respective, anticipated payoffs from these strategies. It has been used not only in modeling competing parties in economics, but also in geopolitical arms and influence pursuits. For our purposes, it can be used to model human behavior, considered here as a threat to a system. Key assumptions in game-theoretic models are as follows: (1) all the possible utilities of the different consequences for each player must be derivable and usable within the model by knowing all the possible goals and aspirations of the different players and (2) players are rational and intelligent enough to work out the consequence of their actions (Ezell et al. 2010). In the case of uncertainty about players' intentions, adaptive methods with branching might be necessary.

Generally, game theory utilizes mathematics, economics, and social and behavioral sciences to model human behavior. Examples of intelligent threats include terrorism and sabotage, which represent an ongoing battle between coordinated opponents participating in a two-party game in which each opponent seeks to achieve his/her own objectives within the system. In the case of terrorism, it is a game of a well-established political system as

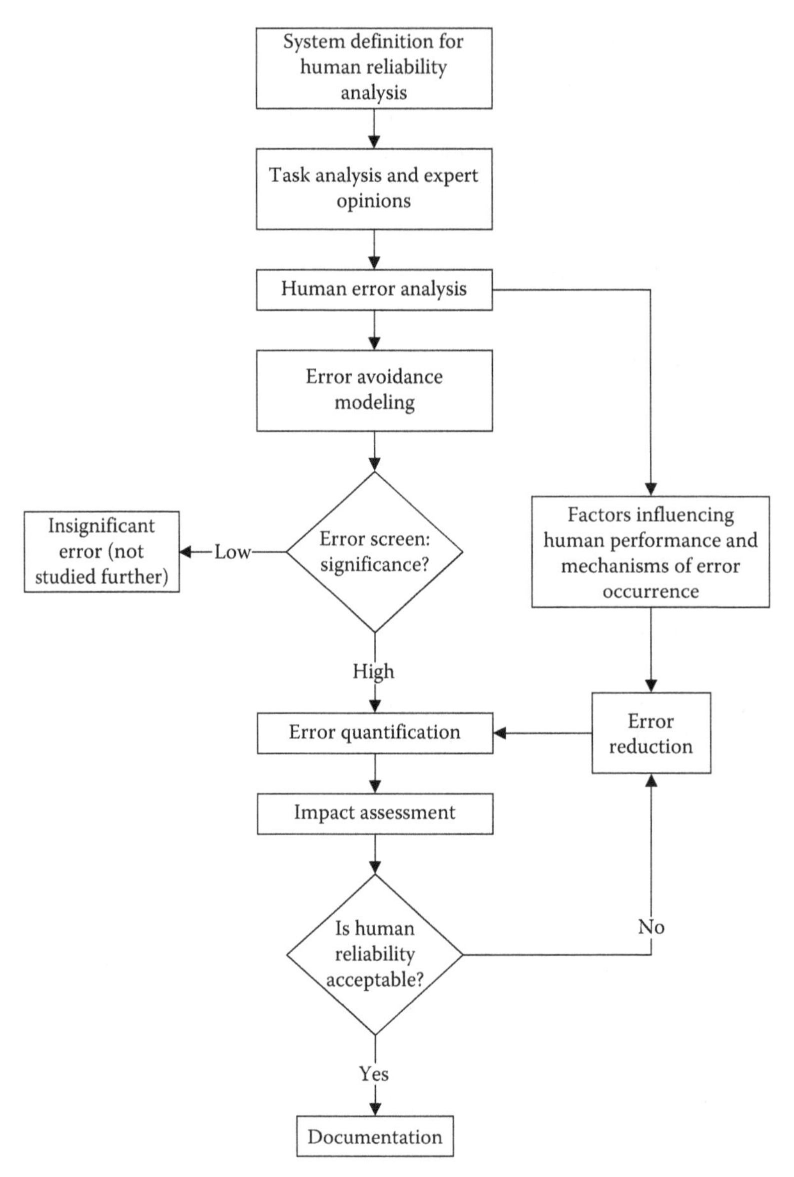

FIGURE 2.35
A procedure for assessing human reliability.

a government versus an emerging organization that uses terrorism to achieve partial or complete dominance. Each player in this game seeks a utility (i.e., benefit) that is a function of the desired state of the system. In the case of terrorism or sabotage, maintaining system survival is the desired state for the government, whereas the opponent seeks a utility based on the failure state of the system. The government, as an opponent, is engaged in risk mitigation by taking actions that seek to reduce the threat, reduce the system vulnerability, and/or mitigate the consequences of any successful attacks. The terrorist, as an opponent, can be viewed as an aggressor who strives to alter or damage the opponent's desired system state. This game involves an intelligent threat and is dynamic. The game continues until the probability of the aggressor being successful in his/her disruptive attempts reaches

TABLE 2.27

Unique Features of Risk Analysis for Asset Protection from Deliberate Human Threats

Features	Unique Characteristics
Risk analysis framework	Should be performed accounting for the perspectives of adversaries as well as the perspectives of defenders; and as a multilevel analysis ranging from an asset to multiassets, to a sector, and to multisectors, to sufficiently account for interdependencies that may affect the risks pertinent to the decision being made.
Asset (target) features	Include attractive assets, critical assets, soft assets, and assets with vulnerabilities that are sufficiently known to adversaries.
Assets (targets) selected by adversaries	Include high-consequence assets (or scenarios) with high probability of success.
Threat features	Include the dynamic nature of threats, threat types, and probabilities; their nonrandomness but deliberateness using design-basis threats; possibly being of unknown or unknowable types.
Threat–asset dependencies	Include dynamically responding to asset protection using countermeasures and consequence mitigation.
Ingenuity of adversaries	Includes converting assets to threats by capitalizing on the efficiency of infrastructures: • Transportation efficiency by converting airplane assets into explosive weapons • Mail efficiency by using mail items for bioagent delivery • Other efficient infrastructure systems including power and information systems
Capabilities of adversaries	Include the ability to select targets and accurately deliver the weapon to them and the ability to adapt to countermeasures to redirect the weapon to another target.
Asset vulnerabilities	Include identifying targets outside the system boundaries to exploit system vulnerabilities through system dependencies.
Consequences	Are broadly defined to include public health, economic loss, loss of vital commodities, interruption of government operation, and national psyche.
Asset and sector interdependencies	Include interdependencies in functionality and subsequently in consequences.
Decision analysis	Includes trade-offs based on national security, safety, and economics.
Information flow	Is a two-way flow of defenders acquiring knowledge about the adversaries; adversaries acquiring knowledge about the assets, countermeasures, and consequence mitigation plans.
Countermeasures	Include countermeasures at the asset level and meta-countermeasures at the multiasset and sector and multisector levels. Countermeasures reduce the probability of selection of an asset as well as the probability of success of an attack.
Consequence mitigation plans	Include mitigation at the local level and meta-mitigations at the state, regional, and national levels. Mitigation actions reduce consequences.
Risk perception and communication	Could include fear factors, hype, psychological aspects, communication effectiveness, and misconceptions.

an acceptable level of risk, a stage where risk is considered under control, and the game is brought to an end. Classical game theory can be used in conjunction with probabilistic risk analysis to determine optimal mitigation actions that maximize benefits. This description of the game in this case is written in a government narrative and can be restated easily in another party narrative.

The objective of this section is to demonstrate the potential of modeling the behavioral aspects of system components within a probabilistic risk analysis framework in an effort to develop suitable measures of risk control for intelligent threats. For a given set of strategies, the behavior of two or more noncooperative (i.e., opposing) players is best modeled using a game-theoretic approach.

A classical example used to introduce game theory is the prisoners' dilemma, which is based on the scenario of two suspects being captured near the scene of a crime. They are questioned separately by a law enforcement agency. Each suspect has to choose whether to confess and implicate the other. If neither person confesses, then both will serve, say, 1 year on a charge of carrying a concealed weapon. If each confesses and implicates the other, both will go to prison for, say, 10 years. However, if one person confesses and implicates the other, and the other person does not confess, the one who has collaborated with the police will go free, while the other person will go to prison for, say, 20 years on the maximum penalty. The strategies in this case are confess or do not confess. The payoffs, herein penalties, are the sentences served. The problem can be expressed compactly in a payoff table of a kind that has become pretty standard in game theory (Table 2.28). The entries of this table mean that each prisoner chooses one of the two strategies: the first suspect chooses a row and the second suspect chooses a column. The two numbers in each cell of the table provide the outcomes for the two suspects for the corresponding pair of strategies chosen by the suspects as an ordered pair, with the first value in the pair being the payoff for the first suspect and the second value in the pair being the payoff for the second suspect. The number to the left of the comma is the payoff to the person who chooses the rows (the first suspect), while the number to the right of the comma is the payoff to the person who chooses the columns (the second suspect). Thus, reading down the first column, if they both confess, each receives a sentence of 10 years, but if the second suspect confesses and the first suspect does not, then the first suspect gets 20 years and the second suspect goes free. This example is not a zero-sum game, as the payoffs are all losses. However, many problems can be cast with losses (negative numbers) and gains (positive numbers), with a total for each cell in the payoff table. A problem in which the payoffs in each cell of the payoff table add up to zero is a *zero-sum game*.

The solution to this problem regarding the suspects should be based on identifying the rational strategies that can be based on both persons wanting to minimize the time they spend in jail. The first suspect might reason as follows:

> Either the other suspect confesses or he/she keeps quiet. If the other suspect confesses and I don't confess, then I will get 20 years, 10 years if I do; therefore, in this case it's best to confess. On the other hand, if the other suspect doesn't confess and I don't either, I get a year, but if I confess I can go free. Either way, it's best if I confess. Therefore, I'll confess.

TABLE 2.28

Payoff Table in Years for the Prisoners' Dilemma Game for S1 and S2

		Second Suspect (S2)	
		Confess (C21)	**Do Not Confess (C22)**
First suspect (S1)	**Confess (C11)**	(10, 10)	(0, 20)
	Do not confess (C12)	(20, 0)	(1, 1)

Note: Underlined values are the respective conditional best payoff, that is, minimum years, for each player.

This logic can be stated in conditional statements as follows:

$$\text{Choice by S1}|\text{C21} = \text{C11 since the prison term is 10 years}$$

$$\text{Choice by S1}|\text{C22} = \text{C11 since the prison term is 0 year}$$

where:
S1 and S2 are suspects 1 and 2, respectively

The C's are as defined in Table 2.28 and are followed by a condition, that is, action by the other noncooperative player. The other player, that is, the second suspect, can and presumably will reason in the same way and come up with the following:

$$\text{Choice by S2}|\text{C11} = \text{C21 since the prison term is 10 years}$$

$$\text{Choice by S2}|\text{C12} = \text{C21 since the prison term is 0 year}$$

In this case, both players would find confessing to be the best rational strategy and go to prison for 10 years each, although if they had acted irrationally and kept quiet they, each could have gotten off with 1 year each.

The rational strategies of the two suspects have fallen into something known as a *dominant strategy equilibrium*, that is, (C11, C21), a term that requires definition. The term *dominant strategy* reflects the fact that an individual player (suspect, in this case) in a game evaluates separately each of the strategy combinations being faced and, for each combination, chooses from these strategies the one that offers the greatest payoff. If the same strategy is chosen for each of the different combinations of strategies the player might face, then that strategy is a dominant strategy for that player in that game. The dominant strategy equilibrium occurs if, in a game, each player has a dominant strategy and each player plays the dominant strategy, then that combination of dominant strategies and the corresponding payoffs is said to constitute the dominant strategy equilibrium for that game. In the prisoners' dilemma game, to confess is a dominant strategy, since both suspects would independently determine confessing to be a dominant strategy, and hence, the dominant strategy equilibrium is reached. The dominant strategy equilibrium is also referred to as the *Nash equilibrium*. When no player can benefit by changing his/her strategy while the other player keeps his/her strategy unchanged, then that set of strategies and the corresponding payoffs constitute the Nash equilibrium.

In summary, we can define the best strategy for a player as the best choice among all cases defined by considering the opponent making a particular choice, one at a time, among all the choices available to the second player, where the best choice is the minimum payoff of all the maximum payoffs corresponding to all the choices available to the second player. A dominant strategy for a player occurs in cases where the *same* best strategy is selected regardless of the opponent's strategy selected. Equilibrium is a state where both players have dominant strategies.

The prisoners' dilemma game is based on two strategies per suspect that can be viewed as deterministic in nature (i.e., nonrandom). In general, many games, especially the games permitting repeatability in choosing strategies by players, can be constructed with strategies that have associated probabilities. For example, strategies can be constructed based on probabilities of 0.4 and 0.6 that add up to one. Such strategies with probabilities are called *mixed strategies*, as opposed to pure strategies that do not involve the probabilities, such as the prisoners' dilemma game. A mixed strategy occurs in a game if a player chooses among two or more strategies at random according to specific probabilities.

Another example is arming/disarming decisions by the United States and the former USSR, the two powerful countries with nuclear arms during the twentieth-century cold war. This example was developed in collaboration with Dr. James Scouras in 2012. Let us start by using the same preference assumptions similar to the prisoner's dilemma in the following order: (1) each country prefers to dominate, (2) each country prefers mutual disarming to mutual arming, and (3) the least liked outcome is to be dominated. In this case, instead of using payoffs, we will use preferences on an ordinal scale of 1 (most preferred), 2, 3, and 4 (least preferred) as shown in Table 2.29(a). According to this table, arming is a dominant strategy, although there is a better outcome of both disarming.

The prisoner's dilemma might not accurately represent the mindset of the United States and the USSR, and it can be argued that the mindset is better represented by the chicken dilemma, also called the hawk-dove preference, that is, similar to two automobile drivers heading toward each other at high speed for a head-on collision and each driver is hoping that the other driver would abandon the course and veer off the collision track. According to the chicken dilemma, the preference assumptions are in the following order: (1) each country prefers to dominate, (2) each country prefers mutual disarming to sole arming, and (3) the least liked is mutual arming. The preferences are shown in Table 2.29(b). In this case, there is no dominant strategy for either country.

A third case can be considered of deadlock, also called the leader's game per Table 2.29(c), in which the preferences are as follows: (1) each country prefers to dominate, (2) each country prefers mutual arming to mutual disarming, and (3) the least liked is being dominated. In this case, there is a dominant strategy of mutual arming.

The last possible case, fourth case, is the stag hunt, also called assurance or reciprocity per Table 2.29(d), in which the preferences are as follows: (1) each country prefers to mutual disarming, (2) each country prefers sole arming to sole disarming, and (3) the least liked is being dominated. In this case, there is no dominant strategy with equilibrium at both (arm, arm) and (disarm, disarm).

It should be noted that generally players might use a mix of the above games, for example, the United States using prisoner's dilemma preferences, whereas the USSR using the

TABLE 2.29

Illustrative Preferences 1 (most preferred), 2, 3, and 4 (least preferred) in Bilateral Nuclear Stability

(a) Prisoner's dilemma preferences

		USSR	
		Disarm	**Arm**
United States	**Disarm**	(2, 2)	(4, 1)
	Arm	(1, 4)	(3, 3)

(b) Chicken dilemma preferences

		USSR	
		Disarm	**Arm**
United States	**Disarm**	(2, 2)	(3, 1)
	Arm	(1, 3)	(4, 4)

(c) Deadlock dilemma preferences

		USSR	
		Disarm	**Arm**
United States	**Disarm**	(3, 3)	(4, 1)
	Arm	(1, 4)	(2, 2)

(d) Stag hunt dilemma preferences

		USSR	
		Disarm	**Arm**
United States	**Disarm**	(1, 1)	(4, 2)
	Arm	(2, 4)	(3, 3)

Note: Underlined values are the respective conditional best preference for each player.

deadlock dilemma preferences. In these cases, the analysis becomes more complex and the probabilities are assigned to account for uncertainties.

In general, gaming could involve more than two players. In the prisoners' dilemma game, a third player that could be identified is the law enforcement agency and its strategies. The solution might change as a result of adding the strategies of this third player. The use of these concepts in risk analysis and mitigation requires further development and exploration. In the nuclear arms example, the addition of other players, such as China, makes a decision situation challenging and perhaps such methods not practical. Several-player games, therefore, might limit the viability of such methods.

Example 2.8: Zero-Sum Payoffs in Pricing Strategy Determination

A simple example in economics is selling a product, such as a microchip processor, in a market with two competing companies at a price of $100 or $200 per processor. The payoffs are profits, after allowing for the costs of all kinds, as shown in Table 2.30. In this example, the two companies are competing for the same market and each firm must choose a high price of $200 per processor or a low price of $100 per processor. At a price of $200, 5,000 processors can be sold for a total revenue of $1,000,000. At a price of $100, 10,000 processors can be sold for a total revenue of $1,000,000. If both companies charge the same price, they split the sales evenly between them; however, if one company charges a higher price, the company with the lower price sells the entire amount and the company with the higher price sells nothing. Payoffs in this case are the profits, or the revenue minus the $500,000 fixed costs. Table 2.30 shows zero-sum payoffs, as the total in each cell is zero.

The solution to this game can be based on the *minimax criterion*, which results in a rational solution where each player chooses the strategy that minimizes the maximum payoff. In this game, the first company's maximum payoff at a price of $100 and $200 is $0 and $500,000, respectively, so the $100 price minimizes the maximum payoffs. The same reasoning applies to the second company; therefore, both companies will choose the $100 price. The reasoning behind the minimax solution in zero-sum games is that the first player (the first company) knows that whatever the company loses, the second player (the second company) gains; therefore, no matter what strategy the first player chooses, the second company will choose the strategy that gives the minimum payoff for that row. The second company reasons conversely, and therefore, the outcome is a dominant strategy equilibrium of ($100, $100) pricing with payoffs of ($0, $0). This outcome is to the best interest of a consumer. The minimax criterion for a two-person, zero-sum game produces a rational solution for each player by choosing the strategy that minimizes the maximum payoffs and the pair of strategies and payoffs such that each player minimizes the maximum payoffs.

Example 2.9: Variable-Sum Game in Price Competition

Continuing the previous sales example of a product, such as a microchip processor, in a market with two competing companies, the product prices are taken as $100, $200, or $300

TABLE 2.30

Zero-Sum Payoff Table (in $1000) for Unit Price Competition for C1 and C2

		Second Company (C2)	
		Price = $100	**Price = $200**
First company (C1)	**Price = $100**	(0, 0)	(500, –500)
	Price = $200	(–500, 500)	(0, 0)

Note: Underlined values are the respective conditional best payoffs for each company.

TABLE 2.31

Payoff Table (in Million Dollars) for Unit-Price ($) Competition for C1 and C2

		Second Company (C2)		
		Price = $100	**Price = $200**	**Price = $300**
First Company (C1)	**Price = $100**	(0, 0)	(50, −10)	(40, −20)
	Price = $200	(−10, 50)	(20, 20)	(90, 10)
	Price = $300	(−20, 40)	(10, 90)	(50, 50)

Note: Underlined values are the respective conditional best payoffs for each company.

per processor. The payoffs are profits, after allowing for the costs of all kinds, as shown in Table 2.31. In this example, the company that charges a lower price will receive more customers, and thus more profits than the competitor. The payoffs in this case do not sum to zero (in million US dollars) and do not sum to a constant value. In this case, payoff may add up to varied amounts in millions of US dollars, depending on the strategies that the two competitors choose. Thus, the minimax solution applies in this case, and the dominant strategy equilibrium is at the ($100, $100) pricing with payoffs of ($0, $0).

2.7 Political, Economic, and Financial Risks

Political, economic, and financial risks can be grouped into categories: (1) political and country risks, (2) economic and financial risks including market and credit risks, (3) operational risks, and (4) reputation risks. These four categories are described in subsequent sections. Additional aspects of political, economic, and financial risk concepts are presented in detail in Chapters 6 and 7.

2.7.1 Political and Country Risks

The increasingly global nature of markets, competition, and events make projects undertaken, investment portfolios constructed, and strategies enacted susceptible to political and country risks. The complexity and interdependencies are intractable, necessitating the deployment of judgment and opinions in analytical frameworks. The commonly used frameworks can be characterized as index based and include the following (Erb et al. 1996):

- Bank of America World Information Services
- Business Environment Risk Intelligence (BERI) S.A.
- Control Risks Information Services (CRIS)
- Economist Intelligence Unit (EIU)
- Euromoney
- Institutional Investor
- Standard and Poor's Rating Group
- Political Risk Services: International Country Risk Guide (ICRG), and Coplin–O'Leary Rating System
- Moody's Investor Services

Using regression and time series analysis, Erb et al. (1996) determined that financial risk measures contain the most information about future equity returns and country risk measures are highly correlated with country equity valuation measures.

Political Risk Services (http://www.prsgroup.com/) is a source to assess the country risk. The underlying framework is based on the methodology developed by Coplin and O'Leary (1994). It provides a series of risk factor ratings to summarize the forecasts for each country based on the following 17 risk components (broken down into 12 in the 18-month forecast and five in the 5-year forecast):

- *Eighteen-month risk factors.* The following 12 factors are analyzed from an 18-month forecast perspective, including political turmoil, which is included in both the 18-month and the 5-year forecast:
  - Turmoil: actions that can result in threats or harm to people or property by political groups or foreign governments, operating within the country or from an external base including riots and demonstrations, politically motivated strikes, disputes with other countries that may affect business, terrorism and guerrilla activities, civil or international war, street crime that might affect international business personnel, and organized crime having an impact on political stability or foreign business. Legal, work-related labor strikes that do not lead to violence are not included in turmoil.
  - Equity restrictions: limitations on the foreign ownership of businesses, emphasizing sectors where limitations are either especially liberal or especially restrictive
  - Operations restrictions: restrictions on procurement, hiring foreign personnel, or locating business activities, as well as the efficiency and honesty of officials with whom business executives must deal and the effectiveness and integrity of the judicial system
  - Taxation discrimination: the formal and informal tax policies that lead to either bias against or special advantages favoring international business.
  - Repatriation restrictions: formal and informal rules regarding the reparation of profits, dividends, and investment capital
  - Exchange controls: formal policies, informal practices, and financial conditions that either ease or inhibit converting local currency to foreign currency, normally a firm's home currency
  - Tariff barriers: the average and range of financial costs imposed on imports
  - Other import barriers: formal and informal quotas, licensing provisions, or other restrictions on imports.
  - Payment delays: the punctuality (or lack thereof) with which government and private importers pay their foreign creditors, based on government policies, domestic economic conditions, and international financial conditions.
  - Fiscal and monetary expansion: an assessment of the effect of the government's spending, taxing, interest rate, and other monetary policies. The assessment is based on a judgment as to whether the expansion is inadequate for a healthy business climate, acceptably expansionist, or so excessively expansionist as to threaten inflation or other economic disorder.
  - Labor policies: government policies, trade union activity, and productivity of the labor force that create either high or low costs for businesses.

- Foreign debt: the magnitude of all foreign debt relative to the size of the economy and the ability of the country's public and private institutions to repay debt service obligations promptly
- *Five-year risk factors.* Four additional factors are analyzed from a 5-year forecast perspective, noting that turmoil is included in both the 18-month and the 5-year forecast as follows:
  - Investment restrictions: the current base and the likely changes in the general climate for restricting foreign investments
  - Trade restrictions: the current base and the likely changes in the general climate for restricting the entry of foreign trade
  - Domestic economic problems: a country's ranking according to its most recent 5-year performance record in per capita Gross Domestic Product (GDP), GDP growth, inflation, unemployment, capital investment, and budget balance
  - International economic problems: a country's ranking according to its most recent 5-year performance record in current account (as a percentage of GDP), the ratio of debt service to exports, and the annual percentage change in the value of the currency

These 17 factors are used in computing summary risk ratings as they stand at the present and forecasted under several regime scenarios. The rating results based on the current and forecast levels are then used to calculate the risk scores.

The rating system comprises 22 variables in three subcategories of risk: political, financial, and economic according to 2011 version of the model. In this model, a separate index is created for each of the subcategories as follows: (1) the political risk index is based on 100 points, (2) the financial risk index is based on 50 points, and (3) the economic risk index is based on 50 points. The total points from the three indices are divided by 2 to produce the weights for inclusion in the composite country risk score making the composite scores to range from 0 to 100. The composite risk score ranges from very low (a score of 80–100 points) to very high risk (a score of 0–49.9 points). Figure 2.36 provides an example political risk map for Africa based on arbitrarily assumed values for the purpose of illustration. Recent enhancements of the model introduced a probabilistic framework.

2.7.2 Economic and Financial Risks

Catastrophic events regardless of their source can have massive impacts on regional economies of afflicted areas, and lead to adverse and sometimes favorable impacts on other adjacent or far economies through economic substitutions. Extreme losses can cause ripples through the interdependencies and global nature of the economy. Moreover, understanding the processes of economic recovery that is affected by the resilience of the infrastructure and economy is important in managing recovery efforts.

Such catastrophic events not only impact economies but also could destabilize financial markets and trigger economic crises in a manner that is not well understood due to complexity associated with individual, corporate, and governmental behavior and actions. As an example, the credit crunch of 2008 has caused the biggest economic crisis for at least 80 years. Most economists and macroeconomic models failed to predict it and renewed efforts to enhance modeling and manage crises in order to help financial risk managers and macroeconomic planners to assess the likelihood and severity of future downturns.

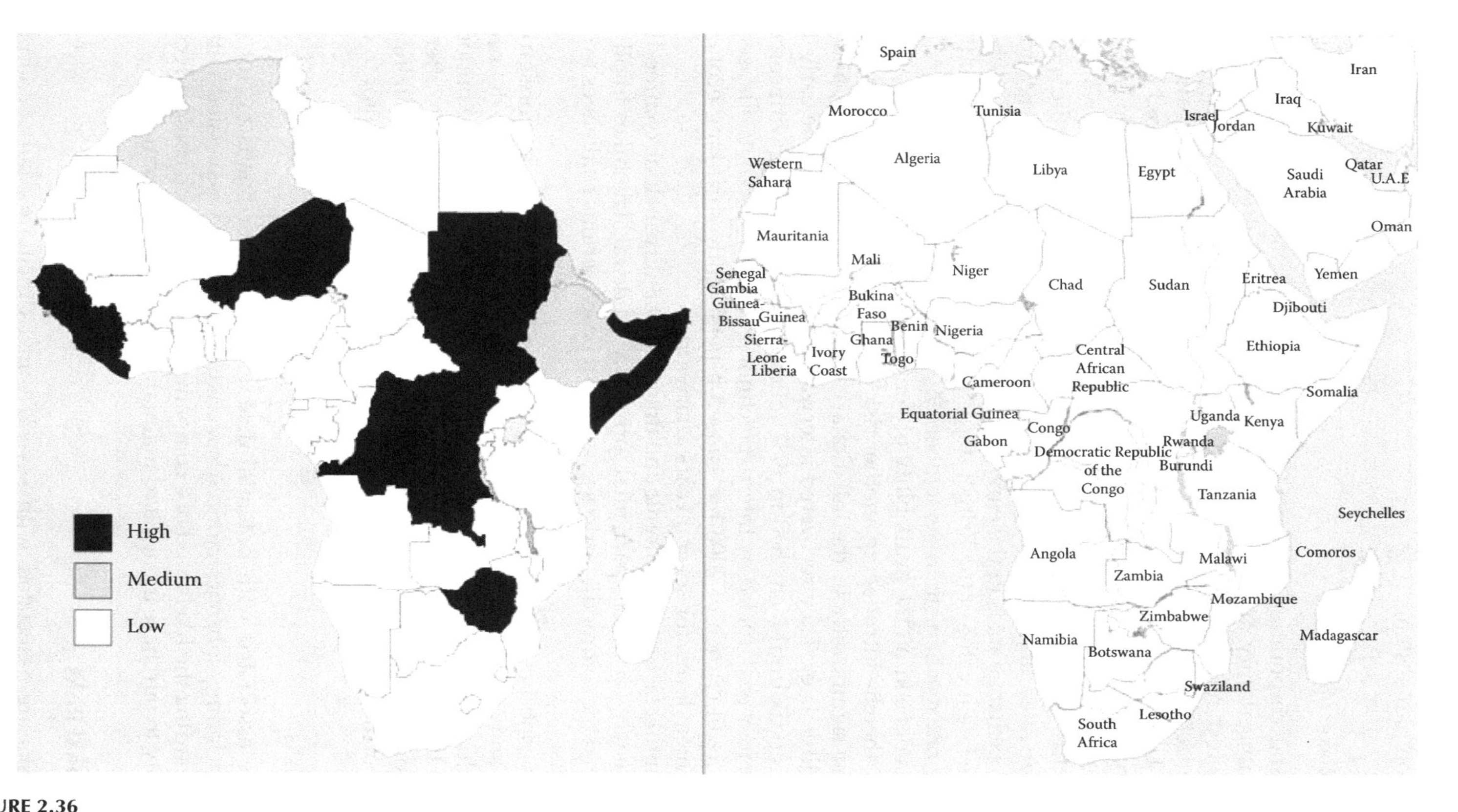

FIGURE 2.36
A hypothetical political risk map for illustration purposes.

Approaches, such as behavioral economics and complexity economics, suggest that business cycles, stock market volatility, and catastrophic collapse are inherent properties of the economy as a complex system.

2.7.3 Market Risks

Governments and corporations operate in economic and financial environments with some levels of uncertainty and instability. A primary contributor to defining this environment is interest rate. Interest rates can have a significant impact on the costs of financing a project and on corporate cash flows and asset values. For example, interest rates in the United States shot up in 1979 and peaked in 1981, followed by a gradual decline with some fluctuations until the credit crush of 2008 that led to a persistently shrinking economy with high unemployment levels for several years afterward.

For projects that target global markets, exchange rate instability can be a major risk source. Exchange rates have been volatile ever since the breakdown of the Bretton Woods system of fixed exchange rates in the early 1970s. An important example of exchange rate instability is the fall in value of the British sterling and Italian lira as a result of the failure of the exchange rate mechanism in September 1992.

Many projects depend on the availability of venture capital and the stock performance of corporations, thereby introducing another risk source related to stock market volatility. Stock prices rose significantly in the inflationary booms of the early 1970s, and then fell considerably a little later. They recovered afterward and fell again in the early 1981. The market rose to a peak until it crashed in 1987, followed by an increase with some swings until reaching a new peak fueled by Internet technologies, after which it collapsed in 2001. The mortgage credit crunch of 2008 has caused a persistently shrinking economy with high unemployment levels for several years afterward.

Other contributing factors to economic and finance instability include commodity prices in general and energy prices in particular, primarily crude oil. The hikes in oil prices in 1973–1974 affected the commodity prices greatly and posed serious challenges to countries and corporations.

Other sources contributing to volatility are derivatives for commodities, foreign currency exchange rates, and stock prices and indices, among others. Derivatives are defined as contracts whose values or payoffs depend on those of other assets, such as the options to buy commodities in the future or options to sell commodities in the future. They offer not only opportunities for hedging positions and managing risks that can be stabilizing but also speculative opportunities to others that can be destabilizing and a contributor to volatility.

2.7.4 Credit Risks

Credit risks are associated with potential defaults on notes or bonds by, for example, corporations, including subcontractors. Also, credit risks can be associated with market perceptions regarding the likelihood of a company defaulting, which could affect its bond rating and ability to purchase money and maintain projects and operations.

2.7.5 Operational Risks

Operational risks are associated with several sources including out-of-control operations risk that could occur when a corporate branch undertakes significant risk exposure that is not accounted for by corporate headquarters, leading potentially to its collapse, for example, the British Barings Bank, which collapsed in 1995 primarily as a result of its failure to control the market exposure created within a small overseas branch of the bank. Another risk source in this category is liquidity risk, in which a corporation requires more funding than it can arrange. Also, such risks could include money transfer risks and agreement breaches. Operational risks include model risks, which are associated with the models and underlying assumptions used to value financial instruments and cash flows incorrectly.

2.7.6 Reputation Risks

The loss of business attributable to a decline in a corporation's reputation can pose another risk source. This risk source can affect a company's credit rating, ability to maintain clients, workforce, and so on. This risk source usually occurs at a slow attrition rate. It can be an outcome of poor management decisions, business practices, and high-profile failures or accidents.

2.8 Risk Treatment and Control

Adding risk treatment and control to risk assessment defines risk management. Risk management involves the coordinated activities to direct and control an organization with regard to risk. For example, risk management is a process by which system operators, managers, and owners make safety decisions and regulatory changes, and choose different system configurations based on the data generated in the risk assessment. Risk management involves using information from the previously described risk assessment stage to make educated decisions about system safety. Risk treatments and control include risk prevention, avoidance, transfer, countermeasures, consequence mitigation, and so on as shown in Figure 2.8.

Risk management seeks an appropriate allocation of available resources in support of stated objectives and any stated constraints; therefore, it requires the definition of acceptable risk and comparative evaluation of options and/or alternatives against risk criteria for decision making. The goals of risk management are to reduce risk to an acceptable level and/or prioritize resources based on comparative analysis. Risk reduction is accomplished by preventing an unfavorable scenario, reducing the frequency, and/or reducing the consequences. A graph of the risk relationship is shown in Figure 2.37a and b as linear and nonlinear contours of constant risk for the risk-neutral and risk-averse stakeholder, respectively. Moreover, the vertical axis is labeled as probability, whereas it is commonly expressed as an annual EP or rate, as shown in Figure 2.4. In cases involving qualitative assessment, a matrix presentation can be used, as shown in Table 2.32. The table shows probability factors, severity factors, and risk (i.e., probability/severity factor) ratings of 0 (lowest) to 3 (highest). The base value of a project is commonly assumed to be 0. Each risk rating value requires a different mitigation plan.

2.8.1 Risk Acceptance

Risk acceptance constitutes a definition of safety as discussed in Section 2.2.8; therefore, risk acceptance is considered a complex subject that often entails much debate. The setting of an acceptable level of risk is important to define the risk performance that a system must achieve to be considered safe. If a system has a risk value above the risk acceptance level,

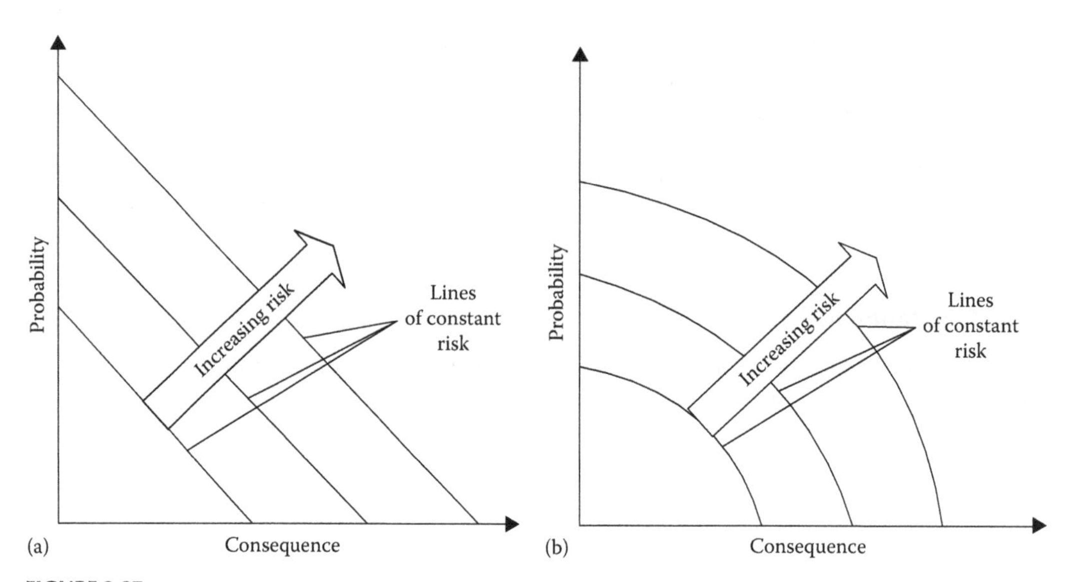

FIGURE 2.37
Risk graph: (a) risk neutral; (b) risk averse.

TABLE 2.32
Qualitative Risk Assessment Using Severity/Probability Factor Rating

		Probability Factor		
		Low	**Medium**	**High**
Severity factor	**High**	2	2	3
	Medium	1	1	2
	Low	0	1	2

Notes: Severity/probability factor rating is as follows: 3, mitigation strategy and detailed contingency plan; 2, mitigation strategy and outlined contingency plan; 1, mitigation strategy; 0, treat as a project base assumption.

actions should be taken to address safety concerns and improve the system through risk reduction measures. One difficulty with this process is defining acceptable safety levels for activities, industries, structures, and so on. Because the acceptance of risk depends on society perceptions, the acceptance criteria do not depend on the risk value alone. This section describes several methods that have been developed to assist in determining acceptable risk values, as summarized in Table 2.33.

Risk managers make decisions based on risk assessment and other considerations, including economical, political, environmental, legal, reliability, producibility, safety, and other factors. The answer to the question "How safe is safe enough?" is a difficult one and is constantly changing due to different perceptions and understandings of risk. To determine an acceptable risk, managers need to analyze the alternatives for the optimal choice. In some industries, an acceptable risk has been defined by consensus. For example, the US Nuclear Regulatory Commission requires that reactors be designed such that the probability of a large radioactive release to the environment from a reactor incident is

TABLE 2.33

Methods for Determining Risk Acceptance

Risk Acceptance Method	Summary
Risk conversion factors	Address the attitudes of the public about risk through comparisons of risk categories; also provide an estimate for converting risk acceptance values between different risk categories.
Farmer's curve	Provides estimated curves for cumulative probability risk profiles for certain consequences (e.g., deaths); demonstrates graphical regions of risk acceptance/nonacceptance.
Revealed preferences	Categorize society preferences for voluntary and involuntary exposure to risk through comparisons of risks and benefits for various activities.
Evaluation of magnitude of consequences	Compares the probability of risks to the magnitude of consequences for different industries to determine acceptable risk levels based on consequences.
Risk effectiveness	Provides a ratio for the comparison of cost to the magnitude of risk reduction. Using cost–benefit decision criteria, a risk reduction effort should not be pursued if the costs outweigh the benefits; this may not coincide with society values about safety.
Risk comparison	Provides a comparison between various activities, industries, etc., and is best suited to compare risks of the same type.

$<1 \times 10^{-6}$ per year. Risk levels for certain carcinogens and pollutants have also been given acceptable concentration levels based on some assessment of acceptable risk. However, risk acceptance for many other activities is not stated.

In some cases, only qualitative implications for risk acceptance are identified, that is, in maritime regulations. The International Maritime Organization High Speed Craft Code and the US Coast Guard Navigation and Vessel Inspection Circular (NVIC) 5-93 for passenger submersible guidance both state that if the end effect is hazardous or catastrophic, then a backup system and a corrective operating procedure are required. These references also state that a single failure must not result in a catastrophic event, unless the likelihood has been determined to be extremely remote.

Often the level of risk acceptance for various activities is implied. Society has reacted to risks through a balance of risk and potential benefits. Measuring this balance of accepted safety levels for various risks provides a means for assessing society values. Threshold values of acceptable risk depend on a variety of issues, including the activity type, industry, users, and society as a whole.

Target safety or reliability levels are required, for example, for developing procedures and rules for buildings and other structures. The selected reliability levels can be related to the probabilities of failure. The following three methods have been used to select target reliability values:

1. Agreeing on a reasonable value in cases of novel structures without prior history
2. Calibrating reliability levels implied in currently and successfully used design codes
3. Choosing target reliability level that minimizes the total expected costs over the service life of the structure including property and other losses

The first approach can be based on expert opinion elicitation, as discussed in Chapter 8. The second approach, code calibration, is the most commonly used approach as it provides the means to build on previous experiences. For example, design codes and rules provided by classification and industry societies can be used to determine the implied reliability and risk levels in respective codes and rules, then target risk levels can be set in a consistent manner, and new codes and rules can be developed to produce future designs and vessels offering similar levels of reliability and/or risk consistency. The third approach can be based on economic and trade-off analysis, as discussed in Chapter 7. In subsequent sections, the methods for determining risk acceptance, summarized in Table 2.33, are discussed.

2.8.1.1 Risk Conversion Factors

Analysis of risks shows that there are different taxonomies that demonstrate the different risk categories, often referred to as *risk factors*. These categories can be used to analyze risks on a dichotomous scale that compares risks by the way they invoke reactions including perceptions by society. For example, the severity category may be used to describe both ordinary and catastrophic events. Grouping events that could be classified as ordinary and comparing the distribution of risk to a similar grouping of catastrophic categories yields a ratio describing the degree of risk acceptance of ordinary events compared to catastrophic events. Comparison of various categories produces the risk conversion values provided in Table 2.34. These factors are useful in comparing the risk acceptance for different activities, industries, and so on. By computing the acceptable risk in one activity, an estimate of acceptable risk in other activities can be calculated based on the risk conversion factors (RCFs). A comparison of several common risks based on origin and volition is shown in Table 2.35. Common risks are classified into voluntary and involuntary groups with immediate and delayed effects or consequences. This grouping can be cross-classified by human-made and natural sources. Example risks in various classification bins are shown in Table 2.35. For example, aviation is a human-made hazard with potentially catastrophic consequences that are voluntary and immediate. Individuals are more willing to accept death due to a voluntary mountain-climbing accident than an involuntary flood-related event. Three hypotheses referred to as the *laws of acceptable risk* can be postulated as provided in Table 2.34:

TABLE 2.34
Risk Conversion Values for Different Risk Factors

Risk Factors	Risk Conversion Factor (RCF)	Computed RCF Value
Origin	Natural/human-made	20
Severity	Ordinary/catastrophic	30
Volition	Voluntary/involuntary	100
Effect	Delayed/immediate	30
Controllability	Controlled/uncontrolled	5–10
Familiarity	Old/new	10
Necessity	Necessary/luxury	1
Costs	Monetary/nonmonetary	NA
Origin	Industrial/regulatory	NA
Media	Low profile/high profile	NA

Note: NA, not available.

TABLE 2.35

Classification of Common Risks

		Voluntary		Involuntary	
Source	**Size**	**Immediate**	**Delayed**	**Immediate**	**Delayed**
Human-made	Catastrophic	Aviation	–	Dam failure Fire in a building Nuclear accident	Pollution Building fire
	Ordinary	Sports Boating Automobiles	Smoking Occupation Carcinogens	Homicide	–
Natural	Catastrophic	–	–	Earthquakes Hurricanes Tornadoes Epidemics	–
	Ordinary	–	–	Lightning Animal bites	Disease

1. The public is willing to accept voluntary risks roughly 1000 times greater than those for involuntarily imposed risks.
2. The statistical death rate appears to be a psychological yardstick for establishing the level of acceptability of other risks.
3. The acceptability of risk appears to be crudely proportional to the third power of the benefits, either real or perceived.

In safety studies of new dams, individuals are concerned about their own risks, which are defined as the total risk of death imposed by a dam on a particular person (i.e., an identifiable life), leading to suggested risk level as follows:

- The average risk of death to particular persons, not to exceed 10^{-6} per exposed person per year
- The risk to a specific person, not to exceed 10^{-5} per year

For existing dams, however, a risk up to 10 times higher could be tolerated.

Based on the above hypothesis that the death rate is the yardstick most commonly used to set a level of acceptable risk, various mortality rates were calculated from the available 1994 and 1995 US data collected by the National Center for Health Statistics, as shown in Table 2.36, and from the National Weather Service for natural disasters, as shown in Table 2.37. These two tables parallel the rates, provided here that involuntary risk to an individual is negligible if it is similar to the risk due to a natural hazard (10^{-6} per year) and it is excessive if it is similar to the risk due to disease (10^{-3} for a 30-year-old person). Over a period of 10 years, these values slightly changed, as footnoted in Table 2.36, and the causes of death have not changed significantly. Figures 2.38 and 2.39 show the trends of age-adjusted death rates and life expectancy, respectively, in the United States based on the data from the Centers for Disease Control and Prevention, the National Center for Health Statistics, and the National Vital Statistics System. Significant improvements have been made in health care, diets, and habits that contributed to these trends.

TABLE 2.36
Individual Fatality Rates Using 1994–1995 Data

Fatal Event	Total Number[a]	Fatalities per Year ($\times 10^{-4}$)[a]	Age-Adjusted Rate ($\times 10^{-4}$)
Total deaths	2,312,200	88.0	50.3
Disease			
Cardiovascular	952,500	36.3	17.5
Cancer	538,000	20.5	13.0
Pulmonary	188,300	7.2	3.4
AIDS	31,256	1.2	NA
Accidents			
Motor vehicles	41,800	1.6	1.6
Falls	13,450	0.52	NA
Poisons	8,994	0.35	NA
Fires/electrical	4,547	0.17	NA
Drowning	3,404	0.13	NA
Firearms/handguns	1,356	0.05	NA
Air/space	1,075	0.04	NA
Water transport	723	0.03	NA
Railway	635	0.02	NA
Suicide	30,900	1.2	1.1
Homicide	21,600	0.8	0.8

[a] 2003–2004 data show 2,398,365 deaths per year and 81.7 fatalities per year ($\times 10^{-4}$).
NA, not available.

TABLE 2.37
Natural Disaster Fatality Rates

Disaster	Years	Deaths	Rate ($\times 10^{-7}$)
Lightning	1959–1993	91	4.2
Tornadoes	1995	30	1.1
	1985–1994	48	1.9
Hurricanes/tropical storms	1995	29	1.1
	1985–1994	20	0.8
Floods	1995	103	3.9
	1985–1994	105	4.2

Risk can be categorized additionally by consequence types. Health risk, financial risk, and performance risk are all risk categories that differ by the types of consequence. It is important to be able to categorize the risk for the purpose of performing risk comparisons. For example, health risk would not be compared to financial risk as they are not similar categories, although methods to convert nonmonetary risk to financial risk are available as discussed in Chapter 5.

2.8.1.2 Farmer's Curves

The Farmer's curve is a graph of the cumulative probability versus the consequence for some activity, industry, or design. This curve introduces a probabilistic approach in

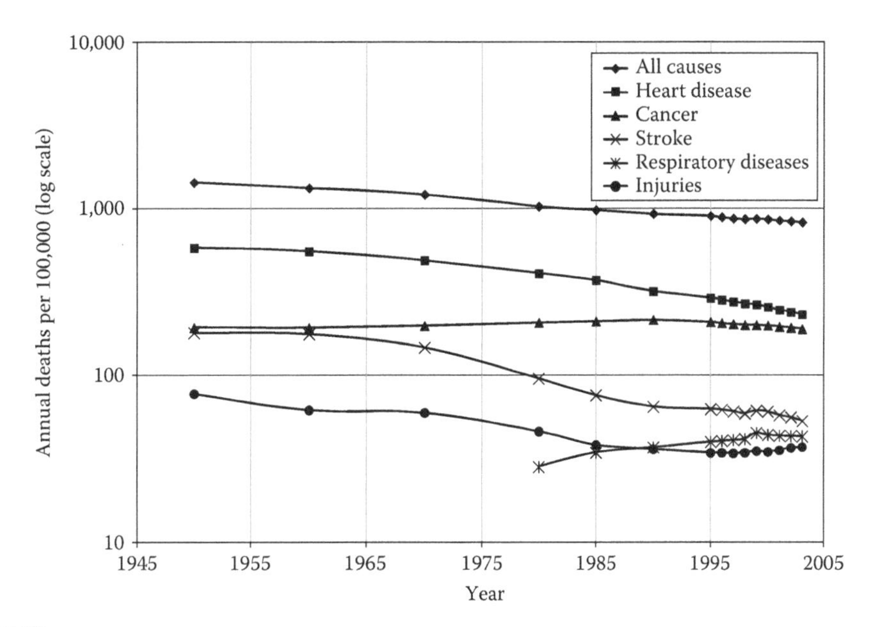

FIGURE 2.38
Trends in age-adjusted death rate.

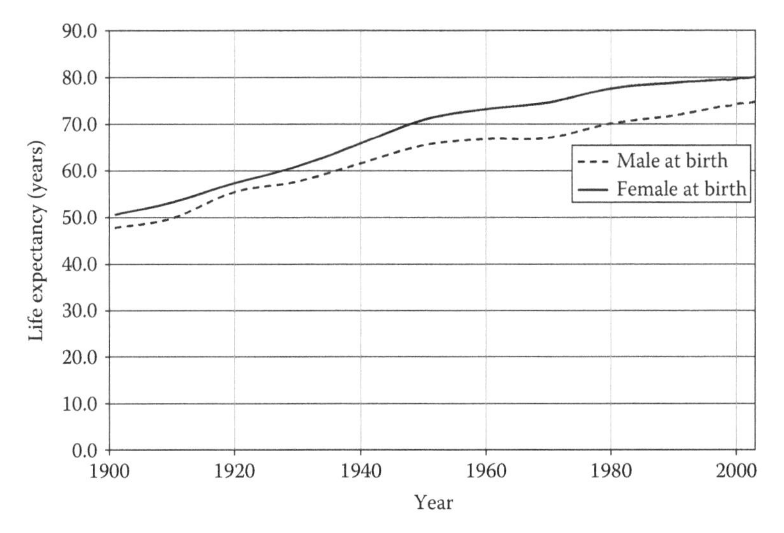

FIGURE 2.39
Trends in life expectancy.

determining the acceptable safety limits. Probability (or frequency) and consequence values are calculated for each level of risk, generating a curve that is unique to the hazard of concern. The area to the right (outside) of the curve is generally considered unacceptable, as the probability and consequence values are higher than the average value delineated by the curve. The area to the left (inside) of the curve is considered acceptable, as the probability and consequence values are less than the estimated value of the curve. An example Farmer's curve for various hazards is provided in Figure 2.40.

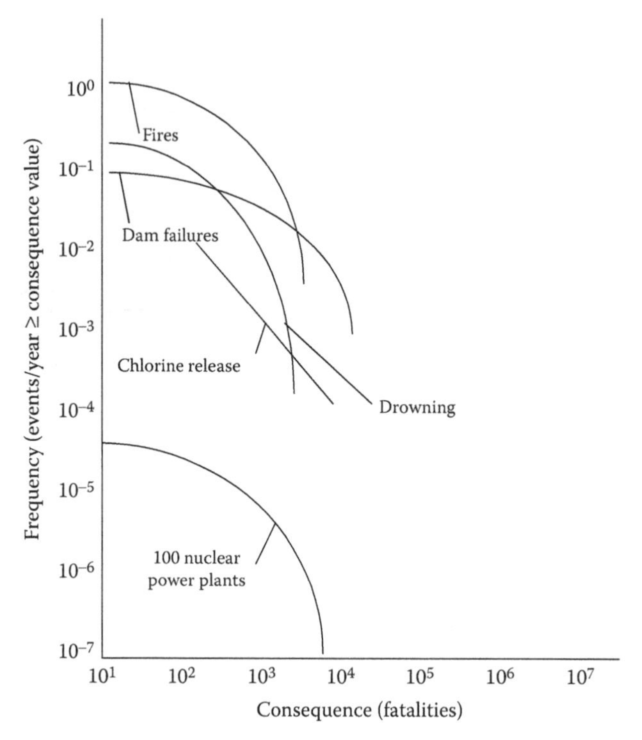

FIGURE 2.40
Farmer's curves comparing risks.

2.8.1.3 Method of Revealed Preferences

The method of revealed preferences provides a comparison of risks versus benefits and categorization of different risk types. The basis for this relationship is that risks are not taken unless there is some form of benefit. Benefit may be monetary or some other item of worth such as pleasure. The different risk types reflect voluntary versus involuntary actions, as shown in Figure 2.41. This technique assumes that the risk acceptance by society is found in the equilibrium generated from historical data on risks versus benefits. The estimated lines for acceptance of different activities are separated by the voluntary/involuntary risk categories. Further analysis of the data leads to estimating the proportionality relationship between risk and benefit as follows:

$$\text{Risk} \sim \text{benefit}^3 \tag{2.14}$$

2.8.1.4 Magnitudes of Risk Consequence

Another factor affecting the acceptance of risk is the magnitude of consequences of the event that can result from some failure. In general, the larger the consequence is, the less the likelihood that this event may occur. This technique has been used in several industries (T.W. Lambe Associates 1982; Whitman 1984) to demonstrate the location of the industry within a society's risk acceptance levels based on consequence magnitude, as shown in Figure 2.42. Further evaluation has resulted in several estimates for the relationship between the accepted probability of failure and the magnitude of consequence for failure

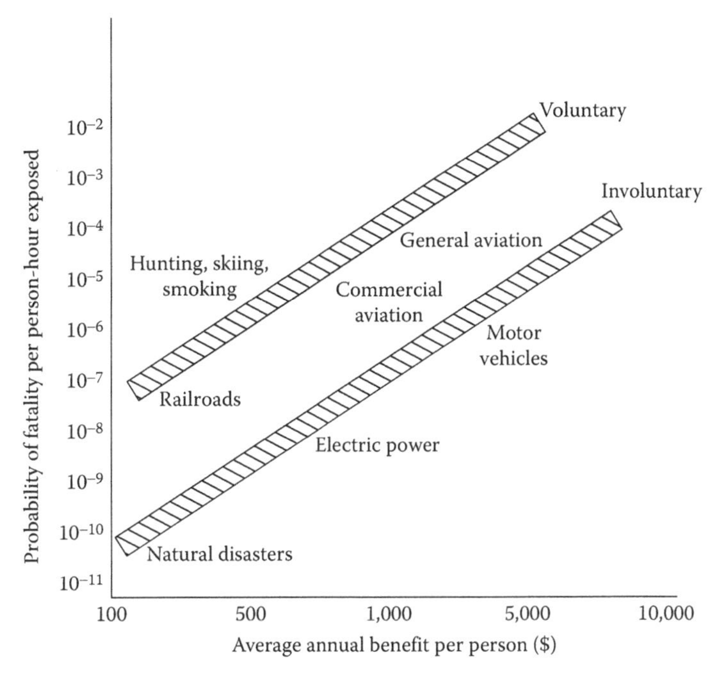

FIGURE 2.41
Accepted risk of voluntary and involuntary activities.

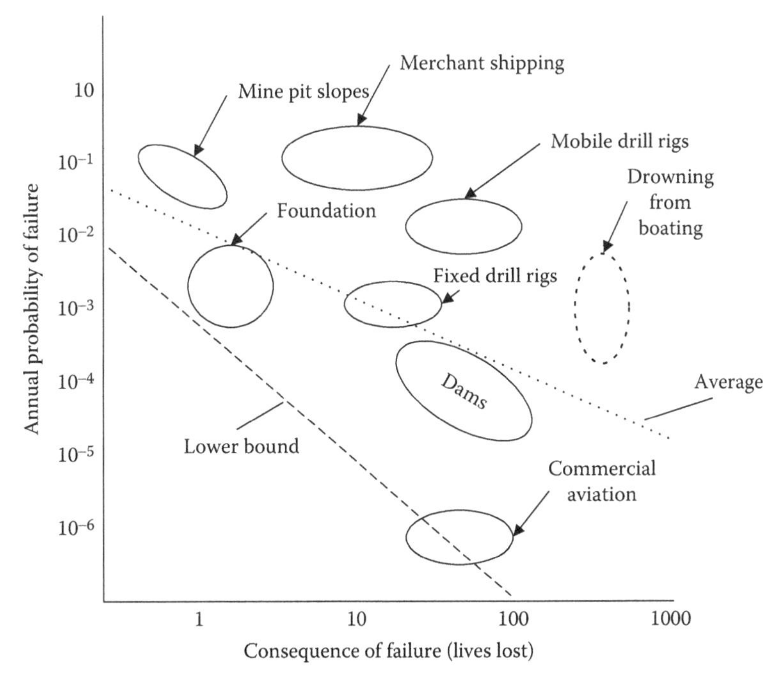

FIGURE 2.42
Target risk based on consequence of failure for industries.

(see Allen 1981) and referred to here as the CIRIA (Construction Industry Research and Information Association) lower bound equation:

$$P_f = 10^{-4}\frac{KT}{n} \tag{2.15}$$

where:
T is the life of the structure
K is a factor regarding the redundancy of the structure
n is the number of people exposed to risk

Another estimate is Allen's (1981) average equation:

$$P_f = 10^{-7}\frac{TA}{W\sqrt{n}} \tag{2.16}$$

where:
A and W are factors regarding the type and redundancy of the structure

Equation 2.15 offers a lower bound, whereas Equation 2.16 offers a central line.

2.8.1.5 Risk Reduction Cost-Effectiveness Ratio

Another measuring tool to assess risk acceptance is the determination of risk reduction effectiveness:

$$\text{Risk reduction cost-effectiveness} = \frac{\text{Cost}}{\Delta\text{Risk}} \tag{2.17}$$

where:
Cost should be attributed to risk reduction
ΔRisk is the level of risk reduction as follows:

$$\Delta\text{Risk} = \text{Risk before mitigation action} - \text{risk after mitigation action} \tag{2.18}$$

The difference in Equation 2.18 is also known as the benefit attributed to a risk reduction action. Risk effectiveness can be used to compare several risk reduction efforts. The initiative with the smallest risk effectiveness provides the most benefit for the cost. Therefore, this measurement may be used to help determine an acceptable level of risk. The inverse of this relationship may also be expressed as cost-effectiveness. This relationship is graphed in Figure 2.43, where the equilibrium value for risk acceptance is shown.

2.8.1.6 Risk Comparisons

This technique uses the frequency of severe incidents to directly compare risks between various areas of interest to assist in justifying risk acceptance. Risks can be presented in ways that can impact how the data are used for decisions. Often, values of risk are manipulated in different forms for comparison, as demonstrated in Table 2.38. Comparison of risk values should be taken in the context of the origin of the values and the uncertainties involved. This technique is most effective for comparing risks that invoke the same human perceptions and consequence categories. Comparing risks of different categories should be done with caution, as the differences between risk and perceived safety may not

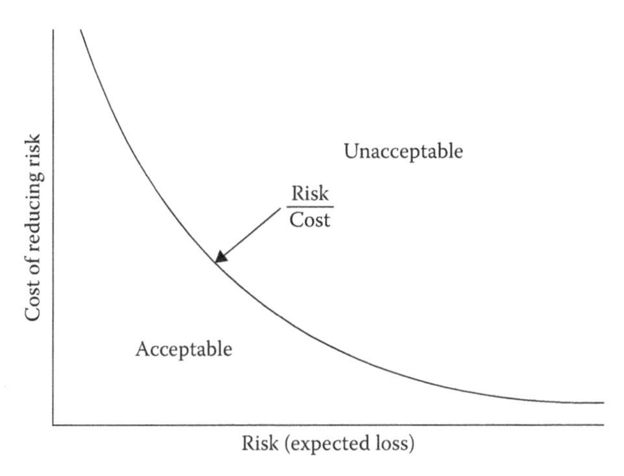

FIGURE 2.43
Cost-effectiveness of risk reduction.

TABLE 2.38
Ways to Express Risk of Death

Ways to Identify Risk of Death	Summary
Number of fatalities	This measure shows the impact in terms of the number of fatalities on society. Comparison of these values is cautioned, as the number of persons exposed to the particular risk may vary. Also, the time spent performing the activity may vary. Different risk category types should also be considered to compare fatality rates.
Annual mortality rate/individual	This measure shows the mortality risk normalized by the exposed population. This measure adds additional information about the number of exposed persons; however, the measure does not include the time spent on the activity.
Annual mortality	This measure provides the most complete risk value, as the risk is normalized by the exposed population and the duration of the exposure.
Loss of life exposure (LLE)	This measure converts a risk into a reduction in the expected life of an individual. It provides a good means of communicating risks beyond probability values.
Odds	This measure is a layman format for communicating probability, e.g., 1 in 4.

provide an objective analysis of risk acceptance. The use of RCFs may assist in transforming different risk categories. Table 2.8 demonstrates various estimates of risk of dying from various activities. Conservative guidelines for determining risk acceptance criteria can be established for voluntary risks using the involuntary risk values of natural causes and conversion factors.

2.8.2 Rankings Based on Risk Results

Another tool for risk management is the development of risk rankings. This ranking may be based on failure probabilities, failure consequences, risks, or other measures relating to risk. Generally, risk items ranked highly should be given high levels of priority; however, risk management decisions may consider other factors such as the costs of actions, their benefits and effectiveness of risk reduction actions, and political and regulatory constraints. The risk ranking results may be presented graphically as needed.

2.8.3 Strategy Tables

A strategy table, sometimes called an *alternative generation table*, is a tabulation of feasible combinations of potential actions to manage risks. Each combination can be viewed as a risk treatment strategy. Such a tool, with origins in systems engineering and decision analysis, offers a structure to identify the key dimensions of the alternatives, to compile lists of the different options available for each dimension, and to generate alternatives by combining one decision options from these dimensions.

Table 2.39 demonstrates a strategy table based on the emergency response planning of a large city to an elevated alert for a potential chemical release by an adversary (Parnell et al. 2005). In this case, the following four necessary functions are identified for each alternative: (1) threat detection, (2) warning, (3) protection, and (4) response. Each function can be performed by several means, with each alternative using one or more of the means to perform each function. For example, an alternative can be defined as the underlined entries in the table, that is, using ground and air sensors for detection, multimedia for warning, gas masks for protection, and the National Guard for response.

Primary advantages of this method are (1) explicitly identifying the dimensions of the alternative; (2) focusing creative thinking on new ways for performing each function and the overall alternative; (3) clearly defining the alternatives; (4) potentially generating many alternatives, for example, $3^4 = 81$ in the example of Table 2.39; (5) offering a comparative representation of the alternatives; and (6) providing a useful tool to communicate with stakeholders. It should be noted that if all dimensions are not identified, unfortunately the set of alternatives will be incomplete.

2.8.4 Decision Analysis

Decision analysis provides a means for systematically dealing with complex problems to arrive at a decision. Information is gathered in a structured manner to support informing the decision-making process. A decision generally deals with three elements: alternatives, impacts, and preferences. The alternatives are possible choices for consideration, and the impacts are the potential outcomes of a decision. Decision analysis provides methods for quantifying preference trade-offs for performance along multiple decision attributes while taking into account risk objectives. Decision attributes are the performance scales that measure the degree to which objectives are satisfied. For example, one possible attribute is reducing lives lost for the objective of increasing safety. Additional examples of objectives may include the following: to minimize the cost, to maximize utility, to maximize reliability, and to maximize profit. The decision outcomes may be affected by uncertainty; however, the goal is to choose the best alternative with the appropriate consideration of uncertainty. The analytical depth and rigor for decision analysis depend on the desired detail for making the decision. Benefit–cost analysis, decision

TABLE 2.39

Alternative Generation Using a Strategy Table for an Elevated Alert for a City

Detect	Warn	Protect	Respond
Patrols	Sirens	Containment	Citizens
Ground and airborne sensors	Television	*Gas masks*	Emergency medical teams
Both	*Multimedia*	Both	*National guard*

trees, influence diagrams, and the analytic hierarchy process are some of the tools to assist in decision analysis. Also, decision analysis should consider constraints such as, in the case of risk-based inspection, availability of a system for inspection, availability of inspectors, preference of certain inspectors, and availability of inspection equipment. Decision trees were briefly introduced in Section 2.5.6.5 and are covered with influence diagrams in Chapter 3.

2.8.5 Benefit–Cost Analysis

Risk managers commonly weigh various factors, including cost and risk. The analysis of three different alternatives is shown graphically in Figure 2.44. The graph shows that alternative C is the best choice, because the levels of residual risk and cost are less than those for alternatives A and B. However, if the only alternatives were A and B, then the decision would be more difficult. Alternative A has higher cost but lower risk than alternative B; alternative B has higher risk but lower cost than alternative A. A risk manager would have to weigh the importance of risk and cost and the availability of resources when making this decision and would also make use of risk-based decision analysis.

Risk–benefit analysis can also be used for risk management. Economic efficiency is important to determine the most effective means of expending resources. At some point, the costs for risk reduction do not provide adequate benefit. This process compares the costs and risk to determine where the optimal risk value is on a cost basis. This optimal value occurs when costs (c) to control risk are equal to the risk cost (r) due to the consequence (loss) (Figure 2.45). The figure shows that for a particular risk r_1 value, the cost is c_1; and for a particular cost c_2 value, the risk is r_2. Investing resources to reduce risks below this equilibrium point would not provide additional financial benefit. This technique can be used when monetary values can be attributed to risks; however, for particular risks, such as risk to human health and environmental risks, monetary values are difficult to estimate for human life and the environment. These valuation-related issues risk attitudes, and the dynamic nature of risk, such as risk homeostasis, are discussed in Chapter 7.

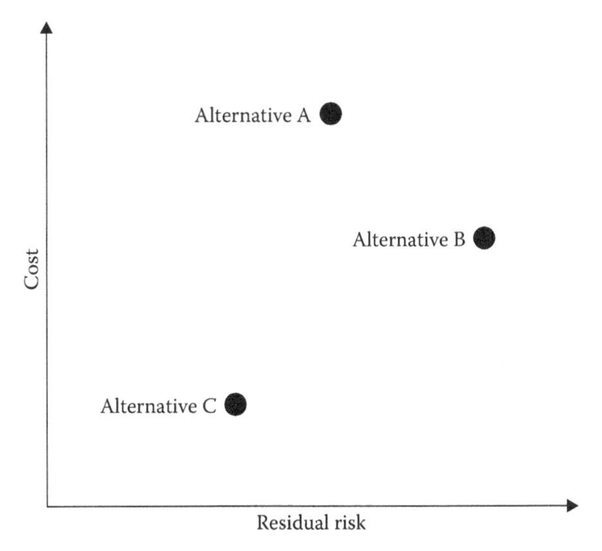

FIGURE 2.44
Risk benefit for three alternatives.

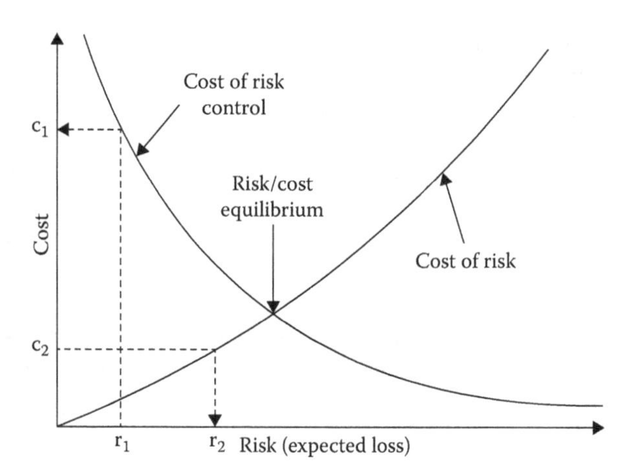

FIGURE 2.45
Comparison of risk and control costs.

2.8.6 Countermeasures and Mitigations

Countermeasures and mitigations are components of risk treatment, and were defined earlier, respectively, as (1) actions taken or a physical capability provided with a principal purpose of reducing or eliminating vulnerabilities or reducing the occurrence of attacks and (2) preplanned and coordinated actions or system features that are designed to reduce or minimize the damage caused by an event; support and complement emergency forces, that is, first responders; facilitate field investigation and crisis management response; and facilitate recovery and reconstitution.

They can be presented in economic and financial terms. Their definitions capture the essence of an effective management process of risk. If implemented correctly, a successful risk mitigation strategy should reduce any adverse (or downside) variations in the financial returns from a project, which are usually measured by (1) net present value (NPV), defined as the difference between the present value of the cash flows generated by a project and its capital cost and calculated as part of the process of assessing and appraising investments, or (2) internal rate of return (IRR), defined as the return that can be earned on the capital invested in the project (i.e., the discount rate that gives an NPV of 0) in the form of the rate that is equivalent to the yield on the investment. These economic concepts are described in Chapters 6 and 7.

Such actions or activities involve direct costs, such as increased capital expenditure or the payment of insurance premiums that might reduce the average overall financial returns from a project. This reduction is often a perfectly acceptable outcome, given the risk aversion of many investors and lenders. A risk mitigation strategy is the replacement of an uncertain and volatile future with one offering a smaller exposure to adverse risks and less variability in the return, although the expected NPV or IRR may be reduced. These two aspects are not necessarily mutually exclusive. Increasing risk efficiency by simultaneously improving the expected NPV or IRR and simultaneously reducing the adverse volatility is sometimes possible and should be sought. Risk mitigation should cover all phases of a project from inception to close down or disposal.

Primary ways to deal with risk within the context of a risk management strategy include the following:

- Risk reduction or elimination
- Risk transfer (e.g., to a contractor or an insurance company)

- Risk avoidance
- Risk absorbance or pooling

These primary ways are described in Chapter 7.

Example 2.10: Benefit–Cost Analysis for Selecting a Transport Method

Table 2.40 shows four transportation methods being considered by the warehouse owner discussed in previous examples in this chapter to supply components from the warehouse to one of its major customers in a foreign country. The available alternatives for the modes of transport are air, sea, road and ferry, or rail and ferry. The company management team has identified four relevant attributes for this decision situation: (1) punctuality, (2) safety of cargo, (3) convenience, and (4) costs. The first three attributes are considered to be benefit parameters, whereas the fourth one is a cost (of transportation). The weights of importance allocated to the three benefit attributes are 30 for punctuality, 60 for safety of cargo, and 10 for convenience. After a brainstorming session by the management team, the performance of each transportation mode was assessed according to the different attributes. The assessment results are shown in Table 2.41, together with the estimated annual cost of using each mode of transport. For the punctuality attribute, alternative A1 is considered to be the best option with a score value of 100; alternative A2 has been assigned a value of 0, indicating that it is the least favorable option. With respect to the other attributes, the same procedure was employed to produce the results summarized in Table 2.41.

TABLE 2.40

Assessment of Modes of Transportation for Delivery to Foreign Clients

		Attributes and Scores (0–100)		
Alternatives	**Cost ($1000)**	**Punctuality**	**Safety**	**Convenience**
A1—air	150	100	70	60
A2—sea	90	0	60	80
A3—road and ferry	40	60	0	100
A4—rail and ferry	70	70	100	0
Weight of importance	–	30	60	10

TABLE 2.41

Benefit-to-Cost Ratio Computations for the Modes of Transportation

		Benefit Scores (0–100)					
Alternatives	**Cost ($1000)**	**Punctuality**	**Safety**	**Convenience**	**Weighted Benefit**	**Weighted Benefit/Cost**	**Rank**
A1—air	150	100	70	60	78	0.52	3
A2—sea	90	0	60	80	44	0.49	4
A3—road and ferry	40	60	0	100	28	0.70	2
A4—rail and ferry	70	70	100	0	81	1.16	1
Weight of importance (normalized weight)	–	30 (0.3)	60 (0.6)	10 (0.1)	100 (1)	–	–

Note: Weighted benefit = (punctuality × 0.3) + (safety × 0.6) + (convenience × 0.1).

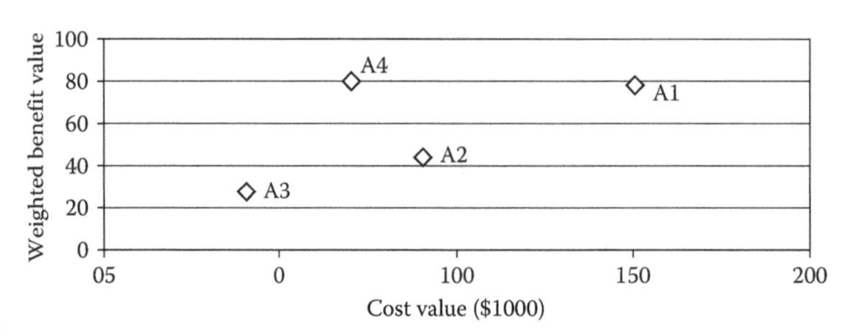

FIGURE 2.46
Cost–benefit analysis of transportation modes.

The optimal alternative can be selected by applying the concept of cost–benefit analysis. The alternatives are ranked according to their weighted benefit-to-cost values; the weight scores were normalized to obtain weight factors that sum up to 1 by dividing each value by the sum of all the weights. Then, each normalized weight was multiplied by the value of each alternative with respect to each attribute, and these values were added horizontally to obtain the total assessment for each alternative. By dividing the weighted assessment (i.e., benefit) by the cost value for each alternative, management can rank the alternatives according to benefit-to-cost ratios. The results are shown in Table 2.41. Inspection of the table reveals that alternative A4 gives the highest ratio (1.16); therefore, the rail and ferry transportation mode can be selected as the best alternative. In Figure 2.46, the plot of the values of benefits versus the values of costs for the alternatives reveals that alternative A4, again, is the best option, with the highest weighted benefit of 81 against a cost of $70,000, confirming previous weighted-benefit-to-cost-ratio computations. Alternative A3 is the second option using the weighted benefit-to-cost ratio, but it is the low-cost value of only $40,000. A cost–benefit trade-off analysis can be made between alternatives A4 and A3. A cost-conscious decision maker might choose alternative A3, whereas a benefit-driven decision maker might select alternative A4. If one is concerned with both, the weighted benefit-to-cost ratio of 1.16 for alternative A4 makes it the optimal choice.

2.9 Data Needs of Risk Methods

In risk assessment, the methods of probability theory are used to represent uncertainties. In this context, uncertainty could refer to event occurrence likelihoods that are characterized by periodicity, such as weather, and to conditions that are existent but unknown, such as the probability of an extreme wave. It can be used to characterize the magnitude of an engineering parameter, yet also the structure of a model. By contrast, probability is a precise concept. It is a mathematical concept with an explicit definition. We use the mathematics of probability theory to represent uncertainties, despite the fact that these uncertainties take many forms. Chapter 1 provides a discussion of types of uncertainty and ignorance and the theories available to model them.

The term *probability* has a precise mathematical definition, but its meaning when applied to the representation of uncertainties is subject to differing interpretations. The *frequentist view* holds that probability is the propensity of a physical system during a theoretically infinite number of repetitions, that is, the frequency of occurrence of an outcome in a long series of similar trials (e.g., the frequency of a coin landing heads up in an infinite number of flips is the probability of that event). In contrast, the *Bayesian view* holds that probability is the rational degree of belief that one holds in the occurrence of an event or the truth of a proposition; probability is manifest in the willingness of an observer to take action based on this belief. This latter view of probability, which has gained wide acceptance in many engineering applications, permits the use of quantified professional judgment in the form of subjective probabilities. Mathematically, such subjective probabilities can be combined or operated on as for any other probability.

Data are required to perform quantitative risk assessment or provide information to support qualitative risk assessment. Information may be available if data have been maintained on the system and components of interest. Information relevant to risk assessment includes the possible failures, failure probabilities, failure rates, failure modes, possible causes, and failure consequences. In the case of a new system, data may be used from similar systems if this information is available. Surveys are a common tool used to provide data. Statistical analysis can be used to assess confidence intervals and uncertainties in estimated parameters of interest. Expert judgment may also be another source of data, as described in Chapter 8. Uncertainty with the quality of the data should be identified to assist in the decision-making process.

Data can be classified as generic and project- or plant-specific data. Generic data include information from similar systems and components. This information may be the only information available in the initial stages of system design; therefore, potential differences due to design or uncertainty may result from using generic data on a specific system. Plant-specific data are specific to the system being analyzed. This information is often developed after the operation of a system. Relevant available data should be identified and evaluated, as data collection can be costly; data collection can be used to update the risk assessment. Bayesian techniques can be used to combine objective and subjective data.

Data can be classified as failure probability data and failure consequence data. The failure probability data can include failure rates, hazard functions, time between failures, results from reliability studies, and any influencing factors and their effects. Failure consequence data include loss reports, damages, litigation outcomes, repair costs, injuries, and human losses, as well as influencing factors and effects of failure prevention and consequence mitigation plans. Areas of deficiency in terms of data availability should be identified, and sometimes failure databases should be constructed. Data deficiency can be used as a basis for data collection and expert opinion elicitation, as described in Chapter 8.

2.10 Risk Representation, Communication, and Documentation

2.10.1 Risk Representation

Risk entails events or sequences of events, called a scenario, with occurrence likelihoods. A scenario can be viewed as a cause and, if it occurs, may result in consequences with severities, called losses. A risk measure accounts for both the probability of occurrence of a scenario and its consequences. Both the probability and its consequences could be uncertain. This section provides the fundamental cases for representing risks.

2.10.1.1 Fundamentals of Risk Representation

As described earlier, the representation or display of risk may include risk matrices (or tables), risk plots (or graphs), and probability distributions of adverse consequences in the form of cumulative probability distributions or EP distributions. The choice of representation techniques depends on the type of analysis (qualitative or quantitative) and stakeholder/decision-maker preferences. The risk display becomes the baseline for comparison of the effectiveness of risk management alternatives. It is important to recognize that the probability of the event is not plotted as a function of its potential adverse consequences. Rather, the two elements of risk are plotted separately on their own axes. Uncertainties in both elements of risk are represented by line segments, which form a cross that depicts the risk of the event.

For a scenario i, the risk pair (p_i, L_i) can be represented in any of the forms provided in Figure 2.47 reflecting the type of data available: (1) point estimates, (2) interval estimates, and/or (3) probability distributions (Ayyub and Kaminskiy 2009). A percentile interval can be used and converted to a probability distribution once a distribution type is assumed.

2.10.1.2 Exceedance Probability Distributions

Risk can be represented using Exceedance Probability (*EP*) *distributions* (or *curves*) as was previously illustrated in Figures 2.4 and 2.6. The EP curve gives the probabilities of specified levels of loss exceedance. The notion "losses" can be expressed in terms of dollars of damage, number of fatalities, casualties, and so on.

The construction of an EP curve begins with data that might be empirically obtained or produced using simulation methods. An example by Kunreuther et al. (2004) illustrates the empirical construction of an EP curve based on a set of loss-producing events. In this example, an EP curve is constructed for a portfolio of residential earthquake policies in Long Beach, California, and based on dollar losses to homes in Long Beach from earthquake

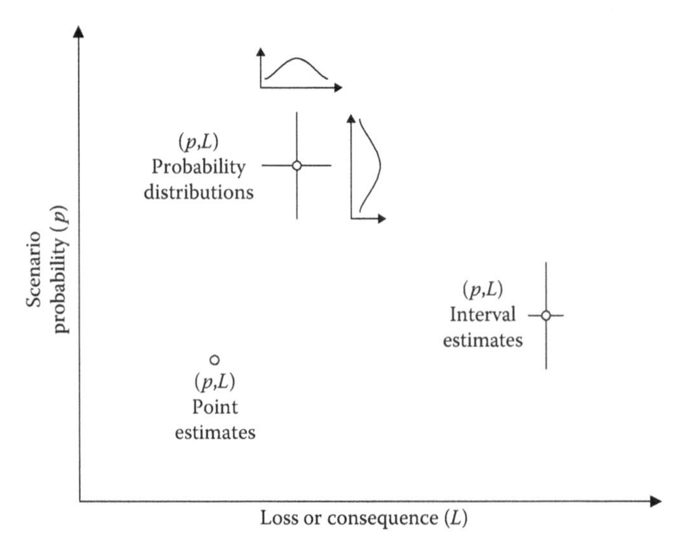

FIGURE 2.47
Risk plots and data types.

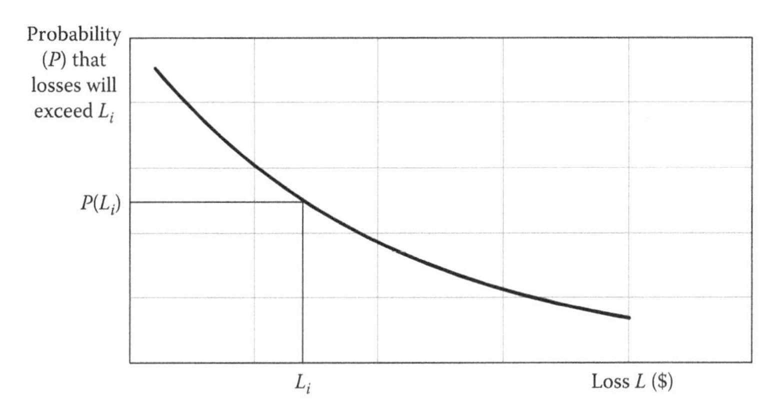

FIGURE 2.48
Sample mean exceedance probability curve.

events. The objective is to combine these loss-producing events that have respective return periods and annual probabilities, with the respective loss estimates obtained after the occurrence of the events to produce probabilities of exceeding losses of different magnitudes. Based on these estimates, the EP, or mean EP, is developed as shown in Figure 2.48. According to this figure with a particular loss L_i, the curve provides the probability that the loss as a random variable exceeds L_i with the respective y-axis value p_i. Thus, the x-axis measures the loss value in particular units of interest and the respective y-axis value is the probability that the loss exceeds a particular value.

Using an EP curve, the effects of countermeasures and mitigation strategies can be examined based on the shifts of the EP curve downward. In other words, the EP curves can be used to estimate benefit–cost effect of these strategies.

The EP curves can also express uncertainty associated with the probability of occurring of undesirable event and the magnitude of the respective loss as a result of uncertainties in specified values as inputs. Such uncertainties can be expressed using the *percentile* EP curves. For example, the 5% and 95% percentile EP curves depict uncertainties associated with losses as well as the uncertainties associated with respective probabilities. In our case, the EP curve depicting uncertainties in losses would show the interval ($L_i^{0.05}$, $L_i^{0.95}$), which can include the loss related to a given mean value L_i associated with probability p_i. Similarly, the EP curve depicting uncertainties in probabilities shows the percentiles ($p_i^{0.05}$, $p_i^{0.95}$) associated with loss mean value L_i.

It should be noted that due to data availability, constructing EP curves is much easier for the problems dealing with natural disasters (such as earthquakes and floods), compared to the risk assessment problems relating to HS where data are limited or nonexistent.

The available data can be assumed to be collected as a set of n disaster events, E_i, $i = 1, 2,\ldots, n$ with respective annual probabilities of occurrence p_i. Also, the respective losses (L_i) associated with these events are estimated after the occurrence of these events. It should be noted that the annual probabilities are defined for the disaster events, not for the losses. For example, the events can be earthquakes with magnitudes that can be physically measured in the form ground accelerations. If one would deal with floods, the respective events can be physically defined as well, for example, water-level elevation. An example of such data of 15 events is given in Table 2.42. Some of the entries in Table 2.42 are for events having equal (or almost equal) values of losses, but with respective annual probabilities that are different, such as for Event_{10} and Event_{11}. These two events correspond to earthquakes of

TABLE 2.42

Constructing Exceedance Probability Curves

Event (E_i)	Annual Probability of Occurrence (p_i)	Loss (L_i)	EP [$EP(L_i)$]	$E(L) = (p_i L_i)$
Event 1	0.002	25,000,000	0.0020	50,000
Event 2	0.005	15,000,000	0.0070	75,000
Event 3	0.010	10,000,000	0.0169	100,000
Event 4	0.020	5,000,000	0.0366	100,000
Event 5	0.030	3,000,000	0.0655	90,000
Event 6	0.040	2,000,000	0.1029	80,000
Event 7	0.050	1,000,000	0.1477	50,000
Event 8	0.050	800,000	0.1903	40,000
Event 9	0.050	700,000	0.2308	35,000
Event 10	0.070	500,000	0.2847	35,000
Event 11	0.090	500,000	0.3490	45,000
Event 12	0.100	300,000	0.4141	30,000
Event 13	0.100	200,000	0.4727	20,000
Event 14	0.100	100,000	0.5255	10,000
Event 15	0.283	0	0.6597	0

different magnitudes occurring, perhaps, at different locations with different populations at risk and producing the same loss estimates.

In order to apply the notions of annual probability and random variable, the disaster events must be, to an extent, *repeatable*, which might be available in the case of natural disasters such as earthquake, floods, and hurricane. It should be noted that identification of events is not necessarily straightforward in the case of terrorist actions. Nevertheless, the respective events can be identified. For example, they might be defined as explosions committed in public sites.

The loss associated with a given disaster event can be treated as a continuous random variable, whereas the number of events occurring in some specified period of time, such as a year, can be treated as a discrete random variable. Table 2.42 provides the loss data estimates for these events. These loss values can be considered as best estimates for the respective events and can be treated as central tendency point estimates of the continuous random variable.

For a set of natural disaster events, E_i, $i = 1,\ldots, n$, each event has an annual probability of occurrence, *pi*, and an associated loss estimate, L_i. The number of events per year is not limited to 1; numerous events can occur in the given year. Fifteen such events are listed in Table 2.42, ranked in descending order of the amount of loss. Event 15 was defined to be encompassing of all other zero-loss events so that the set of all events is collectively exhaustive. Although the probabilities of Events 1–15 in this example add to 1, this contrivance may lead some readers to believe that this is a requirement, whereas it is not.

The events in Table 2.42 are assumed to be independent Bernoulli random variables with the following probability mass functions:

$$P(E_i \text{ occurs}) = p_i \tag{2.19a}$$

$$P(E_i \text{ does not occur}) = 1 - p_i \tag{2.19b}$$

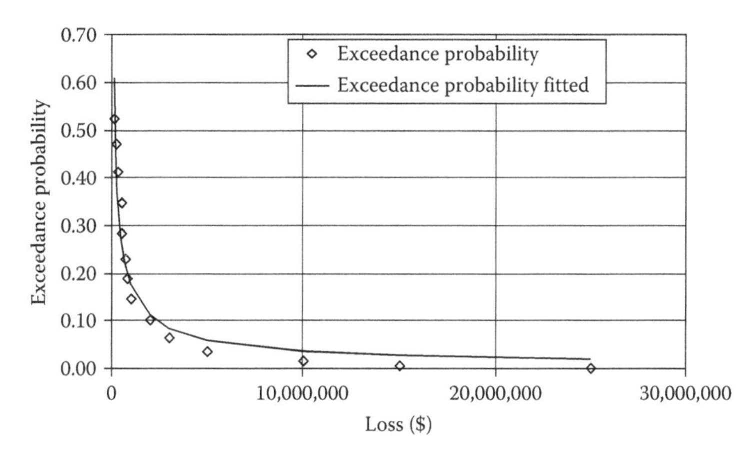

FIGURE 2.49
Mean exceedance probability curve based on the data from Table 2.38.

The expected loss (E) for a given event E_i is

$$E(L) = p_i L_i \tag{2.20}$$

If the events are indexed in reverse order of their losses (i.e., $L_i \geq L_{i+1}$), the mean (expected) EP for a given loss EP(L_i) can be found as

$$\text{EP}(L_i) = P(L > L_i) = 1 - P(L \leq L_i) = 1 - \prod_{j=1}^{i}(1 - p_j) \tag{2.21}$$

The EP curve based on the data from Table 2.42 is shown in Figure 2.49 with a fitted curve using a nonlinear model. In general, the summation of the probabilities of Events 1–15 can exceed 1, since they are independent Bernoulli events.

2.10.2 Risk Communication

Risk communication, defined as an interactive process of exchange of information and opinion among stakeholders such as individuals, groups, and institutions, often involves multiple messages about the nature of risk or expressing concerns, opinions, or reactions to risk managers or to legal and institutional arrangements for risk management. Risk communication greatly affects risk acceptance and defines the acceptance criteria for safety.

Risk communication provides vital links between the risk assessors, the risk managers, and the public for understanding risk; however, this does not necessarily mean that risk communication will always lead to agreement among different parties. An accurate perception of risk provides for rational decision making. The *Titanic* was deemed to be unsinkable yet was lost on its maiden voyage. Space shuttle flights were perceived to be safe enough for civilian travel until the challenger disaster. These disasters obviously had risks that were not perceived as significant until after the disaster. Risk communication is a dynamic process that must be considered prior to management decisions.

The communication process deals with technical information about controversial issues; therefore, it must be skillfully performed by risk managers and communicators who might be viewed as adversaries to the public. Risk communication between risk assessors and risk managers is necessary to apply risk assessments effectively in decision making. Risk managers must participate in determining the criteria for determining what risk is acceptable and unacceptable. This communication between the risk managers and the risk assessors is necessary for a better understanding of risk analysis in making decisions.

Risk communication also provides the means for risk managers to gain acceptance and understanding by the public. Risk managers need to go beyond the risk assessment results and consider other factors in making decisions. One of these concerns is politics, which is largely influenced by the public. Risk managers often fail to convince the public that risks can be kept to acceptable levels, as shown by the public's perception of toxic waste disposal and NPP operation safety. The public's perceived fear can lead to risk managers making conservative decisions to appease the public.

The value of risk calculated from risk assessment is not the only consideration for risk managers. All risks are not created equal, and society has established risk preferences based on public preferences. Decision makers should take these preferences into consideration when making decisions concerning risk.

To establish a means of comparing risks based on the society preferences, RCFs may be used. The RCF expresses the relative importance of different attributes concerning risk. Examples of possible RCFs are shown in Table 2.34. These values were determined by inferences of public preferences from statistical data taking into consideration the consequence of death. For example, the voluntary and involuntary classification depends on whether the events leading to the risk are under the control of the person at risk or not. Society, in general, accepts a higher level of voluntary risk than involuntary risk by an estimated factor of 100, according to Table 2.34, indicating that an individual will accept a voluntary risk that is 100 times greater than an involuntary risk.

The process of risk communication can be enhanced and improved in four aspects: (1) process, (2) channels, (3) message, and (4) target audiences. The risk assessment and management process should have clear goals with transparency, balance, and competence. The communication channels should take full advantage of all channel types available to target audiences, including Web sites, e-mail, and social networks. The contents of the message should account for audience orientation and uncertainty, provide risk comparison, and be complete. Consumer guides should be made available which introduce risks associated with a specific technology, the process of risk assessment and management, acceptable risk, decision making, uncertainty, costs and benefits, and feedback mechanisms. Improving the risk literacy of consumers is an essential component of the risk communication process.

The US Army Corps of Engineers 1992 *Engineering Pamphlet* on risk communication (EP 1110-2-8) provides the following considerations for communicating risk:

- Risk communication must be free of jargon.
- Consensus of experts needs to be established.
- Materials cited and their sources must be credible.
- Materials must be tailored to audience.
- The information must be personalized to the extent possible.
- Motivation discussion should stress a positive approach and the likelihood of success.
- Risk data must be presented in a meaningful manner.

According to the ISO 31010 Standard (2009c), successful risk assessment depends on effective communication and consultation with stakeholders since they assist in the following:

- Developing a communication plan
- Defining the context appropriately
- Ensuring that the interests of stakeholders are understood and considered
- Bringing together different areas of expertise for identifying and analyzing risk
- Ensuring that different views are appropriately considered in evaluating risks
- Ensuring that risks are adequately identified
- Securing endorsement and support for a treatment plan

Stakeholders have an important role of contributing to the interfacing of the risk assessment process with other management disciplines, including change management, project and program management, and financial management.

Fischhoff (2006) offers a strategy for the content of risk communication after severe adverse events as follows:

- Acknowledge the gravity of the events and the tragedy who have suffered.
- Recognize the public's concerns, emotions, and efforts to manage the risk.
- Assure the audience that the relevant officials are doing all that they can.
- Express a coherent, consistent communication philosophy for all risks.
- Provide quantitative risk estimates, including the uncertainties associated with the estimates.
- Provide summary analyses of possible protective actions considering all the expected effects.
- Lead by example, showing possible models for responsible bravery.
- Commit to earning and keeping the public trust.

Fischhoff (2005) provides additional information and explanations of these steps.

2.10.3 Risk Documentation

Any risk assessment process and results should be documented. Risks should be expressed in understandable terms, and the units in which the level of risk is expressed should be clear. The reporting requirement depends on the objectives and scope of the assessment; however, except for very simple assessments, the documentation can include the following (ISO 2009c):

- Objectives and scope
- Description of relevant parts of the system and their functions
- A summary of the external and internal context of the organization and how it relates to the situation, system, or circumstances being assessed
- Risk criteria applied and their justification

- Limitations, assumptions, and justification of hypotheses
- Assessment methodology
- Risk identification results
- Data, assumptions, and their sources and validation
- Risk analysis results and their evaluation
- Sensitivity and uncertainty analysis
- Critical assumptions and other factors that need to be monitored
- Discussion of results
- Conclusions and recommendations
- References

In cases where the risk study leads to a continuing risk management process, the management process should be performed and documented in such a way that it can be maintained throughout the life cycle of the system, organization, equipment, or activity. The assessment and the management process should be updated as significant new information becomes available and/or the context changes.

2.11 Limitations and Pitfalls of Risk Assessment

The 2008 financial crisis, terrorism, hurricane Katrina, BP oil spill, and so on all have something in common—the methods used to assess and manage these risks were fundamentally flawed (Hubbard 2009). In these cases, where risks were not properly assessed and managed, the risk methods themselves can be considered to be the biggest risk. In his treatment of this subject, Hubbard (2009) identifies the challenges to risk management. In the following list, we provide each challenge followed by how it was addressed in this book:

- *Confusion regarding the concept of risk.* In this chapter, we devoted special attention to terminology, units of measurement, and processes in order to make risk management results reproducible and defensible.
- *Completely unavoidable human errors in subjective judgment of risk.* Chapter 7 covers the data needs and expert opinion elicitation as a formal process to collect data in order to reduce the chance of such errors.
- *Entirely ineffectual but popular subjective scoring methods.* The coverage of risk assessment in this book favors probabilistic risk quantification in order to produce risk estimates in meaningful units that are suitable for benefit-to-cost analysis.
- *Misconceptions that block the use of better, existing methods.* Risk assessment methods are not only summarized but also compared among each other to enhance users' understanding of the limitations and applicability of such methods.
- *Recurring errors in even the most sophisticated methods.* The foundational aspects of risk assessment methods are presented with any underlying assumptions and limitations. Simulations methods are covered in Chapter 4 to enable the development of validation methods.

- *Institutional factors.* The coverage and treatment includes risk assessment and management processes including organization matters to reduce such instructional error chances.
- *Unproductive incentive structure.* Separating the risk assessment from the decision making would create the safeguards that are necessary to address this challenge.

2.12 Exercise Problems

Problem 2.1 What is the difference between hazards and threats? Provide examples.

Problem 2.2 Define risk and provide a classification of risk based on its sources. Provide an example for each risk source.

Problem 2.3 What is the difference between risk and uncertainty? How can you identify and differentiate between them in the following cases?

a. ABC Grocery and Supermarket Outlets plans to automate its warehouse by installing a computer-controlled order-packing system and a conveyor system for moving goods from storage to the warehouse shipping area.

b. Starting an automobile by turning the automobile key in the starter switch is based on limiting the system to the following potential failure modes: battery problems, defects in the starting subsystem, defects in the fuel subsystem, defects in the ignition subsystem, engine failure modes, and an act of vandalism that causes the automobile not to start, among the possible failure modes.

Problem 2.4 How would you assess the performance of a transportation system of a city?

Problem 2.5 Define security vulnerabilities of a university campus.

Problem 2.6 Use the ISO 31000 (2009a) definition of risk to define an example objective, uncertainties, and effect.

Problem 2.7 What is risk assessment and its methodologies? Draw a flowchart for risk-based life cycle management for the project described in Problem 2.3a.

Problem 2.8 Tabulate the types of risk events and scenarios that can be developed for the automobile system of Problem 2.3b.

Problem 2.9 Prepare an RBS associated with the project of Problem 2.3a from the point of view of the PM company that represents the owner of ABC Grocery and Supermarket Outlets.

Problem 2.10 Use the information provided in Problem 2.8 to analyze and assess risks associated with automobile subsystems that could lead to not being able to start the automobile. Use the following methods to provide your assessment:

a. Failure mode and effect analysis (FMEA)

b. Fault tree analysis (FTA)

Your model can be limited to the following potential failure modes: battery problems, defects in the starting subsystem, defects in the fuel subsystem, defects in the ignition subsystem, engine failure modes, and an act of vandalism that causes the automobile not to start. The undesirable event is that the car will not start on turning the key.

Problem 2.11 For Problem 2.3a, use the following methods for analyzing and assessing risks encountered by the contractor company constructing the automated warehouse project:

a. Failure mode and effect analysis (FMEA)
b. Fault tree analysis (FTA)

The undesirable event in these models is that the project will not be finished on time.

Problem 2.12 What is the median return on the project's risk profile provided in Figure 2.5? Estimate the standard deviation from the figure. Assuming that you have cash reserve of $100 million, would you consider this investment acceptable to you?

Problem 2.13 Define risk attitude and appetite. What are the differences? Provide examples.

Problem 2.14 Define risk aversion, seeking, and neutrality. Provide examples.

Problem 2.15 Use Table 2.9 to perform rational analytical computations in order to make observations based on the comparison of risk perceptions by members of the League of Women Voters and college students to experts, assuming that the experts have superior perceptions. You may use the mathematics of probability to make observations on the relative position of one group compared to another including comparing each to the opinions of experts. For example, does one group tend to overestimate or underestimate risk by risk groups, perhaps voluntary, involuntary, and so on?

Problem 2.16 Adapt the ISO 31000 (2009a) risk management methodology presented in Figure 2.11 to manage your graduate degree pursuit.

Problem 2.17 Upon graduation, you decided to start a business. Adapt the ISO 31000 (2009a) risk management methodology presented in Figure 2.11 to manage this start-up business. You make any necessary assumptions and clearly state them in a list.

Problem 2.18 Adapt the ISO 31000 (2009a) risk management methodology presented in Figure 2.11 to manage your performance in one of your educational courses or a project at your work or a personal project.

Problem 2.19 Use Figures 2.12 and 2.13 to identify the applicable risk categories to design and build a bridge crossing a water navigation channel to connect two countries with strong economic and political ties.

Problem 2.20 Use Figures 2.12 and 2.13 to identify the applicable risk categories to protect a mininuclear facility from deliberate human threats from outside the facility. The facility is located at the western coast of the Persian Gulf. A mini-nuclear facility has several buried nuclear reactors that are transported to site by cargo ships and require refueling every 10 years in the United States.

Problem 2.21 From a contractor's perspective, develop an RBS for the design and construction of a customized residential structure within 1 year as a lump sum contract.

Problem 2.22 From the perspective of an investor in a start-up company, develop an RBS for the development of a business based on successful social network software.

Problem 2.23 Consider the use of cranes to lift girders in bridge construction by a contractor. This activity has historically resulted in accidents. You are asked to examine the activity using events and causal factor analysis with barrier analysis to suggest barriers of various types. Develop the necessary tabulation structure to justify your suggestions.

Problem 2.24 Consider the use of powered nail guns in building construction by a contractor. This activity has historically resulted in accidents of nail striking workers. You are asked to examine the activity using events and causal factor analysis with barrier analysis to suggest the barriers of various types. Develop the necessary tabulation structure to justify your suggestions.

Problem 2.25 Mining operations are considered to include some of the most hazardous activities in industrial work. Thousands of miners die from mining accidents each year, especially in the processes of coal mining and hard rock mining, particularly in developing countries. Mining accidents can have a variety of causes, including leaks of poisonous gases such as hydrogen sulfide or explosive natural gases, especially methane, dust explosions, collapsing of mines, and flooding. Consider one category of mine accidents and familiarize yourself with it by researching it on the web. You are asked to examine this category using events and causal factor analysis with barrier analysis to suggest barriers of various types. Develop the necessary tabulation structure to justify your suggestions.

Problem 2.26 Use the case of a worker falling from a ladder discussed in Section 2.4.3.1 to develop an Ishikawa diagram similar to Figure 2.15. Discuss the basis for your diagram.

Problem 2.27 Consider the use of powered nail guns in building construction by a contractor. This activity has historically resulted in accidents of nail striking workers. Develop an Ishikawa diagram similar to Figure 2.15 for such accidents. Discuss the basis for your diagram.

Problem 2.28 Accidental deaths of children by guns can occur for various reasons. Over 3000 children were killed by guns in 2006 with 154 of those being accidental. In one case in 2011, a 3-year-old boy was found dead at his home due to a self-inflicted gunshot wound. Develop an events and causal factors diagram as a basis for a forensic analysis to structure an investigation. You may make any necessary assumptions in lieu of real information on this case.

Problem 2.29 Use Pareto analysis to identify the most important causal factors for accidental deaths of children by guns based on the solution of Problem 2.28.

Problem 2.30 Use Pareto analysis to identify the most important causal factors for the child drowning case demonstrated in Figure 2.15.

Problem 2.31 A corporation's server receives hundreds of security hits per year that can be grouped into three types in terms of increasing criticality (type 1, type 2, and type 3). These hits have not resulted in any damage; however, they can be treated as PEs. The conditional probabilities of damage given a hit of a particular type were estimated using event trees to be 0.001, 0.005, and 0.01, respectively. Estimate the annual rate of damaging hits based on the numbers of hits in a year for the three types of 100, 50, and 5, respectively.

Problem 2.32 A mining company classifies safety violations by its workers into three types in terms of increasing criticality (type 1, type 2, and type 3). These hits have not resulted in any accidents; however, they can be treated as PEs. The conditional probabilities of an accident given a violation of a particular type were estimated using event trees to be 0.01, 0.05, and 0.1, respectively. Estimate the annual rate of accident based on the number of violations over a period of 5 years for the three types of 100, 20, and 2, respectively.

Problem 2.33 Use the construction of a highway bridge to illustrate the use of a risk register to identify and manage risks for two objectives: (1) human health and safety and (2) completing a project on time and within budget.

Problem 2.34 Use the sprinkler system event tree of Figure 2.25 to develop a Swiss cheese model for communicating the risks and defenses.

Problem 2.35 Use the sprinkler system event tree of Figure 2.25 to perform FMEA for enhancing the availability of the system. Assume the values to illustrate the computations of the RPN.

Problem 2.36 The operation of a system entails steps A–I as illustrated in the arrow diagram below with the circles defining the termination nodes for completing the steps. The arrows represent the steps needed to operate a system. The operation of the system has some redundancy represented by more than one sequence of arrows to get to a node. A line arrow represents a step that must be performed. The length of the line does not have any significance. A line shows only the logical sequence of steps. Find the minimal cut sets (or construct an FT diagram) as an equivalent logic diagram using appropriate gates to attain the top undesirable event of not reaching node 8 starting from node 1.

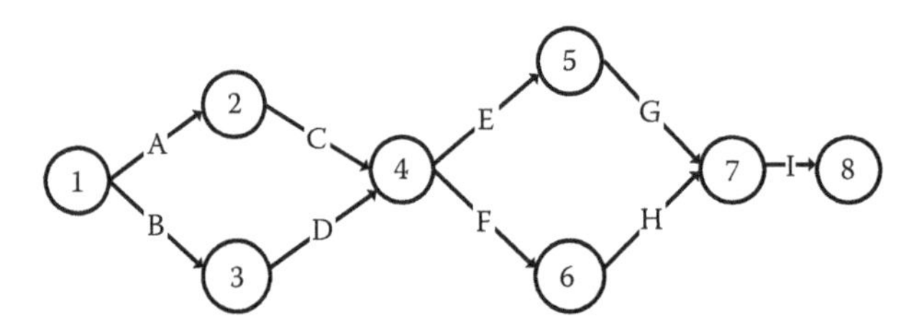

Problem 2.37 The operation of a system entails steps A–G as illustrated in the arrow diagram below with the circles defining the termination nodes for completing steps. Refer to Problem 3.36 for additional information on the meaning of the arrows and circle in the diagram. Find the minimal cut sets (or construct an FT diagram) as an equivalent logic diagram using appropriate gates to attain the top undesirable event of not reaching node 7 starting from node 1.

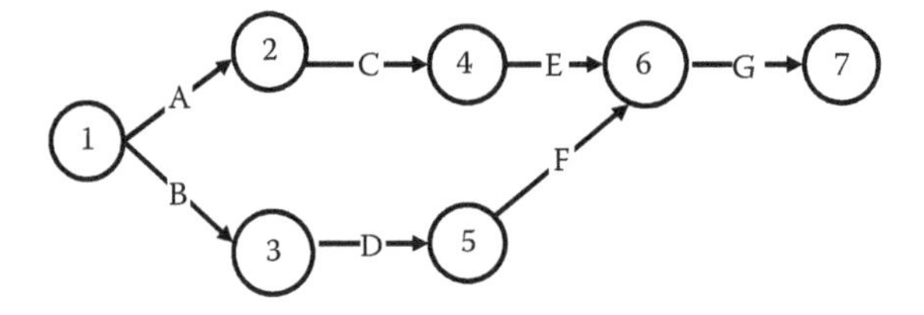

Problem 2.38 Use the FT diagram provided in Figure 2.32, case a, to compute the probability of the top event assuming the following probabilities for the basic events: $P(A) = P(B) = P(C) = P(D) = P(E) = P(F) = 0.01$.

Problem 2.39 Use the FT diagram provided in Figure 2.32, case b, to compute the probability of the top event assuming the following probabilities for the basic events: $P(A) = P(B) = P(C) = P(D) = P(E) = P(F) = P(G) = P(H) = 0.01$.

Problem 2.40 Use the FT diagram provided in Figure 2.32, case c, to compute the probability of the top event assuming the following probabilities for the basic events: $P(A) = P(B) = P(C) = P(D) = P(E) = P(F) = P(G) = P(H) = 0.01$.

Problem 2.41 Use the FT diagram provided in Figure 2.32, case d, to compute the probability of the top event assuming the following probabilities for the basic events: $P(A) = P(B) = P(C) = P(D) = P(E) = P(F) = P(G) = P(H) = 0.01$.

Problem 2.42 Use the FT diagram provided in Figure 2.32, case a, to compute the probability of the top event assuming the following probabilities for the basic events: $P(A) = P(B) = P(C) = 0.01$, $P(D) = 0.02$, and $P(E) = P(F) = 0.03$.

Problem 2.43 Use the FT diagram provided in Figure 2.32, case b, to compute the probability of the top event assuming the following probabilities for the basic events: $P(A) = P(B) = 0.01$, $P(C) = P(D) = 0.02$, $P(E) = 0.03$, and $P(F) = P(G) = P(H) = 0.04$.

Problem 2.44 Use the FT diagram provided in Figure 2.32, case c, to compute the probability of the top event assuming the following probabilities for the basic events: $P(A) = P(B) = 0.01$, $P(C) = P(D) = 0.02$, $P(E) = 0.03$, and $P(F) = P(G) = P(H) = 0.04$.

Problem 2.45 Use the FT provided in Figure 2.32, case d, to compute the probability of the top event assuming the following probabilities for the basic events: $P(A) = P(B) = 0.01$, $P(C) = P(D) = 0.02$, $P(E) = 0.03$, and $P(F) = P(G) = P(H) = 0.04$.

Problem 2.46 Compute the Fussell–Vesely and Birnbaum sensitivity factors for the tree solved in Problem 2.39 based on Figure 2.32, case b. Compare the results from the two methods and discuss.

Problem 2.47 Use Problem 2.4 to define the human errors and factors that could cause the failure of the automobile to start. Assess the significance of these errors by subjectively assigning a probability of occurrence of a major engine failure as a result of human errors.

Problem 2.48 The table below shows the strategies taken by two politicians planning the final two days of campaigning in two key cities in their campaigns to win an election in their state. Strategy S1 is to spend 1 day in each city and strategy S2 is to spend 2 days in the same city. The payoffs below are the total net votes won from the opponent. Is the payoff table a zero-sum game? What is the best alternative for each politician? Provide a justification for the selections. Did you obtain the Nash equilibrium? Why or why not?

		Second Politician	
		S1 (Two Days, Two Cities)	**S2 (Two Days, One City)**
First politician	**S1 (two days, two cities)**	(100, −100)	(200, −200)
	S2 (two days, one city)	(0, 0)	(100, −100)

Problem 2.49 Two contractors are planning to bid on a project. Each contractor can bid one of the following two prices:

1. Bidding price 1 (BP1): \$300,000
2. Bidding price 2 (BP2): \$350,000

The payoffs in the table below are the profits to be yielded from the combinations of strategies by the two contractors. Is the payoff table a zero-sum game? What is the optimal option for each contractor? Use the minimax criterion to obtain the solution. Provide a justification for the selections.

		Second Contractor	
		BP1 ($300,000)	**BP2 ($350,000)**
First contractor	**BP1 ($300,000)**	(0, 0)	(50, −20)
	BP2 ($350,000)	(−10, 40)	(20, 20)

Problem 2.50 Use the structure of Table 2.29 on bilateral nuclear stability and the assumption of deadlock to develop its preference table. Discuss your results.

Problem 2.51 Use the structure of Table 2.29 on bilateral nuclear stability and the assumption of stag hunt to develop its preference table. Discuss your results.

Problem 2.52 Use the framework provided by political risk services to propose a probabilistic framework to assess the risk profile instead of a scoring scheme. Illustrate your proposed framework using hypothetical country cases. The framework is available at http://www.prsgroup.com/ICRG_Methodology.aspx.

Problem 2.53 Use CIRIA lower bound equation, that is, Equation 2.15, to investigate the effects of the life of the structure, the factor regarding the redundancy of the structure, and the number of people exposed to risk. Use Figure 2.42 to estimate the working values for the redundancy factor.

Problem 2.54 Use Allen's (1981) average equation, that is Equation 2.16, to investigate the effects of the life of the structure, the number of persons exposed to risk, and A and W as the factors for the type and redundancy of the structure, respectively. Use Figure 2.42 to estimate the working values for the redundancy factors.

Problem 2.55 The owner and the PM team of the project in Problem 2.3a prepared a list of construction alternatives showing their costs in millions of dollars and the anticipated attributes for the alternatives expressed as (1) the risk levels associated with each alternative, (2) the impact of each alternative on the environment, and (3) the constructability of each alternative. You are asked to help them select the optimal alternative by applying the concept of cost–impact analysis. The table below shows the results of a brainstorming session performed by the owner and the PM team, where they assessed the attributes and scored them against each alternative, taking into account the cost for each alternative. For the risk attribute, alternative A1 is considered to be of very low risk (i.e., risk value = 0, indicating that it is a good alternative with respect to this attribute) and alternative A2 was assigned a high-risk value (i.e., risk value = 100, indicating that it is the worst alternative with respect to risk). The risk values for all alternatives are summarized in the table. With respect to the other attributes, the same procedure is employed. For example, alternative A3 was given a score of 100 in regard to having high environmental impact, whereas alternative A4 was assigned a value of 0. Similarly, alternative A1 was assigned a value of 100 (worst) for difficulty of construction, whereas alternative A2 was assigned a value of 0 (best) for that attribute. The PM team then assigned a weight score for each attribute based on its importance to the project. They assigned importance scores of 100 for risk, 80 for environmental impact, and 50 for constructability. Rank the alternatives based on the weighted benefit-to-cost ratios. Recommend the optimal alternative, that is, the largest benefit-to-cost ratio, to the owner and the PM team. (*Hint:* Define benefit as 100 minus impact and normalize the weight scores by their sum to obtain weight factors that sum up to 1.)

		Attributes as Impact Scores (0–100)		
Alternatives	**Cost ($ Million)**	**Risk**	**Environmental Impact**	**Difficulty of Construction**
A1	90	0	65	100
A2	110	100	90	0
A3	170	80	100	95
A4	60	45	0	50
Weight of importance		100	80	50

Problem 2.56 Use Table 2.39 to compute benefit-to-cost ratios and provide recommendations. Generate alternatives using all possible combinations based on all the underlined items, and assume cost values and risk reduction values. Assume a city population of 500,000.

Problem 2.57 Consider four alternative risk reduction measures for an automobile as provided in the following table:

Alternative	Cost ($)	Mean Risk Reduction ($1000)	Alternative Success Probability
A1	200	1000	0.5
A2	150	2000	0.5
A3	300	4000	0.8
A4	500	6000	1.0

The mean risk reduction can be treated as the benefit. The alternatives are not 100% effective as provided in the table. Compute the risk/benefit ratios and recommend the appropriate alternative. Discuss the alternative success probability and if it has an additional effect on your recommendation beyond its use in the computation.

Problem 2.58 Consider four alternative risk reduction measures for a commercial airplane as provided in the following table:

Alternative	Cost ($)	Mean Risk Reduction ($1,000)	Alternative Success Probability
A1	2,000	100,000	0.2
A2	4,500	200,000	0.3
A3	6,000	400,000	0.8
A4	10,000	600,000	1.0

The mean risk reduction can be treated as the benefit. The alternatives are not 100% effective as provided in the table. Compute the risk/benefit ratios and recommend the appropriate alternative. Discuss the alternative success probability and if it has an additional effect on your recommendation beyond its use in the computation.

Problem 2.59 Use Problem 2.3b to recommend the methods for risk acceptance of the system.

Problem 2.60 Use Problem 2.3a to recommend the risk mitigation strategies for the project. Categorize the strategies as risk reduction or elimination, risk transfer, risk avoidance, or risk absorbance and pooling.

Problem 2.61 Use Problem 2.3b to outline the risk communication plans to users (e.g., operators, automobile mechanics).

Problem 2.62 Ten events with losses are shown in the following table:

Event (E_i)	Loss (L_i)	Computed Annual Probability of Occurrence (p_i)
E_1	1,000,000	0.001
E_2	500,000	0.002
E_3	400,000	0.080
E_4	400,000	0.080
E_5	300,000	0.100
E_6	200,000	0.120
E_7	100,000	0.140
E_8	30,000	0.160
E_9	20,000	0.180
E_{10}	0	0.300

Annual probabilities were computed using logic trees for these events as shown in the table. Construct the loss EP curve.

Problem 2.63 Construct the loss EP curve based on the following six events with losses as shown in the following table along with their corresponding annual probabilities:

Event (E_i)	Loss (L_i)	Computed Annual Probability of Occurrence (p_i)
E_1	10,000,000	0.0001
E_2	8,500,000	0.0002
E_3	5,400,000	0.0080
E_4	3,400,000	0.0080
E_5	1,300,000	0.0100
E_6	0	0.1000

Problem 2.64 Construct the loss EP curve based on the following six events with losses as shown in the following table along with their corresponding annual probabilities:

Event (E_i)	Loss (L_i)	Computed Annual Probability of Occurrence (p_i)
E_1	100,000,000	0.0002
E_2	80,500,000	0.0004
E_3	50,400,000	0.0010
E_4	30,400,000	0.0090
E_5	10,300,000	0.0120
E_6	100,000	0.2000

3

System Definition and Structure

This chapter has the objective of providing definitions and methods for structuring problems and decision situations, and defining systems with special attention on applications. The chapter discusses how to select an appropriate level of detail and coverage in defining a system for supporting risk analysis and management studies. By starting with and focusing on what is called an answer variable, analysts have the tools to design models that are relevant and effective for addressing decision situations.

CONTENTS

3.1 Introduction

Performing risk analysis requires defining the problem at hand, which could span several disciplines or departments in an organization and encompass economic, environmental, technological, societal, and political dimensions. The stakeholders can be diverse, thus posing a challenge to risk analysts to appropriately define the problem. Defining and structuring a problem requires skill and perhaps a specialized facilitator who could work with all stakeholders to effectively achieve this objective. This process is called *system definition* for *structuring a problem* or a decision situation, and is the topic of this chapter.

Risk must be assessed, analyzed, and managed within a systems framework toward the objective of optimizing the utilization of available resources and for the purpose of maximizing benefits and utility to stakeholders in a cost-effective manner. Such a view of risk analysis and management requires structuring and formulating a problem or approaching a design with the following in mind: (1) the structure must be within a systems framework; (2) the approach must be systematic and must capture all critical and relevant aspects of the problem or decision situation; (3) uncertainties must be assessed and considered; and (4) an optimization scheme of the utilization of available resources, including maximizing benefits and utility to stakeholders, should be constructed. The objective of this chapter is to define these dimensions and provide background materials and introduce related methods.

3.2 Perspectives on System Definition

The term "system" originates from the Greek word *systema*, which means an organized whole. Informally, what is a system? According to *Webster's Dictionary*, a system is defined as "a regularly interacting or interdependent group of items forming a unified whole." A system can also be defined as "a set or arrangement of things so related or connected as to form a unity or organic whole," such as a solar system, school system, or system of highways, or as "a set of facts, principles, rules, etc. classified or arranged in a regular, orderly form so as to show a logical plan linking the various parts." The term "system science" is usually associated with observations, identification, description, experimental investigation, and theoretical modeling and explanations that are associated with natural phenomena in fields such as biology, chemistry, and physics. The term "system analysis" includes the ongoing analytical processes of evaluating various alternatives in design and model construction by employing mathematical methods for optimization, reliability assessment, statistics, risk analysis, and operations research, among other tasks.

For scientists and engineers, the definition of a system can be stated as "a regularly interacting or interdependent group of items forming a unified whole that has some attributes of interest." Alternatively, a system can be defined as a group of interacting, interrelated, or interdependent elements that together form a complex whole that can be a complex physical structure, process, or procedure of some attributes of interest. All parts of a system are related to the same overall process, procedure, or structure, yet they are most likely different from one another and often perform different functions.

The discipline of systems engineering establishes the configuration and size of system hardware, software, facilities, and personnel through an interactive process of analysis and design in order to satisfy an operational mission for the system to perform in a cost-effective manner. A system engineering process identifies mission requirements and translates them into design requirements at succeeding lower levels to ensure operational and performance satisfaction. Control of the evolving development process is maintained by a systems engineering organization through a continuing series of reviews and audits of technical documentation produced by systems engineering and other engineering organizations. The essence of systems engineering is structure; therefore, a systems engineer is expected to analyze and define the system as a set of elements or parts connected so as to form a whole with special attention to interfaces among its elements. Systems engineers understand the system by bringing structure to it. Choosing a particular structure requires an understanding of a system's nature and functions, and it leads to determining its constituent elements, associated technologies, costs, schedule, and constraints, among other considerations, for a new or a revised system. No clearly defined guidelines are available for the choice of system elements; however, the definition of these elements leads to interfaces among them that need to be considered in system analysis. Structured approaches provide a mechanistic listing of interactions among the elements. Understanding, controlling, and optimizing interfaces are major tasks for systems engineers, who sometimes spend more time working with the interfaces than on the elements themselves. Systems engineers leverage their understanding of the entire system to determine the various interface requirements of the elements. Seeing and comprehending the big picture offer a basis for identifying interfaces that can affect the chosen elements and impact the structure of the system. Figure 3.1 shows

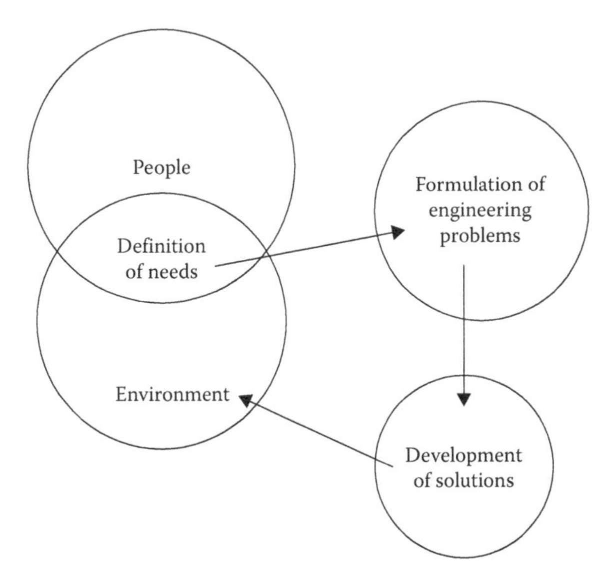

FIGURE 3.1
Engineers and systems.

how systems engineers identify the needs from an environment, structure problems, and provide solutions that feed into the environment through a dynamic process.

Systems can be grouped into various categories: (1) natural systems, such as river systems and energy systems; (2) human-made systems that can be imbedded in the natural systems, such as hydroelectric power systems and navigation systems; (3) physical systems, which are made of real components occupying space, such as automobiles and computers; (4) conceptual systems that could lead to physical systems; (5) static systems, which are without any activity, such as bridges subjected to dead loads; (6) dynamic systems, such as transportation systems; and (7) closed or open-loop systems, such as a chemical equilibrium process and logistic system, respectively. Blanchard (1998) provides additional information on these categories.

The analysis of systems requires the development of models that represent system behavior by focusing on selected attributes of interest or particular objectives of interest. Models for various categories, including natural or human-made systems, can be viewed as abstractions of their respective real systems. Systems scientists or engineers play a major role in defining the level of detail for such an abstraction, as well as the type and extent of information required in order to model these attributes properly and adequately and to predict system behavior. In general, a model can be viewed as an assemblage of knowledge and information regarding the most relevant system behavior and attributes. The availability of knowledge (or lack thereof, i.e., ignorance) and information (or its deficiencies, i.e., uncertainty) play major roles in defining these models as discussed in Chapter 1.

System definition commonly entails abstraction through discretization since most systems present themselves in continuums, spatially, and temporally. The discretization process is subjective and builds on the underlying analytical objective of a study. It involves uncertainty; however, it is necessary to facilitate the development and execution of a model.

Example 3.1: Safety of Flood Control Dams

The primary purposes of most flood control dams are to protect life, property, and environment, and grade stabilization for navigation. Other functions that can also be as important are supplying drinking water, hydropower generation, sediment control, and recreational utilities. Flood control dams are designed and constructed to provide sufficient capacity to store runoffs from a 10- to 100-year storm. A principal spillway is commonly used to pass floodwater from the storage pool (i.e., the reservoir of a dam) by means of a pipe through the dam over a period of several days. Any excess runoff passes immediately over an emergency spillway, which is usually a grassy waterway. Some flood control dams in dry and windy areas rarely contain any water but must have large capacities to control flash floods. Figures 3.2 and 3.3 show a flooded dam and a dam failure, respectively. Figure 3.2 shows workers trying to cross a flooded dam. Figure 3.3 shows a segment of the failed reservoir of the dam.

The US Army Corps of Engineers (USACE) has the responsibility of planning, designing, constructing, and maintaining a large number of US flood control dams. The safety of these dams is of great interest to the USACE. The safety assessment of a dam requires defining a dam system to include (1) the dam facility of structures, foundations, spillways, equipment, warning systems, and personnel; (2) the upstream environment that can produce storms and floods; and (3) the downstream environment, including the potential flood consequences. Due to the complexity of storm development and yield, the upstream segment of a system is difficult to define and would require substantial effort to study. Similarly, the downstream segment is complex in its nature and methods

FIGURE 3.2
Workers crossing Lacamas Lake Dam in Camas, WA, during the February 1996 flood. (Courtesy of the Washington State Dam Safety Office, Olympia, WA.)

FIGURE 3.3
Dam failure on the slope of Seminary Hill, Centralia, WA, 1991. (Courtesy of the Washington State Dam Safety Office, Olympia, WA.)

of assessment. The dam facility itself typically receives the bulk of engineering attention. Systems engineers need to define systems with an appropriate level of detail to achieve an intended study goal.

Example 3.2: Protecting a City from Hurricanes

The City of New Orleans is located in a hurricane-prone region. The city is protected by a hurricane protection system (HPS) consisting of levees and floodwalls, pumping stations, and gates. Figure 3.4a shows an example floodwall that is part of the HPS, and Figure 3.4b shows a breach, that is, failure of the HPS after the 2005 Hurricane Katrina. This example illustrates the discretization of the HPS for the purpose of risk quantification.

(a) Floodwall

(b) Breached Floodwall

FIGURE 3.4
The hurricane protection system of New Orleans. (Courtesy of US Army Corps of Engineers, Alexandria, VA.)

According to the American Society of Civil Engineers, the 2005 Hurricane Katrina is one of the strongest storms to hit the US coast with intense winds, high rainfall, waves, and storm surge. It impacted the Gulf of Mexico shores of Louisiana, Mississippi, and Alabama. The City of New Orleans was built on low-lying marshland along the Mississippi River. Levees and floodwalls were built around the city and adjacent parishes to protect against flooding. As a result of the hurricane, 1118 people were confirmed

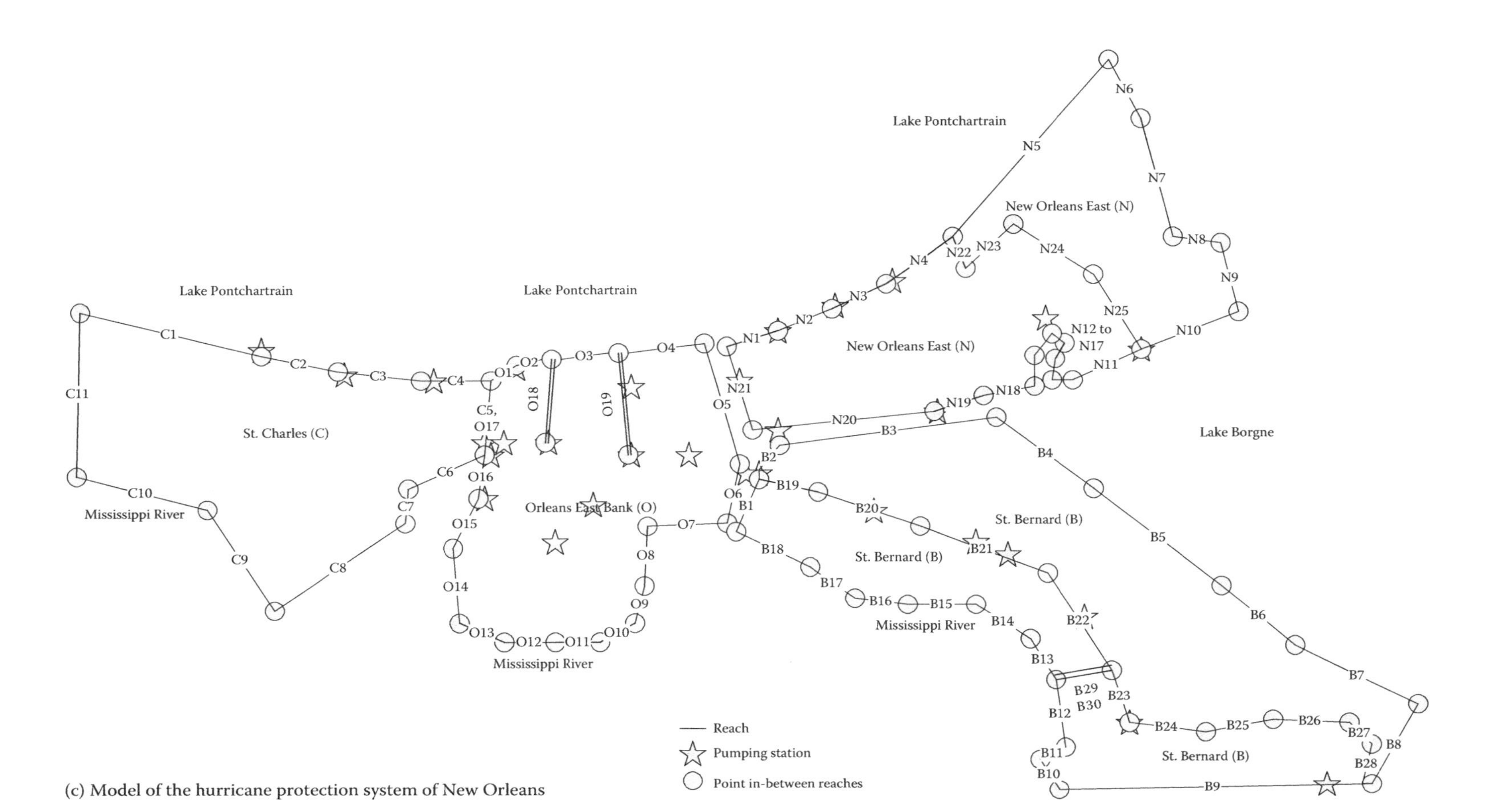

(c) Model of the hurricane protection system of New Orleans

FIGURE 3.4
(Continued) The hurricane protection system of New Orleans. (Courtesy of US Army Corps of Engineers, Alexandria, VA.)

dead in Louisiana and 135 people are still missing and presumed dead. Thousands of homes were destroyed, and the direct damage to residential and nonresidential property is estimated at $21 billion in 2005 and the damage to public infrastructure is another $6.7 billion. Nearly 124,000 jobs were lost, and the region's economy was crippled. The HPS of New Orleans extends several hundred miles with variations in soil conditions, geometry, height, strength, and so on. To analyze the ability of a primary segment of the HPS to protect the city, a simplified model was constructed as shown in Figure 3.4c. Each linear segment in this model represents a levee or a floodwall reach that is treated as a homogeneous segment in the development of the model. This abstraction through discretization is subjective and involves a lot of uncertainty; however, it is necessary to facilitate the development of a model for risk quantification.

3.3 Methods for System Definition

This section introduces methods for developing system models: (1) functional analysis, (2) requirements analysis, (3) work breakdown structure, (4) contributing factor diagrams, (5) decision trees and influence diagrams, (6) Bayesian networks, (7) process modeling method, (8) black-box method, (9) state-based method, and (10) component integration method. It is very common to use a combination of several models to represent a system in order to achieve a study's objectives.

3.3.1 Functional Analysis

Engineers use scientific knowledge to develop and build systems, including products and infrastructure, to meet societal needs. These needs can be represented orderly as an outcome of function analysis and described in a function structure. A function structure is a model of the system without material features such as shape, dimensions, and materials of the elements. The structure offers a hierarchy of the functions intended for the entire system and its parts, and their relations. It offers a logical representation from a limited number of elementary (or general) functions. Functions are abstractions of what a system should do. This type of analysis forces us to think about the system in an abstract manner, and stimulates creativity and prevents us from jumping immediately into solutions.

Functional analysis can be based on an underlying process required to meet a need. The process would offer the basis to identify the key basic functions and their relations. Another basis of functional analysis is to define a collection of elementary, that is, general, functions followed by structuring them in a logical manner. The following general steps can be followed in performing functional analysis:

- Define and describe the main function of the system in the form of a black box to meet the need. The main function is sometimes defined by its underlying parts as defined in the next step.
- Develop a list of subfunctions based on a process typically followed or used to meet the need. The process steps that are viewed as tasks offer clues for defining the corresponding functions. Only the processes that are carried out by the system are functions, whereas processes performed by a user of the system are user tasks. For user tasks, define only functions that support the user in performing the task. For example, for a user task of opening a container, supporting

functions would include an opening (or hole), a cover (or cap or door), and a mechanism for opening.

- Develop a function structure particularly for complex systems. There are three principles of structuring: putting functions in a chronological order, connecting inputs and outputs of flows between functions, or using some other logic applicable to the system. The structure can be visualized in a hierarchy.
- Elaborate the function structure with other secondary and tertiary functions as needed.
- Identify and document any variations of the function structure including moving the system boundary, changing the sequence of subfunctions, and splitting or combining the functions. Exploring these variations is the essence of functional analysis to explore possible solutions for meeting the need.

It is always desirable and advantageous to keep the function structure as simple as possible. Block diagrams of functions should remain conveniently arranged using simple and informative symbols.

The function structure can be loosely assembled into a hierarchy of functional, sequential, communicational, procedural, temporal, and logical attributes as follows:

- Functions with subfunctions that contribute directly to perform a single function
- Sequential breakdowns that show data flow processed sequentially from input to output
- Communicational breakdowns based on information and data needs
- Procedural breakdowns based on logic flow paths
- Temporal breakdowns for differing functions at different times
- Logical breakdowns based on developing logical flows for functions

Multiple functional hierarchies can be based on more than one of these criteria to sort and decompose the functions. Each criterion provides a different way of looking at the information, which can be useful for solving different types of problems. The most common functional hierarchy is a decomposition based on functional grouping, where the lower tier functions taken in total describe the activity of the upper tier function, providing a more detailed description of their top-level functions.

Example 3.3: Functional Analysis of Dams

Dams are intended to meet particular needs that may include protecting life, property, and environment from flooding due to natural and human-made hazards, grade stabilization for navigation, supplying drinking water, hydropower generation, trapping sediment, and recreational. Limiting the model to only the physical system of a dam, a function structure is shown in Figure 3.5 as a block diagram. This function structure can be used to develop a system work breakdown structure as discussed in subsequent sections. The system breakdown structure is the top-down hierarchical division of the dam into its subsystems and components, including people, structure, foundation, floodplain, river and its tributaries, procedures, and equipment. By dividing the dam environment into major subsystems, an organized physical definition for the dam system can be created. This definition allows for a better evaluation of hazards and potential effects of these hazards. Evaluating risk hierarchically (top-down), rather than in a fragmented manner, we can achieve rational, repeatable, and systematic outcomes.

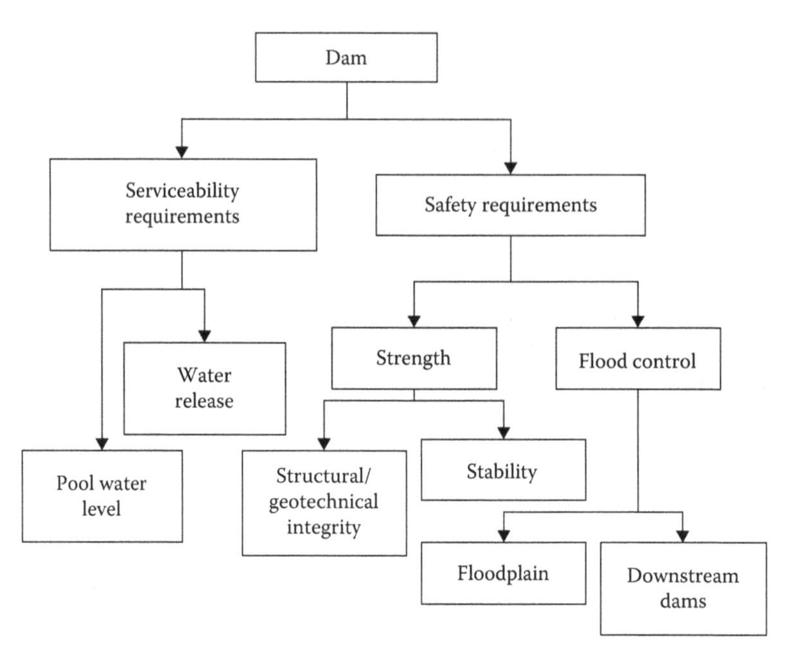

FIGURE 3.5
Functional structure of a dam.

3.3.2 Requirements Analysis

The definition of a system requires a specific goal, which can be determined from either needs identification or problem articulation. The goal statement should then be used to define a hierarchy of objectives that, in turn, can be used to develop a list of functional and performance requirements for the system. These requirements form the basis for system definition methods that are described here.

Requirements analysis can be defined as the detailed study of the performance requirements of a system to ensure that the completed system achieves its intended utility to a user and meets the goal stated. According to this method, the user's needs should be determined, evaluated for their completeness, and translated into quantifiable, verifiable, and documented performance requirements. Requirements analysis feeds directly into functional analysis, as well as allocation of functions, design, and synthesis.

Functional analysis examines the characteristic actions of hardware, software, facilities, or personnel that are necessary to satisfy performance requirements of the system. Functional analysis might establish additional requirements on all supporting elements of the system by examining their detailed operations and interactions. The overall set of system requirements derived by these analyses leads to both functional and performance requirements. Functional requirements define what the system must do and are characterized by verbs, because they imply action on the part of the system. The system gathers, processes, transmits, informs, states, initiates, or ceases. Also, any necessary physical requirements can be included as a part of the performance requirements. Physical requirements define the physical nature of a system, such as mass, volume, power, throughput, memory, and momentum. They may also include details, down to the type and color of paint, location of the ground segment equipment, and specific environmental protection. For example, aerospace company systems, unlike many commercial products, strongly emphasize functional requirements, thus prompting the need for a significant evaluation of the system's functional requirements of a system and allocation of functional requirements to the physical architecture.

The function structure of the example dam in Figure 3.5 might lead to the following functional requirements:

- Water release gates to meet the function of water release as shown in the figure
- Flood warning system to meet the function of safety for the floodplain

The corresponding performance requirements, as examples, for the above two items are as follows:

- Water flow capacity as a performance requirement in cubic meters per second in order to meet the water release function in a timely manner
- Time and means of warning the population of a city for safe evacuation

3.3.3 Work Breakdown Structure

The work breakdown structure as shown in Figure 3.6 for a dam is a hierarchy that defines the hardware, software, processes, and services of a system. The work breakdown structure is a physical-oriented family tree composed of hardware, software, services, processes, and data that result from engineering efforts during the design and development of a system. The sample breakdown of a dam into systems and subsystems in Figure 3.6 focuses on the physical subsystems, components, and human population at risk. The system was divided into subsystems, such as the dam facility subsystem that includes structural members, foundations, gates, turbines, spillway, alarms, and reservoir. The work breakdown structure was developed for the goal of performing risk analysis of dams. Each subsystem can be affected by and can affect other subsystems outside the hierarchy presented. While this breakdown is not complete, it does illustrate the hierarchy of the system and subsystem relations.

3.3.4 Contributing Factor Diagrams

The contributing factor diagrams are used to identify variables and their dependencies that can be used to analytically evaluate quantities, called *answer variables*, selected by a risk analyst to structure a problem entailing risks. Contributing factor diagrams are similar to influence diagrams but are not as formal and detailed. Influence diagrams are covered in this chapter in a subsequent section. A contributing factor diagram consists of variables graphically enclosed in ovals, circles, or rectangles and connected by directed arrows. The selection of a shape does not have any significance other than for convenience. The directed arrows represent the evaluation or computational dependencies among the variables. The construction of a contributing factor diagram should move from the top variable, the *answer variable*, to the basic variables and can be constructed as follows:

1. Identify and select answer variables in consultation with stakeholders and specialists in various areas. Commonly, economic answer variables are selected, such as net present value (NPV) or internal rate of return (IRR), which are discussed further in Chapter 6. Settling on appropriate answer variables can be challenging

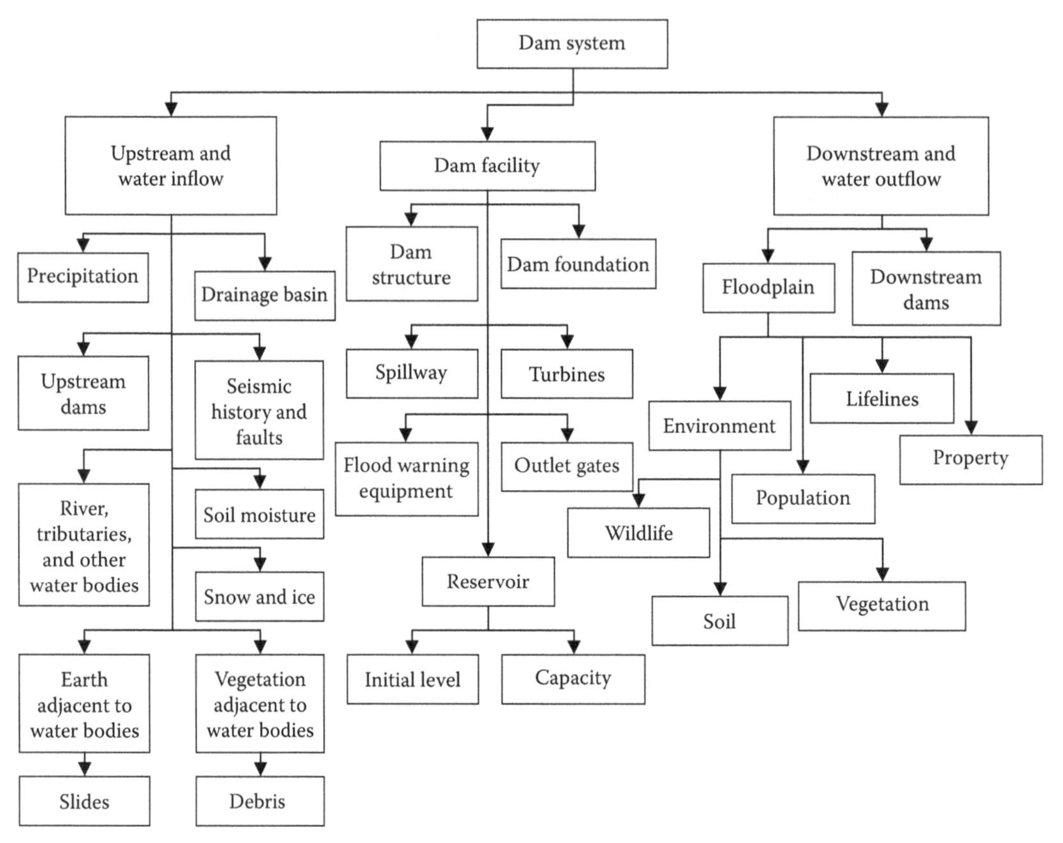

FIGURE 3.6
Work breakdown structure of a dam.

to a team and can result in several answer variables. These variables should be placed at the center of a contributing factor diagram in oval shapes. Ayyub et al. (2009a, 2009b) developed a risk model for the risk-based protection of hurricane-prone regions and used water volume and level entering a protected area by levees and floodwall as an answer variable among other subsequently computed answer variables, such as potential life and property loss.

2. Select the units of measurement for the answer variables, such as dollars per year or tons per year. Ayyub et al. (2009a, 2009b) used water volume, water elevation, monetary units, and life loss count as units of measurements in the risk model for the risk-based protection of hurricane-prone regions.
3. Identify and select primary contributing variables to the answer variables. For example, income and cost variables can be used with directed arrows feeding from them to the answer variables. For each variable, the units of measurement should be identified. Quantitative models are needed to express the dependencies among the variables.
4. Define lower level variables that feed into previously defined variables and their units.
5. Repeat step 4 until sufficient refinement is established for data collection or as defined by data availability.

These steps are presented in general terms to permit their use to solve diverse problems.

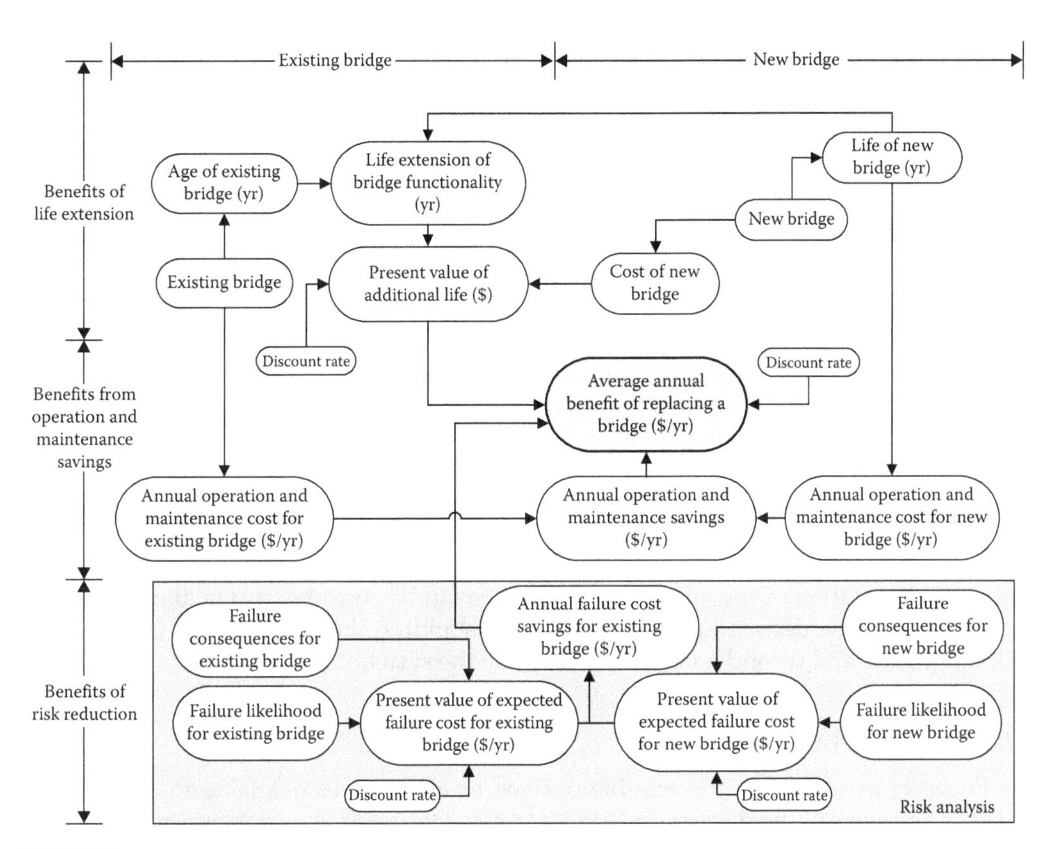

FIGURE 3.7
Contributing factors for risk-informed replacement of an existing bridge.

Example 3.4: Replacement of a Highway Bridge

Infrastructure rehabilitation involves decisions on replacement of major systems such as highway bridges. This bridge replacement need might result from structural (i.e., strength) or functional deficiencies. This decision situation requires the development of an economic model to assess the annual benefit to replace an existing bridge with a new one. Figure 3.7 provides a contributing factor diagram for such a decision situation. The answer variable in this case was identified as the *average annual benefit of replacing the bridge* (expressed in dollars per year). This variable was placed in the middle of the figure as shown by the shaded, bolded shape and was used as the starting point to develop the contributing factor diagram. The determination of this quantity requires three primary computational tracks: (1) the annual benefit generated by extended bridge functionality beyond the age of the existing bridge due to the added life provided by the new bridge; (2) the annual benefit of reduced operation and maintenance costs; and (3) the annual benefit of reduced expected failure costs. The first track is shown in the top portion of the figure, and the annual benefit of reduced operation and maintenance costs is shown in the middle portion of the figure. The risk analysis is shown within a shaded box at the bottom of the figure. The arrows in the figure indicate the computational dependencies among the variables.

3.3.5 Decision Trees and Influence Diagrams

Decision trees and influence diagrams share common features and relationships. They are sometimes employed simultaneously for examining a decision situation. Decision trees use arrows to represent a decision scenario progression as a process covering variables of interest to represent choices available and potential outcomes. Influence diagrams use the arrows to express dependencies among variables that are enclosed in circles, ovals, or other shapes. The two representations of a particular decision situation should be equivalent if the two approaches are properly implemented. The two methods are described in the subsequent sections.

3.3.5.1 Decision Trees

The elements of a decision model must be constructed in a systematic manner based on a decision-making goal or objectives for a decision-making process. One graphical tool for performing an organized decision analysis is a decision tree. A decision tree is constructed by showing the alternatives for decision making and associated uncertainties. The result of choosing one of the alternative paths in the decision tree is the potential consequences of the decision (Ayyub and McCuen 2011). The construction of a decision model requires definition of the following elements: objectives of the decision analysis, decision variables, decision outcomes, and associated probabilities and consequences. The decision analysis leads to identification of the scope of the decisions to be considered. The boundaries for the problem can be determined from first understanding the objectives of the decision-making process and second using them to define the system.

3.3.5.2 Decision Variables

The decision variables are the feasible options or alternatives available to the decision maker at any stage of the decision-making process. The decision variables for the decision model need to be defined. Ranges of values that can be taken by the decision variables should be defined. Decision variables for inspecting mechanical or structural components in an industrial facility can include what components or equipment to inspect and when, which inspection methods to use, assessment of the significance of detected damage, and repair/replace decisions. Therefore, assigning a value to a decision variable means making a decision at a specific point within the process. This point within the decision-making process is referred to as a *decision node*, which is identified in the model by a square as shown in Figure 3.8.

3.3.5.3 Decision Outcomes

The decision outcomes, with the associated occurrence probabilities, for the decision model must also be defined. The decision outcomes are events that can occur as a result of a decision. They are random in nature and their occurrence cannot be fully controlled by the decision maker. Decision outcomes can include the outcomes of an inspection (whether damage is detected) and the outcomes of a repair (whether a repair is satisfactory). The decision outcomes can occur after making a decision at points within the decision-making process called *chance nodes*. The chance nodes are identified in the model using circles, as shown in Figure 3.8.

3.3.5.4 Associated Probabilities and Consequences

The decision outcomes take values that can have associated probabilities and consequences. The probabilities are necessary because of the random (chance) nature of these outcomes.

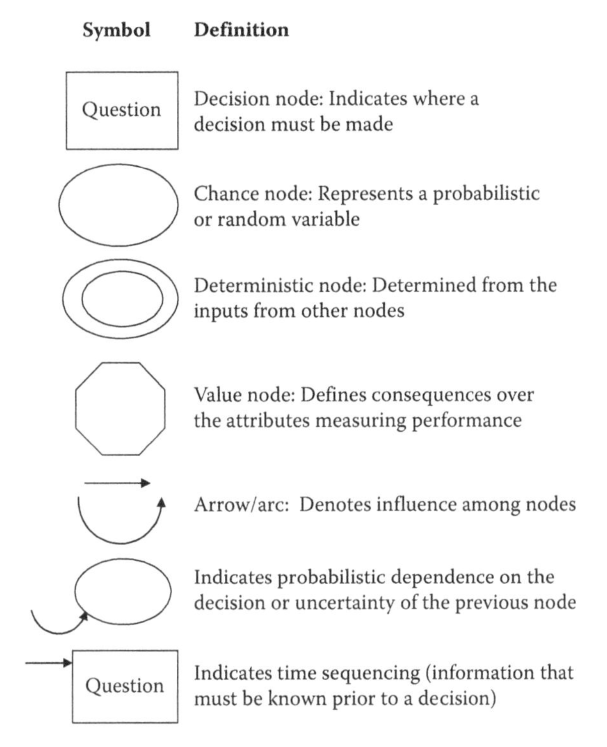

FIGURE 3.8
Symbols and their definitions for influence diagrams and decision trees.

The consequences can include, for example, the cost of failure due to damage that was not detected by an inspection method.

3.3.5.5 Tree Construction

Decision trees are commonly used to structure and analytically examine the available information on a decision situation. Risk profiles could provide some of the necessary information. A decision tree includes the decision and chance nodes. The decision nodes, which are represented by squares in a decision tree, are followed by possible actions (or alternatives, A_i) that can be selected by a decision maker. The chance nodes, which are represented by circles in a decision tree, are followed by outcomes that can occur without the complete control of the decision maker. The outcomes have both probabilities (P) and consequences (C). Here, the consequence can be a cost. Each tree segment followed from the beginning (left end) of the tree to the end (right end) of the tree is called a *branch*. Each branch represents a possible scenario of decisions and possible outcomes, and the total expected consequence (cost) for each branch can be computed. Then, the most suitable decisions can be selected to obtain the minimum cost. In general, utility values can be used and maximized instead of cost values. Also, decisions can be based on risk profiles by considering both the total expected utility value and the standard deviation of the utility value for each alternative. The standard deviation can be critical for decision making as it provides a measure of uncertainty for the utility values of the alternatives as discussed in Chapter 7. Influence diagrams can be constructed to model dependencies among decision variables, outcomes, and system states using the same symbols of Figure 3.8. In case of influence diagrams, arrows are used to represent dependencies among linked items as described in the next section.

Example 3.5: Decision Analysis for Selecting an Inspection Strategy

The objective in this example is to examine inspection methods of welds for quality assurance using a decision tree. This example is for illustration purposes and is based on hypothetical probabilities, costs, and consequences. The first step of the decision analysis for an inspection method selection is to identify a system with a safety concern, using methods such as risk assessment techniques. After performing the risk assessment, available inspection alternatives can be examined to select an appropriate solution. For example, the welds of the hull plating of a ship could be selected as a hull subsystem requiring risk-based inspection. If the welds would fail due to poor weld quality, the adverse consequence could be very significant in terms of economic losses, environmental damages, and potential loss of human life, even vessel loss. An adequate inspection program is necessary to mitigate this risk and keep it at an acceptable level. Previous experiences and knowledge of the system can be used to identify candidate inspection methods. For the purpose of illustration, only four candidate inspection methods are considered in Figure 3.9: visual inspection, dye penetrant inspection, magnetic particle inspection, and ultrasonic testing.

The outcome of an inspection method is either detection or nondetection of a defect, as identified by an occurrence probability, *P*. These outcomes originate from a chance node. The costs or consequences of these outcomes are represented by *C*. The probability and cost estimates are assumed for each inspection method based on its portion of the decision tree.

The computational treatment is limited to expected value computations for illustration purposes. The total expected cost for each branch in Figure 3.9 was computed by summing up the products of the pairs of cost and probability along the branch, and then the total expected cost for the inspection method was obtained by adding up the total expected costs of the branches on that portion of the decision tree. Assuming that the decision objective is to minimize the total expected cost, the "magnetic particle test" alternative should be selected as the optimal method. Although this is not the least

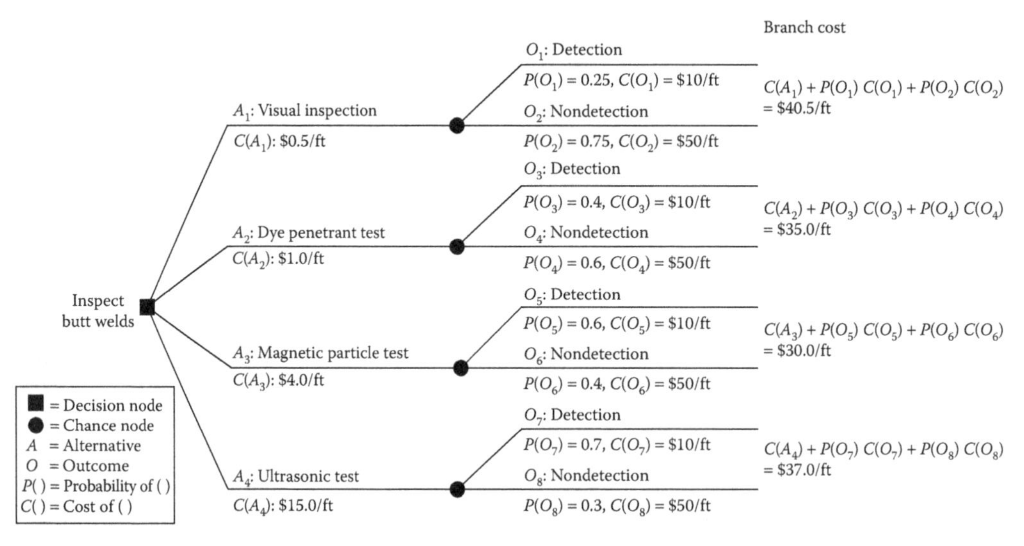

FIGURE 3.9
Decision tree for weld inspection methods.

expensive testing method, its total branch cost is the least. This analysis does not consider the standard deviation of the total cost when making the appropriate selection. Risk profiles of the candidate inspection methods can be constructed as the cumulative distribution functions of the total costs for these methods. Risk dominance can then be identified and an appropriate selection can be made as will be discussed in Chapter 7.

Example 3.6: Decision Analysis for Selection of a Personal Flotation Device Type

Decision analysis may also be applied to engineered consumer products such as personal flotation devices (PFDs). One application is the assessment of alternative PFD designs based on their performances. For this example, the objective of the decision analysis is to select the best PFD type based on a combination of the probability of PFD effectiveness and reliability. Probability values have not been included, as this example is intended only to demonstrate a decision tree as shown in Figure 3.10. The decision criteria could vary based on the performance considerations or concerns of the decision maker. For this example, the alternative with the largest value of combined effectiveness and reliability would be the best alternative.

3.3.5.6 Influence Diagrams

An influence diagram is a graphical method that shows the dependence relationships among the decision elements of a system. Influence diagrams have objectives similar to those for contributing factor diagrams, but they have more detail. Influence diagrams

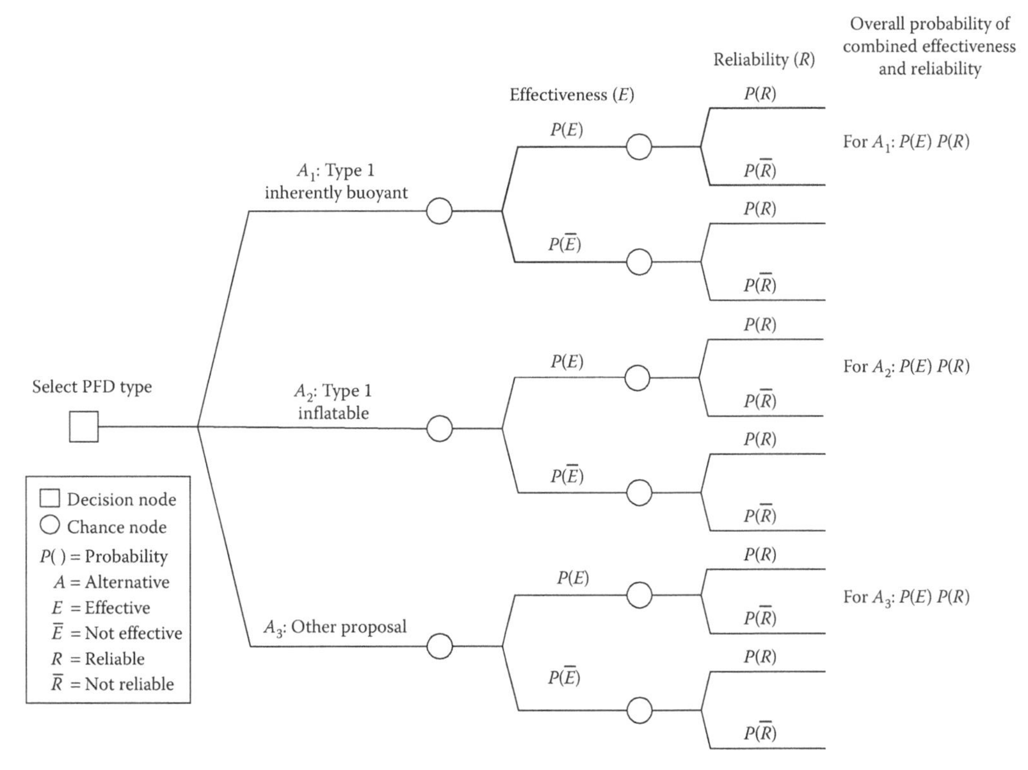

FIGURE 3.10
Selecting a personal flotation device (PFD) based on effectiveness and reliability.

provide compact representations of large decision problems by focusing on dependencies among various decision variables.

Influence diagrams consist of decision nodes, chance nodes, outcomes, and directed arrows indicating dependencies. These compact representations help facilitate the definition and scope of a decision prior to lengthy analysis. They are particularly useful for problems with a single decision variable and a significant number of uncertainties (ASME 1993). Symbols used for creating influence diagrams are shown in Figure 3.8. The first shape in the figure (rectangle) is used to identify a decision node that indicates where a decision must be made. A circular or elliptical shape is used to identify a chance node representing a probabilistic random variable with uncertain outcomes. Double oval shapes are used to identify a deterministic node with a quantity in it that is determined from the inputs from other nodes. The pentagon shape is a *value node*, which is used to define consequences over the attributes that measure performance. The next two symbols (arrows or arcs) are used to represent influence or dependency among nodes. The last shape is used to indicate time sequencing (i.e., information that must be known prior to a decision).

Generally, the process begins with identifying the decision criteria and then further defining what influences the criteria. An example of an influence diagram for selecting weld inspection method is shown in Figure 3.11. An influence diagram showing the relationship of the factors influencing the selection of a PFD type is shown in Figure 3.12. The

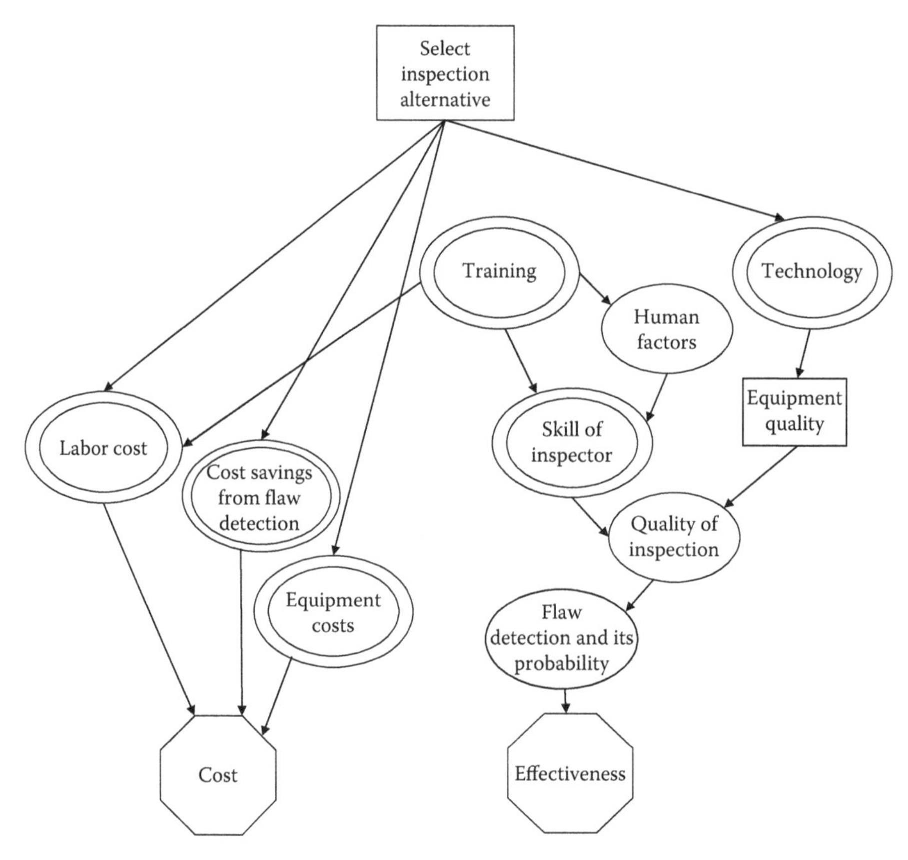

FIGURE 3.11
Influence diagram for selecting a weld inspection method.

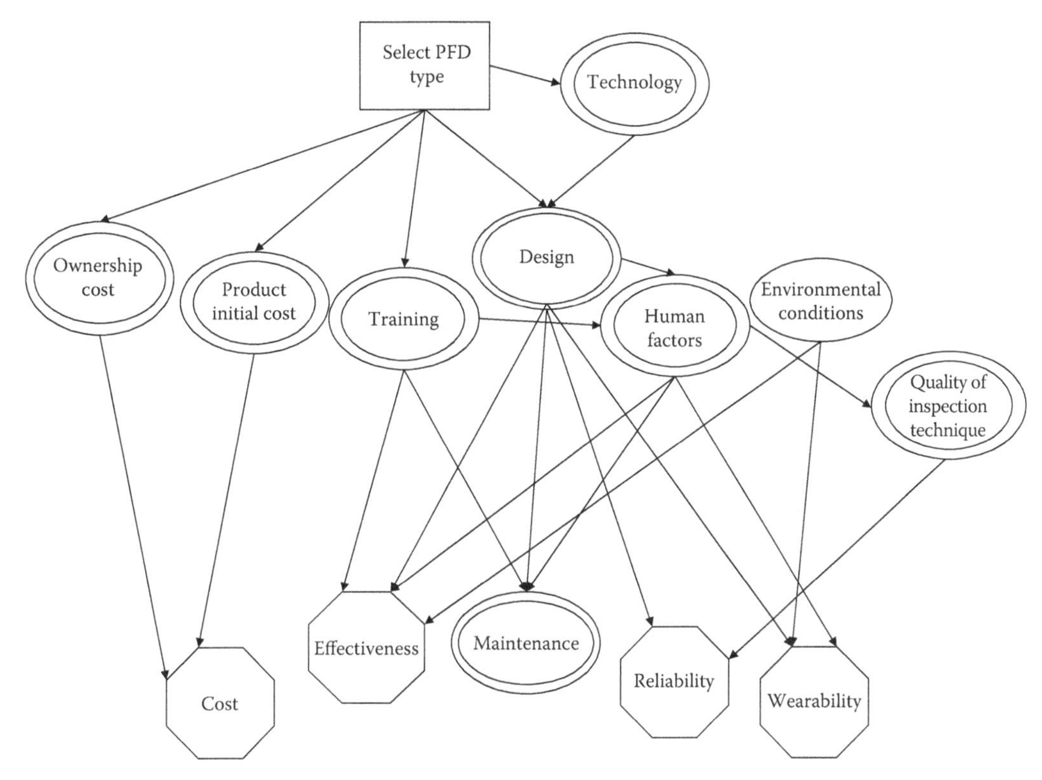

FIGURE 3.12
Influence diagram for selecting a personal flotation device (PFD) design.

example shown in Figure 3.13 is for a protection system of a hurricane-prone region as discussed by Ayyub et al. (2009a).

3.3.6 Bayesian Networks

Bayesian networks constitute a class of probabilistic models for modeling logic and dependency among variables representing a system. A Bayesian network consists of the following:

- Set of variables
- Graphical structure connecting the variables to represent their interdependencies
- Set of conditional distributions defining these interdependencies

A Bayesian network is commonly represented as a graph consisting of a set of nodes and arcs. The nodes represent the variables, and the arcs represent the conditional dependencies in the model. The absence of an arc between two variables indicates conditional independence, that is, the probability of one of the variables does not depend directly on the state of the other.

The construction of a Bayesian network should include all variables that are important in modeling the system. The causal relationships among the variables should be used to guide the connections (i.e., arcs) made in the graph. Prior knowledge should be used

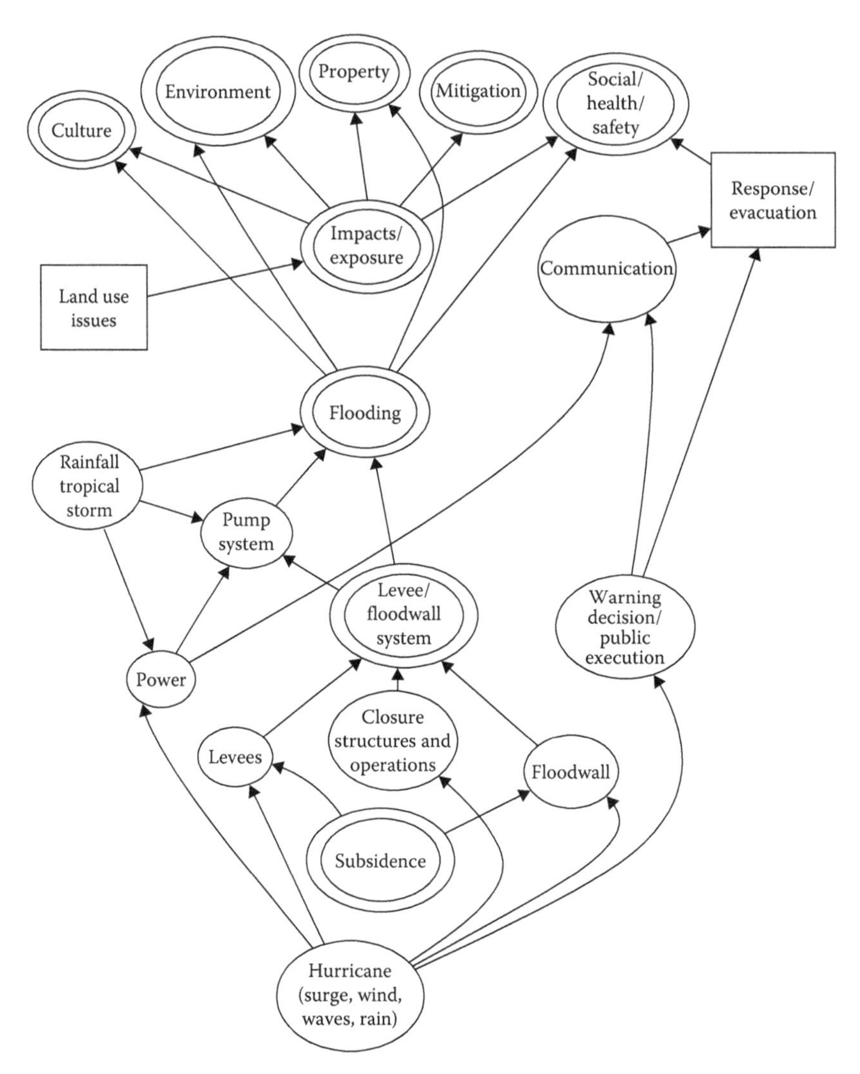

FIGURE 3.13
Influence diagram for a protection system of a hurricane-prone region.

to specify the conditional distributions. Such causal knowledge links variables in the model in such a way that arcs lead from causes to effects. The arcs are considered to be directed arcs (i.e., arcs with arrowheads showing causal directions).

3.3.6.1 Variables

A variable can be viewed as a mapping from the space of possible outcomes to discrete numerical values or continuous ranges of real values. Probability models can be used to assign likelihood values to these outcomes using probability mass functions or density functions, respectively. For example, in a medical experiment, men and women of different ages are studied and the relevant variables would be the sex of the participant, the age of the participant, and the experimental result. The variable of sex has only two possible values: male or female. But the variable of age can take on many values.

3.3.6.2 Relationships in a Bayesian Model

Bayesian models permit analysts to use commonsense and real-world knowledge to eliminate needless complexity in the model of a system. For example, a model builder would be likely to know that the time of day would not normally directly influence an oil leak in a car. Any influence on the leak would be based on other, more direct factors, such as temperature and driving conditions. Meaningless relationships are not explicitly declared in a Bayesian model and are excluded. After establishing all the variables in a model, variables that cause changes in the system should deliberately be associated with those variables that they influence. Only these specified influences are considered in the analysis and are represented by conditioning arcs between nodes. Each arc should represent a causal relationship between a temporal antecedent (known as the *parent*) and its later outcome (known as the *child*). By focusing on significant dependencies, system complexity is reduced in the model, and unnecessary joint probability distributions are not constructed because joint distributions for a real-world model are usually unknown and cannot be quantified.

3.3.6.3 Inference

Inference, also called *model evaluation*, is the process of updating probabilities of outcomes based on the relationships in the model and the evidence known about the situation at hand. Bayesian models apply evidence about recent events or observations to obtain outcomes. The model is exercised by clamping a variable to a state that is consistent with an observation, and the mathematical mechanics are performed to update the probabilities of all the other variables that are connected with the variable representing the new evidence. After an inference evaluation, the updated probabilities reflect the new levels of belief in (or probabilities of) all possible outcomes included in the model. These beliefs are mediated by the original assessment of belief performed by the analyst. The beliefs originally set in the model are known as *prior probabilities*, because they are entered before any evidence is known about the situation. The beliefs computed after evidence is entered are known as *posterior probabilities*, because they reflect the levels of belief computed in light of the new evidence. The computational algorithms follow Bayes' theorem and Bayesian techniques.

3.3.6.4 Network Creation

A Bayesian network can be created according to the following steps:

1. Create a set of variables representing the distinct key elements of the situation being modeled. Every variable in the real-world situation is represented by a Bayesian variable. Each such variable describes a set of states representing all possible distinct situations for the variable.
2. For each such variable, define the set of outcomes or states that each can have. This set is composed of mutually exclusive and collectively exhaustive outcomes and must cover all possibilities for the variable such that no important distinctions are shared between states. The causal relationships among the variables can be constructed by answering such questions as follows: (1) What other variables (if any) directly influence this variable? and (2) What other variables (if any) are directly influenced by this variable? In a standard Bayesian network, each variable

is represented by an ellipse or squares or any other shape, called a *node*. A node is, therefore, a Bayesian variable.

3. Establish the causal dependency relationships among the variables. This step involves creating arcs leading from the parent variable to the child variable. Each causal influence relationship is described by an arc connecting the influencing variable to the influenced variable. The influence arc has a terminating arrowhead pointing to the influenced variable. An arc connects a parent (influencing) node with a child (influenced) node. A directed acyclic graph (DAG) is desirable, in which only one semipath (i.e., sequence of connected nodes, ignoring direction of the arcs) exists between any two nodes.
4. Assess the prior probabilities by supplying the model with numeric probabilities for each variable in light of the number of parents the variable was given in step 3. Use conditional probabilities to represent dependencies, as shown in Figure 3.14. The figure also shows the effect of arc reversal on conditional probability representations. The first case shows that X_2 and X_3 depend on X_1. The joint probability of variables X_2, X_3, and X_1 can be computed using conditional probabilities based on these dependencies as follows (see Figure 3.14):

$$P(X_1, X_2, X_3) = P(X_3 \mid X_1)P(X_2 \mid X_1)P(X_1) \tag{3.1}$$

Case 2 displays different dependencies of X_3 on X_1 and X_2, leading to the following expression for the joint probabilities (see Figure 3.14):

$$P(X_1, X_2, X_3) = P(X_3 \mid X_1, X_2)P(X_2)P(X_1) \tag{3.2}$$

The models for cases 3 and 4 (Figure 3.14) were constructed using the same approach. The reversal of an arc changes the dependencies and conditional probability structure, as illustrated in Figure 3.15. Bayesian tables and probability trees can be used to represent the dependencies among the variables. A Bayesian table is a tabulated representation of the dependencies, whereas a probability tree is a graphical representation of multilevel dependencies using directed arrows similar to Figure 3.14. The examples at the end of this section illustrate the use of Bayesian tables and probability trees for this purpose.

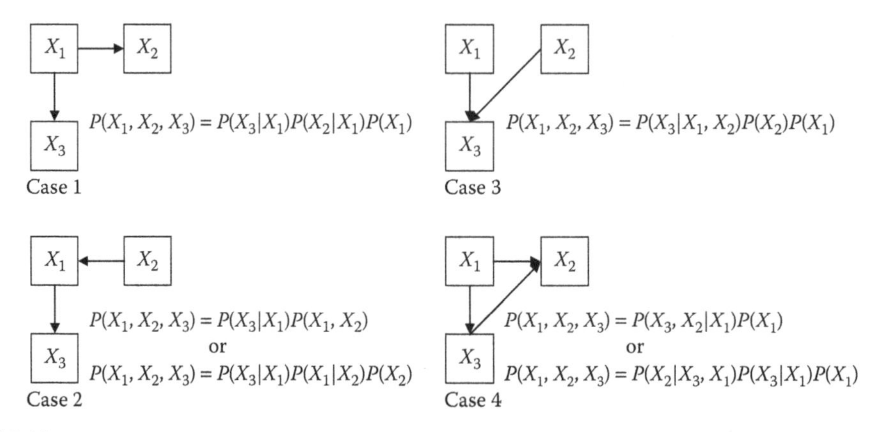

FIGURE 3.14
Conditional probabilities for representing directed arcs.

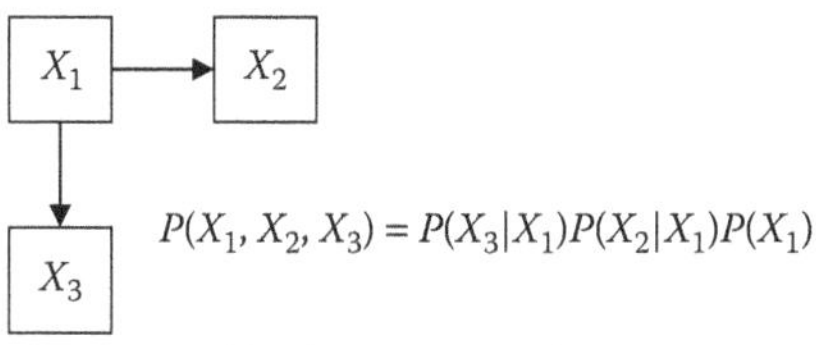

Arc reversal leads to an equivalent representation as follows:

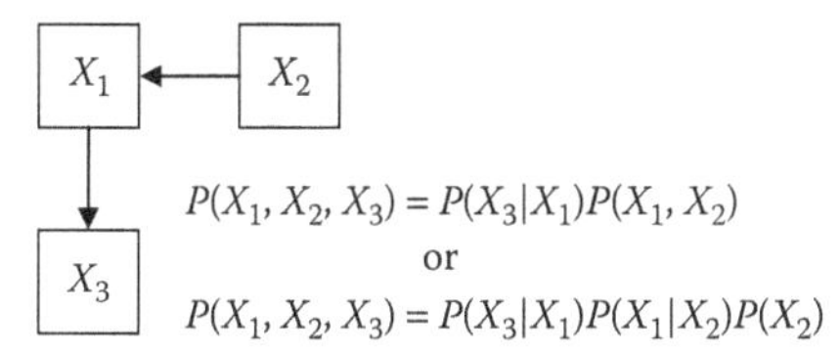

FIGURE 3.15
Arc reversal and effects on conditional probabilities.

5. Bayesian methods, as described in Appendix A, can be used to update the probabilities based on gaining new information, as demonstrated in subsequent examples. By fusing and propagating values of new evidence and beliefs through Bayesian networks, each proposition is eventually assigned a probability value in keep with the axioms of probability theory. The impact of each new piece of evidence is viewed as a perturbation that propagates through the network via the arcs that are message passing among connected variables.

Example 3.7: Bayesian Tables for Two Dependent Variables *A* and *B*

A simple computational example is used here to illustrate the use of Bayesian methods to update the probabilities for the case of two variables *A* and *B* for which a directed arrow runs from *B* to *A*, indicating that *B* affects *A*. The *a priori* probability of *B* is 0.0001. The conditional probability of *A* given *B*, denoted as $P(A|B)$, is given by the following table based on previous experiences:

	Conditional Probability of Events Related to Variable *A* Given the Following	
	Variable *B*	**Variable $\bar{B}$**
Variable *A*	$P(A \mid B) = 0.95$	$P(A \mid \bar{B}) = 0.01$
Variable $\bar{A}$	$P(\bar{A} \mid B) = 0.05$	$P(\bar{A} \mid \bar{B}) = 0.99$

The $P(B|A)$ is of interest and can be computed as follows:

$$P(B \mid A) = \frac{P(A \mid B)P(B)}{P(A)} \tag{3.3}$$

The term $P(A)$ in Equation 3.3 can be computed using the theorem of total probability based on *B* and the complement of *B* as follows:

$$P(B \mid A) = \frac{P(A \mid B)P(B)}{P(A \mid B)P(B) + P(A \mid \bar{B})P(\bar{B})} \tag{3.4}$$

Substituting the probabilities from the table above, the following conditional probability can be computed:

$$P(B \mid A) = \frac{(0.95)(0.0001)}{(0.95)(0.0001) + (0.01)(1 - 0.0001)} = 0.009411 \tag{3.5}$$

The computations of the probability of B for two cases, given an A occurrence and given an $\overline{A}$ occurrence, can be represented using Bayesian tables as follows based on the known prior probability of B:

Prior Probability of Variable B	Conditional Probabilities of Variables A and B	Joint Probabilities of Variables A and B	Posterior Probability of Variable B after Variable A Has Occurred
Case 1: Occurrence of A			
$P(B) = 0.0001$	$P(A \mid B) = 0.95$	$P(B)\,P(A \mid B) = 0.000095$	$P(B \mid A) = P(B)P(A \mid B)/P(A) = 0.009412$
$P(\overline{B}) = 0.9999$	$P(A \mid \overline{B}) = 0.01$	$P(\overline{B})\,P(A \mid \overline{B}) = 0.009999$	$P(\overline{B} \mid A) = P(\overline{B})\,P(A \mid \overline{B})/P(A) = 0.990588$
Total = 1.0000		$P(A) = 0.010094$	Total = $P(B \mid A) + P(\overline{B} \mid A) = 1.000000$
Case 2: Occurrence of $\overline{A}$			
$P(B) = 0.0001$	$P(\overline{A} \mid B) = 0.05$	$P(B)\,P(\overline{A} \mid B) = 0.000005$	$P(B \mid \overline{A}) = P(B)\,P(\overline{A} \mid B)/P(\overline{A}) = 0.000005$
$P(\overline{B}) = 0.9999$	$P(\overline{A} \mid \overline{B}) = 0.99$	$P(\overline{B})\,P(\overline{A} \mid \overline{B}) = 0.989901$	$P(\overline{B} \mid \overline{A}) = P(\overline{B})\,P(\overline{A} \mid \overline{B})/P(\overline{A}) = 0.999995$
Total = 1.0000		$P(\overline{A}) = 0.989906$	Total = $P(B \mid \overline{A}) + P(\overline{B} \mid \overline{A}) = 1.000000$

The tables are structured in a manner to facilitate the evaluation of Bayes' theorem starting with the prior probabilities and the dependence relationship defined. The joint probability can then be computed as the product of the prior and the respective conditional probabilities, and summed up to obtain the probability of the event defining the respective case. The last two columns in each table show the computations of the posterior probabilities based on Bayes' theorem. It can be noted that the total $P(A) + P(\overline{A})$ in the two tables is 1.

Example 3.8: Probability Trees for Two Dependent Variables A and B

Probability trees can be used to express the relationships of dependency among random variables. The Bayesian problem of Example 3.7 can be used to illustrate the use of probability trees; the probability tree for the two cases of Example 3.7 is shown in Figure 3.16.

Example 3.9: Bayesian Tables for Identifying Defective Electric Components

A batch of 1000 electric components was produced in a week at a factory; after exhaustive, time-consuming tests, it was found that 30% of them are defective and 70% are nondefective. Unfortunately, all components are mixed together in a large container. Selecting at random a component from the container has a nondefective prior probability of 0.7. The objective of the company is to screen all the components to identify the defective components. A quick test on each component can be used for this screening. This test has 0.8 detection probability for a nondefective component and a 0.9 detection probability for a defective component. The prior probabilities must be updated using the probabilities associated with this quick test.

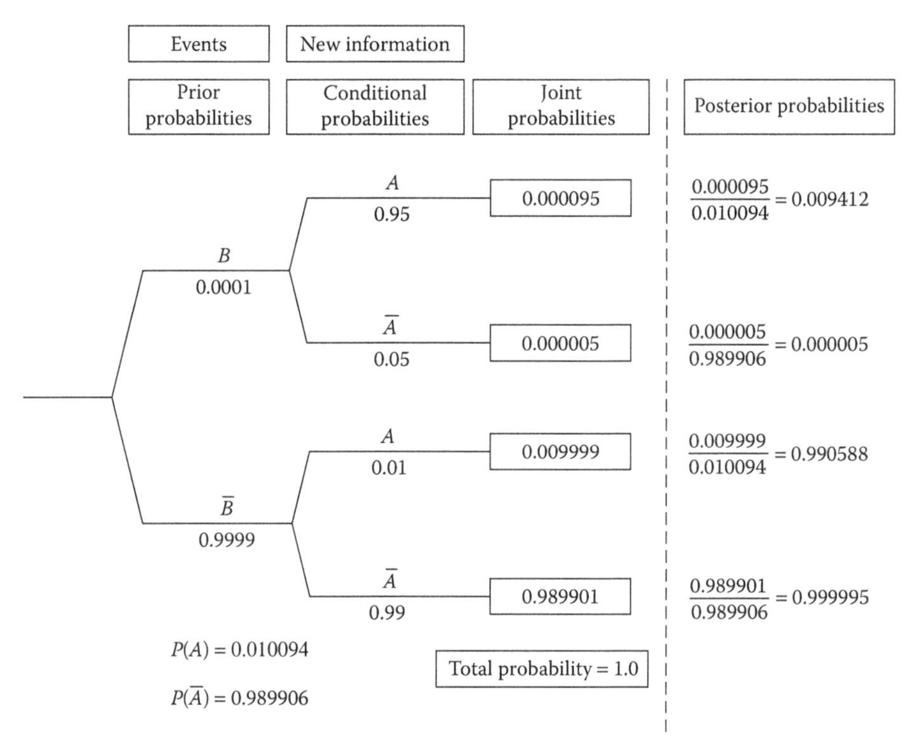

FIGURE 3.16
Probability tree representation of a Bayesian model.

The Bayesian tables can be constructed based on the following definitions of variables:

Nondefective component = B.
Defective component = $\overline{B}$.
Component passing the quick test = A.
Component not passing the quick test = $\overline{A}$.

The Bayesian tables can then be constructed for two cases as follows:

Prior Probability of Variable B	Conditional Probabilities of Variables A and B	Joint Probabilities of Variables A and B	Posterior Probability of Variable B after Variable A Has Occurred
For the case of given the occurrence of A			
$P(B) = 0.7000$	$P(A\mid B) = 0.80$	$P(B)P(A\mid B) = 0.560000$	$P(B\mid A) = P(B)P(A\mid B)/P(A) = 0.949153$
$P(\overline{B}) = 0.3000$	$P(A\mid \overline{B}) = 0.10$	$P(\overline{B})P(A\mid \overline{B}) = 0.030000$	$P(\overline{B}\mid A) = P(\overline{B})P(A\mid \overline{B})/P(A) = 0.050847$
Total = 1.0000		$P(A) = 0.590000$	Total = $P(B\mid A) + P(\overline{B}\mid A) = 1.000000$
For the case of given the occurrence of $\overline{A}$			
$P(B) = 0.7000$	$P(\overline{A}\mid B) = 0.200$	$P(B)\,P(\overline{A}\mid B) = 0.140000$	$P(B\mid \overline{A}) = P(B)P(\overline{A}\mid B)/P(\overline{A}) = 0.341463$
$P(\overline{B}) = 0.3000$	$P(\overline{A}\mid \overline{B}) = 0.900$	$P(\overline{B})P(\overline{A}\mid \overline{B}) = 0.270000$	$P(\overline{B}\mid \overline{A}) = P(\overline{B})P(\overline{A}\mid \overline{B})/P(\overline{A}) = 0.658537$
Total = 1.0000		$P(\overline{A}) = 0.410000$	Total = $P(B\mid \overline{A}) + P(\overline{B}\mid \overline{A}) = 1.000000$

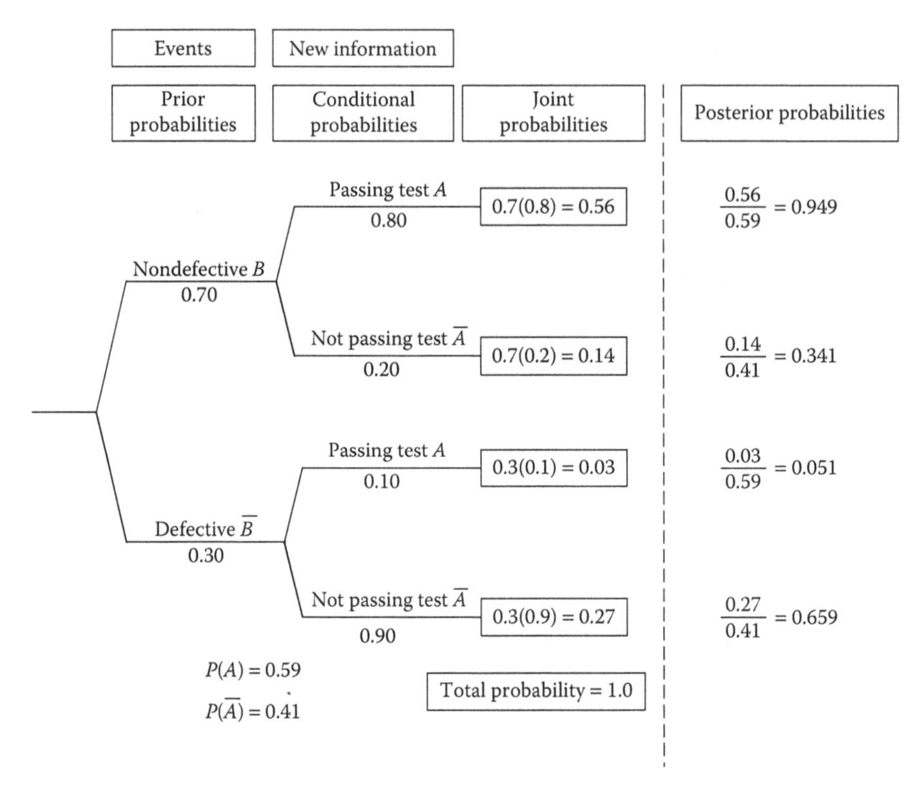

FIGURE 3.17
Probability tree representation of a defective electric component problem.

It can be noted that total $P(A) + P(\overline{A}) = 1$.

The decision situation of this example can be used to illustrate the use of probability trees, as shown in Figure 3.17, which also shows the conditional probabilities obtained from the information of the test. The posterior probabilities calculated using the Bayesian approach are shown at the right side of the tree. From the tree, the probability of a component failing the test can be computed. For example, the probability that a component is nondefective and fails the test can be computed as the joint probability by applying the multiplication rule as follows:

- P(nondefective and failing the test) = 0.7(0.2) = 0.14

The probability that a component is defective and fails the test is as follows:

- P(defective and failing the test) = 0.3(0.9) = 0.27

A component can, therefore, fail the test in two cases of being nondefective and being defective. The probability of failing the test can then be computed by adding the two joint probabilities as follows:

- P(failing the test) = 0.14 + 0.27 = 0.41

Hence, the probability of the component passing the test can be computed as the probability of the complementary event as follows:

- P(passing the test) = 0.56 + 0.03 = 0.59

The posterior probability can be determined by dividing the appropriate joint probability by the respective probability values. For example, to determine the posterior probability that the component is nondefective, the joint probability that comes from the tree branch of a nondefective component of 0.14 can be used as follows:

- Posterior P(component nondefective) = 0.14/0.41 = 0.341

All other posterior probabilities on the tree are calculated similarly. The posterior probabilities of nondefective and defective components must add up to 1: 0.341 + 0.659 = 1.

Example 3.10: Bayesian Network for Diagnostic Analysis

A Bayesian network can be used to represent a knowledge structure that models the relationships among possible medical difficulties, their causes and effects, patient information, and diagnostic tests results. Figure 3.18a provides schematics of key dependencies. The figure shows three diagnostic tests of X-ray, tuberculosis skin test, and dyspnea, that is, shortness of breath. Tuberculosis results in positive X-ray and tuberculosis skin tests, whereas lung cancer results in positive X-ray and dyspnea. Bronchitis results in dyspnea. Tuberculosis vaccination results in positive tuberculosis skin test. Also, the figure shows the dependencies among the patient information and medical difficulties. The problem was simplified by eliminating the tuberculosis vaccination and exposure boxes and the tuberculosis skin test box as provided in Figure 3.18b. The simplified case of Figure 3.18b is used in the rest of the example.

The following table provides conditional probabilities that define arcs in Figure 3.18b:

Event Affected	Causal Event(s) or Condition(s)	Conditional Probability
Tuberculosis (T)	With a visit to Asia (V)	$P(T \mid V) = 0.05$
Tuberculosis (T)	Without a visit to Asia ($\overline{V}$)	$P(T \mid \overline{V}) = 0.01$
Lung cancer (C)	Smoker (S)	$P(C \mid S) = 0.10$
Lung cancer (C)	Nonsmoker ($\overline{S}$)	$P(C \mid \overline{S}) = 0.01$
Bronchitis (B)	Smoker (S)	$P(B \mid S) = 0.60$
Bronchitis (B)	Nonsmoker ($\overline{S}$)	$P(B \mid \overline{S}) = 0.30$
Positive X-ray (X)	Tuberculosis or cancer (TC)	$P(X \mid TC) = 0.9891$
Positive X-ray (X)	No tuberculosis or cancer ($\overline{TC}$)	$P(X \mid \overline{TC}) = 0.04906$
Dyspnea (D)	B and TC	$P(D \mid B \text{ and } TC) = 0.90$
Dyspnea (D)	$\overline{B}$ and TC	$P(D \mid \overline{B} \text{ and } TC) = 0.70$
Dyspnea (D)	B and $\overline{TC}$	$P(D \mid B \text{ and } \overline{TC}) = 0.80$
Dyspnea (D)	$\overline{B}$ and $\overline{TC}$	$P(D \mid \overline{B} \text{ and } \overline{TC}) = 0.10$

The probabilities of having dyspnea (D) are given by the following values:

Combination Cases of Tuberculosis or Cancer (TC) and Bronchitis (B)	Probabilities of Dyspnea (D) Cases	
	$P(D)$	$P(\overline{D})$
(TC, B)	0.9	0.1
(TC, $\overline{B}$)	0.7	0.3
($\overline{TC}$, B)	0.8	0.2
($\overline{TC}$, $\overline{B}$)	0.1	0.9

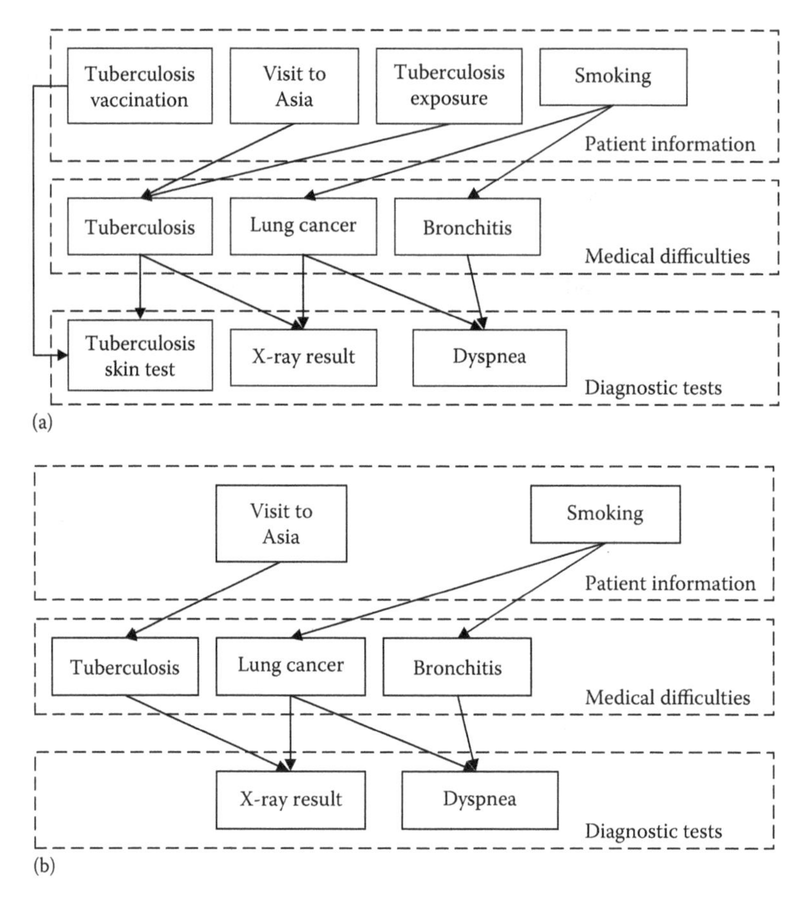

FIGURE 3.18
Bayesian networks for diagnostic analysis of medical tests. (a) Extended model; (b) simplified model.

The true and false states in the first column of the above table are constructed from the following logic table:

Tuberculosis	Lung Cancer	Tuberculosis or Cancer Cases
Present	Present	*TC*
Present	Absent	*TC*
Absent	Present	*TC*
Absent	Absent	$\overline{TC}$

We can now assign unconditional probabilities to the antecedent variables, that is, the variables that do not have any arcs coming to them. In this example, they are the visit to Asia (*V*) and smoking (*S*) variables as shown in Figure 3.19. Figure 3.19 is based on Figure 3.18b with an added intermediate event defining the union of tuberculosis (*T*) and cancer (*C*) to match the conditional probabilities available. The unconditional probabilities assigned herein have no impact on the utility of the resulting model and can be based on statistical information or relative frequency of visiting Asia and smoking in the general population, respectively. Using the unconditional and conditional probabilities provided so far, Figure 3.19 was populated with probabilities using computations

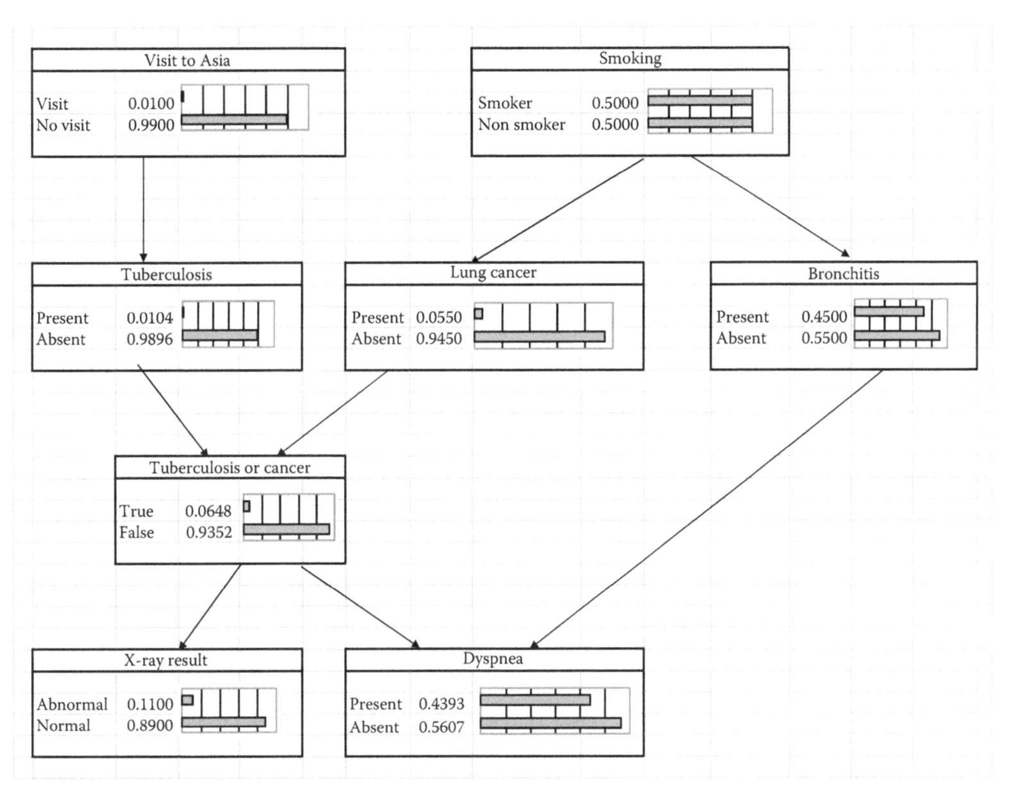

FIGURE 3.19
Propagation of probabilities in percentages in a Bayesian network.

based on Bayesian methods and probability mathematics, as discussed in Appendix A. For example, $P(T)$ can be computed as follows:

$$\begin{aligned} P(T) &= P(T \mid V)P(V) + P(T \mid \overline{V})P(\overline{V}) \\ &= (0.05)(0.01) + (0.01)(0.99) = 0.0104 \end{aligned} \tag{3.6}$$

In Figure 3.19, the P(Absent) is computed as 1 − P(Present). All other values can be computed similarly, except for the case of the union of $T \cup C = TC$ that is computed as follows based on the assumption of independence for the variables T and C:

$$\begin{aligned} P(T \cup C) &= P(T) + P(C) - P(T)P(C) \\ &= 0.0104 + 0.055 - 0.0104(0.055) = 0.064828 \end{aligned} \tag{3.7}$$

This probability can alternately be computed as follows to obtain the same result:

$$\begin{aligned} P(T \cup C) &= 1 - P(\overline{T})P(\overline{C}) \\ &= 1 - 0.9896(0.9450) = 0.064828 \end{aligned} \tag{3.8}$$

For the case of $P(D)$, the value is computed as follows based on the two arcs feeding into the dyspnea variable (D):

$$P(D) = P(D \mid B \cap TC)P(B \cap TC) + P(D \mid \bar{B} \cap TC)P(\bar{B} \cap TC) + P(D \mid B \cap \overline{TC})P(B \cap \overline{TC}) + P(D \mid \bar{B} \cap \overline{TC})P(\bar{B} \cap \overline{TC}) \tag{3.9}$$

Substituting the corresponding probabilities produces,

$$\begin{aligned} P(D) &= 0.9(0.45)(0.064828) + 0.7(0.55)(0.064828) \\ &\quad + 0.8(0.45)(0.935172) + 0.10(0.55)(0.935172) \\ &= 0.0262553 + 0.0249587 + 0.3366619 + 0.0514344 = 0.4393103 \end{aligned} \tag{3.10}$$

A Bayesian table can be used instead to compute such probabilities, for example, the first directed arrow of Figure 3.19 from V to T can be evaluated as follows:

For the case of given the occurrence of V (occurrence of a visit)			
Prior Probability of Variable V	**Conditional Probabilities of Variables T and V**	**Joint Probabilities of Variables T and V**	**Posterior Probability of Variable V after Variable T Has Occurred**
$P(V) = 0.0100$	$P(T \mid V) = 0.05$	$P(V)P(T \mid V) = 0.0005$	$P(V \mid T) = P(V)P(T \mid V)/P(T) = 0.04808$
$P(\bar{V}) = 0.9900$	$P(T \mid \bar{V}) = 0.01$	$P(\bar{V})P(T \mid \bar{V}) = 0.0099$	$P(\bar{V} \mid T) = P(\bar{V})P(T \mid \bar{V})/P(T) = 0.95192$
Total = 1.0000		$P(T) = 0.0104$	Total = $P(V \mid T) + P(\bar{V} \mid T) = 1.00000$

Probability trees can be used to express the relationships of dependency among random variables in this case and are shown in Figure 3.20. Similar treatments can be developed for all the relationships (i.e., directed arrows) of Figure 3.19.

An algorithm for propagating probabilities in the Bayesian network can be developed based on the above computational procedure and used to update the beliefs attached to each relevant node in the network. Such an implementation can be achieved in spreadsheets. The computations of the probabilities presented in Figure 3.19 are summarized in Table 3.1.

The next step is to lock some of the variables. For example, interviewing a patient can produce the information for the box of visiting Asia to lock it with certainty as the patient has visited Asia [i.e., $P(A) = 1$], as shown in Table 3.2. Such a finding propagates through the network, and the probability functions of several nodes are updated. Further updates can be made based on knowing whether a patient is a smoker and the results are summarized in Table 3.3.

Table 3.3 shows that the patient has a probability of 0.185367 for a positive X-ray test and a probability of 0.563500 for a positive dyspnea test. Let us carry this hypothetical example further and assume that an X-ray test was performed and was found negative, that is, normal. In this case, the variable (X) should be locked, and the probabilities require updating. The results for this case are shown in Table 3.4. The computations for this case are significantly different from the previous cases. We could assume that the case presented in Table 3.3 offers the prior probabilities as shown in the Bayesian Table 3.5a–c. These tables also show the computations for the following cases:

- Posterior probabilities of TC with $\bar{X}$ locked (Table 3.5a)
- Posterior probabilities of T with TC updated (Table 3.5b)
- Posterior probabilities of C with TC updated (Table 3.5c)

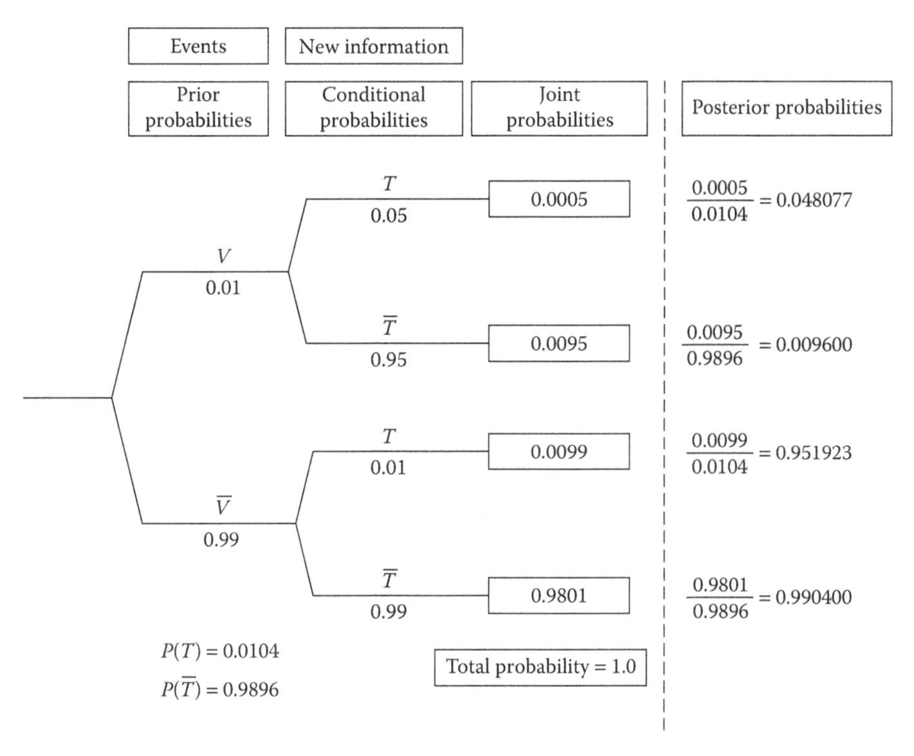

FIGURE 3.20
Probability tree representation of a diagnostic analysis problem.

TABLE 3.1
Propagation of Probabilities for the Bayesian Network of Figure 3.19

Variable	Source	Probability Expression	Probability Value	Probability of the Complement
Asia visit (V)	Input	Not applicable	0.010000	0.990000
Smoking (S)	Input	Not applicable	0.500000	0.500000
Tuberculosis (T)	Computed	$P(T \mid V)P(T) + P(T \mid \overline{V})P(\overline{V})$	0.010400	0.989600
Cancer (C)	Computed	$P(C \mid S)P(S) + P(C \mid \overline{S})P(\overline{S})$	0.055000	0.945000
T or C (TC)	Computed	$1 - P(\overline{T})P(\overline{C})$	0.064828	0.935172
Bronchitis (B)	Computed	$P(B \mid S)P(S) + P(B \mid \overline{S})P(\overline{S})$	0.450000	0.550000
X-ray (X)	Computed	$P(X \mid TC)P(TC) + P(X \mid \overline{TC})P(\overline{TC})$	0.110002	0.889998
Dyspnea (D)	Computed	$P(D \mid B \cap TC)P(B \cap TC) + P(D \mid \overline{B} \cap TC)P(\overline{B} \cap TC) +$ $P(D \mid B \cap \overline{TC})P(B \cap \overline{TC}) + P(D \mid \overline{B} \cap \overline{TC})P(\overline{B} \cap \overline{TC})$	0.439311	0.560690

Propagating further up the network, the probability of T based on TC can be computed as follows using results from Table 3.5a and 3.5b:

$$P(T) = P(T \mid TC)P(TC) = 0.3448(0.001938) = 0.0006682 \tag{3.11}$$

Also, the probability of C based on TC can be computed as follows using the results from Table 3.5a–c:

$$P(C) = P(C \mid TC)P(TC) = 0.6897(0.001938) = 0.0013366 \tag{3.12}$$

TABLE 3.2

Propagation of Probabilities for the Bayesian Network of Figure 3.19 with V Locked

Variable	Source	Probability Expression	Probability Value	Probability of the Complement
Asia visit (V)	Locked	Not applicable	1.000000	0.000000
Smoking (S)	Input	Not applicable	0.500000	0.500000
Tuberculosis (T)	Computed	$P(T\mid V)P(T)+P(T\mid \overline{V})P(\overline{V})$	0.050000	0.950000
Cancer (C)	Computed	$P(C\mid S)P(S)+P(C\mid \overline{S})P(\overline{S})$	0.055000	0.945000
T or C (TC)	Computed	$1-P(\overline{T})P(\overline{C})$	0.102250	0.897750
Bronchitis (B)	Computed	$P(B\mid S)P(S)+P(B\mid \overline{S})P(\overline{S})$	0.450000	0.550000
X-ray (X)	Computed	$P(X\mid TC)P(TC)+P(X\mid \overline{TC})P(\overline{TC})$	0.145180	0.854820
Dyspnea (D)	Computed	$P(D\mid B\cap TC)P(B\cap TC)+P(D\mid \overline{B}\cap TC)P(\overline{B}\cap TC)+$ $P(D\mid B\cap \overline{TC})P(B\cap \overline{TC})+P(D\mid \overline{B}\cap \overline{TC})P(\overline{B}\cap \overline{TC})$	0.453344	0.546656

TABLE 3.3

Propagation of Probabilities for the Bayesian Network of Figure 3.19 with V and S Locked

Variable	Source	Probability Expression	Probability Value	Probability of the Complement
Asia visit (V)	Locked	Not applicable	1.000000	0.000000
Smoking (S)	Locked	Not applicable	1.000000	0.000000
Tuberculosis (T)	Computed	$P(T\mid V)P(T)+P(T\mid \overline{V})P(\overline{V})$	0.050000	0.950000
Cancer (C)	Computed	$P(C\mid S)P(S)+P(C\mid \overline{S})P(\overline{S})$	0.100000	0.900000
T or C (TC)	Computed	$1-P(\overline{T})P(\overline{C})$	0.145000	0.855000
Bronchitis (B)	Computed	$P(B\mid S)P(S)+P(B\mid \overline{S})P(\overline{S})$	0.600000	0.400000
X-ray (X)	Computed	$P(X\mid TC)P(TC)+P(X\mid \overline{TC})P(\overline{TC})$	0.185367	0.814633
Dyspnea (D)	Computed	$P(D\mid B\cap TC)P(B\cap TC)+P(D\mid \overline{B}\cap TC)P(\overline{B}\cap TC)+$ $P(D\mid B\cap \overline{TC})P(B\cap \overline{TC})+P(D\mid \overline{B}\cap \overline{TC})P(\overline{B}\cap \overline{TC})$	0.563500	0.436500

TABLE 3.4

Propagation of Probabilities for the Bayesian Network of Figure 3.19 with V, S, and X Locked

Variable	Source	Probability Expression	Probability Value	Probability of the Complement
Asia visit (V)	Locked	Not applicable	1.000000	0.000000
Smoking (S)	Locked	Not applicable	1.000000	0.000000
Tuberculosis (T)	Computed	See Table 3.5b	0.000668	0.999332
Cancer (C)	Computed	See Table 3.5c	0.001337	0.998663
T or C (TC)	Computed	See Table 3.5a	0.001938	0.998062
Bronchitis (B)	Computed	$P(B\mid S)P(S)+P(B\mid \overline{S})P(\overline{S})$	0.600000	0.400000
X-ray (X)	Locked	Not applicable	0.000000	1.000000
Dyspnea (D)	Computed	$P(D\mid B\cap TC)P(B\cap TC)+P(D\mid \overline{B}\cap TC)P(\overline{B}\cap TC)+$ $P(D\mid B\cap \overline{TC})P(B\cap \overline{TC})+P(D\mid \overline{B}\cap \overline{TC})P(\overline{B}\cap \overline{TC})$	0.520582	0.479418

TABLE 3.5

Propagation of Probabilities for the Bayesian Network of Figure 3.19 with *V*, *S*, and $\overline{X}$ Locked

(a) Posterior Probabilities of *TC* with $\overline{X}$ Locked

Prior Probability for Variable (*TC*) Based on Table 3.3	Data on Conditional Probabilities of $\overline{X}$	Joint Probabilities Relating to *TC* and $\overline{X}$	Posterior Probabilities of *TC* with $\overline{X}$ Locked
$P(TC) = 0.1450$	$P(\overline{X} \mid TC) = 0.01089$	$P(TC)P(\overline{X} \mid TC) = 0.0016$	$P(TC \mid \overline{X}) = 0.0019380$
$P(\overline{TC}) = 0.8550$	$P(\overline{X} \mid \overline{TC}) = 0.95094$	$P(\overline{TC})P(\overline{X} \mid \overline{TC}) = 0.8131$	$P(\overline{TC} \mid \overline{X}) = 0.9988062$
Totals = 1.0000		$P(\overline{X}) = 0.8146$	$P(TC \mid \overline{X}) + P(\overline{TC} \mid \overline{X}) = 1.0000000$

(b) Posterior Probabilities of *T* with *TC* Updated

Prior Probability for Variable (*T*) Based on Table 3.3	Conditional Probabilities of *TC* Based on Table 3.3	Joint Probabilities Relating to *T* and *TC*	Posterior Probabilities of *T* with *TC* Locked
$P(T) = 0.0500$	$P(TC \mid T) = 1.0000$	$P(T)\,P(TC \mid T) = 0.0500$	$P(T)\,P(TC \mid T)/P(TC) = 0.3448$
$P(\overline{T}) = 0.9500$	$P(TC \mid \overline{T}) = P(C) = 0.1000$	$P(\overline{T})P(TC \mid \overline{T}) = 0.0950$	$P(\overline{T})P(TC \mid \overline{T})/P(TC) = 0.6552$
Totals = 1.0000		$P(TC) = 0.1450$	$P(T \mid TC) + P(\overline{T} \mid TC) = 1.0000$

(c) Posterior Probabilities of *C* with *TC* Updated

Prior Probability for Variable (*C*) Based on Table 3.3	Conditional Probabilities of *TC* Based on Table 3.3	Joint Probabilities Relating to *C* and *TC*	Posterior Probabilities of *C* with *TC* Locked
$P(C) = 0.1000$	$P(TC \mid C) = 1.0000$	$P(C)P(TC \mid C) = 0.1000$	$P(C)P(TC \mid C)/P(TC) = 0.6897$
$P(\overline{C}) = 0.9000$	$P(TC \mid \overline{C}) = P(T) = 0.0500$	$P(\overline{C})P(TC \mid \overline{C}) = 0.0450$	$P(\overline{C})P(TC \mid \overline{C})/P(TC) = 0.3103$
Totals = 1.0000		$P(TC) = 0.1450$	$P(C \mid TC) + P(\overline{C} \mid TC) = 1.0000$

TABLE 3.6

Propagation of Probabilities for the Bayesian Network of Figure 3.19 with *V*, *S*, $\overline{X}$, and *D* Locked

Variable	Source	Probability Expression	Probability Value	Probability of the Complement
Asia visit (*V*)	Locked	Not applicable	1.000000	0.000000
Smoking (*S*)	Locked	Not applicable	1.000000	0.000000
Tuberculosis (*T*)	Computed	See Table 3.7b	0.001053	0.998947
Cancer (*C*)	Computed	See Table 3.7c	0.002106	0.997894
T or *C* (*TC*)	Computed	See Table 3.7a	0.003053	0.996947
Bronchitis (*B*)	Computed	See Table 3.7d	0.922269	0.077731
X-ray (*X*)	Locked	Not applicable	0.000000	1.000000
Dyspnea (*D*)	Locked	Not applicable	1.000000	0.000000

Similarly, the probabilities are propagated after locking *D* as shown in Table 3.6 with the Bayesian table (Table 3.7). Propagating further up the network, the probability of *T* based on *TC* can be computed as follows using the results from Table 3.7a and 3.7b:

$$P(T) = P(T \mid TC)P(TC) = 0.3448(0.003053) = 0.0010526 \quad (3.13a)$$

TABLE 3.7

Bayesian Tables for Probability Propagation for the Bayesian Network of Figure 3.19 with V, S, $\overline{X}$, and D Locked

(a) Posterior Probabilities of TC with D Locked

Prior Probability for Variable (TC) Based on Table 3.5	Conditional Probabilities of D	Joint Probabilities Relating to TC and D	Posterior Probabilities of TC with D Locked
$P(TC) = 0.0019$	$P(D \mid TC) = P(DB \bot TC)P(DB) + P(D\overline{B} \mid TC)\, P(D\overline{B}) = 0.8200$	$P(TC)P(D \mid TC) = 0.0016$	$P(TC \mid D) = 0.003053$
$P(\overline{TC}) = 0.9981$	$P(D \mid \overline{TC}) = 0.6(0.8) + 0.4(0.1) = 0.5200$	$P(\overline{TC})P(D \mid \overline{TC}) = 0.5190$	$P(\overline{TC} \mid D) = 0.996974$
Totals = 1.0000		$P(D) = 0.5206$	$P(TC \mid D) + P(\overline{TC} \mid D) = 1.000000$

(b) Posterior Probabilities of T with TC Updated

Prior Probability for Variable (T) Based on Table 3.3	Conditional Probabilities of TC Based on Table 3.3	Joint Probabilities Relating to T and TC	Posterior Probabilities of T with TC Updated
$P(T) = 0.0500$	$P(TC \mid T) = 1.0000$	$P(T)P(TC \mid T) = 0.0500$	$P(T)P(TC \mid T)/P(TC) = 0.3448$
$P(\overline{T}) = 0.9500$	$P(TC \mid \overline{T}) = P(C) = 0.1000$	$P(\overline{T})P(TC \mid \overline{T}) = 0.0950$	$P(\overline{T})P(TC \mid \overline{T})/P(TC) = 0.6552$
Totals = 1.0000		$P(TC) = 0.1450$	$P(T \mid TC) + P(\overline{T} \mid TC) = 1.0000$

(c) Posterior Probabilities of C with TC Updated

Prior Probability for Variable (C) Based on Table 3.3	Conditional Probabilities of TC Based on Table 3.3	Joint Probabilities Relating to C and TC	Posterior Probabilities of C with TC Updated
$P(C) = 0.1000$	$P(TC \mid C) = 1.0000$	$P(C)P(TC \mid C) = 0.1000$	$P(C)P(TC \mid C)/P(TC) = 0.6897$
$P(\overline{C}) = 0.9000$	$P(TC \mid \overline{C}) = P(T) = 0.0500$	$P(\overline{C})P(TC \mid \overline{C}) = 0.0450$	$P(\overline{C})P(TC \mid \overline{C})/P(TC) = 0.3103$
Totals = 1.0000		$P(TC) = 0.1450$	$P(C \mid TC) + P(\overline{C} \mid TC) = 1.0000$

(d) Posterior Probabilities of B with TC Updated

Prior Probability for Variable (B) Based on Table 3.3	Conditional Probabilities of D Based on Table 3.3	Joint Probabilities Relating to B and D	Posterior Probabilities of B with TC Updated
$P(B) = 0.6000$	$P(D \mid B) = 0.80019$	$P(B)P(D \mid B) = 0.48012$	$P(D)P(D \mid B)/P(D) = 0.9223$
$P(\overline{B}) = 0.4000$	$P(D \mid \overline{B}) = 0.10116$	$P(\overline{B})P(D \mid \overline{B}) = 0.04047$	$P(\overline{B})P(D \mid \overline{B})/P(D) = 0.0777$
Totals = 1.0000		$P(D) = 0.52058$	$P(B \mid D) + P(\overline{B} \mid D) = 1.0000$

Also, the probability of C based on TC can be computed as follows using the results from Table 3.7a–c:

$$P(C) = P(C \mid TC)P(TC) = 0.6897(0.003053) = 0.0021056 \tag{3.13b}$$

The probability of $(D|B)$ in Table 3.7d can be computed as follows using the results from Tables 3.1 through 3.4:

$$\begin{aligned} P(D \mid B) &= P(D \mid B \cap TC)P(TC) + P(D \mid B \cap \overline{TC})P(\overline{TC}) \\ &= 0.90(0.001938) + 0.80(1 - 0.001938) = 0.8001938 \end{aligned} \tag{3.14a}$$

It should be noted that $P(D \mid B \cap TC) = 0.90$, $P(TC) = 0.001938$, and $P(D \mid B \cap \overline{TC}) = 0.80$ based on Tables 3.1 and 3.4. In this table, we are using values from the previous case as given in Table 3.4. Similarly, the probability of $(D \mid \bar{B})$ in Table 3.7d can be computed as follows using the results from Tables 3.1 through 3.4:

$$\begin{aligned} P(D \mid \bar{B}) &= P(D \mid \bar{B} \cap TC)P(TC) + P(D \mid \bar{B} \cap \overline{TC})P(\overline{TC}) \\ &= 0.70(0.001938) + 0.10(1 - 0.001938) = 0.101163 \end{aligned} \tag{3.14b}$$

It should be noted that $P(D \mid \bar{B} \cap TC) = 0.70$, $P(TC) = 0.001938$, and $P(D \mid B \cap \overline{TC}) = 0.10$ based on Tables 3.1 and 3.4.

3.3.7 Process Modeling Methods

The definition of a system in this case can be viewed as a process that emphasizes an attribute of the system. The steps involved in this process form a spiral of system definitions with a hierarchical structure and solutions of problems through decision analysis by learning, abstraction, modeling, and refinement. Example processes include engineering systems as products to meet user demands, engineering systems with life cycles, and engineering systems defined by a technical maturity process. These three example processes are described in subsequent sections for demonstration purposes.

3.3.7.1 Systems Engineering Process

The systems engineering process focuses on the interaction between humans and their environment as shown in Figure 3.1. The steps involved in a systems engineering process can be viewed as constituting a spiral hierarchy. A systems engineering process has the following steps (Figure 3.21):

1. *Recognition of need or opportunity.* The recognition of need or opportunity results from the interaction of humans with various environments, so this step can be considered as being not a part of the spiral but its first cause. The step can be viewed as an entrepreneurial activity, rather than an engineering task. The discovery of a need can be articulated in the form of a goal for a proposed system with a hierarchical breakdown into objectives. The delineation of the goals of the system should form the basis for and produce the requirements desired by eventual users of the system. For a government, the goals should also include the long-term interests of the public.
2. *Identification and qualification of the goal, objectives, and functional and performance requirements.* The goal or mission of the system must be stated and delineated. This statement should then be used to define a hierarchy of objectives that can be used to develop a list of performance requirements for the systems. These definitions of the goal, objectives, and performance requirements can be used to compare the cost-effectiveness of alternative system design concepts. The objectives and performance requirements should include relevant aspects of effectiveness, cost, schedule, and risk, and should be traceable to the goal. To facilitate tradeoff analyses, they should be stated in quantifiable and verifiable terms to some

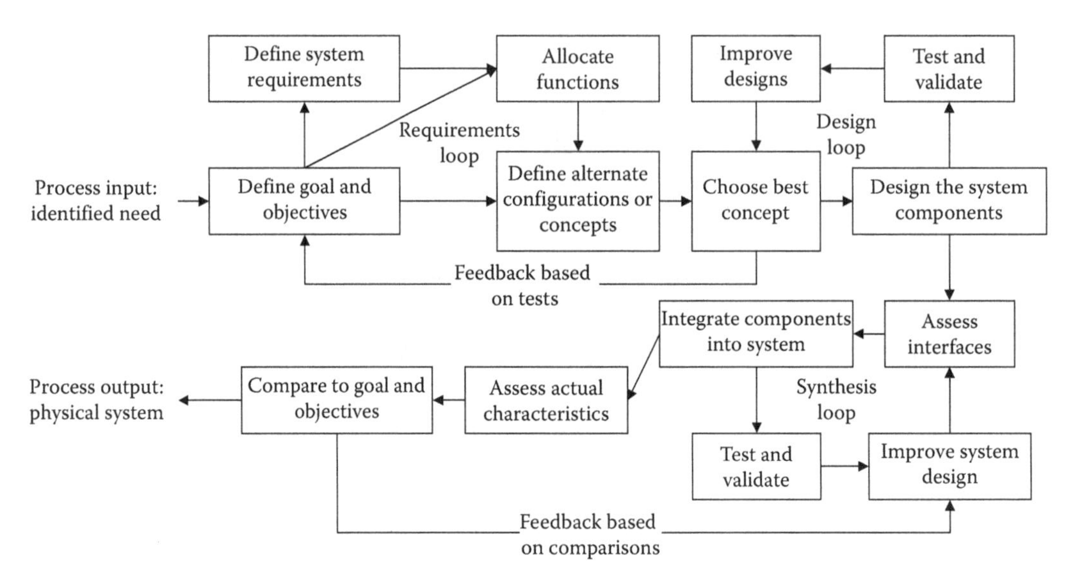

FIGURE 3.21
Systems engineering process.

meaningful extent. At each turn of a loop or spiral, the objectives and performance requirements should be documented for tractability and tracing them to various system components. As the systems engineering process continues, the performance requirements should be translated into a functional hierarchy for the system, allocated to components of the system. The performance and functional requirements should be quantitatively described.

3. *Creation of alternative design concepts.* Establishing a clear understanding of what the system should accomplish is a prerequisite to devising a variety of ways in which the goal, objectives, and requirements can be met. Sometimes, the alternatives can come about as a consequence of integrating the available component design options. Using a bottom-up alternative creation, various concept designs can be developed. It is essential to maintain objectivity in the process to not be drawn to a specific option that would limit or obscure the examination of other options. An analyst or designer must stay an outsider in order to maintain objectivity. This detachment would allow the analyst or designer to avoid premature focus on a single design and would permit the discovery of a truly superior design.
4. *Testing and validation.* At this stage, some testing and validation of the concepts might be necessary in order to establish an understanding of the limitations, capabilities, and characteristics of various concepts. The testing and validation can be experimentally, analytically, or numerically performed using laboratory tests, analytical models, or simulation, respectively. The insight gained from this step might be crucial for subsequent steps of this process.
5. *Performance of trade-off studies and selection of a design.* Trade-off studies start by assessing how well each design concept meets the goals, objectives, and requirements of the system, including effectiveness, cost, schedule, and risk, both quantitatively and otherwise. This assessment can utilize the testing and validation results of the previous step. These studies can be performed using system

models that analytically relate various concept characteristics to performance and functional requirements. An outcome of these studies can be determination of the bounds of the relative cost-effectiveness of the design concepts. Selection among the alternative design concepts must take into account subjective factors that are not quantifiable and were not incorporated in the studies. When possible, mathematical expressions, called objective functions, should be developed and used to express the values of combinations of possible outcomes as a single measure of cost-effectiveness. The outcome of this step identification of the best concept is to be advanced to next steps.

6. *Development of a detailed design.* One of the first issues to be addressed is how the system should be subdivided into subsystems and components in order to represent accurately an engineering product of interest. The partitioning process stops when the subsystems or components are simple enough to be managed holistically. Also, the system might reside within a program that has well-established activities or groups. The program activities might drive the definitions of the system hierarchy of subsystems and components. These program activities should be minimized in number and complexity, as they define various interfaces and could have a strong influence on the overall system cost and schedules. Partitioning is more of an art than a science; however, experiences from other related systems and judgment should be utilized. Interfaces can be simplified by grouping similar functions, designs, and technologies. The designs for the components and subsystems should be tested, verified, and validated. The components and subsystems should map conveniently onto an organizational structure, if applicable. Some of the functions that are needed throughout the system, such as electrical power availability, or throughout the organization, such as purchasing, can be centralized. Standardization of such things as part lists or reporting formats is often desirable. The accounting system should follow, not lead, the system architecture. Partitioning should be done essentially all at once, broadly covering the entire system. Similar to system design choices, alternative partitioning plans should be considered and compared before selecting the optimal plan and its implementation.
7. *Implementing the selected design decisions.* The design spiral or loop of successive refinement should proceed until reaching diminishing returns. The next step is to reverse the partitioning process by *unwinding* the process. This unwinding phase is called *system integration.* Conceptual system integration takes place in all steps of the process, that is, when a concept has been selected, the approach is verified by unwinding the process to test whether the concept at each physical level meets the expectations and requirements. The physical integration phase is accomplished during fabrication or manufacturing of the system. The subsystem integration should be verified and validated to ensure that the subsystems conform to design requirements individually and at the interfaces, such as mechanical connections, power consumption, and data flow. System verification and validation consist of ensuring that interfaced subsystems achieve their intended results collectively as one system.
8. *Performance of missions.* In this step, the physical system is called upon to meet the need for which it was designed and built. During this step, the system effectiveness at the operational site should be validated. Also, the step includes maintenance and logistics documentation, definition of sustaining engineering activities,

compilation of development and operations *lessons-learned* documents, and, with the help of specialty engineering disciplines, identification of improvement opportunities for quantifiable system objectives. Sometimes only bounds, rather than final values, are possible in this step. The spread between any upper and lower bound estimates of system attributes or performances can be reduced as a result of increasing the level of validation and testing, and continually improving and enhancing the design.

3.3.7.2 Life Cycle of Engineering Systems

Engineering products can be treated as systems that have a life cycle. A generic life cycle of a system begins with the initial identification of a need and extends through planning, research, design, production or construction, evaluation, consumer use, field support, and ultimately product phase out or disposal, as shown in Figure 3.22. A system life cycle is sometimes known as the *consumer-to-consumer cycle*, which has major activities applicable to each phase of the life cycle, as illustrated in Table 3.8. The steps illustrated show a logical flow and associated functions for each step or effort. Although the generic steps are the same, various systems might require different specific details in terms of what has to be done. A large system requiring new development, such as a satellite or major ground system, may evolve through all the steps, whereas a relatively small item, such as an element of a space segment or the maintenance phase of a software contract, may not. In considering the life cycle of a system, each of the steps identified should be addressed even though all steps may not be applicable. The life cycle of a product is a general concept that needs to be tailored for each user or customer. The life cycle of systems according to the Department of Defense (DoD) and the National Aeronautics and Space Administration (NASA) is tied to the government procurement process as discussed in Example 3.11, but the general applicability of the concept of a system life cycle is independent of the user and the procurement process.

Example 3.11: Life Cycle of NASA Engineering Systems

The NASA uses the concept of life cycle for a program (program life cycle). The program life cycle consists of distinct phases separated by control gates. The NASA uses its life cycle model not only to describe how a program evolves over time but also to

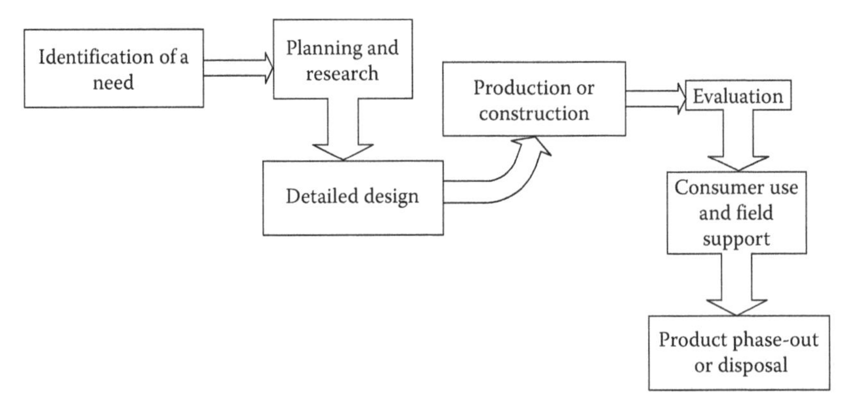

FIGURE 3.22
Life cycle of engineering systems.

TABLE 3.8

System Life Cycle vs. Consumer-to-Consumer Cycle

Phases of System Life Cycle	Phases of Consumer-to-Consumer Cycle	Activities
Identification of need	Consumer	"Wants or desires" for systems due to obvious deficiencies/problems or made evident through basic research results
System planning function	Producer	Marketing analysis; feasibility study; advanced system planning through system selection; specifications and plans; acquisition plan research, design, and production; evaluation plan; system use and logistic support plan; planning review; proposal
System research function		Basic research; applied research based on needs; research methods; results of research; evolution from basic research to system design and development
System design function		Design requirements; conceptual design; preliminary system design; detailed design; design support; engineering model or prototype development; transition from design to production
Production and/or construction function		Production and/or construction requirements; industrial engineering and operations analysis such as plant engineering, manufacturing engineering, methods engineering, and production control; quality control; production operations
System evaluation function	Consumer	Evaluation requirements; categories of test and evaluation; test preparation phase including planning and resource requirements; formal test and evaluation; data collection, analysis, reporting, and corrective action; retesting
System use and logistic support function		System distribution and operational use; elements of logistics and life cycle maintenance support; system evaluation; modifications; product phase-out; material disposal, reclamation, and recycling

aid management in program control. The boundaries between phases are defined so that they precede decisions. Decisions to proceed may be qualified by liens that must be removed within a reasonable time. A program that fails to pass a control gate and has enough resources may be allowed to readdress the deficiencies or may be terminated. The governmental agency operates within a fiscal budget and annual funding that lead to implicit funding control gates at the beginning of fiscal years. While these gates place planning requirements on the project and can make significant replanning necessary, they are not part of an orderly systems engineering process; rather, they constitute one of the sources of uncertainty that affect project risks and should be included in project risk considerations. The NASA model can generally be defined to include the following phases that are provided under separate headings.

PRE-PHASE A—ADVANCED STUDIES

The objective of this phase is to produce a broad spectrum of ideas and alternatives for missions from which new projects or programs can be selected. Major activities and their products in this phase are intended to (1) identify missions consistent with the NASA charter, (2) identify and involve users, and (3) perform preliminary evaluations of possible missions. Typically, this phase consists of loosely structured examinations of new ideas, usually without central control and mostly oriented toward small studies. Also, program or project proposals are prepared which include mission justification and objectives, possible operations concepts, possible system architectures, and cost, schedule, and risk estimates. The phase also produces master plans for existing program areas. The control gates for this phase are informal proposal reviews. Descriptions of projects suggested generally include initial system design and operational concepts, preliminary project organization, schedule, testing and review structure, and documentation requirements. This phase is of an ongoing nature because technological progress makes possible missions that were previously impossible. Manned trips to the moon and the taking of high-resolution pictures of planets and other objects in the universe illustrate past responses to this kind of opportunity. New opportunities will continue to become available as our technological capabilities grow.

PHASE A—CONCEPTUAL DESIGN STUDIES

The objective of this phase is to determine the feasibility and desirability of a suggested new major system in preparation for seeking funding. This phase includes major activities such as (1) preparation of mission needs statements, (2) development of preliminary system requirements, (3) identification of alternative operations and logistics concepts, (4) identification of project constraints and system boundaries, (5) consideration of alternative design concepts, and (6) demonstration of the credibility and the feasibility of designs. System validation plans are initiated in this phase. Also, systems engineering tools and models are acquired, environmental impact studies are initiated, and program implementation plans are prepared. The control gates are conceptual design review and pre-phase B nonadvocate review. This phase is frequently described as a structured version of the previous phase.

PHASE B—CONCEPT DEFINITION

The objective of this phase is to define the project in enough detail to establish an initial baseline. This phase includes major such activities as: (1) reaffirmation of the mission needs statement, (2) preparation of a program initiation agreement, (3) preparation of a systems engineering management plan, (4) preparation of a risk management plan, (5) initiation of configuration management, (6) development of a system-level cost-effectiveness model, (7) restatement of the mission needs as system requirements, (8) establishment of the initial requirement traceability matrix, (9) selection of a baseline system architecture at some level of resolution and concept of operation, (10) identification of science payload, (11) definition of internal and external interface requirements, (12) definition of the work breakdown structure, (13) definition of verification approach and policies, (14) preparation of preliminary manufacturing plans, (15) identification of government resource requirements, (16) identification of ground test and facility requirements, (17) development of statement of work, (18) revision and publication of project implementation plans, and (19) initiation of advanced technology development programs. The control gates include project definition and cost review, program and project requirements review, and safety review. Trade-off studies in this phase should precede rather than follow system design decisions. A feasible system design can be defined as a design that can be implemented as designed, and can then accomplish the goal of the system within the constraints imposed by the fiscal and operating environment. To be credible, a design must not depend on the occurrence of unforeseen breakthroughs in the state of the art. While a credible design may assume likely improvements in the state of the art, it is nonetheless riskier than the one that does not.

PHASE C—DESIGN AND DEVELOPMENT

The objective of this phase is to design a system and its associated subsystems, including its operations systems, so that it will be able to meet its requirements. This phase has primary tasks and activities that include (1) addition of subsystem design specifications to the system architecture; (2) publication of subsystem requirements documents; (3) preparation of subsystem verification plans; (4) preparation of interface documents; (5) repetition of the process of successive refinement to get "design-to" and "build-to" specifications and drawings, verification plans, and interface documents at all levels; (6) augmentation of documents to reflect the growing maturity of the system; (7) monitoring of the project progress against project plans; (8) development of the system integration plan and the system operations plans; (9) documentation of trade-off studies performed; (10) development of the end-to-end information system design and the system deployment approach; (11) identification of opportunities for preplanned product improvement; and (12) confirmation of science payload selection. Control gates include system-level preliminary design reviews, subsystem (and lower level) preliminary design reviews, subsystem (and lower level) critical design reviews, and system-level critical design reviews. The purpose of this phase is to unfold system requirements into system and subsystem designs. Several popular approaches can be used in the *unfolding process* such as code-and-fix, the waterfall, requirements-driven design, and/or evolutionary development.

PHASE D—FABRICATION, INTEGRATION, TEST, AND CERTIFICATION

The purpose of this phase is to build the system designed in the previous phase. Activities include a fabrication system for hardware and coding of software, integration, verification and validation, and certified acceptance of the system.

PHASE E—PREOPERATIONS

The purpose of this phase is to prepare the certified system for operations by performing activities that include initial training of operating personnel and finalization of the integrated logistics support plan. For flight projects, the focus of activities then shifts to prelaunch integration and launch, whereas for large flight projects, extended periods of orbit insertion, assembly, and shakedown operations are necessary. In some projects, these activities can be treated as minor items, allowing this phase to be combined with either its predecessor or its successor. The control gates are launch readiness reviews, operational readiness reviews, and safety reviews.

PHASE F—OPERATIONS AND DISPOSAL

The objective of this phase is to actually meet the initially identified need and then to dispose of the system in a responsible manner. This phase includes major activities such as (1) training replacement operators, (2) conducting the mission, (3) maintaining the operating system, and (4) disposing of the system. The control gates are operational acceptance reviews, regular system operations reviews, and system upgrade reviews. This phase encompasses the problem of dealing with the system when it has completed its mission. The end of life depends on many factors. For example, the disposal of a flight system with short mission duration, such as a space lab payload, may require little more than deintegration of the hardware and return to its owner; the disposal of a large flight project with long duration may proceed according to long-established plans or may begin as a result of unplanned events, such as accidents. In addition to uncertainty as to when this part of the phase begins, the activities associated with safely deactivating and disposing of a system may be long and complex. As a result, the costs and risks associated with different designs should be considered during the planning process.

3.3.7.3 Technical Maturity Model

The technical maturity model is another view of the life cycle of a project. According to this model, the life cycle considers a program as an interaction between society and engineering. The model concentrates on the engineering aspects of the program and not on the technology development through research. The program must come to fruition by meeting both the needs of the customer and the technical requirements. Therefore, by keeping distinctions among technical requirements, needs, and technology development, the motivations, wants, and desires of the customer are differentiated from the technology issues during the course of the project.

3.3.7.4 Spiral Development Process

A product or a system can be developed using a spiral process as shown in Figure 3.23. Spiral development is used for designing marine, aerospace, and other advanced systems. Figure 3.23 shows phases similar to those included in previously presented process

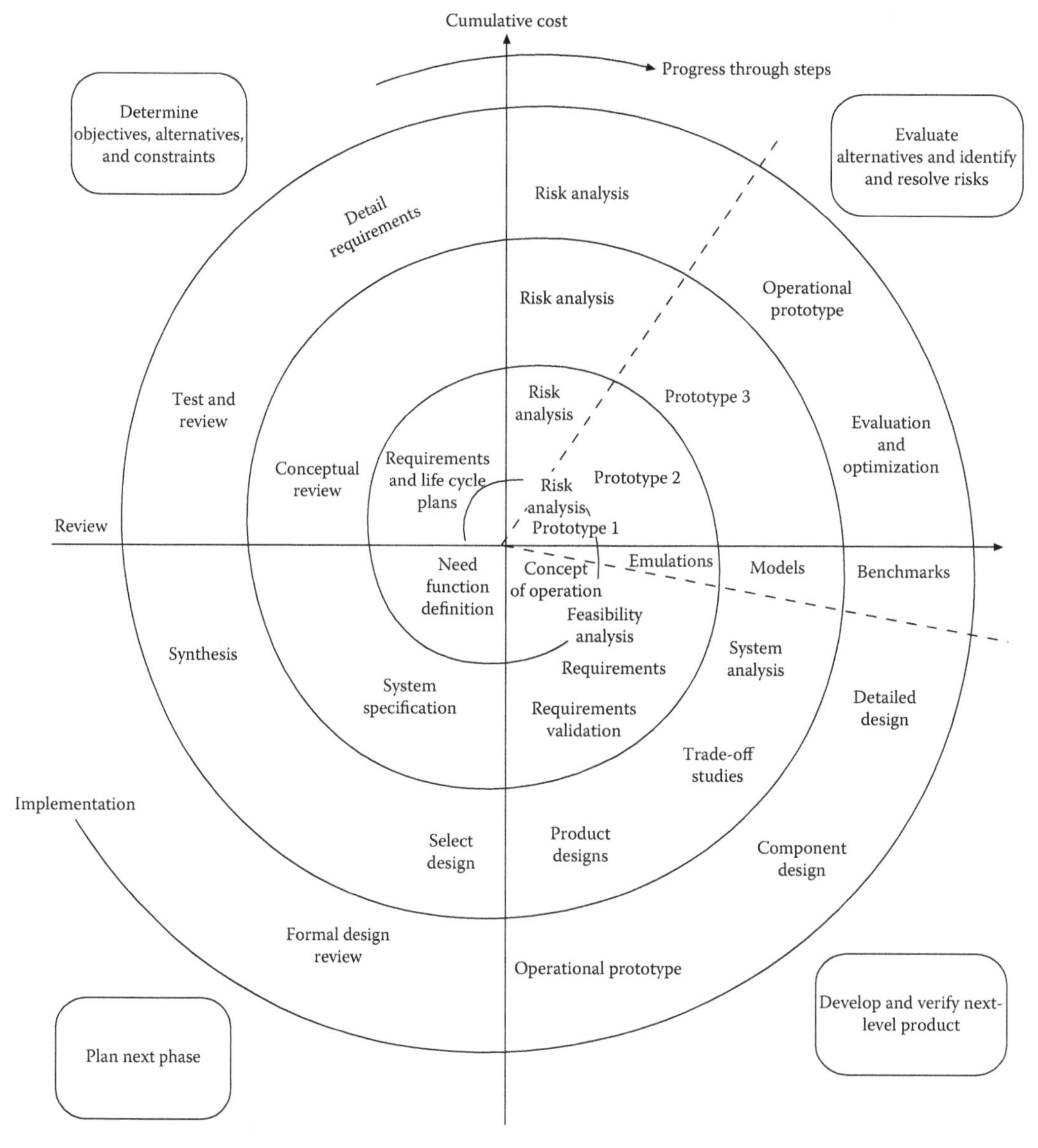

FIGURE 3.23
Spiral development process.

modeling methods in this chapter, with an added spiral organization and risk review and analysis at various levels of development.

3.3.8 Black-Box Method

Historically, engineers have built analytical models to represent natural and human-made systems using empirical tools for observing system attributes of interest (*system output variables*) and trying to relate them to some other controllable or uncontrollable input variables. For example, a structural engineer might observe the deflection of a bridge as an output of an input such as a load at the middle of its span. Varying the intensity of the load changes the deflection. Empirical test methods would vary the load incrementally and the corresponding deflections are measured, thereby producing a relationship as follows:

$$y = f(x) \tag{3.15}$$

where:
x is an input variable
y is an output variable
f is a function that relates input to output

In general, a system might have several input variables that can be represented as a vector **X**, and several output variables that can be represented by a vector **Y**. A schematic representation of this model is shown in Figure 3.24. According to this model, the system is viewed as a whole entity without any knowledge on how the input variables are processed within the system to produce the output variables. This black-box view of the system has the advantage of shielding an analyst from the physics governing the system and providing the analyst with an opportunity to focus on relating the output to the input within some range of interest for the underlying variables. The primary assumptions according to this model are (1) the existence of causal relationships between input and output variables as defined by the function f and (2) the effect of time (i.e., time lag or time prolongation within the system), which are accounted for by methods of measurement of input and output variables.

For complex engineering systems or natural systems, the numbers of input and output variables might be large with varying levels of importance. In such cases, a systems engineer would be faced with the challenge of identifying the most significant variables and how they should be measured. Establishing a short list of variables might be a most difficult task, especially for novel systems. Some knowledge of the physics of the system might help in this task of system identification. Then, the analyst needs to decide on the nature of the time relation between input and output by addressing the following questions:

- Is the output instantaneous as a result of the input?
- If the output lags behind the input, what is the lag time? Are the lag times for the input and output related (e.g., exhibiting nonlinear behavior)?

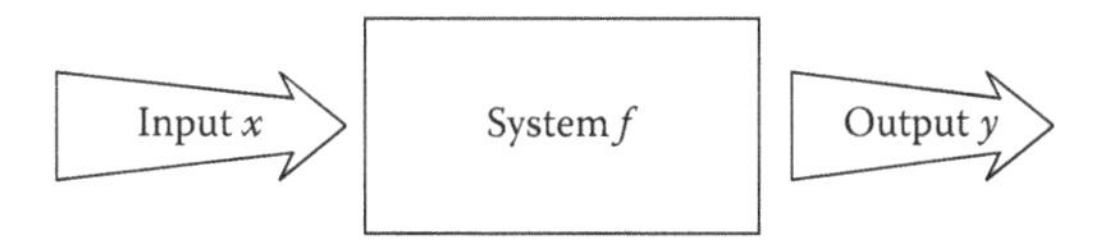

FIGURE 3.24
Black-box system model.

- Does the function f depend on time, number of input applications, or magnitude of input?
- Does the input produce an output and linger within the system, affecting future outputs?

Answering these questions is important for the purposes of defining the model, its applicability range, and validity.

Example 3.12: Probable Maximum Flood

The USACE classes dams according to both size and hazard, where hazard is defined in terms of loss of life and economic loss (Committee on Safety Criteria for Dams 1985). Small dams are 25–40 ft high, intermediate dams are 40–100 ft high, and large dams are over 100 ft high. Low-hazard dams are those for which failure of the dam would result in no loss of life and minimal economic loss. A significant hazard is one that would cause some loss of life and appreciable economic loss, and a high hazard would result in the loss of more than a few lives and excessive economic loss.

The USACE (1965) uses three methods of determining extreme floods, depending on the return period and the intended use. Frequency analyses are used when the project requires defining a storm event with a relatively common return period and are based on gauge records. This type of analyses is used for low-hazard dams, small to intermediate dams, or small dams with significant hazard classifications. A standard project flood (SPF) is used when some risk can be tolerated but where an unusually high degree of protection is justified because of risk to life and property (Ponce 1989). The SPF includes several combinations of meteorological and hydrological conditions but does not include extremely rare combinations. The SPF is typically used for dams that are classed as a significant hazard and are intermediate to large in size. For projects requiring a substantial reduction in risk, such as dams classed as high hazard, the probable maximum flood (PMF) is used. The PMF is the most severe and extreme combination of meteorological and hydrological events that could possibly occur in an area. Flood prediction can be based on black-box models as shown in Figure 3.25. For river systems, time can play a major role in the form of time lag, time prolongation, and system nonlinearity.

Frequency analyses of gauge data conducted by the USACE are based on recommendations in *Bulletin 17B* (US Interagency Advisory Committee on Water Data 1982). The SPF is developed from a standard project storm (SPS). The PMF is based on an index rainfall and a depth–area–duration relationship. A hydrograph is then developed based on this rainfall minus hydrologic extractions. For basins <1000 mi^2 (2590 km^2), the storms are usually based on localized thunderstorms; for basins >1000 mi^2 (2590 km^2), the storms are usually a combination of events. Due to these differences, the PMF for the smaller basins is based on a 6-hr or 12-hr time increment. For large basins, this procedure is considerably more complex. The SPF is developed very similarly to the PMF except that the index flood is decreased by ~50%.

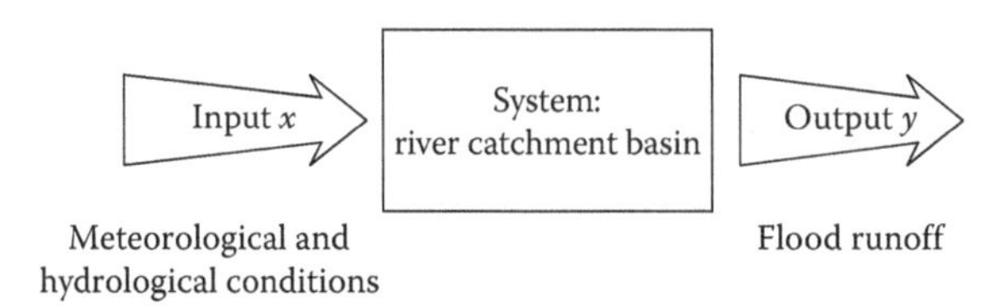

FIGURE 3.25
Black-box system model for flood prediction.

The use of the PMF has often been questioned because rainfalls and floods of that magnitude have not been experienced in a lifetime. However, studies conducted by the USACE have shown that dozens of storms across the United States have exceeded one-half of the probable maximum precipitation (PMP) for those particular areas (Committee on the Safety of Existing Dams 1983; USACE 1965). Based on these data, the USACE assumes that the PMP is a reasonable basis from which to estimate the maximum likely hydrological event, although it continues to be debated by its engineers.

3.3.9 State-Based Method

A convenient modeling method of systems can be based on identifying state variables that would be monitored either continuously or at discrete times. The values of these state variables over time provide a description of the model required for the system. The state variables should be selected such that each one provides unique information. Redundant state variables are not desirable. The challenge faced by systems engineers is to identify the minimum number of state variables that would accurately represent the behavior of the system over time.

Although it is common that the components of a system are modeled to have one of two possible states—a functioning state or a failed state—in general, component models can have more than two states. Such models provide the tools necessary to model repairable systems. For example, a method used to develop reliability models is the *state-space method for system reliability evaluation.* A system according to this method is described by its states and the possible transitions between these states. The system states and the possible transitions are illustrated by a state-space diagram, which is also known as a *Markov diagram.* The case of a two-component parallel system of Figure 3.26a has an illustrative diagram as shown in Figure 3.26b. The various states of the system can be defined as the combination of all possible states of the underlying components, as summarized in Table 3.9. According to the state-space method, the components are not restricted to having only two possible states and may have a number of different states, such as functioning, derated, on standby, completely failed, and under maintenance. Various failure modes may also be defined as separate states. The transitions between the states are caused by various mechanisms and activities such as failures, repairs, replacements, and switching operations. Common cause failures may also be modeled by the state-space method. The number of system states, however,

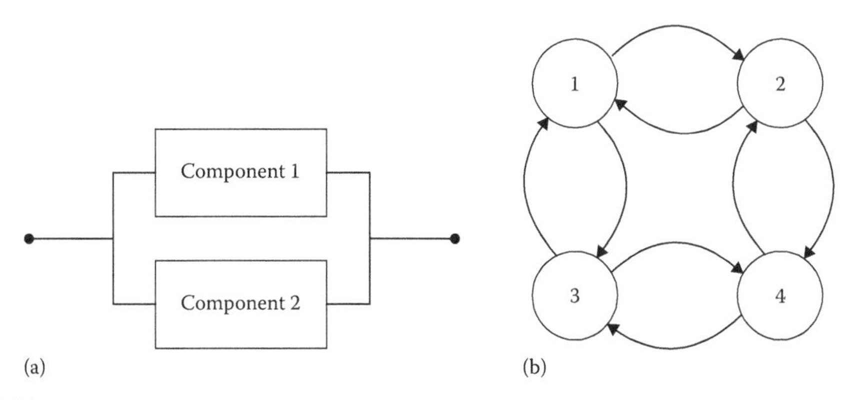

FIGURE 3.26
(a) Parallel system of two components; (b) state-space diagram for the parallel system.

TABLE 3.9

Definition of States of a System Based on the States of Its Components

System State According to Figure 3.26b	State of Component 1 of Figure 3.26a	State of Component 2 of Figure 3.26a	Description of the State of the System
1	Functioning	Functioning	System survival is based on both components functioning.
2	Failed	Functioning	System survival is based on one component functioning and one component failed.
3	Functioning	Failed	System survival is based on one component functioning and one component failed.
4	Failed	Failed	System failure is based on both components failed.

increases rapidly with the size and complexity of the system, making it suitable only for relatively small systems.

The methods described here require developing models that describe the transitions of state variables from some set of values to another set of values. It is common that these transitions are not predictable due to uncertainty and can only be characterized probabilistically. The state transition probabilities are of interest and can be empirically assessed and modeled as described in the subsequent examples. Also, Markov models are applied in Chapter 9 to illustrate the performance of risk-based maintenance planning for marine structure.

Example 3.13: Markov Modeling of a Three-State System

Consider a system that can exist only in three states: functioning state (S_1), degraded state (S_2), and failed state (S_3). If we assume that the states have time-invariant transition probabilities among themselves as shown in Figure 3.27 and provided by the following transition probability matrix (P) from an originating state in row i to a destination state in column j:

$$\text{Originating state } S_i \quad \overset{\text{Destination state } S_j}{\begin{bmatrix} 0.80 & 0.10 & 0.10 \\ 0.30 & 0.20 & 0.50 \\ 0.20 & 0.10 & 0.7 \end{bmatrix}} \tag{3.16}$$

The probabilities in each row add up to one since they correspond to arrows originating from the same state. Defining P_1, P_2, and P_3 as the fraction of the times, that is, probabilities, of finding the system in the respective states, the following state equations should be met:

$$P_1 = 0.80P_1 + 0.30P_2 + 0.20P_3 \tag{3.17a}$$

$$P_2 = 0.10P_1 + 0.20P_2 + 0.10P_3 \tag{3.17b}$$

$$P_3 = 0.10P_1 + 0.50P_2 + 0.70P_3 \tag{3.17c}$$

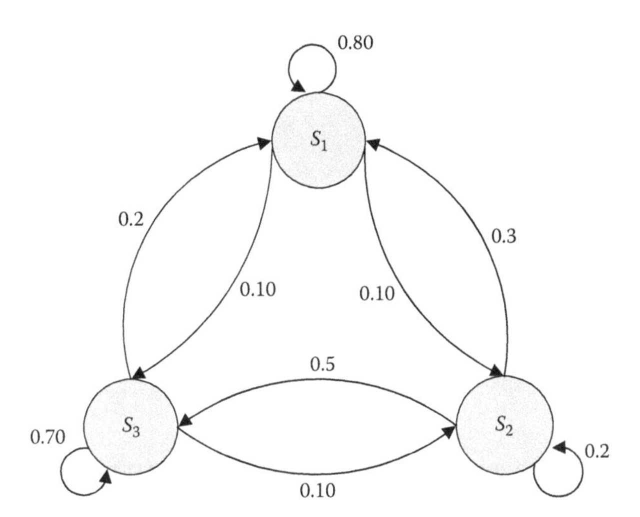

FIGURE 3.27
Markov transition probabilities for a three-state system.

Moreover, the following requirement is necessary since the system can exist only in one of the three states:

$$1 = P_1 + P_2 + P_3 \tag{3.17d}$$

Using any two state requirements from Equation 3.17a–c in Equation 3.17d, we can solve for the values of P_1, P_2, and P_3 as 19/36, 4/36, and 13/36, respectively.

Example 3.14: Markov Modeling of a Single Repairable System

Repairable systems can be assumed for the purpose of demonstration to exist in either a normal (operating) or a failed state, as shown in Figure 3.28. A system in a normal state either stays in the normal state as governed by its reliability level (i.e., continues to be normal) or makes a transition to the failed state through a failure mode. Once it is in a failed state, the system either stays in the failed state as governed by its suitability to be repaired (i.e., it continues to be failed until repaired) or makes a transition to a normal state through repair (or replacement), and if repaired it is assumed to be restored to "as good as new." Therefore, four parameters needed to characterize the situation as follows (see Figure 3.28):

- Fractional time spent in the normal state S_1, interpreted as a probability and denoted P_1
- Fractional time spent in the failed state S_2, interpreted as a probability and denoted P_2
- Average failure rate λ that is related to the mean time between failures (MTBFs) as $\lambda = 1/\text{MTBF}$
- Average repair rate μ that is related to the mean time to repair (MTTR) as $\mu = 1/\text{MTTR}$

These parameters can be used to define the following state equations:

$$\lambda P_1 - \mu P_2 = 0 \tag{3.18}$$

$$P_1 + P_2 = 1 \tag{3.19}$$

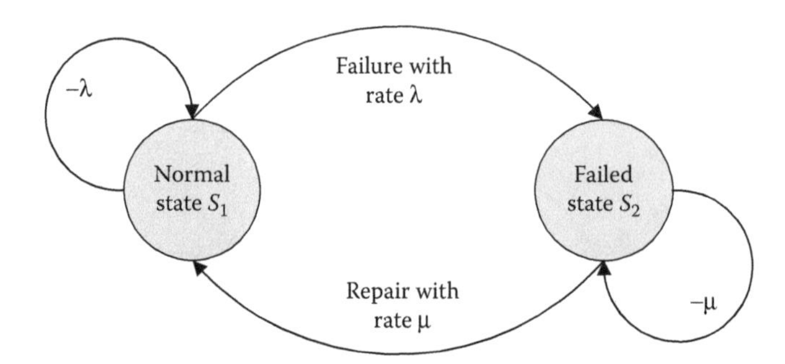

FIGURE 3.28
Markov transition diagram for a repairable system.

Equation 3.18 states that on average we should have the same number of repairs as failures, whereas Equation 3.19 states that the system exists in one of two states. The transition rate matrix (R) from an originating state in row i to a destination state in column j is assumed as follows:

$$\text{Originating state } S_i \quad \overset{\text{Destination state } S_j}{\begin{bmatrix} -10 & 10 \\ 90 & -90 \end{bmatrix}} \tag{3.20}$$

In this case, the rows should add up to zero, and the negative rates are for mathematical convenience to express the rate deficiency from a rate of zero that corresponds to a system staying in this zero-rate state. Equations 3.18 and 3.19 can be rearranged by substitution to give an expression for system *availability* which is P_1:

$$P_1 = \frac{\mu}{\mu + \lambda} \tag{3.21}$$

The concept of availability as provided in Equation 3.21 is discussed in Chapter 4. In cases where the repair rate (μ) is significantly greater than the failure rate (λ):

$$P_1 = \frac{\mu}{\lambda} \tag{3.22}$$

It should be noted that this result is the same as that of MTTR/MTBF.

The probabilities and rates in this case can be summarized as provided in Table 3.10. In this case, the MTBF = 1/10 = 0.10 year and the MTTR = 1/90 = 0.011 year. Such a model can be used to examine, for example, the reliability of electronic devices.

3.3.10 Component Integration Method

Systems can be viewed as assemblages of components. For example, in structural engineering, a roof truss can be viewed as a multiple-component system. The truss in Figure 3.29 has 13 members. The principles of statics can be used to determine the member forces and reactions for a given set of joint loads. By knowing the internal forces and material properties, other system attributes, such as deformations, can be evaluated. In this case, the physical connectivity of the real components can be defined as the connectivity of the

TABLE 3.10
Daily Transition Probabilities

Parameter	Notation	Value	Comments
Normal probability	P_1	0.90	$P_1 + P_2 = 1$, i.e., $0.90 + 0.10 = 1.0$
Failed probability	P_2	0.10	
Failure rate (1/year)	λ	10	$\lambda P_1 - \mu P_2 = 0$, i.e., $10(0.90) - 90(0.10) = 0$
Repair rate (1/year)	μ	90	

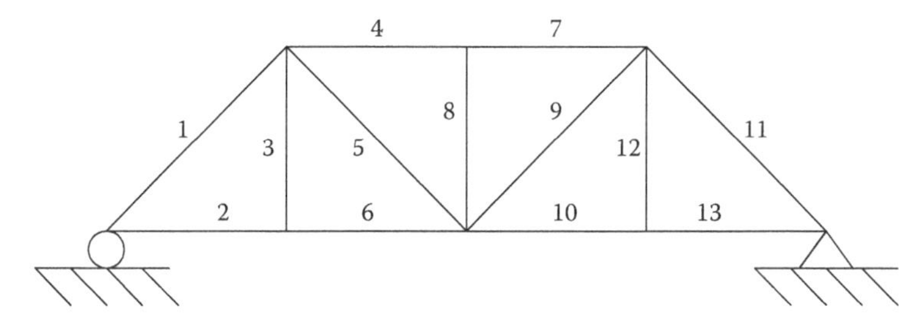

FIGURE 3.29
Truss structural system.

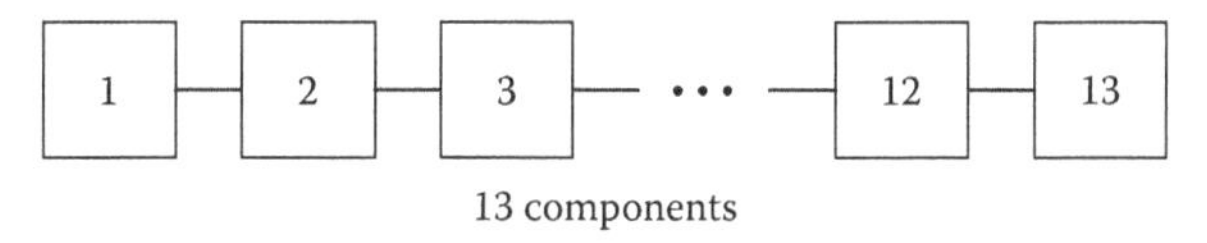

FIGURE 3.30
Truss series system as a reliability block diagram.

components in the structural analysis model. However, if we were interested in the reliability and/or redundancy of the truss, a more appropriate model would be as shown in Figure 3.30, called a *reliability block diagram*. The representation of the truss in Figure 3.30 emphasizes the attributes of reliability or redundancy. According to this model, the failure of one component would result in the failure of the truss system. Ayyub and McCuen (2011), Ang and Tang (2007), and Kumamoto and Henley (1996) provide the details on reliability modeling of systems.

3.4 Hierarchical Definitions of Systems

3.4.1 Introduction

Using one of the perspectives and models of Section 3.2 to define a system, information then needs to be gathered to develop an *information-based system definition*. The information can be structured in a hierarchical manner to facilitate its construction, completeness, and accuracy of representation, although the resulting hierarchy might not achieve all these requirements. The resulting information structure can be used to construct the knowledge levels of the system for the purpose of analyzing and interpreting system behavior. Also, the resulting hierarchy can be used to develop a *generalized system definition* that can generically be used in representing other systems and problems.

A generalized system formulation allows researchers and engineers to develop a complete and comprehensive understanding of human-made products, natural systems, processes, and services. In a system formulation, an *image* or a *model* of an object that emphasizes certain important and critical properties is defined. Systems are usually identified based on the *level of knowledge* and/or information that this level contains. Based on their knowledge levels, systems can be classified into five consecutive hierarchical levels. The higher levels include all the information and knowledge introduced in the lower ones, in addition to more specific information. System definition is usually the first step in an overall methodology formulated for achieving a set of objectives that define a goal. For example, in construction management, real-time control of construction or production activities can be one of these objectives; however, in order to develop a control system for a construction activity, this activity has to be suitably defined depending on its nature and methods of control. *Hierarchical control systems* were determined to be suitable for construction activities (Abraham et al. 1989). Thus, the hierarchical nature of a construction activity must be emphasized. The generalized system definition as discussed in this section can be used for this purpose. The hierarchical system classification enables the decomposition of the overall construction activity into subsystems that represent the different processes involved in each activity. Then, each process can be decomposed into tasks that are involved in performing the process, and the breakdown required for a hierarchical control system is obtained. In this section, the basic concepts of system identification and definitions are introduced, together with some additional concepts that could be used in modeling and solving problems in engineering and sciences. Construction activities are modeled and discussed using the methods presented in this section in a systems framework for the purpose of demonstration. The knowledge system is upgraded throughout the course of the coverage in this section from one system level to the next level in order to illustrate the use of the developed concepts for controlling the construction activities.

3.4.2 Knowledge and Information Hierarchy

The definition of a system is commonly considered as the first step in an overall methodology formulated for achieving a set of objectives (Chestnut 1965; Hall 1962, 1989; Klir 1969, 1985; Wilson 1984). A system can be defined in many ways, as discussed in Section 3.2; however, here we use the common definition of a system as "an arrangement of elements with some important properties and interrelations among them." In order to introduce a comprehensive definition of a system, a more specific description is required based on several main knowledge levels (Klir 1969, 1985). Further classifications of systems are possible within each level using methodological distinctions based on, for example, their nature (i.e., natural or designed), human activity, or social and cultural factors (Wilson 1984). Chestnut (1965) and Hall (1962, 1989) provided hierarchical formulations of systems based on the available information and the degree of detail. Klir (1969, 1985) introduced a set approach for the system definition problem that was criticized by Hall (1989) because of its inability to express the properties of the overall system, knowing the qualities of its elements. However, for construction activities, the set approach is suitable for representing the variables of the problem. The ability to infer information about the overall system, knowing the behavior of its components, can be dealt with using special techniques as discussed by Klir (1985). Once a system is defined, the next step is to define its environment (Chestnut 1965; Hall 1962, 1989; Klir 1969, 1985; Wilson 1984). The environment is defined as "everything within a particular universe that is not included in the system." Hall (1989) introduced an interesting notion within systems thinking that allows a change in boundaries between a defined system and its environment. For the purposes of this section, the formation and structuring of systems are based on the concepts and approaches introduced by Klir (1969, 1985). In the following sections, knowledge and an example control system are gradually built up in successive levels. Each knowledge level is described briefly.

3.4.2.1 Source Systems

At the first level of knowledge, which is usually referred to as level 0, the system is known as a *source system*. Source systems have three different components: object systems, specific image systems, and general image systems. The object system, a model of the original object, is composed of an object, attributes, and a backdrop. The object represents the specific problem under consideration. The attributes are the important and critical properties or variables selected for measurement or observation as a model of the original object. The backdrop is the domain or space within which the attributes are observed. The specific image system is developed based on the object. This image is built through observation channels that measure the attribute variation within the backdrop. The attributes when measured by these channels correspond to the variables in the specific image system. The attributes are measured within a support set that corresponds to the backdrop. The support can be time, space, or population. Combinations of two or more of these supports are also possible. Before upgrading the system to a higher knowledge level, the specific image system can be abstracted into a general format. For this purpose, a mapping function is utilized from different states of the variables to a general state set that is used for all the variables. Some methodological distinctions can be defined in this level. Ordering is one of the distinctions realized within state or support sets. Any set can be either ordered or not ordered, and those that are ordered may be partially ordered or linearly ordered. An ordered set has elements that can take, for example, real values or values on an interval or ratio scale. A partially ordered set has elements that take values on an ordinal scale. A non-ordered set has components that take values on a nominal scale. Distance is another form of distinction, where the distance is a measure between pairs of elements of an underlying set. It is obvious that if the set is not ordered, the concept of distance is not valid. Continuity is another form of distinction, where variables or support could be discrete or continuous. The classification of variables as input or output variables forms another distinction. Those systems that have classified input/output variables are referred to as *directed systems;* otherwise, they are referred to as *neutral systems*. The last distinctions that could be realized in this level are related to the observation channels, which could be classified as crisp or fuzzy.

3.4.2.2 Data Systems

The second level of a hierarchical system classification is the data system. The data system includes a source system together with actual data introduced in the form of states of variables for each attribute. The actual states of the variables at different support instances yield the overall states of the attributes. Special functions and techniques are used to infer information regarding an attribute, based on the states of the variables representing it.

3.4.2.3 Generative Systems

At the generative knowledge level, support-independent relations are defined to describe the constraints among the variables. These relations could be utilized in generating states of the basic variables for a prescribed initial or boundary condition. The set of basic variables includes those defined by the source system and possibly some additional variables that are defined in terms of the basic variables. There are two main approaches for expressing these constraints. The first approach consists of a support-independent function that describes the *behavior* of the system. A function defined as such is known as a *behavior function*. The second approach consists of relating successive states of different variables. In other words, this function describes a relationship between the current overall state of the basic variables and the next overall state of the same variables. A function defined as such is known as a *state-transition function*. An example state-transition function is provided in Example 3.13 using Markov chains. A generative system defined by a behavior function is referred to as a *behavior system*; if it is defined by a state-transition function, it is known as a *state-transition system*. State-transition systems can always be converted into equivalent behavior systems that make the behavior systems more general.

Most engineering and scientific models—such as Newton's basic law of force, computed as the product of mass of an object and its acceleration, or computed the stress in a rod under axial loading as the applied force divided by the cross-sectional area of the rod—can be considered as generative systems that relate basic variables such as mass and acceleration to force or axial force and area to stress, respectively. In these examples, these models are behavior systems.

3.4.2.4 Structure Systems

Structure systems are sets of other systems or subsystems. The subsystems could be source, data, or generative systems. These subsystems may be coupled because they have common variables or due to interaction of some other forms.

3.4.2.5 Metasystems

Metasystems are introduced for the purpose of describing changes within a given support set. The metasystem consists of a set of systems defined at some lower knowledge level and some support-independent relations. Referred to as a replacement procedure, this relation defines the changes in the lower level systems. All the lower level systems should share the same source system. A metasystem can be viewed in relation to the structure system using two different approaches. The first approach is introduced by defining the system as a structure metasystem. The second approach consists of defining a metasystem of a structure system whose elements are behavior systems.

Example 3.15: System Definition of a Structure

A structure, such as a building, can be defined using a hierarchy of information levels to assess the structural adequacy resulting from loads applied to the structure. The system levels for this case are provided for demonstration purposes as follows:

- Goal: assess the structural adequacy of the building
- Source system objects: columns, beams, slabs, footings, dead load, live load, etc.
- Data system: dimensions, material properties, load intensities, etc.

- Generative system: prediction models of stress, such as stiffness analysis, stress computation, ultimate strength assessment of components in flexure, shear, and buckling
- Structure system: performance functions, defined as strength of components minus their respective load effects and used to assess the reliability of each component
- Metasystem: overall structural adequacy assessment of the system based on its components using system reliability concepts

3.5 System Complexity

Our most troubling long-range problems, such as economic forecasting and trade balances, climate change, defense systems, and genetic modeling, center on systems of extraordinary complexity. The systems that host the problems, such as computer networks, economics, ecologies, and immune systems, appear to be as diverse as the problems. Humans as complex, intelligent systems have the ability to anticipate the future and learn and adapt in ways that are not yet fully understood. Engineers and scientists, who study or design systems, have to deal with complexity, which is the interest in the field of complexity. Understanding and modeling system complexity can be viewed as a pretext for solving complex scientific and technological problems, such as finding a cure for acquired immunodeficiency syndrome (AIDS), solving long-term environmental issues, or using genetic engineering safely in agricultural products. The study of complexity has led to, for example, chaos and catastrophe theories. Even if complexity theories would not produce solutions to problems, they can still help us to understand complex systems and perhaps direct experimental studies. Theory and experiment go hand in hand, thereby providing opportunities to make major contributions.

The science of complexity was founded at the Santa Fe Institute by a group of physicists, economists, mathematicians, and computer scientists that included Nobel Laureates in physics and economics (Murray Gell-Mann and Kenneth Arrow, respectively). They noted that scientific modeling and discovery tend to emphasize linearity and reductionism, and consequently developed the science of complexity, which is based on assumed interconnectivity, coevolution, chaos, structure, and order to model nature, human social behavior, life, and the universe in a unified manner (Waldrop 1992).

Complexity can be classified into two broad categories: (1) complexity with structure or (2) complexity without structure. The complexity with structure was termed *organized complexity* by Weaver (1948). Organized complexity can be observed in a system that involves nonlinear differential equations with many interactions among a large number of components and variables that define the system, such as in life, behavioral, social, and environmental sciences. Such systems are usually nondeterministic in their nature. Problem solutions related to such models of organized complexity tend to converge to statistically meaningful averages (Klir and Wierman 1999). Advancements in computer technology and numerical methods have enhanced our ability to obtain such solutions effectively and inexpensively. As a result, engineers design complex systems, such as a space mission to a distant planet, in simulated environments and operations, and scientists can conduct numerical experiments involving, for example, nuclear blasts. In the area of simulation-based design, engineers are using parallel computing and physics-based modeling to simulate fire propagation in engineering systems or the turbulent flow of a jet engine with molecular motion and modeling. These computer and numerical advancements are not limitless, as the increasing computational requirements lead to what is termed *transcomputational problems* capped by *Bremermann's limit* (Bremermann 1962). The nature of such transcomputational problems can be studied by the theory of computational complexity (Garey and Johnson 1979). Bremermann's limit was estimated based on quantum theory using the following proposition (Bremermann 1962):

> No data processing systems, whether artificial or living, can process more than 2×10^{47} bits per second per gram of its mass.

Data processing here is defined as transmitting bits over one or several channels of a system. Klir and Folger (1988) provide additional information on the theoretical basis for this proposition. Consider a hypothetical computer that has the entire mass of the Earth (6×10^{27} g) and operates for a time period equal to the estimated age of the Earth (3.14×10^{17} sec). This imaginary computer would be able to process 2.56×10^{92} bits or, rounded to the nearest power of 10, 10^{93} bits, defining Bremermann's limit. Many scientific and engineering problems defined with more details can exceed this limit. Klir and Folger (1988) provide the examples of pattern recognition and human vision that can easily reach transcomputational levels. In pattern recognition, consider a square $q \times q$ spatial array defining $n = q^2$ cells that partition the recognition space. Pattern recognition often involves color. Using k colors, for example, the number of possible color patterns within the space is k^n. In order to stay within Bremermann's limit, the following inequality must be met:

$$k^n \leq 10^{93} \tag{3.23}$$

Figure 3.31 shows a plot of this inequality for values of $k = 2$–10 colors. For example, using only two colors, a transcomputational state is reached at $q \geq 18$ colors. These computations in pattern recognition can be directly related to human vision and the complexity associated with processing information by the retina of a human eye. According to Klir and Folger (1988), if we consider a retina of about 1 million cells, with each cell having only two states of *active* and *inactive* in recognizing an object, modeling the retina in its entirety would require the processing of the following bits of information, far beyond Bremermann's limit:

$$2^{1,000,000} = 10^{300} \tag{3.24}$$

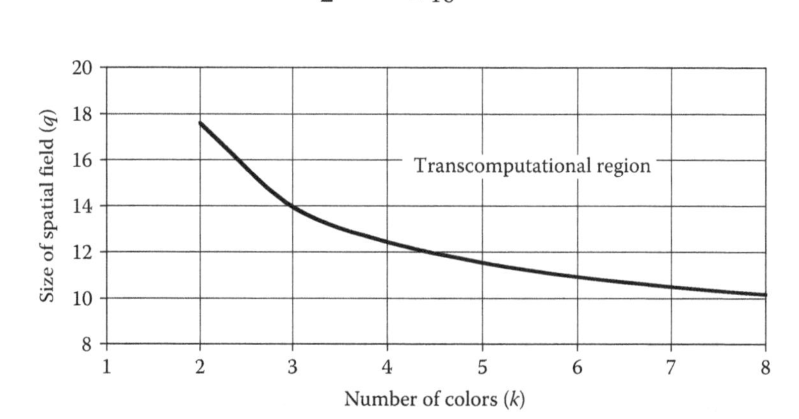

FIGURE 3.31
Bremermann's limit for pattern recognition.

Generally, an engineering system should be modeled with portions of its environment that interact significantly with it in order to assess some system attributes of interest. The level of interaction with the environment can only be subjectively assessed. The complexity of the system model increases along with the size of the environment and level of detail, possibly in a manner that does not have a recognizable or observable structure. This complexity without structure is difficult to model and deal with in engineering and sciences. By increasing the complexity of the system model, our ability to make *relevant* assessments regarding the attributes of the system can diminish, thus presenting a trade-off between relevance and precision in system modeling. Our goal should be to model a system with a level of detail sufficient to result in adequate precision that will lead to relevant decisions in order to meet the objectives of the system assessment.

Living systems show signs of these trade-offs between precision and relevance in order to deal with complexity. The survival instincts of living systems have evolved and manifest themselves as processes to cope with complexity and information overload. The ability of a living system to make relevant assessments diminishes with increases in information input, as discussed by Miller (1978). Living systems commonly need to process information in a continuous manner in order to survive. For example, a fish needs to process visual information constantly in order to avoid being eaten by another fish. When a school of larger fish rushes toward the fish, presenting it with multiple images of threats and dangers, the fish might not be able to process all of the information and can become confused. By considering the information processing capabilities of living systems to be input–output black boxes, the input and output to such systems can be measured and plotted in order to examine such relationships and any nonlinear characteristics that they might exhibit. Miller (1978) described these relationships for living systems using the following hypothesis, which was analytically modeled and experimentally validated:

> As the information input to a single channel of a living system—measured in bits per second—increases, the information output—measured similarly—increases almost identically at first but gradually falls behind as it approaches a certain output rate, the channel capacity, which cannot be exceeded. The output then levels off at that rate, and, finally, as the information input rate continues to go up, the output decreases gradually towards zero as breakdown or the confusion state occurs under overload.

This hypothesis was used to construct families of curves to represent the effects of information input overload, as shown in Figure 3.32. Once the input overload is removed, most living systems recover instantly from the overload and the process is completely reversible; however, if the energy level of the input is much larger than the channel capacity, a living system might not fully recover from this input overload. Living systems also adjust the way they process information in order to deal with an information input overload using one or more of the following processes to varying degrees, depending on the complexity of the living system: (1) *omission*, by failing to transmit information; (2) *error*, by transmitting information incorrectly; (3) *queuing*, by delaying transmission; (4) *filtering*, by giving priority in processing; (5) *abstracting*, by processing messages with less than complete details; (6) *multiple channel processing*, by simultaneously transmitting messages over several parallel channels; (7) *escape*, by acting to cut off information input; and (8) *chunking*, by grouping information in meaningful chunks. These actions can also be viewed as simplifying means to cope with complexity and/or an information input overload.

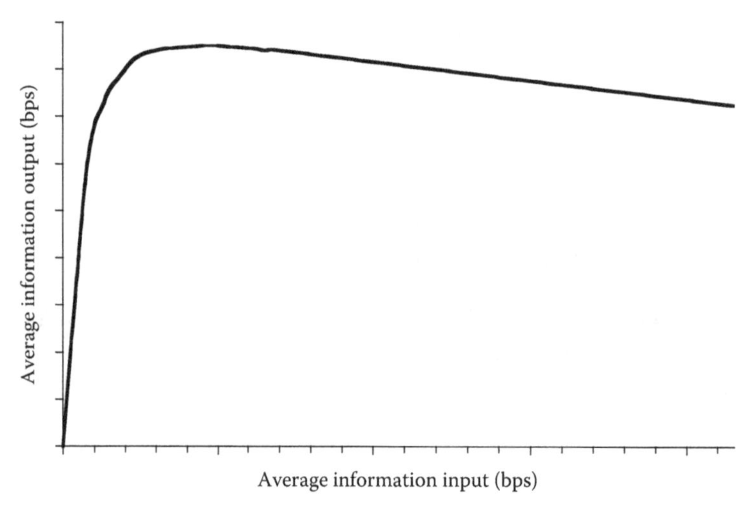

FIGURE 3.32
Relationship of input and output information transmission rates for living systems.

3.6 Exercise Problems

Problem 3.1 Provide example performance and functional requirements for an office building. Develop portions of a work breakdown structure for an office building.

Problem 3.2 Provide example performance and functional requirements for a residential house. Develop portions of a work breakdown structure for a house.

Problem 3.3 Develop and discuss a systems engineering process for a low-income townhouse as an engineering product.

Problem 3.4 Construct a contributing factor diagram based on the influence diagram of a protection system of a hurricane-prone region provided in Figure 3.13. You may make any necessary assumptions.

Problem 3.5 Construct an influence diagram based on the contributing factor diagram for bridge replacement in Figure 3.7. You may make any necessary assumptions.

Problem 3.6 Develop and discuss the life cycle of a major highway bridge as an engineering system.

Problem 3.7 Describe three engineering systems that can be modeled using the black-box method. What are the inputs and outputs for each system?

Problem 3.8 Describe three natural systems that can be modeled using the black-box method. What are the inputs and outputs for each system?

Problem 3.9 Describe three engineering systems that can be modeled using the state-based method. What are the states for each system?

Problem 3.10 Describe three natural systems that can be modeled using the state-based method. What are the states for each system?

Problem 3.11 A textile company is considering three options for managing its sales operation in the textile business:

D1: local or national production facilities

D2: international or foreign production facilities

D3: combination of local and internationally production facilities

The cost of the decision depends on future demand on its textile products. The annual costs for each decision alternative for three levels of demand (in US$) are as follows:

	Future Demand State		
Decision Alternative	**High (S_1)**	**Medium (S_2)**	**Low (S_3)**
D1	$500,000	$550,000	$450,000
D2	$450,000	$300,000	$800,000
D3	$350,000	$400,000	$650,000

The company estimated the probability of S_3 to be three times that of S_2, and the probability of S_1 to be equal to that of S_2.

1. Construct a decision tree for this decision situation showing the probability values and cost values in a graphical representation.
2. What is the recommended strategy using the expected value approach?

Problem 3.12 A computer company is in the process of selecting the best location for its headquarters in Cairo. After careful research and study, the company decision makers developed four decision alternatives based on four locations as follows:

D1: location A

D2: location B

D3: location C

D4: location D

The success of an alternative depends on the economic and market situation. Three market states yield the following profits (in US$) to the company:

	Market State		
Strategy	**Weak (S_1)**	**Average (S_2)**	**Strong (S_3)**
D1	$10,000	$15,000	$14,000
D2	$8,000	$18,000	$12,000
D3	$6,000	$16,000	$21,000
D4	$9,000	$16,000	$14,000

The computer company estimated the probability of S_1 to be the same as that of S_3 and to be twice that of S_2.

1. Construct a decision tree for this decision situation showing the probability values and profit values in a graphical representation.
2. What is the recommended strategy using the expected value approach?

Problem 3.13 An organization is in the process of restructuring its management systems. The top managers asked the systems manager to help in choosing the best design for the new structure, which should improve the performance and increase the success likelihood of the organization. After careful research and study, two systems designs were proposed as follows:

D1: flat organizational structure

D2: matrix organizational structure

The success of the selection process depends on determining employee satisfaction, which can be related to smoothness of work within the organization. The two possible satisfaction levels yield the following costs (in £) to the organization:

	Satisfaction Levels	
Strategy	**High (S_1)**	**Low (S_2)**
D1	£25,000	£45,000
D2	£30,000	£30,000

The organization assessed the probabilities of satisfaction and found the probability of S_1 to be 0.35.

1. Construct a decision tree for this decision situation showing the probability values and costs values in a graphical representation.
2. Why should the systems manager recommend using the expected value approach?

Problem 3.14 An engineer inspects a piece of equipment and estimates the probability of the equipment running at peak efficiency to be 75%. He/she then receives a report that the operating temperature of the machine has exceeded an 80°C critical level. Past records of operating performance suggest that the probability of exceeding the temperature of 80°C when the machine is working at peak efficiency is 0.3. Also, the probability of the temperature being exceeded if the machine is not working at peak efficiency is 0.8.

1. Revise the engineer's initial probability estimate based on this additional information from past records.
2. Draw a probability tree for this situation.

Problem 3.15 A company's sales manager estimates that, for the coming year, the probability of having a high sales level is 0.2, the probability of having a medium sales level is 0.7, and the probability of having a low sales level is 0.1. The manager requested and received a sales forecast report from the company's forecasting unit suggesting a high sales level next year. The track record of the forecasting unit of the company was used to assess the following probabilities: the probability of high sales forecast given that the market will generate high sales is 0.9; the probability of high sales forecast given that the market will generate medium sales is 0.6; and the probability of high sales forecast given that the market will generate low sales is 0.3. Revise the sales manager's initial estimates of the probability of

1. High sales
2. Medium sales
3. Low sales

Draw a probability tree associated with this situation.

Problem 3.16 Use the following state-transition diagram to construct a transition rate matrix and solve for the fractions of times for the system being in the two states:

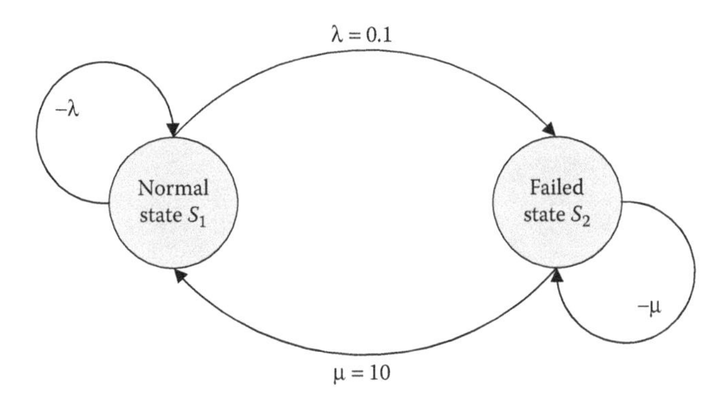

Problem 3.17 Use Table 3.9 to define two system states of S_1 (both components functioning properly) and S_2 (one or more components not functioning properly) to construct a state-transition diagram, assuming that the failure rates (λ) for both components are the same and the repair rates (μ) for both components are also the same. Construct a rate transition matrix assuming $\lambda = 10$ and $\mu = 90$. Compute the system availability based on the assumption that the system is available if one or more components are functioning properly.

Problem 3.18 Use Table 3.9 to define three system states of S_1 (both components functioning properly), S_2 (one of the components functioning properly), and S_3 (both components not functioning properly) to construct a state-transition diagram, assuming that the failure rates (λ) for both components are the same and the repair rates (μ) for both components are also the same. Construct a rate transition matrix assuming $\lambda = 10$ and $\mu = 90$. Compute the system availability based on the assumption that the system is available if one or more components are functioning properly.

Problem 3.19 Use the following state-transition diagram to construct a transition rate matrix and solve for the fractions of times for the system being in the four states:

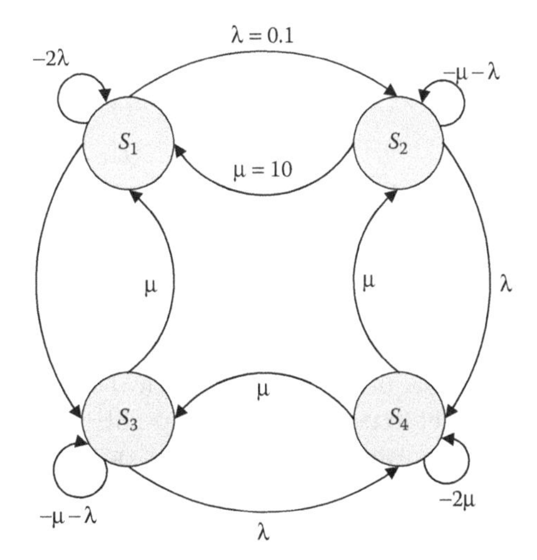

Problem 3.20 Use the following state-transition diagram to construct a transition probability matrix and solve for the fractions of times for the system being in the four states:

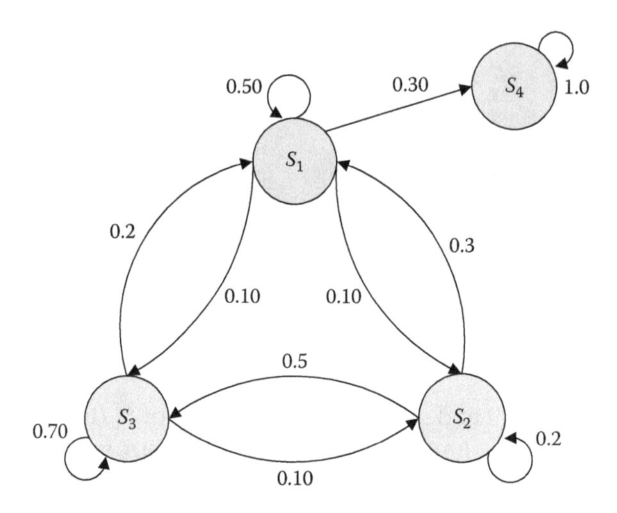

Problem 3.21 Use the following state-transition diagram to construct a transition probability matrix and solve for the fractions of times for the system being in the four states:

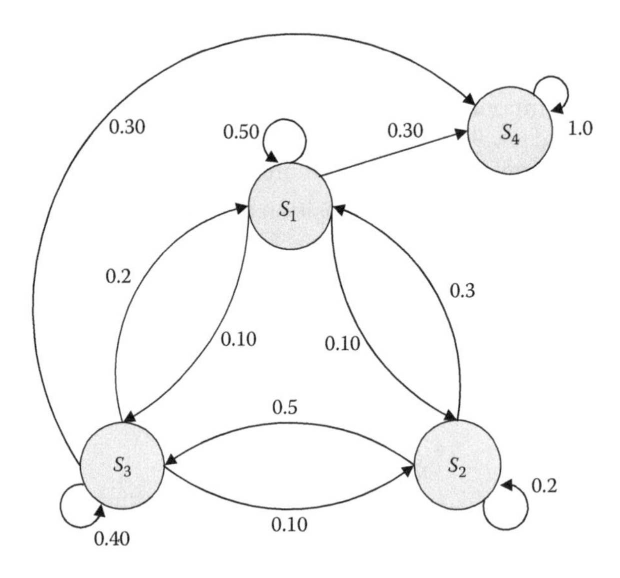

Problem 3.22 Reproduce the sequence of mathematical calculations of Example 3.10 and construct a spreadsheet to perform the computations for all the cases.

Problem 3.23 Reevaluate Example 3.10 by locking the variables as follows for a patient: has not visited Asia, non-smoker, negative X-ray, and negative dyspnea. Compare the results of the example and discuss.

Problem 3.24 Reevaluate Example 3.10 by locking the variables as follows for a patient: visited Asia, smoker, positive X-ray, and positive dyspnea. Compare the results of the example and discuss.

Problem 3.25 Reevaluate Example 3.10 using different probability values as follows:

		Probability of Dyspnea	
Tuberculosis or Cancer	**Bronchitis**	**Present**	**Absent**
True (TC)	Present (B)	0.95	0.05
True (TC)	Absent ($\bar{B}$)	0.80	0.20
False ($\overline{TC}$)	Present (B)	0.90	0.10
False ($\overline{TC}$)	Absent ($\bar{B}$)	0.05	0.95

Problem 3.26 Reevaluate Example 3.10 using different probability values as follows:

		Probability of Dyspnea	
Tuberculosis or Cancer	**Bronchitis**	**Present**	**Absent**
True (TC)	Present (B)	0.80	0.20
True (TC)	Absent ($\bar{B}$)	0.60	0.40
False ($\overline{TC}$)	Present (B)	0.70	0.30
False ($\overline{TC}$)	Absent ($\bar{B}$)	0.25	0.75

Problem 3.27 Build an information-based hierarchical system definition for an office building by defining the source system, data system, generative system, structure system, and metasystem.

Problem 3.28 Repeat Problem 3.15 for a highway bridge.

Problem 3.29 Repeat Problem 3.15 for a residential house.

Problem 3.30 Provide engineering examples of structured and unstructured complexity.

Problem 3.31 Provide science examples of structured and unstructured complexity.

Problem 3.32 Provide two cases of transcomputational problems. Explain why they are transcomputational in nature.

4

Failure Probability Assessment

This chapter covers the methods for assessing the failure probability of components and systems. The failure probability can be computed based on the survival events defined as the complements of the respective underlying failure events using survival probability computations or reliability analysis. Both analytical and empirical methods can be used for this purpose at the component level or the multicomponent discretized system level. This chapter discusses how to select an appropriate method, compute the probabilities of interest, and model time-variant reliability and correlations among basic random variables, and provides illustrative examples. It has a primary learning objective to develop basic skills for modeling the performance of components and systems as a basis to quantify nonperformance probabilities.

CONTENTS

4.1 Introduction

The reliability of an engineering system can be defined as its ability to fulfill its design purpose and as performance requirements, for some time period and under particular environmental conditions. The theory of probability provides the fundamental bases by which to quantify or measure this ability and for the development of reliability and hazard functions and computing failure or nonperformance probabilities. The reliability assessment methods can be based on analytical strength-and-load performance functions, generally described as supply-and-demand performance functions or empirical life data, and can be used to compute the reliability for a given set of conditions that are either time invariant or time dependent. For qualitative and/or preliminary risk analysis, reliability data reported in the literature for similar systems can be used as discussed in Chapter 8.

The reliability of a component or system can be assessed by the probability of meeting satisfactory performance requirements according to some performance functions under specific service and extreme conditions within a stated time period. In estimating this probability, component and system uncertainties are modeled using random variables with mean values, variances, and probability distribution functions or using nonparametric methods.

The objective of this chapter is to introduce failure probability and reliability assessment methods for components and systems that are based on either analytical models or empirical data with some discussion of Bayesian methods. These methods are needed to determine failure rates and hazard functions, which can be applied in decision and problem-solving techniques, such as economic and trade-off studies. Also, such assessments can be fed into the risk analysis and management process as the failure probabilities for defining and quantifying risks.

4.2 Analytical Performance-Based Assessment Methods

Several methods are available for reliability and failure probability assessment purposes that are based on strength-and-load performance functions, such as moment methods including the first-order reliability moment (FORM) method, the first-order second moment (FOSM) method, the advanced second moment (ASM) method, and the computer-based Monte Carlo simulation (MCS) methods. In this section, the two types of probabilistic methods for performance-based assessment are described: (1) the moment methods and (2) the MCS methods using direct, conditional expectation (CE) and importance sampling (IS).

4.2.1 Performance Function and Integration

The reliability of an element of a system can be determined based on the performance function that can be expressed in terms of basic random variables (X_i) for relevant loads and structural strength. Mathematically, the performance function Z can be described as follows using consistent units:

$$Z = Z(X_1, X_2, \ldots, X_n) = \text{supply} - \text{demand} \tag{4.1a}$$

or

$$Z = Z(X_1, X_2, \ldots, X_n) = \text{structural strength} - \text{load effect} \tag{4.1b}$$

or

$$Z = Z(X_1, X_2, \ldots, X_n) = S - L \tag{4.1c}$$

where:

Z is called the *performance function* of interest
S is the strength or supply
L is the load or demand, as illustrated in Figure 4.1

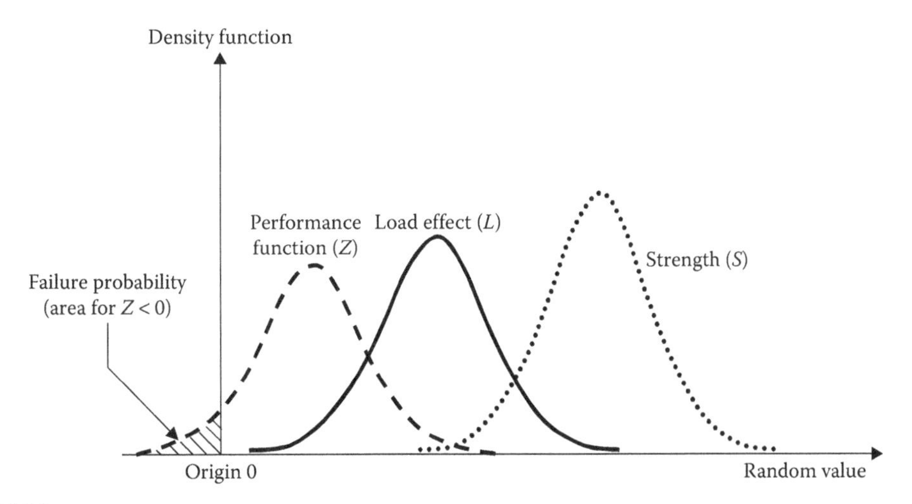

FIGURE 4.1
Performance function for reliability assessment.

The failure surface (or the *limit state*) of interest can be defined for the case of $Z = 0$. Accordingly, when $Z < 0$, the element or component or system is in the failure state according to the performance function, and when $Z > 0$, it is in the survival state. If the joint probability density function (PDF) for the basic random variables (X_i) is $f_{X_1, X_2, \ldots, X_n}(x_1, x_2, \ldots, x_n)$, then the failure probability, P_f, of the element can be obtained by the integral:

$$P_f = \int \cdots \int f_{X_1, X_2, \ldots, X_n}(x_1, x_2, \ldots, x_n) dx_1\, dx_2 \ldots dx_n \tag{4.2}$$

where the integration is performed over the region in which $Z < 0$. In general, the joint PDF is not explicitly or fully known, and the integral is a formidable task. For practical purposes, alternate methods for evaluating P_f are necessary. Reliability is assessed as 1 minus the failure probability.

4.2.2 Moment Methods and Failure Probability

This section covers the moment methods for computing reliability according to a performance function. The section starts with the fundamental case that provides the reliability in terms of a reliability index and a probability, and progressively introduces complexities to the performance function, distribution types and correlations, and the corresponding methods to deal with these complexities for computing the reliability index and the probability.

4.2.2.1 Reliability Index

Instead of using direct integration, as given by Equation 4.2, the performance function Z of Equation 4.1 can be expanded using a Taylor series about the mean value of the basic random variables and then truncated at the linear terms (Ayyub and McCuen 2011). Therefore, the first-order approximate mean and variance of Z can be shown, respectively, as

$$\mu_Z \approx Z(\mu_{X_1}, \mu_{X_2}, \ldots, \mu_{X_n}) \tag{4.3}$$

and

$$\sigma_Z^2 \approx \sum_{i=1}^{n} \sum_{j=1}^{n} \frac{\partial Z}{\partial X_i} \frac{\partial Z}{\partial X_j} \mathrm{Cov}(X_i, X_j) \tag{4.4a}$$

where:

μ is the mean of a random variable
$\mathrm{Cov}(X_i, X_j)$ is the covariance of X_i and X_j
μ_Z is the mean of Z
σ_Z^2 is the variance of Z

The partial derivatives of $\partial Z/\partial X_i$ are evaluated at the mean values of the basic random variables. For uncorrelated random variables, the variance expression can be simplified as

$$\sigma_Z^2 = \sum_{i=1}^{n} \sigma_{X_i}^2 \left(\frac{\partial Z}{\partial X_j} \right)^2 \tag{4.4b}$$

A measure of reliability can be estimated by introducing the *reliability index*, β, which is based on the mean and standard deviation of Z as

$$\beta = \frac{\mu_Z}{\sigma_Z} \tag{4.5}$$

The reliability index according to Equation 4.5 is a measure of the mean margin of safety in units of σ_Z. If Z is assumed to be normally distributed, then it can be shown that the failure probability P_f is

$$P_f = 1 - \Phi(\beta) \tag{4.6}$$

where Φ is the cumulative distribution function (CDF) of the standard normal variate as provided in Appendix A. The procedure of Equations 4.3 through 4.6 produces exact results when performance function Z is linear and normally distributed. For the performance function of Equation 4.1c, the limit state of $Z = 0$ can be expressed as shown in Figure 4.2, and the reliability index for uncorrelated random variables is given by

$$\beta = \frac{\mu_Z}{\sigma_Z} = \frac{\mu_S - \mu_L}{\sqrt{\sigma_S^2 + \sigma_L^2}} \tag{4.7a}$$

For the case of lognormally distributed S and L, the failure probability in terms of the reliability index can be derived as

$$P_f = \Phi(-\beta) = \Phi\left\{\frac{-\ln\left(\dfrac{\mu_L}{\mu_S}\sqrt{\dfrac{\delta_S^2 + 1}{\delta_L^2 + 1}}\right)}{\sqrt{\ln\left[(\delta_S^2 + 1)(\delta_L^2 + 1)\right]}}\right\} \tag{4.7b}$$

where:

 δ is the coefficient of variation (COV) = σ/μ

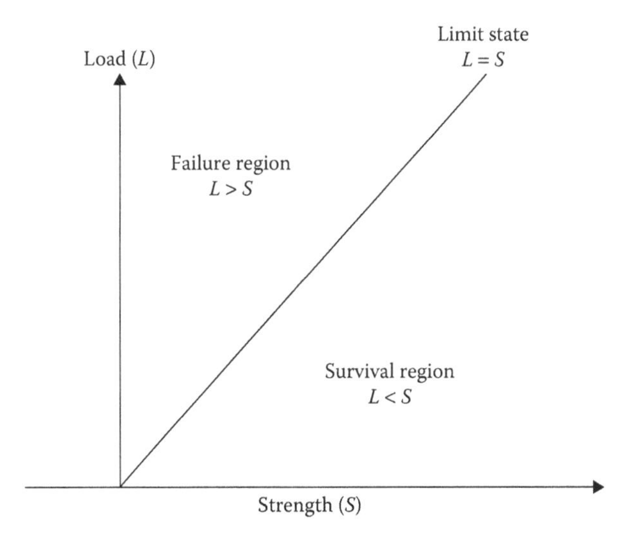

FIGURE 4.2
Performance function for a linear, two-random variable case.

The derivation of this case of lognormally distributed S and L is best achieved by defining Z as S/L and failure when $Z \leq 1$.

4.2.2.2 Nonlinear Performance Functions and the Advanced Second Moment Method

For nonlinear performance functions, the Taylor series expansion of Z is linearized at some point on the failure surface referred to as the *design point* or *checking point* or *the most likely failure point* rather than at the mean which was the case for linear Z. Assuming that the basic random variables (X_i) are uncorrelated, the following transformation to reduced or normalized coordinates can be used:

$$Y_i = \frac{X_i - \mu_{X_i}}{\sigma_{X_i}} \tag{4.8}$$

If the X_i are correlated, they must be transformed to uncorrelated random variables, that is, an orthogonal coordinate system, as described in Section 4.2.2.4. It can be shown that the reliability index, β, is the shortest distance to the failure surface from the origin in the reduced Y-coordinate system, that is, the normalized coordinate system according to Equation 4.8. The shortest distance is illustrated in Figure 4.3 using the performance function of Equation 4.1c, which, in the reduced coordinates, becomes

$$Y_L = \frac{\sigma_S}{\sigma_L} Y_S + \frac{\mu_S - \mu_L}{\sigma_L} \tag{4.9}$$

where:

Y is the reduced coordinate of a random variable according to Equation 4.8

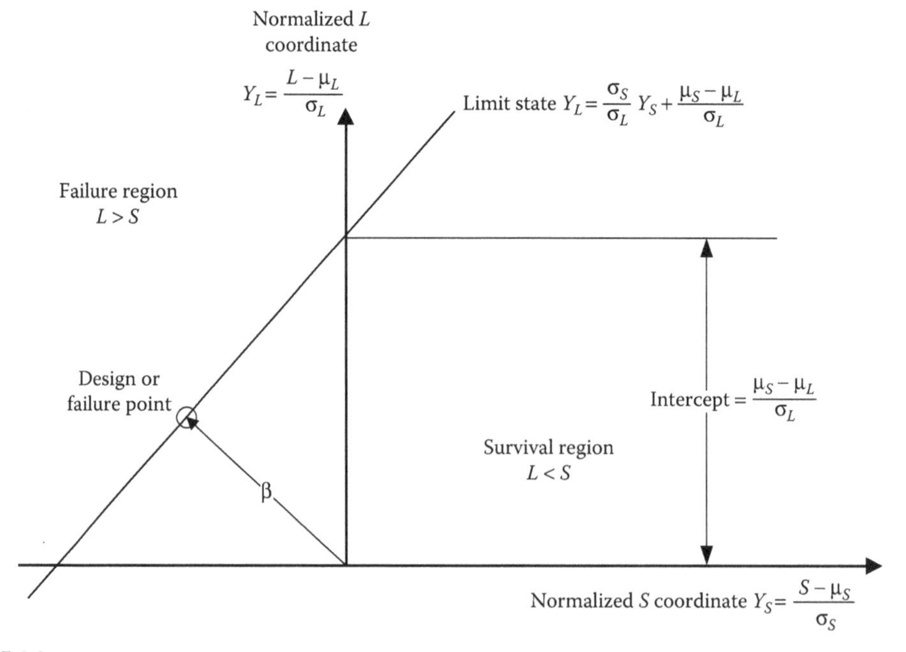

FIGURE 4.3
Performance function for a linear, two-random variable case in normalized coordinates. L, load; S, strength.

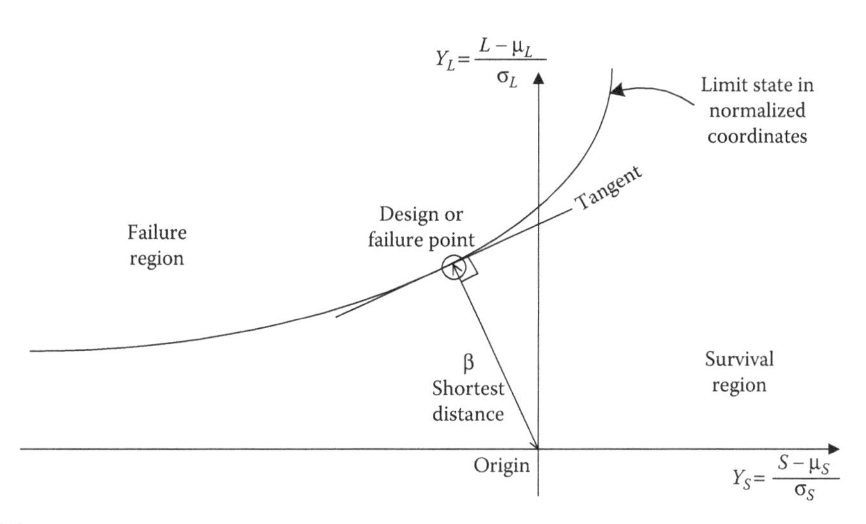

FIGURE 4.4
Performance function for a nonlinear, two-random variable case in normalized coordinates. *L*, load; *S*, strength.

The shortest distance from the origin to the line of Equation 4.9 is shown in Figure 4.3. The point on the failure surface that corresponds to the shortest distance is the most likely failure point. The concept of the shortest distance applies also to a nonlinear performance function, as shown in Figure 4.4. Using the original *X*-coordinate system, the reliability index, β, and the design point $(x_1^*, x_2^*, \ldots, x_n^*)$ can be determined by solving the following system of nonlinear equations iteratively for β called the ASM method:

$$\alpha_i = \frac{\left(\frac{\partial Z}{\partial X_i}\right)\sigma_{X_i}}{\left[\sum_{i=1}^{n}\left(\frac{\partial Z}{\partial X_i}\right)^2 \sigma_{X_i}^2\right]^{1/2}} \tag{4.10}$$

$$x_i^* = \mu_{X_i} - \alpha_i \beta \sigma_{X_i} \tag{4.11}$$

$$Z(x_1^*, x_2^*, \ldots, x_n^*) = 0 \tag{4.12a}$$

or

$$Z(\mu_{X_1} - \alpha_1\beta\sigma_{X_1}, \mu_{X_2} - \alpha_2\beta\sigma_{X_2}, \ldots, \mu_{X_n} - \alpha_n\beta\sigma_{X_n}) = 0 \tag{4.12b}$$

where α_i is the directional cosine and $0 \leq |\alpha_i| \leq 1$, and the partial directives are evaluated at the design point. Once β is obtained, Equation 4.6 can be used to evaluate P_f. The above formulation, however, is limited to normally distributed random variables. In reliability assessment, the directional cosines can be viewed as measures of the importance of the corresponding random variables in determining the reliability index β; the greater the value of $|\alpha_i|$, the greater the importance. Also, partial safety factors (γ) applied to respective mean values that are used in load and resistance factor design (LRFD) can be computed as follows:

$$\gamma = \frac{x^*}{\mu_X} \tag{4.13}$$

In general, partial safety factors take on values >1 for the load variables (in this case, they are called load amplification factors) and values <1 for strength variables (in this case, they are called strength reduction factors).

4.2.2.3 Equivalent Normal Distributions

If a random variable X is not normally distributed, the use of Equations 4.10 through 4.12 requires transforming this variable to an equivalent normally distributed random variable. The parameters of the equivalent normal distribution, $\mu_{X_i}^N$ and σ_X^N, can be estimated by imposing two conditions: the CDFs and PDFs of a non-normal random variable and its equivalent normal variable should be equal at the design point on the failure surface. The first condition can be expressed as

$$\Phi\left(\frac{x_i^* - \mu_{X_i}^N}{\sigma_{X_i}^N}\right) = F_i(x_i^*) \tag{4.14a}$$

The second condition is

$$\phi\left(\frac{x_i^* - \mu_{X_i}^N}{\sigma_{X_i}^N}\right) = f_i(x_i^*) \tag{4.14b}$$

where:

F_i is the non-normal CDF
f_i is the non-normal PDF
Φ is the CDF of standard normal variate
ϕ is the PDF of the standard normal variate

The standard deviation and mean of equivalent normal distributions can be shown, respectively:

$$\sigma_{X_i}^N = \frac{\phi\{\Phi^{-1}[F_i(x_i^*)]\}}{f_i(x_i^*)} \tag{4.15a}$$

and

$$\mu_{X_i}^N = x_i^* - \Phi^{-1}[F_i(x_i^*)]\sigma_{X_i}^N \tag{4.15b}$$

Having determined $\sigma_{X_i}^N$ and $\mu_{X_i}^N$ for each random variable, β can be solved using the same procedure of Equations 4.10 through 4.12. It should be noted that the values of $\sigma_{X_i}^N$ and $\mu_{X_i}^N$ are computed at the failure point (x_i^*) using Equation 4.15a and 4.15b. In an iterative solution, as the failure point is updated in each iteration, these $\sigma_{X_i}^N$ and $\mu_{X_i}^N$ values should be reevaluated.

For example, let us consider the case of the lognormal distribution where f_i is provided by Equation A.65, and F_i can be evaluated as follows:

$$F_i(x_i^*) = \Phi\left[\frac{\ln(x_i^*) - \mu_{Y_i}}{\sigma_{Y_i}}\right] \tag{4.16}$$

where σ_{Y_i} and μ_{Y_i} are the parameters of the lognormal distribution that can be computed based on its mean and variance as provided by Equations A.66 and A.67, respectively. Substituting Equation 4.16 into Equation 4.15a and 4.15b produces the following respective equations noting that Φ^{-1} cancels Φ:

$$\sigma_{X_i}^N = \frac{\phi\left[\dfrac{\ln(x_i^*) - \mu_{Y_i}}{\sigma_{Y_i}}\right]}{f_i(x_i^*)} \tag{4.17a}$$

and

$$\mu_{X_i}^N = x_i^* - \left[\frac{\ln(x_i^*) - \mu_{Y_i}}{\sigma_{Y_i}}\right]\sigma_{X_i}^N \tag{4.17b}$$

The expression of the standard normal density function (φ) is

$$\phi(z) = \frac{1}{\sqrt{2\pi}}\exp(-0.5z^2) \tag{4.18a}$$

or

$$\phi\left[\frac{\ln(x_i^*) - \mu_{Y_i}}{\sigma_{Y_i}}\right] = \frac{1}{\sqrt{2\pi}}\exp\left[-0.5\left(\frac{\ln(x_i^*) - \mu_{Y_i}}{\sigma_{Y_i}}\right)^2\right] \tag{4.18b}$$

and

$$\sigma_{X_i}^N = \frac{\dfrac{1}{\sqrt{2\pi}}\exp\left[-0.5\left(\dfrac{\ln(x_i^*) - \mu_{Y_i}}{\sigma_{Y_i}}\right)^2\right]}{\dfrac{1}{x_i^*\sigma_{Y_i}\sqrt{2\pi}}\exp\left[-0.5\left(\dfrac{\ln(x_i^*) - \mu_{Y_i}}{\sigma_{Y_i}}\right)^2\right]} = x_i^*\sigma_{Y_i} \tag{4.18c}$$

The ASM method is capable of dealing with nonlinear performance functions and non-normal probability distributions; however, the accuracy of the solution and the convergence of the procedure depend on the nonlinearity of the performance function in the vicinity of the design point and the origin. If there are several local minimum distances to the origin, the solution process may not converge onto the global minimum. The probability of failure is calculated from the reliability index β using Equation 4.6, which is based on normally distributed performance functions. Therefore, the resulting failure probability, P_f, based on the ASM, is approximate except for linear performance functions because it does not account for any nonlinearity in the performance functions.

4.2.2.4 Correlated Random Variables

Reliability analysis of some components and systems require the use of correlated basic random variables, such as angle of internal friction and cohesion for soil layers, when assessing the reliability of gravity structures. In this section, this correlation is assumed to be characterized in terms of bivariate correlation, that is, between pairs of random variables. Also, correlated random variables are assumed to be normally distributed because non-normal and correlated random variables require additional information, such as their joint PDF or conditional distributions, for their unique and full definition. Such joint probability information is usually unavailable and is difficult to obtain. A correlated (and normal) pair of random variables X_1 and X_2 with a correlation coefficient ρ can be transformed into noncorrelated, that is, orthogonal, pair Y_1 and Y_2 by solving for two eigenvalues and the corresponding eigenvectors, as follows:

$$Y_1 = \frac{1}{2t}\left(\frac{X_1 - \mu_{X_1}}{\sigma_{X_1}} + \frac{X_2 - \mu_{X_2}}{\sigma_{X_2}}\right) \tag{4.19a}$$

$$Y_2 = \frac{1}{2t}\left(\frac{X_1 - \mu_{X_1}}{\sigma_{X_1}} - \frac{X_2 - \mu_{X_2}}{\sigma_{X_2}}\right) \tag{4.19b}$$

where:
$t = \sqrt{0.5}$

The resulting Y variables are not correlated with respective variances that are equal to the eigenvalues (λ) as follows:

$$\sigma_{Y_1}^2 = \lambda_1 = 1 + \rho \tag{4.20a}$$

$$\sigma_{Y_2}^2 = \lambda_2 = 1 - \rho \tag{4.20b}$$

For a correlated pair of random variables, Equations 4.10 and 4.11 have to be revised, respectively, to the following:

$$\alpha_{Y_1} = \frac{\left[\left(\frac{\partial Z}{\partial X_1}\right)t\sigma_{X_1} + \left(\frac{\partial Z}{\partial X_2}\right)t\sigma_{X_2}\right]\sqrt{1+\rho}}{\left[\left(\frac{\partial Z}{\partial X_1}\right)^2 \sigma_{X_1}^2 + \left(\frac{\partial Z}{\partial X_2}\right)^2 \sigma_{X_2}^2 + 2\rho\left(\frac{\partial Z}{\partial X_1}\right)\left(\frac{\partial Z}{\partial X_2}\right)\sigma_{X_1}\sigma_{X_2}\right]^{1/2}} \tag{4.21a}$$

$$\alpha_{Y_2} = \frac{\left[\left(\frac{\partial Z}{\partial X_1}\right)t\sigma_{X_1} - \left(\frac{\partial Z}{\partial X_2}\right)t\sigma_{X_2}\right]\sqrt{1-\rho}}{\left[\left(\frac{\partial Z}{\partial X_1}\right)^2 \sigma_{X_1}^2 + \left(\frac{\partial Z}{\partial X_2}\right)^2 \sigma_{X_2}^2 + 2\rho\left(\frac{\partial Z}{\partial X_1}\right)\left(\frac{\partial Z}{\partial X_2}\right)\sigma_{X_1}\sigma_{X_2}\right]^{1/2}} \tag{4.21b}$$

and

$$x_1^* = \mu_{X_1} - \sigma_{X_1} t\beta(\alpha_{Y_1}\sqrt{\lambda_1} + \alpha_{Y_2}\sqrt{\lambda_2}) \tag{4.22a}$$

$$x_2^* = \mu_{X_2} - \sigma_{X_2} t\beta(\alpha_{Y_1}\sqrt{\lambda_1} - \alpha_{Y_2}\sqrt{\lambda_2}) \tag{4.22b}$$

where the partial derivatives are evaluated at the design point.

4.2.2.5 Numerical Algorithms

A numerical algorithm presented in this section for the ASM method can be used to assess the reliability of a component or a system according to a linear or nonlinear performance function that may include non-normal random variables. Also, correlated random variables can be handled using this algorithm. Moreover, the performance function can be a closed or nonclosed expression. For nonclosed forms of the performance functions, the implementation of this method requires the use of efficient and accurate numerical algorithms in order to deal with these nonclosed forms. The ASM algorithm can be summarized by the following steps using two cases:

Case (a): noncorrelated random variables

1. Assign the mean value for each random variable as a starting design point value:

$$(x_1^*, x_2^*, \ldots, x_n^*) = (\mu_{X_1}, \mu_{X_2}, \ldots, \mu_{X_n})$$

2. Compute the standard deviation and mean of the equivalent normal distribution for each non-normal random variable using Equations 4.14 and 4.15.
3. Compute the partial derivative $\partial Z/\partial X_i$ of the performance function with respect to each random variable evaluated at the design point to use in Equation 4.10.
4. Compute the directional cosine, α_i, for each random variable, as given in Equation 4.10 at the design point.
5. Compute the reliability index, β, by substituting Equation 4.11 into Equation 4.12a and satisfying the limit state $Z = 0$ in Equation 4.12b using a numerical root-finding method such as the bisection method.
6. Compute a new estimate of the design point by substituting the resulting reliability index, β, obtained in step 5, into Equation 4.11.
7. Repeat steps 2–6 until the reliability index, β, converges within an acceptable tolerance, say 1%–5%.

Case (b): correlated random variables

1. Assign the mean value for each random variable as a starting design point value:

$$(x_1^*, x_2^*, \ldots, x_n^*) = (\mu_{X_1}, \mu_{X_2}, \ldots, \mu_{X_n})$$

2. Compute the standard deviation and mean of the equivalent normal distribution for each non-normal random variable using Equations 4.14 and 4.15.
3. Compute the partial derivative $\partial Z/\partial X_i$ of the performance function with respect to each noncorrelated random variable evaluated at the design point to use in Equation 4.10.
4. Compute the directional cosine, α_i, for each noncorrelated random variable as given in Equation 4.10 at the design point. For correlated pairs of random variables, Equation 4.21a and 4.21b should be used instead of Equation 4.10.
5. Compute the reliability index, β, by substituting Equations 4.11 (for noncorrelated random variables) and 4.22a and 4.22b (for correlated random variables) into Equation 4.12a and satisfying the limit state $Z = 0$ in Equation 4.12b using a numerical root-finding method such as the bisection method.

6. Compute a new estimate of the design point by substituting the resulting reliability index, β, obtained in step 5 into Equations 4.11 (for noncorrelated random variables) and 4.22a and 4.22b (for correlated random variables).
7. Repeat steps 2–6 until the reliability index, β, converges within an acceptable tolerance, say 1%–5%.

Example 4.1: Reliability Assessment Using a Nonlinear Performance Function

The strength–load performance function for a component is assumed to have the following form:

$$Z = X_1 X_2 - \sqrt{X_3}$$

where the X_1, X_2, and X_3 are basic random variables with the probabilistic characteristics in the following table:

Random Variable	Mean Value (μ)	Standard Deviation (σ)	COV	Distribution Type	
				Case (a)	Case (b)
X_1	1	0.25	0.25	Normal	Lognormal
X_2	5	0.25	0.05	Normal	Lognormal
X_3	4	0.80	0.20	Normal	Lognormal

Using the first-order reliability analysis based on the first-order Taylor series, the following can be obtained from Equations 4.3 through 4.5:

$$\mu_Z \cong 1 \times 5 - \sqrt{4} = 5 - 2 = 3$$

$$\sigma_Z \cong \sqrt{5^2(0.25)^2 + 1^2(0.25)^2 + (-0.5/\sqrt{4})^2(0.8)^2}$$
$$= \sqrt{1.5625 + 0.0625 + 0.04} = 1.2903$$

$$\beta \cong \frac{\mu_Z}{\sigma_Z} = \frac{3}{1.2903} = 2.325$$

These estimates are applicable to both cases (a) and (b). Using advanced second-moment reliability analysis, the following solutions can be developed for cases (a) and (b).

CASE (a)

Case (a): Iteration 1

Random Variable	Failure Point Set at the Mean Values	$\frac{\partial Z}{\partial X_i}\sigma_{X_i}$	Directional Cosines (α)
X_1	1.000E+00	1.250E+00	9.687E−01
X_2	5.000E+00	2.500E−01	1.937E−01
X_3	4.000E+00	−2.000E−01	−1.550E−01

The derivatives in the above table are evaluated at the failure point. The failure point in the first iteration is assumed to be the mean values of the random variables. The reliability index can be determined by solving for the root according to Equation 4.12a for the limit state of this example as follows:

$$Z = (\mu_{X_1} - \alpha_1\beta\sigma_{X_1})(\mu_{X_2} - \alpha_2\beta\sigma_{X_2}) - \sqrt{\mu_{X_3} - \alpha_3\beta\sigma_{X_3}} = 0$$

Therefore, solving for the root produces $\beta = 2.37735$ for this iteration that is used to update the failure point using Equation 4.11 as provided in the table below. Once the failure point is updated, the derivatives and directional cosines are updated, and a new root, that is, β, is determined.

Case (a): Iteration 2

Random Variable	Updated Failure Point per Equation 4.11	$\frac{\partial Z}{\partial X_i}\sigma_{X_i}$	Directional Cosines (α)
X_1	4.242E−01	1.221E+00	9.841E−01
X_2	4.885E+00	1.061E−01	8.547E−02
X_3	4.295E+00	−1.930E−01	−1.555E−01

Therefore, the second iteration produces $\beta = 2.3628$.

Case (a): Iteration 3

Random Variable	Updated Failure Point per Equation 4.11	$\frac{\partial Z}{\partial X_i}\sigma_{X_i}$	Directional Cosines (α)
X_1	4.187E−01	1.237E+00	9.846E−01
X_2	4.950E+00	1.047E−01	8.329E−02
X_3	4.294E+00	−1.930E−01	−1.536E−01

The process is repeated again to produce a third estimate of $\beta = 2.3628$, which indicates that β has converged to 2.3628. The failure probability is $1 - \Phi(\beta) = 0.009068$. The partial safety factors can be computed using Equation 4.13 as follows:

Random Variable	Updated Failure Point per Equation 4.13	Partial Safety Factors
X_1	0.418378	0.418378
X_2	4.950849	0.99017
X_3	4.290389	1.072597

CASE (b)

The parameters of the lognormal distribution can be computed for the three random variables based on their respective means (μ) and deviations (σ) using Equations A.66 and A.67 as follows:

$$\sigma_Y^2 = \ln\left[1 + \left(\frac{\sigma_X}{\mu_X}\right)^2\right] \text{ and } \mu_Y = \ln(\mu_X) - \frac{1}{2}\sigma_Y^2$$

The results of these computations are summarized as follows:

Random Variable	Distribution Type	First Parameter (μ_Y)	Second Parameter (σ_Y)
X_1	Lognormal	−0.03031231	0.24622068
X_2	Lognormal	1.608189472	0.04996879
X_3	Lognormal	1.366684005	0.20

Case (b): Iteration 1

Random Variable	Failure Point Set at the Mean Values	Equivalent Normal Standard Deviation	Equivalent Normal Mean Value	$\frac{\partial Z}{\partial X_i}\sigma_{X_i}^N$	Directional Cosines (α)
X_1	1.000E+00	2.462E−01	9.697E−01	1.231E+00	9.681E−01
X_2	5.000E+00	2.498E−01	4.994E+00	2.498E−01	1.965E−01
X_3	4.000E+00	7.922E−01	3.922E+00	−1.980E−01	−1.557E−01

The derivatives in the above table are evaluated at the failure point. The failure point in the first iteration is assumed to be the mean values of the random variables. The reliability index can be determined by solving for the root according to Equation 4.12b for the limit state of this example using the following equation:

$$Z = (\mu_{X_1}^N - \alpha_1\beta\sigma_{X_1}^N)(\mu_{X_2}^N - \alpha_2\beta\sigma_{X_2}^N) - \sqrt{\mu_{X_3}^N - \alpha_3\beta\sigma_{X_3}^N} = 0$$

Therefore, $\beta = 2.30530$ for this iteration that is used to update the failure point using Equation 4.11 as provided in the table below. Once the failure point is updated, the derivatives and directional cosines are updated, and a new root, that is, β, is determined.

Case (b): Iteration 2

Random Variable	Updated Failure Point per Equation (4.11)	Equivalent Normal Standard Deviation	Equivalent Normal Mean Value	$\frac{\partial Z}{\partial X_i}\sigma_{X_i}^N$	Directional Cosines (α)
X_1	4.202E−01	1.035E−01	7.718E−01	5.050E−01	9.118E−01
X_2	4.881E+00	2.439E−01	4.992E+00	1.025E−01	1.850E−01
X_3	4.206E+00	8.330E−01	3.912E+00	−2.031E−01	−3.667E−01

Therefore, the second iteration produces $\beta = 3.3224$.

Case (b): Iteration 3

Random Variable	Updated Failure Point per Equation 4.11	Equivalent Normal Standard Deviation	Equivalent Normal Mean Value	$\frac{\partial Z}{\partial X_i}\sigma_{X_i}^N$	Directional Cosines (α)
X_1	4.584E−01	1.129E−01	8.020E−01	5.465E−01	9.118E−01
X_2	4.843E+00	2.420E−01	4.991E+00	1.109E−01	1.850E−01
X_3	4.927E+00	9.758E−01	3.803E+00	−2.198E−01	−3.667E−01

The process is repeated again to produce a third estimate of $\beta = 3.3126$.

Case (b): Iteration 4

Random Variable	Updated Failure Point per Equation 4.11	Equivalent Normal Standard Deviation	Equivalent Normal Mean Value	$\frac{\partial Z}{\partial X_i}\sigma_{X_i}^N$	Directional Cosines (α)
X_1	4.612E−01	1.136E−01	8.041E−01	5.499E−01	9.118E−01
X_2	4.843E+00	2.420E−01	4.991E+00	1.116E−01	1.850E−01
X_3	4.989E+00	9.880E−01	3.789E+00	−2.212E−01	−3.667E−01

The process is repeated again to produce a fourth estimate of $\beta = 3.3125$.

Case (b): Iteration 5

Random Variable	Updated Failure Point per Equation 4.11	Equivalent Normal Standard Deviation	Equivalent Normal Mean Value	$\frac{\partial Z}{\partial X_i}\sigma_{X_i}^N$	Directional Cosines (α)
X_2	4.843E+00	2.420E−01	4.991E+00	1.116E−01	1.850E−01
X_1	4.612E−01	1.136E−01	8.041E−01	5.500E−01	9.118E−01
X_3	4.989E+00	9.880E−01	3.789E+00	−2.212E−01	−3.667E−01

The process is repeated again to produce a fifth estimate of $\beta = 3.3125$, which indicates that β has converged to 3.3125. The failure probability is $1 - \Phi(\beta) = 0.0004619$. The partial safety factors can be computed using Equation 4.13 as follows:

Random Variable	Updated Failure Point per Equation 4.13	Partial Safety Factors
X_1	0.461189	0.461189
X_2	4.843135	0.968627
X_3	4.988968	1.247242

It is evident from this example that selecting the distribution type can have a significant effect on the resulting failure probabilities.

4.2.3 Monte Carlo Simulation Methods

Monte Carlo Simulation (MCS) techniques are basically sampling processes that are used to estimate the failure probability of a component or system. The basic random variables in Equation 4.1 are randomly generated and substituted into Equation 4.1, and then the fraction of cases that resulted in failure are determined and used to estimate the failure probability. Three methods are described in this section: the direct MCS (DMCS), the CE method, and the IS variance reduction method.

4.2.3.1 Direct Monte Carlo Simulation Method

In the direct simulation method (also called simply MCS or brute force MCS), samples of the basic noncorrelated or correlated variables are randomly drawn according to their corresponding probabilistic characteristics and fed into the performance function Z as given by Equation 4.1. Assuming that N_f is the number of simulation cycles (called also trials, repeats, or iterations) for which $Z < 0$ in N simulation cycles, an estimate of the failure probability can be expressed as:

$$\overline{P}_f = \frac{N_f}{N} \tag{4.23}$$

The estimated failure (or unsatisfactory performance) probability $\overline{P}_f$ should approach the true value when N approaches infinity. The variance of the estimated failure probability can be approximately computed using the variance expression for a binomial distribution as follows (see Appendix A):

$$\text{Var}(\overline{P}_f) \approx \frac{(1 - \overline{P}_f)\overline{P}_f}{N} \tag{4.24}$$

Therefore, an estimate of the COV of the estimate failure probability is

$$\mathrm{COV}(\overline{P}_f) \approx \frac{1}{\overline{P}_f}\sqrt{\frac{(1-\overline{P}_f)\overline{P}_f}{N}} \tag{4.25}$$

These equations show that direct simulation can be computationally inefficient in some cases, especially for small failure probabilities, by observing that the smaller the failure probability, the greater the average number of cycles needed to result in a failure. In subsequent sections, other methods are described for the purpose of increasing the efficiency of simulation methods.

4.2.3.2 Conditional Expectation Simulation Method

The Conditional Expectation (CE) simulation method can be used to estimate the failure probability according to the performance function Z as described in Equation 4.1. The CE method requires generating all the basic random variables in Equation 4.1 except a random variable with high variability (i.e., high COV or large standard deviation) or that offers computational advantages, called a control variable, X_k. Sometimes the control variable is selected on the basis of being able to reduce the performance function to an analytically manageable form as required by the CE method. The CE is computed as the CDF of the control variable evaluated at the generated values of the remaining variables based on the condition $Z < 0$. The resulting CE is an estimate of the failure probability in each simulation cycle.

For the following fundamental performance function, two cases of control variables are examined concurrently:

$$Z = S - L \tag{4.26}$$

For a randomly generated value of L or S, the failure probability for each cycle is given, respectively, for the two cases of control variables, by the following equation:

$$P_{f_i} = F_S(l_i) \tag{4.27}$$

or

$$P_{f_i} = 1 - F_L(s_i) \tag{4.28}$$

In Equations 4.27 and 4.28, S and L are the control variables, respectively. The failure probability, P_f, can be estimated by the following equation for either case:

$$\overline{P}_f = \frac{\sum_{i=1}^{N} P_{f_i}}{N} \tag{4.29}$$

where:

N is the number of simulation cycles

The accuracy of Equation 4.29 can be estimated by the variance and the COV as given by the following equations:

$$\mathrm{Var}(\overline{P}_f) = \frac{\sum_{i=1}^{N} (P_{f_i} - \overline{P}_f)^2}{N(N-1)} \tag{4.30}$$

and

$$\mathrm{COV}(\bar{P}_\mathrm{f}) = \frac{\sqrt{\mathrm{Var}(\bar{P}_\mathrm{f})}}{\bar{P}_\mathrm{f}} \tag{4.31}$$

For the general performance function of Equation 4.1, the conditional failure probability based on a control variable X_k is given by

$$\begin{aligned} P_{\mathrm{f}_i} &= P(X_k < z_{k_i}) \\ &= F_{X_k}(z_{k_i}) \end{aligned} \tag{4.32}$$

where F_{X_k} is the CDF of X_k and z_{k_i} is the rearranged performance function of Equation 4.1 such that the failure domain is redefined from $Z < 0$ to $X_k < Z_k$ and the generated values are substituted in Z_k.

According to the generalized CE (GCE), this CE concept can be extended by having more than one carefully selected control variables in order to produce a simpler expression of the performance function by treating the remaining, randomly generated variables as constants in the simulation cycles. For example, the failure probability based on a nonlinear performance function, such as

$$Z = X_1X_2 - X_3X_4 \tag{4.33}$$

with four normally distributed uncorrelated random variables, can be assessed by randomly generating X_2 and X_3, and computing the conditional probability based on the fundamental concept covered in Equations 4.5, 4.6, and 4.7a as follows:

$$\begin{aligned} P_{\mathrm{f}_i} &= 1 - \Phi(\beta_i) \\ &= 1 - \Phi\left[\frac{x_{2_i}\mu_{X_1} - x_{4_i}\mu_{X_3}}{\sqrt{(x_{2_i})^2\sigma_{X_1}^2 - (x_{4_i})^2\sigma_{X_3}^2}}\right] \end{aligned} \tag{4.34}$$

where a lower case x is a generated value of the corresponding random variable X that is treated in the computations as a constant, a real value.

4.2.3.3 Importance Sampling

The probability of failure of a structure according to the performance function of Equation 4.1 is provided by the integral of Equation 4.2. In evaluating this integral with direct simulation, the efficiency of the simulation process depends on the magnitude of the probability of unsatisfactory performance (i.e., the location of the most likely failure point or design point). The larger the margin of safety (Z) and the smaller its variance, the larger the simulation effort required to obtain sufficient simulation runs with unsatisfactory performances; in other words, smaller failure probabilities require larger numbers of simulation cycles. This deficiency can be addressed by using IS. In this method, the basic random variables are generated according to some carefully selected probability distributions [the *importance density function*, $h_{\underline{X}}(\underline{x})$] with mean values that are closer to the design point than their original (actual) probability distributions. It should be noted that the design point is not known in advance. The analyst can only guess such that simulation runs with failures are obtained more frequently and the simulation efficiency is increased. To compensate for the change in the probability distributions to $h_{\underline{X}}(\underline{x})$, the results of the simulation cycles should be corrected. The fundamental equation for this method is given by

$$\bar{P}_f = \frac{1}{N}\sum_{i=1}^{N} I_{f_i} \frac{f_{\underline{X}}(x_{1i}, x_{2i}, \ldots, x_{ni})}{h_{\underline{X}}(x_{1i}, x_{2i}, \ldots, x_{ni})} \quad (4.35)$$

where:

N is the number of simulation cycles

$f_{\underline{X}}$ $(x_{1i}, x_{2i}, \ldots, x_{ni})$ is the original joint density function of the basic random variables evaluated at the *ith* generated values of the basic random variables

$h_{\underline{X}}$ $(x_{1i}, x_{2i}, \ldots, x_{ni})$ is the selected joint density function of the basic random variables evaluated at the *i*th generated value of the basic random variables

I_f is the performance indicator function that takes values of either 0 for failure and 1 for survival

For noncorrelated basic random variables, the joint density function $f_{\underline{X}}(x_{1i}, x_{2i}, \ldots, x_{ni})$ can be replaced by the product of the density functions of the individual random variables. Similarly, the joint density function $h_{\underline{X}}(x_{1i}, x_{2i}, \ldots, x_{ni})$ can be replaced by the product of the corresponding importance density functions. In Equation 4.35, $h_{\underline{X}}(\underline{x})$ is the sampling (or weighting) density function or the importance function. Efficiency (and thus the required number of simulation cycles) depends on the choice of this sampling density function. The COV of the estimate failure probability can be based on the variance of a sample mean as follows:

$$\text{COV}(\bar{P}_f) = \frac{\sqrt{\frac{1}{N(N-1)}\sum_{i=1}^{N}\left[I_{f_i}\frac{f_{\underline{X}}(x_{1i}, x_{2i}, \ldots, x_{ni})}{h_{\underline{X}}(x_{1i}, x_{2i}, \ldots, x_{ni})} - \bar{P}_f\right]^2}}{\bar{P}_f} \quad (4.36)$$

4.2.3.4 Correlated Random Variables

In this section, correlation between pairs of random variables is treated for simulation purposes. Correlated random variables are assumed to be normally distributed, as non-normal and correlated random variables require additional information such as marginal probability distribution for their unique and full definition. Such information is commonly not available and is difficult to obtain. A correlated (and normal) pair of random variables X_1 and X_2 with a correlation coefficient ρ can be transformed using linear regression as follows:

$$X_2 = b_0 + b_1 X_1 + \varepsilon \quad (4.37)$$

where:

b_0 is the intercept of a regression line between X_1 and X_2

b_1 is the slope of the regression line

ε is the random (standard) error with a mean of zero and a standard deviation as given in Equation 4.38c

These regression model parameters can be determined in terms of the probabilistic characteristics of X_1 and X_2 as follows:

$$b_1 = \frac{\rho\sigma_{X_2}}{\sigma_{X_1}} \quad (4.38a)$$

$$b_0 = \mu_{X_2} - b_1\mu_{X_1} \quad (4.38b)$$

$$\sigma_\varepsilon = \sigma_{X_2}\sqrt{1-\rho^2} \quad (4.38c)$$

The simulation procedure for a correlated pair of random variables (X_1 and X_2) can then be summarized as follows:

1. Compute the intercept (b_0) of a regression line between X_1 and X_2, the slope of the regression line (b_1), and the standard deviation of the random (standard) error (ε) using Equation 4.38a–c.
2. Generate a random (standard) error using a zero mean and a standard deviation as given by Equation 4.38c.
3. Generate a random value for X_1 using its probabilistic characteristics (i.e., mean and variance).
4. Compute the corresponding value of X_2 as follows (based on Equation 4.37):

$$x_2 = b_0 + b_1 x_1 + \varepsilon \tag{4.39}$$

   where:
   b_0 and b_1 are computed in step 1
   ε is the generated random (standard) error from step 2
   x_1 is the generated value from step 3

5. Use the resulting random (but correlated) values of x_1 and x_2 in the simulation-based reliability assessment methods.

The above procedure is applicable for both the direct simulation method and the IS method. In the case of IS, correlated random variables should not be selected for defining the sampling (or importance) density function ($h_{\underline{X}}$) in order to keep the method valid in its present form.

Example 4.2: Simulation-based Reliability Assessment Using a Nonlinear Performance Function

This example uses the same performance function of Example 4.1 and demonstrates the application of simulation methods for case (a) of the example that deals with normally distributed random variables. Table 4.1 show 30 simulation cycles of random numbers (u) in three columns followed by generated random variables (x) in three columns computed as $x = u\sigma + \mu$ for each cycle and each random variable.

The DMC results are shown in two columns of a performance function evaluation (z) and a binary value of a failure indicator (I_f) function that takes a value of either 0 in case of failure ($z < 0$) or 1 in case of survival ($z \geq 0$). In the 30 simulation cycles, no failure was observed, that is, the failure probability estimate in this DMC case is 0.

The IS method was implemented by using an importance function (h) based on changing the mean value of X_3 to 6 and retaining the same standard deviation and distribution type for it. The generated values of x_3 were updated as shown in Table 4.1 using the same random numbers of u_3. The performance value (z) was reevaluated as shown in the IS columns. For all failure cases, that is, $z < 0$, $I_{f_i}[f_{X_3}(x_{3_i})/h_{X_3}(x_{3_i})]$ was computed per Equation 4.35 at the respectively generated x_3 value. The table shows only one cycle with $z < 0$ at $i = 30$ where $I_{f_i} = 1$, $f_{X_3}(x_{3_i}) = 0.006475806$ and $h_{X_3}(x_{3_i}) = 0.45116395$ producing $I_{f_i}[f_{X_3}(x_{3_i})/h_{X_3}(x_{3_i})] = 0.014354$.

As for the CE method, three cases were examined using one at a time the three random variables X_1, X_2, and X_3 as control variables, and computing the respective probabilities in each cycle as follows:

- $P_{f_i} = F_{X_1}(\sqrt{x_{3_i}}/x_{2_i})$ for X_1 as a control variable
- $P_{f_i} = F_{X_2}(\sqrt{x_{3_i}}/x_{1_i})$ for X_2 as a control variable
- $P_{f_i} = 1 - F_{X_3}(x_{1_i}x_{2_i})^2$ for X_3 as a control variable

TABLE 4.1

Assessing Failure Probability Using Simulation Methods (Example 4.2)

	Random Numbers			Random Values			DMC		IS			CE		
Cycle (i)	u_{1i}	u_{2i}	u_{3i}	x_{1i}	x_{2i}	x_{3i}	z_i	I_{fi}	x_{3i}	z_i	$I_{f_i} f_{X3}/h_{X3}$	$P_{f_i}(CE, X_1)$	$P_{f_i}(CE, X_2)$	$P_{f_i}(CE, X_3)$
1	0.376518	0.601560	0.494390	0.921340	5.064347	3.988750	2.668801	0	5.988750	2.218795	0	0.007706	4.70209E−30	0
2	0.464271	0.689917	0.113790	0.977580	5.123904	3.034707	3.266987	0	5.034707	2.765213	0	0.004144	3.23402E−38	0
3	0.448521	0.639832	0.020745	0.967651	5.089502	2.369118	3.385666	0	4.369118	2.834616	0	0.002633	1.19987E−42	0
4	0.472658	0.978586	0.357154	0.982852	5.506347	3.707139	3.486533	0	5.707139	3.022965	0	0.004643	2.41488E−34	0
5	0.505367	0.667774	0.233599	1.003363	5.108444	3.418364	3.276743	0	5.418364	2.797887	0	0.005351	7.28179E−37	0
6	0.428679	0.011367	0.741982	0.955065	4.430531	4.519575	2.105517	0	6.519575	1.678101	0	0.018733	6.54503E−29	0
7	0.339809	0.632782	0.899083	0.896754	5.084808	5.021073	2.319046	0	7.021073	1.910090	0	0.012634	7.24837E−24	0
8	0.844216	0.929123	0.434228	1.252984	5.367323	3.867504	4.758573	0	5.867504	4.302877	0	0.005632	3.75483E−43	0
9	0.740141	0.010239	0.524155	1.160945	4.420631	4.048467	3.120029	0	6.048467	2.672746	0	0.014652	2.52689E−39	0
10	0.949507	0.646730	0.931645	1.410023	5.094127	5.190521	4.904567	0	7.190521	4.501324	0	0.013516	4.73343E−42	0
11	0.021760	0.775842	0.137190	0.495328	5.189557	3.125574	0.802603	0	5.125574	0.306559	0	0.004178	5.22803E−09	0.000558
12	0.441400	0.197682	0.013203	0.963145	4.787518	2.223858	3.119811	0	4.223858	2.555869	0	0.002943	1.16112E−43	0
13	0.399552	0.044745	0.784594	0.936373	4.575478	4.630241	2.132555	0	6.630241	1.709429	0	0.017052	1.57839E−27	0
14	0.650482	0.328875	0.417781	1.096656	4.889245	3.833940	3.403774	0	5.833940	2.946463	0	0.008241	3.87167E−38	0
15	0.220090	0.250189	0.837036	0.807027	4.831526	4.785879	1.711508	0	6.785879	1.294202	0	0.014304	2.66993E−20	0
16	0.930782	0.219434	0.278541	1.370410	4.806473	3.530255	4.707942	0	5.530255	4.235190	0	0.007418	4.81071E−48	0
17	0.219738	0.071137	0.535137	0.806730	4.633156	4.070553	1.720146	0	6.070553	1.273858	0	0.011968	7.90562E−24	0
18	0.995358	0.640109	0.516311	1.650359	5.089688	4.032717	6.391651	0	6.032717	5.943655	0	0.007722	4.92183E−52	0
19	0.346492	0.642253	0.612315	0.901297	5.091122	4.228287	2.532336	0	6.228287	2.092962	0	0.008553	7.65509E−28	0
20	0.076248	0.426935	0.781609	0.642308	4.953954	4.622110	1.032053	0	6.622110	0.608616	0	0.011784	1.90429E−11	9.55E−15
21	0.737512	0.697514	0.344810	1.158923	5.129316	3.680503	4.026020	0	5.680503	3.561103	0	0.006141	4.04012E−41	0
22	0.197226	0.570277	0.304809	0.787107	5.044270	3.591506	2.075255	0	5.591506	1.605745	0	0.006259	1.71267E−25	0
23	0.060445	0.364424	0.589996	0.612238	4.913336	4.182027	0.963130	0	6.182027	0.521762	0	0.009768	1.57719E−11	1.39E−10
24	0.586682	0.527312	0.658814	1.054755	5.017129	4.327382	3.211604	0	6.327382	2.776411	0	0.009603	4.61722E−34	0
25	0.224811	0.863712	0.227707	0.810989	5.274287	3.402865	2.432702	0	5.402865	1.952982	0	0.004648	5.66564E−28	0
26	0.374237	0.582343	0.456018	0.919837	5.051973	3.911623	2.669210	0	5.911623	2.215609	0	0.007465	2.10449E−30	0
27	0.259425	0.071783	0.295543	0.838721	4.634340	3.570190	1.997421	0	5.570190	1.526791	0	0.008915	2.16607E−28	0
28	0.309426	0.373377	0.360101	0.875630	4.919269	3.713450	2.380430	0	5.713450	1.917179	0	0.007486	2.10756E−29	0
29	0.103617	0.463270	0.971569	0.684698	4.976951	5.523494	1.057497	0	7.523494	0.664809	0	0.017381	1.80426E−10	0
30	0.012866	0.244410	0.672744	0.442442	4.826953	4.358002	0.048065	0	6.358002	−0.385860	0.014354	0.011602	0.129927722	0.241576

CE, conditional expectation; DMC, direct Monte Carlo; IS, importance sampling.

In this case, each simulation cycle produces a failure probability and an estimate of the failure probability can be obtained per Equation 4.29.

Figure 4.5 compares the estimates of the DMC and IS as the number of simulation cycles is incrementally increased to 2000 cycles. The DMC shows the first failure at $i = 166$. The failure probability was estimated in Example 4.1 case (a) to be 0.009068 that can be assumed to be close to the true value. We can observe that the DMC converges at about 1250 cycles, whereas the IS requires more than 2000 cycles. The IS was explored further by changing the assumed mean value of X_3 in the importance function as shown in Figure 4.6 in terms of a ratio of the estimated failure probability and the true value of 0.009068. The results in Figure 4.6 are based on 100 simulation cycles and show that the trends are greatly affected by the choice of the importance function (h).

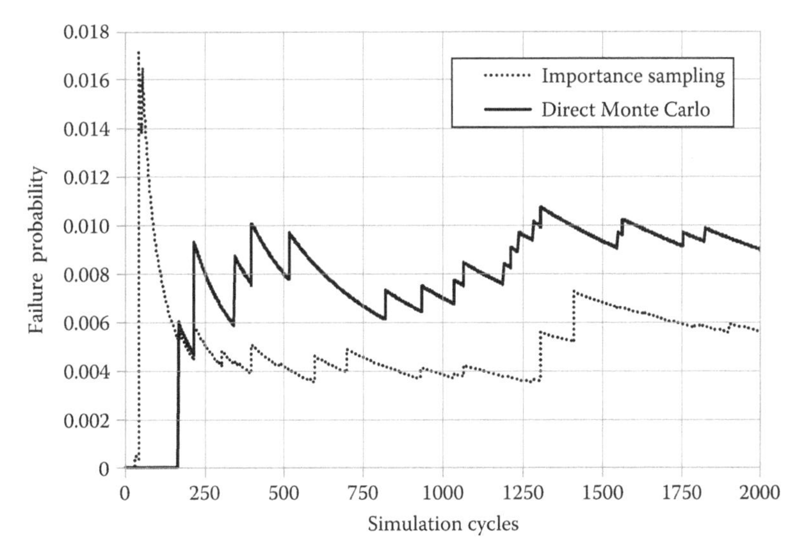

FIGURE 4.5
Direct Monte Carlo simulation (DMCS) and importance sampling (IS) methods.

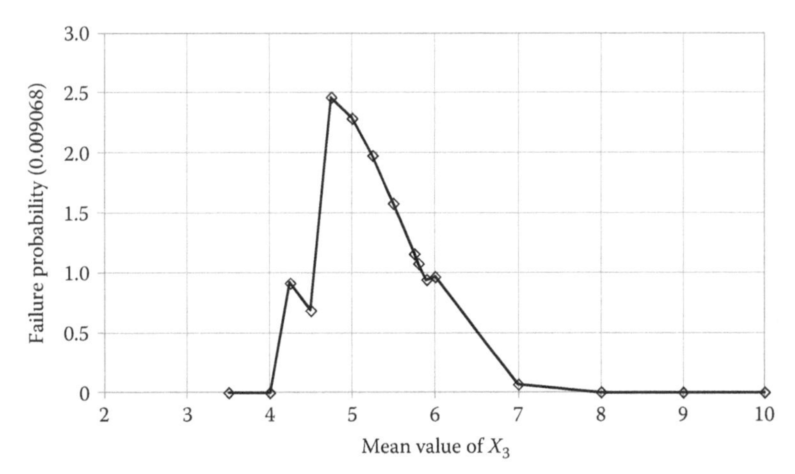

FIGURE 4.6
Effects of changing the mean value of X_3 on failure probability estimated using importance sampling (IS).

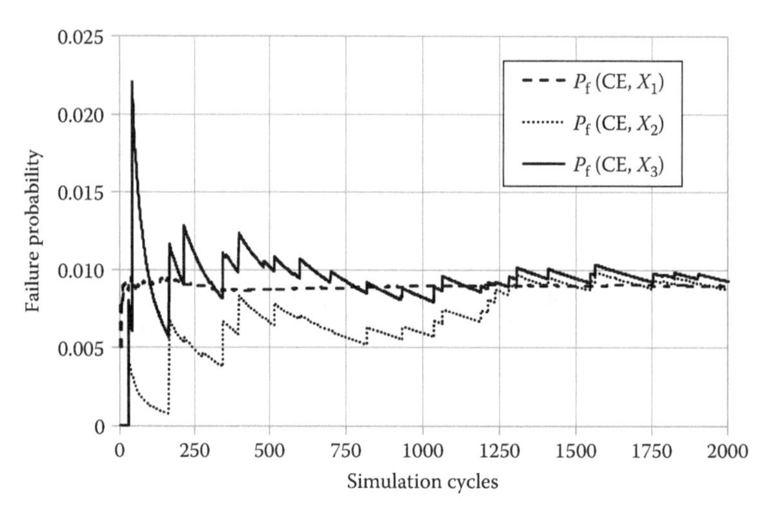

FIGURE 4.7
Effects of control variable selection on failure probability estimated using conditional expectation (CE).

Figure 4.7 compares the estimates of the CE using the three control variables of X_1, X_2, and X_3. It shows that the CE simulation converges to the true value at about 1250 simulation cycles regardless of the control variable selected; however, X_2 shows the poorest trend.

We can observe from these comparisons that the CE method offers the best performance and reliability in this example.

4.2.4 Time-Dependent Reliability Analysis

Several methods for analytical time-dependent reliability assessment are available. In these methods, significant loads as a sequence of pulses can be described by a Poisson process with mean occurrence rate λ, random intensity L, and duration ΔT. The performance function (Z) of a component or system at any instant of time (t) can be defined as

$$Z(t) = S(t) - L(t) \tag{4.40}$$

where:

$S(t)$ is the strength at time t

$L(t)$ is the load at time t

The instantaneous probability of failure at time t can then be defined as the probability of $S(t)$ less than $L(t)$; however, this instantaneous probability treatment does not recognize what has previously happened to the component or system from the start of its life to the present represented by time t. We are usually interested in the first occurrence of L exceeding S, not the instantaneous occurrence, requiring the imposition of a condition on L exceeding S for the first time in its life. Such a conditional probability concept is the basis for computing what is termed *time-dependent reliability* and estimated using the reliability function $R(t)$.

The reliability function, $R(t)$, is defined as the probability that a component or a system survives during an interval of time $(0, t]$ based on a performance function Z. Assuming the load events (pulses) to follow a Poisson process with a rate λ means that the time to a first load occurrence (or time between successive occurrences) is exponentially distributed. We are, however, seeking to characterize the time to failure, that is, not only the time to load occurrence, and recognize that some load occurrences may lead to failure; therefore, we can make the following expressions based on the exponential distribution as covered in Appendix A:

$$R(t) = \exp(-\lambda p_L t) \tag{4.41}$$

where λp_L is the product of load rate (λ) and a function (p_L) that estimates the fraction of loads that produce failure over the time period (0, t] in order to account for any degradation of the strength (S) over this period. The strength degradation, for example, due to corrosion of a structural member, can be modeled by a function $0 \le c(t) \le 1$ and used as a multiplier to an initial strength (S_0 at $t = 0$). This probability, p_L, is taken as the average value over the period (0, t] as follows:

$$p_L = 1 - \frac{1}{t}\int_{\tau=0}^{\tau=t} P[Z(\tau) > 0]\mathrm{d}\tau = 1 - \frac{1}{t}\int_{\tau=0}^{\tau=t} P[c(\tau)S_0 > L]\mathrm{d}\tau \tag{4.42}$$

where $Z = S_0 - L$ as an example performance function, with S_0 being the initial strength at $t = 0$. Substituting Equation 4.42 into Equation 4.41 and accounting for the uncertainty in the initial strength produce the following expression:

$$R(t) = \int_{s=0}^{\infty} \exp\left(-\lambda t\left\{1 - \frac{1}{t}\int_{\tau=0}^{\tau=t} P[c(\tau)s > L]\mathrm{d}\tau\right\}\right) f_{S_0}(s)\mathrm{d}s \tag{4.43a}$$

where $f_{S_0}(s)$ is the PDF of the initial strength (S_0). Noting that the expression $P(c(\tau)s > L)$ in Equation 4.43a is the CDF of L evaluated at $c(\tau)s$, the reliability function can be written as

$$R(t) = \int_{s=0}^{\infty} \exp\left(-\lambda t\left\{1 - \frac{1}{t}\int_{\tau=0}^{\tau=t} F_L[c(\tau)s]\mathrm{d}\tau\right\}\right) f_{S_0}(s)\mathrm{d}s \tag{4.43b}$$

Equation 4.43a results in the expected value of the reliability function based on uncertainties associated with the time-dependent degradation of the strength. Someone can also compute the reliability function based on averaged material degradation at the same time as follows:

$$R(t) = \exp\left(\int_{s=0}^{\infty} -\lambda t\left\{1 - \frac{1}{t}\int_{\tau=0}^{\tau=t} F_L[c(\tau)s]\mathrm{d}\tau\right\} f_{S_0}(s)\mathrm{d}s\right) \tag{4.43c}$$

Engineers are interested in $R(t)$ estimated by Equation 4.43b rather than Equation 4.43c. The reliability can be expressed in terms of the failure rate or hazard function, $h(t)$ as follows:

$$h(t) = -\frac{\mathrm{d}}{\mathrm{d}t}\ln[R(t)] \tag{4.44a}$$

The functions $R(t)$ and $h(t)$ are related as follows:

$$R(t) = \exp\left[-\int_{0}^{t} h(\tau)\mathrm{d}\tau\right] \tag{4.44b}$$

and

$$h(t) = \frac{f(t)}{1 - F(t)} \tag{4.44c}$$

The concept of the hazard function is discussed in detail in Section 4.3.5. The reliability $R(t)$ is based on complete survival during the service life interval $(0, t]$. It means the probability of successful performance during the service life interval $(0, t]$. Therefore, the probability of failure, $P_f(t)$ or $F(t)$, can be computed as the probability of the complementary event, $P_f(t) = 1 - R(t)$, not being equivalent to $P[S(t) < L(t)]$, the latter being just an instantaneous failure at time t without regard to previous performance.

Example 4.3: Time-Dependent Reliability Assessment with Corrosion

This example uses a fundamental performance function of $Z = c(t)S - L$ to demonstrate a computational procedure for estimating $R(t)$ according to Equation 4.43b. The random variables S and L are assumed to be normally distributed with mean values of 6 and 4, and standard deviations of 0.5 and 1, respectively. A load rate (λ) is assumed to be 1. A fundamental corrosion model is used as $c(t) = 1 - at^b$, with the parameter a being varied incrementally to facilitate parametric analysis from 0, 0.01, 0.02, 0.03, 0.05, to 0.10; the parameters $b = 1$, $t > t_p$, t_p = coating life taken as 0 in this example; and the condition $0 < at^b < 1$.

The reliability function $R(t)$ can be evaluated according to the following computational procedure:

- In the ith simulation cycle, randomly generate S as s_i based on a random number u_i as provided in columns 2 and 3 of Table 4.2 that shows demonstrative computations for 10 simulations cycles.
- Evaluate $R_i(t)$ using Equation 4.43b for all the t values of interest, $t = 1, 2, 3, \ldots, 10$ years, for example, for each simulation cycle as follows:
  - For $t = 1$:
    Evaluate $1/t \int_{\tau=0}^{\tau=1} F_L[c(\tau)s_i]d\tau$ using trapezoidal rule based on, say, 100 increments. Table 4.2 shows the use of the trapezoidal rule (with a uniform grid) to calculate the integral for one increment as $\int_0^1 F_L[c(\tau)s_i]d\tau \approx [1 - 0/2(1)]\{F_L[c(0)s_i] + F_L[c(1)s_i]\}$ and then compute $R_i(t = 1) = \exp[-\lambda t(1 - \{1/t \int_{\tau=0}^{\tau=1} F_L[c(\tau)s_i]d\tau\})]$ as shown in columns 4–7.
  - For $t = 2$:
    Evaluate $1/t \int_{\tau=0}^{\tau=2} F_L[c(\tau)s_i]d\tau$ using trapezoidal rule for two increments as $\int_0^2 F_L[c(\tau)s_i]d\tau \approx [2 - 0/2(2)]\{F_L[c(0)s_i] + 2F_L[c(1)s_i] + F_L[c(2)s_i]\}$ and compute $R_i(t = 2) = \exp[-\lambda t(1 - \{1/t \int_{\tau=0}^{\tau=1} F_L[c(\tau)s_i]d\tau\})]$ as shown in columns 8–10.
  - For $t = 3$:
    Evaluate $1/t \int_{\tau=0}^{\tau=3} F_L[c(\tau)s_i]d\tau$ using trapezoidal rule for three increments as $\int_0^3 F_L[c(\tau)s_i]d\tau \approx [3 - 0/2(3)]\{F_L[c(0)s_i] + 2F_L[c(1)s_i] + 2F_L[c(2)s_i] + F_L[c(3)s_i]\}$ and compute $R_i(t = 3) = \exp[-\lambda t(1 - \{1/t \int_{\tau=0}^{\tau=3} F_L[c(\tau)s_i]d\tau\})]$ as shown in columns 11–13.
  - Repeat the process until $t = 10$.
- Repeat the previous step for the next simulation cycle $i + 1$ to obtain $R_{i+1}(t)$ for $t = 1, 2, \ldots, 10$ and until $i = N$ cycles.
- For each t, compute the statistics of $R(t)$ and check for convergence as follows:

$$\overline{R}(t) = \frac{\sum_{i=1}^{N} R_i(t)}{N}$$

where:
 N is the number of simulation cycles

TABLE 4.2

Assessing Time-Dependent Reliability for $t = 1$, 2, and 3 Years for Example 4.3 (with Corrosion Using $a = 0.03$ and $b = 1$)

			$t = 1$				$t = 2$			$t = 3$		
Cycle (i) (1)	u_i (2)	s_i (3)	$F_L(s_i)$ at $\tau = 0$ (4)	$F_L(s_i)$ at $\tau = 1$ (5)	$p_{L_i}(t = 1)$ (6)	$R_i(t = 1)$ (7)	$F_L(s_i)$ at $\tau = 2$ (8)	$p_{L_i}(t = 2)$ (9)	$R_i(t = 2)$ (10)	$F_L(s_i)$ at $\tau = 3$ (11)	$p_{L_i}(t = 3)$ (12)	$R_i(t = 3)$ (13)
1	0.376518	5.842681	0.967312	0.952283	0.040203	0.960595	0.932166	0.048989	0.906669	0.906054	0.059623	0.836216
2	0.464271	5.955161	0.974718	0.962175	0.031553	0.968939	0.944962	0.038992	0.924978	0.922079	0.048155	0.865486
3	0.448521	5.935301	0.973523	0.960562	0.032957	0.967580	0.942853	0.040625	0.921963	0.919411	0.050039	0.860606
4	0.472658	5.965705	0.975334	0.963010	0.030828	0.969642	0.946056	0.038148	0.926543	0.923468	0.047178	0.868026
5	0.505367	6.006727	0.977611	0.966114	0.028138	0.972255	0.950151	0.035002	0.932390	0.928692	0.043528	0.877584
6	0.428679	5.910131	0.971942	0.958437	0.034811	0.965788	0.940089	0.042774	0.918009	0.915932	0.052512	0.854245
7	0.339809	5.793508	0.963554	0.947352	0.044547	0.956431	0.925897	0.053961	0.897697	0.898330	0.065270	0.822169
8	0.844216	6.505968	0.993894	0.989578	0.008264	0.991770	0.982811	0.011035	0.978172	0.972598	0.014788	0.956605
9	0.740141	6.321890	0.989881	0.983506	0.013307	0.986782	0.973966	0.017285	0.966020	0.960192	0.022497	0.934736
10	0.949507	6.820047	0.997599	0.995544	0.003428	0.996578	0.992042	0.004817	0.990411	0.986316	0.006819	0.979752

The accuracy of this estimate can be evaluated through the variance and the COV as given by

$$\mathrm{Var}(\bar{R}(t)) = \frac{\sum_{i=1}^{N} [R_i(t) - \bar{R}(t)]^2}{N(N-1)}$$

and

$$\mathrm{COV}[\bar{R}(t)] = \frac{\sqrt{\mathrm{Var}[\bar{R}(t)]}}{\bar{R}(t)}$$

In this example, we used $N = 100$ cycles and the corrosion parameter varied from 0 to 0.1 in increments as shown in Figures 4.8 and 4.9. The figures show the reliability trend as a function of time and associated statistical uncertainty in the estimates. Also, these figures show the effects of corrosion rate on life estimates and the associated uncertainties.

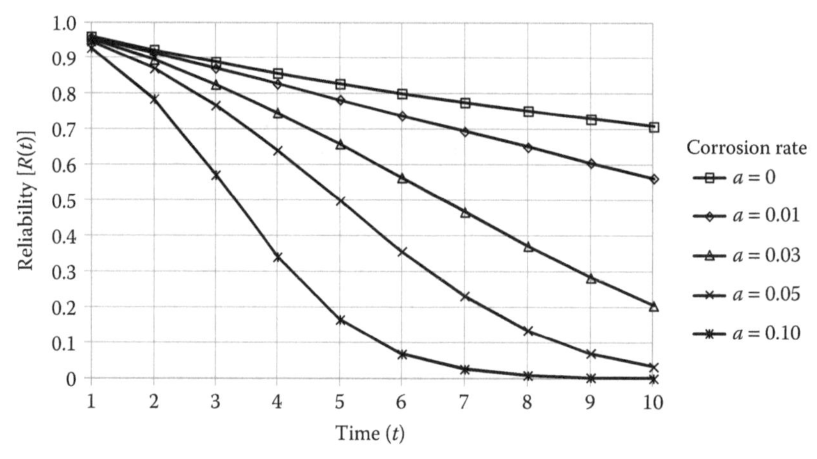

FIGURE 4.8
Effects of corrosion rate on reliability estimates.

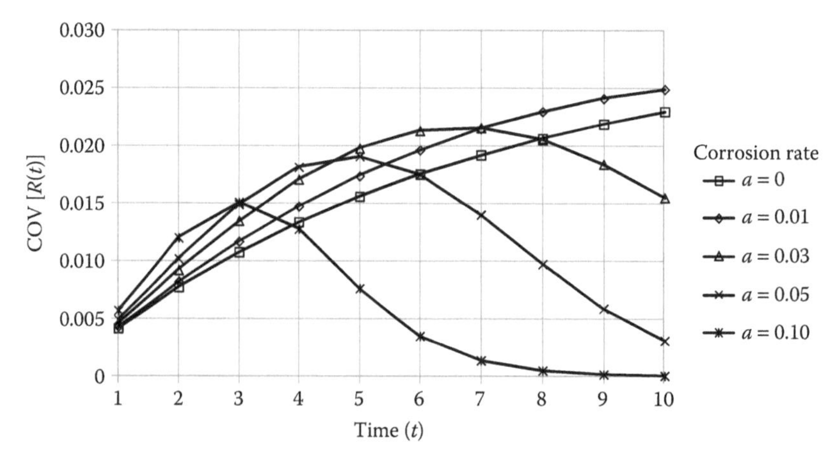

FIGURE 4.9
Effects of corrosion rate on sampling uncertainty in reliability estimates. COV, coefficient of variation.

4.2.5 Resilience

Resilience is defined in Chapter 2 according to the Presidential Policy Directive (PPD)-21 (2013) as the ability to prepare for and adapt to changing conditions and withstand and recover rapidly from disruptions. Resilience includes the ability to withstand and recover from disturbances of the deliberate attack types, accidents, or naturally occurring threats or incidents. Additionally, measuring the resilience of a system's function can be based on the persistence of a corresponding functional performance under uncertainty in the face of disturbances (Ayyub 2013). This definition is consistent with the International Organization for Standardization (ISO 2009a) risk definition of the "effect of uncertainty on objectives." In this section, the focus is on the methods used for measuring resilience and associated metrics.

Ayyub (2013) provided resilience metrics as illustrated in Figure 4.10 that shows a schematic representation of a system performance (Z) with aging effects and an incident occurrence with a rate (λ) according to a Poisson process. At time t_i, it might lead to a failure event with a duration ΔT_f. The failure event concludes at time t_f. The failure event is followed by a recovery event with a duration ΔT_v. The recovery event concludes at time t_v.

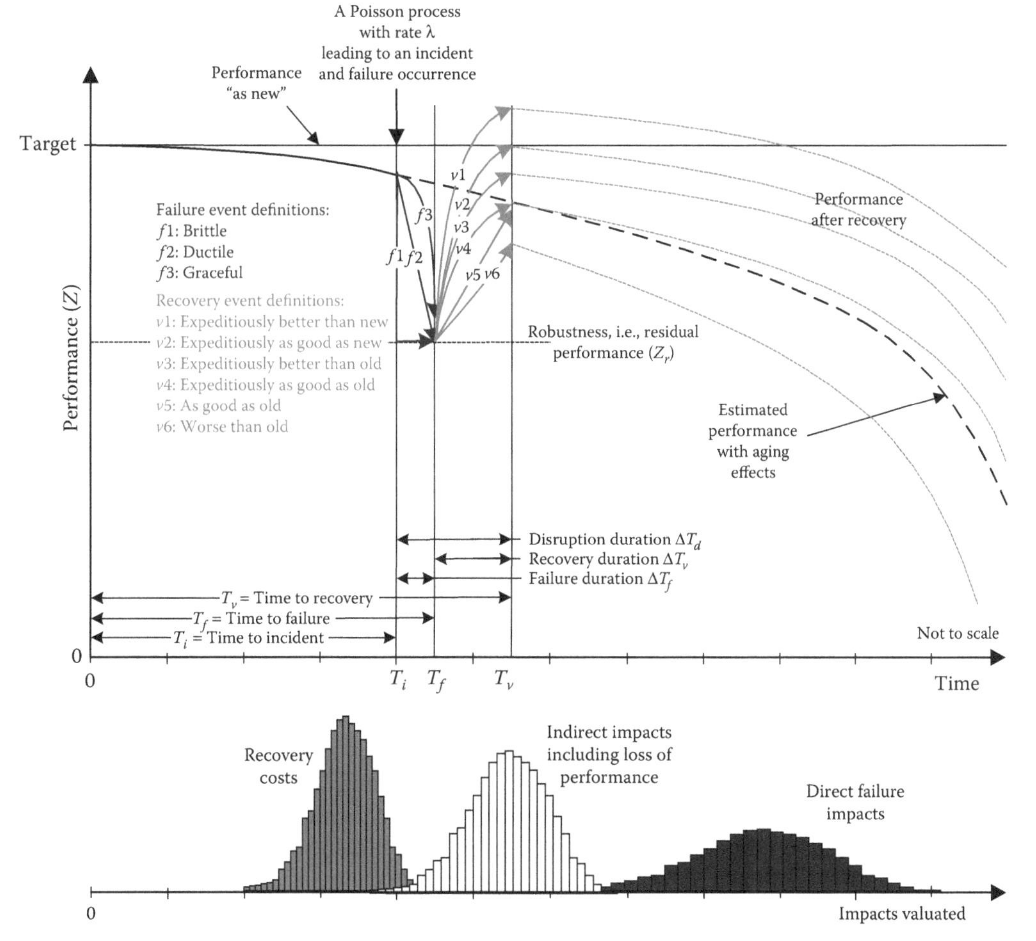

FIGURE 4.10
Resilience metrics.

The total disruption (D) has a duration of $\Delta T_d = \Delta T_f + \Delta T_v$. The figure shows for illustration purposes three failure events—brittle ($f1$), ductile ($f2$), and graceful ($f3$)—and six recovery events—expeditious recovery to better than new ($v1$), expeditious recovery to as good as new ($v2$), expeditious recovery to better than old ($v3$), expeditious recovery to as good as new ($v4$), recovery to as good as old ($v5$), and recovery to worse than old ($v6$). These events define various rates of change of performance of the system. The figure also shows the aging performance trajectory and the estimated trajectory after recovery. The proposed model to measure resilience is

$$\text{Resilience } (R_e) = \frac{T_i + F\Delta T_f + V\Delta T_v}{T_i + \Delta T_f + \Delta T_v} \tag{4.45a}$$

where for any failure event (f) as illustrated in Figure 4.10, the corresponding *failure profile* (F) is measured as follows:

$$\text{Failure } (F) = \frac{\int_{t_i}^{t_f} f dt}{\int_{t_i}^{t_f} Z dt} \tag{4.45b}$$

Similarly, for any recovery event (v) as illustrated in Figure 4.10, the corresponding *recovery profile* (V) is measured as follows:

$$\text{Recovery } (V) = \frac{\int_{t_f}^{t_v} v dt}{\int_{t_f}^{t_v} Z dt} \tag{4.45c}$$

The failure profile value (F) can be considered as a measure of robustness and redundancy, whereas the recovery profile value (V) can be considered as a measure of resourcefulness and rapidity. The time to failure (T_f) can be characterized by its PDF computed as follows:

$$-\frac{d}{dt}\int_{c=0}^{\infty} \exp\left(-\lambda t\left\{1 - \frac{1}{t}\int_{\tau=0}^{t} F_L[\alpha(\tau)s]d\tau\right\}\right) f_{S_0}(s)ds \tag{4.45d}$$

where Z is defined as the system's performance in terms of its strength (S) minus the corresponding load effect (L) in consistent units, that is, $Z = S - L$. Both L and S are treated as random variables, with F_L being the cumulative probability distribution function of L, and f_S being the PDF of S. The aging effects are considered in this model by the term $\alpha(t)$ representing a degradation mechanism as a function of time t. It should be noted that the term $\alpha(t)$ can also represent improvement to the system. Equation 4.45d is based on a Poisson process with an incident occurrence, such as loading, rate of λ, and is based on Ellingwood and Mori (1993). The PDF of T_f as shown in Equation 4.45d is the negative of the derivative of the reliability function.

The units of performance at the system level vary depending on the system type and the objectives of the analysis. Examples of performance types and units of measurement for selected systems for demonstration purposes are as follows:

- For buildings, space availability might be treated as the performance attribute of interest measured using the units of area per day.
- For highway bridges, throughput traffic might be treated as the performance attribute of interest measured using the units of traffic count per day.
- For water treatment plants, available water production capacity might be treated as the performance attribute of interest measured using the units of water volume per day.
- For electric power distribution systems, power delivered might be treated as the performance attribute of interest measured using the units of power delivered per day.
- For communities, economic output might be treated as the performance attribute of interest measured using the units of dollars.

The resilience model of Equation 4.45 can be used for systems, such as buildings, other structures, facilities, infrastructure, networks, and communities. The primary basis for evaluating Equation 4.45 is the definition of performance (Z) at the system level with meaningful and appropriate units, followed by the development of an appropriate breakdown for this performance, using what is termed herein as performance segregation. The performance segregation should be based on some system-level logic that relates the components of the performance breakdown to the overall performance at the system level as the basis for a system model. This model can be used to aggregate the performance of components to assess the system-level performance. Such performance segregation and aggregation analysis are essential for examining the resilience of systems for buildings, other structures, facilities, infrastructure, networks, and communities. The uncertainties associated with the performance components can be modeled as random variables with any necessary performance events in order to use Boolean algebra and the mathematics of probability to characterize the performance Z in Equation 4.45. System analysis for reliability quantification as presented in Section 4.4 may offer a basis. The development of such a system-level model relating components' performances to a system performance is beyond the scope of this book. Such a model is domain specific; however, future studies should set meta-methodological requirements for the development of such models. Figure 4.11 shows an example plot for the case of two identical resilience metrics, that is, resilience components, for the entire range of values of R_{ei} aggregated using the following model:

$$\text{Resilience } (R_{e12}) = \frac{R_{e1}R_{e2}}{R_{e1} + R_{e2} - R_{e1}R_{e2}} \tag{4.45e}$$

The figure also shows the effect of increasing the number of components from 1 to 10. The downward intensification is attributed to the independence assumptions.

Figure 4.10 also shows the economic valuation of resilience. Chapter 5 provides additional information on economic valuation. The figure demonstrates potential direct and indirect losses, and cost of recovery.

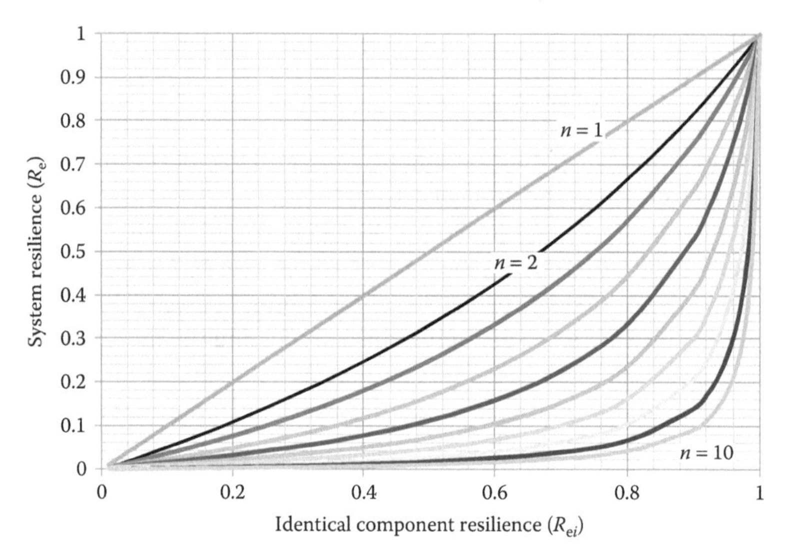

FIGURE 4.11
System resilience aggregate based on two identical resilience components.

4.3 Empirical Reliability Analysis Using Life Data

4.3.1 Failure and Repair

The basic notion of reliability analysis based on life data is *time to failure*. The useful life of a product can be measured in terms of its time to failure. The time to failure can also be viewed as an exposure measure for the product. In addition to time, other possible exposure measures include the number of cycles to failure of mechanical, electrical, temperature, or humidity; the number of demands for standing-by systems; and the number of travel miles. Without loss of generality, the time to failure is mainly used as a measure of exposure in this book. The treatment using other exposure measures is almost identical to the time to failure case.

Products based on the same design and produced by the same production process are expected to have different times to failure due to uncertainties associated with materials used in product manufacturing, uncertainties in manufacturing processes, and variability in exposure and environment during product utilization. Therefore, the time to failure for a product should be treated as a random variable, probabilistically modeled, and statistically characterized.

If the failed product is subject to repair or replacement, it is *repairable* (as opposed to *nonrepairable*). The respective repair or replacement requires some time to get done and is referred to as *time to repair/replace*. The time to repair is another random variable widely used in reliability analysis of repairable systems.

Generally speaking, the time to failure is used for the nonrepairable components or systems. For repairable products, another important characteristic is *time between failures*. This is another random variable or a set of random variables. For example, it can be assumed that the time to the first failure is the same random variable as the time between the first and second failures, the time between the second and third failures, and so on. These times might be the same random variable in the case of perfect repair/replacement. But if the repair/replacement or any maintenance action is not perfect, these times might not be the same, and one needs to consider these times between failures as different random variables.

4.3.2 Types of Data

Reliability estimation requires the respective life (time to failure, time between failures, and/or time to repair) data. Failure data often contain not only times to failure (the so-called *distinct failures*), but also times in use (or exposure length of time) that do not terminate with failures. Such exposure time intervals terminating with nonfailure are *times to censoring* (TTCs). Therefore, life data of equipment can be classified into two types: complete and censored data. *Complete life data* are commonly based on equipment tested to failure or times to failure based on equipment use (i.e., field data). Complete life data consist of available times to failure for the equipment based on these tests or field information. *Censored life data* include some observation results that represent only lower or upper limits on observations of times to failure. For example, if a piece of equipment has not failed at some time t and the equipment is removed from service, then t is considered to be a lower limit on the time to failure and can be used for estimation. The equipment data that produce lower limit values on times to failure are *right-censored data*. In some engineering applications, *left-censored data* with upper limit values on times to failure might also take place. For example, for hydropower equipment, complete data or right-censored data are commonly encountered. In warranty data, left-censored data can be encountered in cases of detecting noncritical failure of components during major inspections of systems per warranty terms, such as for automobiles. Other types of data are possible, such as interval censoring (e.g., in the case of grouped data).

Censored data can be further classified into type I or type II data. Type I data are based on observations of a life test, which for economical or other reasons must be terminated at specified time t_0. As a result, only the lifetimes of those units that have failed before t_0 are known exactly. If, during the time interval $(0, t_0]$, s out of n sample units failed, then the information in the dataset obtained consists of s observed, ordered times to failure as follows:

$$t_1 < t_2 < \cdots < t_s \tag{4.46}$$

as well as the information that $(n - s)$ units have survived for time t_0. The last portion of this information is important and must be used for the reliability and hazard rate function (HRF) estimation. It should be noted that in the case of type I censoring, the number of observed failures (s) is random.

In some life data testing, testing is continued until a specified number of failures (r) is achieved, that is, the respective test or observation is terminated at the rth failure. In this case, r is not random. This type of testing (observation or field data collection) results in type II censoring. The information obtained is similar to the case of type I censoring and includes r observed, ordered times to failure:

$$t_1 < t_2 < \cdots < t_r \tag{4.47}$$

as well as the information that $(n - r)$ units have survived for time t_r. But, as opposed to type I censoring, the test or observation duration t_r is random, which should be taken into account during the respective statistical estimation procedures.

In reliability engineering, type I right-censored data are commonly encountered. Figure 4.12 shows a summary of these data types. Other types of data are possible, such as random censoring. A typical situation where one deals with random censoring is the presence of several failure modes (FMs), such as strength mode of failure (FM1) and fatigue mode of failure (FM2) for structural components, and the problem is to estimate the reliability and/or hazard functions for each FM separately. For instance, if one is interested in estimating the hazard functions for FM1, all times to FM2 must be treated as TTC, which are obviously random.

In engineering, life data of interest are commonly based on failures that result in equipment replacement or major repair or rehabilitation that renders it new; therefore,

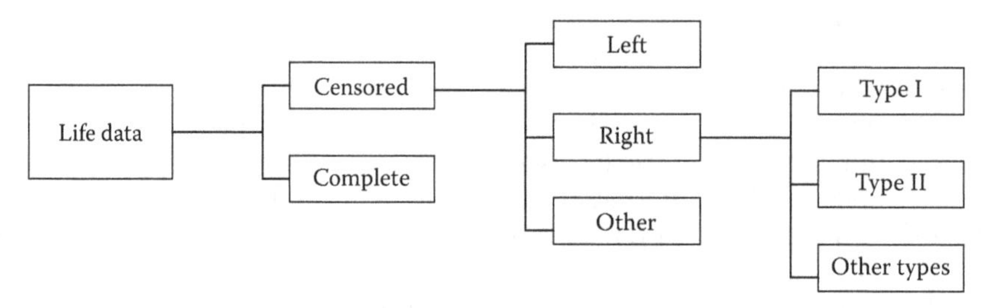

FIGURE 4.12
Types of life data.

such data can be treated just as for nonrepairable equipment. Examples 4.4 through 4.6 provide samples of complete time to failure data, right-censored data, and data based on random censoring, respectively.

Example 4.4: Data of Distinct Failures

The following array provides an example of complete data. In this example, the following sample of 19 times to failure for a structural component given in years to failure is provided for illustration purposes:

26, 27, 28, 29, 30, 31, 32, 33, 34, 35, 36, 37, 38, 39, 40, 42, 43, 50, 56

The time to failure in this case is a random variable because the 19 components show variability in their failure times in spite of being produced based on the same design and manufacturing processes. The same array can be used as an example of a sample of times to repair, if the times are given in, say, hours.

Example 4.5: Right-Censored Data

In this example, tests of equipment are used for demonstration purposes to produce observations in the form of life data as given in Table 4.3. The data in the table provide an example of type I censored data (the sample size is 12), with time to censoring equal to 51 years. If the data collection was assumed to terminate just after the eighth failure, the data would represent a sample of type II right-censored data with the same sample size of 12. The respective data are given in Table 4.4.

TABLE 4.3
Example of Type I Right-Censored Data (in Years) for Equipment

Time order number	1	2	3	4	5	6	7	8	9	10	11	12
Time (years)	7	14	15	18	31	37	40	46	51	51	51	51
TTF or TTC	TTF	TTF	TTF	TTF	TTF	TTF	TTF	TTF	TTC	TTC	TTC	TTC

Note: TTC, time to censoring; TTF, time to failure.

TABLE 4.4
Example of Type II Right-Censored Data (in Years) for Equipment

Time order number	1	2	3	4	5	6	7	8	9	10	11	12
Time (years)	7	14	15	18	31	37	40	46	46	46	46	46
TTF or TTC	TTF	TTF	TTF	TTF	TTF	TTF	TTF	TTF	TTC	TTC	TTC	TTC

Note: TTC, time to censoring; TTF, time to failure.

TABLE 4.5

20,000 Simulation Cycles for the Two Failure Modes (FMs) of Strength and Fatigue for a Structural Member

Year	Time to Failure (Years)	Number of Occurrences of a Given FM	
		Strength (FM1)	Fatigue (FM2)
1984	1	0	0
1985	2	7	0
1986	3	6	0
1987	4	3	0
1988	5	0	0
1989	6	1	7
1990	7	1	12
1991	8	0	20
1992	9	1	36
1993	10	1	47
1994	11	5	61
1995	12	3	33
1996	13	1	74
1997	14	2	65
1998	15	2	58
1999	16	2	44

FM1, strength mode of failure; FM2, fatigue mode of failure.

Example 4.6: Random Censoring

Table 4.5 contains time to failure data in which two FMs were observed in a structural member. The data in this example were generated using MCS. A simulation cycle is terminated once a failure occurs according to one of the modes at time t, making this time t for the other mode as a time to censoring, and the next simulation cycle repeats the process from the start of life (1984) of the structural member. The simulation process is terminated at the end of 1999 in case of no failure. The table shows data for 20,000 simulations and reported as counts of failure by mode in each calendar year based on these repetitions.

4.3.3 Availability

The sum of time to failure and time to repair/replacement including time for any maintenance action resulting in restoration of a failed product to a functioning state can be combined in one measure of probability to find a given product in a functioning state. If the time to failure is characterized by its mean, mean time to failure (MTTF), and the time to repair is characterized by its mean, mean time to repair (MTTR), a definition of this probability of finding a given product in a functioning state can be given by the following ratio for *availability* (A):

$$A = \frac{\text{MTTF}}{\text{MTTF} + \text{MTTR}} \tag{4.48}$$

The above ratio, the availability of the product, is widely used in reliability and risk assessment.

4.3.4 Reliability, Failure Rates, and Hazard Functions

As a random variable, the time to failure (TTF, or T for short) is completely defined by its *reliability function*, $R(t)$, which is traditionally defined as the probability that a unit or a component does not fail in time interval $(0, t]$ or, equivalently, the probability that the unit or the component survives the time interval $(0, t]$ under a specified environment, such as stress conditions (e.g., mechanical and/or electrical load, temperature, humidity). For each product, the allowable stress conditions, as commonly given in the technical specifications, are based on analyzing the uncertainty associated with this time to failure. The probability part of this definition of the TTF can be expressed using the reliability function $R(t)$ as follows:

$$R(t) = P(T > t) \tag{4.49}$$

where:
- P is probability
- T is time to failure
- t is any time period

The reliability function is also called the *survivor* (or *survivorship*) *function.*

Another function that can completely define any random variable (TTF as well as TTR) is the CDF, $F(t)$, which is related to the respective reliability function as follows:

$$F(t) = 1 - R(t) = P(T \leq t) \tag{4.50}$$

The CDF is the probability that the product does not survive the time interval $(0, t]$.

Assuming the TTF to be a random variable, continuous, and positively defined, and $F(t)$ to be differentiable, the CDF can be written as follows:

$$F(t) = \int_0^t f(x)\mathrm{d}x \quad \text{for } t > 0 \tag{4.51}$$

where the function $f(t)$ is the so-called PDF, or unconditional density function of the TTF, which is different from the *hazard* (or *failure*) *rate function*, considered as a conditional PDF. The hazard (or failure) rate function is introduced in Section 4.3.5.

Some examples of commonly used distributions of the TTF for engineering products emphasizing their reliability functions are briefly discussed in Sections 4.3.4.1 through 4.3.4.3. Appendix A provides a summary coverage of these distributions.

4.3.4.1 Exponential Distribution

The exponential distribution has a reliability function $R(t)$ as given by

$$R(t) = \exp(-\lambda t) \tag{4.52}$$

where its parameter λ is the failure rate. The failure rate as a general notion is discussed in a Section 4.3.5. The exponential distribution is characterized by time-invariant failure rate; that is, λ is constant.

4.3.4.2 Weibull Distribution

The reliability function of the two-parameter Weibull distribution is

$$R(t) = \exp\left(-\frac{\beta}{\alpha}t\right) \tag{4.53}$$

where:
- α is the scale parameter
- β is the shape parameter

Comparing Equations 4.52 and 4.53 reveals that the exponential distribution is a specific case of the Weibull distribution, with $\beta = 1$ and $\lambda = 1/\alpha$.

4.3.4.3 Lognormal Distribution

Another widely used probability model for the TTF is the lognormal distribution. This distribution is closely related to the normal distribution because a random variable (T) that is lognormally distributed must have a normally distributed $\ln(T)$. The reliability function of the lognormal distribution is given by

$$R(t) = 1 - \Phi\left(\frac{\ln(t) - \mu}{\sigma}\right) = \Phi\left(-\frac{\ln(t) - \mu}{\sigma}\right) \tag{4.54}$$

where μ and σ are parameters of the lognormal distribution (denoted μ_Y and σ_Y in Equation A.65 and related to the mean and standard deviation in Equation A.66), called the *log mean* and *log standard deviation*, respectively, and

$$\Phi(y) = \frac{1}{\sqrt{2\pi}} \int_{-\infty}^{y} \exp\left(-\frac{x^2}{2}\right) dx \tag{4.55}$$

is the standard normal CDF, that is, for the normal distribution that has a mean of 0 and a standard deviation of 1.

4.3.5 Hazard Functions

The conditional probability $P(t < T \le t + \Delta t | T > t)$ is the failure probability of a product unit in the time interval $(t, t + \Delta t]$, with the condition that the unit is functioning at time t, for small Δt as $\Delta t \to 0$. This conditional probability can be used as a basis for defining the hazard function for the unit by expressing the conditional probability as follows:

$$\begin{aligned} P(t < T \le t + \Delta t | T > t) &= \frac{f(t)}{R(t)}\Delta t \\ &= h(t)\Delta t \end{aligned} \tag{4.56}$$

The function

$$h(t) = \frac{f(t)}{R(t)} \tag{4.57}$$

is the hazard (or failure) rate function (HRF).

The difference between the PDF, $f(t)$, and the HRF, $h(t)$, is clarified using two example situations. The first example situation is based on a new unit that was put to service at time $t = 0$. At time $t = t$, what is the probability that the unit will fail in the interval $(t, t + \Delta t]$ using a small Δt? According to Equation 4.36, this probability is approximately equal to $f(t)$ at time t multiplied by the length of the interval Δt, that is, $f(t)\ \Delta t$. The second situation deals with an identical unit that has survived until time t. What is the probability that the unit will fail in the next small interval $(t, t + \Delta t]$? This conditional probability is approximately equal to the hazard rate $h(t)$ at time t multiplied by the length of the small interval Δt, that is, $h(t)\ \Delta t$.

The CDF, $F(t)$, for the time to failure, T, and the reliability function, $R(t)$, can always be expressed in terms of the so-called *cumulative hazard rate function* (CHRF), $H(t)$, as follows:

$$F(t) = 1 - \exp[-H(t)] \tag{4.58}$$

and

$$R(t) = \exp[-H(t)] \tag{4.59}$$

Based on Equation 4.59, the CHRF can be expressed through the respective reliability function as

$$H(t) = -\ln[R(t)] \tag{4.60}$$

It can be shown that the CHRF and the hazard (failure) rate function are related to each other as

$$h(t) = \frac{\mathrm{d}H(t)}{\mathrm{d}t} \tag{4.61}$$

The CHRF and its estimates must satisfy the following conditions:

$$H(0) = 0 \tag{4.62a}$$

$$\lim_{t \to \infty}[H(t)] = \infty \tag{4.62b}$$

where $H(t)$ is a nondecreasing function that can be expressed as follows:

$$\frac{\mathrm{d}H(t)}{\mathrm{d}t} = h(t) \geq 0 \tag{4.63}$$

For the reliability functions introduced for the exponential, Weibull, and lognormal distributions, the respective hazard functions are given below. For the exponential distribution, the hazard (failure) rate function is constant and is given by

$$h(t) = \lambda \tag{4.64}$$

and the exponential CHRF is

$$H(t) = \lambda t \tag{4.65}$$

The Weibull hazard (failure) rate function is a power law function, which can be written as

$$h(t) = \frac{\beta}{\alpha}\left(\frac{t}{\alpha}\right)^{\beta-1} \tag{4.66}$$

The corresponding Weibull CHRF is

$$H(t) = \frac{t}{\alpha}\beta \tag{4.67}$$

For the lognormal distribution, the cumulative hazard (failure) rate function can be obtained, using Equations 4.60 and 4.54, as

$$H(t) = -\ln\left\{\Phi\left[-\frac{\ln(t) - \mu}{\sigma}\right]\right\} \tag{4.68}$$

for which the function Φ and parameters μ and σ were introduced in a Section 4.3.4.3. The lognormal hazard (failure) rate function can be obtained as the derivative of the corresponding CHRF:

$$\begin{aligned} h(t) &= \frac{\mathrm{d}H(t)}{\mathrm{d}t} \\ &= \frac{\dfrac{1}{t\sigma}\phi\left[\dfrac{\mu - \ln(t)}{\sigma}\right]}{\Phi\left[\dfrac{\mu - \ln(t)}{\sigma}\right]} \end{aligned} \tag{4.69}$$

4.3.6 Selection and Fitting of Reliability Models

In reliability and risk assessment problems, one generally deals with two types of probabilistic models to represent failure and repair time distributions and random processes. In this section, the selection and fitting of distribution functions are introduced.

The best lifetime distribution for a given product is one based on the probabilistic physical model of the product; unfortunately, such models might not be available. Nevertheless, the choice of the appropriate distribution should not be absolutely arbitrary, and at least some physical requirements must be satisfied. For example, the distributions to model time to failure or time to repair must be positively defined. In other words, the probability to observe a negative value of time to failure must be 0. The lognormal, Weibull, and exponential distributions are examples of such distributions. As another example, modeling aging products requires a time to failure distribution having an increasing failure rate, for example, the Weibull distribution has a shape parameter >1.

In some applications or problems, the assessment can be of only the reliability or CDF without parametric estimation based on the chosen distribution function. In such situations, the so-called nonparametric estimation of distribution is sufficient.

4.3.6.1 Complete Data without Censoring

In order to estimate the CHRF and the HRF, as provided in Equations 4.60 and 4.61, respectively, an empirical reliability (survivor) function is needed. The empirical reliability function can be used for parametric fitting of an analytical reliability function. Finally, using Equations 4.60 and 4.61, the hazard functions are evaluated for the time interval of interest.

If the available data are complete (i.e., without censoring), the following empirical reliability (survivor) function (i.e., estimate of the reliability function) can be used:

$$S_n(t) = \begin{cases} 1 & 0 \le t < t_1 \\ \dfrac{n-i}{n} & t_i \le t < t_{i+1} \text{ and } i = 1,\ 2, \ldots, n-1 \\ 0 & t_n \le t < \infty \end{cases} \tag{4.70}$$

where:

t_i is the ith failure time denoted according to their ordered values (order statistics) as $t_1 \le t_2 \le \cdots \le t_k$, where k is the number of failures

n is the sample size

In the case of complete data with distinct failures, $k = n$. The estimate can also be applied to the type I and II right-censored data. In the case of type I censoring, the time interval of $S_n(t)$ estimation is $(0, T]$, where $T = t_0$ is the test (or observation) duration. In the case of type II censoring, the respective time interval is $(0, t_r]$, where t_r is the largest observed failure time. This commonly used estimate $S_n(t)$ is the *empirical survivor function.*

Based on Equation 4.70, an estimate of the CDF of TTF can be obtained as follows:

$$F_n(t) = 1 - S_n(t) \tag{4.71}$$

where:

$F_n(t)$ is an estimate of the CDF of time to failure

Example 4.7: Single-Failure-Mode, Small-Sample Data without Censoring

The single-failure-mode, noncensored data presented in Example 4.4 are used to illustrate the estimation of an empirical reliability function using Equation 4.70. The sample size n in this case is 19. The TTFs and the results of calculations of the empirical survivor function $S_n(t)$ are given in Table 4.6. The results are plotted in Figure 4.13 as points, although sometimes they are plotted as a step function with continuity to the left of the point.

Example 4.8: Single-Failure-Mode, Small-Sample, Type I, Right-Censored Data

Equation 4.70 can be applied to type I and II right-censored data, as noted earlier and which is illustrated in this example. The data for this example are given in Table 4.3, based on single-failure-mode, type I, right-censored data. The TTFs and the calculation results of the empirical survivor function based on Equation 4.70 are given in Table 4.7. The sample size n is 12. Censoring was performed at the end (i.e., without any censoring between failures). The empirical survivor function in the case of right censoring does not reach the 0 value on the right (i.e., at the longest TTF observed). The results are plotted in Figure 4.14 as individual points.

Example 4.9: Single-Failure-Mode, Large-Sample Data

Examples 4.7 and 4.8 illustrated similar treatments for estimating the reliability function for samples with right censoring and samples without censoring. For both cases,

TABLE 4.6

Empirical Survivor Function $S_n(t)$ Based on Data of Example 4.4

Time Order Number	TTF (Years)	Empirical Survivor Function
0	0	19/19 = 1
1	26	18/19 = 0.947368
2	27	17/19 = 0.894737
3	28	16/19 = 0.842105
4	29	15/19 = 0.789474
5	30	14/19 = 140.736842
6	31	13/19 = 0.684211
7	32	12/19 = 0.631579
8	33	11/19 = 0.578947
9	34	10/19 = 0.526316
10	35	9/19 = 0.473684
11	36	8/19 = 0.421053
12	37	7/19 = 0.368421
13	38	6/19 = 0.315789
14	39	5/19 = 0.263158
15	40	4/19 = 0.210526
16	42	3/19 = 0.157895
17	43	2/19 = 0.105263
18	50	1/19 = 0.052632
19	56	0/19 = 0

TTF, time to failure.

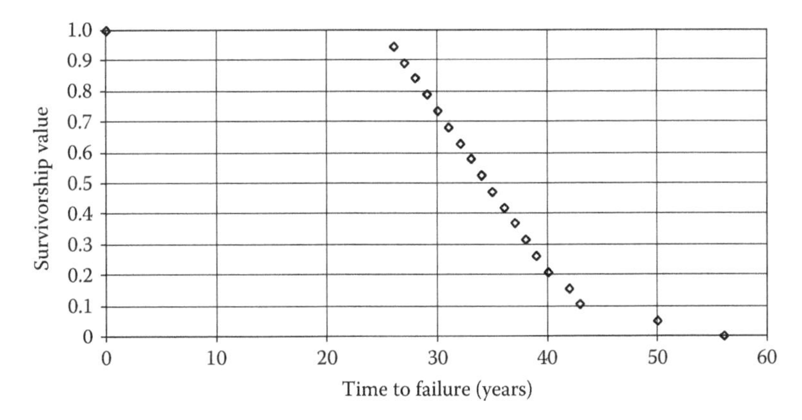

FIGURE 4.13
Survivorship function for single failure mode (FM) without censoring (Example 4.7).

TABLE 4.7

Empirical Survivor Function $S_n(t)$ Based on Data Given in Table 4.3

Time Order Number	Time to Failure (Years)	Time to Censoring (Years)	Empirical Survivor Function
0	0	–	1.000000
1	7	–	0.916667
2	14	–	0.833333
3	15	–	0.750000

(Continued)

TABLE 4.7

(Continued) Empirical Survivor Function $S_n(t)$ Based on Data Given in Table 4.3

Time Order Number	Time to Failure (Years)	Time to Censoring (Years)	Empirical Survivor Function
4	18	–	0.666667
5	31	–	0.583333
6	37	–	0.500000
7	40	–	0.416667
8	46	–	0.333333
9	–	51	0.333333
10	–	51	0.333333
11	–	51	0.333333
12	–	51	0.333333

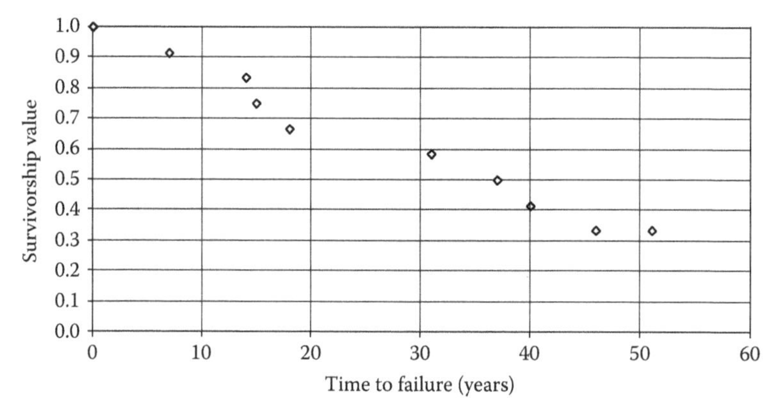

FIGURE 4.14
Survivorship function for single failure mode (FM) with censoring (Example 4.8).

an empirical survivor function was assessed based on Equation 4.70. The data in this example are based on MCS. The TTFs and the estimation results of the empirical survivor function based on Equation 4.70 are given in Table 4.8. The table shows only a portion of data because the simulation process was carried out for 20,000 simulation cycles. The complete dataset covers the years from 1937 to 2060. For example, the survivorship value in the year 1974 is computed as (20,000 − 5)/20,000 = 0.999750. The empirical survivorship values are shown in Figure 4.15. Also, the figure shows the fitted reliability function using loglinear transformation and regression as discussed in Example 4.12.

TABLE 4.8

Example 4.9 Data and Respective Empirical Survivor Function $S_n(t)$

Year	Time to Failure (Years)	Number of Failures	Survivor Function
1937	0	0	1.000000
⋮	⋮	⋮	⋮
1973	36	0	1.000000
1974	37	5	0.999750
1975	38	14	0.999050

(Continued)

TABLE 4.8

(Continued) Example 4.9 Data and Respective Empirical Survivor Function $S_n(t)$

Year	Time to Failure (Years)	Number of Failures	Survivor Function
1976	39	17	0.998200
1977	40	21	0.997150
1978	41	26	0.995850
1979	42	31	0.994300
1980	43	36	0.992500
1981	44	43	0.990350
1982	45	48	0.987950
1983	46	55	0.985200
1984	47	63	0.982050
1985	48	69	0.978600
1986	49	77	0.974750
1987	50	84	0.970550
1988	51	91	0.966000
1989	52	99	0.961050
1990	53	106	0.955750
1991	54	113	0.950100
1992	55	118	0.944200
1993	56	127	0.937850
1994	57	133	0.931200
1995	58	140	0.924200
1996	59	144	0.917000
1997	60	151	0.909450
1998	61	155	0.901700
1999	62	161	0.893650
2000	63	165	0.885400
2001	64	170	0.876900
2002	65	172	0.868300
2003	66	177	0.859450
2004	67	179	0.850500

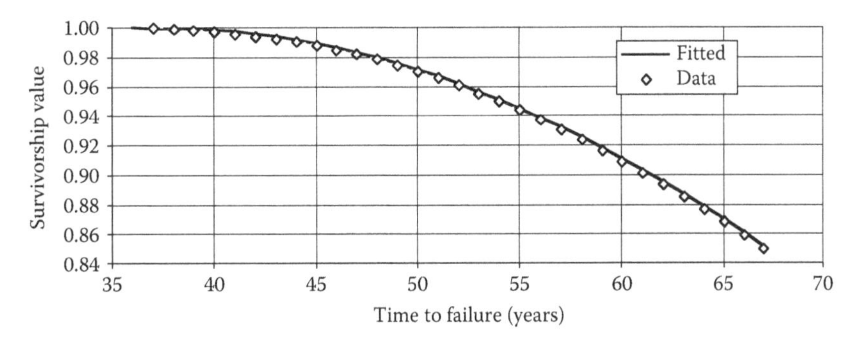

FIGURE 4.15
Empirical survivor function (Example 4.9) and fitted reliability function using loglinear transformation and regression (Example 4.12).

4.3.6.2 Samples with Censoring

In the case of censored data, the Kaplan–Meier (or *product limit*) estimation procedure can be applied to obtain the survivor function that accounts for both TTFs and TTCs. The Kaplan–Meier estimation procedure is based on a sample of n items, among which only k values are distinct failure times with r observed failures. Therefore, $(r - k)$ repeated (nondistinct) failure times exist. The failure times are denoted similar to Equations 4.46 and 4.47, according to their ordered values: $t_1 \leq t_2 \leq \cdots \leq t_k$ and $t_0 = 0$. The number of items under observation (censoring) just before t_j is denoted by n_j. The number of failures at t_j is denoted by d_j. Then, the following relationship holds:

$$n_{j+1} = n_j - d_j \tag{4.72}$$

Under these conditions, the product limit estimate of the reliability function, $S_n(t)$, is given by

$$S_n(t) = \begin{cases} 1 & 0 \leq t < t_1 \\ \prod_{j=1}^{i} \left(\dfrac{n_j - d_j}{n_j} \right) & t_i \leq t < t_{i+1} \text{ for } i = 1, 2, \ldots, k - 1 \\ 0 & t_k \leq t < \infty \end{cases} \tag{4.73}$$

where:
  t is the time to failure for a component

For cases where $d_j = 1$, (i.e., one failure at time t_j), Equation 4.73 becomes

$$S_n(t) = \begin{cases} 1 & 0 \leq t < t_1 \\ \prod_{j=1}^{i} \left(\dfrac{n_j - 1}{n_j} \right) & t_i \leq t < t_{i+1} \text{ for } i = 1, 2, \ldots, k - 1 \\ 0 & t_k \leq t < \infty \end{cases} \tag{4.74}$$

For uncensored (complete) samples with $d_j = 1$, the product limit estimate coincides with the empirical $S_n(t)$ given by Equation 4.70 as follows:

$$\text{For } i = 1: \quad S_n(t) = \prod_{j=1}^{1} \left(\frac{n_j - 1}{n_j} \right) = \left(\frac{n-1}{n} \right) = \left(\frac{n-1}{n} \right)$$

$$\text{For } i = 2: \quad S_n(t) = \prod_{j=1}^{2} \left(\frac{n_j - 1}{n_j} \right) = \left(\frac{n-1}{n} \right)\left(\frac{n-2}{n-1} \right) = \left(\frac{n-2}{n} \right)$$

$$\text{For } i = 3: \quad S_n(t) = \prod_{j=1}^{3} \left(\frac{n_j - 1}{n_j} \right) = \left(\frac{n-1}{n} \right)\left(\frac{n-2}{n-1} \right)\left(\frac{n-3}{n-2} \right) = \left(\frac{n-3}{n} \right)$$

$$\vdots \qquad \vdots \qquad \vdots$$

$$\text{Therefore, for any } i: \quad S_n(t) = \prod_{j=1}^{i} \left(\frac{n_j - 1}{n_j} \right) = \left(\frac{n-i}{n} \right)$$

Example 4.10: A Small Sample with Two Failure Modes

This example illustrates estimating the reliability function based on randomly censored data. In this example, life data consist of times to failure related to multiple FMs. The reliability function corresponding to each FM is estimated using Equation 4.73. For example, two FMs, FM1 and FM2, are considered here. Such a TTF sample can be represented as follows:

$$t_1(\text{FM1}) \le t_2(\text{FM1}) \le t_3(\text{FM2}) \le t_4(\text{FM1}) \le \cdots \le t_k(\text{FM2})$$

If the reliability function related to only FM1 needs to be estimated, all TTFs having the FM2 must be treated as TTCs. For cases involving more than two FMs in a sample, the reliability function for a specific FM*i* can be estimated by treating the TTFs associated with FMs other than FM*i* as TTCs. It should be noted that censoring means that an item survived up to the time of censoring and the item was removed from testing or service.

A sample of 12 TTFs associated with two FMs, FM1 and FM2, is shown in Table 4.9. The calculations of the empirical survivor function based on Equation 4.73 are given in the table. The computational details of the empirical survivorship values for FM1 are provided in Table 4.10, where sample size n is 12 and c_j is the number of items censored at time j. At time order 7 in Tables 4.9 and 4.10, $S_n(16.2) = 1 - 1/6 = 0.8333$. Similarly, at time order number 9 in these tables, $S_n(49.6) = (1 - 1/6)(1 - 1/4) = 0.625$. Other values in the table can be computed in a similar manner.

Example 4.11: Large Sample with Two Failure Modes

The data given in this example were generated by the US Army Corps of Engineers (USACE) for lock and dam gates for the purpose of demonstration. Two FMs, FM1 and FM2, are simulated in this example. A portion of these data related to

TABLE 4.9

Small-Sample Data of Example 4.10 and Respective Empirical Survivor Function for FM1 $S_n(t)$

		Number of Occurrences of a Given FM		
Time Order Number	Time to Failure (Years)	Strength (FM1)	Failure (FM2)	Empirical Survivor Function for M1 (Strength)
0	0	–	–	1.000000
1	0.1	0	1	1.000000
2	1.1	0	1	1.000000
3	1.9	0	1	1.000000
4	6.2	0	1	1.000000
5	9.0	0	1	1.000000
6	11.7	0	1	1.000000
7	16.2	1	0	0.833333
8	21.3	0	1	0.833333
9	49.6	1	0	0.625000
10	51.0	1	0	0.416667
11	51.7	1	0	0.208333
12	68.3	1	0	0.000000

FM, failure mode; FM1, strength mode of failure; FM2, fatigue mode of failure.

TABLE 4.10

Computational Details of Empirical Survivor Function for FM1 $S_n(t)$ (Example 4.10)

Time Order Number (j)	Time to Failure (t_j) (Years)	Number of Failures for FM1 (d_j)	Number of Censorings for FM1 (c_j)	$n_j = n_{j-1} - d_{j-1} - c_{j-1}$	$(1 - d_j/n_j)$	Empirical Survivor Function for FM1
0	0	–	–	–	–	1.000000
1	0.1	0	1	12	–	1.000000
2	1.1	0	1	11	–	1.000000
3	1.9	0	1	10	–	1.000000
4	6.2	0	1	9	–	1.000000
5	9.0	0	1	8	–	1.000000
6	11.7	0	1	7	–	1.000000
7	16.2	1	0	6	1 – 1/6	0.833333
8	21.3	0	1	5	–	0.833333
9	49.6	1	0	4	1 – 1/4	0.625000
10	51.0	1	0	3	1 – 1/3	0.416667
11	51.7	1	0	2	1 – 1/2	0.208333
12	68.3	1	0	1	0	0.000000

FM1, strength mode of failure.

one component is examined here. The full sample size is 20,000. The TTFs and the results of calculations of the empirical survivor function based on Equation 4.73 are given in Table 4.11. The complete dataset covers years from 1984 until 2060. The results are plotted in Figure 4.16 as a step function. The figure also shows the fitted reliability function using loglinear transformation and regression (as discussed in Example 4.13).

TABLE 4.11

Example 4.11 Data and Empirical Survivor Function for FM1 $S_n(t)$

Year	Time to Failure (Years)	Number of Occurrences of a Given FM		Survivor Function for FM1 (Strength)
		Strength (FM1)	Fatigue (FM2)	
1984	0	0	0	1.000000
1985	1	7	0	0.999650
1986	2	6	0	0.999350
1987	3	3	0	0.999200
1988	4	0	0	0.999200
1989	5	1	7	0.999150
1990	6	1	12	0.999100
1991	7	0	20	0.999100
1992	8	1	36	0.999050
1993	9	1	47	0.999000
1994	10	5	61	0.998748
1995	11	3	33	0.998597
1996	12	1	74	0.998546
1997	13	2	65	0.998445
1998	14	2	58	0.998343

(Continued)

TABLE 4.11

(Continued) Example 4.11 Data and Empirical Survivor Function for FM1 $S_n(t)$

Year	Time to Failure (Years)	Number of Occurrences of a Given FM		Survivor Function for FM1 (Strength)
		Strength (FM1)	Fatigue (FM2)	
1999	15	2	44	0.998241
2000	16	1	55	0.998190
2001	17	2	64	0.998087
2002	18	1	73	0.998036
2003	19	1	67	0.997984

FM, failure mode; FM1, strength mode of failure; FM2, fatigue mode of failure.

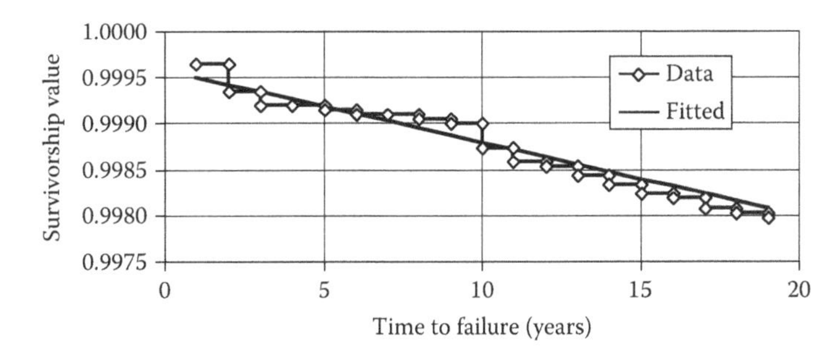

FIGURE 4.16
Empirical survivor function (Example 4.11) and fitted reliability function using loglinear regression and transformation (Example 4.13).

4.3.6.3 Parametric Reliability Functions

Besides the traditional distribution estimation methods, such as the method of moments and maximum likelihood described in Appendix A, the empirical survivor functions can be used to fit analytical reliability functions. After evaluating an empirical reliability function, analytical parametric HRFs, such as those given by Equation 4.59, can be fitted using the empirical survivorship function obtained from life data. The Weibull reliability function was used in studies performed for the USACE as provided in Equation 4.53, including the exponential reliability function as its specific case (Ayyub and Kaminskiy 2001). Also, the reliability function having a polynomial CHRF was used as follows:

$$R(t) = \exp[-H(t)] \tag{4.75}$$

where

$$H(t) = a_0 + a_1 t + a_2 t^2 \tag{4.76}$$

Therefore, the hazard function is given by

$$h(t) = a_1 + 2a_2 t \tag{4.77}$$

For the special case where the parameters a_0 and a_2 equal 0, Equation 4.76 reduces to the exponential distribution. For the special case where the parameters a_0 and a_1 equal 0,

Equation 4.76 reduces to the specific case of the Weibull distribution with the shape parameter of 2. This specific case is known as the *Rayleigh distribution*. The estimation of the parameters in these models can be based on linear or nonlinear curve fitting. Methods involving curve fitting are described in Sections 4.3.6.4 and 4.3.6.5.

4.3.6.4 Parameter Estimation Using Loglinear Transformation

Equations 4.75 through 4.77 provide exponential models with parameters a_0, a_1, and a_2. The logarithmic transformation of a linear and a quadratic polynomial CHRF reliability function leads to the following respective expressions:

$$-\ln[R(t)] = a_0 + a_1 t \tag{4.78}$$

$$-\ln[R(t)] = a_0 + a_1 t + a_2 t^2 \tag{4.79}$$

This loglinear transformation permits the use of linear regression methods to solve for the unknown parameters, a_0, a_1, and a_2, using the least-squares method. Using y to denote the left side of these equations, $y = -\ln[R(t)]$, the following solutions can be obtained for the parameters according to Equation 4.78:

$$a_1 = \frac{\sum t_i y_i - \frac{1}{n}\sum t_i \sum y_i}{\sum t_i^2 - \frac{1}{n}\left(\sum t_i\right)^2} \tag{4.80a}$$

and

$$a_0 = \frac{\sum y_i}{n} - \frac{a_1 \sum x_i}{n} \tag{4.80b}$$

where all summations are performed over all the empirical values of the survivorship function. The parameters of Equation 4.79 can be obtained by solving the following simultaneous equations that can be derived from least-squares optimization:

$$\begin{aligned} na_0 + a_1\sum t_i + a_2\sum t_i^2 &= \sum y_i \\ a_0\sum t_i + a_1\sum t_i^2 + a_2\sum t_i^3 &= \sum t_i y_i \\ a_0\sum t_i^2 + a_1\sum t_i^3 + a_2\sum t_i^4 &= \sum t_i^2 y_i \end{aligned} \tag{4.81}$$

The estimated parameters based on this method are approximate, because applying standard normal linear regression techniques results in violation of some linear regression assumptions, such as the additive normally distributed errors. The violation results from transforming $R(t)$ to $\ln[R(t)]$, producing parameter estimates that are based on the least squares in the $\ln[R(t)]$ space, not the $R(t)$ space. This shortcoming can be alleviated by performing the least-squares estimation using the nonlinear model for $R(t)$ as given in

Equation 4.75 that requires applying numerical optimization methods, as discussed and illustrated in Example 4.14.

Example 4.12: Loglinear Transformation for Parameter Estimation of Example 4.9 Data

The data of Example 4.9 are used to illustrate the use of the loglinear model of Equation 4.79 for parameter estimation as provided in Table 4.12. For Example 4.9 data, the loglinear least-squares estimation gives the following values of the parameter estimates: $a_0 = 0.263018$, $a_1 = -0.013930$ (1/year), and $a_2 = 0.000185$ (1/year^2). All the model parameter estimates

TABLE 4.12

Empirical Survivor Function, $S_n(t)$, and Fitted Reliability Function Using Loglinear Transformation and Regression (Example 4.12)

Year	Time to Failure (Years)	Number of Failures	Survivor Function	Fitted Reliability Function
1937	0	0	1.000000	–
⋮	⋮	⋮	⋮	⋮
1973	36	0	1.000000	–
1974	37	5	0.999750	0.999127
1975	38	14	0.999050	0.999182
1976	39	17	0.998200	0.998868
1977	40	21	0.997150	0.998184
1978	41	26	0.995850	0.997131
1979	42	31	0.994300	0.995711
1980	43	36	0.992500	0.993926
1981	44	43	0.990350	0.991776
1982	45	48	0.987950	0.989265
1983	46	55	0.985200	0.986395
1984	47	63	0.982050	0.983170
1985	48	69	0.978600	0.979593
1986	49	77	0.974750	0.975668
1987	50	84	0.970550	0.971399
1988	51	91	0.966000	0.966791
1989	52	99	0.961050	0.961849
1990	53	106	0.955750	0.956578
1991	54	113	0.950100	0.950984
1992	55	118	0.944200	0.945073
1993	56	127	0.937850	0.938851
1994	57	133	0.931200	0.932326
1995	58	140	0.924200	0.925503
1996	59	144	0.917000	0.918390
1997	60	151	0.909450	0.910995
1998	61	155	0.901700	0.903325
1999	62	161	0.893650	0.895388
2000	63	165	0.885400	0.887193
2001	64	170	0.876900	0.878747
2002	65	172	0.868300	0.870060
2003	66	177	0.859450	0.861140
2004	67	179	0.850500	0.851996

TABLE 4.13

Empirical Survivor Function, $S_n(t)$, and Fitted Reliability Function Using Loglinear Regression and Transformation (Example 4.13)

Year	Time to Failure (Years)	Number of Occurrences of FM1 (Strength)	Survivor Function for FM1 (Strength)	Fitted Reliability Function
1984	0	0	1.000000	–
1985	1	7	0.999650	0.999507
1986	2	6	0.999350	0.999428
1987	3	3	0.999200	0.999349
1988	4	0	0.999200	0.999270
1989	5	1	0.999150	0.999191
1990	6	1	0.999100	0.999112
1991	7	0	0.999100	0.999033
1992	8	1	0.999050	0.998955
1993	9	1	0.999000	0.998876
1994	10	5	0.998748	0.998797
1995	11	3	0.998597	0.998718
1996	12	1	0.998546	0.998639
1997	13	2	0.998445	0.998560
1998	14	2	0.998343	0.998481
1999	15	2	0.998241	0.998402
2000	16	1	0.998190	0.998323
2001	17	2	0.998087	0.998245
2002	18	1	0.998036	0.998166
2003	19	1	0.997984	0.998087

are of high statistical significance. The multiple adjusted correlation coefficient squared (R^2) is 0.999, indicating a good fit. The fitted values of the reliability function and the respective empirical survivor function are given in Table 4.12 and Figure 4.15.

Example 4.13: Loglinear Transformation for Parameter Estimation of Example 4.11 Data

In this example, the reliability function is fitted in a manner similar to that for Example 4.11 for FM1, which corresponds to the strength FM. The loglinear least-squares estimation produced the following values as parameter estimates: a_0 = 0.000414, and a_1 = 0.000079 (1/year). The parameter a_2 turns out to be statistically insignificant; therefore, this parameter was excluded from the model. The multiple adjusted correlation coefficient squared (R^2) = 0.971, which shows a sufficiently good fit. The fitted values of reliability function and the empirical survivor function are given in Table 4.13 and Figure 4.16.

4.3.6.5 Nonlinear Model Estimation

With three parameters, the model provided by Equations 4.75 through 4.77 is nonlinear with respect to time. The parameters can be estimated and the errors analyzed using nonlinear regression analysis procedures. The estimation of nonlinear model parameters can be essentially based on using numerical optimization methods. For this reason, the same dataset treated by different nonlinear estimation procedures might yield different results. The procedure recommended and used in this section is minimization of the sum of the error squared. Most nonlinear estimation procedures require some initial estimates of the parameters in order to start their iterative solution procedures. In the case of loglinear models, or other models that can be transformed to linear ones, the estimates obtained using loglinear transformation can serve as good initial estimates. The examples in this section illustrate a nonlinear estimation procedure.

Example 4.14: Fitting a Nonlinear Model for the Data of Example 4.9

In this example, the nonlinear model of Equations 4.75 through 4.77 is used and its parameters are estimated using nonlinear fitting. For the data of Example 4.9, the nonlinear least-squares estimation gives the following values of the parameter estimates: $a_0 = 0.262649$, $a_1 = -0.013915$ (1/year), and $a_2 = 0.000185$ (1/year^2). These estimates were obtained using the quasi-Newton method of optimization to minimize the sum of the errors, that is, residuals, squared. A numerical algorithm is advised for this purpose, or commercially available software can be used. Such optimization methods require an initial estimate of the solution. The estimates obtained using loglinear estimation from Example 4.12 were used as initial estimates. The estimates obtained using the nonlinear estimation are very close to those obtained using loglinear estimation, with the estimates of a_2 being equal. Both approaches result in good fit, as shown in Table 4.14. Nevertheless, the nonlinear estimates provide better fit based on the sums of the squared residuals; the sum of the squared residuals for the nonlinear model fit is 0.0000046, whereas it is only 0.000962 for the model obtained by loglinear estimation. The fitted reliability function and the empirical survivor function are given in Table 4.14.

TABLE 4.14

Empirical Survivor Function, $S_n(t)$, and Fitted Reliability Function Using Loglinear Regression and Nonlinear Regression (Example 4.14)

Year	Time to Failure (Years)	Number of Failures	Empirical Survivor Function	Fitted Reliability Function	
				Loglinear Regression	Nonlinear Regression
1937	0	0	1.000000	–	–
⋮	⋮	⋮	⋮	⋮	⋮
1973	36	0	1.000000	–	–
1974	37	5	0.999750	0.999127	0.998533
1975	38	14	0.999050	0.999182	0.998551
1976	39	17	0.998200	0.998868	0.998199
1977	40	21	0.997150	0.998184	0.997478
1978	41	26	0.995850	0.997131	0.996388
1979	42	31	0.994300	0.995711	0.994930
1980	43	36	0.992500	0.993926	0.993106
1981	44	43	0.990350	0.991776	0.990918
1982	45	48	0.987950	0.989265	0.988369
1983	46	55	0.985200	0.986395	0.985461

(*Continued*)

TABLE 4.14

(Continued) Empirical Survivor Function, $S_n(t)$, and Fitted Reliability Function Using Loglinear Regression and Nonlinear Regression (Example 4.14)

Year	Time to Failure (Years)	Number of Failures	Empirical Survivor Function	Fitted Reliability Function	
				Loglinear Regression	Nonlinear Regression
1984	47	63	0.982050	0.983170	0.982198
1985	48	69	0.978600	0.979593	0.978582
1986	49	77	0.974750	0.975668	0.974619
1987	50	84	0.970550	0.971399	0.970312
1988	51	91	0.966000	0.966791	0.965667
1989	52	99	0.961050	0.961849	0.960687
1990	53	106	0.955750	0.956578	0.955379
1991	54	113	0.950100	0.950984	0.949748
1992	55	118	0.944200	0.945073	0.943801
1993	56	127	0.937850	0.938851	0.937544
1994	57	133	0.931200	0.932326	0.930983
1995	58	140	0.924200	0.925503	0.924125
1996	59	144	0.917000	0.918390	0.916978
1997	60	151	0.909450	0.910995	0.909549
1998	61	155	0.901700	0.903325	0.901846
1999	62	161	0.893650	0.895388	0.893877
2000	63	165	0.885400	0.887193	0.885650
2001	64	170	0.876900	0.878747	0.877174
2002	65	172	0.868300	0.870060	0.868457
2003	66	177	0.859450	0.861140	0.859508
2004	67	179	0.850500	0.851996	0.850336

Example 4.15: Fitting a Nonlinear Model for the Data of Example 4.11

This example illustrates the fitting of the reliability function similar to Example 4.14 for the strength FM1 described in Example 4.11. The nonlinear least-squares estimation gives the following values of the parameter estimates: $a_0 = 0.000414$, and $a_1 = 0.000086$ (1/year). Similar to the previous example, the estimates obtained using the nonlinear estimation are very close to the respective estimates obtained using loglinear estimation, with the estimates of a_0 being equal. Both approaches result in a good fit, as shown in Table 4.15. Similar to Example 4.14, the nonlinear estimates provide a slightly better fit than the loglinear estimation; the sum of the squared residuals for the nonlinear model fit is 0.000000128, whereas it is 0.000000238 for the model obtained by loglinear estimation in Example 4.13.

4.3.6.6 Probability Plotting

Probability plots are visual representations that show reliability data and preliminary estimation of assumed TTF distribution parameters by graphing the transformed values of an empirical survivor function (or CDF) versus the time (or the transformed time) on a specially constructed probability paper. Reliability data that follow the underlying distribution of a probability paper type will fall on a straight line. Commercial probability papers are available for all the typical life distribution models (e.g., refer to the *2000 Engineering Statistics Handbook* of the National Institute of Standards and Technology). The example that follows illustrates the use of probability plotting of reliability data applied to the Weibull distribution.

TABLE 4.15

Empirical Survivor Function, $S_n(t)$, and Fitted Reliability Function Using Loglinear Regression and Nonlinear Regression (Example 4.15)

Year	Time to Failure (Years)	Number of Occurrences of FM1 (Strength)	Empirical Survivor Function for FM1 (Strength)	Fitted Reliability Function	
				Loglinear Regression	Nonlinear Regression
1984	0	0	1.000000	–	–
1985	1	7	0.999650	0.999507	0.999500
1986	2	6	0.999350	0.999428	0.999414
1987	3	3	0.999200	0.999349	0.999329
1988	4	0	0.999200	0.999270	0.999243
1989	5	1	0.999150	0.999191	0.999158
1990	6	1	0.999100	0.999112	0.999072
1991	7	0	0.999100	0.999033	0.998986
1992	8	1	0.999050	0.998955	0.998901
1993	9	1	0.999000	0.998876	0.998815
1994	10	5	0.998748	0.998797	0.998730
1995	11	3	0.998597	0.998718	0.998644
1996	12	1	0.998546	0.998639	0.998559
1997	13	2	0.998445	0.998560	0.998473
1998	14	2	0.998343	0.998481	0.998387
1999	15	2	0.998241	0.998402	0.998302
2000	16	1	0.998190	0.998323	0.998216
2001	17	2	0.998087	0.998245	0.998131
2002	18	1	0.998036	0.998166	0.998045
2003	19	1	0.997984	0.998087	0.997960

FM1, strength mode of failure.

Example 4.16: Probability Plotting of Weibull Distribution for the Data of Example 4.10

A transformation of the reliability Weibull function can be developed by taking the logarithm of the reliability function of Equation 4.53 twice as follows:

$$\ln\left\{\ln\left[\frac{1}{R(t)}\right]\right\} = \beta\ln(t) - \beta\ln(\alpha) \tag{4.82}$$

By denoting $y = \ln\{\ln[1/R(t)]\}$ and $x = \ln(t)$, y is therefore linear in x with a slope of β. Replacing $R(t)$ by the respective empirical survivor function, $S_n(t)$, a linear regression procedure can be used to fit the following line to the transformed data: $y(x) = bx + a$. The distribution parameters can be estimated as follows: $\beta = b$ and $\alpha = \exp(-a/\beta)$.

The values of these estimates for the data of Example 4.10 are as follows: $\beta = 0.5554$, $\alpha = 1543246.1$, and $a = -7.91411$. The fitted reliability function and the respective empirical survivor function are given in Table 4.16. The respective probability plot is given in Figure 4.17. The sum of the squared residuals for the Weibull distribution fitted using the probability paper is 0.000000271, which is worse than 0.000000128 based on the nonlinear estimation in Example 4.15 and 0.000000238 for the model obtained by loglinear estimation in Example 4.13 for the same data. Nevertheless, the probability paper estimates can be used as initial estimates for the nonlinear estimation.

TABLE 4.16

Empirical Survivor Function, $S_n(t)$, and Fitted Weibull Reliability Function Using Probability Paper (Example 4.16)

Year	Time to Failure (Years)	Number of Occurrences of FM1 (Strength)	Survivor Function for FM1 (Strength)	Probability Paper Fitted Reliability Function
1984	0	0	1.000000	–
1985	1	7	0.999650	0.999635
1986	2	6	0.999350	0.999463
1987	3	3	0.999200	0.999327
1988	4	0	0.999200	0.999211
1989	5	1	0.999150	0.999107
1990	6	1	0.999100	0.999012
1991	7	0	0.999100	0.998923
1992	8	1	0.999050	0.998841
1993	9	1	0.999000	0.998762
1994	10	5	0.998748	0.998688
1995	11	3	0.998597	0.998616
1996	12	1	0.998546	0.998548
1997	13	2	0.998445	0.998482
1998	14	2	0.998343	0.998418
1999	15	2	0.998241	0.998356
2000	16	1	0.998190	0.998297
2001	17	2	0.998087	0.998238
2002	18	1	0.998036	0.998181
2003	19	1	0.997984	0.998126

FM1, strength mode of failure.

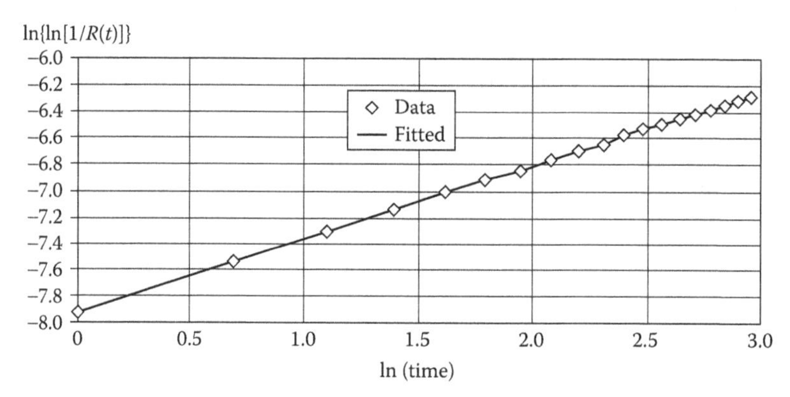

FIGURE 4.17
Weibull probability paper plotting (Example 4.16).

4.3.6.7 Assessment of Hazard Functions

Once the parameters of the underlying life distributions are known (i.e., estimated), the assessment of the CHRF and hazard (failure) rate function is reduced to applying Equations 4.60 and 4.61, respectively. In this section, two examples of the hazard function calculations are provided for demonstration purposes. The first example is based on the reliability function with a polynomial CHRF, as provided by Equations 4.75 through 4.77

and developed in Example 4.14. The second example is based on the Weibull reliability function from Example 4.16.

Example 4.17. Hazard Function Assessment from a Polynomial Cumulative Hazard Function

Example 4.14 demonstrated the development of a polynomial cumulative hazard function from reliability data. The resulting reliability function, expressed according to Equation 4.75 and using the estimated parameters, is

$$R(t) = \exp(-0.262649 + 0.013915t - 0.000185t^2)$$

Using Equation 4.76, the CHRF is

$$H(t) = 0.262649 - 0.013915t + 0.000185t^2$$

where:
t is the time in years

The respective hazard (failure) rate function is the derivative of $H(t)$, as provided by Equation 4.77, which can be written as

$$h(t) = -0.013915 + 0.000370t$$

The results of these calculations are given in Table 4.17 and Figure 4.18. Taking into account that the HRFs are used for projections, the table covers the years from 1990 to 2010. It can be observed from the figure that the hazard (failure) rate function increases with time, which indicates the aging of the equipment.

TABLE 4.17

Hazard Rate Function (HRF) and Cumulative Hazard Rate Function (CHRF) for Reliability Function with a Polynomial CHRF (Example 4.14 Data and Example 4.17 Computations)

Year	Time to Failure (Years)	HRF	CHRF
1980	43	0.001995	0.006369
1981	44	0.002365	0.008549
1982	45	0.002735	0.011099
1983	46	0.003105	0.014019
1984	47	0.003475	0.017309
1985	48	0.003845	0.020969
1986	49	0.004215	0.024999
1987	50	0.004585	0.029399
1988	51	0.004955	0.034169
1989	52	0.005325	0.039309
1990	53	0.005695	0.044819
1991	54	0.006065	0.050699
1992	55	0.006435	0.056949
1993	56	0.006805	0.063569
1994	57	0.007175	0.070559
1995	58	0.007545	0.077919
1996	59	0.007915	0.085649
1997	60	0.008285	0.093749

(Continued)

TABLE 4.17

(Continued) Hazard Rate Function (HRF) and Cumulative Hazard Rate Function (CHRF) for Reliability Function with a Polynomial CHRF (Example 4.14 Data and Example 4.17 Computations)

Year	Time to Failure (Years)	HRF	CHRF
1998	61	0.008655	0.102219
1999	62	0.009025	0.111059
2000	63	0.009395	0.120269
2001	64	0.009765	0.129849
2002	65	0.010135	0.139799
2003	66	0.010505	0.150119
2004	67	0.010875	0.160809
2005	68	0.011245	0.171869
2006	69	0.011615	0.183299
2007	70	0.011985	0.195099
2008	71	0.012355	0.207269
2009	72	0.012725	0.219809
2010	73	0.013095	0.232719

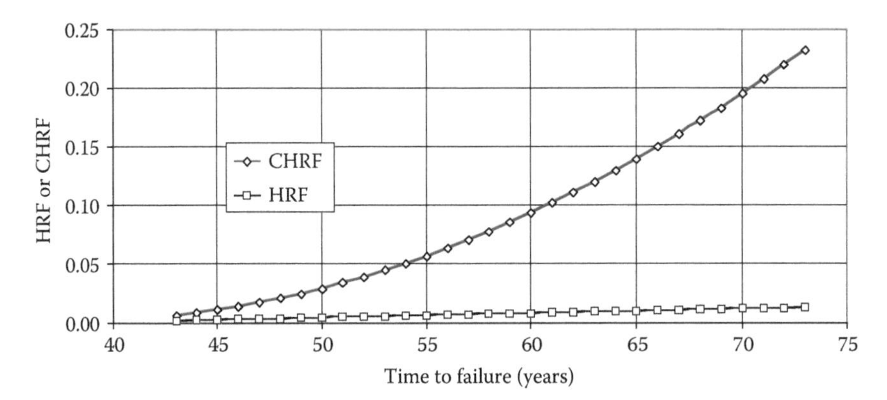

FIGURE 4.18
Cumulative hazard rate function (CHRF) and hazard rate function (HRF) (Example 4.17).

Example 4.18: Assessing the Hazard Function for the Weibull Distribution

This example is based on the Weibull reliability function obtained using the probability plotting from Example 4.16. The Weibull CHRF $H(t)$ is given by Equation 4.67 and the respective hazard (failure) rate function $h(t)$ by Equation 4.66. Using these equations and the estimates of the distribution parameters from Example 4.16, the following expressions for $H(t)$ and $h(t)$ can be obtained:

$$H(t) = \left(\frac{t}{1543246.1}\right)^{0.5554}$$

$$h(t) = \left(\frac{0.5554}{1543246.1}\right)\left(\frac{t}{1543246.1}\right)^{0.5554-1} = 3.60\times10^{-7}\left(\frac{t}{1543246.1}\right)^{-0.4446}$$

TABLE 4.18

Hazard Rate Function (HRF) and Cumulative Hazard Rate Function (CHRF) for Weibull Reliability Function (Example 4.16 Data and Example 4.18 Computations)

Year	Time to Failure (Years)	HRF	CHRF
1985	1	0.000203025	0.000366
1986	2	0.000149180	0.000537
1987	3	0.000124572	0.000673
1988	4	0.000109616	0.000789
1989	5	9.92629E−05	0.000894
1990	6	9.15341E−05	0.000989
1991	7	8.54709E−05	0.001077
1992	8	8.05444E−05	0.00116
1993	9	7.64351E−05	0.001239
1994	10	7.29372E−05	0.001313
1995	11	6.99110E−05	0.001385
1996	12	6.72582E−05	0.001453
1997	13	6.49067E−05	0.001519
1998	14	6.28030E−05	0.001583
1999	15	6.09058E−05	0.001645
2000	16	5.91830E−05	0.001705
2001	17	5.76091E−05	0.001763
2002	18	5.61636E−05	0.00182
2003	19	5.48296E−05	0.001876
2004	20 (prediction)	5.35934E−05	0.00193
2005	21 (prediction)	5.24433E−05	0.001983
2006	22 (prediction)	5.13698E−05	0.002035
2007	23 (prediction)	5.03645E−05	0.002086
2008	24 (prediction)	4.94205E−05	0.002136
2009	25 (prediction)	4.85316E−05	0.002185
2010	26 (prediction)	4.76927E−05	0.002233

The resulting hazard functions are given in Table 4.18. The table covers the years from 1985 to 2010 that includes predictions beyond the data covering the years 2004 to 2010. Contrary to the previous example, the hazard (failure) rate function in this case is decreasing in time, which shows that the given unit is improving with respect to FM1, which might not be realistic, in which case a different probability distribution should be considered.

4.3.7 Case Study: Reliability Data Analysis of Hydropower Equipment

This case study provides a summary of a small portion of a reliability rehabilitation project carried out by the USACE in 1996. The reliability rehabilitation project consisted of structural and mechanical work on USACE-operated facilities, such as locks, dams, and hydropower plants. The objective of reliability rehabilitation projects is to estimate the capital expenditure required to replace features of structural and nonstructural components and systems in a cost-effective manner. Hydropower equipment and plants are included with major rehabilitation programs that are funded by specific US Congressional appropriations. A justification for rehabilitation should include rigorous technical and economic analyses in order to compete successfully for limited appropriation funds, and technical analysis for hydropower equipment, such as generators, must include reliability analysis of equipment. Although the discussion in this section is limited to hydropower generators, approaches used were applied to other types of hydropower equipment. The general objective of this case study is to illustrate the assessment methods of the time-dependent reliability and hazard functions of hydropower generators.

4.3.7.1 Reliability Data

The data used in this study were taken from the 1993 inventory by the USACE of hydropower equipment. The inventory was obtained from the USACE in the form of a database of records for 785 hydropower generators. The inventory was limited to generators with power (P) of more than 5 MW and plant-on-line (POL) dates after 1930. Table 4.19 contains a fragment of the records available in the database. Each record is related to one generator and consists of the following fields: (1) plant name, (2) unit number, (3) POL date, (4) power (kW), (5) rewind date, (6) rewind rating (kW), (7) rewind reason, (8) age at failure (years), and (9) age or exposure time (years).

Analyzing the data, one can conclude that lifetime data are right randomly censored data. In other words, the age of a generator is either the time to failure (for equipment that was repaired or replaced) or the time to censoring (for equipment that was not repaired or replaced). Because the database included equipment that was installed between 1930 and 1993, the generators installed in the 1930s are based on technologies and materials that might be significantly different than those used, for example, in the 1950s or 1990s. Therefore, the POL date (T) was used to stratify the population of generators into groups as follows: (1) $1970 < T \leq 1993$, (2) $1960 < T \leq 1970$, (3) $1950 < T \leq 1960$, and (4) $1930 < T \leq 1950$. Each group spans 10 years, except the first group, which spans 23 years because no failures were reported for generators with $T > 1980$. Combining the last 23 years in one group produces some failure records within this time span to be used for analysis purposes. An implied assumption in this group breakdown is that technologies and materials used in manufacturing generators are strongly correlated

TABLE 4.19

Fragment of Records in Generator Database

Plant Name	Unit Number	POL Date	Power (kW)	Rewind Date	Rewind Rating	Age at Failure (Years)	Age (Years)
Norris	1	September 1, 1936	50,400	November 1, 1990	55,620	54	54
Wheeler	1	November 1, 1936	32,400	September 1, 1984	35,100	48	48
Wheeler	2	April 1, 1937	32,400	June 1, 1986	35,100	49	49
Ontario Power	9	January 1, 1938	8,776	–	–	0	55
Pickwick	2	June 1, 1938	36,000	December 1, 1986	40,400	49	49
Bonneville	2	June 6, 1938	43,200	January 1, 1975	54,200	37	37
Bonneville	1	July 18, 1938	43,200	–	–	0	55
Pickwick	1	August 1, 1938	36,000	May 1, 1986	40,400	48	48
Guntersville	1	August 1, 1939	24,300	October 1, 1978	28,800	39	39
Guntersville	2	October 1, 1939	24,300	July 1, 1979	28,800	40	40

POL, plant-on-line.

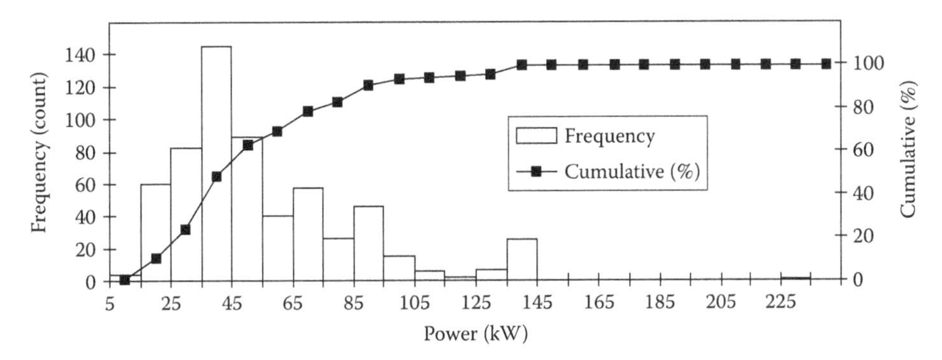

FIGURE 4.19
Power rating of hydropower generators.

TABLE 4.20
Definition of Groups of Hydropower Generators

Group Designation	POL Interval (Years)	Power Capacity (P) (MW)	Number (n) of Units/Number (r) of Failures	Fraction of Surviving Equipment $[(n - r)/n]$
4.1	$1930 < T \leq 1950$	Low power ($P \leq 30$)	63/38	0.396
4.2	$1930 < T \leq 1950$	Medium power ($30 < P \leq 50$)	43/37	0.140
4.3	$1930 < T \leq 1950$	High power ($P > 50$)	17/11	0.353
3.1	$1950 < T \leq 1960$	Low power ($P \leq 30$)	84/17	0.798
3.2	$1950 < T \leq 1960$	Medium power ($30 < P \leq 50$)	62/17	0.726
3.3	$1950 < T \leq 1960$	High power ($P > 50$)	86/29	0.663
2.1	$1960 < T \leq 1970$	Low power ($P \leq 30$)	32/1	0.969
2.2	$1960 < T \leq 1970$	Medium power ($30 < P \leq 50$)	50/9	0.820
2.3	$1960 < T \leq 1970$	High power ($P > 50$)	65/15	0.769
1.1	$1970 < T \leq 1993$	Low power ($P \leq 30$)	85/0	1.000
1.2	$1970 < T \leq 1993$	Medium power ($30 < P \leq 50$)	74/2	0.973
1.3	$1970 < T \leq 1993$	High power ($P > 50$)	124/4	0.968

POL, plant-on-line.

with T; therefore, the variable T can be used to reflect this effect. The second factor used for the stratification is the power rating of generators, P. A histogram of the power ratings of the hydropower generators is shown in Figure 4.19. The data were divided into the following groups based on power capacity P (in megawatts): (1) low power, $P \leq 30$ MW; (2) medium power, $30 < P \leq 50$ MW; and (3) high power, $P > 50$ MW. The simultaneous stratification of the generators population by T and P resulted in 12 groups of low, medium, and high power for each of the four time periods for POL. The number of units in these groups and the fractions of surviving units in each group are provided in Table 4.20.

4.3.7.2 Fitting Reliability Models

The development of reliability assessment models is based on both variables (T and P). If one of them is determined to be insignificant, it can be dropped from the model and the model revised accordingly. The following possible model development scenarios can be considered:

- *Both variables*—power rating P and POL date T—*are significant*. The result in this case consists of 12 reliability models, 1 model for each combination of P and T. Alternatively, one multivariable reliability model can be developed as a function of both P and T.
- *Either* P *or* T *is significant*. The result in this case is three or four reliability models, respectively. Each model in this case is for the different values of the significant variable (P or T). Alternatively, one multivariable reliability model can be developed as a function of either P or T.
- *Both* P *and* T *are insignificant*. The result in this case is one model that is independent of P and T.

4.3.7.2.1 Individual Univariate Models for the 12 POL and Power Combinations

Analyzing Table 4.20, one can notice that one of the 12 groups of POL and power combinations (group 1.1) has no failures. This group without failures was treated using confidence interval estimation for the exponential distribution as discussed at the end of this section. For each of the remaining 11 groups, the reliability model fitting started with constructing the product limit estimates, $S_n(t)$, of the respective reliability functions using Equation 4.74. For example, the reliability function estimates, $S_n(t)$, for group 3 are given in Table 4.21. Then, the following second-order polynomial exponential reliability function defined by Equations 4.75 through 4.77 was fitted to each respective estimate $S_n(t)$:

$$R(t) = \exp[-(a_0 + a_1 t + a_2 t^2)] \tag{4.83}$$

where:
t is TTF in years

TABLE 4.21

Reliability Function Estimate $S_n(t)$ for Groups 3.1–3.3

Years to Failure	Average Power (kW)	POL Year	$S_n(t)$
Group 3.1			
0	18,334.6	1955	1.00000
5	18,334.6	1955	0.98809
22	18,334.6	1955	0.96428
23	18,334.6	1955	0.95238
24	18,334.6	1955	0.94048
25	18,334.6	1955	0.92857
26	18,334.6	1955	0.91667
28	18,334.6	1955	0.90476
30	18,334.6	1955	0.88095
32	18,334.6	1955	0.86904
34	18,334.6	1955	0.85681
38	18,334.6	1955	0.81287
39	18,334.6	1955	0.78665
40	18,334.6	1955	0.75751
41	18,334.6	1955	0.70701

(Continued)

TABLE 4.21

(Continued) Reliability Function Estimate $S_n(t)$ for Groups 3.1–3.3

Years to Failure	Average Power (kW)	POL Year	$S_n(t)$
Group 3.2			
0	40,327.8	1954	1.00000
14	40,327.8	1954	0.98387
19	40,327.8	1954	0.96774
21	40,327.8	1954	0.93548
27	40,327.8	1954	0.91935
29	40,327.8	1954	0.90323
30	40,327.8	1954	0.85484
31	40,327.8	1954	0.79032
33	40,327.8	1954	0.75806
34	40,327.8	1954	0.74159
36	40,327.8	1954	0.72393
Group 3.3			
0	68,929.1	1957	1.00000
16	68,929.1	1957	0.97674
18	68,929.1	1957	0.96512
22	68,929.1	1957	0.94186
25	68,929.1	1957	0.93023
27	68,929.1	1957	0.88372
28	68,929.1	1957	0.84884
29	68,929.1	1957	0.80233
30	68,929.1	1957	0.74419
31	68,929.1	1957	0.68605
32	68,929.1	1957	0.67442
34	68,929.1	1957	0.66169

POL, plant-on-line.

The least-squares estimates of the model parameters were obtained using quasi-Newton and simplex minimization methods. Initial estimates of the model parameters were obtained using loglinear transformation as described in Section 4.3.6.4. The final estimates of model parameters and adjusted squared multiple correlation coefficient R^2 (or multiple R for linear first-order cases) for each group are given in Table 4.22.

For group 1.1, in which no failures were observed, the exponential distribution was used as the model for time to failure distribution. The only possible way to get a rough estimate of the exponential distribution parameter is to construct the following upper confidence limit on the hazard rate parameter a_1 as defined in Equation 4.77 with $a_0 = 0$ and $a_2 = 0$:

$$a_{1u} = \frac{\chi^2_{\alpha,2}}{2T_s} \tag{4.84}$$

where:

a_{1u} is the upper confidence limit on the hazard rate parameter a_1

$\chi^2_{\alpha,2}$ is the lower percentile of the chi-squared distribution at the α level with 2 degrees of freedom

T_s is the total censoring time (i.e., time in service), as given by

TABLE 4.22

Final Estimates of Model Parameters

Group	Model Type	Number of Distinct Failures	a_0	a_1 (year^{-1})	a_2 (year^{-2})	R^2 Value or Adjusted R^2
4.1	Nonlinear (second-order)	17	−1.71776	0.1113	−0.00091	0.98379
4.2	Nonlinear (second-order)	25	0.02563	−0.01068	0.001028	0.99095
4.3	Nonlinear (second-order)	10	−0.0472	0.015172	−1.3E − 05	0.96907
3.1	Nonlinear (second-order)	14	0.04129	−0.00708	0.000323	0.96884
3.2	Nonlinear (second-order)	10	0.27943	−0.03042	0.000895	0.93594
3.3	Nonlinear (second-order)	11	0.71266	−0.0738	0.001965	0.95459
2.1	Loglinear (first-order)	2	0	0.002268	0	1.00000
2.2	Nonlinear (second-order)	5	−0.00049	−0.004	0.00062	0.96568
2.3	Nonlinear (second-order)	9	0.000716	−0.00995	0.000931	0.99464
1.1	Lower limit using the exponential distribution	0	0	0.00061124	0	Not available
1.2	Loglinear (first-order)	2	0	0.001938	0	1.00000
1.3	Loglinear (second-order)	3	0	−0.00062	0.001066	1.00000

$$T_s = \sum_{i=1}^{n} t_{si} \quad (4.85)$$

where t_{si} is the censoring time for the ith equipment unit for $i = 1, 2, \ldots, n$. Using $\alpha = 0.5$ for group 1.1 where $T_s = 1134$ years and $n = 85$, $\chi^2_{\alpha,2}$ was obtained from the tabulated chi-squared distribution tail areas (Ayyub and McCuen 2011) as 1.3863; a_{1u} was calculated as 0.00061124 per year. The resulting a_{1u} for group 1.1 looks reasonable in comparison with other groups, such as group 2.1 in Table 4.22.

4.3.7.2.2 Bivariate Models Using POL Years

In order to study the significance of the power capacity, the following model was fitted for each POL group using the respective average power (P) values in megawatts, as illustrated in Table 4.21:

$$R(t,P) = \exp[-(a_0 + a_1 t + a_2 t^2 + b_1 P + b_2 tP)] \quad (4.86)$$

where:

b_1 and b_2 are power-related model parameters

TABLE 4.23

Bivariate Models Using Average Plant-on-Line (POL) Dates: $R(t, P)$

Group	a_0	a_1 (year^{-1})	a_2 (year^{-2})	b_1 (MW^{-1})	b_2 (year^{-1} MW^{-1})	Adjusted R^2
4	0.02244285	0	0.000484044	0	0	0.561
3	0.09401995	−0.01428786	0.000431951	−0.002186	0.000177	0.878
2	0.00107225	−0.00872835	0.000865890	0	0	0.975
1	−0.00067435	0	0	0	0.00059541	0.981

The significance of each factor included in Equation 4.86 was studied using stepwise regression. The estimated model parameters and adjusted R^2 for each group are given in Table 4.23. Model parameters with zero estimated values are parameters that were determined not to be significant according to stepwise regression. The models in Table 4.23 are less accurate than those in Table 4.22 based on their adjusted R^2. Therefore, the models in Table 4.22 were selected as the final ones. It should be noted that the lower accuracy of the bivariate models can be attributed to the fact that the four bivariate models are based on the same volume of data used for fitting the 12 univariate models of Table 4.22. The bivariate model for group 1 shows that the power capacity might be a significant factor of the equipment aging process.

4.3.7.2.3 *Trivariate Model Using Average Power and POL Years*

By using stepwise regression, the following model was fitted to the entire dataset, using average power values and POL year in the form of two digits, for example, the year 1963 has a T value of 63:

$$R(t,P,T) = \exp[-(a_0 + a_1 t + a_2 t^2 + b_1 P + b_2 T + b_3 PT + b_4 Pt + b_5 PTt)] \tag{4.87}$$

where T is the POL year (in years, counting from 1900) for each average power capacity group (P in megawatts) for each power capacity group. The following factors were determined to be significant: t, t^2, and interaction Pt. Thus, the following model was obtained:

$$R(t,P) = \exp[-(0.030706679 - 0.012733166t + 0.000593775t^2 + 0.000051563Pt)] \tag{4.88}$$

The adjusted R^2 value for this model is 0.765. Thus, again the model of Equation 4.88 turns out to be less accurate than those in Table 4.22 based on their adjusted R^2. Nevertheless, similar to the bivariate model of Equation 4.86, this model of Equation 4.88 shows that the power capacity P is the second (after the unit age t) significant factor of the equipment aging process.

4.4 Software Reliability

Using the reliability definition that was introduced in Section 4.1 as a basis, software reliability can be consistently defined as its ability to provide failure-free operation for a specified time in a specified environment, and this reliability can be consistently quantified as the probability of failure-free operation for a specified time in a specified environment. Three major classes of software reliability analysis methods can be identified:

1. Black-box reliability analysis entails estimating the software reliability based on failure observations from testing or operation of software as a black box without considering or even knowing the internal details or working of the software.
2. Software metric-based reliability analysis focuses on evaluating the software based on its configuration and development process, for example, lines of code, number of statements and/or complexity, or its development process and conditions, such as developer and programmer experience and testing methods used.
3. Architecture-based reliability analysis evaluates the reliability of software as a system from its component reliabilities and the system's architecture, for example, the way the system is constructed from its components. Such approaches are sometimes called component-based reliability estimation, or gray or white-box approaches.

Most software reliability concepts presently in use are adapted from concepts covered in Sections 4.2 and 4.3 for hardware reliability. Adapting and applying hardware reliability concepts to estimate and characterize software reliability should recognize some fundamental differences in the underlying failure processers and nature of hardware compared to software such as the following:

- Hardware failures are generally attributable to physical wearout, deterioration, or degradation, whereas software failures are not generally driven by such mechanisms.
- The so-called independence assumption in modeling hardware failures holds well in the case of physical faults by reducing the complexity of the reliability models and making the use of redundancy very effective for designing hardware with fault tolerance, for example, the use of active redundancy in combination with voting, or standby redundancy through reconfiguration upon failure detection. Therefore, it is feasible to design systems with high hardware reliabilities. These concepts do not generally apply or hold for the case of software reliability.
- The root causes of software failure are sometimes design faults, that is, human error in the development process or maintenance. Such design faults would cause a failure under certain circumstances. The probability of the activation of a design fault is typically usage dependent and time independent. Such software faults could lead to hardware failure, making software and hardware reliability interconnected and interdependent. With an increasing complexity of hardware and software systems, design faults resulting from such an integrated hardware/software system become a difficult-to-predict concern, and the division between hardware and software reliability is somewhat artificial.
- Software failures are generally caused by design faults. In contrast to hardware, software can be perfect, that is, fault-free codes; however, developing complex fault-free software is unfortunately infeasible. Formal software reliability methods can enhance the correctness of software and enables predicting residual errors using reliability growth models, where correctness is a statement in the context of meeting a set of specifications that implicitly presumes such a set to

be correct, that is, correctness does not ensure reliability because the specifications might be faulty. Unfortunately, formal software verification techniques are not effective for large software systems, such as consumer operation systems or word processors. The infeasibility of developing complex, fault-free software systems and the inability of guaranteeing the absence of faults make the assessment of software reliability a necessity to fulfill high-dependability requirements for complex systems.

- Software copies are completely dependent; therefore, many hardware fault tolerance principles ineffective or unsuitable for enhancing software reliability. For example, instead of using redundant copies, software reliability can be improved by using design diversity, for example, using what is termed *N-version programming* for enhancing software reliability.
- Software faults become active and cause failures depending on the usage profile, that is, they depend on the input sequence and operation conditions. The usage profile cannot be predicted with a reasonable degree of certainty since it is driven by the behavior of users, although software is expected to fail in the same way for the same operational conditions and same parameters. Such uncertainty justifies the use of a probabilistic framework.
- User profiles and operation profiles are commonly used in software reliability analysis and measurement. These profiles are used to estimate weight factors for assessing software reliability.

Studies have shown that for complex systems, a significant number of failures are typically caused by software faults.

The concepts of reliability R(t), failure probability $F(t)$, failure density $f(t)$, and hazard rate $h(t)$ are used in software reliability analysis in the same manner as was introduced in Sections 4.2 and 4.3 (see, e.g., Equations 4.57 through 4.63), where t means the execution time that is approximated by the clock time. Additionally, $M(t)$ is introduced as a random process of the number of failures experienced by time t with the mean value function $\mu(t) = E[M(t)]$ and the failure intensity function $\lambda(t) = \mathrm{d}\mu(t)/\mathrm{d}t$, that is, the derivative of the mean value function $\mu(t)$. The term *failure intensity rate* is used herein instead of the failure rate in order to make a distinction between the two rates. Other hardware-related reliability concepts are adopted for use in software reliability including probability of failure per demand taken as $1 - R(t)$, where R is the reliability of a single execution, and availability concepts for combined hardware and software such as downtime, uptime, and reboot time.

Software testing can be used to collect data on the number of faults as a function of time. The resulting empirically constructed cumulative fault count as a function of time can be used to fit a model. This model enables predicting future reliability with additional testing or predicting remaining faults. A suitable model for this purpose that offers the added property of simplicity is

$$M(t) = c[1 - \exp(\lambda t)] \tag{4.89a}$$

where:

c is the initial number of faults that is commonly unknown and should be estimated from curve fitting

$\lambda(t)$ is the time-variant failure intensity rate introduced earlier

Such a model produces a Poisson process for $M(t)$ characterized as

$$P[M(t)=m]=\frac{[cF(t)]^m}{m!}\exp[-cF(t)] \tag{4.89b}$$

This model is based on the assumption that all faults have the same failure intensity function $\lambda(t) = d\mu(t)/dt = f(t)/[1 - F(t)]$. The failure probability function can be determined by integrating the failure intensity function $\lambda(t)$. Other alternate models exist, such as the binomial model and S-shaped model that are not covered herein.

This introductory coverage of software reliability can be enhanced by representing usage profiles that might be different than the testing profiles, accounting for the noninstantaneous removal of faults in actual usage, and including active-standby or cold-standby redundancies if existent. Lyu (1996) and Pham (1999) provide additional information on software reliability.

4.5 Bayesian Methods

The procedures discussed in Section 4.3 are related to the so-called statistical inference. Applying any of such procedures is usually associated with some assumptions, for example, a sample is composed of *uncorrelated identically distributed random variables.* The identically distributed property can be stated according to a specific distribution (e.g., the exponential or Weibull distribution). Such an assumption sometimes is checked using appropriate hypothesis testing procedures. Nevertheless, even if the corresponding hypothesis is not rejected, these characteristics cannot be taken with absolute certainty. In the framework of statistics, data result from observations, tests, measurements, polls, and so on. These data can be viewed as *objective information.*

Bayesian statistical inference is based not only on objective information but also on the so-called subjective information. The subjective information includes sources such as expert opinions, experience based on previously solved problems that are similar to the one under consideration, intuition, and so on. This information is usually used as so-called *prior information,* as opposed to *posterior information* (estimate) regarding the parameters of interest, which is based on the prior information as well as regular statistical samples (objective information). In order to use the prior (subjective) information in Bayesian statistical inference, the subjective information must be expressed in a probabilistic form, which is discussed in Section 4.5.1.

Bayesian statistics is based on Bayes' theorem, which, generally speaking, can be expressed in continuous, discrete, or mixed forms. For the applications considered in this book, the continuous form given below is quite sufficient.

4.5.1 Bayes' Theorem

Bayes' theorem forms the basis for Bayesian methods, as described in Appendix A. Reliability assessment involves estimation of parameters (θ), such as a moment or a probability distribution parameter. It can be any parameter—time to failure or time to repair—or any reliability index, such as the mean time between failures and hazard or failure rate. It is assumed that the parameter θ is a continuous random variable so that the prior and posterior distributions of θ can be represented in the form of continuous PDFs. The continuous prior PDF of θ is denoted $h(\theta)$, and a likelihood function $l(\theta|t)$ can be constructed based on sample data, denoted by t. The likelihood function $l(\theta|t)$ provides an assessment of the occurrence likelihood of the new information given t or as a function of the parameter θ. According to Bayes' theorem, the posterior PDF of θ is given by

$$f(\theta \mid t) = \frac{h(\theta)\, l(\theta \mid t)}{\int_{-\infty}^{\infty} h(\theta)\, l(\theta \mid t)\, d\theta} \tag{4.90}$$

The posterior (Bayes') point estimate of the parameter of interest θ can be computed using the so-called *loss function*. The loss function is a measure of discrepancy between the true value of the parameter θ and its estimate $\hat{\theta}$. Several possible loss functions are available; the most popular one is the squared error loss function, which is given by

$$L(\theta, \hat{\theta}) = (\theta - \hat{\theta})^2 \tag{4.91a}$$

If the likelihood function of Equation 4.91a is used, the corresponding Bayes' point estimate of the posterior mean of θ is

$$\hat{\theta}_{\text{posterior}} = \int_{-\infty}^{\infty} \theta f(\theta|t) d\theta \tag{4.91b}$$

The prior point estimate of θ is

$$\hat{\theta}_{\text{prior}} = \int_{-\infty}^{\infty} \theta h(\theta) d\theta \tag{4.91c}$$

The Bayes' analog of the classical confidence interval is Bayes' probability interval. For constructing the 100(1 − α)% Bayes' probability interval (θ_l, θ_u), the following relationship based on the posterior distribution can be used:

$$P(\theta_l < \theta \leq \theta_u) = \int_{\theta_l}^{\theta_u} f(\theta|t) d\theta = 1 - \alpha \tag{4.92}$$

In reliability and risk analysis, the Bayesian technique is most often applied in estimation of the binomial and exponential (or Poisson) distributions. The respective procedures are briefly discussed in the following sections.

4.5.2 Estimating Binomial Distribution

The binomial distribution plays an important role in reliability and risk analysis. For example, if for a redundant unit, two failures are observed per 12 demands, the probability of failure per demand can be modeled by a binomial probability, and an estimate of this probability is $P = 1/6$. Another example is a situation when n identical units are simultaneously placed in service and observed during a specified time t. The r units failed were not replaced or repaired. In this case, the number of failures, r, can be considered as a discrete random variable having the binomial distribution with parameters n and $p(t)$, where $p(t)$ is the probability of failure of a single unit during time t. The function $p(t)$ is the time to failure CDF, whereas $[1 - p(t)]$ is the reliability or survivor function. An estimate of the failure probability $(\hat{p})$ is $\hat{p} = r/n$, which is also the maximum likelihood estimate.

To obtain the Bayesian estimate for probability p, we can use a binomial test in which the number of units (n) tested is fixed in advance. The probability distribution of the number of failed units (r) during the test is given by the binomial distribution PDF with parameters n and r as follows:

$$f(r;n,p) = \frac{n!}{(n-r)!\,r!} p^r (1-p)^{n-r} \tag{4.93}$$

where:

f is the binomial probability mass function
r is the random variable
n and p are the binomial distribution parameters

The corresponding likelihood function is given by

$$l(p|r) = cp^r (1-p)^{n-r} \tag{4.94}$$

where c is a constant that does not depend on the parameter of interest, p, and can be assigned a value of 1 because constant c drops out from the posterior prediction equation. For any continuous prior distribution of parameter p with PDF $h(p)$, the corresponding posterior PDF can be written as

$$f(p|r) = \frac{p^r (1-p)^{n-r} h(p)}{\int_{-\infty}^{\infty} p^r (1-p)^{n-r} h(p) dp} \tag{4.95}$$

To better understand the difference between statistical inference and Bayes' estimation, the following case of the uniform prior distribution is discussed. The prior distribution in this case is the standard uniform distribution, which is given by

$$h(p) = \begin{cases} 1 & \text{for } 0 < p \le 1 \\ 0 & \text{otherwise} \end{cases} \tag{4.96}$$

Based on Equation 4.95, the respective posterior distribution can be written as

$$f(p \mid r) = \frac{p^{(r+1)-1}(1-p)^{(n-r+1)-1}}{\int_{0}^{1} p^{(r+1)-1}(1-p)^{(n-r+1)-1} dp} \tag{4.97}$$

The posterior PDF of Equation 4.97 is the PDF of the beta distribution that is introduced in Example 4.19. The mean value of this distribution, which is Bayes' estimate of interest $p_{\text{posterior}}$, is given by

$$p_{\text{posterior}} = \frac{r+1}{n+2} \tag{4.98}$$

Example 4.19: Shooting a Target

Assessing the effectiveness of a new weapon system requires life testing. Experience shows that the success rate is ~50%; therefore, a simple test of tossing a coin can be viewed as an accurate representation of this war asset. Tossing a coin three times ($n = 3$) with one success ($r = 1$; e.g., tails up) is considered here. Bayes' estimate of the probability of success according to Equation 4.98 is $p_{\text{posterior}} = 2/5$, which is less than the respective classical estimate (p_C) of r/n, which is equal to 1/3, in this case. The prior and posterior distributions are provided in Figure 4.20. The flat prior distribution used in this example represents, in a sense, a state of equally likely likelihood allocation due to lack of knowledge. As the sample size increases, the classical and Bayes' estimates get closer to each other.

The most widely used prior distribution for parameter p of the binomial distribution is the beta distribution. The PDF of the distribution can be written in the following form:

$$h(b;x_0,n_0) = \begin{cases} \dfrac{\Gamma(n_0)}{\Gamma(x_0)\Gamma(n_0 - x_0)} p^{x_0-1}(1-p)^{n_0-x_0-1} & \text{for} \quad 0 \le p \le 1 \\ 0 & \text{otherwise} \end{cases} \tag{4.99}$$

where $n_0 > x_0 \ge 0$, and $\Gamma(\alpha)$ is the gamma function in terms of α, which is given by

$$\Gamma(\alpha) = \int_0^\infty t^{\alpha-1} e^{-t} dt \tag{4.100}$$

The mean and the variance (var) of the beta distribution (p_{prior}) are given, respectively, by

$$p_{\text{prior}} = \frac{x_0}{n_0} \tag{4.101}$$

$$\text{var}(p_{\text{prior}}) = \frac{x_0(n_0 - x_0)}{n_0^2(n_0 + 1)} \tag{4.102}$$

The mean of Equation 4.101 is the prior mean, if the beta distribution is used as the prior distribution. In the following application, the COV (k) of this distribution is needed:

$$k_{\text{prior}} = \left[\frac{n_0 - x_0}{x_0(n_0 + 1)}\right]^{1/2} \tag{4.103}$$

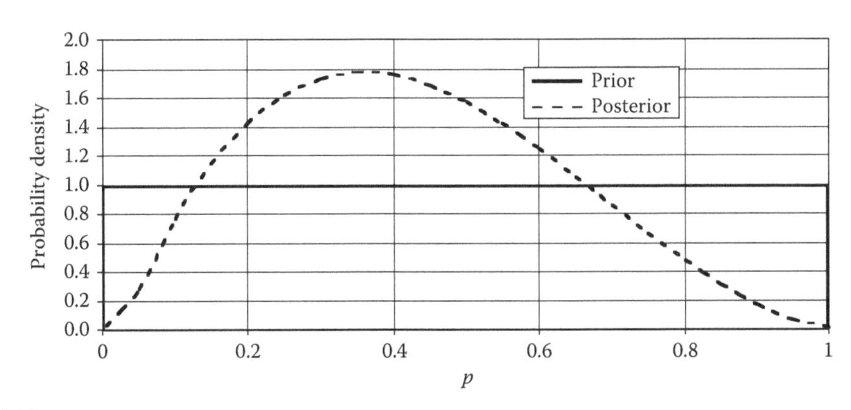

FIGURE 4.20
Prior and posterior personal density functions (PDFs) (Example 4.19).

The PDF of the beta distribution provides a variety of different shapes depending on the values of the distribution parameters. The standard uniform (flat) distribution, used in Example 4.19, is a special case of the beta distribution.

The popularity of the beta distribution, as a prior distribution in estimating the parameter of binomial distribution used as a reliability or survivor function at a given time, stems from having a resulting posterior distribution from the same family of beta distributions. The beta prior distribution belongs to the so-called *conjugate* prior distributions, because, generally speaking, a prior distribution that results in a posterior distribution from the same family as the prior one is referred to as a *conjugate prior distribution.*

Using Bayes' theorem from Equation 4.99 with the binomial likelihood function of Equation 4.94 and the beta prior PDF in the form of Equation 4.99, the posterior PDF can be obtained in the following form:

$$f(p \mid x) = \frac{\Gamma(n + n_0)}{\Gamma(x + x_0)\Gamma(n + n_0 - x - x_0)} p^{(x+x_0)-1}(1 - p)^{(n+n_0-x-x_0)-1} \tag{4.104}$$

which is of course the beta PDF. Therefore, Bayes' point estimate (i.e., the mean of the posterior distribution) is given by

$$p_{\text{posterior}} = \frac{r + x_0}{n + n_0} \tag{4.105}$$

An interpretation of the parameters of prior distribution sometimes is needed. The parameter n_0 can be interpreted as a number of fictitious binomial trials resulting in x_0 fictitious successes. In a reliability context, the same parameters could be interpreted as a number of failures (x_0) observed in a test (or in the field) of n_0 identical units during a fixed time. Assessment of the parameters of the prior distribution is discussed later in this section. The prior distribution parameters can also be estimated based on the real prior data—data collected on similar equipment or data collected on a predecessor of the currently manufactured product, using the respective sample size n_0 and the number of failures observed x_0.

Based on Equation 4.92 and the posterior PDF given by Equation 4.104, the corresponding 100(1 – α)% two-sided Bayesian probability interval for p can be obtained as the simultaneous solutions of the following equations with respect to the respective lower and upper values, p_l and p_u:

$$P(p < p_l) = I_{p_l}(r + x_0, n + n_0 - r - x_0) = \frac{\alpha}{2} \tag{4.106a}$$

$$P(p > p_u) = I_{p_u}(r + x_0, n + n_0 - r - x_0) = 1 - \frac{\alpha}{2} \tag{4.106b}$$

where $I_x(k, m)$ is the incomplete beta function as given by

$$I_x(k,m) = \frac{\Gamma(k + m - 1)}{\Gamma(k)\Gamma(m)} \int_0^x u^{k-1}(1 - u)^{m-1} du \tag{4.107}$$

A practical approach of choosing the parameters of the prior distribution is based on assessing its moments (mean and variance). For example, an expert can provide an estimate of the prior probability p_{prior} of Equation 4.101 and a measure of uncertainty related to this estimate in the form of standard error: the square root of the variance of Equation 4.102 or the COV according to Equation 4.103. Having these estimates and solving a system of two equations, the parameters of interest, n_0 and x_0, can be evaluated.

Example 4.20: Reliability Analysis of Life Rafts

This example illustrates Bayes' reliability estimation process based on the prior subjective information in the form of expert opinion. Life rafts on boats are required for certain types of vessels. An expert has assessed the prior mean (i.e., point estimate) of the reliability function as $p_{\text{prior}} = x_0/n_0 = 0.9$. Selecting the parameters x_0 and n_0 can be based on the value of the COV used as a measure of uncertainty, or accuracy, of the prior point estimate p_{prior}. Some values of the COV and the corresponding values of the parameters x_0 and n_0 obtained as the solutions of Equations 4.101 and 4.103 for $p_{\text{prior}} = x_0/n_0 = 0.9$ are given in Table 4.24.

Example 4.21: Reliability of a New Product

A design engineer assesses the reliability of a new component at the end of its useful life (T = 10,000 hours) as 0.75 with a standard deviation of 0.19. A sample of 100 new components has been tested using an accelerated life technique for 10,000 hours, and 29 failures have been recorded. Given the test results, one needs to find the posterior mean and the 90% Bayesian probability interval for the component reliability. The prior distribution of the component reliability is assumed to have a beta distribution.

The prior mean is subjectively assessed as 0.75 and the COV is 0.19/0.75 = 0.25. Using Equations 4.101 and 4.103, the parameters of the prior distribution are evaluated as $x_0 = 3.15$ and $n_0 = 4.19$. Thus, according to Equation 4.105, the posterior point estimate of the new component reliability is $R(10{,}000) = (3.15 + 71)/(4.19 + 100) = 0.712$. Applying Equation 4.106, the 90% lower and upper confidence limits are found to be 0.637 and 0.782, respectively. Figure 4.21 depicts the respective prior and posterior distributions of estimates of the reliability function at 10,000 hours.

TABLE 4.24

Selection of Parameters for Reliability Estimation of Life Rafts

x_0	n_0	COV
0.9	1	0.236
9	10	0.100
90	100	0.033
900	1000	0.001

COV, coefficient of variance.

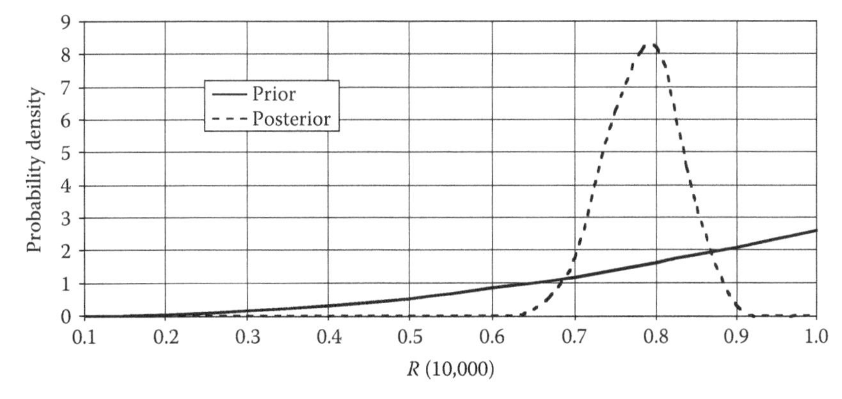

FIGURE 4.21
Prior and posterior personal density functions (PDFs) (Example 4.21).

4.5.3 Parameter Estimation for the Exponential Distribution

In this section, a Bayesian approach for estimation of the parameter λ of the exponential distribution is described. The same approach can be applied to the estimation of the occurrence rate of failures for the homogeneous Poisson process, as well as for the Poisson distribution itself.

A sample of n failure times from the exponential distribution, among which only r are distinct times to failure ($t_1 < t_2 < \cdots < t_r$) and $n - r$ are TTC [$t_{c1}, t_{c2}, \ldots, t_{c(n-r)}$], so that the so-called *total time on test*, T, is given by

$$T = \sum_{i=1}^{r} t_i + \sum_{j=1}^{n-r} t_{ci} \tag{4.108}$$

Based on these data, one needs to estimate the parameter λ for the exponential distribution using Bayes' approach.

Using the gamma distribution as the prior distribution of parameter λ, it is convenient to write the probability density of gamma distribution as a function of λ in the following form:

$$h(\lambda;\delta,\rho) = \frac{1}{\Gamma(\delta)}\rho^{\delta}\lambda^{\delta-1}e^{-\rho\lambda} \tag{4.109}$$

where the parameters $\lambda > 0$, $\rho \leq 0$, and $\delta \leq 0$. These parameters can be interpreted as having δ fictitious failures in p total time, leading to $\lambda = \delta/p$. Selection of the distribution parameters is discussed later, but for the time being, these parameters are assumed to be known. Also, it is assumed that the quadratic likelihood function of Equation 4.91a is used.

For the exponential time-to-failure data, the likelihood function can be written as

$$l(\lambda|\mathrm{t}) = f(t_1)\,f(t_2)\cdots f(t_r)\,R(t_{c,1})\,R(t_{c,2})\cdots R(t_{c,n-r}) \tag{4.110a}$$

where:

$f(t_i)$ is the PDF at the time to failure t_i

$R(t_{c,i})$ is the reliability value at the time to censoring $t_{c,i}$

Therefore, the following likelihood function can be obtained:

$$\begin{aligned} l(\lambda|t) &= \prod_{i=1}^{r}\lambda e^{-\lambda t_i}\prod_{j=1}^{n-r}e^{-\lambda t_{cj}} \\ &= \lambda^r e^{-\lambda T} \end{aligned} \tag{4.110b}$$

where:

T is the total time on test as given by Equation 4.108.

Using Bayes' theorem with the prior distribution given by Equation 4.109 and the likelihood function of Equation 4.100, one can find the posterior density function of the parameter, λ, as

$$f(\lambda \mid T) = \frac{e^{-\lambda(T+\rho)}\lambda^{r+\delta-1}}{\int_0^\infty \lambda^{r+\delta-1}e^{-\lambda(T+\rho)}d\lambda} \tag{4.111}$$

Recalling the definition of the gamma function of Equation 4.80, the integral in the denominator of Equation 4.111 is

$$f(\lambda \mid T) = \frac{(\rho + T)^{\delta+r}}{\Gamma(\delta + r)}\lambda^{r+\delta-1}e^{-\lambda(T+\rho)}$$

or

$$\int_0^\infty \lambda^{r+\delta-1}e^{-\lambda(T+\rho)}d\lambda = \frac{\Gamma(\delta + r)}{(\rho + T)^{\delta+r}}$$

Finally, the posterior PDF of λ can be written as

$$f(\lambda \mid T) = \frac{(\rho + T)^{\delta+r}}{\Gamma(\delta + r)}\lambda^{r+\delta-1}e^{-\lambda(T+\rho)} \tag{4.112}$$

Comparing the above function with the prior one of Equation 4.109 reveals that the posterior distribution is also a gamma distribution with parameters $\rho' = r + \delta$ and $\lambda' = T + \rho$. In other words, the chosen prior gamma distribution turns out to be conjugate one in this case.

Because a quadratic loss function is assumed, the point Bayesian estimate of λ is the mean of the posterior gamma distribution with parameters ρ' and λ'. Therefore, the point Bayesian estimate, $\lambda_{posterior,}$ can be obtained as

$$\lambda_{posterior} = \frac{\rho'}{\lambda'} = \frac{r + \delta}{T + \rho} \tag{4.113}$$

The corresponding probability intervals can be obtained using Equation 4.92. For example, the 100(1 − α)% level, upper, one-sided Bayes' probability interval for λ can be obtained from the following equation based on the posterior distribution (Equation 4.112):

$$P(\lambda < \lambda_u) = 1 - \alpha \tag{4.114}$$

The same upper one-sided probability interval for λ can be expressed in a more convenient form similar to the classical confidence interval (i.e., in terms of the chi-squared distribution) as follows:

$$P\{2\lambda(\rho + T) < \chi^2_{1-\alpha}[2(\delta + r)]\} = 1 - \alpha \tag{4.115}$$

such that

$$\lambda_u = \frac{\chi^2_{1-\alpha,2(\delta+r)}}{2(\rho + T)} \tag{4.116}$$

Contrary to classical estimation, the number of degrees of freedom, $2(\delta + r)$, for Bayes' probability limits is not necessarily integer. The chi-squared value in Equation 4.116 can be obtained from the tables of the chi-squared probability distribution available in probability and statistics textbooks (e.g., Ayyub and McCuen 2011).

Similar to the case of the beta prior distribution, the gamma distribution was selected herein as the prior distribution for illustration purposes. The reliability interpretation of Bayes' estimation of λ can be based on the estimate $\lambda_{posterior}$ of Equation 4.113. The parameter δ can be considered as a prior (fictitious) number of failures observed during a prior (fictitious) test, having ρ as the total time on test. Therefore, one would intuitively choose the prior estimate of λ as the ratio δ/ρ, which coincides with the *mean value* of the prior gamma distribution of Equation 4.109. The respective real-world situation is commonly quite an opposite one. Usually, one has a prior estimate of λ, while the parameters δ and ρ are needed and must be obtained. Having the prior point estimate λ_{prior}, one can only estimate the ratio $\delta/\rho = \lambda_{prior}$. For estimating these parameters separately, some additional information about the degree of belief or accuracy of this prior estimate is required. Because variance of the gamma distribution is δ/ρ^2, the COV of the prior distribution is $1/\delta^{1/2}$ as the ratio of standard deviation to mean. Similar to the case of the beta prior distribution in estimation of binomial probability, the COV can be used as a measure of relative accuracy of the prior point estimate of λ_{prior}. Thus, having an assessment of the prior point estimate, λ_{prior}, and the relative error of this estimate, one can estimate the corresponding parameters of the prior gamma distribution. In order to demonstrate the scale of these errors, the following numerical example is constructed based on a prior point estimate λ_{prior} of 0.01 (in some arbitrary units). The corresponding values of the COV for different values of the parameters δ and ρ, are given in Table 4.25.

Example 4.22: Exponential and Gamma Distributions for Reliability Modeling of Computer Chips

A sample of identical computer chips was tested. Six failures were observed during the test. The total time on the test is 1440 hours. The time-to-failure distribution is assumed to be exponential. The gamma distribution with the mean of 0.01 hour^{-1} and with a COV of 30% was selected as a prior distribution to represent the parameter of interest, λ. The posterior point estimate and the upper 90% probability limit for λ are needed.

Based on the prior mean and COV, the respective parameters of the prior distribution are found as $\delta = 11.1$ and $\rho = 1100$ hours from Table 4.25. Using Equation 4.113, the point posterior estimate of the mean of the hazard rate is evaluated as

$$\lambda_{posterior} = \frac{11.1 + 6}{1110 + 1440} = 6.71 \times 10^{-3} \text{hour}^{-1}$$

TABLE 4.25

Relating the Coefficient of Variance (COV) to Prior Shape and Scale Parameters for the Gamma Distribution

Shape Parameter (δ) as a Prior Number of Failures	Scale Parameter (ρ) as a Prior Total Time on Test	COV
1	100	1.00
5	500	0.45
10	1,000	0.32
100	10,000	0.10

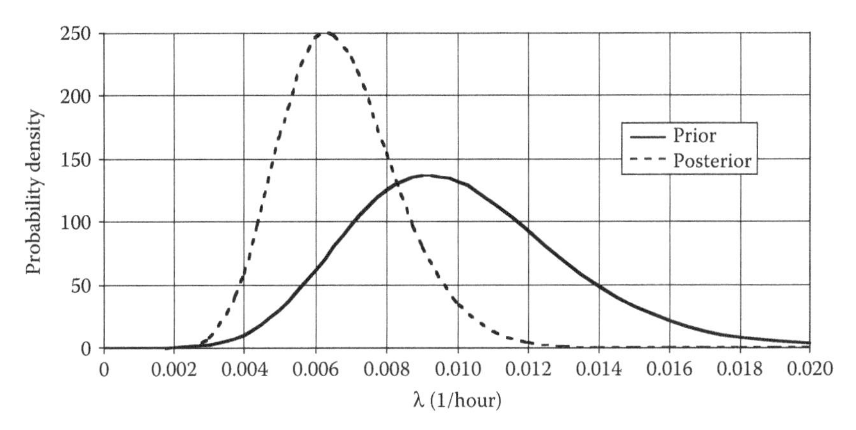

FIGURE 4.22
Prior and posterior personal density functions (PDFs) (Example 4.22).

Using Equation 4.116, the 90% upper limit of the one-sided Bayes probability interval for λ can be computed as follows using $\chi^2_{0.9,17.1} = 9.435$ based on $\alpha = 0.10$ and 17.1 degrees of freedom:

$$\lambda_u = \frac{\chi^2_{0.9,17.1}}{2(2550)} \approx 1.85 \times 10^{-3}\text{hour}^{-1}$$

Figure 4.22 shows the prior and posterior PDFs in this case.

4.6 Reliability Analysis of Systems

The objective of this section is to provide, develop, and demonstrate the methods needed for assessing the hazard functions of the most widely used system models. Systems are assumed to be composed of components that have statistically independent failure events; the reliability functions for these components are defined based on the techniques discussed in the preceding sections of this chapter. Topics involving correlation, ductility, redundancy, and load shedding and redistribution within a system are not discussed in this book.

4.6.1 System Failure Definition

Generally speaking, the problem of assessing system hazard functions can be reduced to the problem of system reliability estimation. As soon as the reliability function of a system is found, the respective hazard functions can be evaluated in the same way as in the case of components. The reliability of a system can be defined based on understanding and modeling the failure of the system. Some systems behave like chains of connected components because a system of this type fails upon the failure of any of the links of its chain-like components. These systems are viewed as being in series with respect to their component connectivity and are termed *weakest link systems*. In parallel systems, the components provide redundancy to each other. A parallel system fails when all its components fail. Redundant systems can be load sharing or nonload sharing. Generally, systems are mixtures of many subsystems, some in series and some in parallel, and can be of a complex nature in terms of connectivity of components and their associated FMs. An analyst must clearly define the failure of a system in the context of failing its components and their associated FMs before computing the reliability and hazard functions of the system.

The so-called *reliability block diagram* (RBD) can be used to represent the structure of a system. An RBD is a success-oriented network describing the function of the system. For most systems considered below, the reliability functions can be evaluated based on their RBD. Reliability assessment at the system starts with fundamental system modeling (i.e., series and parallel systems) and proceeds to more complex systems. Additional information on functional modeling and system definition is provided in Chapter 3.

4.6.2 Series Systems

A series system composed of n components functions if and only if all of its n components are functioning. Figure 4.23 depicts an example of the RBD of a series system consisting of three components. The reliability function of a series system composed of n components, $R_s(t)$, is given by

$$R_s(t) = \prod_{i=1}^{n} R_i(t) \tag{4.117}$$

where:

$R_i(t)$ is the reliability function of the ith component

If a series system is composed of identical components with reliability functions, $R_c(t)$, Equation 4.117 is reduced to

$$R_s(t) = [R_c(t)]^n \tag{4.118}$$

Applying the relationship between a reliability function and its CHRF (i.e., Equations 4.58 and 4.59) to Equation 4.117, the following relationship between the system CHRF, $H_s(t)$, and the CHRFs of its components, $H_i(t)$, can be written as

$$H_s(t) = \sum_{i=1}^{n} H_i(t) \tag{4.119a}$$

By taking the derivative of $H_s(t)$ and applying Equation 4.61, the following relationship between the system hazard (failure) rate function, $h_s(t)$, and the hazard rates of its components, $h_i(t)$, can be obtained:

$$h_s(t) = \sum_{i=1}^{n} h_i(t) \tag{4.119b}$$

For the case of the series system composed of identical components with CHRFs $H_c(t)$ and hazard rates $h_c(t)$, Equation 4.119a and 4.119b is reduced, respectively, to

$$H_s(t) = n\ H_c(t) \tag{4.120a}$$

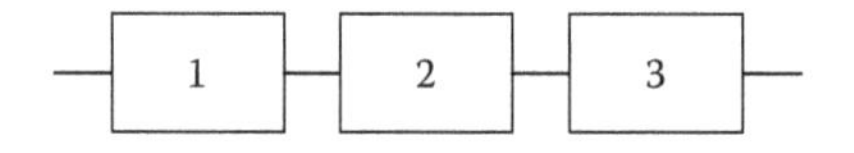

FIGURE 4.23
Series system composed of three components.

$$h_s(t) = n\ h_c(t) \tag{4.120b}$$

Thus, the hazard functions for a series system can be easily evaluated based on the hazard functions of the components of the system.

An examination of Equations 4.119 and 4.120 reveals that the series system composed of components having increasing hazard (failure) rates has an increasing failure rate, as illustrated in Example 4.23.

Example 4.23: Assessing the Hazard Function of a Series System of Three Identical Components

In this example, three identical components with the same hazard function are used to develop the system hazard function. The component hazard functions are given by

$$H_c(t) = 0.262649 - 0.013915t + 0.000185t^2$$

and

$$h_c(t) = -0.013915 + 0.000370t$$

where:
t is the time in years

Applying Equation 4.119a and 4.119b with $n = 3$, the following expressions for the cumulative hazard functions of the series system composed of three identical components with the above given hazard functions can be obtained:

$$H_s(t) = 0.787947 - 0.041745t + 0.000555t^2$$

and

$$h_s(t) = -0.041745 + 0.001110t$$

The resulting hazard functions are given in Table 4.26 and Figures 4.24 and 4.25.

TABLE 4.26

Hazard Rate Function (HRF) and Cumulative Hazard Rate Function (CHRF) for a Series System of Three Identical Components (Example 4.23)

Year	Time to Failure (Years)	System HRF	System CHRF
1980	43	0.005985	0.019107
1981	44	0.007095	0.025647
1982	45	0.008205	0.033297
1983	46	0.009315	0.042057
1984	47	0.010425	0.051927
1985	48	0.011535	0.062907
1986	49	0.012645	0.074997
1987	50	0.013755	0.088197
1988	51	0.014865	0.102507
1989	52	0.015975	0.117927
1990	53	0.017085	0.134457
1991	54	0.018195	0.152097
1992	55	0.019305	0.170847
1993	56	0.020415	0.190707
1994	57	0.021525	0.211677

(Continued)

TABLE 4.26

(Continued) Hazard Rate Function (HRF) and Cumulative Hazard Rate Function (CHRF) for a Series System of Three Identical Components (Example 4.23)

Year	Time to Failure (Years)	System HRF	System CHRF
1995	58	0.022635	0.233757
1996	59	0.023745	0.256947
1997	60	0.024855	0.281247
1998	61	0.025965	0.306657
1999	62	0.027075	0.333177
2000	63	0.028185	0.360807
2001	64	0.029295	0.389547
2002	65	0.030405	0.419397
2003	66	0.031515	0.450357
2004	67	0.032625	0.482427
2005	68	0.033735	0.515607
2006	69	0.034845	0.549897
2007	70	0.035955	0.585297
2008	71	0.037065	0.621807
2009	72	0.038175	0.659427
2010	73	0.039285	0.698157

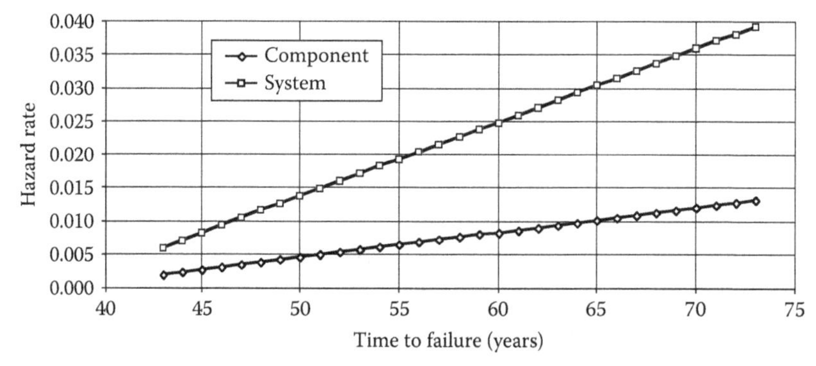

FIGURE 4.24
Hazard rate function (HRF) for a series system of three identical components (Example 4.23).

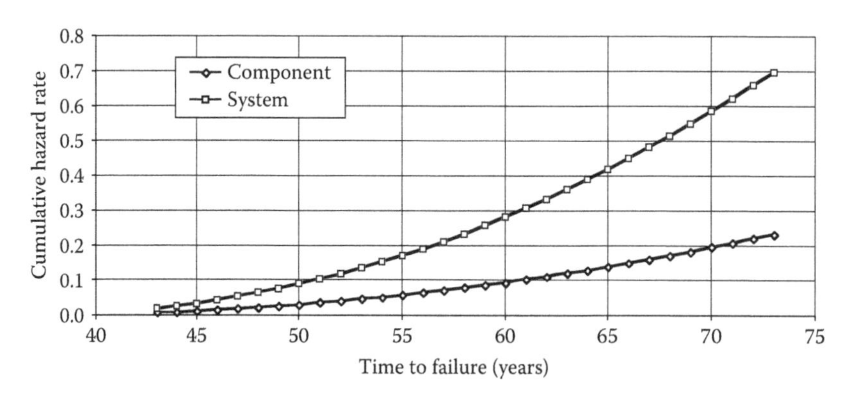

FIGURE 4.25
Cumulative hazard rate function (CHRF) for a series system of three identical components (Example 4.23).

Example 4.24: Assessing the Hazard Functions of a Series System of Four Different Components

The HRFs for one component of this system are from Example 4.23. Additional HRFs for three components are assumed in a similar manner. The failure data and survivor functions for these components are given in Tables 4.27 through 4.29. The parameters of the HRFs based on Equations 4.76 and 4.77 are given in Table 4.30. The parameters of the HRFs of the series system composed of these components were obtained using Equation 4.119a and 4.119b as given in Table 4.30.

TABLE 4.27

Example 4.24 Data and Empirical Survivor Function, $S_n(t)$, for Component 2

Year	Time to Failure (Years)	Number of Failures	Survivor Function
1937	0	0	1.000000
⋮	⋮	⋮	⋮
1972	35	0	1.000000
1973	36	11	0.999450
1974	37	15	0.998700
1975	38	19	0.997750
1976	39	24	0.996550
1977	40	30	0.995050
1978	41	35	0.993300
1979	42	42	0.991200
1980	43	47	0.988850
1981	44	56	0.986050
1982	45	62	0.982950
1983	46	71	0.979400
1984	47	77	0.975550
1985	48	86	0.971250
1986	49	94	0.966550
1987	50	102	0.961450
1988	51	109	0.956000
1989	52	118	0.950100
1990	53	123	0.943950
1991	54	132	0.937350
1992	55	139	0.930400
1993	56	145	0.923150
1994	57	151	0.915600
1995	58	158	0.907700
1996	59	164	0.899500
1997	60	167	0.891150
1998	61	173	0.882500
1999	62	178	0.873600
2000	63	181	0.864550
2001	64	184	0.855350
2002	65	189	0.845900
2003	66	190	0.836400
2004	67	193	0.826750

TABLE 4.28

Example 4.24 Data and Empirical Survivor Function, $S_n(t)$, for Component 3

Year	Time to Failure (Years)	Number of Failures	Survivor Function
1937	0	0	1.000000
⋮	⋮	⋮	⋮
1973	36	0	1.000000
1974	37	9	0.999550
1975	38	16	0.998750
1976	39	18	0.997850
1977	40	24	0.996650
1978	41	27	0.995300
1979	42	33	0.993650
1980	43	40	0.991650
1981	44	45	0.989400
1982	45	52	0.986800
1983	46	58	0.983900
1984	47	67	0.980550
1985	48	73	0.976900
1986	49	80	0.972900
1987	50	89	0.968450
1988	51	95	0.963700
1989	52	103	0.958550
1990	53	110	0.953050
1991	54	118	0.947150
1992	55	124	0.940950
1993	56	131	0.934400
1994	57	137	0.927550
1995	58	144	0.920350
1996	59	150	0.912850
1997	60	155	0.905100
1998	61	159	0.897150
1999	62	165	0.888900
2000	63	169	0.880450
2001	64	174	0.871750
2002	65	176	0.862950
2003	66	181	0.853900
2004	67	182	0.844800

TABLE 4.29

Example 4.24 Data and Empirical Survivor Function, $S_n(t)$, for Component 4

Year	Time to Failure (Years)	Number of Failures	Survivor Function
1937	0	0	1.000000
⋮	⋮	⋮	⋮
1972	35	0	1.000000
1973	36	12	0.999400

(Continued)

TABLE 4.29

(Continued) Example 4.24 Data and Empirical Survivor Function, $S_n(t)$, for Component 4

Year	Time to Failure (Years)	Number of Failures	Survivor Function
1974	37	17	0.998550
1975	38	19	0.997600
1976	39	25	0.996350
1977	40	31	0.994800
1978	41	35	0.993050
1979	42	44	0.990850
1980	43	49	0.988400
1981	44	59	0.985450
1982	45	66	0.982150
1983	46	77	0.978300
1984	47	83	0.974150
1985	48	92	0.969550
1986	49	99	0.964600
1987	50	106	0.959300
1988	51	115	0.953550
1989	52	126	0.947250
1990	53	127	0.940900
1991	54	140	0.933900
1992	55	150	0.926400
1993	56	155	0.918650
1994	57	161	0.910600
1995	58	168	0.902200
1996	59	173	0.893550
1997	60	177	0.884700
1998	61	184	0.875500
1999	62	185	0.866250
2000	63	193	0.856600
2001	64	195	0.846850
2002	65	198	0.836950
2003	66	202	0.826850
2004	67	209	0.816400

TABLE 4.30

Parameters of Hazard Rate Functions (HRFs) for Four Components and a Series System (Example 4.24)

Component Number or System	Parameter a_0	Parameter a_1 (1/year)	Parameter a_2 (1/year^2)
Component 1	0.262649	−0.013915	0.000185
Component 2	0.261022	−0.014371	0.000199
Component 3	0.264099	−0.014097	0.000189
Component 4	0.281940	−0.015469	0.000213
Series system	1.069710	−0.057852	0.000786

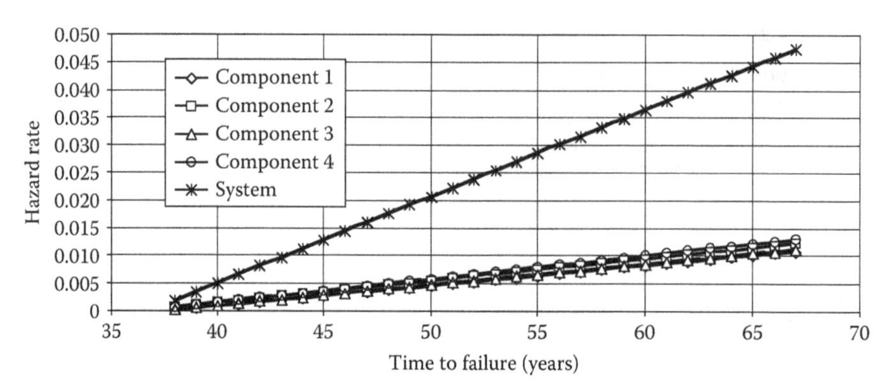

FIGURE 4.26
Hazard rate functions (HRFs) for series system of four different components (Example 4.24).

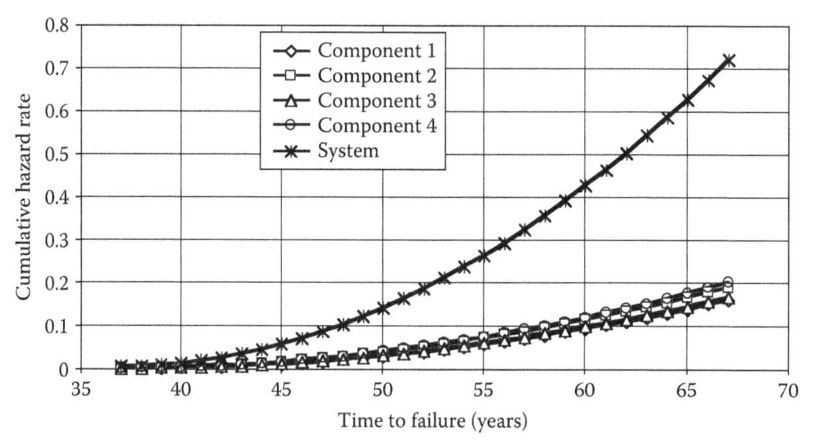

FIGURE 4.27
Cumulative hazard rate functions (CHRFs) for series system of four different components (Example 4.24).

Based on the parameter estimates for the series system, and applying Equations 4.119 and 4.120, the HRFs can be estimated by algebraically summing up the component hazard functions. The resulting system functions are

$$H_s(t) = 1.069710 - 0.057852t + 0.000786t^2$$

and

$$h_s(t) = -0.057852 + 0.001572t$$

Figures 4.26 and 4.27 show the respective HRFs.

4.6.3 Parallel Systems

A parallel system composed of n components can be defined as a system that functions or survives if at least one of its n components functions or survives. Figure 4.28 depicts an example of the RBD for a parallel system consisting of three components.

The reliability function of a parallel system composed of n components, $R_s(t)$, is given by

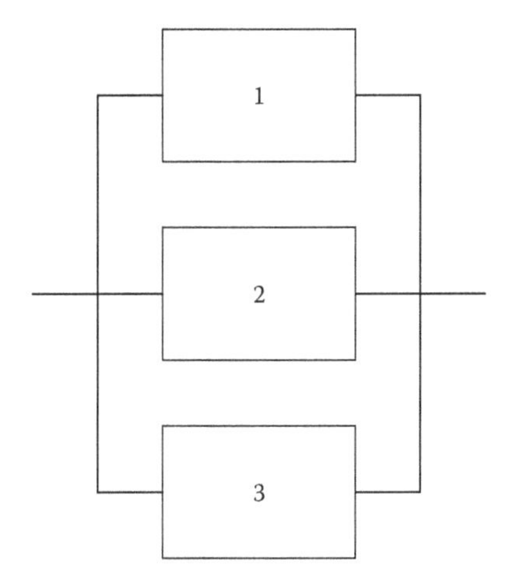

FIGURE 4.28
Parallel system composed of three components.

$$R_s(t) = 1 - \prod_{i=1}^{n} [1 - R_i(t)] \tag{4.121}$$

where:
$R_i(t)$ is the reliability function of the ith component

If a parallel system is composed of identical components with reliability functions, $R_c(t)$, Equation 4.121 is reduced to

$$R_s(t) = 1 - [1 - R_c(t)]^n \tag{4.122}$$

Compared with a series system composed of the same components, the respective parallel system is always more reliable. A parallel system is an example of a redundant system.

Applying the relationship between a reliability function and its CHRF, as provided by Equation 4.60, the following relationship between the parallel system CHRF, $H_s(t)$, and the reliability functions of its components, $R_i(t)$, can be written as

$$H_s(t) = -\ln\left\{1 - \prod_{i=1}^{n} [1 - R_i(t)]\right\} \tag{4.123a}$$

By taking the derivative of $H_s(t)$ and using Equation 4.61, the relationship between the system hazard (failure) rate function, $h_s(t)$, and the reliability functions of its components, $R_i(t)$, can be obtained as follows:

$$h_s(t) = -\frac{\sum_{j=1}^{n}\left\{\prod_{i=1,i\neq j}^{n} [1 - R_i(t)]\frac{\mathrm{d}R_j(t)}{\mathrm{d}t}\right\}}{R_s(t)} \tag{4.123b}$$

where $R_s(t)$ is given by Equation 4.121. For example, for $n = 3$, Equation 4.123b takes on the following form:

$$h_s(t) = -\frac{[1 - R_2(t)][1 - R_3(t)]\frac{dR_1(t)}{dt} + [1 - R_1(t)][1 - R_3(t)]\frac{dR_2(t)}{dt} + [1 - R_1(t)][1 - R_2(t)]\frac{dR_3(t)}{dt}}{1 - [1 - R_1(t)][1 - R_2(t)][1 - R_3(t)]}$$

For practical problems, it might be easier to numerically differentiate Equation 4.123a instead of directly using Equation 4.123b.

For the case of a parallel system composed of identical components with reliability functions $R_c(t)$, Equation 4.123a and 4.123b are reduced to

$$H_s(t) = -\ln\{1 - [1 - R_c(t)]^n\} \tag{4.124a}$$

$$h_s(t) = -\frac{n[1 - R_c(t)]^{n-1}\frac{dR_c(t)}{dt}}{1 - [1 - R_c(t)]^n} \tag{4.124b}$$

Example 4.25: Assessing the Hazard Function of a Parallel System of Three Identical Components

A parallel system composed of the same identical components as used in Example 4.23 is used to demonstrate the assessment of the system hazard functions. Thus, for each component, the hazard functions are

$$H_c(t) = 0.262649 - 0.013915t + 0.000185t^2$$

and

$$h_c(t) = -0.013915 + 0.000370t$$

where:
  t is the time in years

Applying Equation 4.59, the component reliability function is given by

$$R_c(t) = \exp[-(0.262649 - 0.013915t + 0.000185t^2)]$$

To calculate the system CHRF, $H_s(t)$, Equation 4.124a can be used with $n = 3$. For calculating the respective system hazard (failure) rate function, $h_s(t)$, Equation 4.124a requires the derivative $dR_c(t)/dt$ which is given by

$$\frac{dR_c}{dt} = -[R_c(t)]h_c(t)$$

The resulting hazard functions are given in Table 4.31 and illustrated in Figures 4.29 and 4.30.

Example 4.26: Assessing the Hazard Functions of a Parallel System of Four Different Components

The parallel system composed of the four different components of Example 4.24 (shown in Table 4.30) is used in this example to demonstrate the case of components in parallel. The system CHRF, $H_s(t)$, can be evaluated using Equation 4.123a and 4.123b.

TABLE 4.31

Hazard Rate Functions (HRFs) for Parallel System Composed of Three Identical Components (Example 4.25)

Year	Time to Failure (Years)	System HRF	System CHRF
1975	38	3.41962E−10	1.05647E−09
1976	39	2.61345E−09	2.44995E−09
1977	40	1.06621E−08	8.57613E−09
1978	41	3.53837E−08	3.02005E−08
1979	42	9.82423E−08	9.41109E−08
1980	43	2.35499E−07	2.55898E−07
1981	44	5.02210E−07	6.16852E−07
1982	45	9.75927E−07	1.34471E−06
1983	46	1.76005E−06	2.69792E−06
1984	47	2.98677E−06	5.05310E−06
1985	48	4.81957E−06	8.93509E−06
1986	49	7.45518E−06	1.50494E−05
1987	50	1.11251E−05	2.43163E−05
1988	51	1.60965E−05	3.79061E−05
1989	52	2.26724E−05	5.72750E−05
1990	53	3.11919E−05	8.42009E−05
1991	54	4.20287E−05	0.000120819
1992	55	5.55903E−05	0.000169657
1993	56	7.23158E−05	0.000233666
1994	57	9.26734E−05	0.000316250
1995	58	0.000117158	0.000421300
1996	59	0.000146286	0.000553210
1997	60	0.000180596	0.000716903
1998	61	0.000220639	0.00091785
1999	62	0.000266978	0.001162076
2000	63	0.000320181	0.001456177
2001	64	0.000380821	0.001807317
2002	65	0.000449465	0.002223232
2003	66	0.000526673	0.002712222
2004	67	0.000612993	0.003283142

CHRF, cumulative hazard rate function.

The reliability functions of the components of the system, $R_i(t)$, can be determined using Equation 4.59. Instead of using Equation 4.123b, the hazard (failure) rate function can be calculated using the following numerical approximation for the derivative of Equation 4.61:

$$h_s(t_i) = \frac{H_s(t_i) - H_s(t_{i-1})}{t_i - t_{i-1}}$$

where t_i ($i = 1, 2, \ldots, n$) are successive times at which H_s is evaluated. For the data used in the report, the difference $(t_i - t_{i-1})$ is equal to 1 year. The resulting hazard functions are given in Table 4.32 and shown in Figures 4.31 and 4.32.

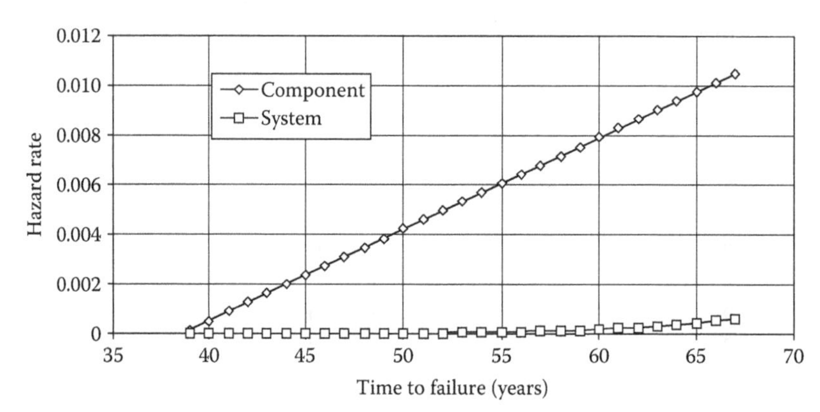

FIGURE 4.29
Hazard rate function (HRF) for parallel system of three identical components (Example 4.25).

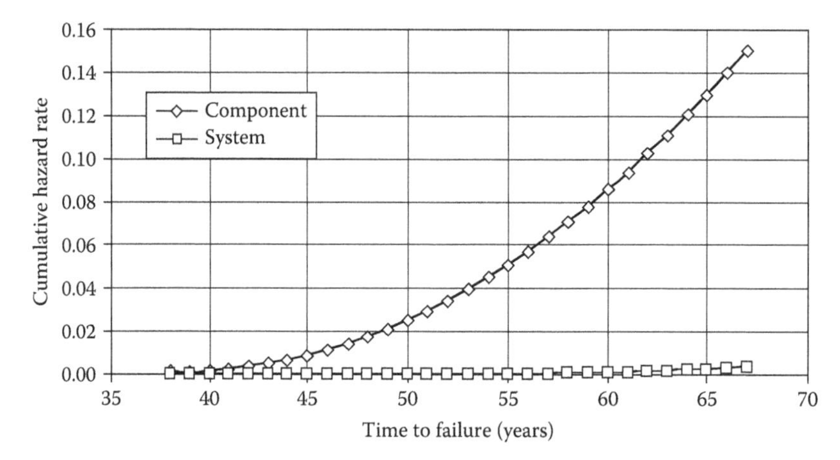

FIGURE 4.30
Cumulative hazard rate function (CHRF) for parallel system of three identical components (Example 4.25).

TABLE 4.32
Hazard Rate Functions (HRFs) for a Parallel System Composed of Four Different Components (Example 4.26)

Year	Time to Failure (Years)	System HRF	System CHRF
1975	38	2.50140E−12	5.20173E−12
1976	39	1.51126E−11	2.03143E−11
1977	40	7.69030E−11	9.72173E−11
1978	41	3.34621E−10	4.31839E−10
1979	42	1.21663E−09	1.64847E−09
1980	43	3.75753E−09	5.40599E−09

(Continued)

TABLE 4.32

(Continued) Hazard Rate Functions (HRFs) for a Parallel System Composed of Four Different Components (Example 4.26)

Year	Time to Failure (Years)	System HRF	System CHRF
1981	44	1.01204E−08	1.55264E−08
1982	45	2.43575E−08	3.98839E−08
1983	46	5.34422E−08	9.33261E−08
1984	47	1.08593E−07	2.01919E−07
1985	48	2.06904E−07	4.08823E−07
1986	49	3.73273E−07	7.82096E−07
1987	50	6.42629E−07	1.42473E−06
1988	51	1.06242E−06	2.48714E−06
1989	52	1.69531E−06	4.18245E−06
1990	53	2.62210E−06	6.80455E−06
1991	54	3.94474E−06	1.07493E−05
1992	55	5.78936E−06	1.65386E−05
1993	56	8.30935E−06	2.48480E−05
1994	57	1.16883E−05	3.65363E−05
1995	58	1.61427E−05	5.26790E−05
1996	59	2.19246E−05	7.46036E−05
1997	60	2.93234E−05	1.03927E−04
1998	61	3.86680E−05	1.42595E−04
1999	62	5.03275E−05	1.92923E−04
2000	63	6.47123E−05	2.57635E−04
2001	64	8.22737E−05	3.39908E−04
2002	65	1.03504E−04	4.43412E−04
2003	66	1.28933E−04	5.72345E−04
2004	67	1.59129E−04	7.31474E−04

CHRF, cumulative hazard rate function.

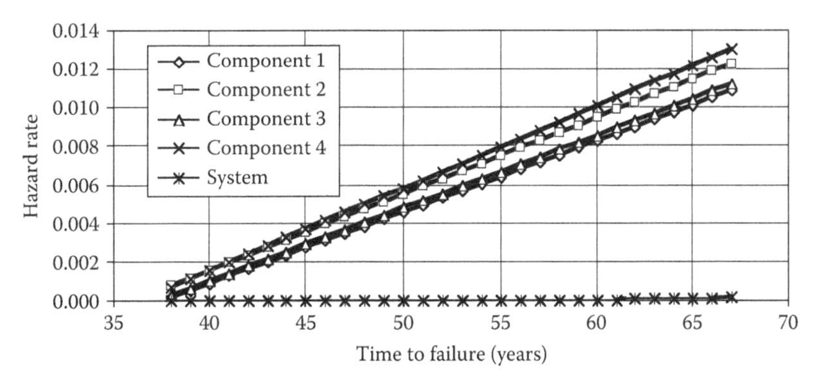

FIGURE 4.31
Hazard rate function (HRF) for parallel system with four different components (Example 4.26).

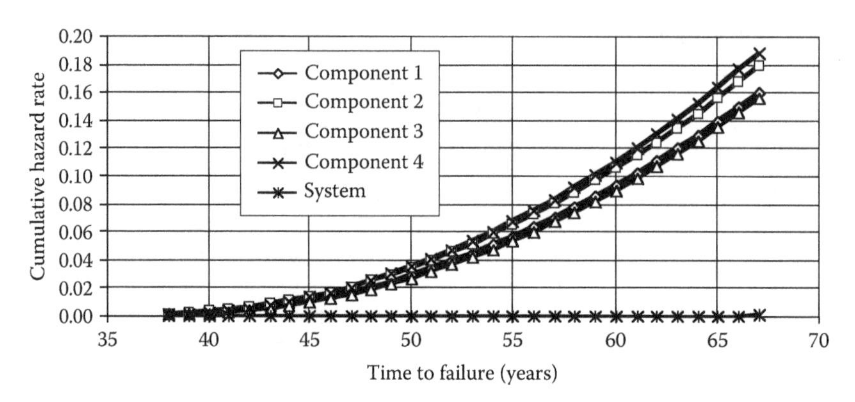

FIGURE 4.32
Cumulative hazard rate function (CHRF) for parallel system with four different components (Example 4.26).

4.6.4 Series–Parallel Systems

Some systems, from the reliability standpoint, can be represented as a series structure of k parallel structures. Figure 4.33 depicts an example RBD of such a system, which is referred to as a *series–parallel* system. These systems are redundant and have alternate loads (or demand) paths.

A series–parallel system similar to the system shown in Figure 4.33 can be analyzed as a simpler system of the composing components. For example, the system in Figure 4.33 can be represented as a series system of two subsystems, called subsystem 1 and subsystem 2. Subsystem 1 is composed of components 1 and 2, connected in parallel, and subsystem 2 is composed of components 3–5, also connected in parallel. Hence, the equivalent structure of the system considered can be represented by the RBD in Figure 4.34.

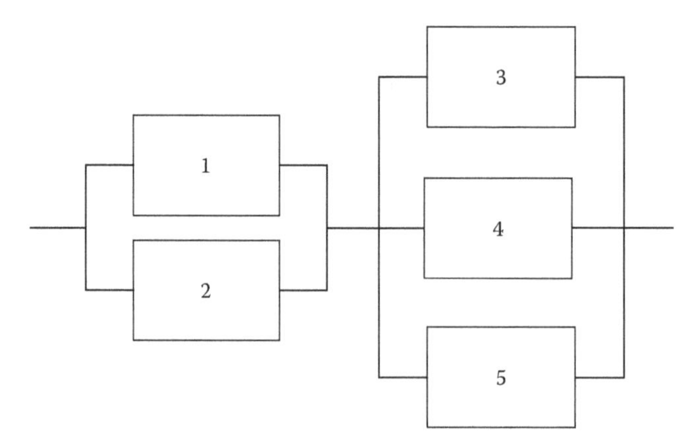

FIGURE 4.33
Series structure of two parallel structures.

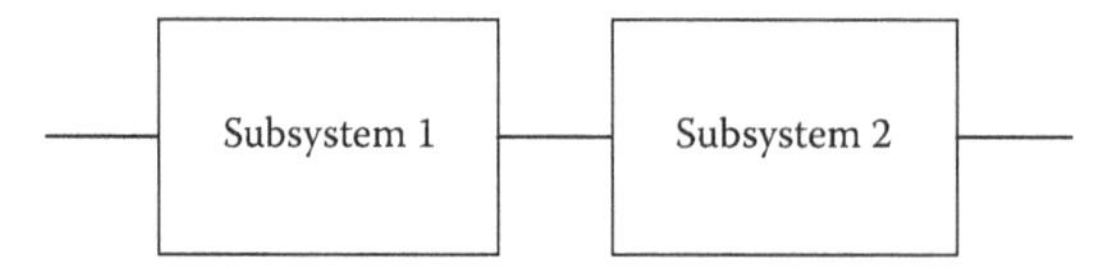

FIGURE 4.34
A system of components equivalent to the system in Figures 4.31 and 4.32.

The following steps can be followed to compute the reliability and hazard functions of the system:

1. Calculate the reliability functions of subsystems 1 and 2 using Equation 4.121 (for parallel systems).
2. Based on the results from the first step, calculate the reliability function of the series system composed of subsystems 1 and 2 using Equation 4.117 (for series systems).
3. Using basic relationships between the reliability function and the hazard functions (Equations 4.59 and 4.61), calculate the CHRF and the hazard (failure) rate function for the system of interest represented in Figure 4.33.

If one is interested in assessing the hazard functions only, the problem can be solved as follows:

1. Calculate the hazard functions for each subsystem as described in Section 4.6.3 for parallel systems.
2. Calculate the system HRF as the HRFs of the series system composed of the subsystems as components of the series system.

Example 4.27: Assessing the Hazard Functions of a Series–Parallel System

In this example, a series–parallel system consisting of two identical subsystems is considered. Each subsystem is composed of the four components connected in parallel, which were considered in Example 4.26. The hazard functions of each subsystem are exactly the same as the respective hazard functions $H_s(t)$ and $h_s(t)$ obtained in Example 4.26. According to Equations 4.119 and 4.120, the hazard function for the series–parallel system can be based on the hazard functions $H_s(t)$ and $h_s(t)$ from Example 4.26. The values of these functions are given in Table 4.33 and depicted in Figures 4.35 and 4.36.

TABLE 4.33

Assessing the Hazard Functions of a Series–Parallel System (Example 4.27)

Year	Time to Failure (Years)	System HRF	System CHRF
1975	38	5.00289E−12	1.040346E−11
1976	39	3.02252E−11	4.062861E−11
1977	40	1.53806E−10	1.944347E−10
1978	41	6.69243E−10	8.636776E−10
1979	42	2.43326E−09	3.296935E−09
1980	43	7.51505E−09	1.081199E−08
1981	44	2.02408E−08	3.105274E−08
1982	45	4.87150E−08	7.976779E−08
1983	46	1.06884E−07	1.866522E−07
1984	47	2.17186E−07	4.038387E−07
1985	48	4.13807E−07	8.176461E−07
1986	49	7.46546E−07	1.564192E−06
1987	50	1.28526E−06	2.849450E−06
1988	51	2.12483E−06	4.974282E−06

(Continued)

TABLE 4.33

(Continued) Assessing the Hazard Functions of a Series–Parallel System (Example 4.27)

Year	Time to Failure (Years)	System HRF	System CHRF
1989	52	3.39061E−06	8.364897E−06
1990	53	5.24420E−06	1.360910E−05
1991	54	7.88948E−06	2.149858E−05
1992	55	1.15787E−05	3.307729E−05
1993	56	1.66187E−05	4.969599E−05
1994	57	2.33766E−05	7.307258E−05
1995	58	3.22855E−05	1.053580E−04
1996	59	4.38492E−05	1.492072E−04
1997	60	5.86468E−05	2.078541E−04
1998	61	7.73360E−05	2.851900E−04
1999	62	1.00655E−04	3.858450E−04
2000	63	1.29425E−04	5.152696E−04
2001	64	1.64547E−04	6.798169E−04
2002	65	2.07007E−04	8.868240E−04
2003	66	2.57865E−04	1.144689E−03
2004	67	3.18258E−04	1.462947E−03

CHRF, cumulative hazard rate function; HRF, hazard rate function.

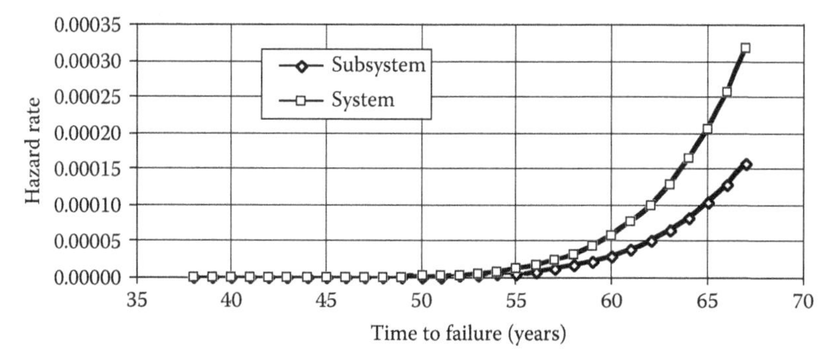

FIGURE 4.35
Hazard rate function (HRF) for a series–parallel system (Example 4.27).

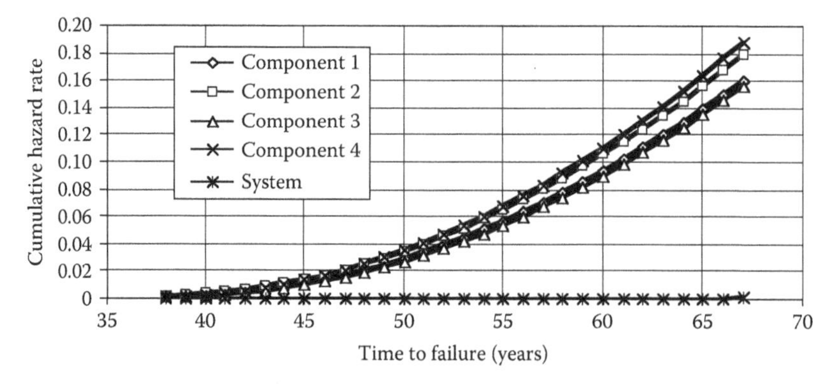

FIGURE 4.36
Cumulative hazard rate function (CHRF) for a series–parallel system (Example 4.27).

4.6.5 *k*-out-of-*n* Systems

Another widely used type of redundant systems is *k*-out-of-*n* systems. Such a system has *n* parallel components; however, at least *k* components must be functioning if the system is to continue operating. An example of this type of redundant system is the cables for a bridge, where a certain minimum number of cables are necessary to support the structure. Another example of *k*-out-of-*n* systems is a three-engine airplane, which can stay in the air if and only if at least two of its three engines are functioning, that is, the plane can be modeled by a two-out-of-three system. The RBD for the two-out-of-three system is given in Figure 4.37. The RBD of Figure 4.37 has more components than the real system, which is why the techniques of system reliability evaluation considered in Sections 4.6.1 through 4.6.4 are not applicable to *k*-out-of-*n* systems.

In engineering practice, parallel systems and *k*-out-of-*n* systems are usually composed of identical components; therefore, this section focuses on *k*-out-of-*n* systems composed of identical components. The reliability function of the *k*-out-of-*n* system, R_s, is given by

$$R_s(t) = \sum_{i=k}^{n} \binom{n}{i} [R_c(t)]^i [1 - R_c(t)]^{n-i} \tag{4.125}$$

Applying the basic relationship between the reliability function and its CHRF (i.e., Equation 4.60 applied to Equation 4.125), the following relationship between the *k*-out-of-*n* system CHRF, $H_s(t)$, and the reliability function of its (identical) components, $R_c(t)$, can be written as

$$H_s(t) = -\ln\left\{ \sum_{i=k}^{n} \binom{n}{i} [R_c(t)]^i [1 - R_c(t)]^{n-i} \right\} \tag{4.126}$$

In order to assess the respective system hazard (failure) rate function, $h_s(t)$, the basic relationship (i.e., Equation 4.61) between the hazard (failure) rate function and the CHRF in the form of Equation 4.126 needs to be applied. Due to the rather complex form of Equation 4.126, numerical differentiation is recommended for practical problems.

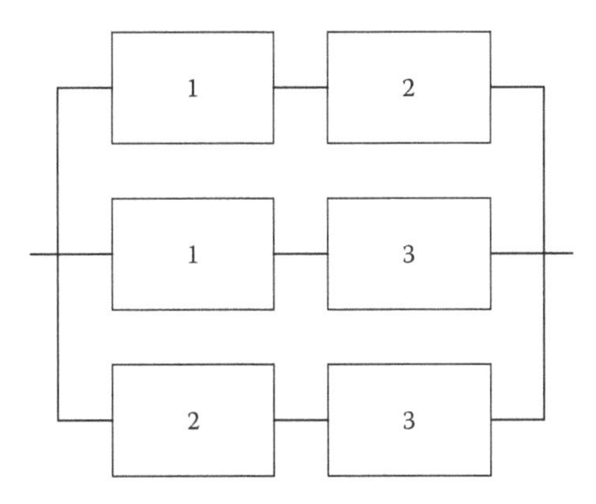

FIGURE 4.37
Two-out-of-three system.

Example 4.28: Assessing the Hazard Functions of a Two-out-of-Three System of Identical Components

A two-out-of-three system composed of identical components having a reliability function as given by:

$$R_c(t) = \exp(-0.262649 + 0.013915t - 0.000185t^2)$$

where:
time, t, is given in years

Equation 4.126 can be used to assess the two-out-of-three system CHRF, $H_s(t)$, which takes the form:

$$H_s(t) = -\ln\left\{\sum_{i=2}^{3}\binom{3}{i}[R_c(t)]^i[1 - R_c(t)]^{n-i}\right\}$$

The above equation can be calculated using the function BINOMDIST in Microsoft's *Excel*. For this example, the hazard (failure) rate function can be calculated using the same approximation as in Example 4.27. The results of the hazard functions calculations are given in Table 4.34 and in Figures 4.38 and 4.39.

TABLE 4.34

Hazard Rate Function (HRF) and Cumulative Hazard Rate Function (CHRF) for a Two-out-of-Three System (Example 4.28)

Year	Time to Failure (Years)	System HRF	System CHRF
1980	43	5.85E−05	1.20E−04
1981	44	9.58E−05	2.16E−04
1982	45	1.47E−04	3.63E−04
1983	46	2.13E−04	5.76E−04
1984	47	2.98E−04	8.74E−04
1985	48	4.01E−04	1.27E−03
1986	49	5.25E−04	1.80E−03
1987	50	6.72E−04	2.47E−03
1988	51	8.43E−04	3.31E−03
1989	52	1.04E−03	4.35E−03
1990	53	1.26E−03	5.61E−03
1991	54	1.50E−03	7.12E−03
1992	55	1.78E−03	8.89E−03
1993	56	2.08E−03	1.10E−02
1994	57	2.41E−03	1.34E−02
1995	58	2.76E−03	1.61E−02
1996	59	3.15E−03	1.93E−02
1997	60	3.56E−03	2.29E−02
1998	61	4.00E−03	2.68E−02
1999	62	4.46E−03	3.13E−02
2000	63	4.95E−03	3.63E−02
2001	64	5.47E−03	4.17E−02
2002	65	6.01E−03	4.77E−02
2003	66	6.58E−03	5.43E−02
2004	67	7.17E−03	6.15E−02

(Continued)

TABLE 4.34
(Continued) Hazard Rate Function (HRF) and Cumulative Hazard Rate Function (CHRF) for a Two-out-of-Three System (Example 4.28)

Year	Time to Failure (Years)	System HRF	System CHRF
2005	68	7.78E−03	6.93E−02
2006	69	8.42E−03	7.77E−02
2007	70	9.07E−03	8.68E−02
2008	71	9.75E−03	9.65E−02
2009	72	1.04E−02	1.07E−01
2010	73	1.12E−02	1.18E−01

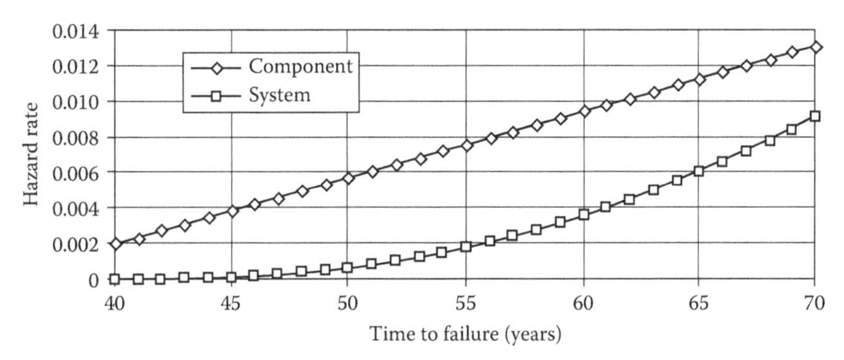

FIGURE 4.38
Hazard rate function (HRF) of a two-out-of-three system of identical components (Example 4.28).

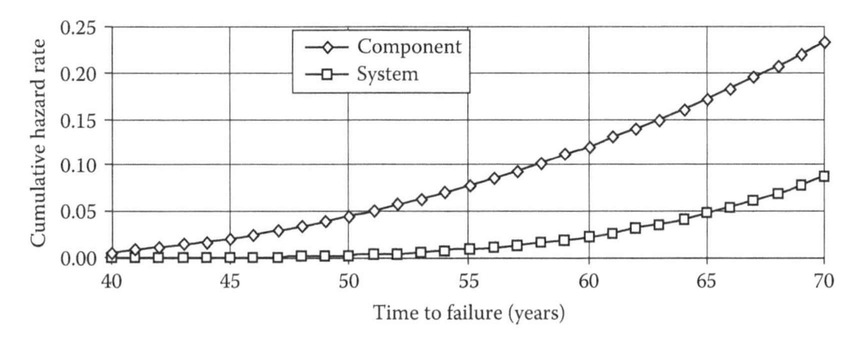

FIGURE 4.39
Cumulative hazard rate function (CHRF) of a two-out-of-three system of identical components (Example 4.28).

Example 4.29: Three-Component Series System as a Three-out-of-Three System and Three-Component Parallel System as a One-out-of-Three System

The difference between the two-out-of-three system and the parallel and series systems composed of the same three identical components is explored in this example. The series system can be treated as a three-out-of-three system, and the parallel system can be treated as a one-out-of-three system. The respective hazard functions for these systems are shown in Figures 4.40 and 4.41. The figures clearly show that the series (three-out-of-three) system is the least reliable, the parallel (one-out-of-three) system is the most reliable, and the HRFs of the three-out-of-three system is somewhere between the HRFs of the series (three-out-of-three) system and the parallel (one-out-of-three) system.

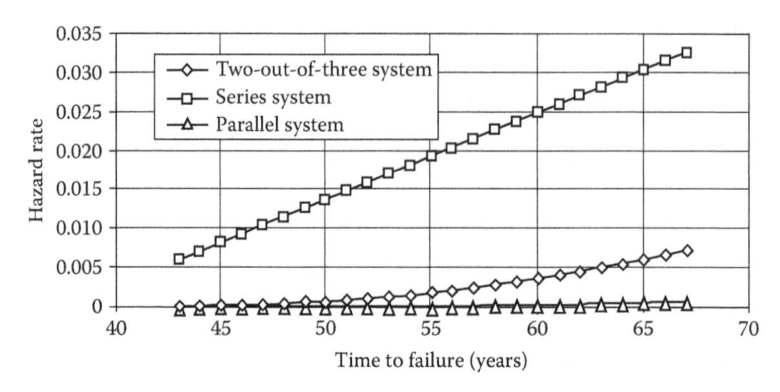

FIGURE 4.40
Hazard rate function (HRF) of a two-out-of-three system, a parallel system, and a series system composed of three identical components (Example 4.29).

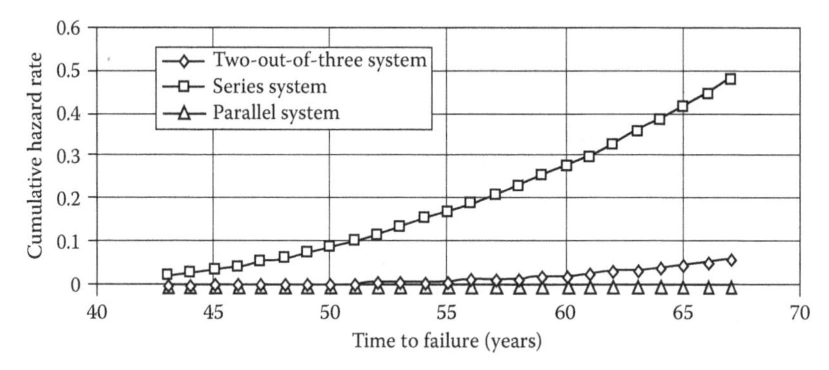

FIGURE 4.41
Cumulative hazard rate function (CHRF) of a two-out-of-three system, a parallel system, and a series system composed of three identical components (Example 4.29).

4.7 Exercise Problems

Problem 4.1 For the following performance function, determine the safety index (β) and the failure probability (P_f) using

a. First-order reliability method
b. Advanced second-moment method

$$Z = X_1 - 2X_2 - \sqrt{X_3}$$

The noncorrelated random variables are assumed to have the following probabilistic characteristics:

Random Variable	Mean Value	Coefficient of Variance (COV)	Distribution Type
X_1	10	0.25	Normal
X_2	4	0.20	Normal
X_3	3	0.40	Normal

Problem 4.2 For the performance function of Problem 4.1, determine the safety index (β) and the failure probability (P_f) using

a. First-order reliability method
b. Advanced second-moment method

The noncorrelated random variables are assumed to have the following probabilistic characteristics:

Random Variable	Mean Value	Coefficient of Variance (COV)	Distribution Type
X_1	10	0.25	Lognormal
X_2	4	0.20	Lognormal
X_3	3	0.40	Lognormal

Problem 4.3 For the following performance function, determine the safety index (β) and the failure probability (P_f) using

a. First-order reliability method
b. Advanced second-moment method

$$Z = X_1 - X_2/X_3$$

The noncorrelated random variables are assumed to have the following probabilistic characteristics:

Random Variable	Mean Value	Coefficient of Variance (COV)	Distribution Type
X_1	5	0.25	Normal
X_2	8	0.40	Normal
X_3	2	0.80	Normal

Problem 4.4 For the performance function of Problem 4.3, determine the safety index (β) and the failure probability (P_f) using

1. First-order reliability method
2. Advanced second-moment method

The noncorrelated random variables are assumed to have the following probabilistic characteristics:

Random Variable	Mean Value	Coefficient of Variance (COV)	Distribution Type
X_1	5	0.25	Lognormal
X_2	8	0.40	Lognormal
X_3	2	0.80	Lognormal

Problem 4.5 For the performance function of Problem 4.3, determine the safety index (β) and the failure probability (P_f) using

1. First-order reliability method
2. Advanced second-moment method

The noncorrelated random variables are assumed to have the following probabilistic characteristics:

Random Variable	Mean Value	Coefficient of Variance (COV)	Distribution Type
X_1	5	0.2	Normal
X_2	8	1.0	Exponential
X_3	2	0.5	Lognormal

Problem 4.6 A project schedule network has two paths of tasks needed to compute the total time to complete the project. They are either T_1 or T_2, as shown by the following time functions:

$$T_1 = t_1 + t_2 + t_5$$

$$T_2 = t_3 + t_4 + t_5$$

Compute the probability of $T_1 > T_2$ by calculating the reliability index (β) and the failure probability (P_f) using

a. First-order reliability method

b. Advanced second-moment method

The noncorrelated random variables are assumed to have the following probabilistic characteristics:

Random Variable	Mean Value	Coefficient of Variance (COV)	Distribution Type
t_1	1	0.25	Normal
t_2	5	0.50	Normal
t_3	4	0.05	Normal
t_4	3	0.20	Normal
t_5	10	0.25	Normal

Problem 4.7 The planning department of a city is considering two structural alternatives to cross a major river in the city by comparing the economics of the two alternatives. The alternatives are to construct either a bridge (B) or a tunnel (T). They estimated the benefit (B)-to-cost (C) ratio $R_i = B_i/C_i$ for each alternative of $i = B$ or T as follows:

$$R_B = \frac{B_B}{C_T}$$

$$R_T = \frac{B_T}{C_T}$$

Compute the probability that $R < 1$ for each alternative by calculating its reliability index (β) and the failure probability (P_f) using

a. First-order reliability method

b. Advanced second-moment method

What would you recommend to the planning department? The noncorrelated random variables are assumed to have the following probabilistic characteristics:

Random Variable	Mean Value	Coefficient of Variance (COV)	Distribution Type
B_B	4	0.35	Normal
C_B	3	0.45	Normal
B_T	5	0.25	Normal
C_T	2	0.05	Normal

Problem 4.8 The planning department of a city is considering two structural alternatives to cross a major river in the city by comparing the economics of the two alternatives.

The alternatives are to construct either a bridge (*B*) or a tunnel (*T*). They estimated the benefit (*B*) to cost (*C*) ratio $R_i = B_i/C_i$ for each alternative of $i = B$ or T as follows:

$$R_B = \frac{B_B}{C_T}$$

$$R_T = \frac{B_T}{C_T}$$

Compute the probability that $R < 1$ for each alternative by calculating its reliability index (β) and the failure probability (P_f) using

a. First-order reliability method

b. Advanced second-moment method

What would you recommend to the planning department? The noncorrelated random variables are assumed to have the following probabilistic characteristics:

Random Variable	Mean Value	Coefficient of Variance (COV)	Distribution Type
B_B	4	0.35	Lognormal
C_B	3	0.45	Lognormal
B_T	5	0.25	Lognormal
C_T	2	0.05	Lognormal

Problem 4.9 The profit from product sales can be computed from the following function of revenue-and-cost relationship, where *R* represents the revenue, *M* the manufacturing cost, *A* the assembly cost, and *T* the transportation cost:

$$P = R - (M + A + T)$$

Determine the reliability index (β) and the failure probability (P_f), that is, the cost exceeding revenue, using

a. First-order reliability method

b. Advanced second-moment method

The noncorrelated random variables are assumed to have the following probabilistic characteristics:

Random Variable	Mean Value	Coefficient of Variance (COV)	Distribution Type
R	18	0.15	Normal
M	4	0.50	Normal
A	6	0.25	Normal
T	5	0.30	Normal

Problem 4.10 Estimate the failure probability for the performance function and random variables defined in Problem 4.1 using the following methods:

a. DMCS

b. CE by trying each variable in the problem as a control variable

Examine the trends using $N = 100$, 1000, 2000, and 5000 cycles.

Problem 4.11 Estimate the failure probability for the performance function and random variables defined in Problem 4.2 using the following methods:

a. DMCS
b. CE by trying each variable in the problem as a control variable

Examine the trends using $N = 100$, 1000, 2000, and 5000 cycles.

Problem 4.12 Estimate the failure probability for the performance function and random variables defined in Problem 4.3 using the following methods:

a. DMCS
b. CE by trying each variable in the problem as a control variable

Examine the trends using $N = 100$, 1000, 2000, and 5000 cycles.

Problem 4.13 Estimate the failure probability for the performance function and random variables defined in Problem 4.4 using the following methods:

a. DMCS
b. CE by trying each variable in the problem as a control variable

Examine the trends using $N = 100$, 1000, 2000, and 5000 cycles.

Problem 4.14 Estimate the failure probability for the performance function and random variables defined in Problem 4.5 using the following methods:

a. DMCS
b. CE by trying each variable in the problem as a control variable

Examine the trends using $N = 100$, 1000, 2000, and 5000 cycles.

Problem 4.15 Estimate the reliability function $R(t)$ according to Equation 4.43b for the performance function and random variables defined in Problem 4.1 using a corrosion model $c(t) = 1 - at$ expressed as follows:

$$Z = c(t)X_1 - 2X_2 - \sqrt{X_3}$$

Examine the trends using $t = 1, 2, \ldots, 20$ and $N = 100$, 1000, 2000, and 5000 cycles for $a = 0$, 0.01, 0.02, 0.05, and 0.1 (*Hint*: see Example 4.2).

Problem 4.16 Estimate the reliability function $R(t)$ according to Equation 4.43b for the performance function and random variables defined in Problem 4.2 using a corrosion model $c(t) = 1 - at$ expressed as follows:

$$Z = c(t)X_1 - 2X_2 - \sqrt{X_3}$$

Examine the trends using $t = 1, 2, \ldots, 20$ and $N = 100$, 1000, 2000, and 5000 cycles for $a = 0$, 0.01, 0.02, 0.05, and 0.1 (*Hint*: see Example 4.2).

Problem 4.17 Estimate the reliability function $R(t)$ according to Equation 4.43b for the performance function and random variables defined in Problem 4.3 using a corrosion model $c(t) = 1 - at$ expressed as follows:

$$Z = c(t)X_1 - \frac{X_2}{X_3}$$

Examine the trends using $t = 1, 2, \ldots, 20$ and $N = 100$, 1000, 2000, and 5000 cycles for $a = 0$, 0.01, 0.02, 0.05, and 0.1 (*Hint*: see Example 4.2).

Problem 4.18 Estimate the reliability function $R(t)$ according to Equation 4.43b for the performance function and random variables defined in Problem 4.4 using a corrosion model $c(t) = 1 - at$ expressed as follows:

$$Z = c(t)X_1 - \frac{X_2}{X_3}$$

Examine the trends using $t = 1, 2, \ldots, 20$, and $N = 100$, 1000, 2000, and 5000 cycles for $a = 0$, 0.01, 0.02, 0.05, and 0.1 (*Hint*: see Example 4.2).

Problem 4.19 Estimate the reliability function $R(t)$ according to Equation 4.43b for the performance function and random variables defined in Problem 4.5 using a corrosion model $c(t) = 1 - at$ expressed as follows:

$$Z = c(t)X_1 - \frac{X_2}{X_3}$$

Examine the trends using $t = 1, 2, \ldots, 20$ and $N = 100$, 1000, 2000, and 5000 cycles for $a = 0$, 0.01, 0.02, 0.05, and 0.1 (*Hint*: see Example 4.2).

Problem 4.20 Give three examples of type I right-censored data.

Problem 4.21 The following tests of identical items were performed:

Test 1: Five items were tested for 10 hours; one failure was observed at 4 hours.

Test 2: 20 items were tested for 40 hours; 5 failures were observed at 3, 7, 11, 15, and 35 hours.

Combine the two datasets into one sample using the following table format:

Order Number	Time to Failure	Time to Censoring
⋮	⋮	⋮
⋮	⋮	⋮

Compute the survivorship function and write the equation for the likelihood function for the combined sample, assuming the exponential failure time distribution.

Problem 4.22 The following array provides an example of a sample of 10 data points that failed at different years. Classify the values as either TTF or TTC. What is the type of data in this array?

Time order number	1	2	3	4	5	6	7	8	9	10
Time (years)	14	18	37	46	55	56	56	56	56	56
TTF or TTC?										

Problem 4.23 The following array provides an example of a sample of 10 data points that failed at different years. Classify the values as either TTF or TTC. If the data collection was assumed to terminate just after the seventh failure, what is the type of data in this array?

Time order number	1	2	3	4	5	6	7	8	9	10
Time (years)	28	36	54	60	64	68	72	72	72	72
TTF or TTC?										

Problem 4.24 Using Equation 4.70, calculate the survivor function for the noncensored sample data of size 10 given as follows:

Time Order Number	Time to Failure (Years)	Provide Survivor Function Value
0	0	
1	28	
2	36	
3	54	
4	60	
5	68	
6	72	
7	75	
8	78	
9	92	
10	95	

Problem 4.25 Use Equation 4.70, calculate the survivor function for the data provided in Problem 4.22 for a sample of size 10 as follows:

Time Order Number	Time to Failure (Years)	Provide Survivor Function Value
0	0	
1	14	
2	18	
3	37	
4	46	
5	55	
6	56	
7	56	
8	56	
9	56	
10	56	

Problem 4.26 Show that the product limit (Kaplan–Meier) estimate reduces to the empirical distribution function for a complete dataset when the failure times are distinct.

Problem 4.27 Use the data provided in Example 4.10 to compute the survivor function for FM2 (fatigue).

Problem 4.28 Use the data provided in Example 4.11 to show the details of computing the survivorship function for FM1, that is, strength, as provided in Table 4.11.

Problem 4.29 Show how the exponential distribution is a specific case of the Weibull distribution presented in Equation 4.53.

Problem 4.30 Use linear regression and logarithmic transformation to determine the coefficients of the following model fitted to the survivor function of Problem 4.24:

$$R(t) = \exp[-H(t)]$$

$$H(t) = a_0 + a_1 t$$

Problem 4.31 Use linear regression and logarithmic transformation to determine the coefficients of the following model fitted to the survivor function of Problem 4.24:

$$R(t) = \exp[-H(t)]$$

$$H(t) = a_0 + a_1 t + a_2 t^2$$

Problem 4.32 Use linear regression and logarithmic transformation to determine the coefficients of the following model fitted to the survivor function of Problem 4.25:

$$R(t) = \exp[-H(t)]$$

$$H(t) = a_0 + a_1 t$$

Problem 4.33 Use linear regression and logarithmic transformation to determine the coefficients of the following model fitted to the survivor function of Problem 4.25:

$$R(t) = \exp[-H(t)]$$

$$H(t) = a_0 + a_1 t + a_2 t^2$$

Problem 4.34 Use linear regression and logarithmic transformation to determine the coefficients of the following model fitted to the survivor function of Problem 4.27:

$$R(t) = \exp[-H(t)]$$

$$H(t) = a_0 + a_1 t + a_2 t^2$$

Problem 4.35 Use linear regression and logarithmic transformation to determine the coefficients of the following model fitted to the survivor function of Problem 4.28:

$$R(t) = \exp[-H(t)]$$

$$H(t) = a_0 + a_1 t + a_2 t^2$$

Problem 4.36 Use nonlinear regression to determine the coefficients of the following model fitted to the survivor function of Problem 4.27:

$$R(t) = \exp[-H(t)]$$

$$H(t) = a_0 + a_1 t + a_2 t^2$$

Problem 4.37 Use nonlinear regression to determine the coefficients of the following model fitted to the survivor function of Problem 4.28:

$$R(t) = \exp[-H(t)]$$

$$H(t) = a_0 + a_1 t + a_2 t^2$$

Problem 4.38 The failure rate functions of two components with independent failure events are $r_1(t) = 10^{-4}$ hour^{-1}, and $r_2(t) = 2 \times 10^{-4}$ hour^{-1}. Find the reliability and failure rate functions for the system when they are arranged (a) in series and (b) in parallel.

Problem 4.39 The probability that an item will survive a 1000-hour mission is 0.4. If the item is operating 800 hours into the mission, the probability of surviving the remaining 200 hours of the mission is 0.85. What is the probability that the item survives the initial 800 hours of the mission?

Problem 4.40 For N identical units observed during 10,000 hours, x failures were observed. Assuming that the number of failures, x, follows a binomial distribution with the probability of failure p, find the mean value and the variance of the statistic $(x - Np)/[Np(1 - p)]^{0.5}$.

Problem 4.41 In assessing the effectiveness of a new brake system for newly designed buses, design engineers have to perform reliability testing to determine the failure probability of the brake system. Since prior information is not available, a uniform distribution for the failure probability is assumed. Using simulation for 500 times, the engineers observed 20 failures. Compute the Bayes mean failure probability.

Problem 4.42 A computer hardware engineer is in the process of assessing the reliability of a new component for a computer system. He found that the reliability of this component at the end of its useful life ($T = 20{,}000$ hours) is given as 0.80 ± 0.20 (mean $\pm$ standard deviation). A sample of 150 new components has been tested using an accelerated life technique for 20,000 hours, and 25 failures have been recorded. Given the test results, find the posterior mean and 90% Bayesian probability interval for the component reliability. The prior distribution of the component reliability is assumed to have a beta distribution.

Problem 4.43 Eight failures were observed during the accelerated life test of a sample of identical computer chips. The total time on test is 1500 hours. The time to failure distribution is assumed to be exponential. The gamma distribution with the mean of 0.02 hour^{-1} and with the COV of 40% was selected as a prior distribution to represent the prior information about the failure rate parameter of interest λ. Estimate the posterior estimate (mean) of λ and the upper 95% probability limit on λ.

Problem 4.44 Activities A–H associated with the operation of a system are illustrated using the arrow diagram below.

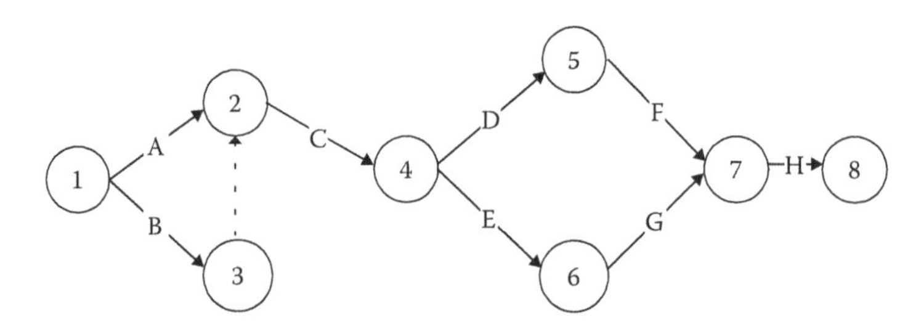

For example, the diagram shows that activity C cannot be started until the completion of either activity A or B. After completing activity C, both D and E can be started. Convert the arrow diagram into an RBD showing series and parallel connections. (*Hint:* A block can replace every arrow. Dotted arrows are dummy activities, which might not appear in the block diagram.) If the timely completion probability of each activity is given in the following table, compute the timely completion probability of the project assuming independent failure events for the activities:

Activity	Reliability
A	0.90
B	0.85
C	0.95
D	0.90
E	0.95
F	0.80
G	0.95
H	0.90

Problem 4.45 Activities A–G associated with the operation of a system are illustrated using the arrow diagram below.

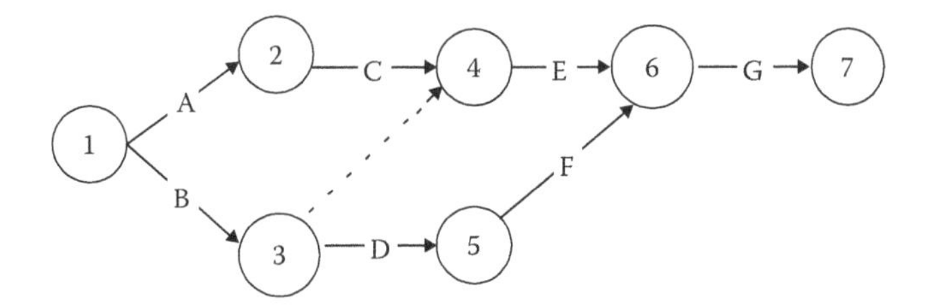

For example, the diagram shows that activity E cannot be started until the completion of either activity B or C. Activity G requires the completion of either E or F. Convert the arrow diagram into a block diagram showing series and parallel connections. (*Hint:* A block can replace every arrow. Dotted arrows are dummy activities, which might not appear in the block diagram.) If the failure probability for each activity is given in the following table, compute the failure probability of the operation assuming independent failure events for the activities:

Activity	Failure
A	0.20
B	0.15
C	0.05
D	0.10
E	0.15
F	0.25
G	0.05

Problem 4.46 The system of computers shown in the figure below consists of four components, connected in a series and parallel arrangement, that is, C1 and C2 that are in series are connected to C3 and C4 that are in parallel.

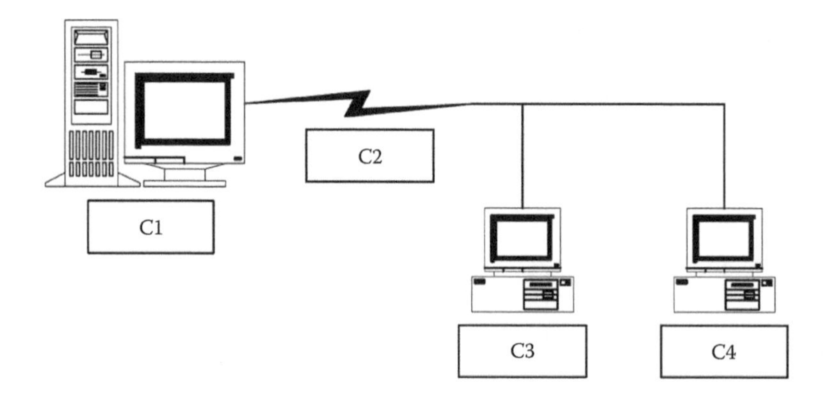

Convert this diagram into a block diagram showing series–parallel connections. If the failure probability for each component is given in the following table, compute the failure probability of the system assuming independent failure events for the components:

Component	Failure
C1	0.10
C2	0.05
C3	0.15
C4	0.20

Problem 4.47 Assuming five identical components connected in series, compute $h_c(t)$, $h_s(t)$, and $H_s(t)$ functions for a system in series. The component cumulative hazard function is given by the following equation:

$$H_c(t) = 0.3 - 0.02t + 0.0002t^2$$

Problem 4.48 Assuming five identical components connected in parallel, compute $h_c(t)$, $h_s(t)$, and $H_s(t)$ functions for a system in parallel. The component cumulative hazard function is given by the following equation:

$$H_c(t) = 0.3 - 0.02t + 0.0002t^2$$

Problem 4.49 Use the data of Problem 4.46 to compute $h_c(t)$, $h_s(t)$, and $H_s(t)$ functions for this series–parallel system using the following functions for each component:

Component	Hazard Function $H_c(t)$
C1	$0.3 - 0.02t + 0.002t^2$
C2	$0.2t$
C3	$0.3 - 0.02t + 0.0002t^2$
C4	$0.3 - 0.02t + 0.0002t^2$

Problem 4.50 Use the data of Problem 4.46 to compute $h_c(t)$, $h_s(t)$, and $H_s(t)$ functions for (1) the case where C1, C2, C3, and C4 are all connected in series, and (2) the case where C1, C2, C3, and C4 are all connected in parallel. Use the following functions for each component:

Component	Hazard Function $H_c(t)$
C1	$0.3 - 0.02t + 0.002t^2$
C2	$0.2t$
C3	$0.3 - 0.02t + 0.0002t^2$
C4	$0.3 - 0.02t + 0.0002t^2$

5

Failure Consequences and Valuations

Risk analysis entails the assessment of event consequences. This chapter starts with analytical methods for consequence assessment, such as cause–consequence diagrams, and follows them with separate sections on valuation including economic valuation, property damage, and human life loss as the primary focus. The chapter also briefly introduces other consequence types such as valuation of disability, health and injuries, tort and professional liability, indirect losses, public health and ecological impacts, dispersion and spread of consequences, and time-delayed consequences.

CONTENTS

5.1 Introduction

The failure of an engineering system could lead to consequences, creating a need to develop and use prediction methods of such potential failure consequences and evaluating their severities. The assessment methods can be based on analytical models or data collection from sources that include accident reports or both. In assessing consequences and severities, uncertainties can be modeled using random variables with probability distribution functions and their parameters or moments.

Consequence is the immediate, short-term, and long-term effects of an event affecting the objectives, for example, an explosion of a chlorine storage tank. These effects may include human and property losses, environmental damages, loss of lifelines, and so on. Broadly stated, events may include successes, and the favorable consequences in this case can be defined as the degree of reward or return or benefits from a success. Such an event could have, for example, beneficial economic outcomes or environmental effects. Failure consequences are the direct outcomes of the action or process of failure. They are the outcomes or effects of failure as a logical result or conclusion. A consequence in this case can be defined as the result of a failure (e.g., gas cloud, fire, explosion, evacuations, injuries, deaths, public and employee health effects, environment damages, or damage to the facility).

Failure or consequence severity is the quality, condition, strictness, impact, harshness, gravity, or intensity of the failure or the consequence. The amount of damage that is (or that may be) inflicted by a loss or catastrophe is a measure of the severity. The severity cannot be assessed with certainty, but it is preferable to try to estimate it with the uncertainty quantified using monetary terms where possible. The uncertain nature of severity necessitates its assessment in probabilistic terms. Failure severity is an assessment of potential losses that could include losses of property, people, wildlife, environment, capability to produce a product, and so on. These losses should also be defined, that is, valuated, in monetary or utility loss terms.

For example, a scenario of events in a chemical plant that lead to release of a chemical has consequences that can be measured, in part, by the amount of chemicals released with the associated uncertainty. As for severity, this chemical release could become a public health hazard as a result of human exposure to the chemicals. Another example is the failure of a dam that produces flooding as a consequence. For example, this flooding leads to a water level at a specific location of, say, average of 5 feet and a standard deviation of 1 foot. The severity of such flooding depends on the interactions of property, humans, and/or the environment with this consequence of the dam failure. For example, the damage to a house at the 5-foot water level can be assessed as a severity in monetary terms in regard to partial loss of the structure and its content.

Severity uncertainty has been recognized in the insurance industry and treated using random variable or stochastic process representations. Measures such as the *maximum possible loss* (MPL) and the *probable maximum loss* (PML) are used to assess, respectively, the worst loss that could occur based on the worst possible combination of circumstances and the loss that is likely based on the most likely combination of circumstances. For example, in the case of fire in a 10-story building, complete loss of the building can be considered as the MPL, whereas a fraction of this total loss can be considered the PML. Since fires are commonly discovered in their incipient stages due to alarms and losses are controlled by systems such as sprinkler systems, the use of PML might meet the needs of an insurance underwriter, especially because an underwriter commonly insures many similar buildings. Also we might be interested to know the exposure that is defined as the extent to which an organization's and/or stakeholder's concerns are subject to an event, and defined by *things at risk* that might include population at risk (PAR), property at risk, and ecological and environmental concerns at risk.

Two of the most difficult consequences to quantify are the loss of human life and the damage to the environment. One way to quantify these consequences is to place different levels of loss in different categories. For example, any event that results in the loss of one to two lives might be labeled as a category 4 loss, three to four lives would be a category 3 loss, five to six lives would be a category 2 loss, and seven or more lives would be a category 1 loss as discussed in Chapter 2. Such approaches are attempts to quantify the consequences that do not easily convert to dollar amounts. It should be noted that different consequence levels can be judged by different groups of people to have different levels of importance. Appropriate valuation should be used as discussed in Section 5.3.

5.2 Analytical Consequence and Severity Assessment

This section presents two methods that are of general applicability to assessing different consequence types: cause–consequence (CS) diagrams and functional modeling. The remaining sections of the chapter provide specific discussions and methods for various consequence types, such as for property loss and life loss.

5.2.1 Cause–Consequence Diagrams

Failure consequences and severities can be assessed using CS diagrams. These diagrams were developed for the purpose of assessing and propagating the conditional effects of a failure using a tree representation to a sufficient level of detail to assess severities as losses. The analysis according to CS starts with selecting a *critical event*, which is commonly selected as a convenient starting point for the purpose of developing a CS diagram. For a given critical event, the consequences are traced using logic trees with event chains and branches. The logic works both backward (similar to fault trees) and forward (similar to event trees). The procedure for developing a CS diagram can be based on answering a set of questions at any stage of the analysis. The questions can include, for example, the following:

- Can this event lead to other failure events?
- What conditions are necessary for this event to lead to other events?
- What are the other components affected by this event?
- What are the other events caused by this event?
- What are the consequences associated with the other (subsequent) events?
- What are the occurrence probabilities of subsequent events or failure probabilities of the components?

Each event in the CS diagram can be analyzed using a fault tree for the purposes of identifying underlying basic events or assessing its probability. Data can be used to assess the probabilities of these events in cases where data are available.

Example 5.1: Failure of Structural Components

In this example, failure scenarios are developed based on the initiating event of buckling of an unstiffened side shell panel of a naval vessel cargo space; these scenarios are used to demonstrate the process of developing CS diagrams. These failure scenarios are classified into two groups: (1) failure scenarios related to the failure of ship systems other than structural failure and (2) failure scenarios involving the ship structural system failure. In this example, only failure scenarios associated with the impact of this failure on the structural system are considered. CS diagrams can be developed based on the procedure shown in Figure 5.1 that presents the sequence of events that should be considered for development of the CS diagram. The consequences associated with the failure scenarios can be grouped as follows:

- Crew—possible injuries and deaths as a result of an overall hull girder failure (hull collapse)
- Cargo—possible loss of cargo, in case of hull failure
- Environment—possible contamination by fuel, lubricant oil, or cargo, in case of hull collapse
- Noncrew—none
- Structure—extensive hull damage, considering the failure of a primary structural member
- Ship—possible loss of ship in case of hull failure
- Other costs—cost of inspection and possible cost of repairs if buckling is detected

The cause–consequence diagram associated with this initiating event is presented in Figure 5.2. The consequences of the possible failure scenarios associated with the buckling of an inner side shell unstiffened panel in the cargo space are presented in Table 5.1. The logic in Figure 5.2 can be followed starting with the left box, *buckling of an inner side shell unstiffened panel*. The six fault trees shown in Figure 5.2 next to each event are provided to demonstrate how to evaluate the probabilities of the respective events; however, these trees are not used in the rest of the example. An explanation of the five-character failure scenarios defined in Table 5.1 is as follows:

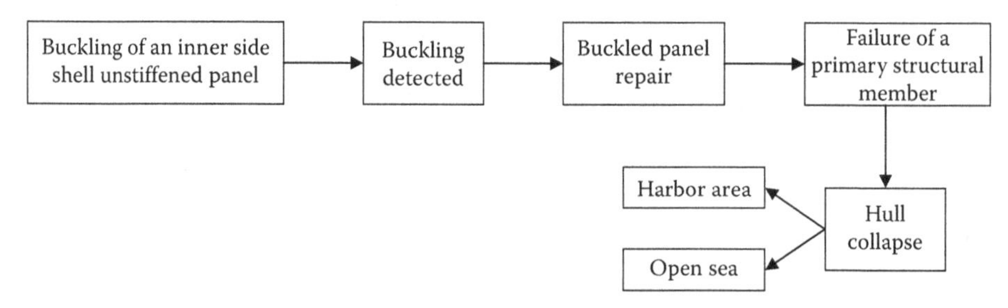

FIGURE 5.1
Buckling of an unstiffened side shell panel and its consequences.

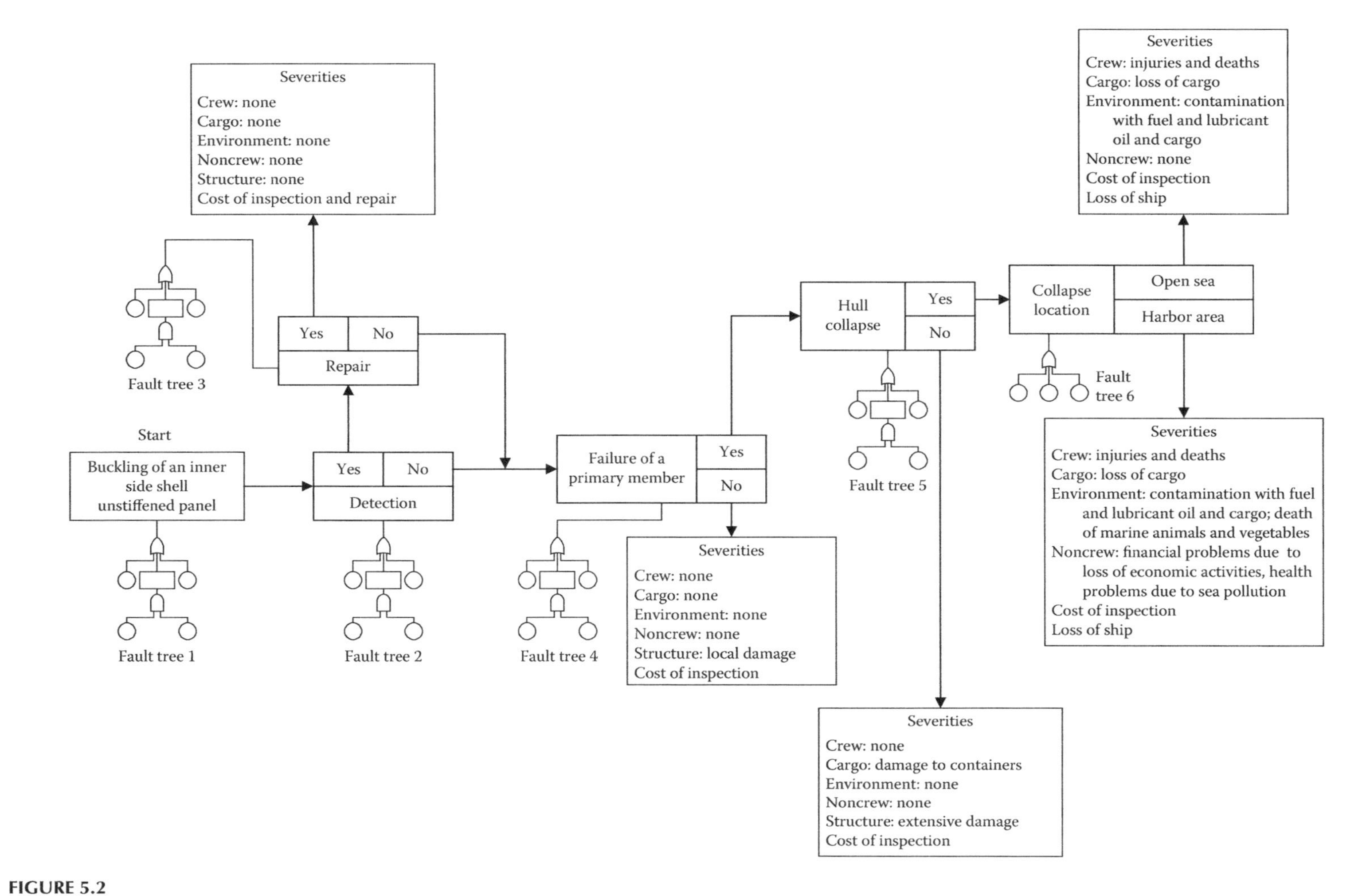

FIGURE 5.2
CS diagram for the buckling of an unstiffened panel.

TABLE 5.1

Structural Consequences Associated with the Buckling of an Unstiffened Panel

Failure Scenario	Severities						
Definition	**Crew**	**Cargo**	**Environment**	**Noncrew**	**Structural System**	**Inspection and Repair**	**Rating**
YYUUU	None	None	None	None	None	Cost of inspection and repair	1
YNYYO NUYYO	Injuries and deaths	Loss of cargo	Contamination with oil (fuel and lubricant) and cargo	None	Loss of ship	Cost of inspection	5
YNYYH NUYYH	Injuries and deaths	Loss of cargo	Contamination with oil (fuel and lubricant) and cargo, death of marine animals and plants	Financial problems due to loss of economic activities, health problems due to sea pollution	Loss of ship	Cost of inspection	5
YNYNU NUYNU	None	Damage to containers	None	None	Extensive damage	Cost of inspection	3
YNNUU NUNUU	None	None	None	None	Local damage		2

H, harbor vicinity; N, no; O, open seas; U, not applicable; Y, yes.

_XXXX: The first character corresponds to the detection of the buckling.
X_XXX: The second character corresponds to the repair of the buckled panel.
XX_XX: The third character corresponds to the failure of a primary structural member.
XXX_X: The fourth character corresponds to the hull collapse.
XXXX_: The fifth character corresponds to the geographical location of the hull failure.

The consequence rating is provided in Table 5.1 using an ordinal scale of 1–5, where 1 is the smallest consequence level and 5 is the greatest consequence level.

5.2.2 Functional Modeling

Assessing the impact of the failure of a system on other systems can be a difficult task. For example, the impact of structural damage on other systems can be assessed using special logic trees and approximate reasoning, pattern recognition, and expert systems based on functional modeling. Prediction of the structural response of the structural components or systems of a ship could require the use of nonlinear structural analysis; therefore, failure definitions must be expressed using deformations rather than forces or stresses. In addition, the recognition and proper classification of failures based on a structural response within the simulation process should be performed based on deformation responses. The process of failure classification and recognition should be automated in order to facilitate its use in simulation algorithms. Failure classification is based on matching a deformation or stress field with a record within a knowledge base of response and failure classes. In cases of no match, a list of approximate matches is provided, with assessed applicability factors. The user can then be prompted to make any changes to the approximate matches and their applicability factors.

Example 5.2: Failure Definition Based on Functional Modeling

Prediction of the structural response of a complex system, such as a floating marine system, could require the use of nonlinear structural analysis. In such cases, failure definitions need to be expressed using deformations rather than forces or stresses. Also, recognition and proper classification of failures based on a structural response within a simulation process should be performed based on deformations. Two failure analysis processes are needed: (1) failure recognition and (2) failure classification. The failure recognition process establishes failure classes based on failure impacts on various ship systems and the failure classification process places the simulation results of a particular simulation cycle expressed as a deformation field in one of the existing classes. The processes of failure classification and recognition should be automated in order to facilitate its use in a simulation algorithm for structural reliability assessment. Figure 5.3 shows a procedure for an automated failure recognition that can be implemented for reliability assessment. Figure 5.4 shows that the failure classification process utilizes matching a deformation or stress field with records within a knowledge base of response and failure classes. In cases of no match, a list of approximate matches is provided, with assessed applicability factors. The user can then be prompted for making any changes to the approximate matches and their applicability factors. In case of poor matches, the user can have the option of activating the failure recognition algorithm shown in

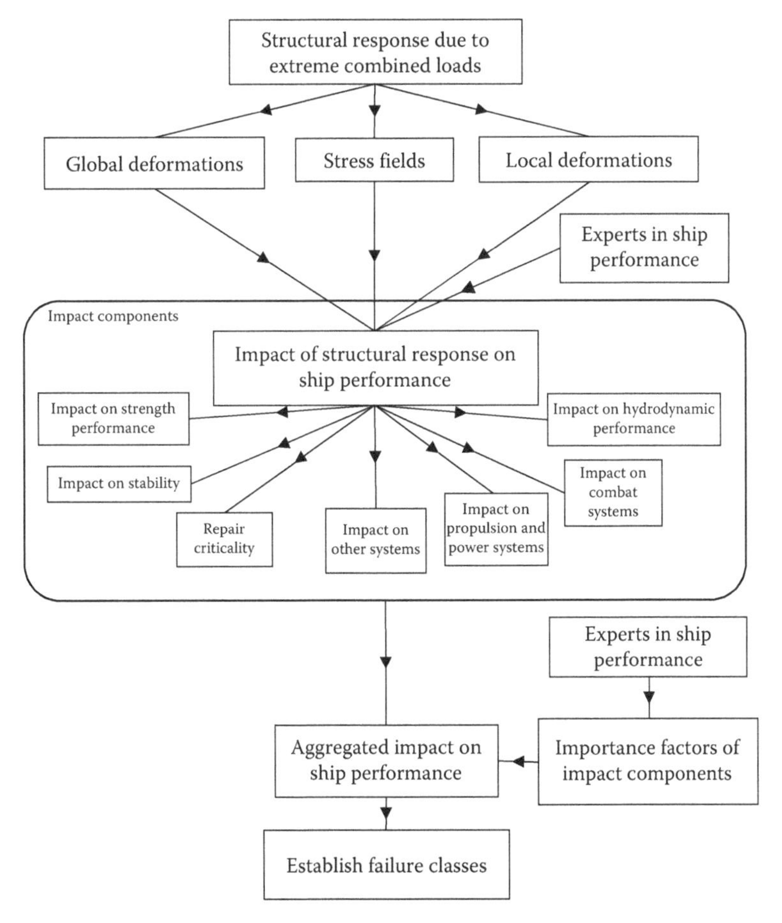

FIGURE 5.3
Failure recognition process for establishing failure classes.

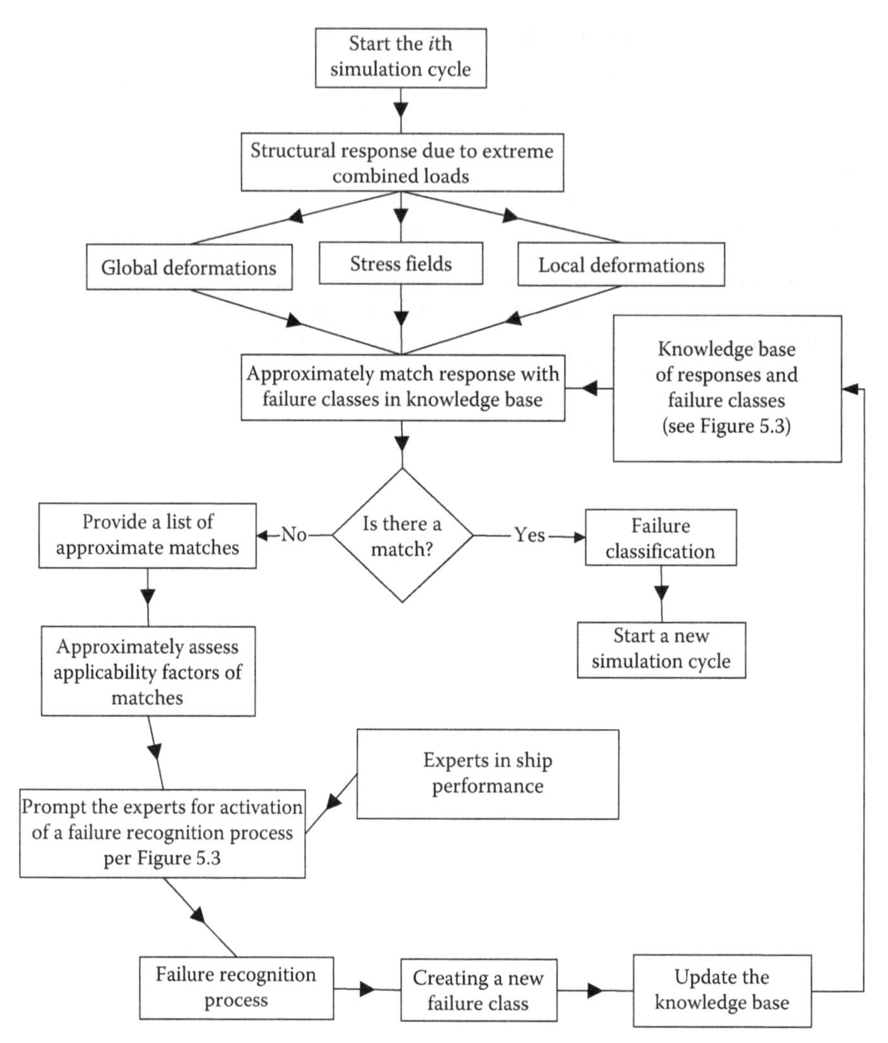

FIGURE 5.4
Failure classification process in simulation.

Figure 5.3 to establish a new failure class in the knowledge base. The adaptive nature of this algorithm allows the updating of the knowledge base based on the new failure class. The failure recognition and classification procedure shown in the figures evaluates the impact of the computed deformation or stress field on several systems of a ship. The severity assessment includes evaluating the remaining strength, stability, repair criticality, propulsion and power systems, combat systems, and hydrodynamic performance. The input of experts in ship performance is necessary to make these evaluations using either numeric or linguistic measures. Then, the assessed impacts are aggregated and combined to obtain overall failure recognition and classification within the established failure classes. The result of this process is then used to update the knowledge base.

A prototype computational methodology for reliability assessment of continuum structures using finite-element analysis with instability failure modes can be developed. A crude simulation procedure can be applied to compare the response with a specified failure definition, and failures can then be counted. By repeating the simulation procedure several times, the failure probability according to the specified failure classes is estimated as the failure fraction of simulation repetitions. Alternatively, conditional expectation can be used to estimate the failure probability in each simulation cycle, and the average failure probability and its statistical error can then be computed.

5.3 Fundamentals of Economic Valuation

5.3.1 Values and Their Distinctions

Risk studies examine the loss of things that we value, such as goods, property, assets, people, and service, for decision-making purposes. Many decision analysis frameworks require their valuations in economic or monetary terms. Approaching broadly from philosophy and particularly from ethics, we can make distinctions among values as (1) instrumental and intrinsic values, (2) anthropocentric and biocentric (or ecocentric) values, (3) existence value, and (4) utilitarian and deontological values (Callicott 2004; NRC 2004). The focus of this section is on economic valuation; however, it is necessary to introduce and discuss these distinctions. An ecosystem is used as an example to discuss these distinctions.

For an ecosystem, the *instrumental value* is derived from its role as a means toward an end other than itself, that is, its value is derived from its usefulness in achieving a goal. In contrast, *intrinsic value*, also called *noninstrumental value*, is its existence independently of any such contribution defined by usefulness. For example, if an animal population provides a source of food for either humans or other species, it has instrumental value that stems from its contribution or *usefulness* to the goal of sustaining the consuming population. If it continues to have value even if it were no longer *useful* to these populations, for example, if an alternative, preferred food source were discovered, such a remaining value would be its intrinsic value. For example, a national park, such as the Grand Canyon, has an intrinsic value component that exists unrelated or independent of direct or indirect use by humans for recreation or investigation. Such an intrinsic value can also stem from cultural sources, such as monuments and burial grounds (NRC 2004).

An *anthropocentric value* system considers humankind as the central focus or final goal of the universe and human beings as the only thing with intrinsic value, and the instrumental value of everything else is derived from its usefulness in meeting human goals. A biocentric value system, that is, nonanthropocentric, assigns intrinsic value to all individual living systems, including but not limited to humans, and assumes that all living systems have value even its usefulness to human beings cannot be determined or can be harmful to human beings.

Existence value reflects the desire of human beings to preserve and ensure the continued existence of certain species or environments to provide for humankind welfare, making it an anthropocentric and utilitarian concept of value and within the domain of instrumental value system. Therefore, utilitarian values are instrumental in that they are viewed as a means toward the end result of increased human welfare as defined by human preferences, without any value judgment about these preferences. The value of certain species or environments comes from generating welfare to human beings, rather than from the intrinsic value of these nonhuman species. This definition permits the potential for substitution or replacement of this source of welfare with an alternative source, that is, the possibility of a welfare-neutral trade-off between continued existence of species or environments and other things that also provide the same utility.

The *deontological value* system is based on an ethical doctrine which holds that the worth of an action is determined by its conformity to some binding rule rather than by its consequences. In this case, deontological value system implies a set of rights that include the right of existence. Something with intrinsic value is irreplaceable and its loss cannot be offset by having more of something else. For example, the death of person is a loss of an intrinsic value because it cannot be offset or compensated by that person having more of something else. The contentious issue is whether this concept should be extended to nonhuman species, for example, animals, either individual animals or species, or all biological creatures, that is, all plant and animal lives, collectively called the biota. In the context of ecosystem valuation, the modern notion of intrinsic value extends the rights beyond human beings. Utilitarian values are based on providing utilities.

This chapter recommends the use of a valuation approach with the following characteristics:

- Anthropocentric in nature based on utilitarian principles
- Consideration of all instrumental values, including existence value
- Its utilitarian basis to permit the potential for substitutability among different sources of values that contribute to human welfare
- Individual's preferences or marginal willingness to trade one good or service for another that can be influenced by culture, income level, and information, making it time and context specific
- Societal values as the aggregation of individual values

This approach is consistent with NRC (2004) and does not capture nonanthropocentric values, for example, biocentric values and intrinsic values as they are related to rights. In some decisions including environmental policy and law, biocentric intrinsic values should be included as was done previously, for example, the Endangered Species Act of 1973.

5.3.2 Total Economic Value

A *total economic value* (TEV) framework can be constructed based on the characteristics defined in Section 5.3.1 and using individual preferences and values. The TEV framework is necessary to ensure that all components of value are recognized and included while avoiding double counting of values (Bishop et al. 1987; Randall 1991). Figure 5.5 provides a classification of TEVs for aquatic ecosystem services with examples (Barbier 1994; NRC 2004).

5.3.3 Economic Valuation

Economic valuation is defined as the worth of a good or service as determined by the market, and used in decision analysis as discussed in Chapters 6 and 7. Economists have dealt with this concept initially by estimating the value of a good to an individual alone, and then extend it broadly as it relates to markets for exchange between buyers and sellers for wealth maximization as discussed in Chapter 6.

Traditionally, the value of a good or service is linked to its price in an open and competitive market determined primarily by the demand relative to supply. Therefore, goods, property, assets, safety of people, service, and so on are treated as commodities, and if there is no market to set the price of a commodity, then it has no economic value. Therefore, the

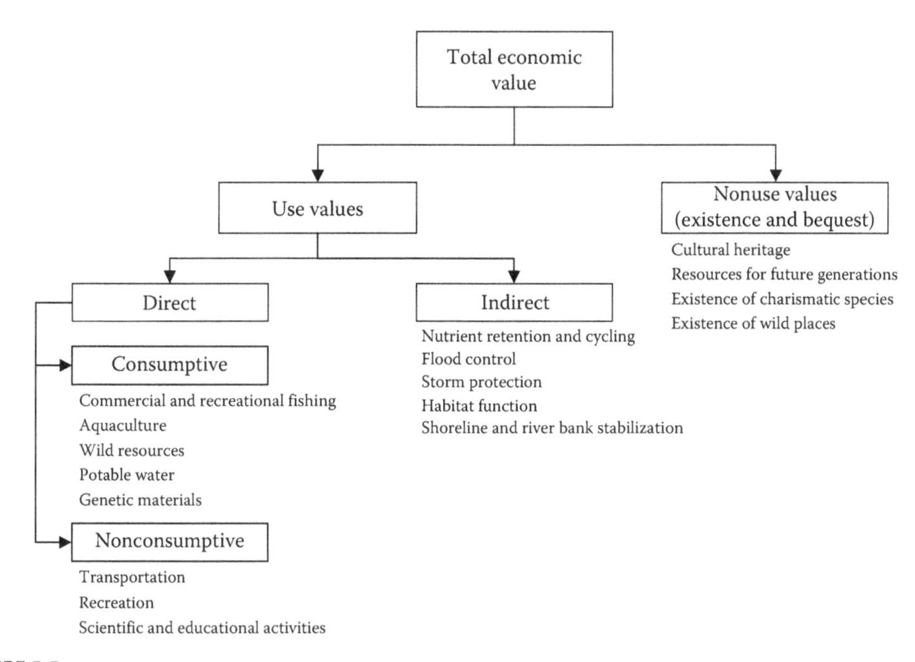

FIGURE 5.5
Classification of total economic value (TEV) for aquatic ecosystem services with examples. [Adapted from Barbier, E.B., *Land Economics*, 70, 155–173, 1994 and National Research Council (NRC), *Valuing Ecosystem Services: Toward Better Environmental Decision-Making*, National Academies Press, Washington, DC, 2004.]

value refers to the market worth of a commodity, which is determined by the equilibrium at which two commodities are exchanged. The limitation herein is in its inability to set a value to things that are not exchanged in markets.

In the *labor theory of value*, a good or service is associated with the amount of discomfort or labor saved through the consumption or use of it. According to this theory, the *exchange value* is recognized without making it equivalent to an economic value, that is, price and value are considered as two different concepts. Accordingly, a value is determined based on the exchange price that does not necessarily represent its true economic value.

An economic measure of the value of a good or the benefit from a service can be defined as the maximum amount a person is willing to pay for this good or service. The concept of *willingness to pay* (WTP) is central to economic valuation. An alternate measure is the *willingness to accept* (WTA) of an amount by the person to forgo taking possession of the good or receiving the service. WTP and WTA produce amounts that are expected to be close; however, generally WTA-generated amounts are greater than WTP-generated amounts due primarily to income levels and affordability factors.

The economic concept of value, including its exchange value, can be criticized as being stripped from moral and ethical considerations. For example, having an exchange value for a good or a service that is harmful in nature, for example, markets of illegal drugs or gambling or prostitution or weaponry, have value in some open markets and in some underground markets, and no value in others. Contrarily not having an exchange value for a good or a service that is good in nature, for example, volunteer work, might not have a market values but this does not necessarily make it without any value. Accounting for such moral and ethical considerations in economic models can be contentious, and commonly such goods or services are ignored. If needed, the concept of TEV should be used in such cases as illustrated in Figure 5.5.

5.4 Real Property Damage

The assessment of real property damage as a result of failure can be quantified in monetary terms using an appropriate mixture of analytical cost estimation models, empirical models, and judgment using expert opinions. The structure and working of the analytical cost estimation models depend on the hazard and properties being investigated. The primary concepts that can be used for assessing property damage are presented in this section using water flooding as a hazard and residential structures and vehicles as the property. Three formulations are provided based on (1) cost estimation models; (2) empirical evidence, data, and simulation; and (3) expert opinion elicitation. In these formulations, damage to residential property as a result of flooding is discussed. Other types of hazard and property might require adaptation or entirely different formulations.

The failure severity in terms of property loss can be assessed as the current replacement value minus depreciation to obtain the *actual cash value* of a property. Sometimes *replacement cost* is used to assess the loss, where replacement cost is defined as the cost of reconstructing the property with like kind and quality. A primary difference between the actual cash value and the replacement cost value is depreciation. The replacement cost is required for both approaches. The replacement cost can be estimated using a work breakdown structure with material and labor estimates, rates, and aggregations. In addition, construction cost indexes can be used to adjust for time and location. Alternatively, the rates per square foot can be used to obtain coarse or rough estimates of the replacement cost. Sometimes size and shape modifiers are used to account for unique variations that are out of the ordinary.

Assessing the content loss of a residential structure can be based on a detailed breakdown of content into structure size, quality, and functions of various spaces in the property. The content loss for each room can then be estimated and aggregated for the entire structure. As for businesses, property loss could include machinery and equipment, furnishings, raw materials, and inventories. Computer programs are commercially available to aid in this type of estimation for both residential and commercial structures. Some aspects of these estimation methods are illustrated in this section.

5.4.1 Analytical Cost Estimation Models

This class of methods is introduced using an example model. The model was developed in 2001 to organize floodplain inventory data and estimate the residential structure and content damage for various depths of flooding on a structure-by-structure basis. This US Army Corps of Engineers (USACE) Floodplain Inventory Tool (CEFIT) estimates residential content values as a function of floodwater depth by factoring in the typical number of rooms, items generally kept in homes of various quality levels, and placement of those items relative to the first floor. CEFIT builds on estimates provided by commercially available software called *Residential Estimator* (RE), developed and marketed by Marshall and Swift, Los Angeles, CA, USA. CEFIT predicts flood damage by assuming that each component or assembly would be cleaned, repaired, replaced, or reset at each given flooding depth. This methodology is depicted in Figure 5.6, which shows how CEFIT uses the *RE* methodology to estimate the structure costs combined with flood–stage (i.e., water-level) data contained in the CEFIT database to provide outputs in the form of flood–damage, that is, flood–stage, relationships for further analysis by engineers or economists.

When a component or assembly is replaced, its full-depreciated replacement costs, as estimated from the RE, are accrued as part of the flood damage. When a component or assembly

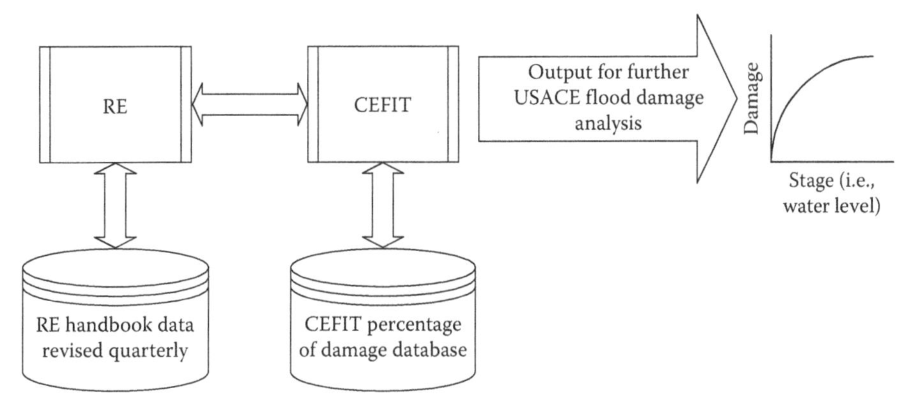

FIGURE 5.6
Corps of Engineers Floodplain Inventory Tool (CEFIT) methodology for computing flood–stage relationships. RE, residential estimator; USACE, US Army Corps of Engineers.

is cleaned or repaired, fractions of the replacement cost are accrued. Thus, the estimated damage at any depth of flooding relies on the assumed response to flooding (clean, repair, replace, or reset) and on the assumed fraction of the replacement cost. CEFIT uses the RE to calculate the replacement cost and applies the technique of aggregating lower level cost information (or component costs) against a listing of quantities, or *bill of quantity*. This modeling technique consists of compiling all the estimates for all the variations of building configurations defined in building the methodology, with all the bills of quantity being a function of the living area. Bills of quantities for 960 building configurations are detailed in the CEFIT database.

Steps for providing key user-defined inputs are given in Figure 5.7. The library of 960 models covers all combinations of key user-defined parameters (eight styles, three building material types, two age periods, five infrastructure types, and four quality types). The user interface of CEFIT permits defining the dwelling type from selections made by the user from pull-down menus. User input data include house configuration, material type, infrastructure type, location, living area, and vertical footage at which water reaches the first floor level. CEFIT selects the model that best fits the user input from the library of 960 models and defines the number of rooms, their size, and location (i.e., which story) in the house.

Next, CEFIT selects the flood level that corresponds to the user input. The model estimates flood damage, including building repair and replacement costs, based on extrapolating to the specified total floor area and updating the remove, clean, replace, and reset operations to the systems and components depending on the predefined flood level. The predefined flood level is accessible for 16 increments of flooding. The flood damage estimate is localized at the price level for any given zip code within the United States.

Example 5.3: Property Loss due to Flooding, Part 1

To illustrate the loss estimation used by CEFIT, a 2000-square-foot home with an effective age of 0 years, located in zip code 22222 (Arlington, VA) was used for illustration purposes. The house has the following characteristics that are needed by CEFIT as input: number of stories = 1, foundation type = slab, construction = standard, style = ranch, quality = average, condition = average, exterior wall = frame, wood siding, and roofing = wood shingle. Table 5.2 shows losses for this residence at flood depths from 1 to 10 feet, as calculated by CEFIT, as a percentage of the RE replacement cost of US$104,747 in 2001. The results are also shown in Figure 5.8.

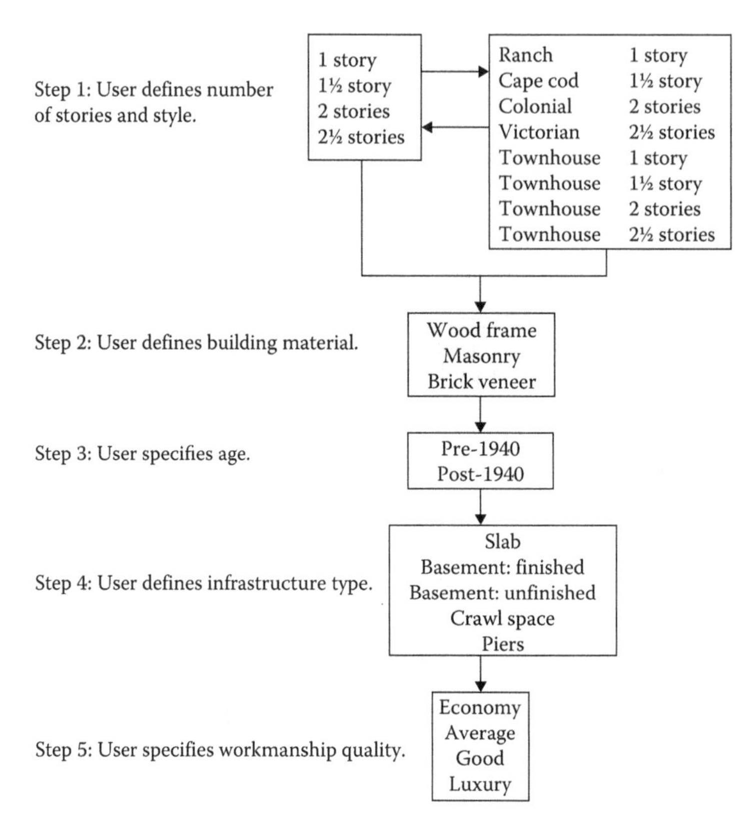

FIGURE 5.7
Steps in providing key Corps of Engineers Floodplain Inventory Tool (CEFIT) user-defined inputs.

TABLE 5.2
Losses as a Function of Water Depth

Water Level (ft)	Damage ($)	Total Replacement Cost (%)
1	24,406	23
2	33,624	32
3	42,004	40
4	49,336	47
5	55,725	53
6	61,382	59
7	66,200	63
8	70,390	67
9	73,847	71
10	76,675	73

5.4.2 Empirical Models and Simulation

Loss and claim data can be used as a basis for constructing regression models. For example, damage survey teams can be assembled and tasked to collect loss and damage data after floods or earthquakes. After flood events, claims are commonly filed with flood insurance programs. Such claims can be analyzed and used to fit regression models. The resulting models can be used for prediction and planning purposes.

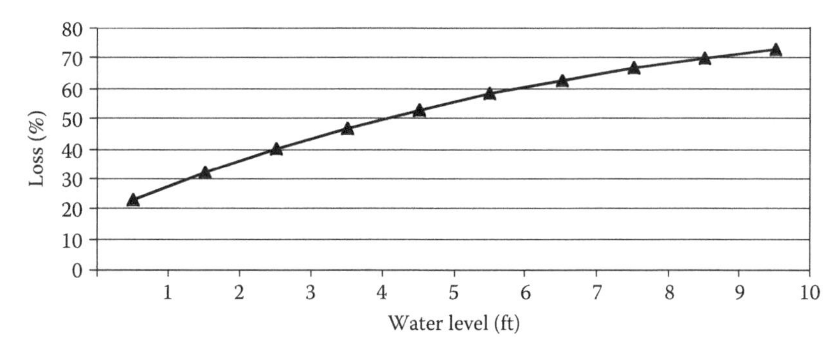

FIGURE 5.8
Damage to a residential structure due to flooding.

Researchers frequently examine empirical evidence of historical events or recent events and infer from the evidence floodwater heights, wind speeds, impact energy, and so on. Also, researchers compute statistics of event rates and losses, and compare them to long-term trends (Clark 2011). Once the hazards are estimated, they are used in simulation to predict potential losses for planning and decision-making purposes.

Simulation is often used to assess the potential losses by examining the impacts of selected scenarios, either analytically constructed or based on actual historic events, an asset, a region, or an inventory of assets. For example, a damaging hurricane that hit a city in the beginning of the previous century can be assumed to impact the same city with its current inventory of assets, and the damages are reassessed and compared to the estimates based on other prediction methods. Such an analysis helps owners, city planners, and insurance companies to evaluate the suitability of first-response capabilities, resource capacity to honor claims, or medical facilities and methods to handle casualties.

The US National Hurricane Center maintains the datasets that provide historical hurricane track information for the Atlantic and Gulf of Mexico. For each hurricane, it generally provides information such as the storm track, central pressure, and wind speed at 6-hour intervals along the track. This information can be used to construct scenarios for loss predictions using current and projected asset inventories.

Example 5.4: Losses due to Hurricanes in the Atlantic

Three modeling companies, AIR Worldwide, EQECAT, and Risk Management Solutions (RMS), developed and used near-term Atlantic hurricane prediction models from 2006 to 2010 to predict the number of hurricanes and losses during this time period (Tables 5.3 through 5.5). AIR modified its near-term model in 2007 to reduce loss predictions relative to the long-term model to ~16%. RMS introduced a modification to its near-term model in 2009 following its annual elicitation of expert opinions that resulted in a reduction in the near-term loss estimates; however, the model still implied a level of loss activity 25% above the long-term historical average. EQECAT made relatively minor adjustments to its near-term model estimates. It should be noted that retrospectively these models initially projected hurricane loss levels at least 35% above the long-term average for the 5-year period (Clark 2011).

Tables 5.3 and 5.4 compare the observed, that is, actual, values with implied projections based on the three near-term models (Clark 2011). For the 5-year period, the aggregate of the actual numbers of hurricanes is greater than that of the values based on the long-term average, but smaller than that of the values based on the near-term models. As for the insured losses in the United States, the three models substantially overpredicted the losses compared to the observed values. Table 5.5 shows hurricane landfall counts and

TABLE 5.3

Number of Atlantic Hurricanes (Number of US Landfalls)

Year	Long-term Average	Observed (Actual)	AIR Predictions	EQECAT Predictions	RMS Prediction
2006	5.9 (1.7)	5 (0)	8.4 (2.4)	8.0 (2.3)	8.4 (2.4)
2007	5.9 (1.7)	6 (1)	6.8 (2.0)	8.0 (2.3)	8.4 (2.4)
2008	5.9 (1.7)	8 (3)	6.8 (2.0)	8.1 (2.3)	8.4 (2.4)
2009	5.9 (1.7)	3 (0)	6.8 (2.0)	8.1 (2.3)	7.6 (2.2)
2010	5.9 (1.7)	12 (0)	6.8 (2.0)	8.1 (2.3)	7.6 (2.2)
Total	29.5 (8.5)	34 (4)	35.6 (10.4)	40.3 (11.5)	40.4 (11.6)

TABLE 5.4

US Insured Losses from Atlantic Hurricanes (2007 US$ in Billion)

Year	Long-term Average	Observed (Actual)	AIR Predictions	EQECAT Predictions	RMS Prediction
2006	10	0	14.0	13.6	14.0
2007	10	0	11.6	13.5	14.0
2008	10	15.2	11.6	13.7	14.0
2009	10	0	11.6	13.7	12.6
2010	10	0	11.6	13.7	12.6
Total	50	15.2	60.4	68.2	67.2

TABLE 5.5

US Hurricane Landfalls and Losses

Year	Number of Landfalls	Losses (2007 US$ in Billion)
2001	0	–
2002	1	0.5
2003	2	2.0
2004	5	25.1
2005	5	61.9
2006	0	–
2007	1	–
2008	3	15.2
2009	0	–
2010	0	–
Decade average	1.7	10.5
Long-term average	1.7	10.0

losses for the first decade of the twenty-first century. We can observe the approximate agreement between the long-term average and the decade average.

5.4.3 Expert Opinions

Expert opinion elicitation can be used to assess property damage. In this section, water flooding is used to illustrate the method. Chapter 8 formally introduces the methods for eliciting expert opinions, which can be defined as a heuristic process of gathering information and data or answering questions on issues or problems of concern.

Example 5.5: Property Loss due to Flooding, Part 2

Expert opinion elicitation is used in this example to illustrate the development of structural and content depth–damage relationships for single-family, one-story homes without basements; residential content-to-structure value ratios (CSVRs); and vehicle depth–damage relationships in a river basin. These damage functions consider the exterior building materials such as brick, brick veneer, wood frame, and metal siding. The resulting consequences can be used in risk studies and in performing risk-based decision making. The expert elicitation was assumed to occur during a face-to-face meeting of members of an expert panel assembled specifically for the issues under consideration. It was further assumed that the meeting of the expert panel was conducted after communicating to the experts in advance of the meeting the background information, objectives, list of issues, and anticipated outcomes from the meeting.

LEVEE FAILURE AND CONSEQUENT FLOODING

In January 1997, the eastern levee of a river failed, causing major flooding near a town. Floodwaters inundated ~12,000 acres and caused damage to over 700 structures. Although the area was primarily agricultural, ~600 residential structures were affected by flooding. This area had a wide range of flooding depths, ranging from maximum depths of about 20 feet (structures totally covered) in the south near the levee break to minimal depths. Residential damage from the flooding was documented. The population of homes within the January 1997 floodplain defines the study area of interest.

FLOOD CHARACTERISTICS

The January 1997 flooding resulted from a trio of subtropical storms. Over a three-day period, warm moist winds from the southwest of the town poured more than 30 inches of rain onto watersheds that were already saturated by one of the wettest Decembers on record. The first of the storms hit on December 29, 1996, with less-than-expected precipitation totals. Only a 0.24-inch rainfall was reported. On December 30, the second storm arrived. The third and most severe storm hit late December 31, 1996, and lasted through January 2, 1997.

Precipitation totals at lower elevations in the central valley were not unusually high, in contrast to the extreme rainfall in the upper watersheds. Downtown, for example, received 3.7 inches of rain from December 26, 1996, through January 2, 1997. However, other locations (elevation 5000 feet) received >30 inches of rainfall, resulting in an orographic ratio of 8:1. A typical storm for this region would yield an orographic ratio of 3:4 between these two locations.

In addition to the trio of subtropical storms, snowmelt also contributed to the already large runoff volumes. Several days before Christmas 1996, a cold storm from the north brought snow to low elevations in foothills. For example, some locations had a snowpack with 5 inches of water content. The snowpack, as well as the snowpack at lower elevations, melted when the trio of warmer storms hit. Not much snowpack loss was observed, however, at snow sensors over 6000 feet in elevation. The effect of the snowmelt was estimated to contribute ~15% to the runoff totals.

Prior to the late December storms, rainfall was already well above the normal in the river basin. In the northern area, the total December precipitation exceeded 28 inches, the second wettest December on record, exceeded only by the 30.8 inches in December 1955.

The available storage in a reservoir was >200% of flood management reservation space on December 1, 1996. By the end of the storm, the available space was ~1% of the flood pool. Another reservoir began in December with just >100% flood management reservation space. At the completion of the storms in early January, ~27% space remained available.

The hydrologic conditions of the January 1997 flooding of the river basin were used as the basis for developing depth–damage relationships and CSVRs. These hydrologic conditions resulted in high-velocity flooding coming from an intense rainfall and a levee failure. This scenario and the flood characteristics were defined and used in the study to assess losses.

BUILDING CHARACTERISTICS

Most of the residential properties affected by the January 1997 flooding were single-story, single-family structures with no basements. The primary construction materials were wood and stucco. Few properties in the study area were two stories, and nearly none had basements. It may be useful to differentiate one story on slab from one story on raised foundations. The study is limited to residential structural types without basements as follows: (1) one story on slab, (2) one story on piers and beams (i.e., raised foundations), and (3) mobile homes.

VEHICLE CHARACTERISTICS

Vehicle classes included in the study are (1) sedans; (2) pickup trucks, sport utility vehicles, and vans; and (3) motorcycles.

STRUCTURAL DEPTH–DAMAGE RELATIONSHIPS

The hydrologic conditions of the January 1997 flooding described earlier were used as the basis for developing these relationships. These hydrologic conditions produced high-velocity flooding due to an intense rainfall and a levee failure. The issues presented to the experts for consideration were (1) the best estimates of the median percentage damage values as a function of flood depth for residential structures of all types and (2) the confidence level for the opinion of the expert (low, medium, or high). The study was limited to residential structural types as follows: (1) type 1, one story on slab without basement; (2) type 2, one story on piers and beams (raised foundation); and (3) type 3, mobile homes.

The experts discussed the issues that produced the assumptions provided in Table 5.6. In this study, structural depth–damage relationships were developed based on expert opinions, and a sample of the results is provided in Table 5.7. The experts provided their best estimates of the median value for percentage damage and their levels of confidence in their estimates. Sample revised depth–damage relationships are shown in Figures 5.9 and 5.10.

TABLE 5.6

Summary of Supportive Reasoning and Assumptions by Experts for Structure Value (2001 US$)

Type 1 and 2 Houses	Type 3 Houses
The median house size is 1400 square feet.	The median size is 24 × 60 feet (1200 square feet).
Houses are wood frame.	Houses are wood frame.
The median house value is $90,000 with land.	The median house value is $30,000 without land.
The median land value is $20,000.	The median house age is 8 years.
The median price without land is about $50 per square foot.	Finished floor is 3 feet above ground level.
The median house age is eight years.	Ceiling height is eight feet.
Heating,Ventilation and Air Conditioning (HVAC) and sewer lines are below finished floor for type 2 houses.	HVAC and sewer lines are below finished floor.
Percentages are of depreciated replacement value of houses.	Percentages are of depreciated replacement value of houses.
Flood without flow velocity was considered.	Flood without flow velocity was considered.
Flood duration was of several days.	Flood duration was of several days.
Flood water was not contaminated but had sediment without large debris.	Flood water was not contaminated but had sediment without large debris.
No septic field damages are included.	No septic field damages are included.
Allowances were made for cleanup costs.	Allowances were made for cleanup costs.

TABLE 5.7

Percentage of Damage to a Type 1 Residential Structure (One Story on Slab without Basement)

	Initial Estimate: Damage by Expert (%)							Aggregated Opinions as Percentiles				
Depth	**1**	**2**	**3**	**4**	**5**	**6**	**7**	**Minimum**	**25%**	**50%**	**75%**	**Maximum**
−1.0	4.0	0.0	3.0	0.0	0.0	0.0	0.0	0.0	0.0	0.0	1.5	4.0
−0.5	4.0	0.0	5.0	0.0	0.0	0.0	0.0	0.0	0.0	0.0	2.0	5.0
0.0	5.0	0.0	10.0	5.0	0.0	10.0	0.0	0.0	0.0	5.0	7.5	10.0
0.5	10.0	40.0	12.0	7.0	10.0	13.0	45.0	7.0	10.0	12.0	26.5	45.0
1.0	15.0	40.0	25.0	9.0	20.0	15.0	55.0	9.0	15.0	20.0	32.5	55.0
1.5	20.0	40.0	28.0	11.0	30.0	20.0	55.0	11.0	20.0	28.0	35.0	55.0
2.0	30.0	40.0	35.0	13.0	30.0	20.0	60.0	13.0	25.0	30.0	37.5	60.0
3.0	40.0	40.0	35.0	15.0	40.0	30.0	60.0	15.0	32.5	40.0	40.0	60.0
4.0	48.0	40.0	40.0	25.0	70.0	50.0	65.0	25.0	40.0	48.0	57.5	70.0
5.0	53.0	65.0	40.0	40.0	70.0	85.0	70.0	40.0	46.5	65.0	70.0	85.0
6.0	65.0	65.0	45.0	50.0	70.0	85.0	75.0	45.0	57.5	65.0	72.5	85.0
7.0	68.0	70.0	75.0	70.0	80.0	90.0	75.0	68.0	70.0	75.0	77.5	90.0
8.0	70.0	75.0	80.0	90.0	80.0	90.0	75.0	70.0	75.0	80.0	85.0	90.0
9.0	73.0	85.0	95.0	100.0	95.0	90.0	75.0	73.0	80.0	90.0	95.0	100.0
10.0	80.0	85.0	100.0	100.0	100.0	100.0	80.0	80.0	82.5	100.0	100.0	100.0
11.0	83.0	85.0	100.0	100.0	100.0	100.0	80.0	80.0	84.0	100.0	100.0	100.0
12.0	85.0	85.0	100.0	100.0	100.0	100.0	80.0	80.0	85.0	100.0	100.0	100.0
	Revised Estimate: Damage by Expert (%)							**Aggregated Opinions as Percentiles**				
Depth	**1**	**2**	**3**	**4**	**5**	**6**	**7**	**Minimum**	**25%**	**50%**	**75%**	**Maximum**
−1.0	1.0	0.0	3.0	0.0	0.0	0.0	5.0	0.0	0.0	0.0	2.0	5.0
−0.5	1.0	0.0	5.0	0.0	0.0	0.0	10.0	0.0	0.0	0.0	3.0	10.0
0.0	10.0	15.0	10.0	5.0	5.0	15.0	35.0	5.0	7.5	10.0	15.0	35.0
0.5	10.0	40.0	25.0	40.0	20.0	45.0	45.0	10.0	22.5	40.0	42.5	45.0
1.0	25.0	40.0	30.0	40.0	20.0	45.0	45.0	20.0	27.5	40.0	42.5	45.0
1.5	25.0	40.0	40.0	40.0	30.0	45.0	45.0	25.0	35.0	40.0	42.5	45.0
2.0	35.0	40.0	45.0	40.0	30.0	45.0	45.0	30.0	37.5	40.0	45.0	45.0
3.0	40.0	40.0	45.0	40.0	40.0	70.0	45.0	40.0	40.0	40.0	45.0	70.0
4.0	48.0	40.0	55.0	40.0	70.0	80.0	55.0	40.0	44.0	55.0	62.5	80.0
5.0	53.0	65.0	55.0	50.0	70.0	85.0	60.0	50.0	54.0	60.0	67.5	85.0
6.0	65.0	65.0	70.0	60.0	70.0	85.0	65.0	60.0	65.0	65.0	70.0	85.0
7.0	68.0	65.0	75.0	85.0	80.0	95.0	75.0	65.0	71.5	75.0	82.5	95.0
8.0	70.0	65.0	80.0	85.0	85.0	95.0	75.0	65.0	72.5	80.0	85.0	95.0
9.0	73.0	85.0	95.0	85.0	85.0	95.0	75.0	73.0	80.0	85.0	90.0	95.0
10.0	80.0	85.0	100.0	85.0	85.0	95.0	80.0	80.0	82.5	85.0	90.0	100.0
11.0	83.0	85.0	100.0	85.0	85.0	95.0	80.0	80.0	84.0	85.0	90.0	100.0
12.0	85.0	85.0	100.0	85.0	85.0	95.0	80.0	80.0	85.0	85.0	90.0	100.0
Confidence	High	High	High	High	High	High	High					

CONTENT DEPTH–DAMAGE RELATIONSHIPS

The hydrologic conditions of the January 1997 flooding described earlier were used as the basis for developing these relationships. These hydrologic conditions produced high-velocity flooding due to an intense rainfall and a levee failure. The issues presented to the experts for consideration were (1) the best estimates of the median

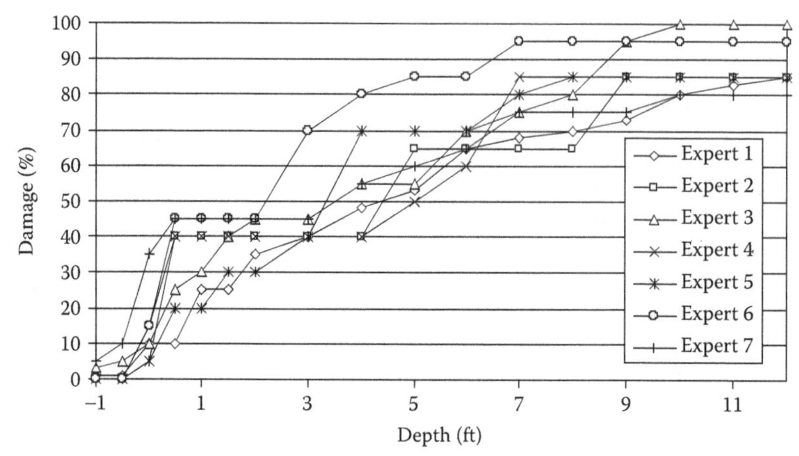

FIGURE 5.9
Percentage of damage to a type 1 residential structure (one story on slab without basement).

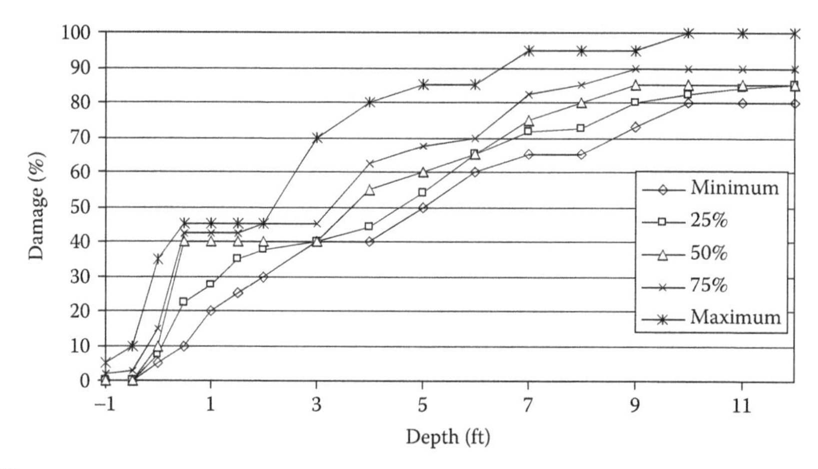

FIGURE 5.10
Aggregated (as percentiles) percentage of damage to a type 1 residential structure (one story on slab without basement).

percentage damage values as a function of flood depth for residential structures of all types and (2) the confidence level for the opinion of the expert (low, medium, or high). The study was limited to residential structural types as follows: (1) types 1 and 2, one story on slab without basement or one story on piers and beams (raised foundation); and (2) type 3, mobile homes. The experts discussed the issues that produced the assumptions provided in Table 5.8. In this study, content depth–damage relationships were developed based on expert opinions (see the sample provided in Table 5.9). Sample revised depth–damage relationships are shown in Figures 5.11 and 5.12.

CONTENT-TO-STRUCTURE VALUE RATIOS

The hydrologic conditions of the January 1997 flooding described earlier were used as the basis for developing these relationships. These hydrologic conditions produced high-velocity flooding due to an intense rainfall and a levee failure. The

TABLE 5.8

Summary of Supportive Reasoning and Assumptions by Experts for Content Value

Houses Types 1–3 (2001 US$)
As a guide, the insurance industry uses 70% ratio of the content to structure value.
The median house value is $90,000 with land.
The median land value is $20,000.
Garage or shed contents are included.
The median content age is eight years.
Percentages are of depreciated replacement value of contents.
Flood without flow velocity was considered.
Flood duration was for several days.
Flood water is not contaminated but has sediment without large debris.
Allowance is made for cleanup costs.
Insufficient time was allowed to remove (or protect) contents.

issues presented to the experts were (1) the best estimates of the median values of a residential structure, its content, and their content-to-structure-value ratios (CSVRs) for all types and (2) the confidence level for the opinion of the expert (low, medium, or high). The study was limited to residential structural types as follows: (1) types 1 and 2, one story on slab without basement or one story on piers and beams (raised foundation) and (2) type 3, mobile homes. The experts discussed the issues that produced the assumptions provided in Table 5.10. In this study, the best estimates of the median value of structures, the median value of contents, and the ratio of content to structure value were developed based on the expert opinions, a sample of which is provided in Table 5.11. The table provides the initial and revised expert opinions of median structure value, median content value, and the CVSR. Each expert provided best estimate value, low value estimate, and high value estimate. Also, the experts provided an assessment of their individual confidence levels for their opinions. Sample CVSRs are shown in Figure 5.13.

VEHICLE DEPTH–DAMAGE RELATIONSHIPS

The hydrologic conditions of the January 1997 flooding described earlier were used as the basis for developing these relationships. These hydrologic conditions produced high-velocity flooding due to an intense rainfall and a levee failure. The issues presented to the experts were (1) the best estimates of the median percentage damage values as a function of flood depth for vehicles of all types and (2) the confidence level for the opinion of the expert (low, medium, or high). The study was limited to residential vehicle classes as follows: (1) type 1, sedans; (2) type 2, pickup trucks, sport utility vehicles, and vans; and (3) type 3, motorcycles. The experts discussed the issues that produced the assumptions provided in Table 5.12. In this study, the best estimates of the median value of vehicle depth–damage relationships were developed based on expert opinions, a sample of which is provided in Table 5.13. Sample relationships are shown in Figures 5.14 and 5.15.

Example 5.6: Property Loss due to the 1906 San Francisco Earthquake

This earthquake of April 18, 1906, ranks as one of the most significant earthquakes of all time according to the US Geological Survey (USGS). The rupture of the northernmost 296 miles (477 km) of the San Andreas fault from northwest of San Juan Bautista to the triple junction at Cape Mendocino resulted in this large earthquake in terms of horizontal displacements and great rupture length.

TABLE 5.9

Percentage of Damage to Contents of Type 1 and 2 Residential Structures (One Story on Slab or One Story on Piers and Beams)

	Initial Estimate: Damage by Expert (%)							Aggregated Opinions as Percentiles				
Depth	**1**	**2**	**3**	**4**	**5**	**6**	**7**	**Minimum**	**25%**	**50%**	**75%**	**Maximum**
−1.0	0.5	0.0	3.0	0.0	0.0	10.0	0.0	0.0	0.0	0.0	1.8	10.0
−0.5	0.5	0.0	5.0	0.0	0.0	20.0	0.0	0.0	0.0	0.0	2.8	20.0
0.0	2.0	30.0	15.0	0.0	0.0	40.0	5.0	0.0	1.0	5.0	22.5	40.0
0.5	2.0	40.0	35.0	20.0	50.0	40.0	10.0	2.0	15.0	35.0	40.0	50.0
1.0	15.0	50.0	35.0	40.0	50.0	40.0	20.0	15.0	27.5	40.0	45.0	50.0
1.5	27.0	60.0	40.0	50.0	60.0	40.0	20.0	20.0	33.5	40.0	55.0	60.0
2.0	35.0	70.0	40.0	60.0	70.0	60.0	40.0	35.0	40.0	60.0	65.0	70.0
3.0	47.0	80.0	70.0	70.0	80.0	80.0	40.0	40.0	85.5	70.0	80.0	80.0
4.0	55.0	80.0	70.0	80.0	80.0	90.0	60.0	55.0	65.0	80.0	80.0	90.0
5.0	80.0	80.0	70.0	90.0	90.0	90.0	60.0	60.0	75.0	80.0	90.0	90.0
6.0	90.0	80.0	70.0	100.0	100.0	90.0	85.0	70.0	82.5	90.0	95.0	100.0
7.0	90.0	80.0	75.0	100.0	100.0	95.0	95.0	75.0	85.0	95.0	97.5	100.0
8.0	90.0	85.0	85.0	100.0	100.0	100.0	100.0	85.0	87.5	100.0	100.0	100.0
9.0	90.0	85.0	90.0	100.0	100.0	100.0	100.0	85.0	90.0	100.0	100.0	100.0
10.0	90.0	85.0	90.0	100.0	100.0	100.0	100.0	85.0	90.0	100.0	100.0	100.0
11.0	90.0	85.0	90.0	100.0	100.0	100.0	100.0	85.0	90.0	100.0	100.0	100.0
12.0	90.0	90.0	90.0	100.0	100.0	100.0	100.0	90.0	90.0	100.0	100.0	100.0
	Revised Estimate: Damage by Expert (%)							**Aggregated Opinions as Percentiles**				
Depth	**1**	**2**	**3**	**4**	**5**	**6**	**7**	**Minimum**	**25%**	**50%**	**75%**	**Maximum**
−1.0	2.0	0.0	3.0	0.0	0.0	2.0	0.0	0.0	0.0	0.0	2.0	3.0
−0.5	2.0	0.0	5.0	5.0	0.0	5.0	0.0	0.0	0.0	2.0	5.0	5.0
0.0	15.0	20.0	15.0	10.0	10.0	30.0	5.0	5.0	10.0	15.0	17.5	30.0
0.5	20.0	30.0	35.0	20.0	30.0	40.0	20.0	20.0	20.0	30.0	32.5	40.0
1.0	25.0	50.0	35.0	40.0	45.0	40.0	20.0	20.0	30.0	40.0	42.5	50.0
1.5	25.0	60.0	40.0	50.0	60.0	40.0	30.0	25.0	35.0	40.0	55.0	60.0
2.0	30.0	70.0	40.0	60.0	70.0	60.0	40.0	30.0	40.0	60.0	65.0	70.0
3.0	40.0	80.0	70.0	70.0	75.0	80.0	40.0	40.0	55.0	70.0	77.5	80.0
4.0	50.0	80.0	70.0	80.0	80.0	90.0	60.0	50.0	65.0	80.0	80.0	90.0
5.0	50.0	80.0	70.0	90.0	90.0	90.0	60.0	50.0	65.0	80.0	90.0	90.0
6.0	85.0	80.0	70.0	95.0	90.0	90.0	70.0	70.0	75.0	85.0	90.0	95.0
7.0	90.0	80.0	75.0	95.0	90.0	95.0	100.0	75.0	85.0	90.0	95.0	100.0
8.0	90.0	85.0	85.0	95.0	90.0	95.0	100.0	85.0	87.5	90.0	95.0	100.0
9.0	90.0	85.0	90.0	95.0	90.0	95.0	100.0	85.0	90.0	90.0	95.0	100.0
10.0	90.0	85.0	90.0	95.0	90.0	95.0	100.0	85.0	90.0	90.0	95.0	100.0
11.0	90.0	85.0	90.0	95.0	90.0	95.0	100.0	85.0	90.0	90.0	95.0	100.0
12.0	90.0	85.0	90.0	95.0	90.0	95.0	100.0	85.0	90.0	90.0	95.0	100.0
Confidence	High	High	High	High	High	High	High					

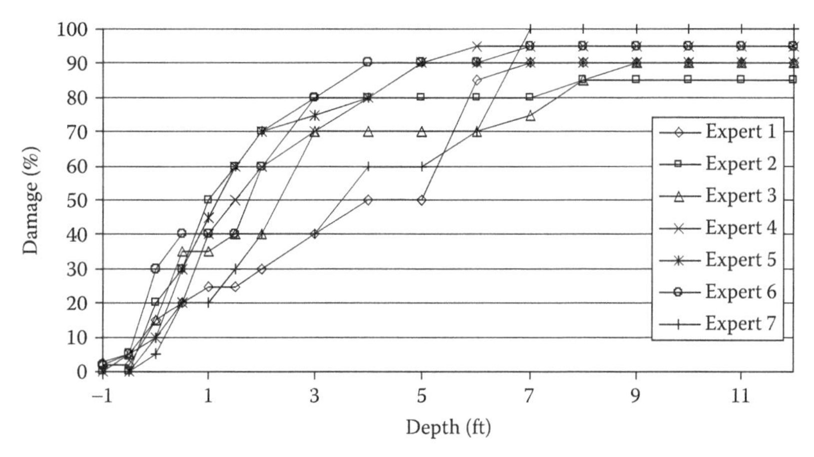

FIGURE 5.11
Percentage of damage to contents of type 1 and 2 residential structures (one story on slab or one story on piers and beams).

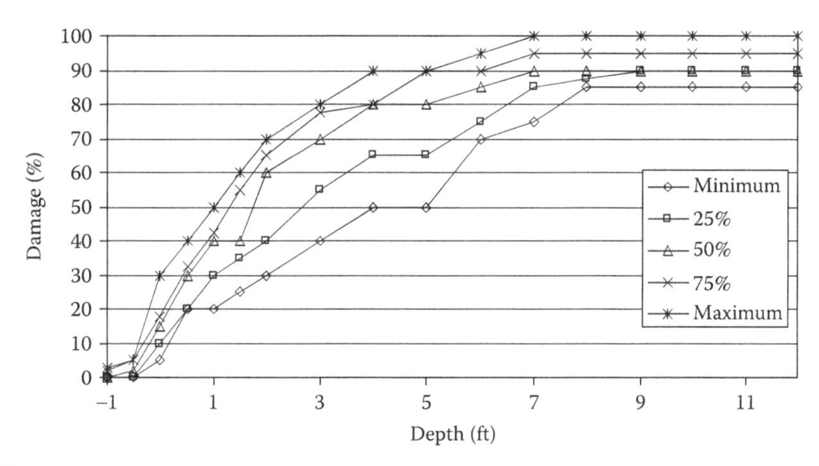

FIGURE 5.12
Aggregated (as percentiles) percentage of damage to contents of types 1 and 2 residential structures (one story on slab or one story on piers and beams).

TABLE 5.10

Summary of Supportive Reasoning and Assumptions by Experts for Content-to-Structure Value Ratio (CSVR)

Type 1–3 Houses (2001 US$)
As a guide, the insurance industry uses 70% for the CSVR.
The median house value is $90,000 with land.
The median land value is $20,000.
Garage or shed contents are included.
The median content age is eight years.
Depreciated replacement value of structure and contents was used.
Insufficient time was allowed to remove (or protect) contents.

TABLE 5.11

Value of Residential Structures, Contents, and Their Ratios for Type 1 and 2 Houses (One Story on Slab or One Story on Piers and Beams)

	Initial Estimate: Damage by Expert (%)							Aggregated Opinions as Percentiles				
Issue	**1**	**2**	**3**	**4**	**5**	**6**	**7**	**Minimum**	**25%**	**50%**	**75%**	**Maximum**
Median structure (1000$)												
Low	70.0	70.0	65.0	50.0	60.0	50.0	40.0	40.0	50.0	60.0	67.5	70.0
Best	90.0	110.0	106.0	70.0	70.0	60.0	70.0	60.0	70.0	70.0	98.0	110.0
High	110.0	250.0	175.0	90.0	80.0	80.0	90.0	80.0	85.0	90.0	142.5	250.0
Median content (1000$)												
Low	35.0	49.0	35.0	25.0	35.0	15.0	10.0	10.0	20.0	35.0	35.0	49.0
Best	50.0	77.0	41.0	50.0	40.0	20.0	20.0	20.0	30.0	41.0	50.0	77.0
High	65.0	175.0	70.0	80.0	45.0	25.0	25.0	25.0	35.0	65.0	75.0	175.0
Content-to-structure-values ratio (CSVR)												
Low	0.50	0.70	0.54	0.50	0.58	0.30	0.25	0.25	0.40	0.58	0.52	0.70
Best	0.56	0.70	0.39	0.71	0.57	0.33	0.29	0.33	0.43	0.59	0.51	0.70
High	0.59	0.70	0.40	0.89	0.56	0.31	0.28	0.31	0.41	0.72	0.53	0.70
	Revised Estimate: Damage by Expert (%)							**Aggregated Opinions as Percentiles**				
Depth	**1**	**2**	**3**	**4**	**5**	**6**	**7**	**Minimum**	**25%**	**50%**	**75%**	**Maximum**
Median structure (1000$)												
Low	70.0	70.0	77.0	50.0	60.0	50.0	50.0	50.0	50.0	60.0	70.0	77.0
Best	90.0	80.0	82.0	70.0	70.0	60.0	70.0	60.0	70.0	70.0	81.0	90.0
High	110.0	90.0	94.0	90.0	80.0	75.0	90.0	75.0	85.0	90.0	92.0	110.0
Median content (1000$)												
Low	35.0	79.0	40.0	25.0	35.0	15.0	10.0	10.0	20.0	35.0	37.5	49.0
Best	50.0	50.0	42.0	50.0	40.0	20.0	20.0	20.0	30.0	42.0	50.0	50.0
High	65.0	51.0	50.0	80.0	45.0	25.0	30.0	25.0	37.5	50.0	58.0	80.0
Content-to-structure-values ratio (CSVR)												
Low	0.50	0.70	0.52	0.50	0.58	0.30	0.20	0.20	0.40	0.50	0.55	0.70
Best	0.56	0.63	0.51	0.71	0.57	0.33	0.29	0.29	0.42	0.56	0.60	0.71
High	0.59	0.57	0.53	0.89	0.56	0.33	0.33	0.33	0.43	0.56	0.58	0.89
Confidence	High	High	Medium	High	High	High	High					

The earthquake resulted in >3,000 direct and indirect deaths from the 1906 San Francisco population of ~400,000, with 225,000 being homeless. It also destroyed 28,000 buildings of the city's 1906 inventory and caused monetary loss of >$400 million in 1906 from earthquake and fire, with $80 million from the earthquake alone (http://earthquake.usgs.gov). Figure 5.16 shows damage to buildings.

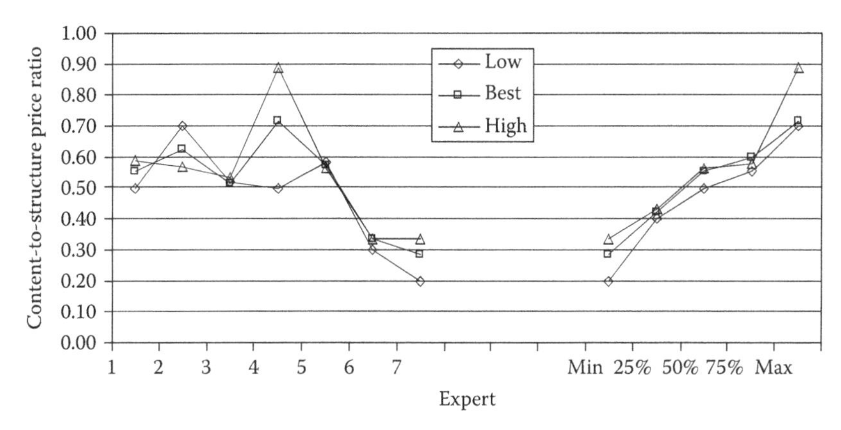

FIGURE 5.13
Content-to-structure-values ratios (CSVRs) for type 1 and 2 houses (one story on slab or one story on piers and beams).

TABLE 5.12

Summary of Supportive Reasoning and Assumptions by Experts for Vehicle Damage

Vehicle Types 1 and 2
The median vehicle age is five years.
Percentages are of the depreciated replacement values of vehicles.
Flood without flow velocity was considered.
Flood duration was for several days.
Flood water is not contaminated but has sediment without large debris.
Allowance was made for cleanup costs.

5.5 Loss of Human Life

Generally, the well-being of a person comprises the quantity and quality of life (QOL). Valuing human life requires the definition of a unit of measurement for human life. Suitable measurement units are counts of individual life lost or counts of life years of these individuals. The latter permits to account for injuries and the QOL as discussed in Section 5.6.

Public policy decisions commonly involve risks to human health and safety. In such policy-related deliberations, we commonly find two competing sides of industry representatives and public interest groups trying to influence policy makers. For example, industry representatives in the occupational safety area might claim that their industry exposure standards are too stringent to levels exceeding any rational limits determined from reasonable cost–benefit analysis. However, public interest groups might claim that the cost–benefit analysis is fundamentally biased and is intended to further the ends of industry to the detriment of workers health and safety. The valuation of human life is central to this debate, although some claim that the value of human life cannot be expressed in monetary terms; however, the competing demands on scarce public funds make it a necessity. Refusal to place an explicit value on life merely forces us to implicitly value it through decisions to fund or not to fund public projects or decisions to take some regulatory actions. Since failures sometimes lead to deaths and designing systems often require trade-off analyses to

TABLE 5.13

Percent Damage to a Type 1 Vehicle (Sedans)

	Initial Estimate: Damage by Expert (%)							Aggregated Opinions as Percentiles				
Depth	**1**	**2**	**3**	**4**	**5**	**6**	**7**	**Minimum**	**25%**	**50%**	**75%**	**Maximum**
0.0	0.0	0.0	0.0	0.0	0.0	0.0	0.0	0.0	0.0	0.0	0.0	0.0
0.5	5.0	0.0	5.0	0.0	0.0	0.0	0.0	0.0	0.0	0.0	2.5	5.0
1.0	20.0	0.0	30.0	10.0	25.0	5.0	10.0	0.0	7.5	10.0	22.5	30.0
1.5	25.0	0.0	30.0	10.0	25.0	5.0	10.0	0.0	15.0	25.0	37.5	50.0
2.0	35.0	30.0	80.0	20.0	30.0	20.0	60.0	20.0	25.0	30.0	47.5	80.0
2.5	50.0	35.0	100.0	40.0	70.0	40.0	70.0	35.0	40.0	50.0	70.0	100.0
3.0	60.0	40.0	100.0	50.0	70.0	60.0	90.0	40.0	55.0	60.0	80.0	100.0
4.0	100.0	40.0	100.0	100.0	80.0	80.0	100.0	40.0	80.0	100.0	100.0	100.0
5.0	100.0	50.0	100.0	100.0	95.0	80.0	100.0	50.0	87.5	100.0	100.0	100.0
	Revised Estimate: Damage by Expert (%)							**Aggregated Opinions as Percentiles**				
Depth	**1**	**2**	**3**	**4**	**5**	**6**	**7**	**Minimum**	**25%**	**50%**	**75%**	**Maximum**
0.0	0.0	0.0	0.0	0.0	0.0	0.0	0.0	0.0	0.0	0.0	0.0	0.0
0.5	10.0	0.0	5.0	0.0	0.0	2.0	0.0	0.0	0.0	0.0	3.5	10.0
1.0	25.0	10.0	20.0	20.0	20.0	10.0	20.0	10.0	15.0	20.0	20.0	25.0
1.5	35.0	30.0	50.0	25.0	25.0	40.0	30.0	25.0	27.5	30.0	37.5	50.0
2.0	40.0	40.0	80.0	30.0	30.0	50.0	50.0	30.0	35.0	40.0	50.0	80.0
2.5	50.0	50.0	100.0	40.0	60.0	60.0	70.0	40.0	50.0	60.0	65.0	100.0
3.0	60.0	100.0	100.0	50.0	70.0	80.0	80.0	50.0	65.0	80.0	90.0	100.0
4.0	100.0	100.0	100.0	100.0	100.0	80.0	100.0	80.0	100.0	100.0	100.0	100.0
5.0	100.0	100.0	100.0	100.0	100.0	80.0	100.0	80.0	100.0	100.0	100.0	100.0
Confidence	High	High	High	High	High	High	High					

maximize the benefits to society, including reducing the likelihood of fatalities, the value of life enters in these analyses, often in an implicit manner. If implicitly done, it undermines consistency among designs, projects, disciplines, and decision makers.

The value of life can be viewed in two different perspectives as either the value of a particular life (VL) or the value of statistical life (VSL). The difference between the VL and the VSL is that the former provides an assessment of a particular life, for example, John Smith's life as an identified person, using methods for determining the limits on life insurance; whereas the latter is based on assessing the implicit value of life using data from compensation premiums paid to workers for risky occupations and for insurance purposes. Figure 5.17 shows a hierarchal listing of these methods. To adjust for the age of affected population, sometimes the VSL year (VSLY) or *life years saved* (LYS) is considered.

Most economists agree that a conceptually appropriate method to value human life in cost–benefit analyses should be consistent with other economic concepts and should be based on individuals' WTP (or on individuals' WTA compensation) for small changes in their probability of survival. Despite this agreement, however, controversy continues on the appropriate technique for actually producing estimates for valuing life. Efforts to assess the value of human life have been based on WTP concepts, earning potential, and assessments of the implicit values in currently accepted and used regulations.

The value of life is sometimes used as an anchor concept or a basis for other estimates relating to health such as injuries and illness. These linkages highlight its importance.

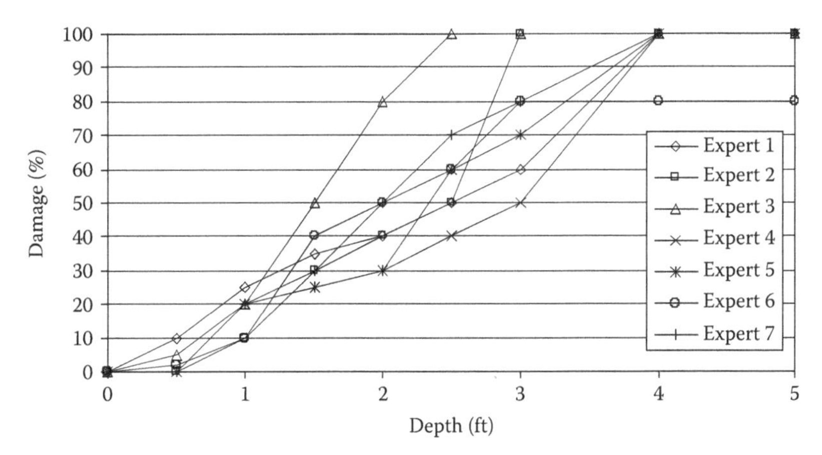

FIGURE 5.14
Percentage of damage to a type 1 vehicle (sedans).

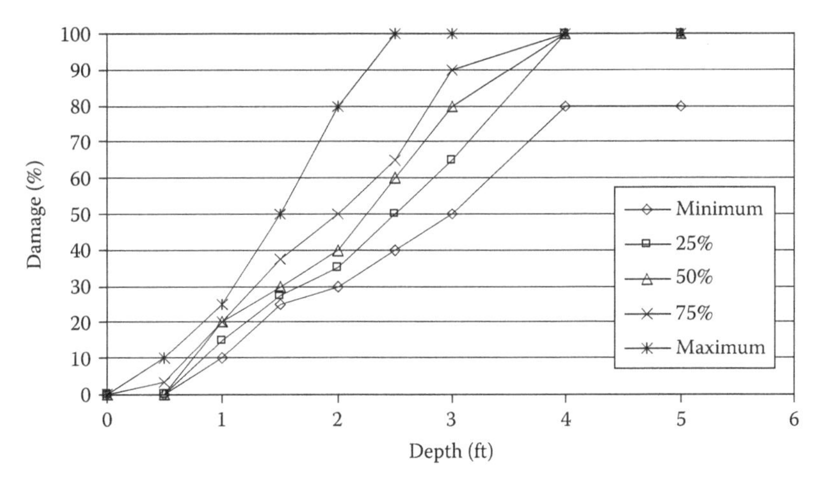

FIGURE 5.15
Aggregated (as percentiles) percentage of damage to a type 1 vehicle (sedans).

This section focuses on two classes of methods for assessing the value of life and provides examples: the WTP method and the human capital (HC) method.

5.5.1 Willingness to Pay Method

The concept of Willingness to Pay (WTP) for assessing the VSL can be examined based on several human behavior types to serve different analytical purposes of which the following are most relevant (Figure 5.17; Viscusi and Aldy 2002):

- WTP based on labor markets for goods and services
- WTP based on consumption markets for goods and services
- WTP based on policy and regulatory decisions

These methods vary in the extent of the direct involvement of individuals in related transactions and the role of groups of individuals in such decisions.

(a)

(b)

FIGURE 5.16
Damage to buildings resulting from the 1906 San Francisco Earthquake (http://earthquake.usgs.gov): (a) Stanford University 1906 Earthquake Damage and (b) financial district.

(c)

FIGURE 5.16
(Continued) Damage to buildings resulting from the 1906 San Francisco Earthquake (http://earthquake.usgs.gov): (c) City Hall.

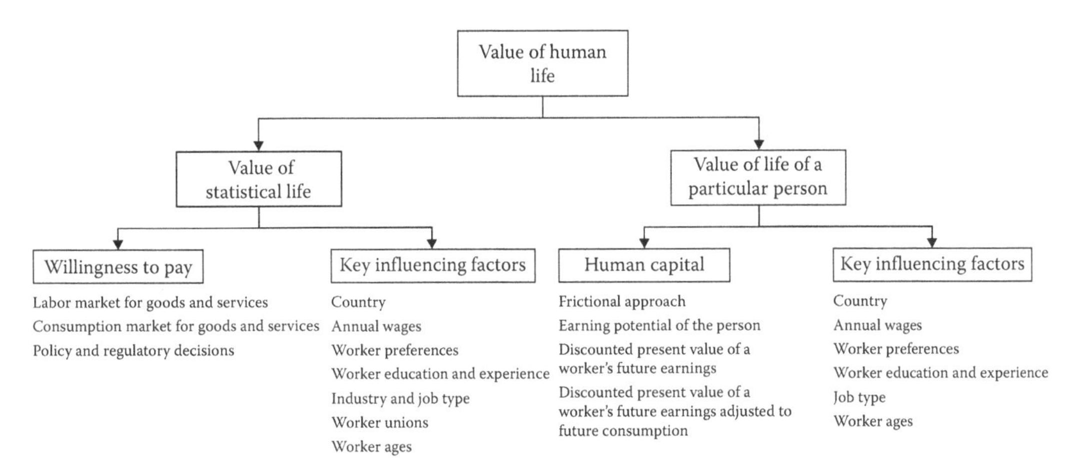

FIGURE 5.17
The value of human life.

Smith (1776) noted "the wages of labor vary with the ease or hardship, the cleanliness or dirtiness, the honorableness or dishonorableness of the employment" that has challenged economists pursuing valuation of labor and examining trade-offs between wages and risks. A logically consistent basis for the valuation of life in safety decisions should be the same criterion used by welfare economists in other areas of cost–benefit analysis, namely, the *potential Pareto improvement principle* (Mishan 1971). A potential Pareto improvement exists when individuals who gain from a social change are able to compensate those who stand to lose from the change and still leave a net gain. Thus, we can conclude that the relevant question can be stated as what individuals are *willing to pay* (or *willing to accept as compensation*) for a change that will affect the loss of life (LOL). Most public safety decisions do not entail the value of an identified individual's life, rather the value of a reduction in the probability of death for a given population, that is, it is the aggregate value a PAR places on programs that save *statistical lives* or the sum of the amounts individuals are willing to pay *ex ante* to *buy* small reductions in the probability of their death (Landfeeld and Seskin 1982). It is compatible with the notion that, if there were a market for "buying" safety, this approach would yield the price that consumers would be willing to pay. It produces a VSL based on a social welfare maximization notion.

Two approaches are commonly followed to quantify the VSL using the WTP methods: (1) analyses of direct survey responses by individuals and (2) statistical estimation of individuals' revealed preferences. Each approach has problems associated with it. Landfeeld and Seskin (1982) provide the results of survey estimates of individuals' WTP for reductions in risk of death (1977 US$). It was observed that the question statement could influence the outcome. For example, asking individuals open-ended questions about their WTP for a coronary care unit that would reduce risk of death from heart attack by 0.002 produced an average WTP of $76 for the unit. The aggregate WTP in this case for a community of N such individuals would be $\$76N/0.002N$ = $38,000 per statistical life saved. Surveys concerning safety and airline travel, employing similar methods, found a value per statistical life of $8.4 million and another survey on the WTP for reducing cancer mortality calculated a value of $1.2 million per statistical life saved. For example, if a population of 100,000 persons was willing to pay an average of $50 each to reduce deaths from 4 per 100,000 to 2 per 100,000, the total WTP can be computed as $5 million and the value per statistical life is estimated to be $2.5 million, since two lives could be saved. The WTP approach yields a substantially higher value than do other approaches. The WTP method does recognize an individual's desire to live longer. Moreover, the WTP has no actuarial base, and someone could argue that it is based on dubious logic since subjects do not have an appropriate appreciation to small-risk values and effects particularly in the case of workers at jobs with greater risks using a wage-risk approach. For example, two jobs, A and B, are similar except that A has one more job-related death per year for every 10,000 workers than does B. The workers in job A earn $500 more per year than those in job B, or $5 million for the 10,000 workers. The value of life of workers in job B who are willing to forgo the money for the lower risk is $5 million. This inferred value is based on perhaps a simplistic understanding of the issues by the workers, their perceptions, biases, and sometimes youth and enthusiasm.

We can observe widespread in the outcomes of such surveys. Moreover, what individuals say that they will do may vary considerably from what they will actually do when confronted with a true case of the situation. Furthermore, the belief in public good might drive some of the answers. Interestingly, social psychologists raised questions about the ability of individuals to respond rationally and consistently to abstract and complex questions involving a hypothetical risk expressed in small numbers, and whether the risk change is communicated as an absolute difference or as a percentage of the starting basis. Finally, the answers of respondents could depend on how the compensation differentials are stated, for example, in labor market terms (e.g., compensation increases) or consumption activity terms (e.g., outfitting a house with smoke detector or a car with lane-keeping sensors). Generally, labor market terms result in greater spreads in the VSL compared to the VSL generated from consumption activity terms. The primary reasons of the wider spreads are lack of information of responders, self-selection, that is, those who work in risky jobs exhibit less risk aversion than the population as a whole, difficulty to separate the risk of death from risk of injury, and so on.

For cases involving public policy regulatory decisions, the maximum WTP can be estimated for individual stakeholders and averaged over all the people involved. These decisions provide a logical information source for estimating the VSL using the concept of WTP by a society represented by its policy makers and regulators. Viscusi and Aldy (2002) examined and summarized the VSL in the United States starting with the Reagan Administration's executive orders that vested the US Office of Management and Budget (OMB) with the responsibility of overseeing and coordinating the review of regulatory impact analyses. OMB recommended as best practices the use of a VSL to monetize the benefits associated with rules that change the population's mortality risk. Until then it was the practice to use the HC method because it was viewed that life is sacred to value. OMB rejected regulations on the HC basis by arguing that the costs exceeded the benefits. The adoption of the VSL methodology resulted in boosting benefits by roughly an order of magnitude, improving the attractiveness of agencies' regulatory efforts. OMB does not specify VSLs in order to provide flexibility to the US agencies in choosing a VSL appropriate to the population affected by their specific rules. Such flexibility has resulted in significant variations in the selected VSL both across agencies and through time (Viscusi and Aldy 2002). In addition, some regulatory impact analyses have included the age of the affected population by using a VSL adjusted by the number of life-years saved [e.g., the Food and Drug Administration (FDA) rule restricting tobacco sales to children, 61 FR 44396, and the Environmental Protection Agency (EPA) rule regulating the sulfur content of gasoline, 65 FR 6698]. The EPA guidelines, as an example, recommend a VSL of $6.2 million (2000 US$). The Federal Aviation Administration (FAA) recommends a VSL of $3 million in its 2002 economic analyses of regulations that is smaller than the EPA value due in part to an anchoring effect based on the earlier Department of Transportation (DOT) valuation work using HC lost earnings, and therefore has about closed this gap.

Leung (2009) reported on efforts at the New Zealand Ministry of Transport to set its VSL by surveying the worldwide values as shown in Table 5.14. The ratio of VSL to the gross domestic product (GDP) per capita is quite revealing in that VSL is on the average 85 times the GDP per capita with a range of 37–123 and the United States has the highest ratio. The data in Table 5.14 can be combined with other predictor variables of interest to develop regression models to help set values of VSL for other countries or populations. McMahon and Dahdah (2008) examined the effects of GDP per capita on WTP-based VSL and developed the following relationship:

$$\ln(\text{VSL}) = 3.015 + 1.125\ \ln(\text{GDP per capita}) \tag{5.1}$$

This logarithmic regression model was based on the data adjusted for purchasing power parity expressed in 2004 US$. This model is plotted in Figure 5.18.

Other considerations relating to VSL that are noteworthy are the trade-offs made by individuals among risks and benefits based on the following:

1. Risk–income (i.e., risk–wage) trade-offs
2. Risk–benefit trade-offs
3. Risk–time trade-off
4. Risk–risk trade-offs

TABLE 5.14

International Comparison of Value of Statistical Life (VSL) in Domestic Currency

Country (Year Currency)	VSL (million)	VSL-to-GDP per Capita Ratio
Austria (2006 Euros)	2.68	87
Belgium (2006 Euros)	5.60 (estimated)	186
Canada (2007 CAD$)	4.60	99
Denmark (2009 DKK kr)	12.20	–
France (2000 Euros)	1.00	42
Germany (2004 Euros)	1.16	–
Netherlands (2003 Euros)	2.40	82
New Zealand (2008 NZ$)	3.35	–
Norway (2005 NOK)	26.50	63
Singapore (2008 SG$)	1.87	37
Sweden (2006 SEK kr)	21.00	66
UK (2007 GBP)	1.64	71
USA (2008 US$)	5.80	123
Average		85.6
Standard deviation		43.5

Source: Adapter from Leung, J., Understanding transport costs and charges: Phase 2—Value of statistical life: A meta analysis, is the current value of safety for New Zealand too low? Technical Report, Financial and Economic Analysis Team, New Zealand Ministry of Transport, 2009.

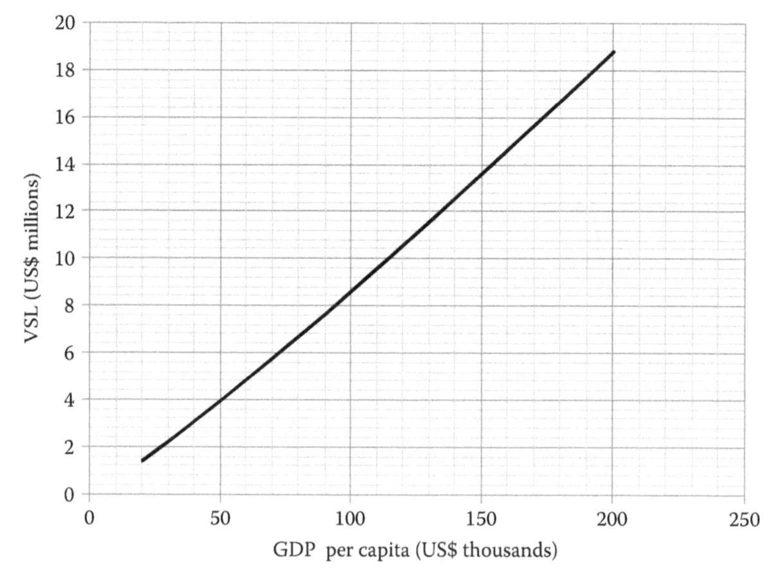

FIGURE 5.18
A model of gross domestic product (GDP) and the value of statistical life (VSL).

The first and second items are discussed earlier. As for the third item, individuals tend to value the effects of delayed risks less than the immediate ones, posing a challenge for modeling discounting and interest rate effects, whereas the fourth item addresses individual or in this case group preferences in replacing a risk type by another as also discussed in Chapter 2. The third item also brings in the need to consider the age of the population affected as discussed earlier. The fourth item manifests itself in the following example forms (Viscusi and Aldy 2002):

- Policies may reduce risks of one type while increasing risks of another type, for example, banning saccharin, the artificial sweetener, in response to an animal study finding that it may be a potential human carcinogen that could result in increasing the risks associated with obesity.
- Policies may reduce risks and create incentives for individuals to undertake less individual effort to reduce their exposure to risks, for example, drivers of *safer cars* drove more recklessly than before, as discussed in Chapter 7.
- Risk reduction policies may result in regulatory expenditures that directly increase fatalities, for example, policies to remove asbestos from buildings may increase the exposure of asbestos by workers.
- The costs of risk reduction policies decrease income available to finance other health and safety expenditures, for example, the costs of risk reduction policies reduce national income, some of which would otherwise be used to promote health and safety with a negative net impact on life.

Section 5.5.3 provides an example VSL generated using the WTP method.

5.5.2 Human Capital Method

The Human Capital (HC) method assumes that a society values an individual's life by future production potential and is usually calculated as the discounted present value of expected labor earnings. It uses the discounted present value of a worker's future earnings as a proxy for the cost of premature death, injury, or illness to a society. Basically, it treats humans as a labor source and an input to the production process. The society's incentive of preventing an incident is the saving in potential output or productivity capacity. Some analysts have employed expected earnings minus consumption, based on the notion that the death of an individual leads to not only cessation of productive contribution but also cessation of claims on future consumption. Whether the gross approach or the net approach is employed, each is implicitly based on the same notion of maximization of society's present and future production. Labor earnings are evaluated before taxes as representing the actual component of the GDP. In addition, nonlabor income, such as investment income from stocks, is excluded since individual capital holdings (and associated earnings) are not materially affected by the death of the individual. Moreover, the method assigns a zero value for persons without labor income such as retired individuals with only investment or pension income and ignores other dimensions that may be more important to an individual than economic loss such as health, being alive, happiness, and nonmarket activities. The only adjustment for nonmarket activities is the value for housekeeping activities based on relevant market values of such household activities. The use of this method requires the selection of a discount rate (i) to account for the time value of money as discussed in Chapter 6. The selection of the discount rate can be difficult and challenging with a great impact on any results from risk studies or benefit–cost analyses.

The results of this method are age specific, and many economists consider it to be based on dubious logic because it ignores an individual's desire to live. The HC of a society values safety because of their aversion to death and injury, not because they want to save productive resources and enhance the GDP. Some *ad hoc* methods have been suggested

TABLE 5.15

Example Earning Potential Multipliers as a Function of Age

Age Range (Years)	Income Multiplier to Annual Income (x)
20–35	30
35–40	20
41–45	14
46–50	12
51–59	10
60–65	7
>66	5

to deal with this criticism by multiplying the present value of future outputs by a factor that takes into account pain, grief, and suffering.

Life insurance utilizes methods that are founded in the HC concepts to determine the maximum life insurance amount for a person. This maximum amount is taken as the larger of (1) the total asset value of the person or (2) the earning potential of the person. The earning potential is a function of the age of the person and is capped in order to limit the insurance premium to a prescribed percentage of the annual income of the person for affordability reasons. The earning potential is computed as the annual income (x) times an income multiplier as demonstrated in Table 5.15. The affordability caps are typically on a sliding scale of about 5% for incomes from $80,000 to $120,000 that grows to about 8% for incomes up to $250,000 and about 10% for incomes >$250,000.

Despite the conceptual problems associated with the HC approach, the technique is widely used out of the perceived relative ease in computation using the objective numbers based on life expectancy, labor force participation, and projected earnings (Landfeeld and Seskin 1982).

A variation of the HC method is called the frictional approach that is used in situations of high unemployment or in short-term situations to estimate only the production lost (or additional costs incurred) between when a worker leaves a job due to an accident and returns or is replaced by another worker or, alternatively, for the time period required to restore production to its preincident state. Moreover, some calculations truncate the income stream at an average retirement age, and other calculations estimate the probability of employment each year and multiply it by the expected income stream in employment in each year (Leung 2009).

5.5.3 Comparison the Willingness to Pay and Human Capital Methods

The HC method offers simplicity and straightforwardness by estimating the discounted present value of future output. But the WTP method offers a conceptually compatible and complete economic measure by assessing the premium that people place on pain, grief, and suffering rather than merely evaluating the lost output or income. The WTP method enables analysts to ask those directly affected by a problem what they consider to be the value of safety. In asking such questions, analysts might be faced with the difficulty of ensuring that both the scope and the content of the questions are understandable. Comparing the advantages and disadvantages of each method does not produce a preferred one, although in recent years the WTP method has gained popularity among risk analysts and economists.

TABLE 5.16

Earthquakes in the United States Leading to 10 Deaths or More (http://earthquake.usgs.gov/)

Year	Earthquake	Deaths	Comments
1812	San Juan Capistrano, California	40	
1868	Hawaii Island, Hawaii	77	Landslides: 31, tsunami: 46
1868	Hayward, California	30	
1872	Owens Valley, California	27	
1886	Charleston, South Carolina	60	
1906	San Francisco, California	3000	Deaths (approximate) from earthquake and fire
1933	Long Beach, California	115	
1946	Aleutian Islands, Alaska	165	Tsunami: 159 Hawaii, 5 Alaska, 1 California
1952	Kern County, California	12	
1959	Hebgen Lake, Montana	28	
1960	Chile, South America	61	Tsunami in Hawaii
1964	Prince William Sound, Alaska	128	Tsunami: 98 Alaska, 11 California, 4 Oregon Earthquake: 15 Alaska
1971	San Fernando, California	65	
1989	Santa Cruz County, California	63	
1994	Northridge, California	60	

Example 5.7: Life Loss due to Earthquakes

Earthquakes can result in significant life loss. The life loss varies as a result of earthquake intensity, regional building practices, first-response preparedness and practices, and other factors. Tables 5.16 and 5.17 list earthquakes resulting in deaths in the Unites States and worldwide. Table 5.17 lists the earthquakes in ascending order by deaths.

5.5.4 War Fatality and Other Compensations

The September 11, 2001, attack on the United States by four planes commanded by 19 hijackers resulted in the death of 2974 people, not including the hijackers, with an additional 24 people missing, and presumed to be dead. The US government created the September 11 Victims Compensation Fund of 2001 to reach an agreement with the families of the victims not to sue the airlines involved. Three elements of compensation were applicable to each victim as follows (Feinberg 2001):

- Economic-based compensation varied for each victim based on annual income, remaining years in work life expectancy, benefits received from the employer or insurance settlements, victim's effective tax rate, and the household size of the victim.
- Non-economic-based compensation has less variation compared to economic-based compensation and account for the trauma. This element is roughly equivalent to the amounts received under existing federal programs by public safety officers who were killed while on duty.
- Additional amounts dispensed on an individual basis for medical expenses or burial and memorial costs.

The summation of the three compensation elements were checked against a $300,000 minimum for a single, deceased victim before subtracting the amounts from other

TABLE 5.17
Earthquakes Worldwide Leading to 50,000 Deaths or More (http://earthquake.usgs.gov/)

Year	Country	Deaths	Magnitude
1556	Shaanxi (Shensi), China	830,000	8
2010	Haiti region	316,000	7.0
1976	Tangshan, China	242,769	7.5
1138	Syria, Aleppo	230,000	
2004	Sumatra	227,898	9.1
856	Damghan, Iran	200,000	
1920	Haiyuan, Ningxia (Ning-hsia), China	200,000	7.8
893	Ardabil, Iran	150,000	
1923	Kanto (Kwanto), Japan	142,800	7.9
1948	Ashgabat (Ashkhabad), Turkmenistan (Turkmeniya, USSR)	110,000	7.3
1290	Chihli, China	100,000	
2008	Eastern Sichuan, China	87,587	7.9
2005	Pakistan	86,000	7.6
1667	Caucasia, Shemakha	80,000	
1727	Tabriz, Iran	77,000	
1908	Messina, Italy	72,000	7.2
1970	Chimbote, Peru	70,000	7.9
1755	Portugal, Lisbon	70,000	8.7
1693	Sicily, Italy	60,000	7.5
1268	Asia Minor, Cilicia	60,000	
1990	Western Iran	50,000	7.4
1783	Italy, Calabria	50,000	

sources, or a $500,000 minimum for a married victim or a victim with dependants before subtracting the amounts from other sources. The other sources include government programs or life insurance, and the benefits received from charities are not considered to be other sources. The 2,880 claims made by families of the victims resulted in compensation totaling $5,996,261,002.08 (2001 US$). The average deceased victim's award, after deductions, was $2,082,128, and the median deceased victim's award was $1,677,633 with the minimum and maximum payments made $250,000 and $7.1 million, respectively.

According to the news reports, the families of US soldiers killed in action or in accidents during the war are eligible for death benefits that could range from $250,000 to >$800,000 (http://articles.latimes.com/2003/apr/05/news/war-benefits5). As for compensation for the deaths of non-Americans, condolence payments were extended to families of Iraqi and Afghani victims as a fixed lump sum of $2000 each (2003 US$).

5.5.5 Typical Values of Statistical Life

The results of studies estimating the VSL based on decisions by policy makers have varied greatly, depending on the data sources, the methodologies used, and the assumptions made. A compilation of the data in 1990 US$ resulted in the following values ($ million) based on WTP concepts: 0.8, 0.9, 1.4, 1.5, 1.6, 1.6, 2, 2.4, 2.4, 2.6, 2.6, 2.8, 2.9, 3, 4.1, 4.6, 5.2, 6.5, 9.7, and 10.3. The median is $2.6 million. A histogram of the value of life based on these 20 values is shown in Figure 5.19. The following examples are based on specific regulatory agencies:

- The DOT regulates and sets overall national transportation policy including railroads, aviation, and safety of waterways, ports, highways, and oil and gas pipelines. VSLs converted to 1990 US$ ranged from $50,000 to $29,000,000, with a median of $312,000. Transportation studies have used $1,400,000 (1990 US$). Another similar plot is provided in Figure 5.19 based on the DOT data. The 2001 Office of the Secretary of Transportation (OST) guidance establishes a minimum value of $3 million per fatality averted. This value and the injury values based on it presented in Section 5.5.6 were used in all FAA analyses until updated in future years.
- The Consumer Product Safety Commission (CPSC) is an independent federal regulatory agency established by the Consumer Product Safety Act. CPSC data on the VSL for various items ranged from $80,000 to $1,400,000 (1990 US$). The median value is not reliable (Figure 5.20).
- The Department of Labor Occupational Safety and Health Administration (OSHA) data on the VSL for various items ranged from $130,000 to $91 billion (1990 US$). The median value might not be reliable at $6.7 million. Other OSHA data were examined and analyzed and the VSL values ranged from $12,000 to $85 million

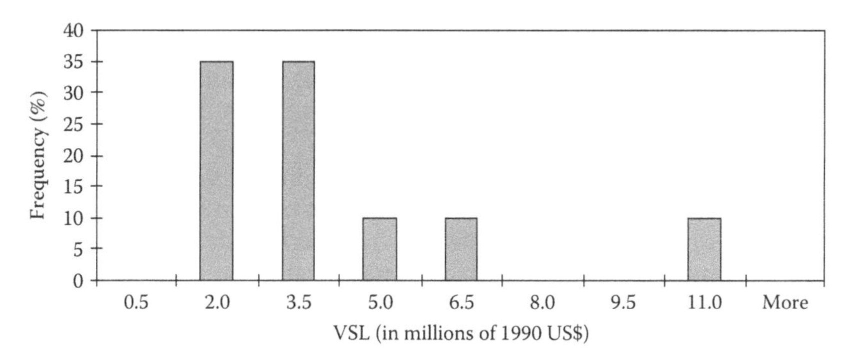

FIGURE 5.19
Value of statistical life (VSL) in wage-risk studies based on the willingness to pay (WTP) method.

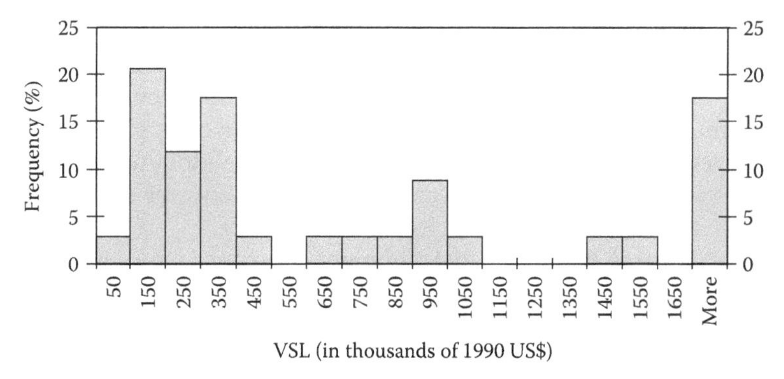

FIGURE 5.20
Value of statistical life.

(1990 US$). The median value in this case, which might not be reliable, was $265,000. Analyzing VSL for five OSHA regulations (relating to asbestos, coke ovens, benzene, arsenic, and acrylnitrile) produced VSL ranging from $200,000 to $20 million (1985 US$) per death avoided.

- The EPA data on the VSL ranged from $9000 to $4.4 billion (1990 US$). The median value, which might not be reliable, was $21.5 million. Analyzing one EPA regulation (relating to benzene) produced VSL of $100 million (1985 US$) for the EPA. Environmental studies on the risks from residential radon exposures resulted in the estimated fatalities from radon and the costs of measures to minimize radon seeping into homes. The computed values of life ranged from $400,000 to $7,000,000 (1989 US$). Spending $4000 (1989 US$) for a picocurie/liter reduction (using a $400,000 VSL) was considered to be cost effective over a 50-year period.
- When various VSLs directly relevant to various governmental agencies were examined and analyzed, the VSL values ranged from $300,000 to $6.5 million (1990 US$). The median value, which might not be reliable, was $1.4 million.

The US agencies have flexibility in choosing a VSL appropriate to the population affected by their specific rules, and therefore, this flexibility has resulted in significant variations in the selected VSL both across agencies and through time as shown in Table 5.18 (Viscusi and Aldy 2002).

5.5.6 Human Life Loss due to Floods Resulting from Dam Failure

5.5.6.1 Introduction

This section focuses on loss of human life resulting from the failure of dams. Dam failure can have various consequences, some of which can be significant, such as LOL. Each system failure that can arise has consequences. This section deals with the definition of floodplains, PAR, dam breach inundation, and fatality rates.

5.5.6.2 Floodplains

A floodplain is defined by the American Geological Institute as the portion of a river valley adjacent to the river channel that is built of sediments during the current regimen of the stream and covered with water when the river overflows its banks at flood stages. The floodplain is a level area near the river channel. Clearly, the floodplain is an integral and necessary component of the river system. If a climate change or land-use change occurs, then the existing floodplain may be abandoned and new floodplain construction begins. Sediment is deposited when the stream flow overtops the banks; this occurs approximately every 1.5–2 years in stable streams. The floodplain extends to the valley walls. In engineering, floodplains are often defined by the water surface elevation for a design flood, such as a 100- or 200-year flood.

Changes in the natural floodplain development are caused by changes in sediment loads or water discharge. Increases in both the sediment and the water discharge are often caused by land-use changes, typically urbanization. Other causes include changes to the channel itself, such as straightening or relocating. Climatic changes can cause the current floodplain to be abandoned; however, this is seldom a concern for engineering, as the time scale is geologic.

TABLE 5.18

Values of Statistical Life Used by US Regulatory Agencies, 1985–2000

Year	Agency	Regulation	VSL (in millions of 2000 US$)
1985	FAA	Protective Breathing Equipment (50 FR 41452)	1.0
	EPA	Regulation of Fuels and Fuel Additives; Gasoline Lead Content (50 FR 9400)	1.7
1988	FAA	Improved Survival Equipment for Inadvertent Water Landings (53 FR 24890)	1.5
	EPA	Protection of Stratospheric Ozone (53 FR 30566)	4.8
1990	FAA	Proposed Establishment of the Harlingen Airport Radar Service Area, TX (55 FR 32064)	2.0
1994	Food and Nutrition Service (USDA)	National School Lunch Program and School Breakfast Program (59 FR 30218)	1.7, 3.5
1995	CPSC	Multiple Tube Mine and Shell Fireworks Devices (60 FR 34922)	5.6
1996	Food Safety Inspection Service (USDA)	Pathogen Reduction; Hazard Analysis and Critical Control Point Systems (61 FR 38806)	1.9
	FDA	Regulations Restricting the Sale and Distribution of Cigarettes and Smokeless Tobacco to Protect Children and Adolescents (61 FR 44396)	2.7
	FAA	Aircraft Flight Simulator Use in Pilot Training, Testing, and Checking and at Training Centers (61 FR 34508)	3.0
	EPA	Requirements for Lead-Based Paint Activities in Target Housing and Child-Occupied Facilities (61 FR 45778)	6.3
	FDA	Medical Devices; Current Good Manufacturing Practice Final Rule; Quality System Regulation (61 FR 52602)	5.5
1997	EPA	National Ambient Air Quality Standards for Ozone (62 FR 38856)	6.3
1999	EPA	Radon in Drinking Water Health Risk Reduction and Cost Analysis (64 FR 9560)	6.3
	EPA	Control of Air Pollution from New Motor Vehicles: Tier 2 Motor Vehicle Emissions Standards and Gasoline Sulfur Control Requirements (65 FR 6698)	3.9, 6.3
2000	CPSC	Portable Bed Rails; Advance Notice of Proposed Rulemaking (65 FR 58968)	5.0

CPSC, Consumer Product Safety Commission; EPA, Environmental Protection Agency; FAA, Federal Aviation Administration; FR, Federal Register; USDA, US Department of Agriculture.

Source: Adapted from Viscusi, W.K. and Aldy, J.E., 2002. The value of a statistical life: A critical review of market estimates throughout the world, John M. Olin Center for Law, Economics, and Business at Harvard Law School Discussion Paper Series. *Paper No. 392*, http://lsr.nellco.org/harvard_olin/392.

5.5.6.3 Demographics

The number of people at risk in the event of capacity exceedance or other uncontrolled release depends on the population within the inundation area and the conditions of release. The planning team defines a variety of scenarios to represent a range of modes of failure, given overtopping and other potential conditions of breaching. For each scenario, specific characteristics of the release are defined, and quantitative characteristics of downstream effects are estimated for economic cost and LOL. Probabilities are associated with each scenario based on reliability analyses of the type discussed in

Chapter 4, and the resulting probability–consequence combinations are used as the basis for risk assessment.

For estimating the characteristics of downstream effects, a fluvial hydraulics model possibly combined with a dam breach analysis is used to forecast depths and extents of flooding. With this information, the economic effect on structures and facilities can be estimated, as can the environmental effect on downstream ecosystems. The number of people at risk, however, depends on additional considerations. These include the time of the day and the season of the year at which the release occurs, the rate of water rise, the available warning time (WT) and the effectiveness of evacuation plans, and the changes in downstream land use. An empirical review of uncontrolled releases at other dams and of levee overtoppings provides an initial basis for estimating the PAR under the various scenarios. Nevertheless, the quantitative historical record of dam failures is small, and any particular project will have characteristics that differ in important ways from those of the database.

A quantitative expression for estimating LOL in dam failures, based on statistical analysis of empirical data related to severe flooding, can be expressed as follows (Ayyub et al. 1998c):

$$\text{LOL} = \frac{\text{PAR}}{1+13.277(\text{PAR}^{0.44})\exp\left[0.750(\text{WT})-3.790(\text{Force})+2.223(\text{WT})(\text{Force})\right]} \tag{5.2}$$

where:

LOL is the potential LOL
PAR is the PAR
WT is the warning time in hours
Force is the forcefulness of the flood water (1 for high force, 0 for low force)

The PAR is defined as the number of people within 3 hours' travel time of the flood wave and includes not just those exposed to treacherous flood waters but all at risk of getting their feet wet. The empirical equation is statistically valid only for PARs <100,000. An example calculation is shown in Figure 5.21. For this example dam, the following values are assumed: PAR = 100,000, WT = 1 hour, and Force = 0 and 1. The resulting values for LOL are 0.3 and 5 persons, respectively, for Force = 0 and 1.

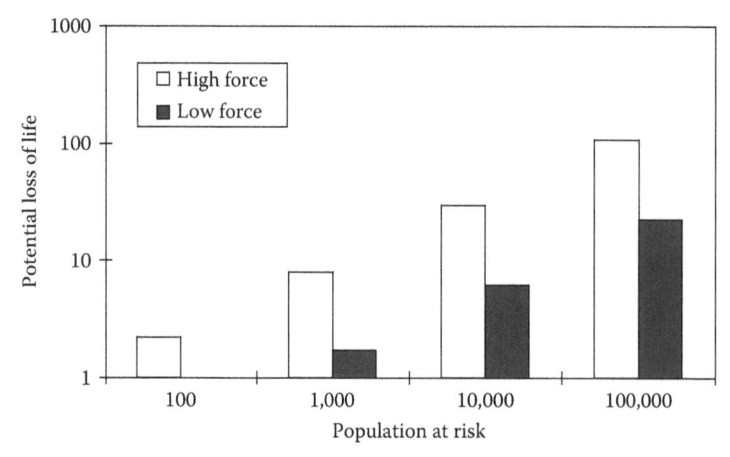

FIGURE 5.21
Example calculation of potential loss of life (LOL) for a warning time (WT) of one hour.

The US Bureau of Reclamation (USBR) suggested in 1989 estimating the PAR by applying an annual exposure factor to the number of residents in the floodplain. The annual exposure factor is the fraction of the year a typical individual spends at home. This factor ranges from about 0.6 to 0.8. The number of residents in the floodplain is estimated from census data, interviews with local planning officials, the number of homes in the area multiplied by the average number of residents per home, planning or cadastral maps, and house-to-house surveys. In most cases, the analysis must be augmented by considering facilities other than homes, such as schools, factories, and shopping centers.

The WT in Equation 5.2 depends on the existing warning system. This is the time in hours before the arrival of flooding by which the "first individuals for each *PAR* are being warned to evacuate," according to the USBR. As a lower bound, the WT is sometimes taken as just the flood travel time (i.e., no warning is issued prior to loss of containment). This seems to be appropriate for events such as earthquake-induced failures but conservative for hydrologically caused failures. The effect of the WT on LOL also depends on the warning procedure (e.g., telephone chain calls vs. siren) and on the evacuation plan. Neither of these factors enters the above equation.

The forcefulness of floodwaters in the above equation is treated as a dichotomous variable with a value of 1 for high force and 0 for low force. High force means that waters are swift and very deep, typical of narrow valleys; low force means that waters are slow and shallow, typical of broad plains. For cases in which the population resides in both topographies, the PAR is subdivided. The PAR should not be divided into any more than two subgroups because nonlinearity in the above equation causes overestimation of LOL when the PAR is subdivided.

5.5.6.4 Simulating Dam Breach Inundation

A number of mathematical models simulate a dam breach of an earthen dam by overtopping. Simulation of a breach requires flow over the dam, flow through the breach, and flow down the dam face. The flow over the dam is typically modeled as weir flow. The breach shape is assumed in all models, either as a regular geometric shape or as the most efficient breach channel shape where the hydraulic radius of the breach channel is maximized similar to stable channel design. The initial breach grows due to collapse of the breach slopes, gravity and hydrodynamic forces, and erosion of the soil, typically modeled using sediment transport equations developed for alluvial river channels (Wahl 1997).

The 1984 National Weather Service breach simulation model uses breach shape and erosion rate as inputs. The increase in erosion of the breach is assumed to be linear. The errors encountered in handling breach morphology in such a simplistic way are quickly overshadowed as the flood wave moves downstream. A more rigorous simulation of the breach morphology can be based on including both gradual erosion of the breach and sudden enlargement. The breach shape can be approximated by a triangle or trapezoid, although many other shapes are possible. Failure time is selected as a small value to maximize outflow. For earthen dams, this should be less than about two hours; for concrete dams, the failure time should be on the order of 0.5 hour.

The outflow from a breach can be modeled based on an implicit finite difference solution of the complete one-dimensional unsteady flow equations. The flow downstream from a breached or breaching dam is modeled using one-dimensional, unsteady St. Venant's equations with proper treatment of parameter uncertainty. The length of the river downstream of the dam can be divided into at least three reaches to differentiate between different flow types. The flow is modeled over the entire downstream river reach at 1-km increments. In addition, the model`accounts for bridges and other structures failing as the breach outflow travels downstream.

Inundation mapping is generally carried out by determining the extent of the flooding over the current topography. The water surface elevation or stage, as determined by breach outflow modeling, is extended to all topographic points with the same elevation to determine the extent of inundation. The most effective way to develop these maps is to use a geographic information system (GIS) based on reliable topographic maps, such as the US Geological Survey quadrangle series for the United States.

5.5.6.5 Dam Failure and Flood Fatalities

Evidence from ancient Babylonia, Egypt, India, Persia, and the Far East shows that dams have served the public for at least 5000 years. The total number of dams in the world that represent a hazard in the event of failure may exceed 150,000. Since the twelfth century, ~2000 dams have failed, although most of these failures were not major dams. About 200 reservoirs in the world failed in the twentieth century, and >8000 people died in these reservoir failures. The reasons behind these numbers of failures and fatalities should be used to improve the safety of dams.

Table 5.19 shows the calculated failure rates for dams based on failure. An estimated failure rate for dams based on this table is $10-^4$, without an indication of fatality rates for the associated failures. The rate is provided as the number of failures per dam per year, that is, per dam-year. Consequences of notable failure dams in the United States for the 1963–1983 period are summarized in Table 5.20.

To calculate the estimated fatality rates for US dam incidents, historical data were collected from a variety of sources including but not limited to the US Committee on Large Dams (USCOLD 1988), the International Commission on Large Dams (ICOLD 1974, 1983), *Engineering News Record* and *American Society for Civil Engineers Journal* articles, the National Performance of Dams Program (NPDP) files and records, the National Inventory of Dams (NID) database, the National Program of Inspection of Dams (USACE 1975), and other sources. Information was collected on the following items: (1) name or names of the dam; (2) state in which it is located; (3) year of completion; (4) year incident occurred; (5) age at time of incident, usually calculated from year of completion minus the incident year; (6) height of dam, in both meters and feet; (7) type of incident from the USCOLD records; (8) number of fatalities; (9) PAR, if available; (10) structure type, classified primarily as earth, gravity, rockfill, timber crib, masonry, arch, or buttress or miscellaneous, cofferdams, and tailing dams; (11) reference source; and (12) additional

TABLE 5.19

Referenced Dam Failure Rates

Area	Failures	Total Dams	Period (Years)	Rate per Dam-Year
United States	33	1764	41	4.5×10^{-4}
	12	3100	14	2.8×10^{-4}
	74	4974	23	6.5×10^{-4}
	1	(dam-year = 4500)		2.2×10^{-4}
World	125	7500	40	4.2×10^{-4}
	9	7833	6	1.9×10^{-4}
Japan	1046	276,971	16	2.4×10^{-4}
Spain	150	1620	145	6.6×10^{-4}
Great Britain	20	2000	150	0.7×10^{-4}

TABLE 5.20

Dam Failure Consequences from Notable US Dam Failures, 1963–1983

Name and Location of Dam	Failure Date	Fatalities	Property Damages
Mohegan Park, Connecticut	March 1963	6	Three million dollars
Little Deer Creek, Utah	June 1963	1	Summer cabins damaged
Baldwin Hills, California	December 1963	5	Forty-one houses destroyed; 986 houses damaged; 100 apartment buildings damaged
Swift, Montana	June 1964	19	Unknown
Lower Two Medicine, Montana	June 1968	9	Unknown
Lee Lake, Massachusetts	March 1968	2	Six houses destroyed; 20 houses damaged; 1 manufacturing plant partially destroyed
Buffalo Creek, West Virginia	February 1972	125	Five hundred and forty-six houses destroyed; 538 houses damaged
Lake 'O Hills, Arkansas	April 1972	1	Unknown
Canyon Lake, South Dakota	June 1972	33	Unable to assess damage; dam failure accompanied by damage caused by natural flooding
Bear Wallow, North Carolina	February 1976	4	One house destroyed
Teton, Idaho	June 1976	11	Seven hundred and seventy-one houses destroyed; 19 houses damaged
Laurel Run, Pennsylvania	July 1977	39	Six houses destroyed; 19 houses damaged
Sandy Run and five others in Pennsylvania	July 1977	5	Unknown
Kelly Barnes, Georgia	November 1977	39	Nine houses, 18 house trailers, and several (but unknown number) college buildings destroyed; 6 houses and 5 college buildings damaged
Swimming Pool, New York	1979	4	Unknown
About 20 dams in Connecticut	June 1982	0	Unknown
Lawn Lake, Colorado	July 1982	3	Eighteen bridges destroyed; 117 businesses and 108 houses damaged; campgrounds, fisheries, and power plant damaged
DMAD, Utah	June 1983	1	Unknown

notes, such as owner, NID number, and data differences between sources. This information was collected for 1337 dam incidents occurring from the late 1880s to 1997. Although the NPDP houses the most extensive collection of the US dam incident information at a single location, it is worthwhile to note the scarcity of available dam information, particularly with respect to the number of fatalities. Additional dam records, even those that contained fatality information, that could not be verified were not included in the database.

The NID data, consisting of records for 75,187 dams existing from 1995 to 1996, were analyzed to compute the age of each dam in 1997 and record its structural type and purpose. Total dam-years and the incident rate were calculated as follows:

$$\begin{aligned}\text{Total dam-years} = &\sum(\text{NID age computed values})\\ &+\sum(\text{age values from incident file})\end{aligned} \tag{5.3}$$

$$\text{Incident rate} = \frac{\text{Total number of incidents occurring}}{\text{Total dam-years}} \tag{5.4}$$

The number of incidents at which fatalities occurred and the total number of fatalities for these incidents were also recorded for the purpose of calculating the number of fatalities per incident and used to compute a fatality rate as follows:

$$\text{Fatality rate} = \frac{\text{Number of fatalities}}{\text{Number of incidents with fatalities}} \times \text{incident rate} \quad (5.5)$$

A dam incident with no LOL was recorded as a fatality number of 0. Where the description of the incident appears to be one in which no LOL would have occurred, but this fact could not be verified, these incidents were recorded as probable 0 fatalities and were separately included in the final results.

After the 1928 St. Francis dam failure and with the development of modern soil mechanics, dam design and construction underwent a dramatic revision. Prior to this time, most dams were not designed or supervised during construction by engineers. To account for these technological changes in dam design and construction, incidents occurring at dams completed after 1940 were additionally analyzed separately. Tables 5.21 and 5.22 show the calculated results for cases where sufficient and significant data are available. The tables show the numbers of accidents and fatalities that occurred in dam-years (defined as the cumulative sum of numbers of dams multiplied by respective years in service) and the corresponding rates.

Approximately 56% of the incidents occurred during the first 5 years after completion of the structure. If dam survival age is plotted, the resulting curve displays the typical hazard rate curve as a bathtub-shaped curve, with high failure rates early, then a uniform rate, and a higher rate again as age increases. Therefore, computations are made for dams over 5 years of age.

Earthen dams account for over 67% of the dam incidents; therefore, the data were subdivided by structural type. Data for earthen dams both with and without the inclusion of tailing dams are shown. Although rockfill dams completed after 1940 had 16 incidents (incident rate = 1.8×10^{-3}), no known fatalities occurred at this type of structure. In addition, no known fatalities were reported for the 35 incidents (incident rate = 1.9×10^{-3}) at timber crib dams or for 12 buttress dam incidents (incident rate = 6.2×10^{-4}). Only one known fatality occurred during an arch dam incident (in 1984). Arch dams accounted for 26 incidents (incident rate = 1.1×10^{-3}), with 5 occurring in post-1940 completed dams.

Dams >15 m are classified by the ICOLD as large dams; therefore, data were calculated for both large and small dams using height as the classifying factor. Most of the small dam incidents occurred at earthen structures. The 1889 incident at Johnstown, with 2209 fatalities, raises the fatality rate for large dams, earthen dams, and dams over 5 years of age. When these situations are eliminated, most of the fatality rates for dams are in the 10^{-4} range, which is less than those for the two major disease categories of cardiovascular and cancer, but higher than those for other US natural disasters as provided in Chapter 2.

When modeling LOL, additional factors should be taken into consideration. The number of fatalities depends on the amount of time the PAR has to evacuate. This was demonstrated in the 1976 Teton Dam incident, where seven fatalities occurred in a PAR of 2000 with <1.5 hours of warning, but only four fatalities occurred in a PAR of 23,000 with >1.5 hours of warning. Another example is Hurricane Georges, which hit the Mississippi Gulf Coast. Emergency operations officials attributed the lack of area fatalities to the fact that the PAR heeded the evacuation warning, which was not the case when Hurricane Camille hit the same area in 1969. The cost of expensive structural changes should be balanced

TABLE 5.21

Calculated Dam Incidents and Fatality Rates

	Number of Incidents		Dam-Years		Incident Rate per Dam-Year		Fatalities		Number of Incidents Involving Fatalities		Fatalities per Incident		Fatality Rate per Dam-Year	
	Total	Post-1940	Total	Post-1940	Total	Post-1940	Total	Post-1940	Total	Post-1940	Total	Post-1940	Total	Post-1940
Total	1337	420	2,877,755	1,724,062	4.6E−04	2.4E−04	3563	208	389	118	9.2	1.8	4.3E−03	4.3E−04
Earth	905	352	2,519,434	1,660,160	3.6E−04	2.1E−04	2632	73	261	91	10.1	0.8	3.6E−03	1.7E−04
Earth, including tailings	928	360	2,527,246	1,667,039	3.7E−04	2.2E−04	2758	199	267	95	10.3	2.1	3.8E−03	4.5E−04
Gravity	155	30	222,254	28,954	7.0E−04	1.0E−03	588	9	62	14	9.5	0.6	6.6E−03	6.7E−04
Rockfill	54	–	33,445	–	1.6E−03	–	199	–	19	–	0.5	–	1.7E−02	–

TABLE 5.22

Calculated Dam Incidents and Fatality Rates

	Number of Incidents		Dam-Years		Incident Rate per Dam-Year		Fatalities		Number of Incidents Involving Fatalities		Fatalities per Incident		Fatality Rate per Dam-Year	
	Total	Post-1940	Total	Post-1940	Total	Post-1940	Total	Post-1940	Total	Post-1940	Total	Post-1940	Total	Post-1940
Small dams (<15 m)	700	209	2,637,019	1,577,325	2.7E−04	1.3E−04	526	183	234	61	2.2	3.0	6.0E−04	4.0E−04
Large dams (≥15 m)	439	202	239,528	146,611	1.8E−03	1.4E−03	2999	24	120	54	25.0	0.4	4.6E−02	6.1E−04
Over 5 years of age	591	205	2,875,394	1,721,982	2.1E−04	1.2E−04	2668	64	197	65	13.5	1.0	2.8E−03	1.2E−04

with the cost of an upgraded warning system if a long WT will be available and evacuation can reasonably be accomplished.

The depth and velocity of the floodwaters can also be included with proper consideration of the structural type in the path of the floodwaters. A flood fatality model similar to the following model for fatalities from an earthquake can be developed:

$$\text{Log}(N(D)) = a(D) + b(D)M \tag{5.6}$$

where:

N is the number of casualties as a function of the magnitude (M)
D is the population density in the area affected
a and b are regression parameters that depend on density ranges

5.6 Disability, Illness, and Injury

Life loss and longevity is not the only aspect valued in life. The QOL is also valued by assessing the health states of individuals as a result of an event. Two methods are commonly used: the Abbreviated Injury Scale (AIS) and the Quality Adjusted Life Year (QALY) where *quality* accounts for disabilities. The World Health Organization (www.who.int) provides death estimates and the Disability-Adjusted Life Year (DALY) provides estimates of member states. The DALY is similar to the QALY concept.

The QALY concept assigns a unit value for a year of perfect health, that is, 1, and a year of less than perfect health as less than unity with death assigned a zero value. This concept provides a standard unit for measuring health gain across diseases, disabilities and population groups. In the case of illness, the concept of utility can be used to measure quality degradation on a cardinal scale to represent the strength of an individual's preferences for specific health states under conditions of uncertainty with primary limitations that include subjectivity and emotional factors by individuals affected by these health states (ASCC 2008). In the case of disability, the QALY is treated as the sum of two components: premature mortality, that is, years of life lost, and morbidity, equivalent years of life lost due to disability depending on the disability significance that is subjectively assessed.

The AIS is an anatomical scoring system first introduced in 1969 that since then has been revised and updated against survival so that it now provides a reasonably accurate ranking of the severity of injury. The AIS is monitored by scaling committees, such as the scaling committee of the Association for the Advancement of Automotive Medicine, and updated as needed.

Injuries are ranked on a scale of 1–6, with 1 being minor, 5 severe, and 6 unsurvivable. This scale represents the threat to life associated with an injury and is not meant to represent a comprehensive measure of severity. The AIS is not an arithmetic injury scale, in that the difference between the AIS levels 1 and 2 is not the same as that between the AIS levels 4 and 5, that is, it is on an ordinal scale. This scale has many similarities with other injury scales, such as the Organ Injury Scale of the American Association for the Surgery of Trauma.

Table 5.23 shows the relationship between the AIS and a fraction of the WTP value, for example, $3,000,000 based on the FAA guidance documents. These percentages reflect the loss of quality and quantity of life resulting from an injury typical of that level.

TABLE 5.23

Abbreviated Injury Scale and Willingness to Pay Value (2001 US$)

Abbreviated Injury Scale Code	Injury Severity	Definition	Multiplier (%)	Willingness to Pay Value ($)
1	Minor	Superficial abrasion or laceration of skin; digit sprain; first-degree burn; head trauma with headache or dizziness (no other neurological signs)	0.2	6000
2	Moderate	Major abrasion or laceration of skin; cerebral concussion (unconscious <15 minutes); finger or toe crush/amputation; closed pelvic fracture with or without dislocation	1.55	46,400
3	Serious	Major nerve laceration; multiple rib fracture (but without flail chest); abdominal organ contusion; hand, foot, or arm crush/amputation	5.75	172,500
4	Severe	Spleen rupture; leg crush; chest wall perforation; cerebral concussion with other neurological signs (unconscious <24 hours)	18.75	562,500
5	Critical	Spinal cord injury (with cord transection); extensive second- or third-degree burns; cerebral concussion with severe neurological signs (unconscious >24 hours)	76.25	2,287,500
6	Fatal	Injuries that, although not fatal within the first 30 days after an accident, ultimately result in death	100.00	3,000,000

TABLE 5.24

Per-Victim Medical and Legal Costs Associated with Injuries (2001 US$)

Abbreviated Injury Scale Code (see Table 5.20)	Injury Severity ($)	Emergency and Medical ($)	Legal and Court ($)	Total Direct Cost ($)
1	Minor	600	1900	2500
2	Moderate	4600	3100	7700
3	Serious	16,500	4700	21,200
4	Severe	72,500	39,100	111,600
5	Critical	219,900	80,100	300,000
6	Fatal	52,600	80,100	132,700

In addition to the WTP values, the DOT identifies other costs associated with fatalities and injuries related to transportation, including the costs of emergency services, medical care, and legal and court services, such as the cost of carrying out the court proceedings but not the cost of settlements. Because medical and legal costs of separate injuries to the same victim are not necessarily additive, the Office of Aviation Policy and Plans (APO) advises that medical and legal costs be valued on a per-victim basis, as provided in Table 5.24. The table provides direct per victim medical and legal costs classified according to the worst AIS injury sustained by each aviation accident victim. The values in Table 5.24 should be added only once to the aggregated sum of the WTP values for injuries suffered by any particular individual.

5.7 Tort and Professional Liability

Liability refers to an entity bringing a claim against one of the parties or more as a result of an exchange based on a contract or otherwise as a result of the claimant's assertion of damages resulting from this exchange. An entity can be a person or a group or an organization. In addition, a claim can be of the tort type that enables a victim with some injury to seek compensation. Criminal cases are initiated and managed by governments, whereas tort cases are initiated by the victim or the victim's survivors. A successful liability case results not in a sentence of punishment but in a judgment that entitles the plaintiffs' financial compensations from defendants. An award of compensatory damages principally shifts all of the plaintiffs' legally cognizable costs to the defendants. On rare occasions, a plaintiff may also be awarded punitive damages, defined as damages in excess of compensatory relief, and in other cases, a plaintiff may obtain an injunction, that is, a court order preventing the defendant from injuring the plaintiff or from invading property rights. In the tort cases, the claimant can be one of the parties of the exchange.

In this section, these two types are briefly discussed. The valuation of liability is beyond the scope of this chapter although some of the general principles covered under valuation and WTP apply.

5.7.1 Tort Liability

Tort stems from the claim of faults and duties, and distinguishes between two general classes of duties (Coleman and Mendlow 2010): (1) duties not to injure *full stop* and (2) duties not to injure negligently, recklessly, or intentionally. Engaging in an activity the law regards as extremely hazardous, for example, blasting with dynamite, a duty of the first sort, that is, a duty not to injure *full stop*, comes with it; however, engaging in an activity of ordinary risk level, for example, driving a car, duty of the second sort, that is, a duty not to injure negligently, recklessly, or intentionally, comes with it.

According to the strict liability, absent some prior agreement, an entity causing damage or harm is responsible for the damages and outcomes of the harm, not the affected entity. According to the fault liability, an entity is expected to reasonably take the interests of others into account and moderate behavior accordingly by particularly taking precautions not to injure and to avoid being careless with respect to the interests of others including not injuring or causing damages intentionally.

It should be noted that strict liability is not defeasible by excuse, for example, the blasting activity traditionally governed by strict liability with the blaster having a duty not to injure by blasting, which means that regardless of the care level the blaster cannot discharge his duty in case of an injury. On the other hand, fault liability is defeasible in cases of a reasonable person of ordinary prudence acting reasonably or justifiably, for example, driving a car as an activity traditionally governed by fault liability with the driver having a duty not to injure by driving faultily, which means that the driver cannot discharge his duty only in cases of injures from negligence, recklessness, or intentional acts.

For particular activities, the cost of any actions needed to manage tort liability risk requires defining the legal duty of parties as a basis to allocate the costs to the activities. Suppose a construction company is developing a site in the center of a city that might affect the adjacent properties, causing the adjacent dwellers financial loss. There is a need to protect, clean, and accommodate the activities of the dwellers. Is it a cost of construction or a cost to the dwellers? Answering this question requires determining whether the company owes the dwellers a duty to prevent impacts to the dwellers, then any cost of protection or damage thereof is a cost of construction. But if the company has no such legal duty, it is the dwellers' responsibility.

Relevant liability-related questions to risk studies are as follows: (1) how much to spend on precautions and (2) what hazards that can be justifiably forgone. Economists offer an answer to the first question based on the notion that a precaution is reasonable when it is rational; a precaution is rational when it is cost justified; and a precaution is cost justified when the cost of the precaution is less than the expected injury cost defined as the cost of the anticipated injury discounted by the probability of the injury's occurrence (Coleman and Mendlow 2010). As for the second question, the same logic can be used by examining the cost of taking precautions, and if found costlier than the expected injury cost, not having these precautions is rational and hence not negligent; in this case, these hazards can be justifiably forgone on this basis. Therefore, this logic induces all rational persons including both injurers and victims alike to take all and only cost-justified precautions. If potential injurers behave rationally, losses will always lie where they fall, that is, with victims. Similarly, rational victims will therefore approach all accidents, assuming that they would have to bear the costs, and will take all and only cost-justified precautions. As a result, fault liability is economically efficient by producing an optimal level of risk taking by both injurers and victims. This principle also applies to strict liability producing efficient outcomes. Under ideal conditions, efficiency requires that individuals, both injurers and victims, take all and only cost-justified precautions, and as a result, strict and fault liability can both be efficient. The primary differences is in the cost allocation of precautions that (1) make the costs of the defendant's conduct in strict liability higher than a rule of fault liability would and (2) make the costs of the plaintiff's conduct higher by the rule of fault liability than a rule of strict liability would. The economic analysis of tort cases offers insights that are useful to risk analysts, but is lacking and incomplete, since the analysis does not account for legal duty, justice, and rights. These considerations are beyond the scope of this section and are provided by Coleman and Mendlow (2010).

5.7.2 Professional Liability

Statutory bases for liability commonly govern the profession of engineering; however, compliance with them does not relieve a business organization of responsibility for the conduct of its agents, employees, or officers, and are subject to tort liability. On an individual basis, a professional is not relieved of responsibility for professional services performed by virtue of being employed with or a partner in or officer of a business organization, that is, the professional is personally accountable and liable only for misconduct, negligent acts, or wrongful acts committed by the person or by any person under his or her direct control and supervision. The business organization is liable up to the full value of its property. The liability can extend beyond the contract clauses governing the professional services, that is, the existence of a contract is not determinative of tort liability and will not necessarily preclude an action in tort by a third party. The rationale in this case is that public policy dictates that liability should not be limited to the terms of the contract. In the case of defects, the nature of the defect and the issue of control are key items to a determination of liability sometimes subject to statute of limitations on time after delivery and acceptance of professional services or goods.

The remainder of this section is used to focus on nuclear liability as an example and in light of the 2011 Fukushima nuclear disaster. This example illustrates the legal complexities associated with liability. It should be noted that even with the best risk-informed planning and guidelines, accidents at nuclear power plants (NPPs) could still occur. The 1990 report from a US Presidential Commission estimates that the catastrophic nuclear accident probability in the United States (about 100 nuclear reactors) in the remaining lifetime of 40 years per plant is 1 in 250,000 years. There are currently 438 NPP units worldwide (predicted to increase to 500); extrapolating the US figure with some uncertainty considerations to obtain the worldwide average time to an accident yields an estimate of 1 in 5000–50,000 years for remaining lifetimes. Given the possibility of another accident, in addition to strengthening safety measures, we should develop dependable liability coverage that can be tapped in an emergency.

In 1957, the United States enacted the Price–Anderson nuclear liability regime for managing the risk of nuclear accidents. The legislation aimed to establish a mechanism for compensating the public for losses and to encourage the private development of nuclear power. With 104 operating reactors, the United States has a total of $11.975 billion in coverage (as of 2011) before congressional authorization for additional funding. The US Department of Energy provides similar liability coverage for its activities.

Internationally, three conventions are available with similar goals: the 1968 Convention on Third Party Liability in the Field of Nuclear Energy, called the Paris Convention; the 1977 Vienna Convention on Civil Liability for Nuclear Damage; and the Convention on Supplementary Compensation (CSC) for Nuclear Damage, which will enter into force when ratified by at least five countries with at least 400 GW of installed nuclear capacity.

Estimates of the damage due to a catastrophic accident range from $110 billion to as much as $7 trillion (2011 US$). Accidents do not recognize political borders and could lead to disputes. Achieving adequate nuclear liability coverage requires an efficient and cost-effective system with adequate funds to pay damages. Starting with the premise of a worldwide need to mitigate the consequences of one catastrophic nuclear accident, each NPP unit can be assessed for a cost share secured by international legal instruments, subject to adjustments based on, among other metrics, a safety rating system to create the incentive to reduce accident rates. To succeed, financing will be essential, perhaps via securities and hedge funds.

5.8 Indirect Losses

Indirect losses, including consequential damage, are second order in that they are induced by the direct losses. They can be classified as time-independent or time-dependent losses. For example, the loss of a building includes the direct loss of its value and indirect losses such as loss of use of the building, which is time dependent. Time-independent losses include, for example, the loss in value of clothing due to a loss of part of the clothing. Indirect losses also include business interruptions due to shutdown or reduced operations. Such losses could include depreciation; an inability to pay mortgages and other indebtedness, salaries of personnel, and maintenance, advertising, and utility expenses; and failure to meet subcontract obligations. The total loss also depends on the period of interruption. Some businesses must continue operation, leading to additional losses due to higher operating rates for space, people, and materials. Indirect losses could also include contingent business interruption due to other contributing properties that are not owned by the loss bearer but are essential for operations, such as an essential supplier of materials. Still other indirect losses could include losing favorable lease terms as a result of loss of leased premises, criminal loss due to dishonesty of employees, and legal liability losses.

An important category of indirect losses as a result of contract breach is *consequential damages* defined as loss of profit or revenue if determined being reasonably foreseeable at the time of contract formation. Such a factual determination could lead to significant loss as a result of the contract breach. A similar concept is the *loss of use* or *loss of function*.

For example, the total loss of an uninsured automobile in a crash could entail not only the monetary amount to replace the car and losses associated with other impacted properties but also at least the monetary amount necessary to maintain the functionality during the period until the care is replaced by perhaps renting a car—this does not include any other altered opportunities and utilities as a result of the car loss and its replacement, noting that the altered states might be favorable or adverse. It is common that risk analysts account only for direct property losses.

5.9 Public Health and Ecological Damages

Assessing health impact to the public requires performing exposure assessment. Failure consequences are used to determine, for example, the effects of varying levels of exposure to particular chemicals or biological agents of interest. People must come in contact with the chemicals in order to become at risk, but the amount of exposure depends greatly on how much of each chemical is present, who might be exposed, and how they are exposed. For instance, because children might play in a polluted stream or people might drink polluted well water or eat polluted fish, these activities must be defined in order to identify everyone who could be exposed. The exposure assessment is followed by toxicity assessment to determine which illnesses or other health effects may be caused by exposure to chemicals. It will also include determining the dose that can cause harmful health effects (i.e., how much of each chemical it takes to cause harm). Generally, the higher the dose, the more likely a chemical will cause harm. These harms need then to be translated into reduced longevity or equivalent life loss.

Ecological risk assessment evaluates the potential adverse effects that human activities have on the plants and animals that make up ecosystems or an environment. When risk assessment is conducted for a particular place such as a watershed, the ecological risk assessment process can be used to identify vulnerable and valued resources, prioritize data collection activity, and link human activities with their potential effects. The assessment of ecological impacts of an event is not treated explicitly in this section, but some of the concepts presented in previous sections can be used for this purpose. Some analytical and modeling methods are described in the remainder of this section. Ecological risk assessment is a process by which scientific information is used to evaluate the likelihood that adverse ecological effects are occurring or may occur as a result of exposure to physical (e.g., site cleanup activities) or chemical (e.g., release of hazardous substances) stressors at a site. These assessments often contain detailed information regarding the interaction of these stressors with the biological community at the site. Part of the assessment process includes creating exposure profiles that describe the sources and distribution of harmful entities, identify sensitive organisms or populations, characterize potential exposure pathways, and estimate the intensity and extent of exposures at a site. For example, toxicity (i.e., effects data) and exposure estimates (i.e., environmental concentrations) are evaluated for the likelihood that the intended use of a pesticide will adversely affect terrestrial and aquatic wildlife, plants, and other organisms. Data required to conduct an ecological risk assessment may include the following:

- Toxicity to wildlife, aquatic organisms, plants, and nontarget insects
- Environmental changes
- Environmental transport
- Estimated environmental concentrations
- Where and how the pesticide will be used
- What animals and plants will be exposed
- Climatologic, metrologic, and soil information

In addition, ecological methods may be used for detecting the patterns of disease occurrence across space and time and relating the rates of disease frequency to environmental, behavioral, and constitutional factors. Several unique sources of bias in ecological data must be considered when designing studies and interpreting their findings. The risk assessment process involves multiple steps, beginning with an appraisal of toxicity and exposure and concluding with a characterization of risk. Risk characterization defines the likelihood that humans or wildlife will be exposed to hazardous concentrations. Thus, risk characterization describes the relationship between exposure and toxicity. Risk assessors identify the species likely to be exposed, the probability of such exposure occurring, and the effects that might be expected. With the use of environmental modeling, scientists can evaluate the environmental and health consequences of operational and accidental chemical releases. The following modeling methods can be used depending on the situation and analysis objectives:

- *Source modeling.* Determining the quantity and the nature of a chemical release is the first step in modeling its transport, fate, human health, and ecological impacts.
- *Emissions modeling.* This modeling method can be used to estimate air emissions from point or area sources such as from waste management and wastewater treatment operations.
- *Air dispersion modeling.* For chemicals that are emitted from sources such as industrial facilities or mobile sources, this modeling method determines both the air concentration and the amount of chemical constituent deposited on surfaces at specified locations.
- *Groundwater and surface water modeling.* This modeling method enables effective and cost-saving management of groundwater resources. It helps decision makers to determine the optimal solutions for pollution control at local, regional, and

national levels. It uses a variety of water quality models and databases for many situations, including point and nonpoint sources and in-stream kinetics.

- *Food web modeling.* This modeling method predicts biological uptake and accumulation of chemicals in aquatic and terrestrial food webs. It uses data and regression methods to estimate chemical concentrations in produce and animal products. The focus is on characterizing the variability in tissue concentration estimates associated with dietary preferences and chemical-specific behavior in biological systems.
- *Ecological modeling.* Risk assessors use a holistic approach to predict ecological risks associated with chemical releases in terrestrial, freshwater, and wetland habitats, recognizing the importance of characterizing the variability and uncertainty inherent in ecological simulations.
- *Stochastic modeling.* Environmental models often provide deterministic results, although the input data include both uncertainty and variability. This method provides a distribution of risks that reflect variability in the input parameters and can provide either a quantitative evaluation or a qualitative discussion of the uncertainty. A statistical method based on response surface methodology can also be used to determine the most sensitive input variables in a Monte Carlo analysis.
- *GIS-based modeling.* This modeling method allows scientists to develop complex, interactive, and flexible applications using geospatial data to simulate and predict real-world events. They may be used to (1) predict the amounts and effects of nonpoint source runoff, (2) evaluate the effects and dangers of pollutants as they travel through the environment, and (3) simulate the effects of environmental policies. Such capabilities provide flexibility to examine what-if scenarios to better understand environmental processes and the effects of environmental policy.
- *Life cycle modeling.* This modeling method might be necessary to assess ecological risk. For example, life cycle emissions for the production and combustion of fuels to produce electricity using electrical energy distribution grids might require modeling many processes that consume fuel or electricity in order to calculate the trade-offs among alternative energy sources.

The Food Safety and Inspection Service (FSIS) of the US Department of Agriculture (USDA) relies greatly on risk assessments as a means of guiding food safety policy decisions. The agency has conducted risk assessments for *Salmonella enteritidis* in eggs and egg products and in ground beef and, with the FDA, it has developed a risk ranking for *Listeria monocytogenes* in a variety of foods. Risk assessment has been used for determining the risks associated with any type of hazard, including biological, chemical, or physical. Having the objective of ensuring that the public is protected from health risks of unsafe foods, exposure assessment in this case must differentiate between short-term exposure for acute hazards and long-term exposure for chronic hazards. For acute hazards, such as pathogens, data on levels of pathogens causing illness in vulnerable population groups are important. For chronic hazards, such as chemicals that may cause cumulative damage, a lifetime averaged exposure is relevant.

The valuation of health and ecological impacts requires the use of concepts covered in Section 5.3.

5.9.1 Toxicity Assessment

Toxicity assessment is frequently dealt with in environmental risk studies of toxins resulting from exposure to a range of environmental hazards on plants, animals, and humans. Toxicity assessment frequently entails performing the following steps (ISO 31010 2009):

- Problem formulation including defining objectives and the scope of the assessment in terms of the range of target populations and hazard types of interest
- Hazard identification of all possible sources of harm to the target population from hazards identified in the previous step based on reviewing literature and expert opinions
- Hazard analysis to understand their nature most importantly in terms of interacting with the target, for example, human exposure to chemical effects might include acute and chronic toxicity, the potential to damage DNA, or the potential to cause cancer or birth defects. For each effect identified, an observed response of a person and the associated dose with the response along with wherever possible the mechanism by which the effect is produced are noted. Two thresholds are of interest: (1) the no observable effect level (NOEL) threshold and (2) the no observable adverse effect level (NOAEL) threshold, for the purpose of defining criteria for acceptability of the risk associated with the hazard. Figure 5.22 shows a dose–response curve for the purpose of illustration with the two NOEL and NOAEL identified. Such curves are usually derived from tests on animals and from experimental systems based on cultured tissues or cells, or the effects of other hazards such as microorganisms or introduced species from field data and epidemiological studies.

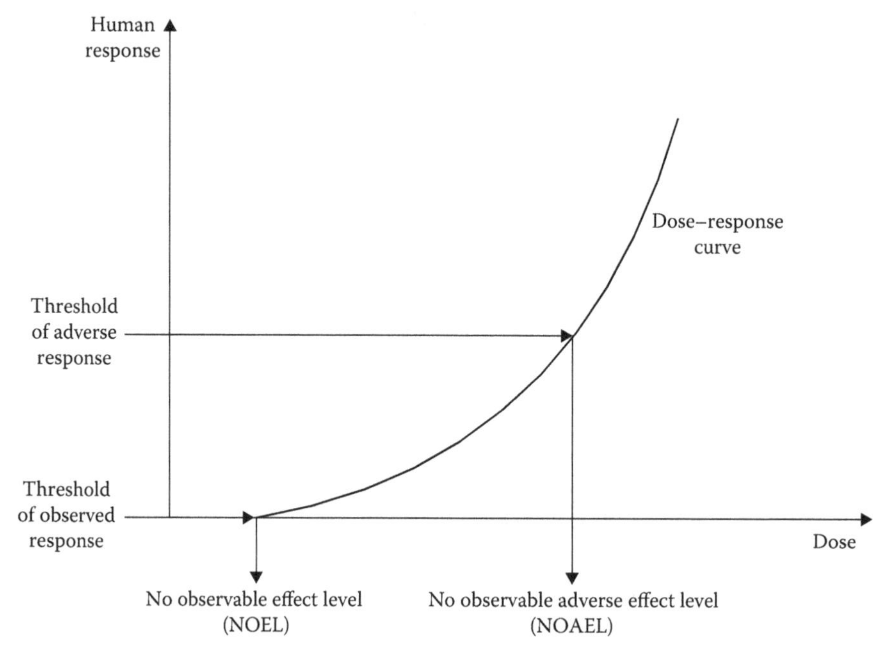

FIGURE 5.22
A dose–response curve.

- Exposure analysis to examine how a hazardous substance or its residues might reach a susceptible target population and in what amount. Such analysis examines pathway analysis and examination of protection and barrier layers.
- Risk characterization to bring together the above elements to estimate the probabilities of particular consequences when effects from all pathways are combined. The primary output is typically an indication of the level of risk from exposure of a particular target to a particular hazard in the context concerned.

An important attribute of this analysis is enhancing the understanding of the nature of the problem and the factors that increase risk by its pathway analysis, protection, and barriers. Moreover, the dose–response curves derived from exposing animals to high levels of a hazard can be extrapolated to estimate the effects of very low levels of a toxin to humans with applicability considerations.

5.9.2 Frequency–Population Curves

Frequency–population (*FN*) curves are a graphical representation of the probability expressed as a cumulative frequency (*F*) of events causing a particular level of harm to a specified population expressed in terms of the number of people affected (*N*) (ISO 31010 2009). This method focuses the attention on high *N* values that may occur with a high frequency *F* because they may be socially or politically unacceptable. Such *FN* curves can be constructed using observed data based on historical losses or calculated from simulation model estimates. Sometimes, a mixture of both is used.

FN curves are effective in communicating risk information to managers and policy makers for setting safety levels. They are also appropriate for comparing risks from similar situations to compensate for lack of data or new situations; however, they should not be used to compare risks of different types. A primary limitation not preserving or communicating the range of effects or outcomes of incidents other than the number of people impacted, how many ways a particular value of *N* is achieved.

A related concept is the *as low as reasonably practicable* (ALARP) in setting safety levels for safety-critical and safety-involved systems. The concept is used as a basis to define what is termed the ALARP principle by requiring a residual risk to be ALARP. This principle and the term are similar to stating *so far as is reasonably practicable* (SFAIRP) in the UK Health and Safety law. A basis for defining the ALARP risk threshold is to demonstrate that the cost involved in reducing the risk further would be grossly disproportionate to the benefit gained. The ALARP principle arises from the belief that infinite time, effort, and money could be spent to reduce a risk to zero.

5.10 Exercise Problems

Problem 5.1 Define failure consequences and severities. Describe the differences between them using your own examples.

Problem 5.2 Demonstrate the differences between failure consequences and severities using examples related to the following fields:

a. Structure engineering

b. Public health

Problem 5.3 What do maximum possible loss (MPL) and probable maximum loss (PML) mean? Show the difference between them using examples from the following fields:

a. Nuclear engineering

b. Environmental engineering

Problem 5.4 What is the purpose of cause–consequence (CS) diagrams and what are their uses?

Problem 5.5 Example 5.1 deals with consequences associated with the structural failure of a component of a ship structural system. Use the information provided in the example to perform the following:

a. Define the sequence of events that can be used to develop the cause–consequence (CS) diagram for failure scenarios related to the failure of ship systems other than structural failure.

b. Draw the cause–consequence (CS) diagram for failure scenarios related to the failure of ship systems other than structural failure. Limit the consequences to five items.

c. Derive a consequence-rating table using the same five character notations and ordinal scale rating as used in Example 5.1.

Problem 5.6 A factory uses a power generator that is located in a generator room. Use the following sequence of events to construct and draw the cause–consequence (CS) diagram for the failure scenarios related to the initiating event of generator overheating:

1. Generator overheating is sufficient to cause fire.
2. Local fire in generator room occurs (or does not occur).
3. Operator fails (or does not fail) to extinguish fire.
4. Fire spreads (or does not spread) to the factory.
5. Factory fire system fails (or does not fail) to extinguish fire.
6. Fire alarm fails (or does not fail) to sound.

Problem 5.7 In the case of a fire in an apartment that is equipped with a smoke detector, the potential consequences of the fire to occupants may be analyzed using the CS diagram method. You may limit the scope of the cause–consequence (CS) diagram development to considering only the following events:

a. The smoke detector operates (or fails to operate) during the fire.

b. The occupants are able (or unable) to escape.

Construct and draw the cause–consequence (CS) diagram based on all possible event occurrences and nonoccurrences.

Problem 5.8 What are the distinctions of value? Define and provide examples.

Problem 5.9 Define the components of the total economic value (TEV) in the context of a dam facility. Make any necessary assumptions.

Problem 5.10 What are the types of formulations used in assessing real property damage? What are the characteristics and differences between these types? Give examples for both types in the engineering field.

Problem 5.11 Based on reviewing the literature, summarize the direct property damage estimate from the 2005 Hurricane Katrina that impacted the City of New Orleans area.

Problem 5.12 Based on reviewing the literature, summarize the impacts to infrastructure from the 2005 Hurricane Katrina that impacted the City of New Orleans area.

Problem 5.13 What are the methods normally used to assess the loss of human life? What are the differences between the methods?

Problem 5.14 If a group of 1000 employees working at a nuclear waste site are willing to pay an average amount of $70 each to reduce causes of deaths from 2 per 1000 to 1 per 1000, what is the total willingness to pay (WTP) value and what is the value of statiscal life (VSL)?

Problem 5.15 If a group of 10,000 employees working in a chemical plant are willing to pay an average of $700 each to reduce causes of deaths from 3 per 10,000 to 1 per 10,000, what is the total WTP value and what is the value of statiscal life (VSL)?

Problem 5.16 For Problem 5.12, the workers at the nuclear waste site were divided into two equal groups that correspond to two types of jobs, A and B. If job B has two more job-related deaths per year for every 1000 employees than does job A, and if the workers of job B earn $400 per year more than those of job A, use the HC method to calculate the value of life for workers in job A who are willing to forgo the additional money for a lower risk level.

Problem 5.17 For Problem 5.12, the workers at the nuclear waste site were divided into two equal groups that correspond to two types of jobs, A and B. If job A has three more job-related deaths per year for every 10,000 employees than does job B, and if the workers of job A earn $600 per year more than those of job B, use the HC method to calculate the value of life of workers in job B who are willing to forgo the additional money for a lower risk level.

Problem 5.18 Use Equation 5.1 and data from http://data.un.org/on GDP per capita to estimate the VSL for the following countries: the United States, the United Kingdom, Israel, Saudi Arabia, Uganda, Malaysia, Japan, China, India, Brazil, and Russia. Provide the results for 2004 and 2010. Discuss the reasonableness of your findings and provide observations.

Problem 5.19 A flood control dam, if overtopped, would lead to flooding without a floodwater force. The WT to the affected population is 6 hours. The size of the PAR is 100,000. Estimate the LOL for this situation as a result of flooding. Plot the trend of LOL as a function of WT.

Problem 5.20 A flood control dam, if overtopped, would lead to flooding without a floodwater force. The WT to the affected population is 4 hours. The size of the PAR is 90,000. Estimate the LOL for this situation as a result of flooding. Plot the trend of LOL as a function of size of the PAR.

Problem 5.21 An initiating event could lead to failure scenarios A and B that involve human injuries. The injuries are estimated for both scenarios as follows:

Scenario A	Scenario B
One injury at AIS = 6 (fatality)	Two injuries at AIS = 6 (fatality)
Ten injuries at AIS = 3	Twelve injuries at AIS = 2
Fifteen injuries at AIS = 4	Fifteen injuries at AIS = 5

Determine the total costs, including medical and legal expenses, associated with each scenario in 2001 US$.

Problem 5.22 An initiating event could lead to failure scenarios A and B that involve human injuries. The injuries are estimated for both scenarios as follows:

Scenario A	Scenario B
Two injuries at AIS = 6 (fatality)	Four injuries at AIS = 6 (fatality)
Five injuries at AIS = 3	Thirty injuries at AIS = 2
Ten injuries at AIS = 4	One injury at AIS = 3

Determine the total costs, including medical and legal expenses, associated with each scenario in 2001 US$.

Problem 5.23 An initiating event could lead to failure scenarios A, B, and C that involve human injuries. The injuries are estimated for the scenarios as follows:

Scenario A	Scenario B	Scenario C
Three injuries at AIS = 6 (fatality)	Two injuries at AIS = 6 (fatality)	One injury at AIS = 6 (fatality)
Twenty injuries at AIS = 2	Ten injuries at AIS = 3	Five injuries at AIS = 3
Ten injuries at AIS = 4	Twelve injuries at AIS = 5	Twenty injuries at AIS = 5
Seven injuries at AIS = 1	Five injuries at AIS = 2	Three injuries at AIS = 2

Determine the total costs, including medical and legal expenses, associated with each scenario in 2001 US$.

6

Engineering Economics and Finance

Decision analysis using results of risk studies commonly entails evaluating alternatives with costs and effects spanning several years. Considering time and the time value of money in engineering economics and financial analysis is the focus of this chapter. Discount rates, cash flows, equivalence, and inflation indices are essential for rational decision making. This chapter covers and illustrates these concepts using practical examples.

CONTENTS

6.1 Introduction

6.1.1 Need for Economics

Present-day engineers are commonly faced with nontechnological, in addition to technological, barriers that limit what can be done to solve a problem or meet a need. Technological barriers limit what engineers can do because they might simply lack the know-how or have not yet developed tools required to solve a problem. However, engineers commonly encounter barriers that are not technological; that is, in addition to designing and building systems, they must meet other constraints, such as budgets and regulations. For example, natural resources necessary to build systems are becoming scarcer and more expensive than ever before. This trend is expected to continue. Also, engineers and economists are aware of the potential negative side effects of engineering innovations, such as air pollution from automobiles. For these reasons, they are often asked to place their project ideas within the larger framework of the environment of a specific planet, country, or region. They must ask themselves if a particular project would offer some net benefit to individuals or a society as a whole. The net benefit assessment requires considering the inherent benefits of the project, plus any negative side effects, including severities associated with failure consequences due to hazards, plus the cost of consuming natural resources, considering both the price that must be paid for them and the realization that once they are used for that project, they will no longer be available for other projects.

Risk analysis requires engineers and economists to work closely together to develop new systems, to solve problems that face society, and to meet the societal needs. They must decide if the benefits of a project exceed its costs and must make this comparison in a unified, systems framework. Results from risk assessment, therefore, should feed into economic models, and economic models might drive technological innovations and solutions. The development of such an economic framework is as important as the physical laws and sciences defining technologies that determine what can be accomplished with engineering. Figure 6.1 shows how problem solving is composed of physical and economic components.

A systems framework is divided into physical and economic environments. The physical environment involves producing physical systems and services depending on physical laws such as Ohm's and Newton's laws. However, much less of a quantitative nature is known about economic environments, as economics is involved more with the actions of people and the structure of organizations. Risk analysis draws from both environments.

Satisfying the sets of requirements for both the physical and economic environments is achieved by linking design and product- and service-producing processes. Engineers and economists need to manipulate systems to achieve a balance in attributes within both the physical and economic environments and within the constraints of limited resources.

This mix of engineering and economics is traditionally termed *engineering economics*. In this book, its use involves the added economics of risk. It plays a crucial and central role with diverse application potentials, such as selecting from design alternatives to increase the

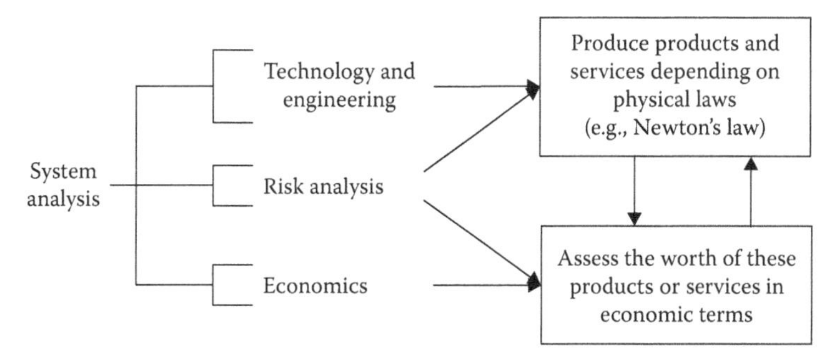

FIGURE 6.1
Systems framework for risk analysis.

capacity of a set of navigational locks and gates, choosing the best design for a high-efficiency gas furnace, selecting the most suitable robot for a welding operation on an automotive assembly line, making a recommendation about whether jet airplanes for an overnight delivery service should be purchased or leased, or considering the choice between reusable and disposable bottles for high-demand beverages. For the second and third examples in particular, engineering knowledge should provide sufficient means to determine a good design for a furnace or a suitable robot for an assembly line, but it is the economic evaluation that allows further definition of the best design or the most suitable robot.

Engineers and economists are concerned with two types of efficiency: (1) physical and (2) economic. Physical efficiency takes the following form:

$$\text{Physical efficiency} = \frac{\text{System output(s)}}{\text{System input(s)}} \tag{6.1}$$

In the furnace example, the system outputs might be measured in units of heat energy and the inputs in units of electrical energy, and if these units are consistent, then physical efficiency is measured as a ratio between 0 and 1. Certain laws of physics (e.g., conservation of energy) dictate that the output from a system can never exceed the input to a system, if it is measured in consistent units. A particular system can only change from one form of energy (e.g., electrical) to another (e.g., heat). Losses incurred along the way due to electrical resistance, friction, and so on always yield efficiencies <1. In an automobile engine, for example, 10%–15% of the energy supplied by the fuel might be consumed simply to overcome the internal friction of the engine. A perfectly efficient system would be the theoretical perpetual motion machine.

The other form of efficiency of interest here is economic efficiency, which takes the following form:

$$\text{Economic efficiency} = \frac{\text{System worth}}{\text{System cost}} \tag{6.2}$$

This ratio is also commonly known as the *benefit–cost ratio*. Both terms of this ratio are assumed to be of monetary units, such as dollars. In contrast to physical efficiency, economic efficiency can exceed unity, and in fact it should if a project is to be deemed economically desirable or feasible. The most difficult part of determining economic efficiency is accounting for all the factors that might be considered benefits or costs of a particular system and converting these benefits or costs into monetary equivalents.

For example, for a transportation construction project that promises to reduce people's travel times to work, how do we place a value on that travel time savings? In addition, if this transportation project introduces new risks while eliminating others, what is the net benefit of these risk-related changes? A systems framework of analysis must provide a means for proper accounting of benefits and risks.

In the final evaluation of most ventures, economic efficiency takes precedence over physical efficiency because projects cannot be approved, regardless of their physical efficiency, if there is no conceived demand for them among the public, if they are economically infeasible, or if they do not constitute a wise use of those resources that they require.

Numerous examples can be cited of engineering systems that have an adequate physical design but little economic worth; that is, such designs may simply be too expensive to produce. For example, a proposal to purify water needed by a large city by boiling it and collecting it again through condensation is such a case. This type of a water purification experiment is done in junior physical science laboratories every day, but at the scale required by a large city, it is simply too costly.

6.1.2 Role of Uncertainty and Risk in Engineering Economics

Engineering economic analyses might require, for simplicity, the assumption of knowing the benefits, costs, and physical quantities with a high degree of confidence. This degree of confidence is sometimes called *assumed certainty*. In virtually all situations, however, there was some doubt as to whether the ultimate values of various quantities exist. Both risk and uncertainty in decision-making activities are caused by a lack of precise knowledge, incomplete knowledge, or a fallacy in knowledge regarding future conditions, technological developments, synergies among funded projects, and so on. Decisions under risk are decisions in which the analyst models the decision problem in terms of assumed possible future outcomes, or scenarios, whose probabilities of occurrence and severities can be estimated. This type of analysis builds on the concepts covered in Chapters 1 through 4. Decisions under uncertainty, by contrast, could also include decision problems characterized by several unknown outcomes or outcomes for which probabilities of occurrence cannot be estimated. Because engineering is concerned with actions to be taken in the future, an important part of the engineering process is improving the level of certainty of decisions with respect to satisfying the objectives of engineering applications. By presenting the concepts relating to ignorance and uncertainty, hierarchy, systems analysis, risk methods, and economics (see Chapters 1 through 6), analysts may combine them in many forms to obtain creative solutions to problems.

6.1.3 Engineering and Economic Studies

Engineering activities dealing with elements of the physical environment are intended to meet human needs that could arise in an economic setting. The engineering process employed from the time a particular need is recognized until it is satisfied may be divided into the following five phases: (1) determination of objectives, (2) identification of strategic factors, (3) determination of means (engineering proposals), (4) evaluation of engineering proposals, and (5) assistance in decision making. These elements of an engineering process are discussed in Chapter 3. These steps can also be presented within an economic framework. The *creative step* involves people with vision and initiative adopting the premise that better opportunities exist than do now. This leads to research, exploration, and investigation of potential opportunities. The *definition step* involves developing system alternatives with specific economic and physical requirements for particular inputs and outputs. The *conversion step* involves converting the attributes of system alternatives to a common measure so that systems can be compared. Future cash flows are assigned to each alternative to account for the time value of money. The *decision step* involves evaluating the qualitative and quantitative inputs and outputs to and from each system as the basis for system comparison and decision making. Decisions among system alternatives should be made on the basis of their differences in regard to accounting for uncertainties and risks.

6.2 Fundamental Economic Concepts

Economics as a field can be defined as the science that deals with the production, distribution, and consumption of wealth, and with the various related problems of labor, finance, and taxation. It is the study of how human beings allocate scarce resources to produce various commodities and how those commodities are distributed for consumption among the people in a society. The essence of economics lies in the fact that resources are scarce, or at least limited, and that not all human needs and desires can be met. Economics deals with the behavior of people; as such, economic concepts have an important qualitative nature that might not be subject to universal interpretation. The principal concern of economists is how to distribute these resources in the most efficient and equitable way. The field of economics has undergone a significant expansion, as the world economy has grown increasingly large and complex, and economists are currently employed in large numbers in private industry, government, and educational institutions. This section introduces a number of important economic concepts.

Utility is the power of a good or service to satisfy human needs. Value designates the worth that a person attaches to an object or service. It is also a measure or appraisal of utility in some medium of exchange and is not the same as cost or price. Consumer goods are the goods and services that directly satisfy human needs, for example, television sets, shoes, and houses. Producer goods are the goods and services that satisfy human needs indirectly as part of the production or construction processes, for example, factory equipment and industrial chemicals and materials.

Economy of exchange occurs when two or more people exchange utilities, where consumers evaluate utilities subjectively in regard to their mutual benefit. *Economy of organization* can be attained more economically by labor savings and efficiency in manufacturing or capital use.

A key objective in engineering applications is the satisfaction of human needs, which nearly always implies a cost. Economic analyses may be based on a number of cost classifications. The *first* (or *initial*) *cost* is the cost to get an activity started, such as property improvement, transportation, installation, and initial expenditures. *Operation and maintenance costs* are experienced continuously over the useful life of an activity. *Fixed costs* arise from making preparations for the future and include costs associated with ongoing activities throughout the operational lifetime of that concern. Fixed costs are relatively constant and can be decoupled from the system input/output. *Variable costs* are related to the level of operational activity. For example, the cost of fuel for construction equipment is a function of the number of days of use. *Incremental or marginal costs* are the additional expenses incurred from increased output in one or more system units (i.e., production increase); they are determined from the variable costs. *Sunk costs* cannot be recovered or altered by future actions and are usually not considered a part of engineering economic analysis.

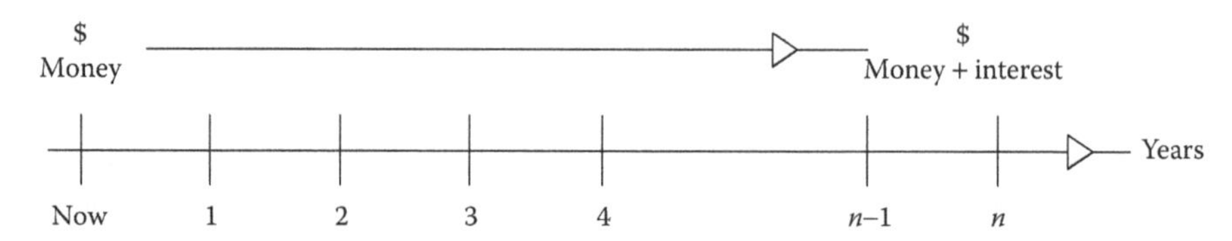

FIGURE 6.2
Time value of money.

Finally, *life cycle costs* are the costs over the entire life cycle of a product, including feasibility, design, construction, operation, and disposal costs.

Economy of exchange is also greatly affected by *supply* and *demand*, which, respectively, express the available number of units in a market for meeting some utility or need and the number of units that a market demands of such units. The supply and demand can be expressed using curves. For example, a demand curve shows the number of units that people are willing to buy and the cost per unit as a decreasing curve, whereas a supply curve shows the number of units that vendors will offer for sale and the unit price as an increasing curve. The *exchange price* is defined by the intersection of the two curves. Elasticity of demand involves price changes and their effect on demand changes. It depends on whether the consumer product is a necessity or a luxury.

The *law of diminishing returns* for a process states that the process can be improved at a rate with a diminishing return, for example, the cost of inspection to reduce the costs of repair and lost production.

Interest is a rental amount, expressed on an annual basis and charged by financial institutions for the use of money. It is also called the *rate of capital growth* or the *rate of gain* received from an investment. For the lender, it consists, for convenience, of (1) risk of loss, (2) administrative expenses, and (3) profit or pure gain. For borrowers, it is the cost of using capital for immediately meeting their needs.

The *time value of money* reflects the relationship between interest and time; that is, money has time value because the purchasing power of a dollar changes with time. Figure 6.2 illustrates the time value of money.

The *earning power of money* represents the funds borrowed for the prospect of gain. Often these funds will be exchanged for goods, services, or production tools, which in turn can be employed to generate an economic gain. The earning power of money involves prices of goods and services that can move upward or downward, where the purchasing power of money can change with time. Both price reductions and price increases can occur where reductions are caused by increases in productivity and availability of goods, and increases are caused by government policies, price support schemes, and deficit financing.

6.3 Cash Flow Diagrams

Cash flow diagrams are a means of visualizing and/or simplifying the flow of receipts and disbursements for the acquisition and operation of items in an enterprise. A cash flow diagram normally has a horizontal axis that is marked off in equal increments, one per period, up to the duration of the project. It also addresses revenues and disbursements, where revenues or receipts are represented by upward arrows and disbursements or payments are represented by downward arrows.

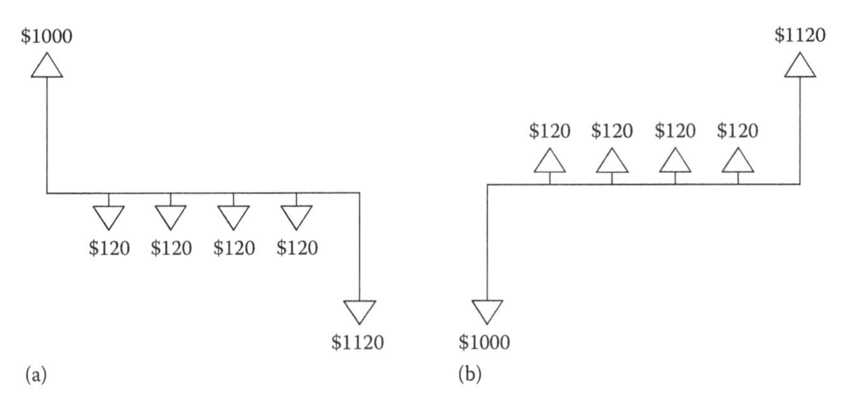

FIGURE 6.3
Typical cash flow diagram: (a) borrower point of view; (b) lender point of view.

All disbursements and receipts (i.e., cash flows) are assumed to take place at the end of the year in which they occur. This is known as the "end-of-year" convention. Arrow lengths are approximately proportional to the magnitude of the cash flow. Expenses incurred before time = 0 are sunk costs and are not relevant to the problem. Because there are two parties to every transaction, it is important to note that cash flow directions in cash flow diagrams depend on the point of view taken. A net cash flow is defined by the arithmetic sum of receipts (+) and disbursements (–) that occur at the same point in time.

Example 6.1: Cash Flow Diagrams

Figure 6.3 shows cash flow diagrams for a transaction spanning 5 years. The transaction begins with a $1000 loan. For years 2, 3, and 4, the borrower pays the lender $120 interest. At year 5, the borrower pays the lender $120 interest plus the $1000 principal. The figures show two types of cash flow arrows. A cash flow over time is represented by an upward arrow, indicating a positive flow, whereas a downward arrow indicates a negative flow. Any cash flow diagram problem will have two cash flows: one for the borrower and the other for the lender.

6.4 Interest Formulae

Interest formulae play a central role in the economic evaluation of engineering alternatives. The objective of this section is to introduce and demonstrate key interest formulae after discussing interest types.

6.4.1 Types of Interest

A payment that is due at the end of a time period in return for using a borrowed amount for this period is called *simple interest*. For fractions of a time period, the interest should be multiplied by the fraction. Simple interest (I) is calculated by the following formula:

$$I = Pni \tag{6.3}$$

where:

P is the principal in dollars or other currency
i is the interest rate expressed as a fraction per unit time
n is the number of years or time periods, that is, consistent in units with the interest rate

The *compound interest* can be computed as

$$I = P\left[(i+1)^n - 1\right] \quad (6.4)$$

Compound interest is a type of interest that results from computing interest on an interest payment due at the end of a time period. If an interest payment is due at the end of a time period that has not been paid, this interest payment is treated as an additional borrowed amount over the next time period, producing an additional interest amount called *compound interest.*

Example 6.2: Simple Interest

A contractor borrows $50,000 to finance the purchase of a truck at a simple interest rate of 8% per annum. At the end of 2 years, the interest owed would be

$$I = \$50,000 \times 0.08 \times 2 = \$8000$$

Example 6.3: Simple Interest over Multiple Years

A loan of $1000 is made at an interest rate of 12% for 5 years. The interest is due at the end of each year and the principal is due at the end of the fifth year. In this case, the principal (P) is $1000, the interest rate (i) is 0.12, and the number of years or periods (n) is 5. Table 6.1 shows the payment schedule based on using Equation 6.3. The amount at the start of each year is the same because, according to the terms of the loan, interest due is payable at the end of the year.

Example 6.4: Compound Interest

A loan of $1000 is made at an interest rate of 12% compounded annually for 5 years. The interest and the principal are due at the end of the fifth year. In this case, the principal (P) is $1000, the interest rate (i) is 0.12, and the number of years or periods (n) is 5. Table 6.2 shows the resulting payment schedule. The amount at the start of each year is not the same because, according to the terms of the loan, the interest due is added to the amount borrowed until the end of the 5 years, when the loan matures.

TABLE 6.1

Resulting Payment Schedule (Example 6.3)

Year	Amount at the Start of Year ($)	Interest at the End of Year ($)	Amount Owed at the End of Year ($)	Payment ($)
1	1000	120	1120	120
2	1000	120	1120	120
3	1000	120	1120	120
4	1000	120	1120	120
5	1000	120	1120	1120

TABLE 6.2

Resulting Payment Schedule (Example 6.4)

Year	Amount at the Start of Year ($)	Interest at the End of Year ($)	Amount Owed at the End of Year ($)	Payment ($)
1	1000.00	120.00	1120.00	0.00
2	1120.00	134.40	1254.40	0.00
3	1254.40	150.53	1404.93	0.00
4	1404.93	168.59	1573.52	0.00
5	1573.52	188.82	1762.34	1762.34

6.4.2 Discrete Compounding and Discrete Payments

Interest formulae presented in this section cover variations of computing various interest types and payment schedules for a loan. The interest formulae are provided in the form of factors. For example, Equation 6.3 includes the factor (ni), which is used as a multiplier to obtain I from P. Seven factors are presented in this section as follows: (1) single-payment, compound-amount factor; (2) single-payment, present-worth factor; (3) equal-payment-series, compound-amount factor; (4) equal-payment-series, sinking-fund factor; (5) equal-payment-series, capital-recovery factor; (6) equal-payment-series, present-worth factor; and (7) uniform-gradient-series factor. In presenting these formulae, the following notations are presented: i is the annual interest rate; n, the number of annual interest periods; P, a present principal sum; A, a single payment in a series of n equal payments made at the end of each annual interest period; and F, a future sum of n annual interest periods. Each case is illustrated with a computational example. Instead of using an annual period, other periods can be used, such as quarters, months, or days; for other periods, the interest (i) should correspond to the period (i.e., interest for a quarter, month, or day). The compounding frequency is discussed in Section 6.4.3.

6.4.2.1 Single-Payment, Compound-Amount Factor

The single-payment, compound-amount factor is used to compute a future payment (F) for an amount borrowed at the present (P) for n years at an interest of i. The future sum is calculated by applying the following formula:

$$F = P(1+i)^n \tag{6.5}$$

Example 6.5: Single-Payment, Compound-Amount Factor

A loan of \$1000 is made at an interest rate of 12% compounded annually for 4 years. The interest is due at the end of each year and the principal is due at the end of the fourth year. The principal (P) is \$1000, the interest rate (i) is 0.12, and the number of years or periods (n) is 4. Therefore,

$$F = \$1000(1 + 0.12)^4 = \$1573.50 \tag{6.6}$$

Figure 6.4 shows the cash flow for the single present amount (P = \$1000) and the single future amount (F = \$1573.50).

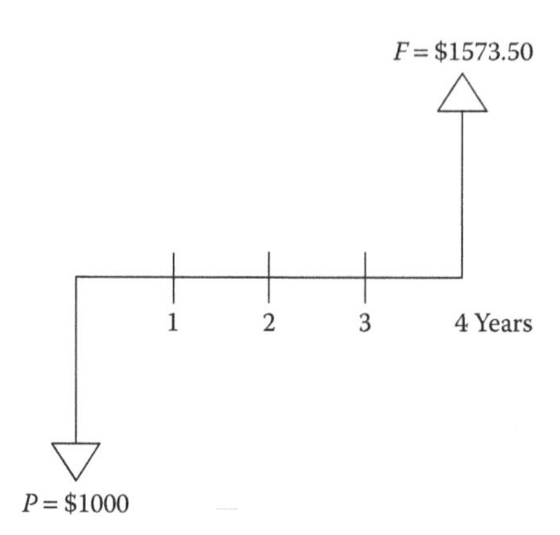

FIGURE 6.4
Cash flow for single-payment compound amount from the perspective of a lender (Example 6.5).

6.4.2.2 *Single-Payment, Present-Worth Factor*

The single-payment, present-worth factor provides the present amount (P) for a future payment (F) for n periods at an interest rate i as follows:

$$P = \frac{F}{(1+i)^n} \tag{6.7}$$

The factor $1/(1+i)^n$ is known as the single-payment, present-worth factor and may be used to find the present worth (P) of a future amount (F).

Example 6.6: Single-Payment, Present-Worth Factor for Construction Equipment

A construction company wants to set aside enough money today in an interest-bearing account in order to have $100,000 4 years from now for the purchase of a replacement piece of equipment. If the company can receive 12% interest on its investment, the single-payment, present-worth factor is calculated as follows:

$$P = \frac{\$100{,}000}{(1+0.12)^4} = \$63{,}550 \tag{6.8}$$

Example 6.7: Single-Payment, Present-Worth Factor for Software Purchase

A construction company wants to set aside enough money today in an interest-bearing account in order to have $1573.5 4 years from now for the purchase of a replacement piece of software. If the company can receive 12% interest on its investment, the single-payment, present-worth factor is

$$P = \frac{\$1573.5}{(1+0.12)^4} = \$1000 \tag{6.9}$$

Example 6.8: Single-Payment, Present-Worth Factor for Bridge Replacement

A town plans to replace an existing bridge that costs $5000 annually in operation and maintenance and has a remaining useful life of 20 years. The new bridge will cost

\$500,000 for construction and an additional \$2000 for annual operation and maintenance. The new bridge is expected to have a useful life of 50 years, thus extending the life of the bridge 30 years (i.e., extending it from the 21st year to the 50th year). If the interest rate is 8%, the single-payment, present-worth factor for 20 years is

$$\frac{1}{(1+i)^n} = \frac{1}{(1+0.08)^{20}} = 0.2145 \tag{6.10}$$

This factor can be used to bring a future expense to its present value. For example, a maintenance payment (\$2000) in the 20th year has a present value of 21.45% of \$2000, or \$429.

Example 6.9: Calculating the Interest Rate for Savings

A construction company wants to set aside \$1000 today in an interest-bearing account in order to have \$1200 4 years from now. The required interest rate must satisfy the following condition:

$$F = \$1000(1+i)^4 = \$1200 \tag{6.11}$$

Solving for i produces the following:

$$i = \sqrt[4]{\frac{\$1200}{\$1000}} - 1 = 0.046635 \tag{6.12}$$

The interest rate i needed is 0.046635, or ~4.7%.

Example 6.10: Calculating the Number of Years

A construction company wants to set aside \$1000 today at an annual interest rate of 10% in order to have \$1200. The number of years required to yield this amount can be computed based on the following condition:

$$F = 1000(1+0.1)^n = 1200 \tag{6.13}$$

Solving for n produces the following:

$$(1+0.1)^n = \frac{1200}{1000} \quad \text{or} \quad n = \frac{\ln(1.2)}{\ln(1.1)} = 1.9129285 \tag{6.14}$$

Therefore, the number of years n is ~2.

6.4.2.3 Equal-Payment-Series, Compound-Amount Factor

The equal-payment-series, compound amount factor is used in economic studies that require the computation of a single-factor value that accumulates from a series of payments occurring at the end of succeeding interest periods. Figure 6.5 represents this cash flow scenario as a graph. At the end of year 1, a payment of \$$A$ begins the accumulation of interest at rate i for $(n - 1)$ years. At the end of year 2, a payment of \$$A$ begins the accumulation of interest at rate i for $(n - 2)$ years. End-of-year payments of \$$A$ continue until year n. The total accumulation of funds at year n is simply the sum of \$$A$ payments multiplied by the appropriate single-payment, present-worth factors. The results are illustrated in Table 6.3.

The total compound amount is simply the sum of the compound amounts for years 1 through n. This summation is a geometric series as follows:

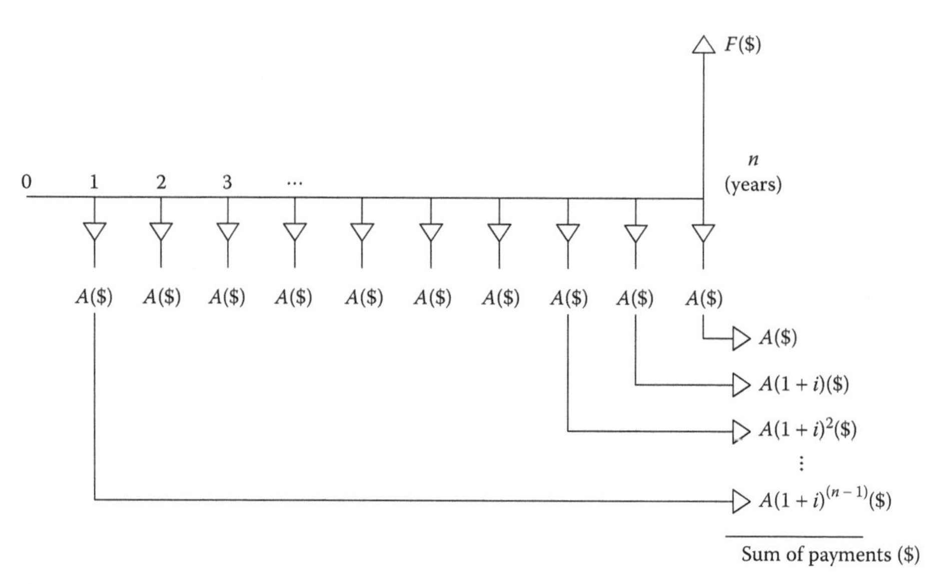

FIGURE 6.5
Equal-payment-series compound amounts.

TABLE 6.3
Total Accumulation of Funds

End of Year	Compound Amount at the End of n Years
1	$\$A(1+i)^{(n-1)}$
2	$\$A(1+i)^{(n-2)}$
3	$\$A(1+i)^{(n-3)}$
⋮	⋮
$n-1$	$\$A(1+i)$
n	$\$A$

$$F = A + A(1+i) + A(1+i)^2 + \cdots + A(1+i)^{n-1} \tag{6.15}$$

With some mathematical manipulation, it can be expressed as

$$F = A\frac{(1+i)^n - 1}{i} \tag{6.16}$$

Example 6.11: Equal-Payment-Series, Compound-Amount Factor for Total Savings

A contractor makes four equal annual deposits of \$100 each into a bank account paying 12% interest per year. The first deposit will be made 1 year from today. The money that can be withdrawn from the bank account immediately after the fourth deposit is

$$F = \$100\left[\frac{(1+0.12)^4 - 1}{0.12}\right] = \$477.9 \tag{6.17}$$

6.4.2.4 Equal-Payment-Series, Sinking-Fund Factor

For an annual interest rate *i* over *n* years, the equal end-of-year amount to accomplish a financial goal of having a future amount of *F* at the end of the *n*th year can be computed from Equation 6.16 as follows:

$$A = F\left[\frac{i}{(1+i)^n - 1}\right] \tag{6.18}$$

where:

A is the required end-of-year payments to accumulate a future amount *F*

Example 6.12: Equal-Payment-Series, Sinking-Fund Factor for Future Savings

A student is planning to have personal savings totaling $1000 4 years from now. If the annual interest rate will average 12% over the next 4 years, the equal end-of-year amount to accomplish this goal is calculated as

$$A = \$1000\left[\frac{0.12}{(1+0.12)^4 - 1}\right] = \$209.2 \tag{6.19}$$

6.4.2.5 Equal-Payment-Series, Capital-Recovery Factor

The equal-payment-series, capital-recovery factor is defined based on a deposit of amount *P* that is made now at an interest rate *i*. The depositor wishes to withdraw the principal plus the earned interest in a series of year-end equal payments over *n* years such that when the last withdrawal is made, no funds should be left in the account. Figure 6.6 summarizes the flow of disbursements and receipts from the depositor's point of view for this scenario. Equating the principal $*P* plus the accumulated interest of Equation 6.5 with the accumulation of equal payments $*A* plus their corresponding interests of Equation 6.16 gives

$$P(1+i)^n = A\frac{(1+i)^n - 1}{i} \tag{6.20}$$

which can be rearranged to give

$$A = P\left[\frac{i(1+i)^n}{(1+i)^n - 1}\right] \tag{6.21}$$

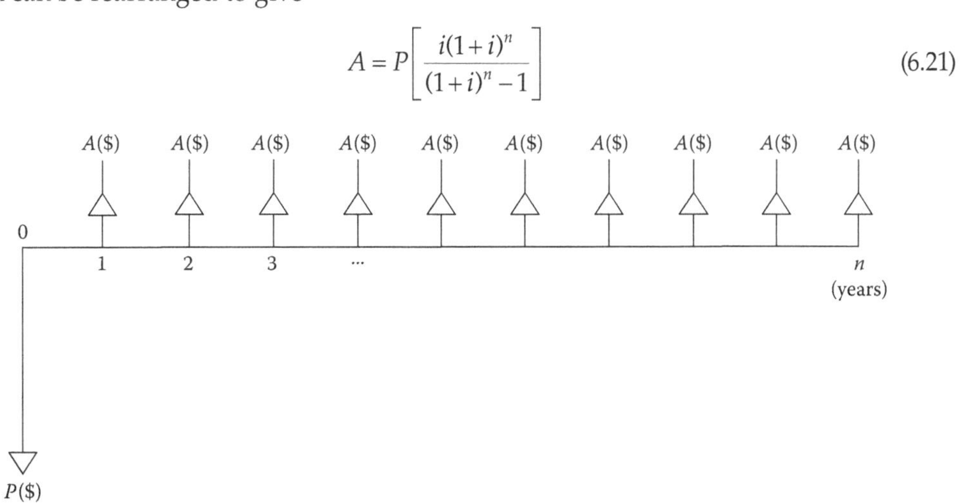

FIGURE 6.6
Equal-payment-series capital recovery.

Example 6.13: Equal-Payment-Series, Capital-Recovery Factor for a Loan

A contractor borrows $1000 and agrees to repay it in 4 years at an interest rate of 12% per year. The payment in four equal end-of-year payments is calculated by applying Equation 6.21 as follows:

$$A = \$1000\left[\frac{0.12(1+0.12)^4}{(1+0.12)^4 - 1}\right] = \$329.2 \tag{6.22}$$

6.4.2.6 *Equal-Payment-Series, Present-Worth Factor*

The present worth P of an equal-payment series A over n periods at an interest rate i is

$$P = A\left[\frac{(1+i)^n - 1}{i(1+i)^n}\right] \tag{6.23}$$

Example 6.14: Equal-Payment-Series, Present-Worth Factor for Investing in a Machine

If a certain machine undergoes a major overhaul now, its output can be increased by 5%, which translates into an additional cash flow of $100 at the end of each year for 4 years. If the annual interest rate is 12%, the amount that could be invested in order to overhaul this machine is calculated by applying Equation 6.23 as follows:

$$P = \$100\left[\frac{(1+0.12)^4 - 1}{0.12(1+0.12)^4}\right] = \$303.7 \tag{6.24}$$

Example 6.15: Present Worth of Annuity Factor for Bridge Replacement

In Example 6.8, a town was planning to replace an existing bridge that costs $5000 annually in operation and maintenance and has a remaining useful life of 20 years. The new bridge will cost $500,000 for construction and an additional $2,000 for annual operation and maintenance. The new bridge will have a useful life of 50 years, thus extending the life of the bridge by 30 years. If the interest rate is 8%, the present worth of annuity factor for 20 years, according to Equation 6.23, is

$$\frac{(1+i)^n - 1}{i(1+i)^n} = \frac{(1+0.08)^{20} - 1}{0.08(1+0.08)^{20}} = 9.818 \tag{6.25}$$

and for 30 years is

$$\frac{(1+i)^n - 1}{i(1+i)^n} = \frac{(1+0.08)^{30} - 1}{0.08(1+0.08)^{30}} = 11.258 \tag{6.26}$$

Example 6.16: Capital-Recovery Factor for Bridge Replacement

Examples 6.8 and 6.15 presented the case of a town replacing an existing bridge. For an interest rate of 8%, the capital recovery factor (to compute equal payments) for 50 years of the cost of the new bridge ($500,000) according to Equation 6.21 is

$$\frac{i(1+i)^n}{(1+i)^n - 1} = \frac{0.08(1+0.08)^{50}}{(1+0.08)^{50} - 1} = 0.08174 \tag{6.27}$$

The annual cost of the new bridge can be taken as the total cost of the bridge multiplied by the capital recovery factor, producing the following amount:

$$\text{Annual cost of new bridge} = \$500{,}000(0.08174) = \$40{,}900 \tag{6.28}$$

6.4.2.7 *Uniform-Gradient-Series Factor*

Often periodic payments do not occur in equal amounts and may increase or decrease by constant amounts (e.g., \$100, \$120, \$140, \$160, \$180, and \$200). The uniform-gradient-series factor (G) is a value of $(n - 1)G$ at the end of year n, that is, 0 at the end of year 1, G at the end of year 2, $2G$ at the end of year 3, and so on. An equivalent equal payment A can be computed as follows:

$$A = G\left[\frac{1}{i} - \frac{n}{(1+i)^n - 1}\right] \tag{6.29}$$

Example 6.17: Uniform-Gradient-Series Factor for Payments

If the uniform-gradient amount is \$100 and the interest rate is 12%, the uniform annual equivalent value at the end of the fourth year is calculated by applying Equation 6.29 as follows:

$$A = \$100\left[\frac{1}{0.12} - \frac{4}{(1 + 0.12)^4 - 1}\right] = \$135.9 \tag{6.30}$$

Example 6.18: Computation of Bridge Replacement Benefits

Examples 6.8, 6.15, and 6.16 presented the case of a town replacing an existing bridge. The existing bridge has annual operation and maintenance costs of \$5000 and has a remaining useful life of 20 years. The new bridge will cost \$500,000 for construction and an additional \$2,000 for annual operation and maintenance. The new bridge will have a useful life of 50 years, thus extending the life of the bridge by 30 years. The applicable interest rate is 8%.

This example demonstrates the computation of the annual benefit gained from replacing the bridge. The benefits of the new bridge include the additional function availability for an additional 30 years and the reduction in operation and maintenance costs by \$2000 per year over the next 20 years. This example does not analyze the costs of replacing the bridge; rather, the focus is only on the benefits. The values calculated in this example were rounded to the nearest \$100.

The benefits credited to the bridge life extension can be assessed based on the annual amount the town is willing to pay for having this functionality available in the future. A willingness-to-pay approach is used here instead of direct benefit assessment, where benefits could be assessed based on the reduced travel time, convenience, increased safety, and so on. The willingness-to-pay approach equates the annual benefits to the annual payments the town would make in these future years as a result of replacing the bridge. The benefits in the 20th year credited to the extended life of the bridge are equal to the annual costs of the new bridge, as calculated in Equation 6.28, multiplied by the present worth of annuity factor for 30 years, as calculated in Equation 6.26. Therefore, the benefit is

$$\text{Benefits in the 20th year} = \$40{,}900(11.258) = \$460{,}500 \tag{6.31}$$

The present worth for the first year of the extended bridge life is equal to the benefits in the 20th year, as calculated in Equation 6.31, multiplied by the single-payment, present-worth factor for 20 years, as calculated in Equation 6.10. Therefore, the present worth is

$$\text{Present worth in the first year} = \$460{,}500(0.2145) = \$98{,}800 \tag{6.32}$$

The annual savings in operation and maintenance costs between the 1st and 20th years are equal to the difference in the operation and maintenance costs of the existing bridge and the new bridge. Therefore, the annual savings are

$$\text{Annual saving in operation and maintenance costs} = \$5000 - \$2000 = \$3000 \tag{6.33}$$

The present worth for the first year of operation and maintenance savings is equal to the annual savings in operation and maintenance costs between the 1st and 20th years, as calculated in Equation 6.33, multiplied by the present worth of annuity factor for 20 years, as calculated in Equation 6.25. Therefore, its present worth is

$$\text{Present worth in the first year} = \$3000(9.818) = \$29{,}500 \tag{6.34}$$

The present worth of the total credit is the sum of the present worth in the first year of bridge extension, as calculated in Equation 6.32, and the present worth in the first year of operation and maintenance savings, as calculated in Equation 6.34. Therefore, the present worth of total credit is given by

$$\text{Present worth of total credit} = \$98{,}800 + \$29{,}500 = \$128{,}300 \tag{6.35}$$

Finally, the average annual credit or benefit spread over 50 years is equal to the present worth of the total credit, as calculated in Equation 6.35, multiplied by the capital recovery factor, as calculated in Equation 6.27. Therefore, the average annual credit, or benefit, is

$$\text{Average annual credit or benefit} = \$128{,}300(0.08174) = \$10{,}500 \tag{6.36}$$

6.4.3 Compounding Frequency and Continuous Compounding

6.4.3.1 Compounding Frequency

The *effective interest rate* is defined as an interest rate that is compounded using a time period less than a year. The *nominal interest rate* is defined as the effective rate times the number of compounding periods in a year. The nominal interest rate is expressed on an annual basis, and financial institutions refer to this rate as the annual percentage rate (APR), also referred to as the nominal rate compounded at a period less than a year. For example, if the effective rate is 1% per month, it follows that the nominal rate is 12% compounded monthly.

The effective interest rate (i) for any time interval (l), which can be different from the compounding period, is given by

$$i = \left(1 + \frac{r}{m}\right)^{l(m)} - 1 \tag{6.37a}$$

where:

i is the effective interest rate in the time interval
r is the nominal interest rate per year
l is the length of the time interval (in years)
m is the reciprocal of the length of the compounding period (in years)

TABLE 6.4

Example Illustrating the Concept of Continuous Compounding

Compounding Frequency	Number of Periods	Effective Interest Rate per Period (%)	Effective Annual Interest Rate (%)
Annually	1.0	18	18
Semiannually	2.0	9	18.81
Quarterly	4.0	4.5	19.25186
Monthly	12.0	1.5	19.56182
Weekly	52.0	0.3462	19.68453
Daily	365.0	0.0493	19.71642
Continuously	∞	0	19.72174

Clearly, if $l(m) = 1$, then $i = r/m$. The product $l(m)$ is called c, which corresponds to the number of compounding periods in the time interval l. It should be noted that c should be ≥1. For the special case of $l = 1$, the effective interest rate (i) for a year is given by

$$i = \left(1 + \frac{r}{m}\right)^m - 1 \tag{6.37b}$$

6.4.3.2 Continuous Compounding

The limiting case for the effective rate is when compounding is performed infinite times in a year. Using $l = 1$, the following limit produces the continuously compounded interest rate (i_a):

$$i_a = \lim_{m \to \infty} \left(1 + \frac{r}{m}\right)^m - 1 \tag{6.38a}$$

This limit produces the following effective interest rate:

$$i_a = \mathrm{e}^r - 1 \tag{6.38b}$$

The concept of continuous compounding is illustrated in Table 6.4.

The presentation of continuous compounding is limited to the case of Equation 6.38a and 6.38b. Extensions of these concepts, such as interest formulae for continuous compounding and discrete payments and interest formulae for continuous compounding and continuous payments, are beyond the scope of this chapter.

6.4.4 Summary of Interest Formulae

The following table provides a summary of the interest formulae presented in the previous sections.

To Find	Given	Multiply by	Notation	Factor Name
For single cash flows				
F	P	$(1+i)^n$	$(F/P, i, n)$	Single-payment, compound amount
P	F	$\frac{1}{(1+i)^n}$	$(P/F, i, n)$	Single-payment, present worth

(*Continued*)

To Find	Given	Multiply by	Notation	Factor Name
For uniform series (annuities)				
F	A	$\frac{(1+i)^n - 1}{i}$	$(F/A, i, n)$	Equal-payment-series, compound amount
A	F	$\frac{i}{(1+i)^n - 1}$	$(A/F, i, n)$	Equal-payment-series, sinking fund
A	P	$\frac{i(1+i)^n}{(1+i)^n - 1}$	$(A/P, i, n)$	Capital recovery
P	A	$\frac{(1+i)^n - 1}{i(1+i)^n}$	$(P/A, i, n)$	Equal-payment-series, present worth
A	G	$\frac{1}{i} - \frac{n}{(1+i)^n - 1}$	$(A/G, i, n)$	Uniform-gradient series

6.4.5 Choosing a Discount Rate

Choosing an appropriate discount rate in risk studies should be based on the situation under consideration. A discount rate to manage the risks associated with a highway system might be different than that to a discount rate to manage the risks associated with growing the energy generation capacity with climate change considerations. In both example cases, benefit-cost analysis can be used and two different rates can be justified.

The discount rate is a fundamental assumption for estimating the value, for example, net present value, of developing and constructing highway systems, power plants, schools, environmental protections, and so on. Decision or policy makers must quantify the social marginal cost and the social marginal benefit for each project and compare these projects in order to allocate limited resources. The discount rate appears in both sides of a benefit-cost analysis, that is, future costs such as maintenance and future benefits such as reduced pollution emissions. Generally, calculating the marginal cost is easier than measuring the marginal benefit. Also, the uncertainty is smaller in the former than in the latter. The examination of the effects including benefits require valuating time of people affected, human health and safety, ecological impacts, and so on that have differing time periods associated with the respective effects. A primary issue arises also in decisions spanning multigenerations creating many situations of mismatch between generations bearing the costs from generations reaping the benefits.

In risk studies that do not have significant social or society-wide impacts, economic efficiency dictates the use of a discount rate representing the opportunity cost of what else an entity, for example, a decision maker, could accomplish with those same funds used to cover the costs of an alternative selected. For example, if the funds could be instead used to invest in the private sector yielding 3% as the next best alternative for using the funds, then 3% would be the discount rate.

In the case of social project funding, justifiably choosing the discount rates requires making ethically subtle choices about the benefits to others. For example, nowadays consumptions could most likely impact future generations due to global change in temperature. In this case, choosing a discount rate for the costs and benefits of reducing CO_2 emissions and other harmful greenhouse gases is very important and could drive alternative considered and decision made. The discount rate for benefit-cost analysis ranges from 1.4 to about 3% based on various debates. The small discount rate is from the Stern Review on the Economics of Climate Change (http://www.webcitation.org/5nCeyEYJr). The US Office of Management and Budget (OMB) provides guidance on this matter and uses a pretax discount rate of 7% as an example in its Circular No. A-94 for benefit–cost analysis of federal programs (http://www.whitehouse.gov/omb/circulars_a094).

6.5 Economic Equivalence Involving Interest

6.5.1 Meaning of Equivalence

Economic equivalence is commonly used in engineering to compare alternatives. In engineering economy, two things are said to be equivalent if they have the same effect. Unlike most individuals involved with personal finances, corporate and government decision makers using engineering economics might not be so much concerned with the timing of a project's cash flows as with the profitability of the project. Therefore, analytical tools are needed to compare projects involving receipts and disbursements occurring at different times, with the goal of identifying an alternative having the largest eventual profitability.

6.5.2 Equivalence Calculations

Several equivalence calculations are presented in this section, for which the calculations involve the following: (1) cash flows, (2) interest rates, (3) bond prices, and (4) loans. Two cash flows have to be presented for the same time period using a similar format to facilitate comparison. When interest is earned, monetary amounts can be directly added only if they occur at the same point in time. Equivalent cash flows are those that have the same value. For loans, the effective interest rate for the loan, also called the *internal rate of return* (IRR), is defined as the rate that sets the receipts equal to the disbursements on an equivalent basis. The equivalence of two cash flows can be assessed at any point in time, as illustrated in Example 6.19.

Example 6.19: Equivalence between Cash Flows

Two equivalent cash flows are presented in Table 6.5. The equivalence can be established at any point in time for an interest rate of 12% compounded annually. For example, if 8 years was selected, $F = \$1000(1 + 0.12)^8 = \$2475.96$ for cash flow 1, whereas $F = \$1000(1 + 0.12)^4 = \$1573.50$ for cash flow 2. It should be noted that two or more distinct cash flows are equivalent if they result in the same amount at the same point in time. In this case, the two cash flows are not equivalent.

Example 6.20: Internal Rate of Return

According to the equivalence principle, the actual interest rate earned on an investment can be defined as the interest rate that sets the equivalent receipts to the equivalent disbursements. This interest rate is called the IRR. In Table 6.6, the following equality can be set as:

$$\$1000 + \$500(P/F,i,1) + \$250(P/F,i,5) = \$482(P/A,i,3)(P/F,i,1) + \$482(P/A,i,2)(P/F,i,5) \tag{6.39}$$

TABLE 6.5

Two Equivalent Cash Flows

Year	Cash Flow 1 ($)	Cash Flow 2 ($)
1	1000.00	0.00
2	0.00	0.00
3	0.00	0.00
4	0.00	1000.00
5	0.00	0.00
6	0.00	0.00
7	0.00	0.00
8	2475.96	1573.50

TABLE 6.6

Converting Cash Flow to Its Present Value

Time (Year End)	Receipts ($)	Disbursements ($)
0	0	−1000
1	0	−500
2	482	0
3	482	0
4	482	0
5	0	−250
6	482	0
7	482	0

By trial and error, $i = 10\%$ makes the above equation valid. The equivalence can be made at any point of reference in time; it does not need to be the origin (time = 0) to produce the same answer.

If the receipts and disbursement of an investment cash flow are equivalent for some interest rate, the cash flows of any two portions of the investment have equal absolute equivalent values at that interest rate; that is, the negative (−) of the equivalent amount of one cash flow portion is equal to the equivalent of the remaining portion on the investment. For example, breaking up the above cash flow (Table 6.6) between years 4 and 5 and performing the equivalence at the fourth year produce the following:

$$-\$1000(F/P,10,4) - \$500(F/P,10,3) + \$482(F/A,10,3) = -(-\$250(P/F,10,1) + \$482(P/A,10,2)(P/F,10,1))$$

or

$$\begin{aligned} -\$1000(1.464) - \$500(1.331) + \$482(3.310) &= -\left[-\$250(0.9091) + \$482(1.7355)(0.9091)\right] \\ -\$534 &= -\$534 \end{aligned} \tag{6.40}$$

Example 6.21: Bond Prices

A bond is bought for $900 and has a face value of $1000 with 6% annual interest that is paid semiannually. The bond matures in 7 years. The yield to maturity is defined as the rate of return on the investment unit its maturity date. Using equivalence, the following equality can be developed:

$$\$900 = \$30(P / A,i,14) + \$1000(P / F,i,14) \tag{6.41}$$

By trial and error, $i = 3.94\%$ per semiannual period. The nominal rate is 2(3.94) = 7.88%, whereas the effective rate is 8.04%.

Example 6.22: Equivalence Calculations for Loans

Suppose a 5-year loan of $10,000 (with interest of 16% compounded quarterly with quarterly payments) is to be paid off after the 13th payment. The quarterly payment is:

$$\$10,000(A / P,4,20) = \$10,000(0.0736) = \$736 \tag{6.42}$$

The balance can be based on the remaining payments as follows:

$$\$736(P / A,4,7) = \$736(6.0021) = \$4418 \tag{6.43}$$

6.5.3 Amortization Schedule for Loans

For calculations involving principal and interest payments, the case of a loan with fixed rate (i) and constant payment A is considered. An amortization schedule for a loan is defined as a breakdown of each loan payment (A) into two portions: an interest payment (I_t) and a payment toward the principal balance (B_t). The following terms are defined: I_t is the interest payment of A at time t and B_t is the portion of payment of A to reduce the balance at time t. The payment can be expressed as follows:

$$A = I_t + B_t, \text{ for } t = 1, 2,\ldots,n \tag{6.44}$$

The balance (B_t) at the end of $t - 1$ is given by

$$B_t = A[P/A,i,n - (t - 1)] \tag{6.45}$$

Therefore, the following relationships can be obtained:

$$I_t = A[P/A,i,n - (t - 1)](i) \tag{6.46}$$

and

$$B_t = A - I_t = A\{1 - [P/A,i,n - (t - 1)](i)\} \tag{6.47}$$

from the following conditions:

$$(P/F,i,n) = 1 - (P/A,i,n)(i) \tag{6.48}$$

and

$$B_t = A(P/F,i,n - t + 1) \tag{6.49}$$

TABLE 6.7

Amortization Calculations for Example 6.23

Year End	Loan Payment ($)	Payment toward Principal (B_t)	Interest Payment (I_t) ($)
1	350.265	$350.265(P/F, 15, 4) = $200.27	150.00
2	350.265	$350.265(P/F, 15, 3) = $230.30	119.97
3	350.265	$350.265(P/F, 15, 2) = $264.85	85.42
4	350.265	$350.265(P/F, 15, 1) = $304.58	45.69
Total	1401.06	$1000.00	401.06

Example 6.23: Principal and Interest Payments

Suppose a 4-year loan of $1000 (at 15% interest compounded annually with annual payments) is to be paid off. The payment is $A = \$1000(A/P, 15, 4) = \$1000(0.3503) = \$350.3$. The results are illustrated in Table 6.7 based on Equation 6.49 and using $I_t = A - B_t$.

6.6 Economic Equivalence and Inflation

6.6.1 Price Indexes

For the purposes of calculating the effect of inflation on equivalence, price indexes are used. A *price index* is defined as the ratio between the current price of a commodity or service and the price at some earlier reference time.

Example 6.24: Economic Equivalence and Inflation

The base year, with an index of 100, is 1967 and the commodity price is $1.46/lb. If the price in 1993 is $5.74/lb, the 1993 index is ($5.74/1.46) × 100 = $3.9315. The actual consumer price index (CPI) and the annual inflation rates are published and can be used for these computations.

6.6.2 Annual Inflation Rate

The annual inflation rate at $t + 1$ can be computed as

$$\text{Annual inflation rate at } t+1 = \frac{\text{CPI}_{t+1} - \text{CPI}_t}{\text{CPI}_t} \tag{6.50}$$

The average inflation rate ($\bar{f}$) can be computed based on the following condition:

$$\text{CPI}_t(1+\bar{f})^n = \text{CPI}_{t+n} \tag{6.51}$$

Therefore, the average inflation rate is

$$\bar{f} = \sqrt[n]{\frac{\text{CPI}_{t+n}}{\text{CPI}_t}} - 1 \tag{6.52}$$

Example 6.25: Annual Inflation Rate

If the CPI for 1966 = 97.2 and the CPI for 1980 = 246.80, the average rate of inflation over the 14-year interval can be obtained by applying Equation 6.52 as follows:

$$\bar{f} = \sqrt[14]{\frac{246.80}{97.2}} - 1 = \left(\frac{246.80}{97.2}\right)^{1/14} - 1 = 6.882\% \tag{6.53}$$

6.6.3 Purchasing Power of Money

The purchasing power at time t in reference to time period $t - n$ is defined as follows:

$$\text{Purchasing power at time } t = \frac{\text{CPI}_{t-n}}{\text{CPI}_t} \tag{6.54}$$

Denoting the annual rate of loss in purchasing power as k, the average rate of loss of purchasing power $(\bar{k})$ can be computed as follows:

$$\frac{\text{CPI}_{\text{base year}}}{\text{CPI}_t}(1-\bar{k})^n = \frac{\text{CPI}_{\text{base year}}}{\text{CPI}_{t+n}} \tag{6.55}$$

Solving for CPI_t produces the following:

$$\text{CPI}_t = (1-\bar{k})^n \text{CPI}_{t+n} \tag{6.56}$$

Therefore,

$$(1+\bar{f})^n = \frac{1}{(1-\bar{k})^n} \tag{6.57}$$

Equation 6.57 relates the average inflation rate $(\bar{f})$ and the annual rate of loss in purchasing power $(\bar{k})$.

6.6.4 Constant Dollars

The constant dollar is defined as follows:

$$\text{Constant dollars} = \frac{1}{(1+\bar{f})^n}(\text{actual dollars}) \tag{6.58}$$

When using actual dollars, the market interest rate (i) is used. When using constant dollars, the inflation-free interest rate (i^*) is used, which is defined as follows for 1 year:

$$i^* = \frac{1+i}{1+f} - 1 \tag{6.59}$$

For multiple years, it is defined as follows:

$$i^* = \frac{(1+i)^n}{(1+\bar{f})^n} - 1 \tag{6.60}$$

6.7 Economic Analysis of Alternatives

6.7.1 Present-, Annual-, and Future-Worth Amounts

The present-worth amount is the difference between the equivalent receipts and the disbursements at present. If F_t is a net cash flow at time t, the present worth (PW) as a function of i is as follows:

$$\text{PW}(i) = \sum_{t=0}^{n} F_t(P/F, i, t) = \sum_{t=0}^{n} F_t(1+i)^{-t} \tag{6.61}$$

The net cash flow (F_t) is defined as the sum of all disbursements and receipts at time t. The annual equivalent amount is the annual equivalent receipts minus the annual equivalent disbursements of a cash flow. It is used for repeated cash flows per year and is calculated by applying the following equation:

$$\text{AE}(i) = \text{PW}(i)(A/P, i, n) = \left[\sum_{t=0}^{n} F_t(1+i)^{-t}\right]\left[\frac{i(1+i)^n}{(1+i)^n - 1}\right] \tag{6.62}$$

The future-worth amount is the difference between the equivalent receipts and the disbursements at some common point in the future:

$$\text{FW}(i) = \sum_{t=0}^{n} F_t(F/P, i, n-t) = \sum_{t=0}^{n} F_t(1+i)^{n-t} \tag{6.63}$$

The amounts PW, AE, and FW differ in the point of time used to compare the equivalent amounts.

Example 6.26: Annual Equivalent Amount

The cash flow illustrated in Table 6.8 is used to compute the annual equivalent amount based on an interest rate of 10% for a segment of the cash flow that repeats as follows:

$$\text{AE}(10) = [-\$1000 + \$400(P/F,10,1) + \$900(P/F,10,2)](A/P,10,2) \tag{6.64}$$

or

$$\text{AE}(10) = [-\$1000 + \$400(0.9091) + \$900(0.8265)](0.5762) = \$61.93 \tag{6.65}$$

TABLE 6.8

Cash Flow for Example 6.26

Year End	Receipts ($)	Disbursements ($)
0	0	−1000
1	400	0
2	900	−1000
3	400	0
4	900	−1000
⋮	⋮	⋮
$n-2$	900	−1000
$n-1$	400	0
N	900	0

6.7.2 Internal Rate of Return

The IRR is the interest rate that causes the equivalent receipts of a cash flow to be equal to the equivalent disbursements of the cash flow. We can solve for i such that the following condition is satisfied:

$$0 = \text{PW}(i) = \sum_{t=0}^{n} F_t (1+i)^{-t} \quad (6.66)$$

which represents the rate of return on the unrecovered balance of an investment (or loan). The following equation can be developed for loans:

$$U_t = U_{t-1}(1+i) + F_t \quad (6.67)$$

where:
U_0 is the initial amount of loan or first cost of an asset (F_0)
F_t is the amount received at the end of the period t
i is the IRR

For example, the following expressions can be provided:

$$U_1 = U_0(1 + i) + F_1$$

$$U_2 = U_1(1 + i) + F_2$$

$$\vdots$$

etc.

The basic equation for i requires the solution of the roots of a nonlinear (polynomial) function; therefore, more than one root might exist. The following three conditions can be used to obtain one root (i.e., single i) as needed: (1) $F_0 = 0$ (the first nonzero cash flow is a disbursement), (2) one change in sign in the cash flow (from disbursements to receipts), and (3) $\text{PW}(0) > 0$ (the sum of all receipts is greater than the sum of all disbursements). In case of multiple IRRs, other methods should be used for economic analyses that are beyond the scope of this chapter.

Example 6.27: Internal Rate of Return

The cash flow illustrated in Table 6.9 is used to solve for i by trial and error using the net cash flow and Equation 6.66. The IRR was determined to be $i^* = 12.8\%$.

6.7.3 Payback Period

The payback period *without* interest is the length of time required to recover the first cost of an investment from the cash flow produced by the investment for an interest rate of 0. It can be computed as the smallest n that produces the following:

$$\sum_{t=0}^{n} F_t \geq 0 \quad (6.68)$$

TABLE 6.9

Cash Flow for Example 6.27

Year End	Receipts ($)	Disbursements ($)
0	0	−1000
1	0	−800
2	500	0
3	500	0
4	500	0
5	1200	0

The payback period *with* interest is the length of time required to recover the first cost of an investment from the cash flow produced by the investment for a given interest rate i. It can be computed as the smallest n that produces the following:

$$\sum_{t=0}^{n} F_t(1+i)^{-t} \geq 0 \tag{6.69}$$

Example 6.28: Payback Period

According to Table 6.9, the payback period for only the $1000 disbursement without interest is 3 years. The payback period for only the $1800 disbursement without interest is 5 years.

6.8 Exercise Problems

Problem 6.1 Define physical efficiency and economic efficiency. Describe the differences between them using your own examples for each.

Problem 6.2 What is engineering economics as a field of study? What is the role of uncertainty in engineering economics?

Problem 6.3 What are the types of costs associated with economic analyses? Classify them with simple examples using engineering applications.

Problem 6.4 What is meant by the time value of money? What is the meaning and use of cash flow diagrams?

Problem 6.5 A person purchased a car at year 2000 (consider it year 0) for $5000. The maintenance costs are $300 per year. The car is sold at the end of the fourth year for $2000. Draw the cash flow diagram for this car from the perspective of the purchaser.

Problem 6.6 In January 1996, a company purchased a used computer system for $10,000. No repair costs were incurred in 1997 and 1998; however, subsequent repair costs were incurred as follows: $1700 in 1999, $2600 in 2000, and $2800 in 2001. The computer was sold in 2001 for $1000. Draw the cash flow diagram for this machine from the perspective of the purchaser.

Problem 6.7 If the amount to be deposited in a bank is $10,000, and the bank is offering 3% per year simple interest, compute the interest at the end of the first year payable by the bank.

Problem 6.8 A contractor borrows $15,000 from a bank. If a simple interest loan for 4 months yields $975 interest, what is the annual interest rate that the bank offers?

Problem 6.9 A construction company borrows a sum of $100,000 at a simple interest rate of 10% for 4 years. If the contract conditions state that the interest is due at the end of each year and the principal is due at the end of the fourth year, prepare a schedule of payments for this 4-year loan.

Problem 6.10 An investor borrows $100,000 from a bank for a 5-year period at a yearly interest rate of 14%. The investor signs a contract to make a simple interest payment each year and to repay the loan after 5 years. Prepare a schedule of payments for the investor for this 5-year loan period.

Problem 6.11 For Problem 6.9, if the interest is compounded and the conditions of the loan state that the interest due each year is added to the amount borrowed until the end of the 4 years, provide a revised schedule of payments to accommodate the new changes in the loan terms.

Problem 6.12 For Problem 6.10, if the interest is compounded and the conditions of the loan state that the interest due each year is added to the amount borrowed until the end of the five years, provide a revised schedule of payments to accommodate the changes in the loan terms.

Problem 6.13 A company wants to know the value of the future sum of money if they deposit the principal amount of $50,000 for 3 years in a bank at a yearly interest rate of 10%.

Problem 6.14 An investor deposits $200,000 in a national bank; if the bank pays 8% interest, how much will the investor have in his account at the end of 10 years?

Problem 6.15 To raise money for a new business, an investor asks a financial institution to loan him some money. He offers to pay the institution $3000 at the end of 4 years. How much should the institution give him if it wants a return of 12% interest per year on the investor's money?

Problem 6.16 How much should a contractor invest in a fund that will pay 9% compound interest if he wishes to have $600,000 in the fund at the end of 10 years?

Problem 6.17 An engineering company would like to have $20,012 after 12 years based on $10,000 deposit. How much interest should the company seek to achieve this sum?

Problem 6.18 In Problem 6.15, the investor finds that he cannot pay more than $2000 at the end of a certain period. Assuming that the same 12% interest is paid on his money, compute the period necessary to satisfy this change in his payment.

Problem 6.19 If a student deposits $500 at the end of each year in a savings account that pays 6% interest per year, how much will be in the account at the end of 5 years?

Problem 6.20 A construction company is considering making a uniform annual investment in a fund with a view toward providing capital at the end of 7 years to replace an excavator. An interest rate of 6% is available; what is the annual investment required to produce $50,000 at the end of the period?

Problem 6.21 A contractor is considering purchasing a used tractor for $6200, with $1240 due as down payment and the balance paid in 48 equal monthly payments at an interest rate of 1% per month. The payments are due at the end of each month. Compute the monthly payments.

Problem 6.22 A student wants to deposit an amount of money in a bank so that he/she can make five equal annual withdrawals of $1000, the first of which will be made 1 year after the deposit. If the fund pays 9% interest, what amount must he/she deposit?

Problem 6.23 The plant manager of a construction company estimates that the maintenance cost of a bulldozer will be $2000 at the end of the first year of its service, $2500 at the end of the second year, and $3000, $3500, and $4000 at the end of the third, fourth, and fifth years, respectively. Knowing that the interest is set at 5%, find the equivalent uniform-series cost each year over a period of 5 years.

Problem 6.24 An investor calculated his end-of-year cash flows to be $1000 for the second year, $2000 for the third year, and $3000 for the fourth year. If the interest rate is 15% per year, find the uniform annual worth at the end of each of the first 4 years. Notice that there is no cash flow at the end of year 1.

Problem 6.25 An engineer is considering two building design alternatives A and B that produce the following cash flows:

Cash Flow	Design A	Design B
Investment	$10,000	$20,000
Annual maintenance costs	$1,000 per year	$400 per year
Salvage value at the end of useful life	$1,200	$2,000
Useful life (years)	5	15

For an interest rate of 8%, which alternative would you select?

Problem 6.26 A company wants to buy a new machine for its new development. Two possible machines have been identified. The following table shows the cash flow for both machines:

Cash Flow	Machine X	Machine Y
Investment	$10,000	$20,000
Annual maintenance costs	$500 per year	$100 in the second year with an increase of $100 per year in subsequent years
Salvage value at the end of useful life	0	$5,000
Useful life (years)	4	12

At an interest rate of 8%, which machine would you select?

Problem 6.27 An investor bought a bond for $100. It has a face value of $95 with 5% annual interest that is paid every 6 months. The bond matures after 25 years.

a. What is the rate of return on this investment?

b. What is the effective rate of return on this investment?

Problem 6.28 A company that invests in bonds bought a bond for $85,000 and incurred costs of $5,000. The bond has a face value of $100,000, with 5% annual interest paid every 6 months. The bond matures after 25 years.

a. What is the rate of return on this investment?

b. What is the effective rate of return on this investment?

Problem 6.29 Consider a 5-year loan given to an investor in the amount of $2000, with an interest rate of 16% compounded quarterly with quarterly payments. What is the schedule of payments for the principal sum and the interest? Prepare a payment schedule for your calculations.

Problem 6.30 Consider a 6-year loan given to an investor in the amount of $4000, with interest of 20% compounded semiannually with semiannual payments. What is the schedule of payments for the principal sum and the interest? Prepare a payment schedule for your calculations.

Problem 6.31 If an index representing the price of cement increases from 231 to 287 over a period of 3 years, compute the average rate of inflation.

Problem 6.32 If an index representing the price of a commodity increases from 46.2 in the year 1998 to 57.4 in the year 2001, compute the average rate of inflation.

Problem 6.33 Two alternatives are considered for implementing an office automation plan in an engineering design firm. The following cash flow table is produced:

Cash Flow	Alternative A	Alternative B
Investment first cost ($)	180,000	460,000
Net annual receipts less expenses ($)	35,000	84,000
Useful life (years)	10	10
Interest rate (%)	10	10

Which alternative should be selected using the annual equivalent amount method?

Problem 6.34 Three alternatives are considered for execution by a construction firm. The following cash flow table is produced:

Cash Flow	Alternative A	Alternative B	Alternative C
Investment first cost in $	390,000	920,000	660,000
Net annual receipts less expenses ($)	69,000	167,000	133,500
Useful life (years)	10	10	10
Interest rate (%)	10	10	10

Which alternative should be selected using the annual equivalent amount method?

Problem 6.35 A small contractor calculated the company's cash flow for a project and found it to be as follows:

Year	Receipts ($)	Disbursements ($)
0	0	−2000
1	+800	0
2	+800	0
3	+800	0

Find the interest rate value that makes the receipts and disbursements equivalent.

Problem 6.36 A small business venture calculated the company's cash flow for a project and found it to be as follows:

Year	Receipts ($)	Disbursements ($)
0	0	−600
1	+500	−250
2	+200	0
3	+150	0
4	+100	0
5	+50	0

Find the interest rate value that makes the receipts and disbursements equivalent.

Problem 6.37 Which of the following two alternatives has the shortest payback period?

Cash Flow	Alternative A	Alternative B
First cost ($)	20,000	10,000
Annual maintenance costs	$2,000 in year 1, increasing by $500 per year	$500 in year 1, increasing by $200 per year
Salvage value at the end of useful life ($)	2,000	4,000
Benefits	$8,000 per year	$3,000 per year
Useful life (years)	10	10

Problem 6.38 Determine the payback period to the nearest year for the following project:

Cash Flow	Values
First cost	$22,000
Annual maintenance costs	$1,000 per year
Overhaul costs	$7,000 every 4 years
Salvage value at the end of useful life	$2,500
Uniform benefits	$6,000 per year
Useful life (years)	12

7

Risk Treatment and Control Methods

Controlling risks effectively requires the use of decision analysis in an economic framework within political and regulatory constraints. Risk treatments needed for risk control should be based on an underlying philosophy for risk management. This chapter introduces fundamental concepts for risk treatment and control within an economic framework, including risk aversion, risk homeostasis, discounting procedures, decision analysis, trade-off analysis, insurance models, and risk financing. The chapter also briefly covers the concept of exposure and residual risk.

CONTENTS

7.1 Introduction

Risk treatment is a component of risk management, as illustrated in Figure 2.8 and discussed in Section 2.8. Treating risk is required to control risk by operators, managers, and owners who can make effective safety decisions and regulatory changes and can choose different system configurations based on the data generated in the risk assessment stage. Risk control involves using information from the previously described risk assessment stage to make rational decisions related to system risks. Risk treatments include failure prevention, threat reduction, vulnerability reduction, failure probability reduction, and consequence mitigation.

Generally, risk management is performed within an economic framework with an objective of optimizing the allocation of available resources in support of a broader goal; therefore, it requires the definition of acceptable risk and comparative evaluation of options and/or alternatives for decision making. Risk treatments have an objective to reduce risk to an acceptable level and/or prioritize resources based on comparative analysis. Section 2.8 provides information on defining acceptable risks and describes the methods for reducing risk by preventing an unfavorable scenario, reducing the rate, and/or reducing the consequences. Also, it describes four primary methods for risk mitigation: (1) risk reduction or elimination, (2) risk transfer to others (e.g., to a contractor or an insurance company), (3) risk avoidance, and (4) risk absorbance or pooling.

Risk control requires expending resources in the present to prevent potential losses in the future. This requirement creates complex decision and trade-off situations. Using a strict economic framework for risk control might produce outcomes that are economically efficient and satisfactory to some stakeholders but not to others, creating ethical and legal dilemmas that could require governmental interventions through regulations for risk control. Examples of governmental regulatory bodies that deal regularly with risk control include the Occupational Safety and Health Administration (OSHA), the Nuclear Regulatory Commission (NuRC), the Environmental Protection Agency (EPA), the National Highway Traffic Safety Administration (NHTSA), and the Food and Drug Administration (FDA). The regulatory efforts of government are necessary in these cases and others, but they might not be needed or preferred in some industries where voluntary or consensus standards can be developed to control risks, such as those of the Underwriters Laboratories (ULs) for various general consumer products (e.g., personal flotation devices [PFDs]).

The objective of this chapter is to introduce fundamental concepts for risk treatment and control within an economic framework, including risk aversion, risk homeostasis, discounting procedures, decision analysis, trade-off analysis, insurance models, and risk finance.

7.2 Philosophies of Risk Control

Risk control can be approached by an organization within a strategic, system-wide, or organization-wide plan. A philosophy for risk control might be constructed based on recognizing that the occurrence of a consequence-inducing event is the tip of an iceberg representing a scenario; therefore, risk control should target the entire scenario to produce an early intervention that could result in reducing the likelihood or elimination of this event. Such a philosophy can be referred to as the *domino theory for risk control* and could apply to cases involving complex scenarios. For example, the domino theory for risk control has been used in industrial accident prevention to eliminate injury-producing events through construction of a domino sequence of events as demonstrated by the following:

- A personal injury as the final domino occurs only as a result of an accident.
- An accident occurs only as a result of a human-related or mechanical hazard.

- A human-related or mechanical hazard exists only as a result of human errors or degradation of equipment.
- Human errors or degradation are inherited or acquired as a result of their environment.
- An environment is defined by conditions into which individuals or processes are placed.

This approach might be suitable for such applications as manufacturing, construction, production, and material handling. A related approach to risk control is the *cascading failure theory* for risk control, according to which control strategies are identified by investigating cascading failures; for example, loss of electric power to a facility might lead to the failure of other systems, which in turn leads to the failure of additional systems, and so on. In this case, risk control can target increasing power availability as a solution. Risk control can be achieved for similar applications through *energy release control* by adopting the following strategies:

- The creation of the hazard can be prevented in the first place during the concept development and design stages. For example, having no-smoking rules can be adopted to reduce the risk associated with fires, and pressure relief valves can be used to reduce risks associated with overpressurizing vessels and tanks.
- The impact of the hazard can be reduced through design and production, such as limiting power and reducing speed limits on highways.
- The release of a hazard that already exists in the design and utilization stages can be prevented. For example, electric fuses can be used to eliminate the release of electrical energy beyond some limits.
- The rate or spatial distribution of release of the hazard from its source can be controlled during the design and utilization stages, for example, brakes of vehicles control the energy in the wheels of vehicles.
- The hazard can be separated from what needs to be protected in time or space in the design, utilization, modification, and accident mitigation stages; for example, traffic lights are designed to keep vehicles and pedestrians from meeting.
- The hazard can be separated from what needs to be protected by interposing a material barrier during the design, utilization, modification, and accident mitigation stages; for example, firewalls can be used to separate a fire in a building within a compartment from other spaces.
- Relevant qualities of the hazard can be modified during the design and utilization stages, such as using fat-free food ingredients.
- What needs to be protected can be made more resistant to damage from hazard during the design, utilization modification, and accident mitigation stages, such as by designing fire- and earthquake-resistant buildings.
- The damage already done by the hazard can be countered and contained; for example, fire sprinkler systems and emergency response teams can be used to protect a facility.
- The object of damage can be repaired and rehabilitated; for example, injured workers and salvage operations can be rehabilitated after an accident.

A risk control philosophy must also define the control measures, time of application, and target of the risk control measures. The control measures can include pressure relief valves, firewalls, and emergency response teams. The time of application identifies when the measure is needed, such as before an event, at the time of an event, or after an event occurs. The targets of the risk control measures could include workers, visitors, machinery, assets, or a population outside a plant.

7.3 Risk Aversion in Investment Decisions

Treating risk within an economic framework enables the identification of economically efficient solutions. Such an approach starts by constructing cash flows for available alternatives as investments. The concepts discussed in Chapter 6 can be used to compute the net present value (NPV) for each alternative. Selecting an optimal alternative can be based on the expected or average NPV, as was demonstrated in the decision tree analyses in Chapter 3; however, this selection criterion might not reflect the complexities involved in real decision situations. This section utilizes an example situation of investment decisions under uncertainty to introduce some key concepts and related complexities based on risk attitudes and appetites of decision makers.

Consider a decision situation involving three alternatives A, B, and C, which could lead to several scenarios each. The scenarios for each alternative are identified by the magnitude of their respective NPVs—extremely low, very low, low, good, high, very high, and extremely high. These scenarios and their NPV values are shown in Table 7.1. The table demonstrates that alternatives A and B have generally smaller returns and smaller spreads than alternative C. The table also shows three cases of probability distributions (p) for the scenarios of equal likelihood, increasing likelihood, and decreasing likelihood. These probability distributions are used to introduce various concepts and cases.

Table 7.2 shows the descriptive statistics of the NPV of alternatives A, B, and C using the three probability distributions for the scenarios of equally likelihood, increasing likelihood, and decreasing likelihood (p). The descriptive statistics were computed as follows:

TABLE 7.1

Seven scenarios for Three Alternatives

Quantity	Extremely Low	Very Low	Low	Good	High	Very High	Extremely High
NPVs ($)							
Alternative A	100	200	300	400	500	600	700
Alternative B	300	400	500	600	700	800	900
Alternative C	0	200	400	600	800	1000	1200
Probabilities (p)							
Equal likelihood	1/7	1/7	1/7	1/7	1/7	1/7	1/7
Increasing likelihood	1/28	2/28	3/28	4/28	5/28	6/28	7/28
Decreasing likelihood	7/28	6/28	5/28	4/28	3/28	2/28	1/28

NPV, net present value.

TABLE 7.2

Descriptive Statistics of the Net Present Values (NPVs) of Alternatives A–C of Table 7.1

Quantity	Alternative A	Alternative B	Alternative C
Equal likelihood			
Expected NPV ($)	400	600	600
Standard deviation of NPV ($)	200	200	400
Coefficient of variation of NPV	0.5	0.333	0.667
Increasing likelihood			
Expected NPV ($)	500	700	800
Standard deviation of NPV ($)	173.21	173.21	346.41
Coefficient of variation of NPV	0.346	0.247	0.433
Decreasing likelihood			
Expected NPV ($)	300	500	400
Standard deviation of NPV ($)	173.21	173.21	346.41
Coefficient of variation of NPV	0.577	0.346	0.866

$$E(\text{NPV}) = \sum_{i=1}^{N=7} \text{NPV}_i p_i \tag{7.1}$$

$$\sigma(\text{NPV}) = \sqrt{\sum_{i=1}^{N=7} p_i[\text{NPV}_i - E(\text{NPV})]^2} \tag{7.2}$$

$$\text{COV}(\text{NPV}) = \frac{\sigma(\text{NPV})}{E(\text{NPV})} \tag{7.3}$$

where:

E is the expected value or mean value

NPV_i is the NPV of scenario i of the seven ($N = 7$) scenarios

p_i is the respective occurrence probability of a scenario

σ is the standard deviation

COV is the coefficient of variation

The expected value measures the average return for an alternative, whereas the standard deviation measures the dispersion in the NPV, reflecting uncertainty associated with the outcome of an alternative. The coefficient of variation (COV) is a measure of dispersion in a normalized or unit-free form. The COV can be interpreted as the standard deviation of NPV, that is, it is a measure of risk per unit value of the expected NPV. In this example, alternatives A and B produce smaller NPVs than alternative C; however, they have less dispersion or uncertainty. But alternative C produces a greater NPV and has a larger dispersion than alternatives A and B. For a decision maker or an investor, this situation might not be clear-cut; one investor might be willing to take on larger dispersion for a potentially larger NPV, while another investor might prefer the reverse.

The inconclusive decision situation in this example can be attributed to the level of satisfaction that an investor, that is, decision maker, might reach based on each alternative. The level of satisfaction for each level of NPV (or wealth, W) that corresponds to each scenario is the *utility* (U), which represents the risk attitude of an investor. The risk attitude of an investor or decision maker may be thought of as a decision maker's preference of taking a chance on an uncertain money payout of known probability versus accepting a sure money amount (i.e., with certainty). For example, suppose a person is given a choice between (1) accepting the outcome of a fair coin toss (where heads means winning $20,000 and tails means losing $10,000) and (2) accepting a certain cash amount of $4000. The expected value in this case is $5000, which is $1000 more than the certain money amount. A risk-neutral decision maker should prefer the coin toss because it has a higher expected value, whereas a risk-averse investor should prefer the $4000 certain amount. If the certain amount were raised to $6000 and the decision maker still preferred the coin toss, he or she would be demonstrating a risk-seeking attitude. Such trade-offs can be used to derive a utility function that represents a decision maker's risk attitude. The risk attitude of a given decision maker is typically a function of the amount at risk. Many people who are risk averse when faced with the possibility of significant loss become risk neutral, and sometimes risk taking when potential losses are relatively small. Because decision makers vary substantially in their risk attitudes, it is necessary to assess both the risk exposure (i.e., the degree of risk inherent in the decision) and the risk attitude of the decision maker using a utility function. Generally, the larger the NPV, the greater the utility, and vice versa. The concept of utility under uncertainty is based on the following axioms:

- Decision making is always rational,
- Decision making takes into considerations all available alternatives, and
- Decision makers prefer more consumption or wealth to less.

These axioms define what is termed *cardinal utility*. The utility for each NPV level is a subjective measure that depends on the nature, personality, and character of a decision maker and sometimes on the environment and timing of the decision situation. For the purpose of illustration, a subjectively constructed utility function was used to produce the utility values shown in Table 7.3 for alternatives A, B, and C. Decision making can be viewed as all about maximizing utility rather than maximizing wealth, because maximizing utility leads to maximizing satisfaction. Commonly, an alternative with the highest expected utility, $E(U)$, is identified and selected. The descriptive statistics of the utility for alternatives A, B, and C using the three probability distributions for the scenarios of equal likelihood, increasing likelihood, and decreasing likelihood (p) are shown in Table 7.4. Alternative C has still a larger expected utility value compared to alternatives A and B with a larger, respective dispersion value.

TABLE 7.3

Utility Values for Net Present Values (NPVs)

Quantity	Extremely Low	Very Low	Low	Good	High	Very High	Extremely High
Alternative A							
NPV ($)	100	200	300	400	500	600	700
Utility	77	148	213	272	325	372	413
Alternative B							
NPV ($)	300	400	500	600	700	800	900
Utility	213	272	325	372	413	448	477
Alternative C							
NPV ($)	0	200	400	600	800	1000	1200
Utility	0	148	272	372	448	500	528

TABLE 7.4

Descriptive Statistics for the Utility of Alternatives A–C

Quantity	Alternative A	Alternative B	Alternative C
Equal likelihood			
Expected utility	260	360	324
Standard deviation of utility	112.48	88.61	180.84
Coefficient of variation of utility	0.433	0.246	0.558
Increasing likelihood			
Expected utility	316	404	412
Standard deviation of utility	92.24	71.58	136.47
Coefficient of variation of utility	0.292	0.177	0.331
Decreasing likelihood			
Expected utility	204	316	236
Standard deviation of utility	102.59	81.90	176.91
Coefficient of variation of utility	0.503	0.259	0.750

To appreciate the impact of utility values on a decision, the different NPVs and utilities for these alternatives are shown in Table 7.3. The expected utility values of Table 7.4 show different preferences compared to the expected NPVs of Table 7.2; therefore, the assignment of utilities results in changing preferences and decisions. For example, the reason for the change in preference for alternative B compared to alternative A is due to the fact that the utility values attributed by an investor or a decision maker to the NPV for alternative B reflect a cautious investor, compared to alternative A [i.e., preferring lower E(NPV) to a large dispersion].

The utility function of Table 7.4 reflects the cautiousness of an investor or a decision maker. The values in Tables 7.2 and 7.4 reveal impeded cautiousness of the investor based on the utility function. Considering alternative B for the equal likelihood scenarios as an example, the respective expected values of NPV and utility value [E(NPV) and $E(U)$, respectively] are as follows:

$$E(\text{NPV}) = \$600 \tag{7.4a}$$

$$E(U) = 360 \tag{7.4b}$$

The result of Equation 7.4a and the utility function presented in Table 7.3 can be used to compute the utility of E(NPV) as follows:

$$U[E(\text{NPV})] = U(600) = 372 \tag{7.5}$$

Because $U[E(\text{NPV})] > E(U)$ for alternative B, based on Equations 7.4b and 7.5, the investor in this case is cautious or risk averse. The meaning of risk aversion in this case is that a certain NPV of \$600 has a utility of 372, which is larger than the weighted utility of a risky project with an E(NPV) of \$600 based on its $E(U)$ of 360. An investor who could receive a certain NPV of \$600 instead of an expected NPV with the same value would be always more satisfied with the higher utility. Therefore, in this case, $U[E(\text{NPV})]$ is larger than $E(U)$ as any incremental increase in NPV results in a nonproportionally smaller increase in utility. Humans generally have an attitude toward risk where small stimuli over time and space are ignored, while the sum of these stimuli, if exerted instantly and locally, could cause a significant response.

In general, risk aversion can be defined by the following relationship:

$$U[E(\text{NPV})] > E[U(\text{NPV})] \tag{7.6a}$$

or

$$U[E(W)] > E[U(W)] \tag{7.6b}$$

The utility function used in the previous example is for a risk-averse investor, as shown in Figure 7.1. The equation used to construct the utility function in Figure 7.1 for illustration purposes is given by

$$U(W) = 0.8W - 0.0003W^2 \tag{7.7}$$

The figure also shows two points that have the coordinates (NPV, U) of ($200, 148) and ($1000, 500). These two points represent two scenarios with, say, equal probabilities of 0.5 each. Therefore, for these two scenarios, the following quantities can be computed:

$$E(\text{NPV}) = 0.5(\$200) + 0.5(\$1000) = \$600 \tag{7.8a}$$

The utility of this E(NPV) is

$$U[E(\text{NPV})] = 0.8(600) - 0.0003(600)^2 = 372 \tag{7.8b}$$

The expected utility of the two points is

$$E[U(\text{NPV})] = 0.5(148) + 0.5(500) = 324 \tag{7.8c}$$

Cases in which utility grows slower than wealth represent risk-averse investors. The intensity of risk aversion depends on the amount of curvature in the curve. The larger the curvature for this concave curve, the higher the risk aversion.

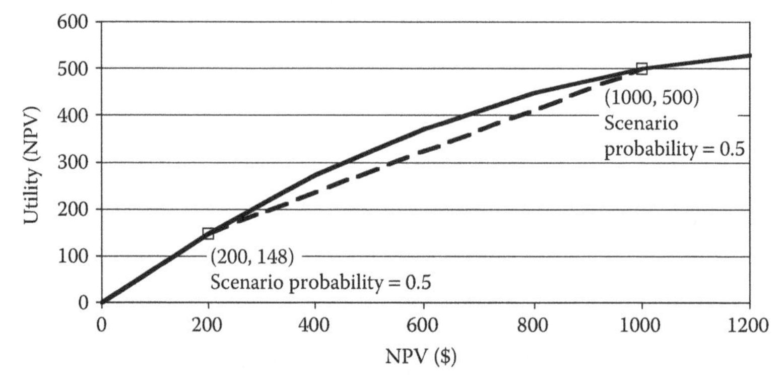

FIGURE 7.1
Utility function for a risk-averse investor. NPV, net present value.

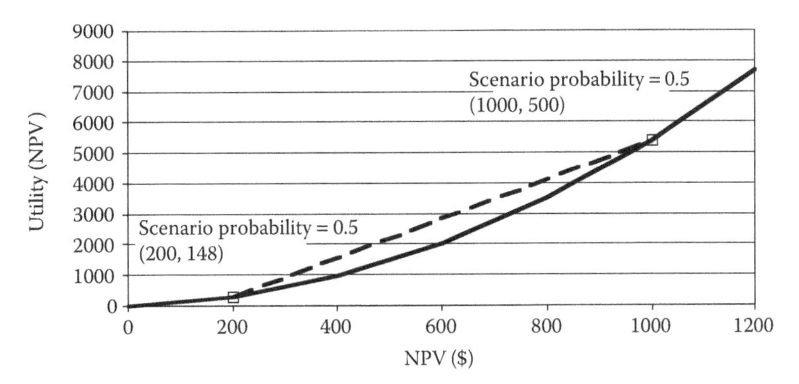

FIGURE 7.2
Utility function for a risk-seeking investor. NPV, net present value.

Although not as common, risk-seeking investors display a risk propensity. In this case, the utility function is convex, as shown in Figure 7.2, and meets the following conditions:

$$U[E(\text{NPV})] < E[U(\text{NPV})] \tag{7.9a}$$

or

$$U[E(W)] < E[U(W)] \tag{7.9b}$$

The utility function for the risk-seeking investor shown in Figure 7.2 was constructed using the following utility function for illustration purposes:

$$U(W) = 0.4W + 0.005W^2 \tag{7.10}$$

The figure also shows two points that have the coordinates (NPV, U) of (\$200, 280) and (\$1000, 5400). These two points represent two scenarios with, say, equal probabilities of 0.5 each. Therefore, for these two scenarios, the following quantities can be computed:

$$E(\text{NPV}) = 0.5(\$200) + 0.5(\$1000) = \$600 \tag{7.11a}$$

The utility of this E(NPV) is

$$U[E(\text{NPV})] = 0.4(600) + 0.005(600)^2 = 2040 \tag{7.11b}$$

The expected utility of the two points is

$$E[U(\text{NPV})] = 0.5(280) + 0.5(54000) = 2840 \tag{7.11c}$$

Cases in which utility grows faster than wealth represent risk-seeking investors. The intensity of risk propensity depends on the amount of curvature in the curve. The greater the curvature for this convex curve, the higher the risk propensity.

The case of risk neutrality is another possibility and is common for governments and large corporations with relatively sizable resources. A risk-neutral investor has a utility function without curvature, as shown in Figure 7.3. In this case, the utility function is linear and meets the following conditions:

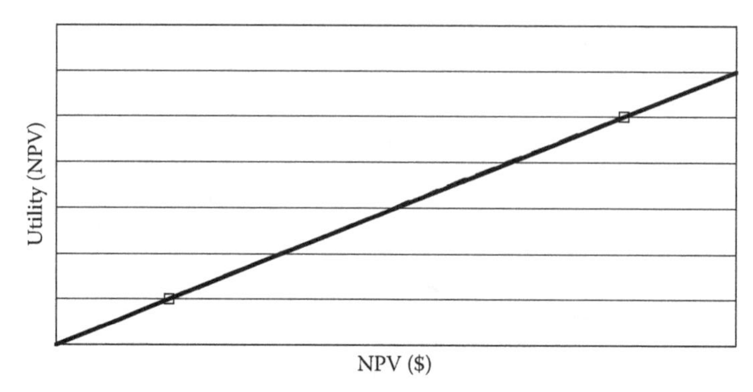

FIGURE 7.3
Utility function for a risk-neutral investor. NPV, net present value.

$$U[E(\text{NPV})] = E[U(\text{NPV})] \tag{7.12a}$$

or

$$U[E(W)] = E[U(W)] \tag{7.12b}$$

The use of NPV is appropriate for most applications; however, it should be noted that the size of an initial investment might need to be considered when selecting among available alternatives. The larger the size of an initial investment, the smaller the rate of return for the same NPV. For this reason, the use of the rate of return might be needed in some applications. Despite this shortcoming of using NPV, it offers a unique representation of the risk-taking willingness of investors through utility functions.

In addition to expected values of NPV or U, the standard deviations of NPV and U should also be considered in investment decision making (the standard deviations of NPV and U are computed for the examples in Tables 7.2 and 7.4); however, the COV of NPV and U can also be used as a normalized, unit-free measure of dispersion. The expected values and standard deviations of NPV and U for investment alternatives can be graphically displayed as shown in Figure 7.4. The figure shows indifference curves for a risk-averse investor that were subjectively constructed and drawn. Each curve represents a line that connects pairs of expected values and standard deviations of return that are judged by an investor to have the same utility level. The utility value assigned to each curve increases in the direction indicated in the figure. The larger the risk aversion, the steeper the indifference curves. In this case, alternative B is the most desirable investment, because it offers the largest return along the same indifference curve.

This section has dealt so far only with a single investment, not a portfolio of investments. Investment decisions about a portfolio might require treating the investments as multiple random variables that can be combined through a sum as follows for a portfolio of two investments:

$$\text{NPV} = \text{NPV}_1 + \text{NPV}_2 \tag{7.13}$$

The concepts covered in Appendix A on multiple random variables can be used herein to compute the mean and standard deviation of the total NPV as follows:

$$E(\text{NPV}) = E(\text{NPV}_1) + E(\text{NPV}_2) \tag{7.14a}$$

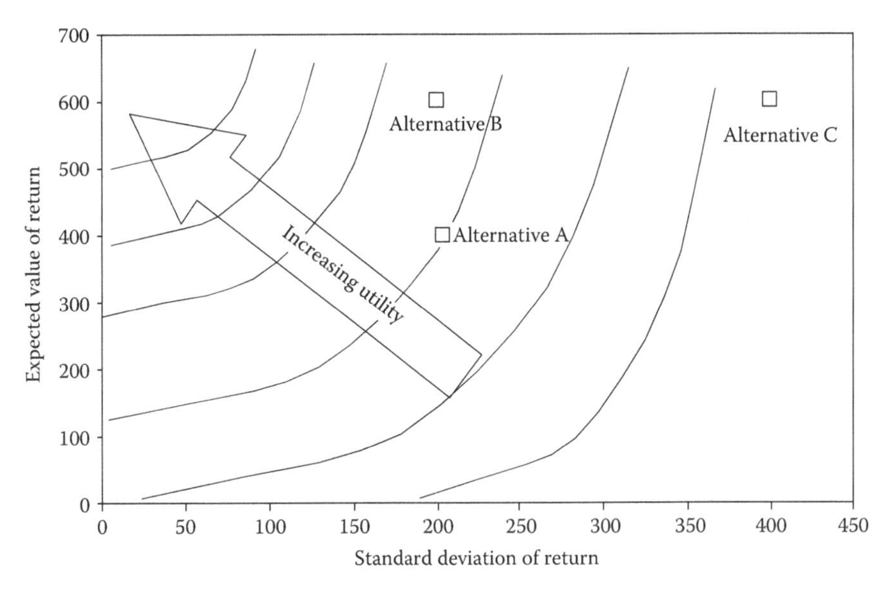

FIGURE 7.4
Indifference curves for risk aversion.

$$\sigma(\text{NPV}) = \sqrt{[\sigma(\text{NPV}_1)]^2 + [\sigma(\text{NPV}_2)]^2 + 2\text{Cov}(\text{NPV}_1, \text{NPV}_2)} \tag{7.14b}$$

where:

$\text{Cov}(\text{NPV}_1, \text{NPV}_2)$ is the covariance of NPV_1 and NPV_2 as a measure of association between NPV_1 and NPV_2 that is given by

$$\text{Cov}(\text{NPV}_1, \text{NPV}_2) = \sum_i \sum_j [\text{NPV}_{1i} - E(\text{NPV}_1)][\text{NPV}_{2j} - E(\text{NPV}_2)]p_{ij} \tag{7.15}$$

where:

p_{ij} is the joint probability of NPV_{1i} and NPV_{2j}

Sometimes, an approximate joint probability can be computed from the marginal probabilities as follows based on the assumption of independence:

$$p_{ij} = p_{1i}p_{2j} \tag{7.16}$$

Covariance, as a measure of correlation, can take negative values, positive values, and a zero value. A zero value for the covariance indicates that the investments are uncorrelated. The sign of the covariance indicates negative or positive correlation corresponding to a direct linear, proportional relationship or an inverse relationship, respectively. A negative correlation according to Equation 7.14b leads to reducing the standard deviation of the NPV of the portfolio, which means reducing the risk, and vice versa for the positive correlation case. Introducing a negative correlation among investments is commonly known as *investment diversification*. Using these concepts, an investor could construct a diagram similar to Figure 7.4 for the entire portfolio. Available investment funds could be allocated to produce an optimal solution that maximizes

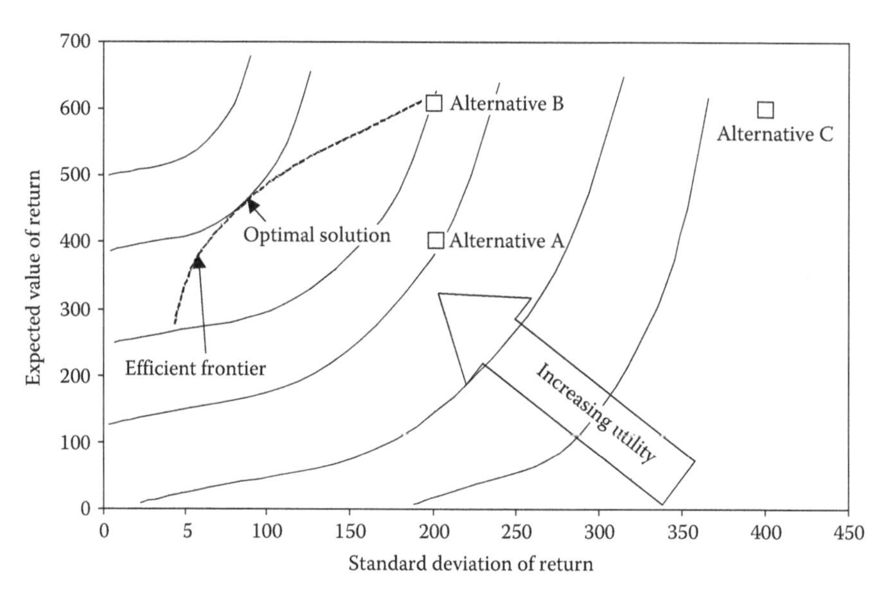

FIGURE 7.5
Minimum variance frontier with an indifference curves for an optimal solution.

the returns and minimizes the standard deviation of the returns. The result is a curve known as the *minimum variance frontier*, which usually has two expected values of return for any value of the standard deviation. The *efficient frontier*, as shown in Figure 7.5, considers only the larger (i.e., upper) expected values of the minimum variance frontier. The efficient frontier can be viewed as an envelope of points that have maximum return values among all available alternatives corresponding to respective standard deviation values; that is, for a specific standard deviation, the alternative that provides maximum return is identified, and the line that connects all the alternatives that maximize the return for a range of standard deviations defines the efficient frontier. The intersection of the efficient frontier with an indifference curve would offer the optimal solution shown in Figure 7.5. The hypothetical efficient frontier shown as a dotted line in Figure 7.5 is for illustration purposes that could be observed in case of having many investment alternatives where the dotted line defines the outer maximum return with minimum standard deviation envelope. In the case of the three alternatives A, B, and C, the efficient frontier is defined by A and B, and since these two alternatives have the same standard deviation, the optimal solution is alternative B since it has the larger expected value between the two alternatives.

Example 7.1: Construction of Utility Functions for Investment Decisions

Investors or decision makers commonly construct utility functions for investment decisions subjectively. Alternative A of Table 7.1 is used in this example to demonstrate the construction of utility functions. Two utility functions are provided which represent the preference or risk attitudes of two investors: a risk-averse investor and a risk-seeking investor, respectively, as follows:

$$U_1(\text{NPV}) = 0.8\text{NPV} - 0.0003\text{NPV}^2 \tag{7.17a}$$

$$U_2(\text{NPV}) = 0.4\text{NPV} - 0.0002\text{NPV}^2 \tag{7.17b}$$

TABLE 7.5

Utility Values for Alternative A Based on Equation 7.17a and 7.17b

	NPV ($)						
	100	200	300	400	500	600	700
U_1(NPV)[a]	77	148	213	272	325	372	413
U_2(NPV)[b]	42	88	138	192	250	312	378

NPV, net present value.
[a] See Equation 7.17a.
[b] See Equation 7.17b.

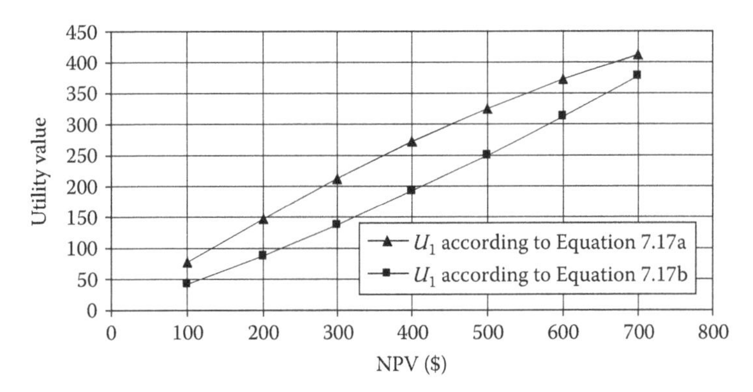

FIGURE 7.6
Utility and net present value (NPV) for alternative A based on Equation 7.17a and 7.17b.

where the NPV ($) values are provided in Table 7.1 for alternative A. The utility functions are evaluated in Table 7.5. Figure 7.6 shows the different slope characteristics for the two utility curves. The curve for U_1, which is concave in shape, represents the risk-averse attitude of the investor, whereas the curve for U_2, which is convex in shape, represents the risk-seeking attitude. These curves relates NPV to U inorder to represent the risk attitude of an investor.

Example 7.2: Efficient Frontier for Screening Design Alternatives

An architectural company has developed six design alternatives for a new commercial structure, denoted as D_1–D_6. The company's management decided to identify the optimal alternative for implementation using economic-based efficient frontier analysis. The expected value and standard deviation of the NPV were assessed for the six alternatives. The standard deviation is viewed herein as a measure of risk associated with each alternative. The statistics of the NPV are presented in Table 7.6. The efficient frontier can be identified based on the results of the six alternatives by plotting them as shown in Figure 7.7. The figure clearly shows the efficient frontier as the alternatives that offer the largest expected NPV for any given standard deviation. As can be observed from the figure, designs D_1, D_2, and D_6 fall on the efficient frontier. The other design alternatives, D_3, D_4, and D_5, are said to be dominated by those three design alternatives that are on the efficient frontier. The management of the company must now decide which design alternative is more economical for implementation among the short list of alternatives that are on the efficient frontier. Based on the expected NPV return only, D_6 can be identified as the optimal alternative; however, with risk reduction considerations, D_2 could also be the optimal design alternative. In addition, if management

TABLE 7.6

Expected and Standard Deviation of Net Present Value (NPV) for Design Alternatives

	Design					
	D1	**D2**	**D3**	**D4**	**D5**	**D6**
Expected NPV ($1000)	100	42	66	66	88	118
Standard deviation of NPV ($1000)	25	4	48	25	65	65
Expected NPV/standard deviation of NPV	4.00	10.50	1.375	2.64	1.35	1.82

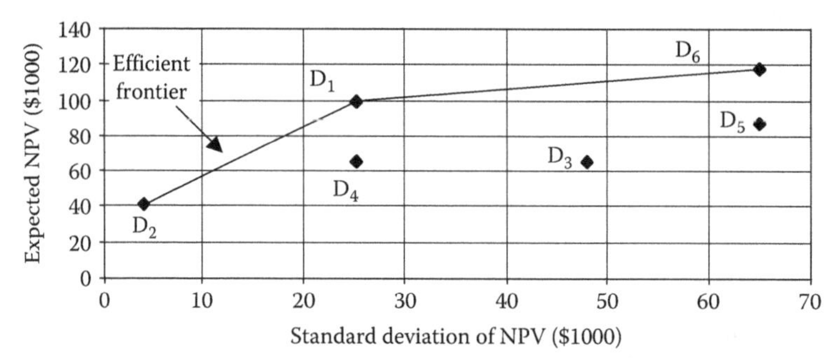

FIGURE 7.7
Efficient frontier for design alternatives. NPV, net present value.

would accept less returns than those offered by D_6 and higher risks than those offered by D_2, then they would prefer D_1. As demonstrated, a trade-off between risk and return can be made among the alternatives falling on the efficient frontier. Such a trade-off requires assessing the attitude of management toward risk as discussed in Example 7.3.

Example 7.3: Selecting Optimal Design Alternative Based on Different Risk Attitudes

Example 7.2 presented the case of selecting an optimal design alternative and discussed the possible trade-offs among alternatives falling on the efficient frontier. Figure 7.8 shows the cases of risk-averse management and risk-seeking management of a company. Utility curves for risk-averse management subjectively assigned in the space of the expected and

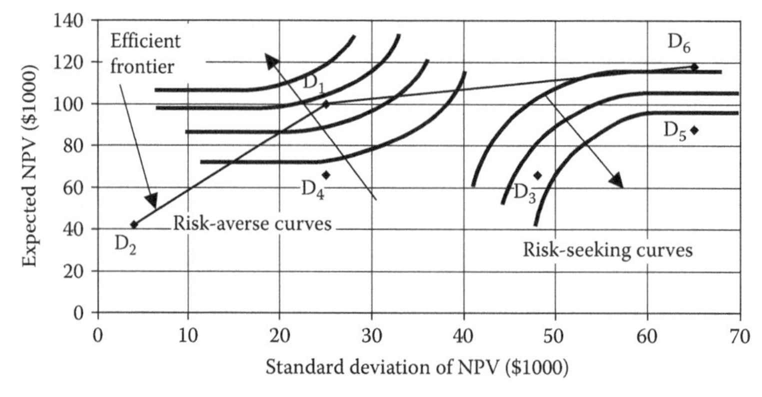

FIGURE 7.8
Efficient frontier and utilities for design alternatives. NPV, net present value.

standard deviation of NPV are shown on the left side of the figure. These risk-averse curves lead the management to select the alternative design D_1. For risk seekers, as shown on the right side of the figure, designs D_3, D_5, or D_6 are among the appealing alternatives. In the risk-seeking case, the alternatives chosen might not all fall on the efficient frontier, that is, the alternatives could include risky ones. Hence, management might choose alternative D_6 despite its high level of risk because it offers a large standard deviation with a high upward return potential, noting that the downward return potential is the associated risk. Finally, if the management is risk neutral, design D_2 would be identified as one that gives the highest value of return in terms of expected NPV per standard deviation of NPV, as shown in Table 7.6, with a ratio of 10.5, followed by D_1 with a ratio of 4.

Example 7.4: Efficient Frontier and Utility Values for Screening Car Product Alternatives

An automobile manufacturer is considering five alternative product designs for its new generation of sedans. The alternatives are denoted as A–E. For each design option, an analytical simulation was carried out to obtain the mean and standard deviation of the marginal profits of each design based on the selling price, expected sales, design reliability, and associated warranty repairs. The simulation results are presented in Table 7.7, which shows the expected profit and standard deviation for the five design alternatives. The production manager of the company would like to maximize the expected return and, being risk averse, would like to minimize the risk for the company. Comparing designs A and B, as shown in Table 7.7, reveals that they offer the same expected return; however, with a larger standard deviation of \$225,000, design B is much riskier than design A. Design A is therefore said to dominate design B. In addition, design B is dominated by design E, which for the same level of standard deviation offers a higher expected profit with an expected return of \$800,000. Similarly, design D dominates design C. The nondominated designs are A, D, and E, which lie on the efficient frontier as shown in Figure 7.9. The axes of Figure 7.9 are switched around compared to Figure 7.8 for the purpose of illustration since both styles are used in the literature. The choice among designs A, E, and D can be made based on the risk attitude of the decision maker. Design A offers a low expected return with a low level of standard deviation, whereas design E offers a high expected return with a high level of standard deviation. Design D offers medium values for both the return and the standard deviation.

To model the risk attitude of the decision maker, utility curves need to be constructed to identify the optimal choice among the alternative designs. Assuming that the risk attitude of the manager can be expressed using the following utility function:

$$U(P) = 0.3P - 0.00015P^2$$

where:

U is the utility
P is the profit

TABLE 7.7

Expected Value and Standard Deviation of Profits for Car Product Designs

Alternative	Expected Profit (\$1000)	Standard Deviation of Profit (\$1000)
A	50	30
B	50	225
C	300	120
D	550	120
E	800	225

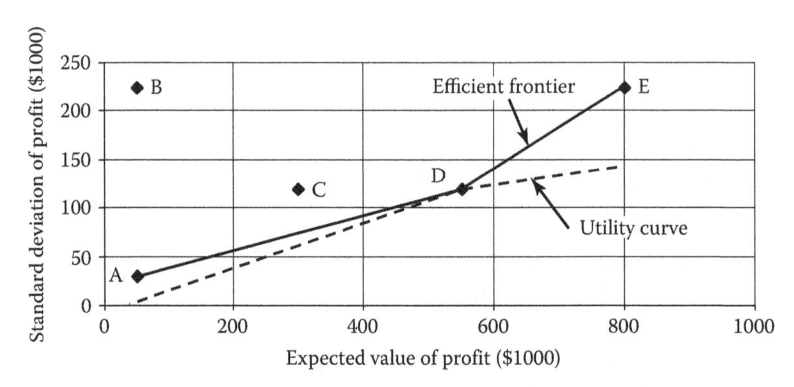

FIGURE 7.9
Efficient frontier and utility curve for design alternatives.

The utility curve takes a concave shape, as shown in Figure 7.9. The utility function is tangent to the efficient frontier at design D; hence, product D with an expected profit of $550,000 and a standard deviation of $120,000 is the optimal solution that maximizes profit and satisfies the risk level accepted by the decision maker.

7.4 Types of Risk Treatments

Risk treatments include countermeasures and mitigations by technological, logical, or user behavioral means. They were defined earlier to include (1) the actions taken or a physical capability provided with a principal purpose of reducing or eliminating vulnerabilities or reducing the occurrence of attacks or threats by technological, logical, or user behavioral means and (2) preplanned and coordinated actions or system features that are designed to reduce or minimize the damage caused by an event; support and complement emergency forces, that is, first responders; facilitate field investigation and crisis management response; and facilitate recovery and reconstitution. The latter form defines mitigations that also include technological, logical, or user behavioral means.

Such treatments involve direct costs, such as increased capital expenditure or the payment of insurance premiums that might reduce the average overall financial returns from a project. Primary ways to deal with risk within the context of a risk management strategy include risk reduction or elimination, risk transfer (e.g., to a contractor or an insurance company), risk avoidance, and risk absorbance or pooling as discussed in subsequent sections.

7.4.1 Risk Reduction or Elimination

Risk reduction or elimination is often the most fruitful approach. For example, could the design of a system be amended so as to reduce or eliminate either the probability of occurrence of a particular risk event or the adverse consequences if it occurs? Alternatively, could the risks be reduced or eliminated by retaining the same design but using different materials or a different method of assembly? Other possible risk mitigation options in this category include, as examples, a more attractive labor relations policy to minimize the risk of stoppages, training of staff to avoid hazards, improved site security to prevent theft and vandalism, preliminary investigation of possible site pollution, advance ordering of key components, noise abatement measures, effective signage, and liaisons with the local community.

7.4.2 Risk Transfer

A general principle of an effective risk management strategy is that commercial risks in projects and other business ventures should be borne wherever possible by the party that is best able to manage them and thus mitigate the risks. Most often, contracts and financial agreements are used to transfer risks. Companies specializing in risk transfer can be consulted for procedures necessary to meet the needs of a project. Risks can also be transferred to an insurance company, which, in return for a payment (i.e., premium) linked to the probability of occurrence and severity associated with the risk, is obliged by the contract to offer compensation to the party affected by the risk. Insurance coverage can include straight insurance for expensive risks with a low probability, such as fire; performance bonds, which ensure that the project will be completed if the contractor defaults; and sophisticated financial derivatives, such as hedge contracts, to avoid such risks as unanticipated losses in foreign exchange markets.

7.4.3 Risk Avoidance

A most intuitive way of avoiding a risk is not to undertake a project in such a way that involves that risk. Consider, for example, the objective to generate electricity. A nuclear power source, although cost efficient, is considered to have a high risk due to potentially catastrophic consequences, so, even though all reasonable precautions would be taken, still the practical solution is to turn to other forms of fuel to avoid that risk. Another example would be the risk that a particularly small contractor would file bankruptcy. In this case, the risk could be avoided by using a well-established contractor instead for that particular job.

7.4.4 Risk Absorbance and Pooling

In cases where risks cannot (or cannot economically) be eliminated, transferred, or avoided, they must be absorbed if the project is to proceed. Normally, a sufficient margin in the finances of a project should be created to cover the risk event should it occur; however, it is not always essential for one party alone to bear all these absorbed risks. Risks can be reduced through pooling, possibly through participation in a consortium of contractors, when two or more parties are able to exercise partial control over the incidence and impact of risk. Joint ventures and partnerships are other examples of pooling risks.

7.4.5 Characterizing Uncertainty for Risk Reduction

Risk can be mitigated through proper uncertainty characterization. The presence of improperly characterized uncertainty can lead to greater estimates of likelihood of adverse events, as well as increased estimated cost margins as a means of compensating for these risks. Risk can be reduced by a proper characterization of uncertainty, which can be achieved through data collection and knowledge construction.

7.5 Risk Homeostasis

According to risk homeostasis concepts as described by Pitz (1992), people accept a certain level of risk in any activity. This risk level is subjectively estimated and accepted in regard to their health, safety, and other things they value in exchange for the benefits or satisfaction they hope to receive from that activity, such as transportation, work, eating, drinking, drug use, recreation, romance, and sports (Wilde 1988). *Homeostasis* is broadly defined as the tendency to maintain, or the maintenance of, normal, internal stability in a living species by coordinated responses of its relevant internal systems that automatically compensate for environmental changes. Risk homeostasis can be defined in a similar manner as an ongoing activity of people of continuously assessing the amount of their risk exposure, comparing it with the amount of risk they are willing to accept, and trying to eliminate any difference between the two risk levels. Thus, if an individual's exposure to risk is subjectively assessed by the individual to be lower than an acceptable level, the individual might tend to engage in actions that increase his or her exposure to risk. If a subjectively experienced risk is higher than an acceptable level, people attempt to exercise greater caution. This balancing act of bringing risk exposures to acceptable levels is continuous; consequently, people choose their future actions in an adaptive manner so that subjectively assessed risk exposures match acceptable risk levels. Each particular adjustment action carries an objective probability of risk of accident or illness; therefore, the aggregation of these adjustment actions across the entire population over an extended period of time of several years yields the temporal rate of accidents or of lifestyle-dependent diseases for the population.

Resulting accident and disease rates, as well as more direct and frequent personal experiences of danger, in turn influence the amount of risk people associated with various activities and lifestyles over the next period of time. Accordingly, people decide on their future actions, and these actions in turn produce the subsequent rate of human-caused mishaps. Such a closed loop representation between the past and the present and between the present and the future produces, over the long run, human-made mishap rates reflecting risk acceptance.

The implication for risk homeostasis concepts is that people alter their behavior in response to implementing health and safety measures to increase their risk exposures to bring them to the same levels as acceptable levels of risk. Reducing the cumulative or total risk level requires motivating people to alter the amount of risk they are willing to undertake. Such an implication can be used to explain the fact that technological efforts toward flood control in the United States have failed to reduce the number of flood victims. Improved impoundment and levee construction have made certain areas less prone to flooding, but, as a consequence, more people have settled in the fertile plains because they were now safer than before, leading to the same end result in terms of the number of flood victims. Subsequent floods, although fewer in number, have caused more human loss and more property damage. Understanding risk homeostasis, then, might affect the choice of risk mitigation actions. For example, reducing the problem of excessive flow of water and flooding might be more effectively mitigated upstream in the form of reforestation or the careful maintenance of wetlands so that more-than-normal precipitation is contained and does not run downhill.

Risk homeostasis can also explain the fact that a random selection of cigarette smokers who were advised to quit by their physician did indeed reduce their cigarette consumption to a much greater extent than a comparison group (Wilde 1988). These former smokers had a lower rate of smoking-related disease; however, they did not live any longer. Also, it could explain why the number of traffic deaths per capita has remained the same or even increased despite the construction of modern, multilane highways. These highways have contributed to a reduction in the number of road deaths per unit distance driven but have maintained or even increased the number of traffic deaths per capita. A sure way to reduce the accident rate on a particular road to zero is to simply close down that road to all traffic. However, road users would move to other roads, and the accidents would migrate with them to other locations (Wilde 1988).

Risk homeostasis could have a great implication for selecting risk mitigation actions. Traditional risk mitigation practices can therefore be called into question, such as prohibiting drinking and driving, closing borders to illicit drug trade, relying on enforcement of laws traditionally, informing the public of certain dangers, and engineering the physical aspects of the built environment. Risk mitigation actions that depend on human conduct might not work or might not be effective in general. These conclusions emphasize the need to account for human behavior within risk mitigation actions and to devote efforts to changing the behavior of humans, aimed at increasing people's desires to be safe and live a healthy lifestyle. Thus, in addition to enforcement, educational, and engineering approaches, a motivational approach to prevention is necessary.

7.6 Insurance for Loss Control and Risk Transfer

Risk management, including loss control, is of central importance for insurers. Insurers typically perform rigorous studies and reviews, followed by periodic site visits and specialized studies. Some insurers utilize specialized methods and protocols for performance measurement and verification.

7.6.1 Loss Control

Loss control for risk management in insurance practices is central to the business of insurance. If insurers and insured systems are able to limit the rate and/or intensity of losses, or at least quantify the risk, pure premiums can be calculated with known distributions and uncertainty. Potentially, the cost of insurance can be lowered, although a variety of market consideration might weigh heavily on determining financial premium and deductible rates. Loss control measures can range from requiring fire sprinklers in buildings to computer ergonomics training in workplaces. The two primary approaches to implementing insurance loss control are contractual and technical. Contractual methods include exclusions on the policy or the ability to shift the loss cost to others, such as in performance surety bonds, for which the insurer can make claims on the contractor in the event of a loss. Insurance providers also limit claims through the use of deductibles and exclusions. Technical methods for loss control include a host of quality assurance techniques used during design, construction, and start-up of a project. These technical methods are captured within the set of tools known as *system commissioning*. Measurement and diagnostics methods can be used to track actual performance and make corrections before claims materialize. Loss control specialists are used to help keep the number of accidents and losses to a minimum. They visit factories, shop floors, and businesses to identify potential hazards and help to eliminate them. In the health insurance area, they might work with an organization to promote preventive health care in the workplace or to limit exposure to certain types of ailments.

7.6.2 Risk Actuaries and Insurance Claim Models

The insurance industry utilizes analytical skills to assess risks and the price of their insurance products. The analytical skills of actuaries are used to assess risks of writing insurance policies on property, businesses, and people's lives and health. For example, the cost of automobile insurance is significantly higher for someone under the age of 25 than for other age groups because actuaries have determined that the risk of insuring automobiles is highly age dependent. Actuaries are a crucial part of the insurance process because they use statistical and mathematical analyses to assess the risks of providing coverage. Actuaries, therefore, need to be aware of general societal trends and legislative developments that may affect risks. Actuaries can work either within insurance companies or for the government, pension-planning organizations, or third-party advisors. The remainder of this section provides an example presentation of an actuary model for assessing risks.

The development of a risk model for insurance purpose requires the assessment of anticipated insurance claims. Several factors can affect the expected loss to insurer as a result of claims, most importantly claim rate (or frequency) and severity. If the uncertainty associated with both can be modeled, a reasonable assessment of claim magnitude may be made. For this purpose, an insurance claim model should be constructed using a combination of analytical skills and expert opinions. Expert opinion elicitation can be used to gather data on claim or accident occurrence rates or frequencies and on claim or accident severities.

The objective of an insurance claim model is to assess the annual magnitude of claims by accounting for uncertainties associated with frequencies, severities, and expert-to-expert variability. Several experts might be used to elicit the necessary information.

The annual rate of events (λ) can be estimated as an interval, such as [0.2, 0.3] or [0.2, 0.9]. The annual rate can be modeled by a Poisson process with an estimated occurrence rate λ. For simplicity, an elicited interval is assumed to be the mean annual rate $\pm k\sigma$, where k is a given real value. The mean (μ) and standard deviation (σ) can be computed based on the interval limits of λ and k. The annual rate based on this model is a random variable distributed according to a continuous distribution with the probability density function $f_\lambda(\lambda)$, which can be represented by such probability distributions as (1) a gamma distribution, (2) a beta distribution, or (3) a negative binomial distribution (or Pascal distribution). In this section, a gamma distribution is used to illustrate computational procedures to assess annual claims. Other distributions could have been used for this assessment. The gamma distribution has two parameters, α and θ, defined as follows:

$$\alpha = \frac{\mu^2}{\sigma^2} \tag{7.18a}$$

and

$$\theta = \frac{\sigma^2}{\mu} \tag{7.18b}$$

where:

μ is the assumed mean of λ
σ is the standard deviation of λ

The probability density function (f_λ) of the gamma distribution is given by

$$f_\lambda(x) = \frac{x^{\alpha-1} e^{\frac{-x}{\theta}}}{\Gamma(\alpha)\theta^\alpha} \tag{7.19}$$

The severity of a claim is the second variable that has to be examined in the assessment of insurance claims. The severity of claims can be modeled using two lognormal distributions representing the lower and upper limits and based on the expert opinion. These two distributions can be treated to have equal likelihood in terms of their representation of future insurance claim severities. Means and standard deviations for both the lower and upper severity limits can be elicited. Therefore, the event occurrence severity is a random variable with the cumulative distribution function (CDF), $F_S(s)$, taking on one of the two lognormally distributed random variables (i.e., low and high estimates of the CDF) with equal probability of 0.5. Each of these distributions is defined by its mean and COV. Other distributions can also be used as $F_S(s)$. The mean and standard deviation of the severity are designated as μ_s and σ_s, respectively. These values are then used in the calculation of equivalent normal mean and standard deviation for the lognormal distribution as follows:

$$\mu_y = \ln(\mu_s) - \frac{1}{2}\sigma_y^2 \tag{7.20a}$$

$$\sigma_y = \sqrt{\ln\left[1 + \left(\frac{\sigma_S^2}{\mu_s}\right)^2\right]} \tag{7.20b}$$

Having defined the normal-equivalent mean and standard deviation, the density function for the lognormal distribution may be shown as

$$F_S(s) = \frac{1}{s\sigma_y\sqrt{2\pi}} e^{-\frac{1}{2}\left(\frac{\ln(s)-\mu_y}{\sigma_y}\right)^2} \tag{7.21}$$

Having identified the major components for modeling the magnitude of the insurance claims, two cases are considered here. The annual rate of claims is regarded first as nonrandom and second as random. Both cases examine the magnitude of claims over a time period t in years (e.g., t = [0, 10]). A stochastic model is therefore gradually constructed in this section as provided under separate headings. Two cases are considered as follows: (1) a fundamental loss accumulation model in which the rate is known either as a nonrandom value or as a random value and the severity is represented by a probability distribution and (2) an extension of the first case, where severity is assessed based on the opinion of several experts. These two cases are developed in the subsequent sections.

7.6.2.1 Modeling Loss Accumulation

The rate (λ) is initially considered to be a nonrandom quantity. Randomness in the rate is added to the model at the end of the section. The severity of each event is modeled using a continuous random variable with the CDF $F_S(s)$. The CDF of the accumulated damage (loss) during a nonrandom time interval [0, t] is given by

$$F(s; t,\lambda) = \sum_{n=0}^{\infty} e^{-\lambda t}\frac{(\lambda t)^n}{n!}F_S^{(n)}(s) \tag{7.22}$$

where:

$F_S^{(n)}(s)$ is the n-fold convolution of $F_S(s)$

In other words, $F_S^{(n)}(s)$ is the probability that the total loss accumulated over n events (during time t) does not exceed s. For $n = 0$, $F_S^{(0)}(s)$ is defined as $F_S^{(0)}(s) = 1$; for $n = 1$, $F_S^{(1)}(s) = F_S$ [i.e., the CDF of S using the mean (μ) and the standard deviation (σ) of S]. For $n = 2$, the twofold convolution $F_S^{(2)}(s)$ can be evaluated using conditional probabilities as

$$F_S^{(2)}(s) = P(S + S < s) = \int_0^{\infty} F_S(s - x) f_S(x) \mathrm{d}x$$

where:

P is the probability

$f_S(s)$ is the density function of severity

This result can be expressed as

$$F_S^{(2)}(s) = \int_0^{\infty} F_S(s - x) \mathrm{d}F_S(x)$$

In the case of a normal probability distribution, the twofold convolution $F_S^{(2)}(s)$ can be evaluated as follows:

$$F_S^{(2)}(s) = P(S + S < s) = F_S(s; 2\mu, \sqrt{2}\sigma)$$

where:

$F_S(s; 2\mu, \sqrt{2}\sigma)$ is the CDF of $(S + S)$ that can be evaluated using the normal CDF of S with a mean value of 2μ and a standard deviation of $\sqrt{2}\sigma$ for uncorrelated and identical severities

For other distribution types, the distribution of the sum $S + S$ needs to be used. In general, for the case of $S + S$, the following special relations can be used:

- $S + S$ is normally distributed if S is normally distributed.
- $S + S$ has a gamma distribution if S has an exponential distribution.
- $S + S$ has a gamma distribution if S has a gamma distribution.

The threefold convolution $F_S^{(3)}(s)$ is obtained as the convolution of the distributions of $F_S^{(2)}(s)$ and $F_S(s)$. For uncorrelated and identical severities represented by a normal probability distribution, the threefold convolution is

$$F_S^{(3)}(s) = P(S + S + S < s) = F_S(s; 3\mu, \sqrt{3}\sigma)$$

Higher order convolution terms can be constructed in a similar manner for n uncorrelated and identical severities represented by a normal probability distribution as follows:

$$F_S^{n}(s) = P(S + S + \cdots + S < s) = F_S(s; n\mu, \sqrt{n}\sigma)$$

The above equation includes the sum of n identical and independent random variables S. Therefore, Equation 7.22 can be written for uncorrelated and identical severities represented by a normal probability distribution as follows:

$$F(s; t, \lambda) = \sum_{n=0}^{\infty} \mathrm{e}^{-\lambda t} \frac{(\lambda t)^n}{n!} F_S(s; n\mu, \sqrt{n}\sigma)$$

If λ is random with the PDF $f_\lambda(\lambda)$, Equation 7.22 can be modified to:

$$F(s;t) = \int_0^\infty \left(\sum_{n=0}^\infty e^{-\lambda t} \frac{(\lambda t)^n}{n!} F_S^{(n)}(s) \right) f_\lambda(\lambda) \mathrm{d}\lambda \tag{7.23}$$

where:
$F_S^{(n)}(s)$ is the n-fold convolution of $F_S(s)$

7.6.2.2 Subjective Severity Assessment

Information on severity might not be available, thus requiring the use of expert opinions as discussed in Chapter 8. If an expert provides two distributions of severity, $F_{S\max}(s)$ and $F_{S\min}(s)$, Equation 7.22 must be replaced by the respective mixture of the two distributions with equal weights. In general, for $j = 1, 2,\ldots, k$ experts, the distribution of accumulated damage (loss) can be represented using one of the following approaches: (1) the respective mixture of the distributions given by Equation 7.23 or (2) the distribution of weighted average with appropriately chosen weights w_j ($j = 1, 2,\ldots, k$). These two approaches are described in this section.

For the mixture of distributions, the CDF of the accumulated damage (loss) during a nonrandom time period $[0, t]$ is given by the following expression based on Equation 7.23:

$$F(s;t) = \sum_{j=1}^{k} w_j F_j(s;t) \tag{7.24}$$

where:
$\sum_{j=1}^{k} w_j = 1$

$$F_j(s;t) = \int_0^\infty \left(\sum_{n=0}^\infty e^{-\lambda t} \frac{(\lambda t)^n}{n!} F_{Sj}{}^{(n)}(s) \right) f_{\lambda j}(\lambda) \mathrm{d}\lambda$$

For the distribution of weighted average, the accumulated damage (loss) distribution is the k-fold weighted convolution of the distributions of Equation 7.23:

$$F(s;t) = F_{w_j}^{(k)}(s;t) \tag{7.25}$$

where:
$F_{wj}(s;t) = F_j(w_j s;t)$ is associated with weight w_j from $\sum_{j=1}^{k} w_j = 1$

For equal weights, each weight is given by

$$w_j = \frac{1}{k} \tag{7.26}$$

In this case, because the distributions $F_j(s;t)$ ($j = 1, 2,\ldots, k$) are assumed to be independent, the mean (μ_S) and the standard deviation (σ_S) of $F(s;t)$ are expressed in terms of the mean (μ_{S_j}) and the standard deviation (σ_{S_j}) of the distributions $F_j(s;t)$ as

$$\mu_S = \sum_{j=1}^{k} \frac{\mu_{S_j}}{k} \tag{7.27a}$$

$$\sigma_S = \frac{1}{k}\left(\sum_{j=1}^{k} \sigma^2{}_{S_j}\right)^{1/2} \tag{7.27b}$$

The closed-form solution of Equation 7.24 can be obtained for some distribution families (e.g., the normal one).

7.6.2.3 Computational Procedures and Illustrations

A computational model based on the above probabilistic model can be developed using some analytical approximations, numerical methods, and/or Monte Carlo simulation approaches including efficient algorithms such as Latin hypercube sampling and importance sampling. The computational procedure has the following features:

- Input data (k experts)
- Distributions of λ and damage (loss)
- Evaluation of Equation 7.23 for each expert
- Combining the results from the previous steps in numerical and graphical forms of a mixed distribution solution based on Equation 7.24 or an averaging distribution solution based on Equation 7.25

Example 7.5: One Expert and Nonrandom Event Occurrence Rate

The numerical example presented in this section illustrates the case of one expert and nonrandom rate λ. The computations for the case of multiple experts can be constructed directly through extension.

The event rate (λ) is assumed to have a value of one event per year, and the loss as a result of one event occurrence is assumed to have the normal distribution with mean (μ_S) of 3 and standard deviation (σ_S) of 0.2 (in \$1000). The model given by Equation 7.22 was evaluated for the following time intervals: $t = 1, 2, 3,$ and 4 years.

To provide the accuracy acceptable for practical applications, the summation of Equation 7.22 includes 11 terms in this case. The normal distribution was used to evaluate the operation of convolution. The n-fold convolution $F_S^{(n)}(s)$ is the normal distribution having a mean equal to the mean of the underlying distribution, $F_S(s)$, multiplied by n, and the respective variance is increased by the same factor n. Selected computational steps of accumulated damage (loss) distributions are illustrated in Table 7.8, where $F(s;t,\lambda)$ is the cumulative probability distribution of the accumulated loss for a time period of t and a rate of occurrence of λ. The table shows sample computations for $t = 1$ year.

The accumulated damage (loss) distributions evaluated for all time intervals are shown in Figure 7.10. The computational steps in each function are associated with the successive convolutions in the sum of Equation 7.22. The figure shows that the median of loss increases as the time exposure increases. Similar statements can be made about other percentiles. The table summarizes the evaluation of the infinite sum of Equation 7.22 using an approximation of 11 terms. The contribution of each term diminishes as n becomes larger. Terms should be accumulated until the contributions become insignificant. The model is evaluated for selected s values as provided in the columns of the tables. Selected s values are provided in the table for demonstration purposes.

TABLE 7.8

Accumulated Damage (Loss) Distribution Based on Equation 7.22 for $t = 1$ year

Number of Events (n) in t	Occurrence Probability of n Events $= e^{-\lambda t}\frac{(\lambda t)^n}{n!}$	$e^{-\lambda t}\frac{(\lambda t)^n}{n!}F_S^{(n)}(s)$ Evaluated at s							
		0.3	0.6	…	3.3	…	6.6	…	36
0	0.367879	3.679E−01	3.679E−01		3.679E−01		3.679E−01		3.679E−01
1	0.367879	0.000E+00	0.000E+00		3.433E−01		3.679E−01		3.679E−01
2	0.18394	0.000E+00	0.000E+00		0.000E+00		1.808E−01		1.839E−01
3	0.061313	0.000E+00	0.000E+00		0.000E+00		1.315E−13		6.131E−02
4	0.015328	0.000E+00	0.000E+00		0.000E+00		0.000E+00		1.533E−02
5	0.003066	0.000E+00	0.000E+00		0.000E+00		0.000E+00		3.066E−03
6	0.000511	0.000E+00	0.000E+00		0.000E+00		0.000E+00		5.109E−04
7	7.3E−05	0.000E+00	0.000E+00		0.000E+00		0.000E+00		7.299E−05
8	9.12E−06	0.000E+00	0.000E+00		0.000E+00		0.000E+00		9.124E−06
9	1.01E−06	0.000E+00	0.000E+00		0.000E+00		0.000E+00		1.014E−06
10	1.01E−07	0.000E+00	0.000E+00		0.000E+00		0.000E+00		1.014E−07
11	9.22E−09	0.000E+00	0.000E+00		0.000E+00		0.000E+00		9.216E−09
	$F(s;t,\lambda) = \sum_{n=0}^{11} e^{-\lambda t}\frac{(\lambda t)^n}{n!}F_S^{(n)}(s) =$	3.68E−01	3.68E−01		7.11E−01		9.17E−01		1.00

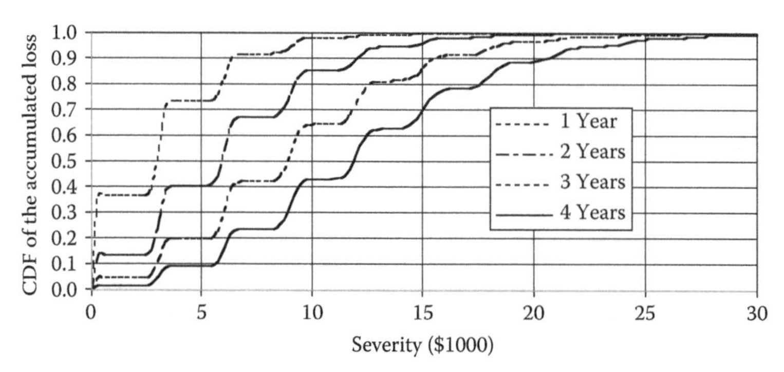

FIGURE 7.10
Cumulative distribution functions (CDFs) of the accumulated loss, $F(s;t,\lambda)$, with nonrandom annual rate.

7.7 Benefit–Cost Analysis

Many decision situations involve multiple hazards and potential failure scenarios. For cases involving several credible consequence scenarios, the risks associated with each can be assessed as the product of the corresponding probabilities and consequences, and the results summed up to obtain the total risk. If the risk is not acceptable, mitigation actions should be considered to reduce it. Justification for these actions can be developed based on benefit–cost analysis. The costs in this case are associated with mitigation actions. The benefits that are associated with mitigation actions can be classified as follows:

- Reduction in the number of severe accidents that lead to reduced fatalities, reduced injuries, and reduced property and environmental loss
- Reduction in the number of incidents (i.e., minor accidents) that lead to reduced injuries and reduced property and environmental losses
- Reduction in the number of incidents and accident precursors leading to reduced errors and deviations, reduced equipment failures, reduced property and environmental losses, and so on
- Secondary and tertiary benefits as a result of intangibles

Benefit assessment sometimes requires the development and use of categories of products and users to obtain meaningful results. An illustrative example of this requirement is the examination of survival data based on the use of PFDs, as provided in the following hypothetically constructed data:

Case	Adults	Children	Adults and Children
Wearing PFDs	$\frac{98}{100}=0.98$	$\frac{320}{400}=0.80$	$\frac{418}{500}=0.836$
Not wearing PFDs	$\frac{950}{1000}=0.95$	$\frac{250}{400}=0.625$	$\frac{1200}{1400}=0.857$

The results for adults in the table show that wearing a PFD reduces drowning risk. Similarly, the results for children also show that wearing a PFD reduces drowning risk. In addition, the data assumed here show that children always wear PFDs, whereas adults do not wear PFDs most of the time. The last column in the table shows the combined results for adults and children without user categories. This combined case produces illogical values—wearing PFDs does not reduce drowning risk. In this case, the large differences between the counts of adults and children, combined with survival rates that depend on respective categories, result in the illogical final results. It is evident from this example that computing frequency reduction as a benefit should be based on properly and carefully constructed categories. The construction of these categories depends on the decision situation.

The present value of incremental costs and benefits can be assessed and compared among alternatives that are available for risk mitigation or system design. Several methods are available to determine which, if any, option is most worth pursuing. In some cases, no alternative will generate a net benefit relative to the base case. Such a finding would be used to argue for pursuit of the base case scenario. The following are the most widely used present value comparison methods (as discussed in Chapter 6):

- NPV
- Benefit-to-cost ratio
- Internal rate of return (IRR)
- Payback period

The NPV method requires that each alternative must meet the following criteria to warrant investment of funds: (1) should have a positive NPV and (2) should have the highest NPV of all alternatives considered. The first condition ensures that the alternative is worth undertaking relative to the base case; that is, it contributes more in incremental benefits than it absorbs in incremental costs. The second condition ensures that maximum benefits are obtained in a situation of unrestricted access to capital funds. The NPV can be calculated as follows:

$$\text{NPV} = \sum_{t=0}^{k} \frac{(B-C)_t}{(1+r)^t} = \sum_{t=0}^{k} \frac{B_t}{(1+r)^t} - \sum_{t=0}^{k} \frac{C_t}{(1+r)^t} \tag{7.28}$$

where:

B is the future annual benefits in constant dollars
C is the future annual costs in constant dollars
r is the annual real discount rate
k is the number of years from the base year over which the project will be evaluated
t is an index running from 0 to k representing the year under consideration

The benefit of a risk mitigation action can be assessed as follows:

$$\text{Benefit} = \text{unmitigated risk} - \text{mitigated risk} \tag{7.29}$$

The cost associated with Equation 7.29 is the cost of the mitigation action. The benefit minus the cost of mitigation can be used to justify the allocation of resources. The benefit-to-cost ratio can be computed as follows and may also be helpful in decision making:

$$\text{Benefit-to-cost ratio}\left(\frac{B}{C}\right) = \frac{\text{Benefit}}{\text{Cost}} = \frac{\text{Unmitigated risk} - \text{mitigated risk}}{\text{Cost of a mitigation action}} \tag{7.30}$$

Ratios >1 are desirable. In general, the larger the ratio, the better the mitigation action.

Accounting for the time value of money would require defining the benefit-to-cost ratio as the present value of benefits divided by the present value of costs. The benefit-to-cost ratio can then be calculated as follows:

$$\frac{B}{C} = \frac{\sum_{t=0}^{k} \frac{B_t}{(1+r)^t}}{\sum_{t=0}^{k} \frac{C_t}{(1+r)^t}} \tag{7.31}$$

where:

B_t is the future annual benefits in constant dollars
C_t is the future annual costs in constant dollars
r is the annual real discount rate
t is an index running from 0 to k representing the year under consideration

A proposed activity with a B/C ratio of discounted benefits to costs of ≥1 is expected to return at least as much in benefits as it costs to undertake, indicating that the activity is worth undertaking.

The IRR is defined as the discount rate that makes the present value of the stream of the expected benefits in excess of the expected costs equal 0 (as discussed in Chapter 6). In other words, it is the highest discount rate at which the project will not have a negative NPV. To apply the IRR criterion, it is necessary to compute the IRR and then compare it with a base rate of, say, a 7% discount rate. If the real IRR is <7%, the project would be worth undertaking relative to the base case. The IRR method is effective in deciding whether a project is superior to the base case; however, it is difficult to utilize it for ranking projects and deciding among mutually exclusive alternatives. Project rankings established by the IRR method might be inconsistent with those of the NPV criterion. Moreover, a project might have more than one IRR value, particularly when a project entails major final costs, such as cleanup costs. Solutions to these limitations exist in capital budgeting procedures and practices that are often complicated or difficult to employ in practice and present opportunities for error.

The payback period measures the number of years required for net undiscounted benefits to recover the initial investment in a project (as discussed in Chapter 6). This evaluation method favors projects with near-term and more certain benefits and fails to consider the benefits beyond the payback period. The method does not provide information on whether an investment is worth undertaking in the first place.

Another issue of interest is the timing to implement an action. The optimal project timing is frequently ignored in economic analysis but is particularly important in the case of large infrastructure projects, such as road improvements. In some cases, benefit–cost analysis may reveal that a greater net benefit can be realized if a project is deferred for several years rather than implemented immediately. Such a situation has a higher likelihood of occurring if the following conditions are met:

- The project benefit stream is heavily weighted to the later years of the project life.
- The project is characterized by large, up-front capital costs.
- Capital and land cost escalation can be contained through land banking or other means.

For example, a project NPV can be calculated for the following two-time scenarios to assess delaying the start of the project by d years: without delay (NPV) and with delay (NPV_d):

$$NPV = \sum_{t=0}^{k} \frac{(B-C)_t}{(1+r)^t} \tag{7.32a}$$

$$NPV_d = \sum_{t=d}^{k+d} \frac{(B-C)_t}{(1+r)^t} \tag{7.32b}$$

To resolve the issues of optimal timing, the NPV for each alternative should be measured for both the current and delayed time scenarios to identify the best alternative and the best starting time.

The models for benefit–cost analysis presented in this section have not accounted for the full probabilistic characteristics of B and C in their treatment. Concepts from reliability assessment of Chapter 4 can be used for this purpose. Assuming B and C to be normally distributed, a benefit–cost index ($\beta_{B/C}$) can be defined similar to Equation 4.7 as follows:

$$\beta_{B/C} = \frac{\mu_B - \mu_C}{\sqrt{\sigma_B^2 + \sigma_C^2}} \tag{7.33}$$

where:
μ and σ are the mean and standard deviation, respectively

The failure probability, interpreted in this case as the probability of realizing the benefit can be computed as

$$P_{f,B/C} = P(C > B) = 1 - \Phi(\beta_{B/C}) \tag{7.34}$$

In the case of lognormally distributed B and C, the benefit–cost index ($\beta_{B/C}$) can be computed as

$$\beta_{B/C} = \frac{\ln\left(\dfrac{\mu_B}{\mu_C}\sqrt{\dfrac{\delta_C^2 + 1}{\delta_B^2 + 1}}\right)}{\sqrt{\ln[(\delta_B^2 + 1)(\delta_C^2 + 1)]}} \tag{7.35}$$

where:
δ is the COV

Equation 7.34 also holds for the case of lognormally distributed B and C. In the case of mixed distributions or cases involving the basic random variables of B and C, the advanced second moment method of Section 4.2.1 or the simulation method of Section 4.2.2 can be used. In cases where benefit is computed as revenue minus cost, benefit might be correlated with cost, requiring the use of the techniques found in Sections 4.2.1.4 and 4.2.2.4.

Example 7.6: Protection of Critical Infrastructure

This example is used to illustrate the cost of benefit–cost analysis using a simplified decision situation. As an illustration, assume that there is a 0.01 probability of an attack on a facility containing hazardous materials during the next year. If the attack occurs, the probability of a serious release to the public is 0.01, with a total consequence of

$100 billion. The total consequence of an unsuccessful attack is negligible. The unmitigated risk can therefore be computed as

$$\text{Unmitigated risk} = 0.01(0.01)(\$100 \times 10^9) = \$10 \times 10^6$$

If armed guards are deployed at each facility, the probability of attack can be reduced to 0.001 and the probability of serious release if an attack occurs can be reduced to 0.001. The cost of the guards for all plants is assumed to be $100 million per year. The mitigated risk can therefore be computed as

$$\text{Mitigated risk} = 0.001(0.001)(\$100 \times 10^9) = \$0.10 \times 10^6$$

The benefit in this case is

$$\text{Benefit} = \$10 \times 10^6 - \$0.1 \times 10^6, \text{ or } {\sim}\$10 \times 10^6$$

The benefit-to-cost ratio is about 0.1; therefore, the $100 million cost might be difficult to justify.

Example 7.7: Efficient Frontier in Benefit–Cost Analysis for a Mode of Transportation

Four transportation modes are being considered by the management of a warehousing company to supply components from the warehouse to one of its major customers in a foreign country. The available alternatives for the modes of transport are (1) road and ferry (A_1), (2) rail and ferry (A_2), (3) air (A_3), and (4) sea (A_4). The management team of the company was not certain of the cost and return values of the alternatives. A brainstorming session by the management team produced probabilistic information for costs and revenues associated with each alternative, as shown in Table 7.9. Table 7.10 shows

TABLE 7.9

Assessments of Modes of Transportation for Delivery to Foreign Clients

Cost		Revenue	
Estimated NPV of Cost ($ millions)	**Probability**	**Estimated NPV of Revenue ($ millions)**	**Probability**
A_1: road and ferry			
100	0.6	300	0.5
90	0.3	250	0.4
80	0.1	200	0.1
A_2: rail and ferry			
80	0.4	210	0.3
70	0.4	225	0.4
35	0.2	240	0.3
A_3: air			
100	0.6	140	0.5
90	0.3	120	0.4
80	0.1	110	0.1
A_4: sea			
140	0.1	200	0.2
120	0.1	100	0.4
10	0.8	80	0.3
–	–	50	0.1

NPV, net present value.

the calculation of the benefits and the benefit-to-cost ratio (B/C) associated with each alternative. From Table 7.10, the alternatives can be ranked based on the B/C ratios to conclude that alternative A_2 is the best choice, with the largest ratio of 2.36, followed in order by alternatives A_4, A_1, and A_3. Figure 7.11 shows the results graphically along with the efficient frontier that includes the most appealing alternatives A_1, A_2, and A_4. Alternative A_3 is considered a risky alternative with low benefit value and high cost value compared to other alternatives. Assuming that the management team is risk averse, from Figure 7.11 alternative A_2 gives the highest benefit ($158 million) with the least cost ($67 million), which is in agreement with the selection based on its greatest B/C ratio of 2.36.

This example can be developed further by computing the standard deviations of the benefits as shown in Table 7.11. The efficient frontier based on the mean and the standard deviation of benefit is shown in Figure 7.12. The figure shows A1 as the most appropriate option.

TABLE 7.10

Benefit-to-Cost Ratios for the Modes of Transportation

Alternatives	Cost ($\$10^6$)	Revenue ($\$10^6$)	Benefits ($\$10^6$)	B/C	Rank
A_1: road and ferry	95	270	175	1.84	3
A_2: rail and ferry	67	225	158	2.36	1
A_3: air	95	129	34	0.36	4
A_4: sea	34	109	75	2.21	2

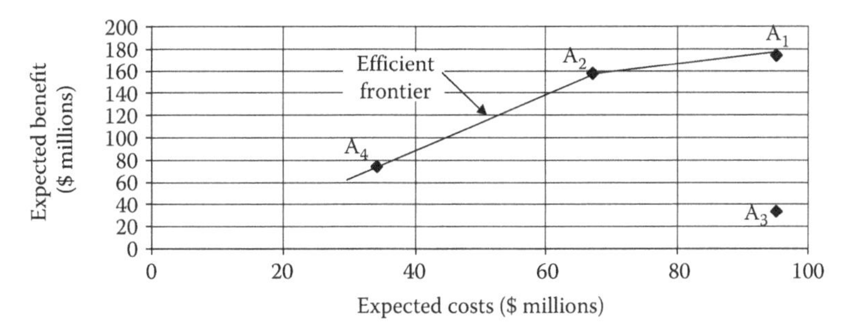

FIGURE 7.11
Efficient frontier for the benefit–cost analysis of transportation modes.

TABLE 7.11

Mean and Standard Deviation of Benefits for the Modes of Transportation

Alternative	Mean Cost ($\$10^6$)	Standard Deviation of Cost ($\$10^6$)	Mean Revenue ($\$10^6$)	Standard Deviation of Revenue ($\$10^6$)	Mean Benefits ($\$10^6$)	Standard Deviation of Benefits ($\$10^6$)
A_1: road and ferry	95	6.71	270	33.17	175	33.84
A_2: rail and ferry	67	16.61	225	12.55	158	20.82
A_3: air	95	6.71	129	11.36	34	13.19
A_4: sea	34	48.21	109	47.84	75	67.92

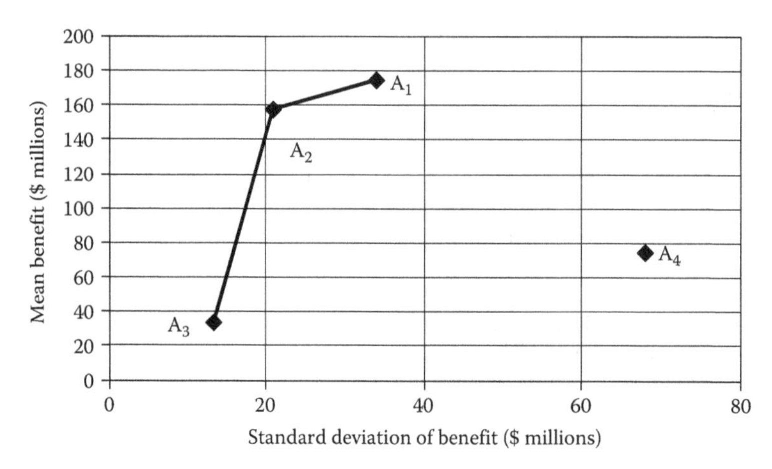

FIGURE 7.12
Efficient frontier using the mean and standard deviation of the benefit of transportation modes.

7.8 Risk Financing

Risk financing can be defined as the activities associated with providing funds to cover the financial effect of unexpected losses experienced by an entity. Risk financing includes the traditional ways of risk transfer by funding reserves for self-insurance and risk pooling. Other contemporary risk finance methods are based on products and solutions offered by banking and insurance industry, such as captive insurance companies and catastrophic bonds.

Risk can be financed using catastrophe bonds (also known as cat bonds) using risk-linked securities that transfer a specified set of risks from a sponsor to investors. Such a risk financing method was created and first used in the mid-1990s in the aftermath of Hurricane Andrew and the Northridge earthquake. The motive behind catastrophe bonds by insurance companies is to alleviate some of the risks they would face if a major catastrophe occurred; that is, they would spread the loss to a larger group of investors in return from their investments using the premiums. For example, an insurance company issues bonds through an investment bank, which are then sold to investors. If no catastrophe occurred, the insurance company would pay a coupon, that is, a fixed percentage return, to the investors; however, on the contrary, if a catastrophe did occur, then the principal would be forgiven and the insurance company would use this money to pay their claim holders. Such investments are usually subscribed by hedge funds, catastrophe-oriented funds, and asset managers to create diversifications in portfolios that are driven by other markets. The catastrophe bonds are often structured as floating rate bonds, the principal of which is lost if specified trigger conditions are met such as major natural catastrophes, for example, an earthquake within an epicenter zone of magnitude that exceeds a particular threshold and with losses in Tokyo. These catastrophe bonds are typically used by insurers as an alternative to traditional catastrophe reinsurance to increase capacity and maintain solvency in case of an event.

For example, an insurer that built up a portfolio of risks by insuring properties in Florida might wish to manage its risk by transferring some of this risk by simply purchasing traditional catastrophe reinsurance or sponsoring a catastrophe bond. The catastrophe bond would pass the risk on to investors. Sponsoring a catastrophe bond requires the consultation with an investment bank to create a special purpose entity that would issue the catastrophe bond. Investors would buy the catastrophe bond that might pay them a coupon based on a market indicator such as the London Interbank Offered Rate (LIBOR) plus a spread of about 3% and 20%. Two possible outcomes for investors are as follows:

1. If no hurricane hit Florida, the investors would make return on the investment.
2. If a hurricane were to hit Florida and trigger the catastrophe bond, the principal initially paid by the investors would be forgiven, that is, lost, and instead used by the sponsor to pay its claims to policy holders.

A captive insurance company is of a special type by being established with the specific objective of insuring risks emanating from their parent group(s), but they sometimes also insure risks of the group's customers. Captive insurance companies, also called captives for short, are licensed by many jurisdictions, called as their domiciles. Most captive insurers are based *offshore,* in places such as Belize, Bermuda, the Cayman Islands, Ireland, and Dubai International Financial Centre.

An example is used in this section to illustrate the use of risk financing in the construction of a large infrastructure project.

Example 7.8: Risk Financing of 2001 Construction of the Rail Link Connecting London with the Channel Tunnel to Paris, France

The UK Department for Transport (DFT) awarded in 1996 a contract to the London and Continental Railways Limited (LCR), a private sector consortium, to build the Channel Tunnel Rail Link (the Link), a high-speed railway linking St. Pancras Station, London, to the Channel Tunnel connecting to Paris, France. According to 2006 report of the House of Commons Committee of Public Accounts on the Channel Tunnel Rail Link, the construction of the Link was to have been funded partly by government grants and LCR borrowing money, secured on future revenue from the operator of the Link, Eurostar, UK. By the end of 1997, Eurostar's revenues were well below LCR's forecasts, and consequently, LCR abandoned its plans to borrow money and approached DFT for an increase in the government grants. The government agreed to finance the project in two sections. The marginal economic justification of the project and changes in the team to build Section 2 resulted in restructuring the deal with LCR backed by DFT and Bechtel and a group of insurers sharing construction risk for Section 2. This 2001 deal included LCR paying Bechtel, also on the design–build team, and the insurers £87 million to bear £315 million of the first £600 million of any cost-construction overrun in a layered liability sharing structure. Additional information on the structure of the deal is available in Pollalis (2006) and the House of Commons Committee of Public Accounts (2006).

7.9 Loss Exposure and Residual Risk

Risk managers are commonly interested in knowing the exposure level of an entity as a result of various activities involving risks. Exposure is defined as the extent to which an organization's and/or stakeholder's concerns are subject to an event, and defined by things at risk that might include population at risk, property at risk, and ecological and environmental concerns at risk. Loss exposure can be defined as the likely maximum loss amount for an entity, also called *probable maximum loss* (PML) or *maximum foreseeable loss* (MFL). Universally accepted quantitative definitions of PML or MFL are unavailable, and someone perhaps could introduce definitions such as the 99th percentile loss value and the maximum loss value, respectively. The importance of knowing the loss exposure levels is to assess their impacts on an enterprise in order to ensure that the enterprise has the capacity or the means to obtain the capacity to meet its obligations. Sometimes insurers are interested to know the *maximum possible loss* (MPL) as the worst loss that could occur based on the worst possible combination of circumstances.

Residual risks are products of accepted risks through self-insurance, contractual and insurance exclusions, risks within time periods to statuary limitations or repose times, and so on. Such residual risks can become cumulative in nature in some cases, and could lead to a significant exposure and the potential for class actions. They can be managed by legal and insurance means.

Example 7.9: Impact of Sea-Level Rise or Potential Storm Surge on Washington, DC

Figure 7.13 shows inundation maps of Washington, DC, as an example, using hypothetical sea-level rise (SLR) of 0.1, 0.4, 1.0, 2.5, and 5.0 m. Each image shows the clipped shape of the river and the data layer of the streets of Washington, DC (Ayyub et al. 2012). The *Washington Post* (2012) reproduced the results of Figure 7.13 as shown in Figure 7.14.

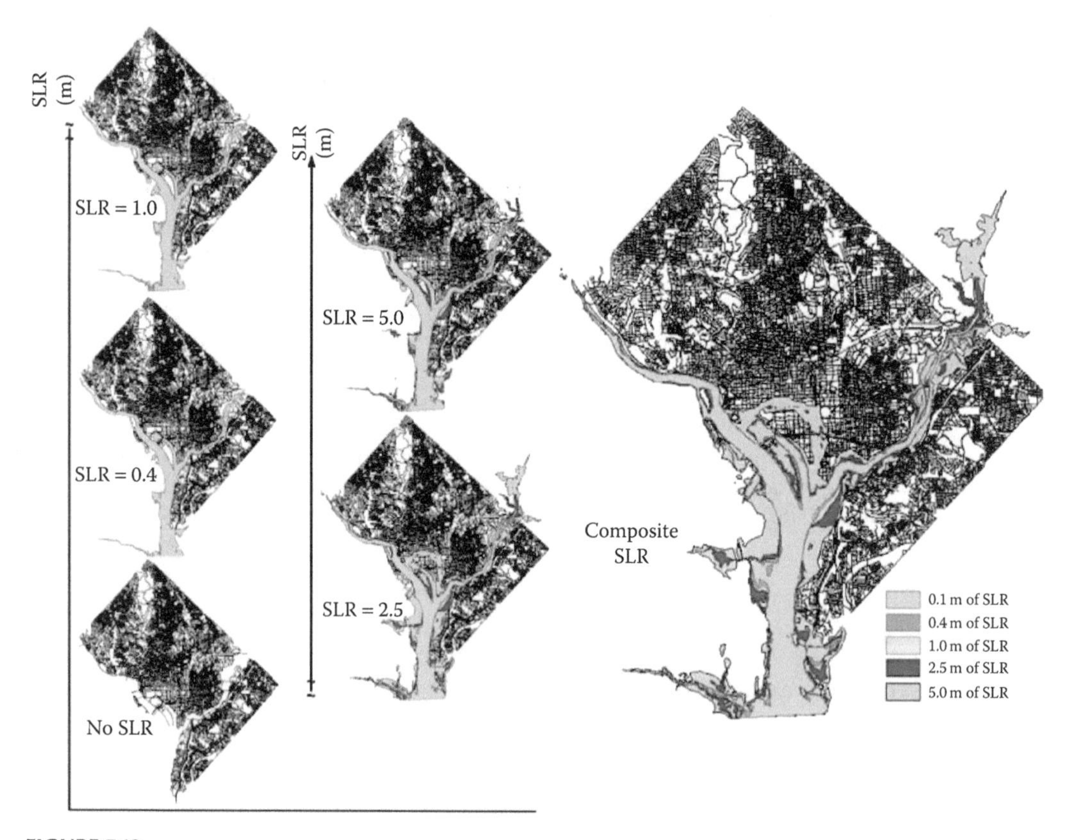

FIGURE 7.13
Impact of sea-level rise (SLR) on Washington, DC, for 0.1, 0.4, 1, 2.5, and 5 m. SLR, sea-level rise. (From Ayyub, B.M., Braileanu, H.G., and Qureshi, N., *Risk Analysis: An International Journal*, 32, 1901–1918, 2012. With permission.)

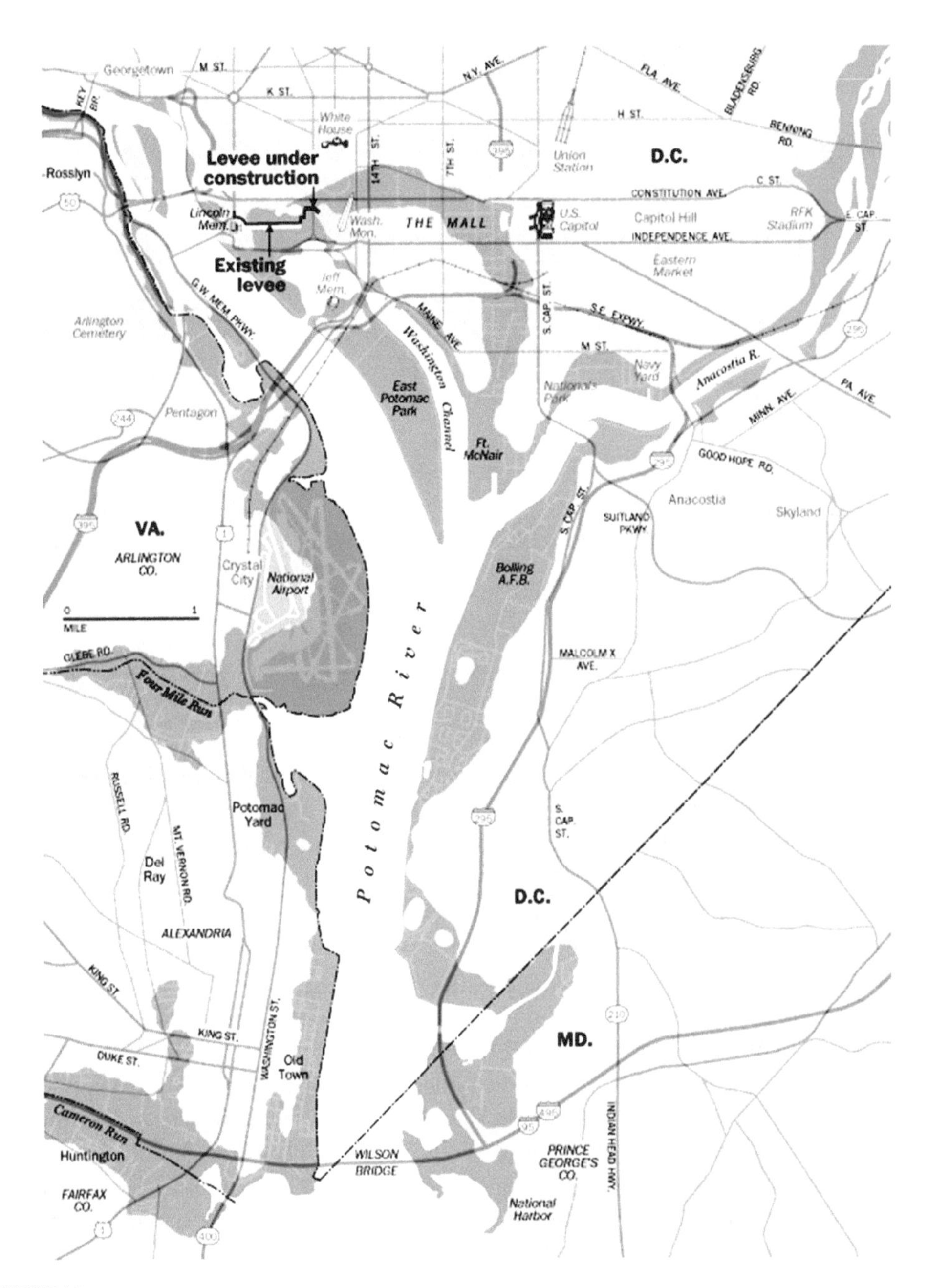

FIGURE 7.14
The impact of a powerful hurricane making landfall around Virginia Beach, Washington, DC. (*Washington Post* 2012 based on results by From Ayyub, B.M., Braileanu, H.G., and Qureshi, N., *Risk Analysis: An International Journal*, 32, 1901–1918, 2012. With permission.)

The *Washington Post* interpreted the results from a powerful hurricane making landfall around Virginia Beach that would push loads of water into the Chesapeake Bay, causing a massive storm surge up the Potomac. Figure 7.15 shows the direct monetary losses of residential and some commercial properties using the city's databases as a function of SLR (Ayyub et al. 2012).

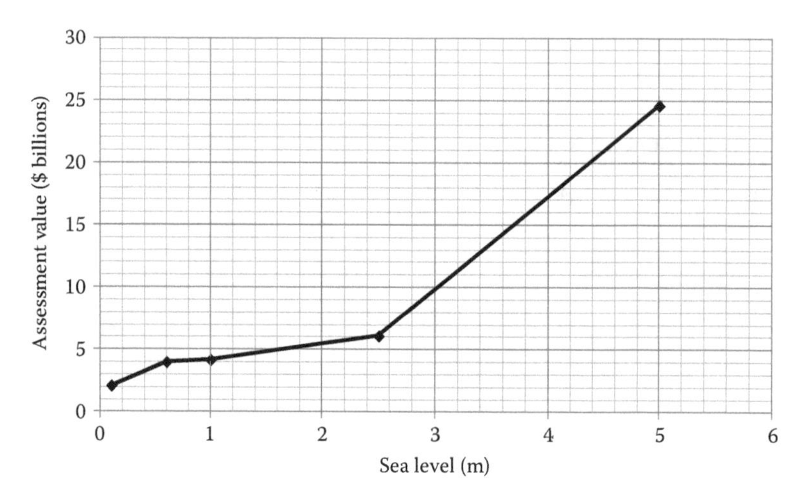

FIGURE 7.15
Assessment value of properties in DC vs. sea-level rise (SLR). (From Ayyub, B.M., Braileanu, H.G., and Qureshi, N., *Risk Analysis: An International Journal*, 32, 1901–1918, 2012. With permission.)

7.10 Exercise Problems

Problem 7.1 What is the meaning of *risk control* and what is its objective? Why is it important to consider in risk assessment studies?

Problem 7.2 What are the different philosophies of risk control? Explain them by developing risk strategies for simple examples.

Problem 7.3 What are the three types of measurements required for defining a risk control philosophy? Give examples for each type of these measurements.

Problem 7.4 How can risk be controlled using economic analysis? What is the meaning of *utility* and what are its axioms? Why is utility important in investment decisions? What are the types of risk–attitude curves?

Problem 7.5 Use the information given in Tables 7.1 and 7.2 to draw the corresponding expected NPV decision tree showing the probability values for equal likelihood, increasing likelihood, and decreasing likelihood for alternatives A and C.

Problem 7.6 Use the information given in Tables 7.1 and 7.2 to draw the corresponding expected NPV decision tree showing the probability values for equal likelihood, increasing likelihood, and decreasing likelihood for alternatives B and C.

Problem 7.7 Use the information given in Tables 7.3 and 7.4 to draw the corresponding utility decision tree showing the probability values for equal likelihood, increasing likelihood, and decreasing likelihood for alternatives A and B.

Problem 7.8 Use the information given in Tables 7.3 and 7.4 to draw the corresponding utility decision tree showing the probability values for equal likelihood, increasing likelihood, and decreasing likelihood for alternative C.

Problem 7.9 Use the information given in Tables 7.1 and 7.3 and utility functions given by Equations 7.7 and 7.10 to draw the corresponding utility curves for decision alternative A. Do the curves that correspond to the two equations differ? Why or why not?

Problem 7.10 Use the information given in Tables 7.1 and 7.3 and utility functions given by Equations 7.7 and 7.10 to draw the corresponding utility curves for decision alternatives B and C. Do the curves that correspond to the two equations differ for each alternative? Why or why not?

Problem 7.11 Using the information given in Table 7.2, plot the expected NPV against the standard deviation of NPV for the three decision alternatives for equal likelihood, increasing likelihood, and decreasing likelihood. What is the optimal alternative based only on the expected NPV information?

Problem 7.12 What do (a) *minimum variance frontier* and (b) *efficient frontier* mean? Using the information given in Table 7.2, plot the efficient frontier curves for the three alternatives based on equal likelihood, increasing likelihood, and decreasing likelihood.

Problem 7.13 Use Equation 7.17a and 7.17b of Example 7.1 to draw the utility curves for alternative B for equal likelihood, increasing likelihood, and decreasing likelihood.

Problem 7.14 A financial services corporation is considering five alternative sites for moving its head office in the near future. Preliminary assessments revealed varying expected and standard deviations of profit for each location as a result of variations in revenues gained from such a move and the associated costs incurred from renting these locations as follows:

Site Alternative	Expected Profit ($1000)	Standard Deviation of Profit ($1000)
A	150	50
B	350	150
C	450	130
D	600	115
E	750	230

Use the following utility (U) function in terms of profit (P) to represent the risk attitude of the corporation:

$$U(P) = 0.25P - 0.0001P^2$$

to plot the efficient frontier and utility curves for this investment situation and to recommend the optimal alternative. (*Hint:* Plot the standard deviation on the vertical axis as shown in Figure 7.9.)

Problem 7.15 Use Equation 7.17a and 7.17b of Example 7.1 to draw the utility curves for alternative C for equal likelihood, increasing likelihood, and decreasing likelihood.

Problem 7.16 A chemical company requested bids from mechanical design companies for equipment that will be installed in a mill they own for the purpose of selecting one of the designs. Five design alternatives were submitted and the chemical company management needs to select the optimal design for implementation. The results of simulating the performance of the designs can be summarized in the form of the expected values and standard deviation of profits as follows:

Equipment Alternative	Expected Profit ($1000)	Standard Deviation of Profit ($1000)
A	120	30
B	100	40
C	220	60
D	315	60
E	350	80

Use the following utility (U) function in terms of profit (P) to represent the risk attitude of the corporation:

$$U(P) = 0.23P - 0.00015P^2$$

to plot the efficient frontier and utility curves for this investment situation and to recommend the optimal alternative. (*Hint:* Plot the standard deviation on the vertical axis as shown in Figure 7.9.)

Problem 7.17 Define risk homeostasis and demonstrate its meaning using simple examples from your own experiences.

Problem 7.18 What are the implications of risk homeostasis and its effect on the risk mitigation process?

Problem 7.19 Reevaluate the accumulated damage (loss) of Example 7.5 by changing the event occurrence rate λ to two events per year. The severity associated with an event occurrence is assumed in this problem to follow a normal probability distribution with mean (μ_S) of 4 and standard deviation (σ_S) of 0.3 (both in $1000). Evaluate the cumulative loss accumulation using the time intervals of 2, 4, 6, and 8 years.

Problem 7.20 Reevaluate the accumulated damage (loss) of Example 7.5 by changing the event occurrence rate λ to three events per year. The severity associated with an event occurrence is assumed in this problem to follow a normal probability distribution with mean (μ_S) of 3 and standard deviation (σ_S) of 0.1 (both in $1000). Evaluate the cumulative loss accumulation using the time intervals of 1, 3, 5, and 7 years.

Problem 7.21 What is meant by benefit–cost analysis? What are the formulae that can be used in benefit–cost analysis?

Problem 7.22 ABC Designs wants to compare design alternatives for a crossing structure. The design alternatives are either over or under a major river crossing the city. The alternatives are a bridge with three possible types of designs, denoted as A_1, A_2, A_3, or A_4, or a tunnel (B). These alternative structures will be operated as toll crossing roads. The designer estimated the different costs of alternatives and their respective lifetime revenues (NPV) in the table below. Perform a benefit–cost analysis to find the alternative that provides the optimal B/C ratio. Plot the five alternatives on an efficient frontier curve and indicate your recommendation for the optimal alternative, assuming the designer to be risk averse.

	Cost		Revenue	
Design Alternative	**Estimated NPV of Cost ($10⁶)**	**Probability**	**Estimated NPV of Revenue ($10⁶)**	**Probability**
A_1: suspension bridge	200	0.5	500	0.4
	170	0.3	250	0.4
	150	0.2	220	0.2
A_2: cast *in situ* bridge	150	0.4	300	0.4
	120	0.3	270	0.5
	100	0.3	200	0.1
A_3: cable-stayed bridge	200	0.6	350	0.5
	150	0.3	250	0.4
	120	0.1	210	0.1
A_4: arched steel girder	160	0.4	280	0.5
	130	0.3	220	0.4
	110	0.3	200	0.1
B: tunnel	250	0.7	450	0.3
	220	0.2	350	0.4
	200	0.1	250	0.2
	–	–	200	0.1

NPV, net present value.

Problem 7.23 What is the definition of *benefit* in benefit–cost analysis? Define the difference between unmitigated and mitigated risks.

Problem 7.24 ABC marketing company is considering launching a new product in the market. The marketing manager and her team have prepared five advertising campaign alternatives for marketing the new product. The alternatives with their estimated costs and their corresponding revenues (NPV) as a result of the advertising campaign are presented in the table below. Perform a benefit–cost analysis to determine the alternative that provides the optimal *B/C* ratio. Plot the five alternatives on an efficient frontier curve showing your recommendation of the optimal alternative, assuming the manager to be risk averse.

	Cost		Revenue	
Design Alternative	**Estimated NPV of Cost ($10⁶)**	**Probability**	**Estimated NPV of Revenue ($10⁶)**	**Probability**
A: advertise on radio	50	0.3	250	0.2
	65	0.3	200	0.5
	75	0.4	125	0.3
B: advertise in newspapers	100	0.5	250	0.3
	120	0.2	230	0.5
	130	0.3	200	0.2
C: advertise on television	150	0.4	450	0.4
	250	0.4	350	0.3
	300	0.2	200	0.3

(Continued)

	Cost		Revenue	
Design Alternative	**Estimated NPV of Cost ($\$10^6$)**	**Probability**	**Estimated NPV of Revenue ($\$10^6$)**	**Probability**
D: advertise on billboards	250	0.4	650	0.6
	270	0.4	500	0.2
	300	0.2	450	0.1
	–	–	300	0.1
E: advertise on company Web site	60	0.5	180	0.4
	75	0.3	160	0.5
	100	0.2	130	0.1

NPV, net present value.

8

Data for Risk Studies

Performing quantitative risk assessment requires information on possible failures, failure probabilities, failure rates, failure modes, possible causes, failure consequences, and uncertainties associated with the information and the underlying system and its environment. This chapter provides guidance on data sources, use of databases, and expert opinion elicitation to support risk studies.

CONTENTS

8.1 Introduction

Risk studies require data for defining event scenarios and assessing occurrence probabilities and consequences. Risk studies require failure data that are commonly not available to risk analysts because they represent products that have not worked as originally envisioned and thus could potentially be used in legal actions against a manufacturer or to gain competitive advantage. Therefore, manufacturers, perhaps at the advice of their legal counsel and marketing departments, do not often reveal such data freely. Due to the scarcity of failure data, efforts have been made on an industry level to pool data sources and protect anonymity of sources. An example of such an effort is the offshore reliability data (OREDA) program for the offshore oil exploration industry.

Data are needed to perform quantitative risk assessment or provide information to support qualitative risk assessment. The relevant information for risk assessment includes possible failures, failure probabilities, failure rates, failure modes, possible causes, failure consequences, and uncertainties associated with the system and its environment. In the case of a new system, data may be used from similar systems if this information is available. Surveys are a common tool used to produce some data. Statistical analysis can be used to assess confidence intervals and uncertainties in estimated parameters of interest. Generally, data can be classified as failure probability data and failure consequence data. The data, if available or existing, provide a history of a system or components of the system. The history is provided through previous system failures, individual component failures, known causes for these failures, maintenance records, and any other information related to the system. In the case of a new system, data could be interpolated or extrapolated from existing information on similar systems or based on the data from known components that comprise the new system. In cases where similar systems are nonexistent, expert opinion elicitation (EE) can be employed.

8.2 Data Sources

Data can be placed in classes with distinct attributes. These class distinctions can come from the source of the information. Figure 8.1 shows a hierarchy of data sources and their usability. Preexisting data can be modified to reflect the stresses of the intended application. Clemens (2002) describes a process depicted in Figure 8.1. If preexisting data provide information needed based on identical items in an identical environment and application, the preexisting data can be transferred into a database for performing risk analyses. Such an exact match is rarely encountered in cutting-edge technology applications, but it does represent the best situation. The next best situation is to find a dataset for similar conditions and then modify the data to make them roughly reflect the new stresses of the intended application. If neither of these scenarios is available, published reliability and consequence data can be used, when applicable. If preexisting data or published data are not available, engineering judgment must be utilized (e.g., use of EE). Another approach is to take preexisting data for a like system tested under differing conditions and compare the stress levels between the application of interest and the test application. The prerecorded failure rates are modified based on the comparative stresses of the two test environments. Bayesian methods can be used to combine the objective and subjective information.

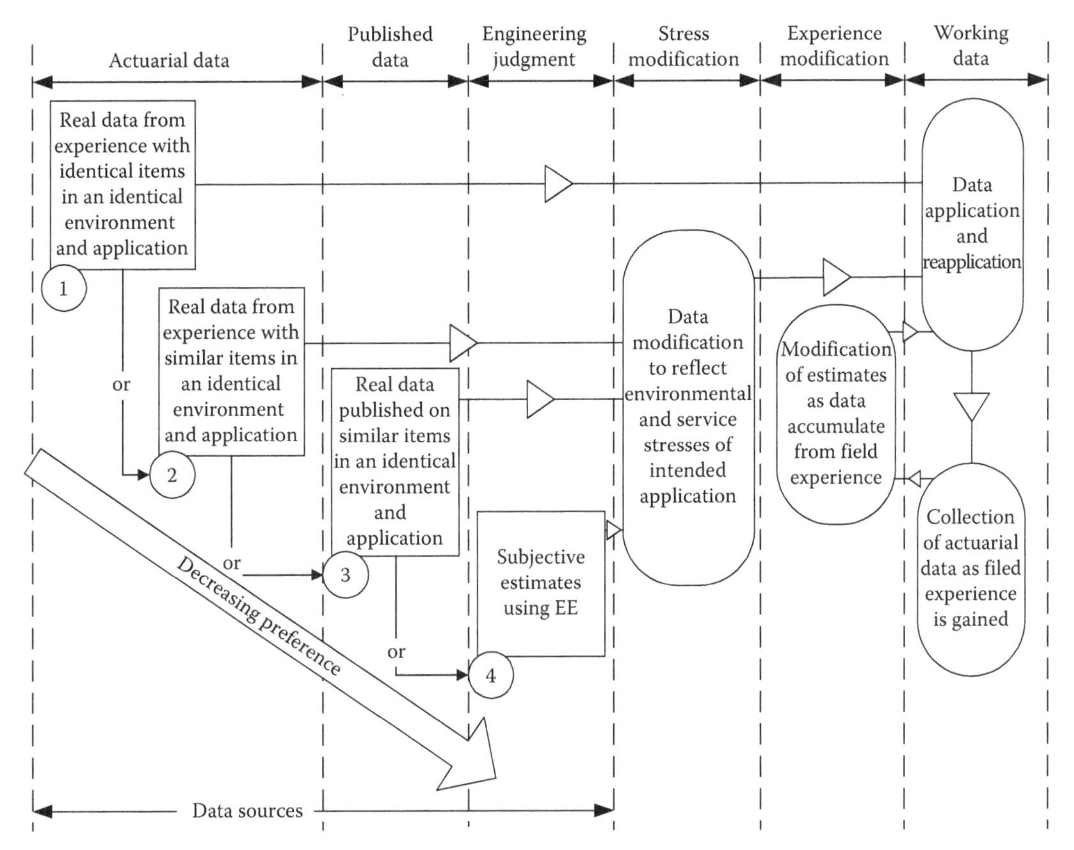

FIGURE 8.1
Data sources. EE, expert opinion elicitation.

Generic data are data that have been generated by looking at machinery or systems that are similar but not necessarily identical to the equipment or systems under consideration. For instance, generic data regarding the failure of a motor-driven pump may come from several different types of pumps used in several different systems. These pumps may be used to pump fuel, lubricating oil, or water. Often, generic data are the only information available in the initial stages of a probabilistic risk assessment (PRA), but these data should be used with care because they are generic and very general in nature. This general information may be used in the beginning stages of a PRA, but more specific data should be acquired for a more thorough analysis.

A thorough PRA must include data that are more specific to the system being analyzed than the generic data used in the early stages of the PRA. The specific data can be data that are collected from identical components and systems or from actual systems similar to the one under consideration. The risk-related data collected for the system are often referred to as *plant-specific data*. If a PRA is conducted in the design stage of a system, plant-specific data are usually not available, and the PRA at this stage must be completed using generic data. A good practice is to perform the PRA using generic data, and when the system enters operation, the PRA is updated using the newly available plant-specific data.

Failure data on different components and systems are usually not available from manufacturers, and generic failure probabilities can be used in these cases. In cases where data are not available, assumed values can be used. Example generic data are provided by Modarres (1993) and Kumamoto and Henley (1996) for mechanical systems, especially nuclear power plants. Another source of failure data is expert judgment provided by chief engineers, systems designers, and systems analysts, as described in subsequent sections.

8.3 Databases

Databases can be classified according to the types and sources of information that they contain; for example, databases can be described as failure databases if they contain information about failure probabilities and consequences. Also, a database can be described as an in-house database, a plant database, a process database, or an industry database, depending on the source and scope of information. This section provides information on databases that can be used in risk studies.

8.3.1 In-House Failure Databases

Risk studies require the knowledge of failure probabilities and consequences. The required information should be current and reflect the condition of the system at the time of the analysis. The development of an in-house database can greatly assist in meeting these requirements of risk-based analysis. The failure database needs to be designed with the proper fields to facilitate the retrieval of information in the desired format in order to compute the failure probabilities and consequences. Data collection forms can be designed to collect information about failures for the purpose of developing a failure database. The various entries in a form should correspond to fields in the database, and completion of a form adds a complete record to the database. Commercial software for developing and managing databases is available. Also, spreadsheet software can be used for this purpose.

8.3.2 Plant Failure Databases

If an in-house failure database is not available, an available system or process database that is similar to the system or process under investigation should be used. The entries of the database should be examined carefully to ensure their applicability to the system or process under investigation. Any entries that are not fully applicable should be examined for possible adjustment based on judgment or other considerations. The sources of the collected information should be documented for future reference or for addressing future inquiries.

8.3.3 Industry Failure Databases and Statistics

Generic information about failures that can be obtained from industry failure databases or statistics should be used after careful examination for its applicability to the system or plant under investigation. Such information is available in the literature or is provided by professional organizations such as the American Society of Mechanical Engineers, the Institute of Electrical and Electronics Engineers, and the American Petroleum Institute. Results from specialized studies are also available, such as for failures during civil construction (Eldukair and Ayyub 1991).

8.3.4 Reliability, Availability, and Maintainability Databases

Various industries have attempted to develop reliability, availability, and maintainability (RAM) databases with varying success. For example, an industry-wide, international marine network was recently formed to develop and collect RAM data, and to share these data at different levels by linking chief engineers, ship operators/managers, regulatory agencies, equipment manufacturers, and shipyards/designers (Inozu and Radovic 1999). Experiences with the development of databases have revealed some difficulty in obtaining failure information from participants due to the legal, insurance, and negative publicity implications and competitiveness and market-share concerns.

8.3.5 Failure Statistics Reported in the Literature

Failure statistics that are reported in the literature can be used after carefully examining them for their applicability to the system or plant under investigation before their use. Eldukair and Ayyub (1991) provide an example of the availability of such information.

8.3.6 Challenges Associated with Data from Other Sources

The definition of failure in most data sources is not clearly stated, particularly in failure-rate summary tables. The lack of standardized recording and reporting methodologies leads to the need to interpret the meaning of the data. For example, the mean is generally considered to be a single figure; however, a range is usually open to interpretation because it is not always clear if it represents the absolute extreme values or a confidence interval, and the corresponding confidence level may not be identified. Some data sources provide probability distribution models, such as normal or lognormal, whereas other sources provide a standard deviation. Methods of recording raw failure data are often not standardized. If the data are only recorded for internal purposes, the data fields could vary considerably from one organization to another. Sometimes government regulatory agencies require that organizations under their purview, such as the Nuclear Regulatory Commission (NuRC) for the US nuclear electrical generating industry, report failures to them in a standardized manner. In these cases, the centralized failure databases can prove to be very valuable for failure analysis and risk studies.

Only data summaries are commonly made available and published, and they can pose a challenge to users. Data summaries show only perspectives constructed by their authors. Often lacking are very important factors such as the size of the original dataset, leading to issues relating to statistical significance. Such summaries might not state if the data are empirically derived from observations or are estimated through some sort of expert judgment. Without these details, the data cannot be fully and properly evaluated.

Failure data might not reveal the underlying technologies of the items that failed. The technological generation can have a significant effect on the relevance of data to various applications. Technological advances usually, but unfortunately not always, bring about an increase in reliability.

The operating environment can significantly impact the causes and definition of failure. If the operating environments differ significantly from the data source, an uninformed user would use the data outside their range of applicability, producing misleading results. How the system is defined is also important. For example, an electrically powered liquid pump could be subdivided into electrical motor failures, mechanical (e.g., rotating) component failures of the bearings and impeller, or mechanical failure of the casing and seals. Defining the system as the pump or the various components can significantly impact the findings.

Example 8.1: Types of Failure Data for an Engine of a Marine Vessel

Failures of components of a system, such as an engine room of a marine vessel, can be categorized as follows: (1) failure on demand (i.e., failure to start), (2) failure during service (i.e., failure during running, also referred to as failure on time), and (3) unavailability due to maintenance and testing, which can also be considered as failure on demand. For marine systems, such as the engine room of a marine vessel, failure probabilities are of the on-demand type. Hence, all failure-on-time rates of components should be converted into failure-on-time probability by multiplying the failure rate by the time of mission for the components. The time of mission is defined as the time of service of a component and can be one of the following types: (1) the expected lifetime for which the components are not subjected to scheduled maintenance and (2) the time interval between scheduled preventive maintenance of the component.

Maintenance can be classified as scheduled or unscheduled. In the first type, maintenance is performed based on a fixed time interval and is intended to prevent failure and its consequences. The scheduled maintenance can be for a component, a subsystem, or a system. The maintenance in this case is intended to occur before the occurrence of failure. The interval of scheduled maintenance can be based on the analysis of failure data of components, subsystems, or systems. In addition, the time interval of scheduled maintenance should account for the failure rate, consequence of failure, ease and accessibility of maintenance, and life cycle cost analysis of the component, such as the expected cost of failure, the expected cost of maintenance, and the total expected cost. The cost of preventive maintenance is commonly less than the cost of failure. Unscheduled maintenance is performed based on indications that failure may occur soon, such as rising temperature readings of lubrication oil or a pressure drop across a valve. In this case, the cost of failure can be insignificant or much less than the cost of preventive maintenance. Section 9.2 includes additional information on modeling and optimizing resources for maintenance.

In this example, the following time intervals for maintenance of components can be used for illustration purposes based on the assumption of perfect maintenance and maintained components becoming as good as new:

- Forty-eight-hour average port-to-port duration for scheduled maintenance of components with failure-on-time rate $\leq$1E−3
- One hundred and sixty-eight-hour scheduled maintenance for components with failure-on-time rate $\leq$1E−4
- Forty-two-day voyage duration for scheduled maintenance of components with failure-on-time rate $\leq$1E−5
- Annual maintenance for scheduled maintenance of components with failure-on-time rate $\leq$1E−6

The above maintenance schedule can be revised based on risk analysis results that provide both failure probabilities and consequences for various failure scenarios. Risk analysis should include all systems and their components, and should assess the importance and effect of each component on the failure rate of the systems and other dependent systems.

The third mode of failure is unavailability, defined as the probability that a system or a component will not work upon demand. In the reliability analysis of each system, two criteria can be calculated: (1) system reliability and (2) system unavailability. These two criteria are different yet of the same importance to measure the risk involved in the design and operation of the system.

8.4 Expert Opinion Elicitation

8.4.1 Introduction

Available or existing data should be used to provide a history of a system or components of the system. In the case of a new system, data could be interpolated or extrapolated from existing information for similar systems or based on the data from known components that comprise the new system. In cases where similar systems are nonexistent, EE can be employed. This section provides background information and guidance on the elicitation of expert opinions.

8.4.2 Theoretical Bases and Terminology

Expert opinion elicitation (EE) can be defined as a heuristic process of gathering information and data or answering questions on issues or problems of concern. In this chapter, a focus on occurrence probabilities and consequences of events was established to demonstrate the process presented in this chapter. For this purpose, the EE process can be defined as a formal process of obtaining information or answers to specific questions about certain quantities referred to as *issues*, such as failure rates, failure consequences, and expected service life. EE should not be used in lieu of rigorous reliability and risk analytical methods but should be used to supplement them and to prepare for them. The EE process presented in this chapter is a variation of the Delphi technique (Helmer 1968) with scenario analysis (Kahn and Wiener 1967) based on uncertainty models (Ayyub 1991, 1992a, 1992b, 1993, 1998; Ayyub and Gupta 1997; Ayyub et al. 1997; Cooke 1991), social research (Bailey 1994), the US Army Corps of Engineers studies (Ayyub et al. 1996), ignorance, knowledge, information, and uncertainty (see Chapter 1), as well as nuclear industry recommendations (NuRC 1997) and Stanford Research Institute protocol (Spetzler and Stael von Holstein 1975). Ayyub (2002) provides additional information on EE.

The terminology of Table 8.1 is used in this chapter for defining and using an EE process. Table 8.1 provides the definitions of terms related to the EE process. The EE process is defined as a formal, heuristic process of gathering information and data or answering questions on issues or problems of concern. The EE process requires the involvement of a *leader* of the EE process who has managerial and technical responsibility for organizing and executing the project, overseeing all participants, and intellectually owning the results.

An *expert* can be defined as a very skillful person with considerable training in and knowledge of a specific field. The expert is the provider of an opinion in the process of EE. An *evaluator* is an expert who has the role of evaluating the relative credibility and plausibility of multiple hypotheses to explain observations. The process involves *evaluators,* who consider available data, become familiar with the views of proponents and other evaluators, question the technical bases of data, and challenge the views of proponents, and *observers,* who can contribute to the discussion but cannot provide expert opinion. The process might require peer reviewers who can provide an unbiased assessment and critical review of the EE process, its technical issues, and results. Some of the experts might be *proponents,* who advocate a particular hypothesis or technical position. In science, a proponent evaluates experimental data and professionally offers a hypothesis that would be challenged by the proponent's peers until proven correct or wrong. *Resource experts* are technical experts with detailed and deep knowledge of particular data, issue aspects, particular methodologies, or use of evaluators.

The *sponsor* of the EE process provides financial support and owns the rights to the results of the EE process. Ownership is in the sense of property ownership. A *subject* is

TABLE 8.1

Terminology and Definitions

Term	Definition
Evaluator	A person who considers available data, becomes familiar with the views of proponents and other evaluators, questions the technical bases of data, and challenges the views of proponents
Expert	A person with related or unique experience with an issue or question of interest for the process
EE process	A formal, heuristic process of gathering information and data or answering questions on issues or problems of concern
Leader of the EE process	An entity having managerial and technical responsibility for organizing and executing the project, overseeing all participants, and intellectually owning the results
Observer	A person who can contribute to the discussion but cannot provide the expert opinion
Peer reviewer	A person who can provide an unbiased assessment and critical review of an EE process, its technical issues, and results
Proponent	A person who is an expert and advocates a particular hypothesis or technical position; in science, a person who evaluates experimental data and offers a hypothesis, which would be challenged by the proponent's peers until proven correct or wrong
Resource expert	A person who is a technical expert with detailed and deep knowledge of particular data, issues, particular methodologies, or use of evaluators
Sponsor of EE process	An entity that provides financial support and owns the rights to the results of the EE process, with ownership being in the sense of property ownership
Subject	A person who might be affected or might affect an issue or question of interest for the process
TF	An entity responsible for structuring and facilitating the discussions and interactions of experts in the EE process, staging effective interactions among experts, ensuring equity in presented views, eliciting formal evaluations from each expert, and creating conditions for direct, noncontroversial integration of expert opinions
TI	An entity responsible for developing the composite representation of issues based on informed members and/or sources of related technical communities and experts; explaining and defending composite results to experts and outside experts, peer reviewers, regulators, and policy makers; and obtaining feedback and revising composite results
TIF	An entity responsible for the functions of both TI and TF

EE, expert opinion elicitation; TF, technical facilitator; TI, technical integrator; TIF, technical integrator and facilitator.

a person who might be affected by or might affect an issue or question of interest for the process. A *technical facilitator* (TF) is an entity responsible for structuring and facilitating the discussions and interactions of experts in the EE process, staging effective interactions among experts, ensuring equity in presented views, eliciting formal evaluations from each expert, and creating conditions for direct, noncontroversial integration of expert opinions. A *technical integrator* (TI) is an entity responsible for developing the composite representation of issues based on informed members and/or sources of related technical communities and experts; explaining and defending composite results to experts, peer reviewers, regulators, and policy makers; and obtaining feedback and revising composite results. A *technical integrator and facilitator* (TIF) is responsible for both functions of TI and TF. TIFs are commonly employed in engineering and economic applications.

8.4.3 Classification of Issues, Study Levels, Experts, and Process Outcomes

The NuRC (1997) classified the issues for EE purposes into three complexity degrees (A, B, or C) with four levels of study in the EE process (I, II, III, and IV), as shown in Table 8.2. A given issue is assigned a complexity degree and a level of study that depend on (1) the significance of the issue to the final goal of the study, (2) the technical complexity and uncertainty level of the issue, (3) the amount of nontechnical contention about the issue in the technical community, and (4) the important nontechnical considerations, such as budgetary, regulatory, scheduling, public perception, or other concerns. Experts can be classified into five types (NuRC 1997): (1) proponents, (2) evaluators, (3) resource experts, (4) observers, and (5) peer reviewers. These types are defined in Table 8.1.

The study level as shown in Table 8.3 involves a TI or a TIF. A TI can be one person or a team (i.e., an entity) that is responsible for developing the composite representation of issues based on informed members and/or sources of related technical communities and experts; explaining and defending composite results to experts and outside experts, peer reviewers, regulators, and policy makers; and obtaining feedback and revising composite results. A TIF can be one person or a team (i.e., an entity) that is responsible for the functions of a TI, and structuring and facilitating the discussions and interactions of experts in the EE process; staging effective interactions among experts; ensuring equity in presented views; eliciting formal evaluations from each expert; and creating conditions for direct, noncontroversial integration of expert opinions. The primary difference between the TI and the TIF is in the intellectual responsibility for the study, which lies with only the TI or the TIF and the experts, respectively. The TIF has

TABLE 8.2

Issue Complexity Degree

Degree of Complexity	Description
A	Noncontroversial; insignificant effect on risk
B	Significant uncertainty; significant diversity; controversial complex
C	Highly contentious; significant effect on risk; highly complex

Source: Nuclear Regulatory Commission (NuRC), *Recommendations for Probabilistic Seismic Hazard Analysis: Guidance on Uncertainty and Expert Use*, Vols. 1 and 2, NuRC, Washington, DC, 1997.

TABLE 8.3

Study Levels

Level	Requirements
I	TI evaluates and weighs the models based on literature review and experience, and estimates the needed quantities.
II	TI interacts with proponents and resource experts, assesses interpretations, and estimates the needed quantities.
III	TI brings together proponents and resource experts for debate and interaction. TI focuses the debate, evaluates interpretations, and estimates needed quantities.
IV	TI and TF (which can be one entity, or TIF) organize a panel of experts to interpret and evaluate, focus discussions, keep the experts' debate orderly, summarize and integrate opinions, and estimate the needed quantities.

Source: Nuclear Regulatory Commission (NuRC), *Recommendations for Probabilistic Seismic Hazard Analysis: Guidance on Uncertainty and Expert Use*, Vols. 1 and 2, NuRC, Washington, DC, 1997.

TF, technical facilitator; TI, technical integrator; TIF, technical integrator and facilitator.

also the added responsibility of maintaining the professional integrity of the process and its implementation.

The TI and TIF processes are required to utilize peer reviewers for quality assurance purposes. Peer review can be classified according to the peer-review method and the peer-review subject. Two methods of peer review can be performed: (1) participatory peer review, which would be conducted as an ongoing review throughout all study stages, and (2) late-stage peer review, which would be performed as the final stage of the study. The former method allows for affecting the course of the study, whereas the latter one might not be able to affect the study without a substantial rework of the study. The second classification of peer review is peer-review subject, which has two types: (1) technical peer review, which focuses on the technical scope, coverage, contents, and results, and (2) process peer review, which focuses on the structure, format, and execution of the EE process. Guidance on the use of peer reviewers is provided in Table 8.4 (NuRC 1997).

The EE process preferably should be conducted to include a face-to-face meeting of experts specifically to address the issues under consideration. The meeting of the experts should be conducted after providing the experts in advance with background information, objectives, a list of issues, and the anticipated outcome of the meeting. The EE based on the TIF concept can result in consensus or disagreement. Consensus can be of four types shown in Figure 8.2. Commonly, the EE process has the objective of achieving consensus type 4, that is, experts agree that a particular probability distribution represents the overall scientific community. The TIF plays a major role in building consensus by acting as a facilitator. Disagreement among experts, whether it is intentional or unintentional, requires the TIF to act as an integrator by using equal or unequal weight factors. Sometimes, expert opinions need to be weighed for appropriateness and relevance rather than being strictly weighted by factors in a mathematical aggregation procedure.

8.4.4 Process Definition

EE has been defined as a formal, heuristic process of obtaining information or answers to specific questions about certain quantities, or issues, such as failure rates, failure consequences, and expected service lives. The suggested steps for an EE process depend on the

TABLE 8.4

Guidance on Use of Peer Reviewers

EE Process	Peer-Review Subject	Peer-Review Method	Recommendation
TIF	Technical	Participatory	Recommended
		Late stage	Can be acceptable
	Process	Participatory	Strongly recommended
		Late stage	Risky, unlikely to be successful
TI	Technical	Participatory	Strongly recommended
		Late stage	Risky, but can be acceptable
	Process	Participatory	Strongly recommended
		Late stage	Risky, but can be acceptable

Source: Nuclear Regulatory Commission (NuRC), *Recommendations for Probabilistic Seismic Hazard Analysis: Guidance on Uncertainty and Expert Use*, Vols. 1 and 2, NuRC, Washington, DC, 1997.

TI, technical integrator; TIF, technical integrator and facilitator.

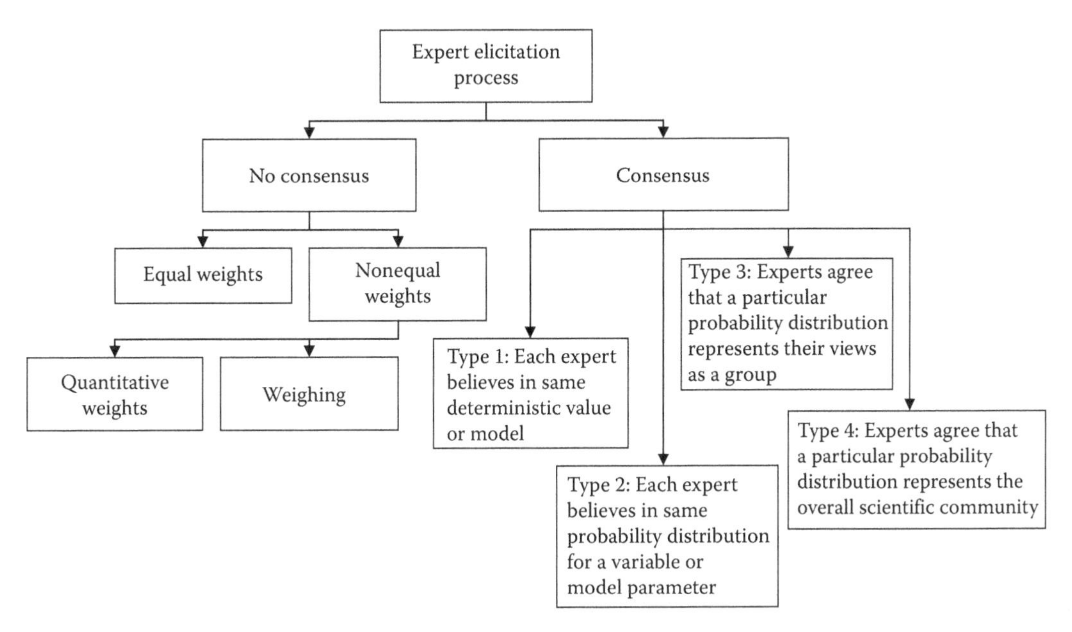

FIGURE 8.2
Outcomes of the expert opinion elicitation (EE) process.

use of a TI or a TIF, as shown in Figure 8.3. The details of the steps involved in these two processes are defined in subsequent subsections.

8.4.5 Need Identification for Expert Opinion Elicitation

The primary reason for using EE is to deal with uncertainty in selected technical issues related to a system of interest. Issues with significant uncertainty, issues that are controversial and/or contentious, issues that are complex, and/or issues that can have a significant effect on risk are most suited for EE. The value of the EE comes from its initial intended uses as a heuristic tool, not a scientific tool, for exploring vague and unknown issues that are otherwise inaccessible. It is not a substitute for scientific, rigorous research.

The identification of need and its communication to experts are essential for the success of the EE process. The need identification and communication should include the definition of the goal of the study and the relevance of issues to this goal. Establishing this relevance makes the experts stakeholders and therefore increases their attention and sincerity levels. Establishing the relevance of each issue or question is essential for enhancing the reliability of data collected from the experts.

8.4.6 Selection of Study Level and Study Leader

The goal of a study and the nature of the issues determine the study level, as shown in Table 8.2. The study leader can be a TI, a TF, or a combined TIF. The leader of the study is an entity having managerial and technical responsibility for organizing and executing the project, overseeing all participants, and intellectually *owning* the results. The primary difference between the TI and the TIF lies in the intellectual responsibility for the study—with only the TI or with both the TIF and the experts. The TIF has also the added

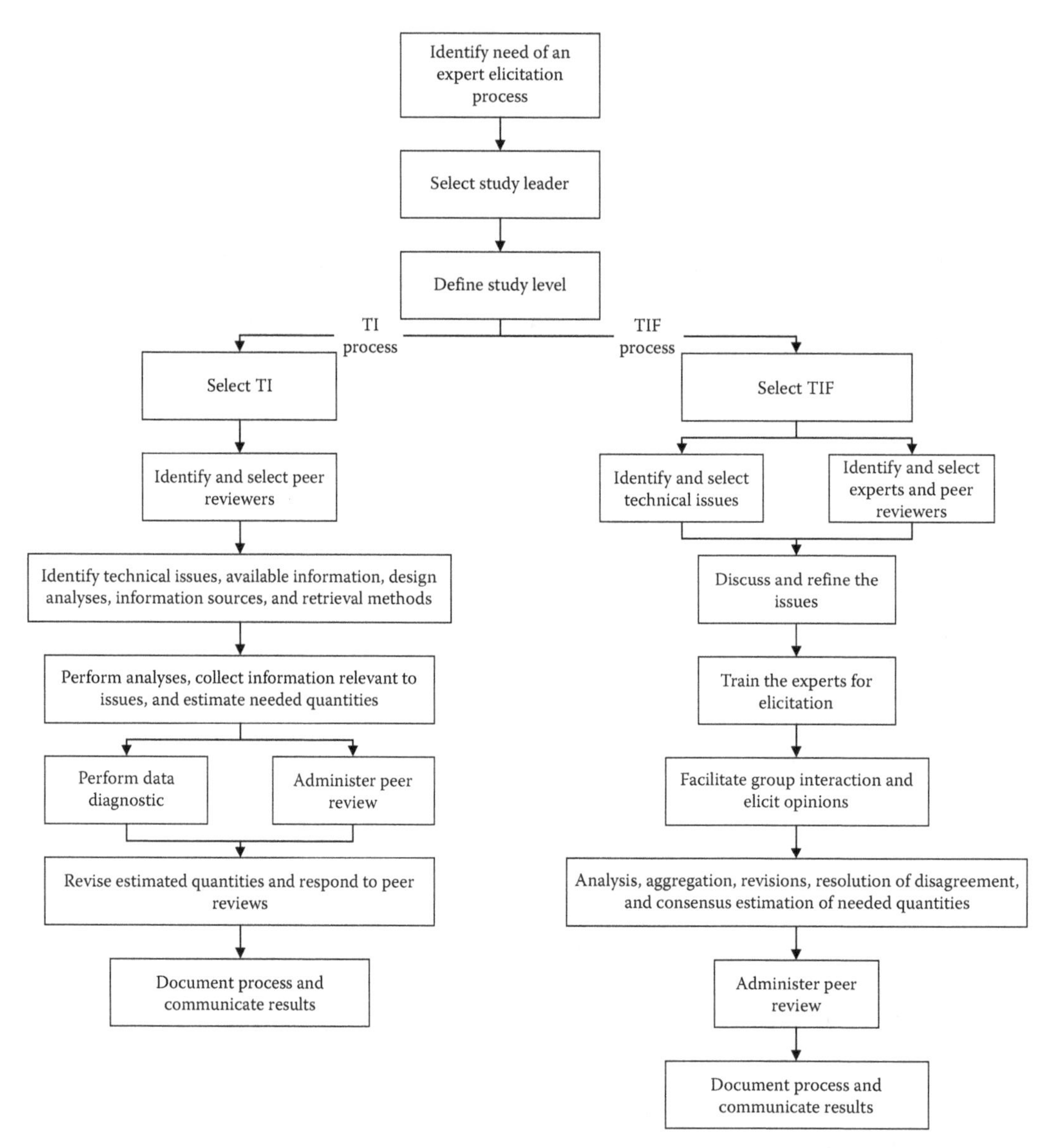

FIGURE 8.3
Expert opinion elicitation (EE) process. TI, technical integrator; TIF, technical integrator and facilitator. [Adapted from Nuclear Regulatory Commission (NuRC), *Recommendations for Probabilistic Seismic Hazard Analysis: Guidance on Uncertainty and Expert Use*, Vols. 1 and 2, NuRC, Washington, DC, 1997.]

responsibility of maintaining the professional integrity of the process and its implementation. The TI is required to utilize peer reviewers for quality assurance purposes.

A study leader should be selected based on the following attributes:

- Outstanding professional reputation and widely recognized competence based on academic training and relevant experience
- Strong communication and interpersonal skills, flexibility, impartiality, and ability to generalize and simplify

- A large contact base of industry leaders, researchers, engineers, scientists, and decision makers
- Ability to build consensus and leadership qualities

The study leader does not need to be a subject expert but should be knowledgeable in the subject matter.

8.4.7 Selection of Peer Reviewers and Experts

8.4.7.1 Selection of Peer Reviewers

Peer review can be classified according to peer-review method and peer-review subject. Two methods of peer review can be performed: (1) participatory peer review, which is conducted as an ongoing review throughout all study stages, and (2) late-stage peer review, which is performed as the final stage of the study. The second classification of peer reviews is by peer-review subject, which can be (1) technical peer review, which focuses on the technical scope, coverage, contents, and results, and (2) process peer review, which focuses on the structure, format, and execution of the EE process.

Peer reviewers are needed for both the TI and TIF processes. The peer reviewers should be selected by the study leader in close consultation with perhaps the study sponsor. Researchers, scientists, and/or engineers who will serve as peer reviewers should have the following:

- An outstanding professional reputation and widely recognized competence based on academic training and relevant experience
- A general understanding of the issues in other related areas or relevant expertise and experiences from other areas
- The availability and willingness to devote the required time and effort
- Strong communication skills, interpersonal skills, flexibility, impartiality, and an ability to generalize and simplify

8.4.7.2 Selection of Experts

The size of an expert panel should be determined on a case-by-case basis. The panel should be large enough to achieve the required diversity of opinion, credibility, and result reliability. In recent EE studies, a nomination process was used to establish a list of candidate experts by consulting the archival literature, technical societies, governmental organization, and other knowledgeable experts (Trauth et al. 1993). Formal nomination and selection processes should establish appropriate criteria for nomination, selection, and removal of experts. For example, the following criteria were used to select experts for an ongoing Yucca Mountain seismic hazard analysis (NuRC 1997):

- Strong relevant expertise through academic training, professional accomplishment and experiences, and peer-reviewed publications
- Familiarity with and knowledge of various aspects related to the issues of interest
- Willingness to act as proponents or impartial evaluators
- Availability and willingness to commit the needed time and effort
- Specific related knowledge and expertise in regard to the issues of interest

- Willingness to participate effectively in debates, to prepare for discussions, and to provide required evaluations and interpretations
- Strong communication and interpersonal skills, flexibility, impartiality, and ability to generalize and simplify

In this NuRC study, the criteria established for expert removal included failure to perform according to commitments and demands as set in the selection criteria and unwillingness to interact with members of the study.

The panel of experts for an EE process should have a balance and broad spectrum of viewpoints, expertise, technical points of view, and organizational representation. The diversity and completeness of the panel of experts are essential for the success of the elicitation process. For example, the panel can include the following:

- Proponents who advocate a particular hypothesis or technical position
- Evaluators who consider available data, become familiar with the views of proponents and other evaluators, question the technical bases of data, and challenge the views of proponents
- Resource experts who are technical experts with detailed and deep knowledge of particular data, issue aspects, particular methodologies, or use of evaluators

The experts should be familiar with the design, construction, operation, inspection, maintenance, reliability, and engineering aspects of the equipment and components of the facility of interest. It is essential for selecting people with basic engineering or technological knowledge; however, they do not necessarily have to be engineers. It might be necessary to include one or two experts from management with engineering knowledge of the equipment and components, consequences, safety aspects, administrative and logistic aspects of operation, EE process, and objectives of this study. One or two experts with a broader knowledge of the equipment and components might be needed. Also, one or two experts with a background in risk analysis and risk-based decision making and their uses in areas related to the facility of interest might be needed.

Observers can be invited to participate in the elicitation process. Observers can contribute to the discussion but cannot provide expert opinion. The observers provide expertise in the elicitation process, probabilistic and statistical analyses, risk analysis, and other support areas. The composition and contribution of the observers are essential for the success of this process. The observers may include the following:

- Individuals with operational, economic, engineering, research, or administrative-related backgrounds from research laboratories or headquarters
- Individuals with expertise in probabilistic analysis, probabilistic computations, consequence computations and assessment, and EE

Biographical sketches about the study leader, TI, TF, experts, observers, and peer reviewers should be assembled. All attendees can participate in discussions during a meeting; however, only the experts can provide answers to questions on the selected issues. The integrators and facilitators are responsible for conducting the EE process. They can be considered to be observers or experts, depending on the circumstances and needs of the process.

8.4.7.3 Items Needed by Experts and Reviewers before the Expert Opinion Elicitation Meeting

The experts and observers should receive the following items before the EE meeting:

- An objective statement of the study
- A list of experts, observers, integrators, facilitators, study leaders, sponsors, and their biographical statements
- A description of the facility, systems, equipment, and components
- Basic terminology and definitions, such as probability, failure rate, average time between unsatisfactory performances, mean (or average) value, median value, and uncertainty
- Failure consequence estimation
- A description of the EE process
- A related example on the EE process and its results, if available
- Aggregation methods of expert opinions such as computations of percentiles
- A description of the issues in the form of a list of questions and background information, with each issue being presented on a separate page with space for recording an expert's judgment, any revisions, and comments
- Clear statements of expectations from the experts in terms of time, effort, responses, communication, and discussion style and format

It might be necessary to personally contact the individual experts for the purpose of ensuring a clear understanding of expectations.

8.4.8 Identification, Selection, and Development of Technical Issues

The technical issues of interest should be carefully selected to achieve certain objectives. The technical issues are related to the quantitative assessment of failure probabilities and consequences for selected components, subsystems, and systems within a facility. The issues should be selected such that they would have a significant impact on the study results. These issues should be structured in a logical sequence starting with a background statement, then the questions, and then selections for answers or the answer format and scales. Personnel with a risk analysis background who are familiar with the construction, design, operation, and maintenance of the facility need to define these issues in the form of specific questions. Also, background materials about these issues should be assembled. The materials will be used to familiarize and train the experts in regard to the issues of interest, as described in subsequent steps.

An introductory statement for the EE process should be developed, which includes the goal of the study and establishes relevance. Instructions should be provided with guidance on expectations, answering the questions, and reporting. The following are guidelines on constructing questions and issues based on social research practices (Bailey 1994):

- Each issue can include several questions; however, each question should address only one answer being sought. It is a poor practice to combine two questions into one.
- Question and issue statements should not be ambiguous, and the use of ambiguous words should be avoided. In EE of failure probabilities, the word *failure* might be vague or ambiguous to some subjects. Special attention should be given to its

definition within the context of each issue or question. The level of language used should be kept to the minimum level possible. Also, be aware that the choice of words can affect the perception of an issue by various subjects.

- The use of factual questions is preferred over abstract questions. Questions that refer to concrete and specific matters result in desirable concrete and specific answers.
- Questions should be carefully structured in order to reduce biases of subjects. Questions should be asked in a neutral format, sometimes more appropriately without lead statements.
- Sensitive topics might require stating questions with lead statements that would establish supposedly accepted social norms in order to encourage subjects to answer the questions truthfully.

Questions can be classified into *open-ended* and *closed-ended questions*. A closed-ended question has the following characteristics: (1) It limits the possible outcomes of response categories; (2) it can provide guidance to subjects, thereby making it easier for a subject to answer; (3) it provides complete answers; (4) it allows for dealing with sensitive or taboo topics; (5) it allows for comparing the responses of subjects; (6) it produces answers that can be easily coded and analyzed; (7) it can be misleading; (8) it allows for guess work by ignorant subjects; (9) it can lead to frustration due to subject perception of inappropriate answer choices; (10) it limits the possible answer choices; (11) it does not allow for detecting variations in question interpretation by subjects; (12) it results in artificially small variations in responses due to the limitation of the possible answers; and (13) it can be prone to clerical errors by subjects who unintentionally select the wrong answer categories.

An open-ended question has the following characteristics: (1) it does not limit the possible outcomes of response categories; (2) it is suitable for questions without known answer categories; (3) it is suitable for dealing with questions with too many answer categories; (4) it is preferred for dealing with complex issues; (5) it allows for creativity and self-expression; (6) it can lead to collecting worthless and irrelevant information; (7) it can lead to nonstandardized data that cannot be easily compared among subjects; (8) it can produce data that are difficult to code and analyze; (9) it requires superior writing skills; (10) it might not communicate properly the dimensions and complexity of the issue; (11) it can be demanding on the time of subjects; and (12) it can be perceived as difficult to answer, thereby discouraging subjects from responding accurately or at all.

The format, scale, and units for the response categories should be selected to best achieve the goal of the study. The minimum number of questions and the question order should be selected with the following guidelines:

- Sensitive questions and open-ended questions should be at the end of the questionnaire.
- The questionnaire should start with simple questions and questions that are easy to answer.
- A logical order of questions should be developed such that questions at the start of the questionnaire feed the needed information into questions at the end of the questionnaire.
- Questions should follow a logical order based on a time sequence or related to a process.
- The order of the questions should not lead to or set a particular response.

- Reliability check questions that are commonly used in pairs (stated positively and negatively) should be separated by other questions.
- Questions should be mixed in terms of format and type in order to maintain the interest of subjects.
- The order of the questions can establish a funnel that starts with general questions followed by more specific questions within several branches of questioning; this funnel technique might not be appropriate in some applications, and its suitability should be assessed on a case-by-case basis.

Some of the difficulties or pitfalls of using questions, with suggested solutions or remedies, include the following (Bailey 1994):

- Subjects might feel that the questionnaire is not legitimate and has a hidden agenda. A cover letter or a proper introduction of the questionnaire is needed.
- Subjects might feel that the results will be used against them. Unnecessary sensitive issues and duplicate issues should be removed, and sometimes assuring a subject's anonymity might provide the needed remedy.
- Subjects might refuse to answer questions on the basis that they have completed their share of questionnaires or are tired of being a guinea pig. Training and education might be needed to create the proper attitude.
- A sophisticated subject who has participated in many studies may begin to question the structure of the questionnaire, test performance, and results. This situation may require sampling around to find a replacement subject.
- A subject might provide normative answers—answers that the subject thinks are being sought. Unnecessary sensitive issues and duplicate issues should be removed, and sometimes assuring a subject's anonymity might provide the needed remedy.
- Subjects might not want to reveal their ignorance and perhaps appear stupid. Emphasizing that there are no correct or wrong answers and assuring a subject's anonymity might provide the needed remedy.
- A subject might think that the questionnaire is a waste of time. Training and education might be needed to create the proper attitude.
- Subjects might feel that a question is too vague and cannot be answered. The question should be restated so that it is very clear.

Once the issues are developed, they should be pretested by administering them to a few subjects for the purpose of identifying and correcting flaws. The results of this pretesting should be used to revise the issues.

8.4.9 Elicitation of Opinions

The elicitation process of opinions should be systematic for all the issues according to the steps presented in this section.

8.4.9.1 Issue Familiarization of Experts

The background materials that were assembled in the previous step should be sent to the experts about 1–2 weeks in advance of the meeting with the objective of providing sufficient time for them to become familiar with the issues. The objective of this step is also to ensure the existence of a common understanding among the experts. The background material should include the objectives of the study; the issues; lists of questions for the issues; descriptions of the systems and processes, the equipment and components, the elicitation process, and the selection methods of experts; and the biographical information on the selected experts. Example results and their meaning, methods of analysis of the results, and lessons learned from previous elicitation processes should also be made available to the experts. It is important to break the questions or issues down into components that can be easily addressed. Preliminary discussion meetings or telephone conversations between the facilitator and the experts might be necessary in some cases to prepare for the elicitation process.

8.4.9.2 Training of Experts

This step is performed during the meeting of the experts, observers, and facilitators. During the training, the facilitator needs to maintain flexibility to refine wording or even change approach based on feedback from experts. For instance, experts may not be comfortable with the term *probability* and may prefer the use of *events per year* or *recurrence interval*. This indirect elicitation should be explored with the experts. The meeting should be started with presentations of background material to establish relevance of the study to the experts and study goals in order to establish a rapport with the experts. Then, information on uncertainty sources and types, occurrence probabilities and consequences, the EE process, technical issues and questions, and aggregation of expert opinions should be presented. Experts need to be trained on providing answers in an acceptable format that can be used in the analytical evaluation of the failure probabilities or consequences. The experts should be trained in certain areas, such as the meaning of probability, central tendency, and dispersion measures, especially experts who are not familiar with the language of probability. Additional training might be required on consequences, subjective assessment, logic trees, problem structuring tools such as influence diagrams, and methods of combining expert evaluations. Sources of bias, including overconfidence and base rate fallacy, and their contribution to bias and error should be discussed. This step should include a search for any motivational bias of experts—as revealed, for example, by previous positions the experts have taken in public; motivational biases could also include wanting to influence decisions and the allocation of funds, believing that they will be evaluated by their superiors as a result of their answers, and/or wanting to be perceived as an authoritative expert. These motivational biases, once identified, can be sometimes overcome by redefining the incentive structure for the experts.

8.4.9.3 Elicitation and Collection of Opinions

The opinion elicitation step starts with a technical presentation of an issue and by decomposing the issue to its components, discussing potential influences, and describing event sequences that might lead to identifying the top events of interest. These top events are the basis for questions related to the issue in the next stage of the EE step. Presentation of the factors, limitations, test results, analytical models, and uncertainty types and sources should allow for questions to eliminate any ambiguity and clarify the scope and conditions for the issue. Discussion of the issue should be encouraged, as such discussion and questions might result in refining the definition of the issue. Then, a form with a statement of the issue should be given to the expert to record their evaluation or input. Each expert's judgment and supportive reasoning should be documented for each issue. It is common to ask to provide several conditional probabilities in order to reduce the complexity of the questions and therefore obtain reliable answers. These conditional probabilities can be based on fault tree and event tree diagrams. Conditioning has the benefit of simplifying the questions by decomposing the problems. Also, it results in a conditional event that has a larger occurrence probability than its underlying events, thus making the elicitation less prone to bias because experts tend to have a better handle on larger probabilities in comparison with very small ones. It is desirable to have the elicited probabilities in the range of 0.1–0.9, if possible. Sometimes it might be desirable to elicit conditional probabilities using linguistic terms (Ayyub 2002). If correlation among variables exists, it should be presented to the experts in great detail and conditional probabilities elicited. Issues should be dealt with one issue at a time, although sometimes similar or related issues might be considered simultaneously.

8.4.9.4 Aggregation and Presentation of Results

The collected assessments from the experts for an issue should be assessed for internal consistency, analyzed, and aggregated to obtain the composite judgments for the issue. The means, medians, percentile values, and standard deviations are computed for each issue. Also, a summary of the reasoning provided during the meeting about the issues should be developed. Uncertainty levels in the assessments should also be quantified. The methods can be classified into consensus methods and mathematical methods. The mathematical methods can be based on assigning equal or different weights to the experts. Percentiles are commonly used to combine expert opinions as shown in Table 8.5. A p percentile value (x_p) for a random variable based on a sample is the value of the parameter such that p% of the

TABLE 8.5

Computations of Percentiles

	25th Percentile		50th Percentile		75th Percentile	
Number of Experts (n)	Arithmetic Average	Geometric Average	Arithmetic Average	Geometric Average	Arithmetic Average	Geometric Average
4	$(X_1 + X_2)/2$	$\sqrt{X_1X_2}$	$(X_2 + X_3)/2$	$\sqrt{X_2X_3}$	$(X_3 + X_4)/2$	$\sqrt{X_3X_4}$
5	X_2	X_2	X_3	X_3	X_4	X_4
6	X_2	X_2	$(X_3 + X_4)/2$	$\sqrt{X_3X_4}$	X_5	X_5
7	$(X_2 + X_3)/2$	$\sqrt{X_2X_3}$	X_4	X_4	$(X_5 + X_6)/2$	$\sqrt{X_5X_6}$
8	$(X_2 + X_3)/2$	$\sqrt{X_2X_3}$	$(X_4 + X_5)/2$	$\sqrt{X_4X_5}$	$(X_6 + X_7)/2$	$\sqrt{X_6X_7}$
9	$(X_2 + X_3)/2$	$\sqrt{X_2X_3}$	X_5	X_5	$(X_7 + X_8)/2$	$\sqrt{X_7X_8}$
10	$(X_2 + X_3)/2$	$\sqrt{X_2X_3}$	$(X_5 + X_6)/2$	$\sqrt{X_4X_5}$	$(X_8 + X_9)/2$	$\sqrt{X_8X_9}$
11	X_3	X_3	X_6	X_6	X_9	X_9
12	X_3	X_3	$(X_6 + X_7)/2$	$\sqrt{X_6X_7}$	X_{10}	X_{10}
13	$(X_3 + X_4)/2$	$\sqrt{X_3X_4}$	X_7	X_7	$(X_{10} + X_{11})/2$	$\sqrt{X_{10}X_{11}}$
14	$(X_3 + X_4)/2$	$\sqrt{X_3X_4}$	$(X_7 + X_8)/2$	$\sqrt{X_7X_8}$	$(X_{11} + X_{12})/2$	$\sqrt{X_{11}X_{12}}$
15	X_4	X_4	X_8	X_8	X_{12}	X_{12}
16	X_4	X_4	$(X_8 + X_9)/2$	$\sqrt{X_8X_9}$	X_{13}	X_{13}
17	$(X_4 + X_5)/2$	$\sqrt{X_4X_5}$	X_9	X_9	$(X_{13} + X_{14})/2$	$\sqrt{X_{13}X_{14}}$
18	$(X_4 + X_5)/2$	$\sqrt{X_4X_5}$	$(X_9 + X_{10})/2$	$\sqrt{X_9X_{10}}$	$(X_{14} + X_{15})/2$	$\sqrt{X_{14}X_{15}}$
19	X_5	X_5	X_{10}	X_{10}	X_{15}	X_{15}
20	X_5	X_5	$(X_{10} + X_{11})/2$	$\sqrt{X_{10}X_{11}}$	X_{15}	X_{15}

data are less than or equal to x_p. Based on this definition, the median value is considered to be the 50th percentile.

Aggregating the opinions of experts requires the computation of the 25th, 50th, and 75th percentiles. The computation of these values depends on the number of experts providing opinions. Table 8.5 provides a summary of the equations needed for 4–20 experts. In the table, X_i indicates the opinion of an expert with the ith smallest value, that is, $X_1 \leq X_2 \leq X_3 \leq \cdots \leq X_n$, where n is the number of experts. As shown in the table, the arithmetic average is commonly used to compute the percentiles. In some cases, where the values of X_i differ by power order of magnitude, the geometric average can be used.

8.4.9.5 Group Interaction, Discussion, and Revision by Experts

The aggregated results need to be presented to the experts for a second round of discussion and revision. The experts should be given the opportunity to revise their assessments of the individual issues at the end of the discussion. Also, the experts should be asked to state the rationale for their statements and revisions. The revised assessments of the experts should be collected for aggregation and analysis. This step can produce either consensus or no consensus, as shown in Figure 8.2. The selected aggregation procedure might require eliciting the weight factors from the experts. In this step, the TF plays a major role in developing a consensus and maintaining the integrity and credibility of the elicitation process. Also, the TI is needed to aggregate the results with reliability measures without biases. The integrator might need to deal with varying expertise levels for the experts, the outliers (i.e., extreme views), the nonindependent experts, and the expert biases.

8.4.9.6 Documentation and Communication

A comprehensive documentation of the process is essential for ensuring the acceptance and credibility of the results. The document should include the complete descriptions of the steps, the initial results, the revised results, the consensus results, and the aggregated result spreads and reliability measures.

Example 8.2: Risk-Based Approval of Personal Flotation Devices

With the introduction of inflatable personal flotation devices (PFDs), the US Coast Guard (USCG) and PFD industry were faced with limitations regarding the current PFD approval practice. Inflatable PFDs perform better than inherently buoyant PFDs in some aspects, but they involve new hazards not present in the traditional, inherently buoyant PFDs. For the approval of inflatable PFDs, it became apparent that in some areas such devices offered performance advantages over inherently buoyant PFDs but also had some disadvantages in other areas. The need to perform equivalency analysis of engineering designs is a common problem for the regulation of engineering systems; therefore, an improved process for evaluating and comparing PFD performance is needed. The introduction of this concept applied to PFD analysis required the use of EE to model the relationships between the performance variables of PFDs and the probability of the PFDs meeting the needs of a person from the population of potential users (i.e., relationships between the performance levels of a PFD and the respective fractions of the population whose needs will be met at these levels).

PFD FREEBOARD

FB is defined as the distance measured perpendicular to the surface of the water to the lowest point where the user's respiration may be impeded. The objective of FB is to minimize the probability of drowning. Greater FB means that user movement and water movement are less likely to cause mouth immersion and water inhalation. Figure 8.4 shows a linear relationship between the FB and the probability of meeting the needs of a PFD user based on EE. Defining this linear relationship requires eliciting two (x, y) points from experts (Table 8.6): FB required to achieve a probability of 1, the absolute minimum FB, and the probability corresponding to the absolute minimum FB.

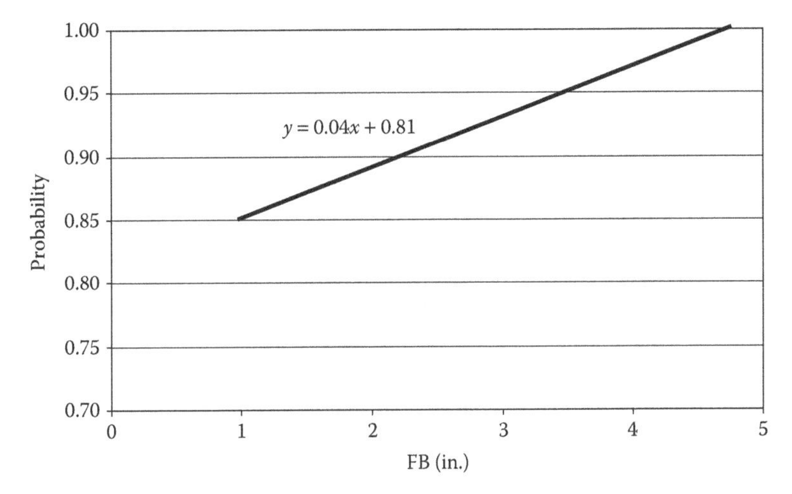

FIGURE 8.4
Probability of meeting the needs of a personal flotation device (PFD) user for freeboard (FB).

TABLE 8.6
EE for FB

Values to Define Model	Expert Opinion Collection								
	Expert 1	**Expert 2**	**Expert 3**	**Expert 4**	**Expert 5**	**Expert 6**	**Expert 7**	**Expert 8**	**Expert 9**
FB required for probability of 1	5	5	3.5	4.5	4	4.75	4.75	5	4.75
Absolute minimum FB	0.5	0.5	1	1	0.5	0.75	1	1	1
Absolute minimum FB probability	0.85	0.8	0.95	0.8	0.8	0.85	0.8	0.9	0.9

	Expert Opinion Aggregation				
	Minimum	**25th Percentile**	**50th Percentile**	**75th Percentile**	**Maximum**
FB required for probability of 1	3.5	4.25	4.75	5	5
Absolute minimum FB	0.5	0.5	1	1	1
Absolute minimum FB probability	0.8	0.8	0.85	0.9	0.95

FB, freeboard.

PFD FACE PLANE ANGLE

FPA is defined as the angle, relative to the surface of the water, of the plane formed by the most forward part of the forehead and chin of a user floating in the attitude of static balance. The objective here is to decrease the probability of drowning. A positive angle is achieved when a user's forehead is higher than his chin. Proper FPA decreases the chances of water inhalation. Figure 8.5 shows a linear relationship between the FPA and the probability of meeting the needs of a PFD user based on EE. Defining this linear relationship requires eliciting two (x, y) points from experts (Table 8.7): FPA required for a probability of 1, the absolute minimum FPA, and the probability corresponding to the absolute minimum FPA.

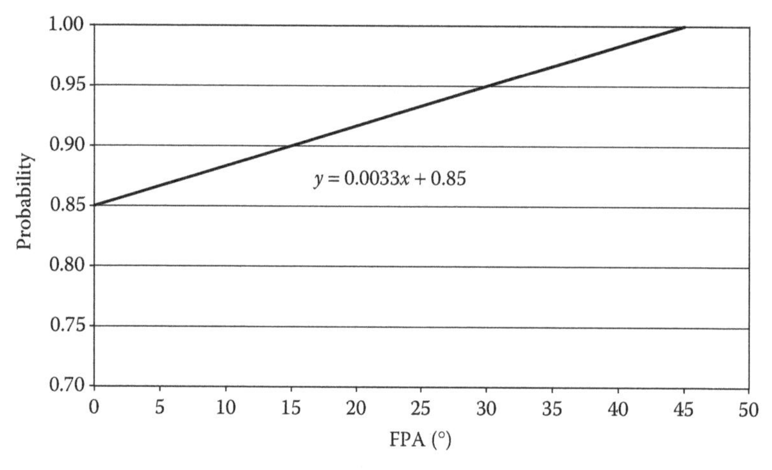

FIGURE 8.5
Probability of meeting the needs of a personal flotation device (PFD) user for face plane angle (FPA).

TABLE 8.7
EE for FPA

	Expert Opinion Collection								
Values to Define Model	**Expert 1**	**Expert 2**	**Expert 3**	**Expert 4**	**Expert 5**	**Expert 6**	**Expert 7**	**Expert 8**	**Expert 9**
FPA required for probability of 1	35	90	30	45	25	60	90	45	45
Absolute minimum FPA	5	−5	−10	0	−5	3	15	0	15
Absolute minimum FPA probability	0.8	0.75	0.9	0.9	0.8	0.9	0.85	0.9	0.5

	Expert Opinion Aggregation				
	Minimum	**25th Percentile**	**50th Percentile**	**75th Percentile**	**Maximum**
FPA required for probability of 1	25	32.5	45	75	90
Absolute minimum FPA	−10	−5	0	10	15
Absolute minimum FPA probability	0.5	0.775	0.85	0.9	0.9

FPA, face plane angle.

PFD CHIN SUPPORT

CS is defined as the PFD device being in direct contact with the jawline while the subject is in either the vertical upright or relaxed face-up position. CS aids the unconscious or exhausted user by preventing the face from falling into the water. CS is considered adequate if the device prevents the subject from touching the chin to the chest while the subject is in the relaxed face-up position of static balance. Figure 8.6 shows CS being provided by the PFD design and not being provided by the PFD design. Defining this relationship requires eliciting one value (Table 8.8): PFD effectiveness without CS.

PFD TORSO ANGLE

TA is the angle between a vertical line and a line passing through the shoulder and the hip. A desirable TA aids in preventing mouth immersions due to waves and the wearer being tipped face down by his or wave movement. A positive TA is achieved when a test participant's hips are forward with respect to his shoulders. Figure 8.7 shows a linear relationship between the TA and the probability of meeting the needs of a PFD user

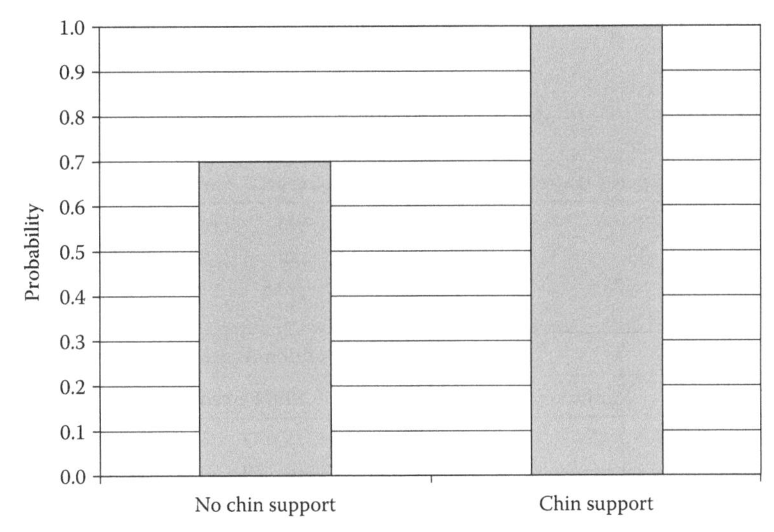

FIGURE 8.6
Probability of meeting the needs of a personal flotation device (PFD) user without chin support (CS).

TABLE 8.8
EE for CS

	Expert Opinion Collection								
Values to Define Model	**Expert 1**	**Expert 2**	**Expert 3**	**Expert 4**	**Expert 5**	**Expert 6**	**Expert 7**	**Expert 8**	**Expert 9**
Probability that the PFD is effective with no CS	0.7	0.6	0.7	0.7	0.5	0.5	0.7	0.7	0.5

	Expert Opinion Aggregation				
	Minimum	**25th Percentile**	**50th Percentile**	**75th Percentile**	**Maximum**
Probability that the PFD is effective with no CS	0.5	0.55	0.7	0.7	0.7

CS, chin support; PFD, personal flotation device.

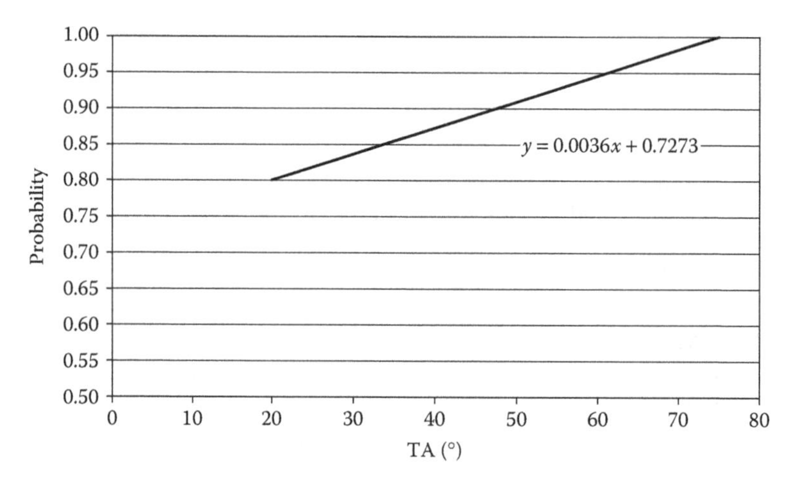

FIGURE 8.7
Probability of meeting the needs of a personal flotation device (PFD) user for face plane angle (FPA). TA, torso angle.

TABLE 8.9
EE for TA

	Expert Opinion Collection								
Values to Define Model	**Expert 1**	**Expert 2**	**Expert 3**	**Expert 4**	**Expert 5**	**Expert 6**	**Expert 7**	**Expert 8**	**Expert 9**
TA at probability of 1	85	75	60	45	45	80	60	80	75
Absolute minimum TA	30	30	20	20	20	10	15	45	15
Absolute minimum TA probability	0.75	0.8	0.85	0.9	0.8	0.8	0.85	0.8	0.5

	Expert Opinion Aggregation				
	Minimum	**25th Percentile**	**50th Percentile**	**75th Percentile**	**Maximum**
TA at probability of 1	45	52.5	75	80	0.7
Absolute minimum TA	10	15	20	30	45
Absolute minimum TA probability	0.5	0.775	0.8	0.85	0.9

TA, torso angle.

based on EE. Defining this linear relationship requires eliciting two (x, y) points from experts (Table 8.9): TA required for a probability of 1, the absolute minimum TA, and the probability corresponding to the absolute minimum.

PFD TURNING TIME FROM FACE DOWN

TT is defined as the average time required for a device to turn a face-down wearer to a position in which the wearer's respiration is not impeded and the majority of test subjects are turned face up. The faster the TT on as large a portion of the population as possible, the more likely it is that the PFD will prevent an unconscious person from drowning. Figure 8.8 shows a linear relationship between the TT and the probability of meeting the needs of a PFD user based on EE. Defining this linear relationship requires eliciting two (x, y) points from experts (Table 8.10): TT required for a probability of 1, the absolute maximum TT, and the probability corresponding to the absolute maximum TT.

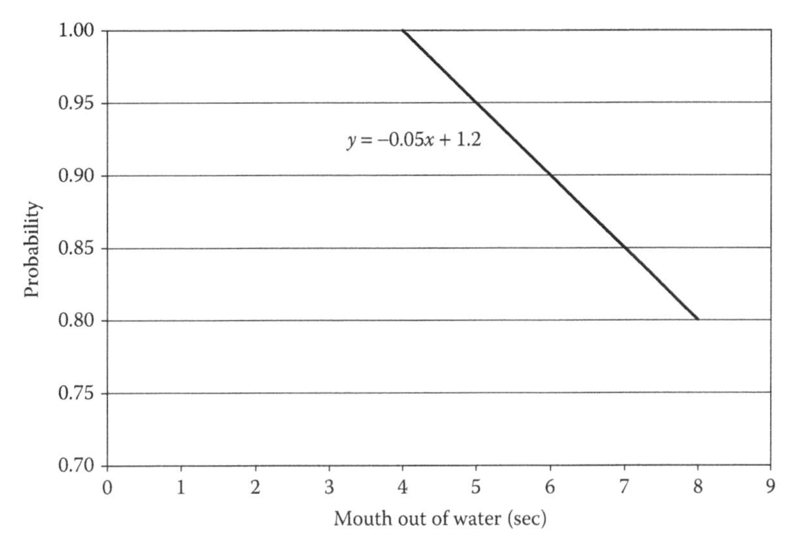

FIGURE 8.8
Probability of meeting the needs of a personal flotation device (PFD) user for turning time (TT).

TABLE 8.10
EE for TT

	Expert Opinion Collection								
Values to Define Model	**Expert 1**	**Expert 2**	**Expert 3**	**Expert 4**	**Expert 5**	**Expert 6**	**Expert 7**	**Expert 8**	**Expert 9**
TT at probability of 1	2.5	3	3	3	5	5	4	5	5
Absolute maximum TT	6	8	6.5	8	10	10	7	10	10
Absolute maximum TT probability	0.85	0.6	0.5	0.8	0.8	0.75	0.8	0.8	0.9

	Expert Opinion Aggregation				
	Minimum	**25th Percentile**	**50th Percentile**	**75th Percentile**	**Maximum**
TT at probability of 1	2.5	3	4	5	5
Absolute maximum TT	6	6.75	8	10	10
Absolute maximum TT probability	0.5	0.675	0.8	0.83	0.9

TT, turning time.

8.5 Model Modification Based on Available Data

Often data are unavailable for some aspects of the model, and adjustments to the model must be made to accommodate this lack of data. For example, a subsystem composed of components with unknown reliability can be modeled by the reliability of the entire subsystem, if that is known. Again, it is of the utmost importance for the model to accurately represent the system being analyzed. The failure probabilities of components and systems can be computed for selected failure modes using reliability methods that are based on the definition of performance functions and limit states. Methods such as the advanced second moment method and simulation with variance reduction techniques can be used for this purpose (Ang and Tang 1984; Ayyub and Haldar 1984; Ayyub and McCuen 2011). Equipment reliability can also be assessed based on statistical and Bayesian analysis of life data, as described in Chapter 4 and Appendix A.

8.6 Failure Data Sources

This section describes the sources of reliability data. These resources were used to construct Appendix B, which provides values for demonstration purposes. These values should not be used in risk studies without a careful examination of their applicability. In addition, this section surveys failure databases that are commonly quoted in the literature. The databases selected here are for illustration purposes.

Anderson and Neri (1990) provide a tabulation of failure rates of mechanical parts. The values were collected for the army aircraft flight safety prediction model and refer to aircraft components. The tabulation provides only part failure rates per hour for broadly categorized components. Some entries are provided as single figures, whereas others are shown as ranges. Supporting information on data sources and/or dates is not provided. Davidson (1994) provides a summary of failure rates for broadly defined systems, equipment, and components. The author uses a logarithmic scale for reporting the data. Modarres (1993) provides the suggested reliability data for the nuclear power industry using a lognormal model. Smith (2001) compiled a versatile and comprehensive list of values; while he covers a wide variety of components, the focus is on instrumentation and telecommunication systems. He provides failure rates per million hours, giving a combination of the lowest and highest failure rates and often the geometric mean.

The Martin Titan handbook, *Procedure and Data for Estimating Reliability and Maintainability*, was a widely distributed source of reliability information in 1959 (Fragola 1996). The handbook contains generic failure rates (per million hours) for a wide range of electrical, electronic, electromechanical, and mechanical parts or assemblies. The US Department of Defense military handbook, *MIL-HDBK-217*, provides consistent and uniform methods for estimating the inherent reliability of military electronic equipment and systems. In this handbook, the failure rate is expressed as a function of a generic failure rate and a set of adjustment factors to modify this generic failure rate by taking into account operating environments. Compared to the Martin Titan handbook, it offers an enormous amount of data; however, its limitations include the following: (1) assuming constant failure rates, (2) taking system failure rate as a summation of part failure rates only, (3) assuming design and manufacturing processes to be perfect, and (4) not accounting for variations in load and environment conditions. The Government/Industry Data Exchange Program (GIDEP 2002), formerly the Failure Rate Databank (FARADA), consists of data from industrial organizations, government laboratories, and repair facilities. This data bank includes both failure rate and replacement rate data collected from field experience, laboratory accelerated life tests, and reliability demonstration tests. It allows the data to be analyzed statistically according to a generic data structure. The Reliability Analysis Center (RAC) Non-Electronic Reliability Notebook (Fragola 1996) of the US Air Force provides a compilation of data from military field operating experiences and test experiences. This database provides failure rates for a variety of component types including mechanical, electromechanical, and discrete electronic parts and assemblies, with the concentration being on items that are not covered by other failure rate sources. Some of the failure rates were derived through syntheses of similar generic part types, with failure rate groupings being made for those of the type that had been subjected to a similar environment. The available data tables in this notebook of about 1,000 pages of data and over 25,000 parts are separated according to the source of information (field, test, and reliability demonstration). The WASH-1400 Reactor Safety Study of the NuRC (1975) used a set of generic failure data for performing PRA for a loss of coolant accident. The OREDA project has offered a collection program for the offshore industry available since the early 1980s (Sandtorv et al. 1996). As an initiative from the Norwegian Petroleum Directorate, this program started with the aim of collecting reliability data for safety important equipment, for example, electric generator, pumps, vessels, and valves. The collected reliability data have included >33,000 data points for 24,000 pieces of offshore equipment. This source includes information on failure rates, failure mode distribution, and repair time with the classification of failure severity. The four severity categories are critical, incipient, degradation, and unknown. Inozu (1993) developed a databank for ships on RAM.

8.7 Exercise Problems

Problem 8.1 What are the differences between TF and TIF in an EE process?

Problem 8.2 What are the success requirements for selecting experts and developing an expert panel? How many experts would you recommend to have? For your range on the number of experts, provide guidance in using the lower and upper ends of the range.

Problem 8.3 Working in teams, select five classmates as a panel of experts and elicit their opinions on five forecasting issues in engineering. Select these issues such that the classmates can pass the tests of experts on these issues. Perform all the steps of EE, and document your process and results as a part of solving this problem.

Problem 8.4 You are asked to form an expert panel and perform EE about the issues provided below that are concerned with current developments by humanity. In addition to obtaining answers to these questions, you are also being asked to assess the confidence of the participants in their answers on a scale from 1 to 7, corresponding to the highest and the smallest confidence, respectively.

- In your opinion, in what year will the median family income (in 2002 or present dollars) reach twice its present amount?
- In what year will the use of electric automobiles, among all automobiles driven, reach 50%?
- In what year will the use of intelligent and autonomous (without a driver) automobiles, among all automobiles driven, reach 50%?
- By what year will the average life expectancy of a human reach more than 120 years?
- By what year will it be possible to have commercial carriers to the outer space?

- In what year will a human for the first time travel to Mars stay at least several days and return to Earth?
- Provide a formal report summarizing the process, listing the experts, and providing answers to these questions.

Problem 8.5 Develop a list of communication forecasting issues and elicit opinions, similar to the exercise in Problem 8.4.

Problem 8.6 Develop a list of bioengineering and health forecasting issues and elicit opinions, similar to the exercise in Problem 8.4.

Problem 8.7 Develop a list of power sources and technologies forecasting issues and elicit opinions, similar to the exercise in Problem 8.4.

Problem 8.8 An optimal clearance between the bottom of an overpass bridge and the water surface of a navigation channel must be determined to permit for safe navigation. A group of seven navigation experts was consulted to offer their opinions about an appropriate design clearance. A formal EE session resulted in the following opinions:

	Expert Opinion Regarding Optimal Clearance						
	Expert 1	Expert 2	Expert 3	Expert 4	Expert 5	Expert 6	Expert 7
Clearance (m)	50	55	65	70	70	75	80

Aggregate the opinions of the experts by computing the minimum, maximum, 25th percentile, 50th percentile, and 75th percentile values.

Problem 8.9 A management consultant is in the process of restructuring the organizational hierarchy of a large corporation. She identified three possible types of organizational structures that are suitable for this large corporation: vertical structure, flat structure, or matrix structure. The selection of a particular type should be based on achieving the highest satisfaction level by employees and their managers. She conducted an EE session using seven experts and received opinions about the best type of structure suitable for the company. The level of satisfaction was measured on a scale of 100 points (lowest, 0; highest, 100) with regard to each structure type as provided in the following table:

Structural Organization Type	Level of Satisfaction (0, lowest; 100, highest)						
	Expert 1	Expert 2	Expert 3	Expert 4	Expert 5	Expert 6	Expert 7
Vertical	65	70	70	75	75	80	75
Flat	70	85	85	60	75	80	85
Matrix	80	70	75	75	90	85	85

Aggregate the opinions of the experts by computing the minimum, maximum, 25th percentile, 50th percentile, and 75th percentile values.

Problem 8.10 The probability of performance failure of a newly designed vertical organizational system of a large corporation needs to be assessed by the research and development department of the corporation. The research and development department identified potential failures at three management levels (top, middle, and lower) as the sources of this organizational system failure. Nine experts in organizational performances were consulted to offer their opinions and provide probability values. The results are summarized in the following table:

Level of Management	Failure Probability of Vertical Structure								
	Expert 1	Expert 2	Expert 3	Expert 4	Expert 5	Expert 6	Expert 7	Expert 8	Expert 9
Top	0.55	0.50	0.45	0.65	0.70	0.65	0.65	0.50	0.65
Middle	0.70	0.65	0.65	0.75	0.80	0.70	0.75	0.65	0.70
Lower	0.85	0.70	0.85	0.85	0.90	0.80	0.80	0.75	0.80

Aggregate the opinions of the experts by computing the minimum, maximum, 25th percentile, 50th percentile, and 75th percentile values.

9

Risk-Based Maintenance of Marine Vessels

A Case Study

The purpose of this chapter is to illustrate using the concepts covered in this book to develop a risk methodology for managing maintenance activities of a marine vessel as a case study. Readers are referred to other case studies, such as the methodology on risk analysis of a protected hurricane-prone region by Ayyub et al. (2009), and the methodology on critical asset and portfolio risk analysis for homeland security using an all-hazards framework by Ayyub et al. (2007) and McGill et al. (2007).

CONTENTS

9.1 Maintenance Methodology

A methodology can be constructed to utilize risk and economic concepts to manage maintenance of a structural system. A marine system is used to illustrate the concepts introduced in the section. The methodology utilizes and builds on previous experiences and addresses the limitations of current maintenance practices. The methodology described here is referred to as risk-based optimal maintenance management of ship structures (ROMMSS) as described by Ayyub et al. (2002). Risk-based methodologies require the use of analytical methods at the system level, which consider subsystems and components in assessing their failure probabilities and consequences. Systematic, quantitative, qualitative, or semiquantitative approaches for assessing the failure probabilities and consequences of engineering systems are used for this purpose. A systematic approach allows an engineer to expediently and easily evaluate complex engineering systems for safety and risk under different operational and extreme conditions. The ability to quantitatively evaluate these systems helps cut the cost of unnecessary and often expensive reengineering, repair, strengthening, or replacement of components, subsystems, and systems. The results of risk analysis can also be utilized in decision analysis methods that are based on the benefit–cost trade-offs.

The ROMMSS is essentially a six-step process that provides a systematic and rational framework for the reduction of total ownership costs for ship structures. This framework combines advanced probabilistic numerical models, optimization algorithms, risk and maintenance cost models, and corrective/preventive maintenance technologies, and directs them toward the cost-effective identification, prioritization, and overall management of ship structure maintenance problems. Such a strategy could lead to the reengineering of ship structure components and system maintenance processes. The basic steps followed for the ROMMSS strategy, as shown in Figure 9.1, are as follows:

1. Selection of ship or fleet system
2. Partitioning of the ship structure into major subsystems and components

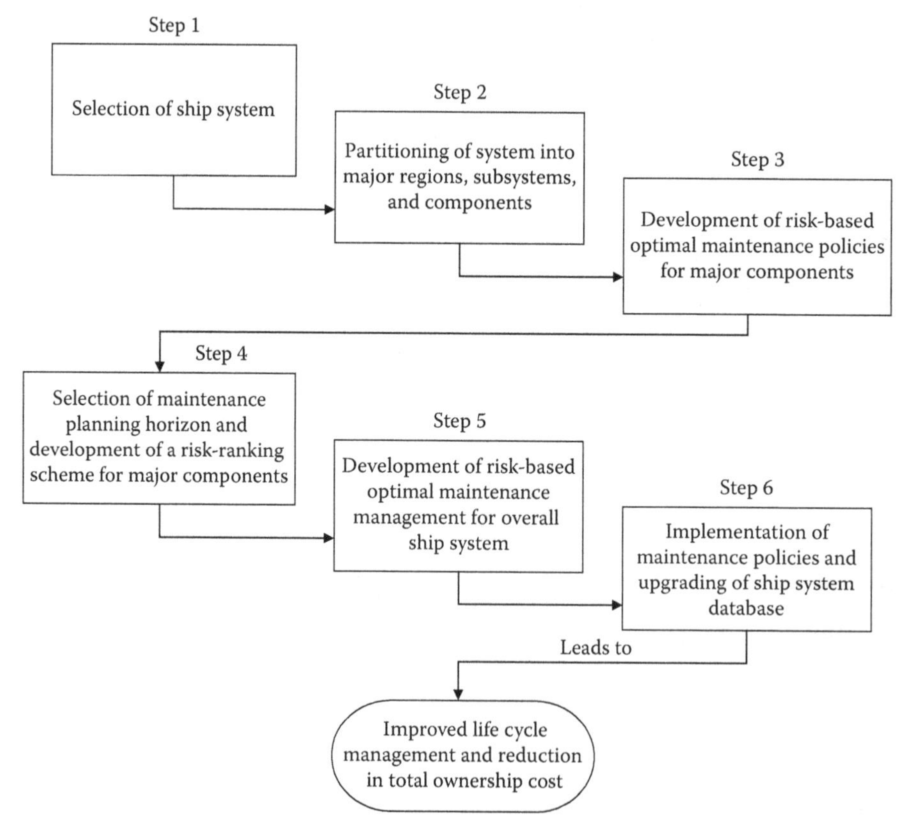

FIGURE 9.1
Flowchart for development of risk-based optimal maintenance management of ship structures (ROMMSS).

3. Development of risk-based optimal maintenance policy for major components within a subsystem
4. Selection of a time frame for maintenance implementation and development of risk-ranking scheme
5. Development of optimal maintenance scheduling for the overall vessel
6. Implementation of optimal maintenance strategies and updating system condition states (CSs) and databases

These steps are described in subsequent sections.

9.2 Selection of Ship or Fleet System

The first task in ROMMSS involves the selection of a ship system for maintenance. This selection could be a single vessel or an entire class of similar ships. The system and its boundaries must first be identified. Although the risk-based methodology advanced in this study is quite general and can be applied to the maintenance of any system within a ship structure, emphasis is placed here on maintenance of the hull structural system. This system includes longitudinals, stringers, frames, beams, bulkheads, plates, coatings, foundations, and tanks. The hull structural system delineates the internal and external shape of the hull, maintains watertight integrity, ensures environmental safety, and provides protection against physical damage. The boundaries of a hull structural system include the hull, its appendages from (and including) the boot topping down to the keel for the exterior surfaces of the ship, the structural coating, and the insulation for the interior and exterior surfaces.

9.3 Partitioning of the System

Components of a typical ship vessel include the main hull form (part of which is below the waterline), single or multiple decks, an engine room, an equipment room, fuel tanks, freshwater tanks, ballast tanks, superstructures, and storage area. These components experience structural deterioration due to loads from a variety of sources—environmental and otherwise. The type, rate, and extent of structural damage are each dependent on the physical location of a component and may be different for different regions of a vessel. Furthermore, the maintenance requirements of various components of a ship structure may differ in terms of frequency, type, and cost, even for components within the same region. The presence of structural damages and the uncertainty associated with its impact pose a risk that can affect the overall safety of a vessel. This risk could manifest itself in terms of loss of watertightness, environmental pollution, or even loss of serviceability.

The basic steps involved in partitioning a ship structural system are demonstrated in Figure 9.2. It should be noted that the major components of some ship structural systems are the basic elements for which the maintenance policies require optimization. As such, partitioning schemes for some vessels might choose to skip steps 2 and 3 of the partitioning process.

An example of a partitioning scheme for a naval vessel is shown in Figure 9.3. The structure is first broken into four artificial regions separated by major transverse bulkheads. For

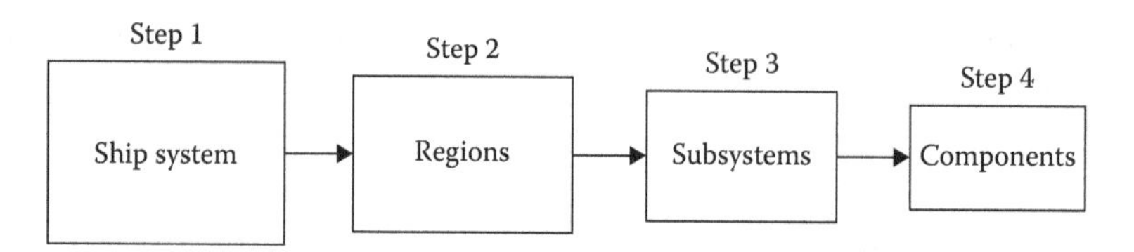

FIGURE 9.2
Basic steps in partitioning a ship structural system.

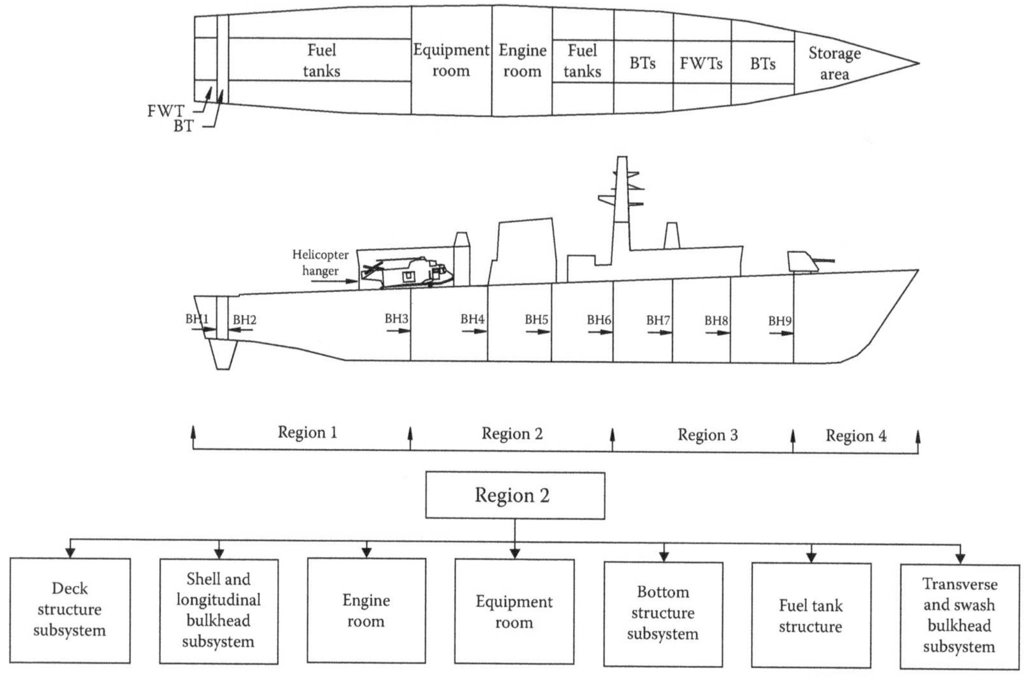

FIGURE 9.3
Demonstration of partitioning scheme for a navy ship. BH, bulkhead; BT, ballast tank; FWT, freshwater tank.

example, region 2, which lies between bulkhead number 3 (BH3) and bulkhead number 6 (BH6), has the major elements such as deck structure, hull plating, longitudinal bulkhead, engine room, equipment room, bottom structure, fuel tank structures, and transverse bulkhead. These subsystems are further broken down into their major components as shown in Figure 9.4.

A partitioning scheme is also demonstrated in Figure 9.5 for a typical tanker ship, where the vessel is broken into fore, midship, and aft regions. The major midship structural subsystems and their components are shown in Figure 9.6.

9.4 Development of Optimal Maintenance Policy for Components

This section discusses the details of step 3 of ROMMSS. Figure 9.7 provides a flowchart for the risk-based optimal maintenance of individual components. Each of the essential steps outlined in the flowchart is discussed in the following subsections.

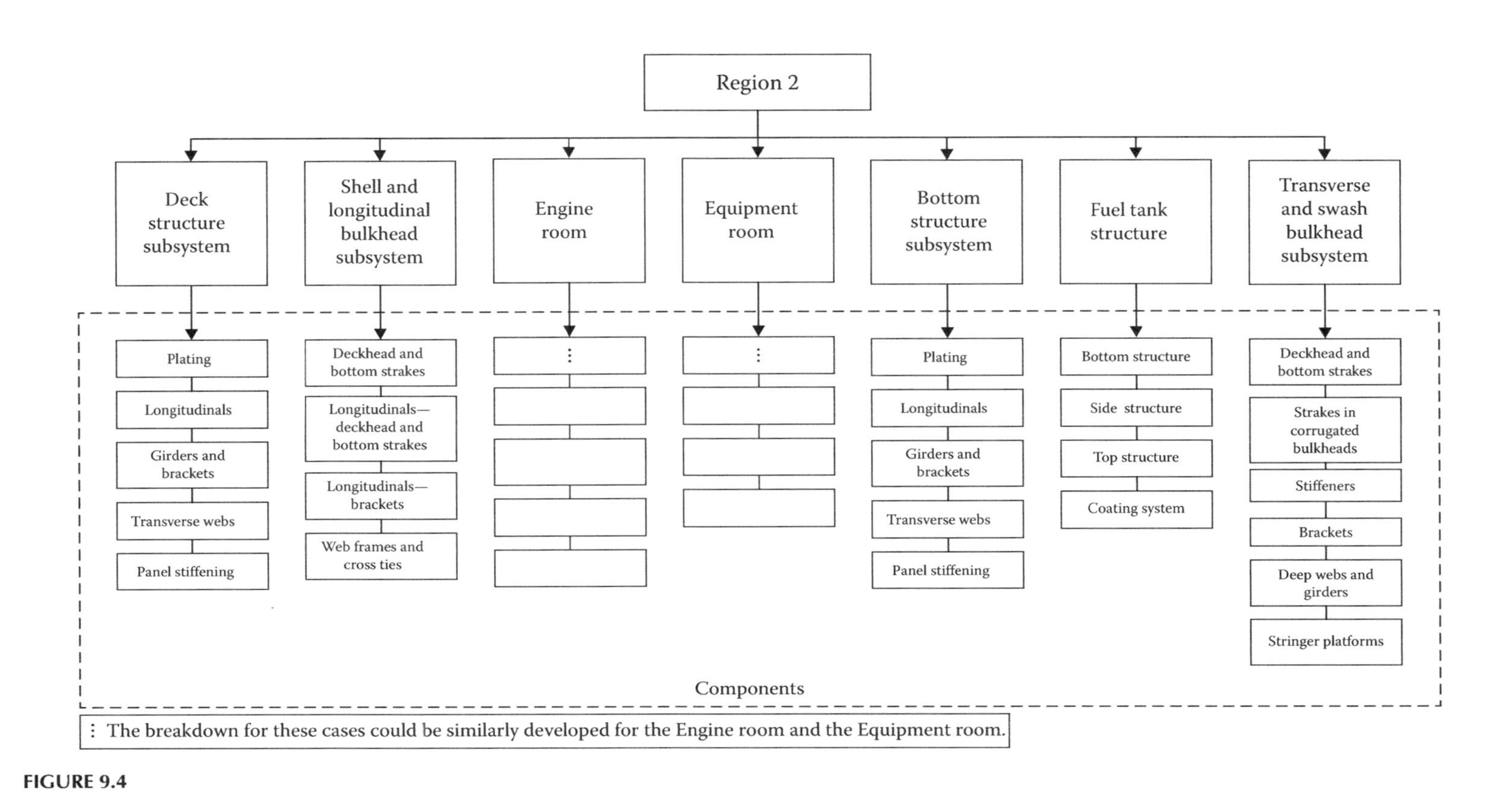

FIGURE 9.4
Demonstration of subsystem partitioning scheme for a navy ship.

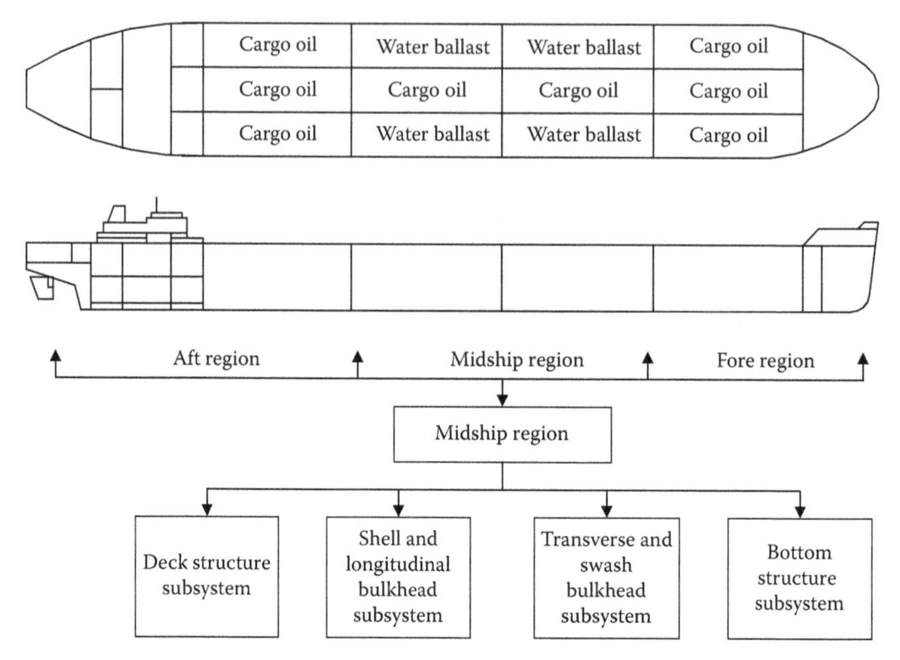

FIGURE 9.5
Demonstration of partitioning scheme for a tanker structure.

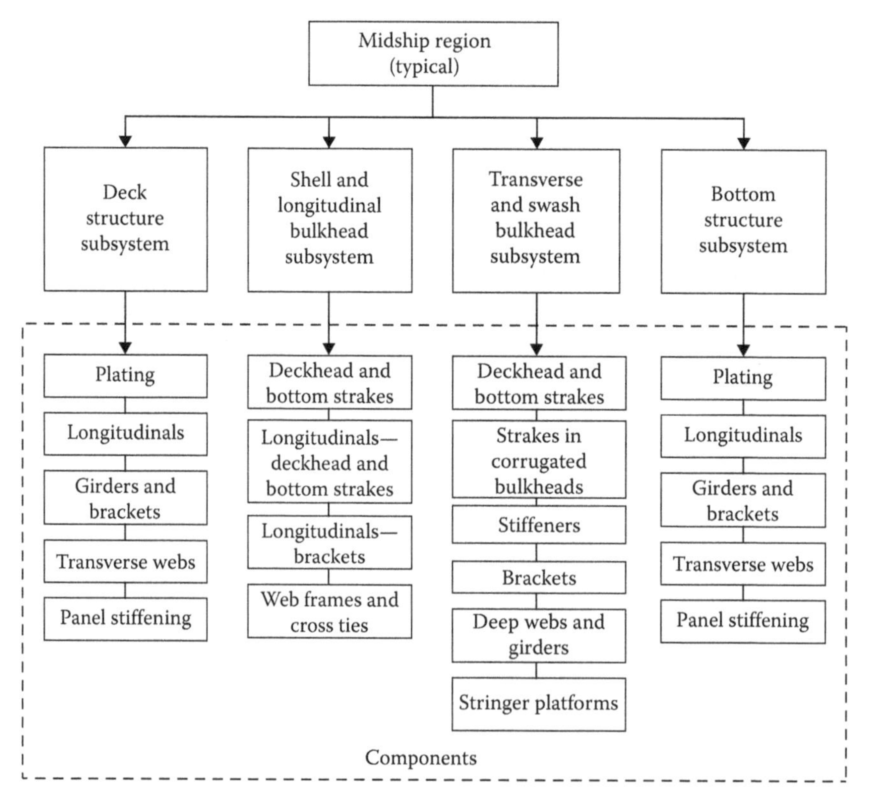

FIGURE 9.6
Typical midship subsystems and components for tanker ship.

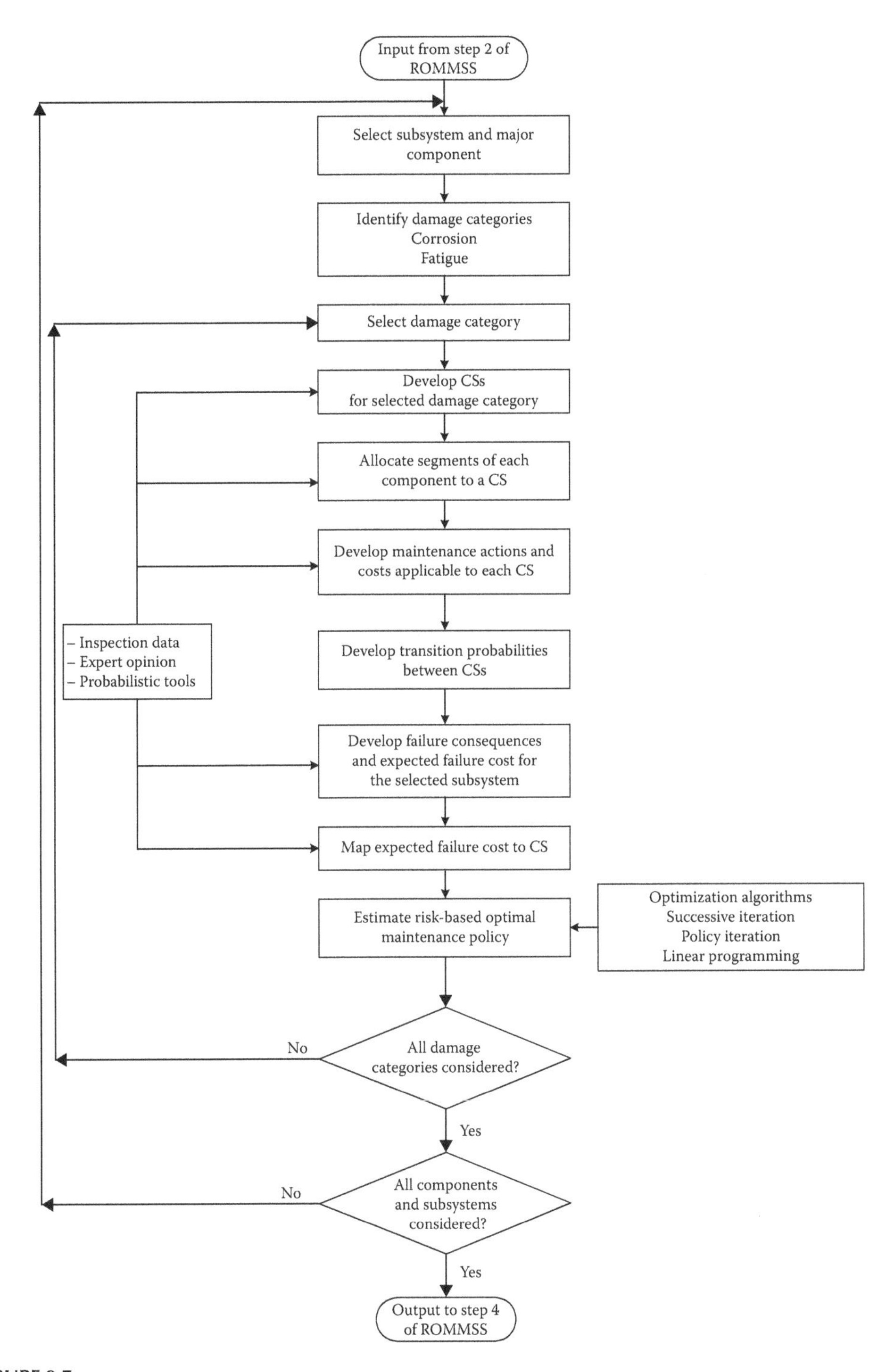

FIGURE 9.7
Flowchart for risk-based optimal maintenance policy for major components. CS, condition state; ROMMSS, risk-based optimal maintenance management of ship structure.

9.4.1 Selection of a Subsystem and Its Major Components

The subsystem must first be identified and then its major component selected. Examples of this process are presented in Figures 9.4 and 9.6.

9.4.2 Identification of Damage Categories

Several damage categories may be applicable to a major component. Identification of these categories must place emphasis on the components that have been known to consume an excessive portion of the overall maintenance budget. A review of ship structure maintenance needs shows that, with respect to budget consumption, the most prominent damage categories for most components include fatigue cracking and corrosion.

Fatigue cracks are the result of repeated application of stress cycles, which gradually weaken the granular structure of a metal. They are typically enhanced by high stresses and are most likely to occur in regions of high stress concentration. Corrosion is the physical deterioration of a metal as a result of chemical or electrochemical reaction with its environment. In steel vessels, corrosion usually starts with breakdown of any protective coating and progresses to rust formation and subsequent metal loss. The rate of corrosion attack depends on many factors, including heat, acidity, salinity, and the presence of oxygen. Although ship surfaces are protected to some degree by paint systems, these systems can fail due to improper application or chipping or simply as a result of aging. Corrosion generally progresses to different degrees in different locations, but the overall result is a gradual reduction in the capacity of a structure for load. As the two aforementioned damage mechanisms are the most common in ship structures, they are the focus of the remainder of this discussion. It should, however, be noted that the proposed methodology is equally applicable to other damage modes. To advance the risk-based methodology, a suitable damage category must be selected.

9.4.3 Development of Condition States

Once a system has been broken down into its major subsystems and components, CSs are employed as a measure of the degree of damage experienced by segments of a given component. CSs serve to rank the level of damage severity among segments. The level of damage could range from "good as new" or "intact" to "failure." The CSs for a particular type of damage have to be defined. Two examples of corrosion-based CSs currently used by various classification societies, naval forces, and inspectors are illustrated in Tables 9.1 and 9.2. Table 9.1 represents an example of CSs allocated based on a visual observation, whereas Table 9.2 represents CSs allocated based on measured values of material thickness.

TABLE 9.1

Condition States (CSs) for Corrosion Damage (Visual Observation)

CS	Name	Description
1	No corrosion	Paint/protection system is sound and functioning as intended.
2	Low corrosion	Surface rust or freckled rust has either formed or is in the process of forming.
3	Medium corrosion	Surface or freckled rust is prevalent and metal is exposed.
4	Active/high corrosion	Corrosion is present and active, and a significant portion of metal is exposed.
5	Section loss	Corrosion has caused section loss sufficient to warrant structural analysis to ascertain the effect of the damage.

TABLE 9.2

Condition States (CSs) for Corrosion Damage (Measured Thickness Loss)

CS	Name	Description
1	No corrosion	Paint/protection system is sound and functioning as intended.
2	Surface corrosion	Less than 10% of metal thickness has been attacked by corrosion.
3	Moderate corrosion	Metal thickness loss is between 10% and 25%.
4	Deep corrosion	Metal thickness loss is between 25% and 50%.
5	Excessive corrosion	Metal thickness is reduced to <50% of original thickness.

In addition, CSs for any damage category can be defined through elicitation of subject matter experts (SMEs).

9.4.4 Allocation of Component Percentages in Each Condition State

Inspections are periodically conducted to ascertain the damaged CSs of the major components of ship structures. These inspections are driven by statutory requirements of Classification Society, Flag Administration Officer requirements, or owner/operator requirements. Generally, the basic defects such as cracking, corrosion, coating breakdown, and buckling are sought for and documented during inspections. An inspection could be conducted either visually or by using more sophisticated equipment such as ultrasonic thickness gauging. The purpose of this step is to allocate the percentage of a major component to the CS corresponding to the damage it has experienced. This task should be performed using the data obtained during the inspection. Exact values of the percentage allocated to each CS are not required for optimal performance of the current methodology. The methodology is robust enough to handle such uncertainties and inexact values. This percentage allocation represents the current distribution of the CSs for a particular component. For example, in a CS allocation scheme consisting of five CSs, the following vector represents the percentage breakdown of the current CSs (i.e., $t = 0$):

$$s^0 = s_1^0, s_2^0, s_3^0, s_4^0, s_5^0 \tag{9.1}$$

The total percentage of components allocated to a CS vector at any time always adds up to 100. Unfortunately, in ship structural systems, current inspection data and records may not be available with which to develop CS distributions. In such instances, the help of SMEs may be elicited to establish current CS distributions. Factors such as the age and travel route of the vessel, as well as the location of the components, must be taken into consideration when eliciting SMEs. A maximum value should be specified for the percentage of the components permitted to be allocated to the worst CS at any time. This threshold or limiting value (s_L) should be based on Flag Administration Officer and Classification Society requirements. Referring to Equation 9.1, s_5^0 must be no greater than s_L (i.e., $s_5^0 \leq s_L$).

9.4.5 Maintenance Actions and Costs

Maintenance and repair actions that can be applied to various segments of a component depend not only on the damage category, but also on the location of the component and the CSs of the component. The cost of these actions can differ significantly. For example, consider the corrosion problem defined previously. Possible maintenance actions include spot blasting, welding, patch coating, addition and maintenance of sacrificial anodes, and section replacement. In general, the cost of maintenance action increases with the severity of a CS.

TABLE 9.3

Demonstrative Maintenance Actions and Associated Costs

CS	Percentage of Component in CS	Possible Maintenance Action	Expected Unit Cost of Maintenance Action ($)
1	s_1^0	1 = No repair	0
		2 = Monitor	C(1,2)
2	s_2^0	3 = No repair	0
		4 = Monitor	C(2,4)
		5 = Spot blast/patch coating	C(2,5)
3	s_3^0	6 = No repair	0
		7 = Spot blast/patch coating	C(3,7)
		8 = Spot blast/weld cover plate/patch coating	C(3,8)
4	s_4^0	9 = No repair	0
		10 = Cut out/weld new plate/spot blast/patch coating	C(4,10)
		11 = Add/maintain sacrificial anode	C(4,11)
5	s_5^0	12 = No repair	0
		13 = Cut out/weld new plate/spot blast/patch coating	C(5,13)
		14 = Replace component	C(5,14)

CS, condition state.

For example, the cost associated with the repair of a level 5 CS is typically much greater than that associated with the repair of a level 1 CS. A risk-based optimal maintenance system must seek to minimize the cost of maintenance. The cost of maintenance actions could include materials, labor costs, and the cost of steel and anode replacement. The unit costs should be based on the dimensions of the component (area, volume, or length). Both the labor costs and the potential maintenance actions should be estimated based on elicitation from SMEs. A summary of potential maintenance actions and the associated costs for the corrosion problem considered previously is shown in Table 9.3. The associated cost designation, $C(a,b)$, reads as "the maintenance cost associated with CS a and maintenance action b." It should be noted from Table 9.3 that every CS has a no-repair maintenance action. An associated expected failure cost is due to the risk of being in a particular CS. This cost is estimated at a subsequent step.

9.4.6 Transition Probabilities for Cases without Maintenance Actions

Ship structural components tend to deteriorate when no maintenance actions are taken. A model must therefore be developed to estimate the deterioration rates of components under such circumstances. The model must have the capability to quantify the uncertainty inherent in such predictions. Furthermore, the prediction model must have the capability to incorporate results from actual experience and to update parameter values when more data become available. A probabilistic Markov chain model, which quantifies uncertainty, is adopted in this study. It estimates the likelihood that a component, in a given CS, would make a transition to an inferior CS within a specified period. An example of the Markov chain model is shown in Figure 9.8. Such Markov chain modeling has been used in bridge management systems for maintenance planning developed by the Federal Highway Administration and utilized by many states.

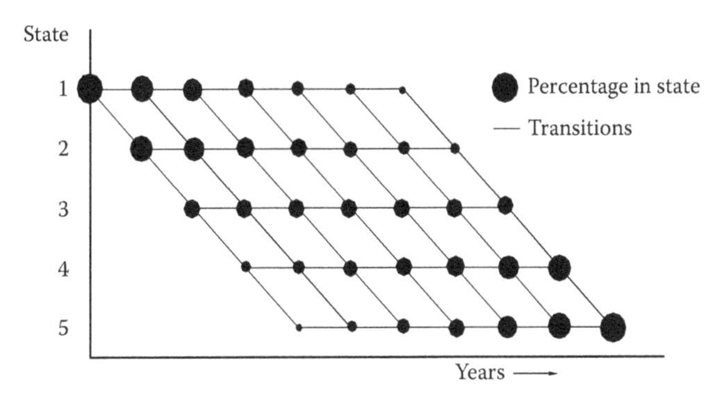

FIGURE 9.8
Demonstration of Markov chain transition between condition states (CSs) for cases without maintenance actions.

For the corrosion problem under consideration, the following assumptions are made in developing Markov chain transition probabilities:

- A one-year time interval for corrosion to progress from one state to an inferior state is assumed. This is a reasonable assumption and consistent with data availability such as the Tanker Structure Cooperative Forum (TSCF) corrosion growth annual rates provided for the components of ship structures.
- CSs are allowed to deteriorate by, at most, one level during a one-year period.
- Aging vessels generally deteriorate faster than new vessels; therefore, transition probabilities between CSs depend on the age of the vessel. Transition probabilities are assumed valid for 5-year intervals. This assumption implies that different corrosion growth rates are assigned depending on the age of the vessel. TSCF, for example, assigns an age-dependent corrosion growth rate for structural components of a tanker vessel.

Based on the above assumptions, transition probabilities between CSs can be estimated using inspection data from two consecutive years. The algorithms for estimating the transition probabilities are given at the end of this section. However, it is expected that such data might not be readily available for some components. Therefore, in such instances, elicitation of SMEs needs to be employed (Ayyub et al. 2002). Users of this system need to elicit opinions of inspectors and engineers about component deterioration, such that the responses could be mathematically converted to the transition probabilities required by the models. An example question could be as follows:

> *Suppose all of the components are in state 1. How long will it take for 50% of them to deteriorate to state 2 if no maintenance action is taken?*

Taking this question as an example, the probability of transition (i.e., deterioration) from CS 1 to CS 2, P_{12} can be computed as

$$P_{12} = 1 - 0.5^{1/T_1} \tag{9.2}$$

where:
T_1 is the number of years used to calculate transition probabilities

Similar questions can be asked about other transition probabilities. It should be noted that a similar approach has been used in bridge management systems.

The optimal maintenance policy selections are based on the theory of discounted dynamic programming. Consider a probabilistic process that is observed to be in a number of states at points in time $t_0, t_1, t_2, \ldots, t_n$. After observing the state of the process, an action must be chosen. The action belongs to a finite set of feasible actions for that state. When the process is in state *i* at time *n* and action *a* is chosen, an expected cost is incurred, denoted by $C(i,a)$. The states for the next time step in the process are determined based on the transition probabilities for action *a*, denoted by $P_{ij}(a)$.

If X_n denotes the state of the process at time *n* and a_n is the action chosen, the previous statement implies that

$$P(x_{n+1} = j \middle| x_0, a_0, \ldots, x_1, a_1, \ldots, x_n = i, a_n = a) = P_{ij}(a) \tag{9.3}$$

Thus, the costs and transition probabilities are functions of only the previous state and subsequent action, assuming that all costs are bounded. To select from the potential actions, some policy must be followed. There are no restrictions on the choice of policies; hence, actions can also be considered random.

An important class of all policies is the class of stationary policies. A policy *f* is called stationary if it is nonrandom, and the action it chooses at time *t* depends only on the state of the process at time *t*; whenever in state *i*, $f(i)$ is chosen. Thus, when a stationary policy is employed, the sequence of states (X_n; $n = 0, 1, 2, \ldots$) forms a Markov chain; hence, such processes are typically termed *Markovian decision processes*.

To find the optimal policy, a criterion for such optimization must be chosen. If we choose as our criterion the total expected return on invested dollars and discount future costs by a discount factor α (such that $0 < \alpha < 1$), then among all policies π, we attempt to minimize the following:

$$V_\pi(i) = E_\pi\left[\sum_{n=0}^{\infty} c(i_n, a_n)\alpha^n \middle| x_0 = i\right] \tag{9.4}$$

where:

E_π is the (conditional) expectation given that policy π is employed

Hence, $V_\pi(i)$ is the total expected discounted cost. A policy π^* is said to be α-optimal if $V_\pi^*(i) \leq V_p(i)$ for all *i* and π.

The main result of dynamic programming (i.e., the optimality equation) is a functional equation satisfied by $V(i)$ as follows:

$$V(i) = \min_a\left[c(i_n, a_n) + \alpha\sum_j P_{ij}(a)V(j)\right] \tag{9.5}$$

An important result of dynamic programming is obtaining the optimality according to Equation 9.5. In other words, if *f* is a stationary policy that, when the process is in state *i*, selects an action that minimizes the right-hand side of Equation 9.5, then

$$V_f(i) = V(i) \text{ for all } i \tag{9.6}$$

It is also true that *V* is the unique bounded solution of the optimality equation.

9.4.7 Failure Consequences and Expected Failure Cost

Deterioration of subsystems of a ship structure poses a risk to operation of the vessel, such as unavailability. The level of risk depends on the consequences of subsystem failure. The consequences of failure could range from unplanned repair, unavailability, and environmental pollution to reduction or loss of serviceability. This task is aimed at identifying and streamlining the consequences of failure associated with a subsystem. Furthermore, it is directed toward estimating the likelihood that being in a particular CS will increase or reduce the realization of these consequences. The approach proposed herein assigns important factors to various components that make up the subsystem. More specifically, this step involves the following:

- Identification and categorization of failure consequence for a subsystem; an example is shown in Table 9.4.
- Development of a rating scheme for the various components of a subsystem; the rating scheme ranks the components of a subsystem in terms of their degree of importance to the overall structural integrity, watertightness, and functional requirements of the subsystem. A rating scheme can be developed as shown in Table 9.5.
- Mapping the cost of failure to the no-repair action that exists within a given CS (Table 9.3). The goal is to estimate the likelihood of whether operating in a particular CS will increase or reduce the chances of incurring a particular failure cost. SMEs can again be called upon to estimate this probability. The probability estimation process must be cast in such a way that experts can supply subjective information that can be translated into numerical values. An example of a probabilistic translation scheme is shown in Table 9.6.

TABLE 9.4

Example of Possible Consequences of Subsystem Failure

Consequence of Failure	Consequence Cost per Incident ($)
1 = Minor structural failure	C_1 = Minor unplanned repair cost
2 = Reduction/loss of serviceability	C_2 = Economic cost due to loss of serviceability
3 = Major structural failure	C_3 = Substantial unplanned repair cost/economic cost
4 = Major oil spill, leak, or other form of environmental pollution	C_4 = Environmental cleaning/litigation cost

TABLE 9.5

Sample Ranking Scheme for a Typical Subsystem

Bottom Structure Components	Level of Importance (1 [low] to 4 [high])
Bottom plating	4
Bottom longitudinals	4
Bottom girders and brackets	4
Bottom transverse webs	3
Panel stiffening	4

TABLE 9.6

An Example of a Probabilistic Translation Scheme

Probability	Value
Low	10^{-6}
Medium	10^{-4}
High	10^{-2}
Very high	10^{-1}

TABLE 9.7

Example of Mapping Condition States (CSs) to Failure Cost

CS	Action	Probability of Failure Consequence	Expected Unit Failure Cost
1	No repair	$P_{1C_1}, P_{1C_2}, P_{1C_3}, P_{1C_4}$	$R_1 = P_{1C_1}C_1 + P_{1C_2}C_2 + P_{1C_3}C_3 + P_{1C_4}C_4$
2	No repair	$P_{2C_1}, P_{2C_2}, P_{2C_3}, P_{2C_4}$	$R_2 = P_{2C_1}C_1 + P_{2C_2}C_2 + P_{2C_3}C_3 + P_{2C_4}C_4$
3	No repair	$P_{3C_1}, P_{3C_2}, P_{3C_3}, P_{3C_4}$	$R_3 = P_{3C_1}C_1 + P_{3C_2}C_2 + P_{3C_3}C_3 + P_{3C_4}C_4$
4	No repair	$P_{4C_1}, P_{4C_2}, P_{4C_3}, P_{4C_4}$	$R_4 = P_{4C_1}C_1 + P_{4C_2}C_2 + P_{4C_3}C_3 + P_{4C_4}C_4$
5	No repair	$P_{5C_1}, P_{5C_2}, P_{5C_3}, P_{5C_4}$	$R_5 = P_{5C_1}C_1 + P_{5C_2}C_2 + P_{5C_3}C_3 + P_{5C_4}C_4$

To perform such mapping operations, an appropriate list of questions must be developed. An example question could be as follows:

Suppose a component is in state 1 (new state). What is the likelihood that it will experience an unplanned repair during its first year of service?

Similar questions can address all failure consequence categories and CSs. The findings can then be summarized to arrive at an expected failure cost, as shown in Table 9.7. It is evident that the procedure can become quite involved and must therefore be computerized to achieve cost-effectiveness.

9.4.8 Transition Probabilities for Cases with Maintenance Actions

Implementation of maintenance actions generally moves a component toward better CSs. Inherent uncertainty is associated with the degree of improvement afforded by a particular maintenance action. Assessing the quality of repair is highly subjective, as it depends on not only the personnel involved but also the shipyard that is used. Therefore, a model must be developed not only to estimate the improvement of a component after a maintenance action has been taken but also to quantify the uncertainty inherent in such improvements. The prediction model must have the capability to incorporate results from actual experience and also update its parameters when more data become available. A Markov chain transition probability model, which quantifies uncertainty, is again adopted in this section. The prediction model quantifies the likelihood that a component in a particular CS would improve from one CS to a superior CS when a specific maintenance action is taken. Elicitation of SMEs is currently the only approach to estimating transition among states when maintenance actions are taken. A suitable list of SME questions should be compiled such that expert opinions can easily be translated into transition probabilities. An example question could be as follows:

Suppose a group of components are operating in state 3 and a particular maintenance action is taken. What, then, are the percentages of components that, as a result, improve to either state 1 or state 2 immediately after the action?

TABLE 9.8

Implementation of Maintenance Actions to Estimate Failure Cost

CS	Percentage of Component in CS	Maintenance Action Number	Transition Probabilities among States					Expected Unit Maintenance Cost	Expected Failure Cost
			1	2	3	4	5		
1	s_1^0	1	$P_{11}(1)$	$P_{12}(1)$	$P_{13}(1)$	$P_{14}(1)$	$P_{15}(1)$	0	R_1
		2	$P_{11}(2)$	$P_{12}(2)$	$P_{13}(2)$	$P_{14}(2)$	$P_{15}(2)$	C(1,2)	R_1
2	s_2^0	3	$P_{21}(3)$	$P_{22}(3)$	$P_{23}(3)$	$P_{24}(3)$	$P_{25}(3)$	C(2,3)	R_2
		4	$P_{21}(4)$	$P_{22}(4)$	$P_{23}(4)$	$P_{24}(4)$	$P_{25}(4)$	C(2,4)	R_2
		5	$P_{21}(5)$	$P_{22}(5)$	$P_{23}(5)$	$P_{24}(5)$	$P_{25}(5)$	C(2,5)	R_2
3	s_3^0	6	$P_{21}(6)$	$P_{22}(6)$	$P_{23}(6)$	$P_{24}(6)$	$P_{25}(6)$	C(3,6)	R_3
		7	$P_{31}(7)$	$P_{32}(7)$	$P_{33}(7)$	$P_{34}(7)$	$P_{35}(7)$	C(3,7)	R_3
		8	$P_{31}(8)$	$P_{32}(8)$	$P_{33}(8)$	$P_{34}(8)$	$P_{35}(8)$	C(3,8)	R_3
4	s_4^0	9	$P_{41}(9)$	$P_{42}(9)$	$P_{43}(9)$	$P_{44}(9)$	$P_{45}(9)$	C(4,9)	R_4
		10	$P_{41}(10)$	$P_{42}(10)$	$P_{43}(10)$	$P_{44}(10)$	$P_{45}(10)$	C(4,10)	R_4
		11	$P_{41}11)$	$P_{42}(11)$	$P_{43}(11)$	$P_{44}(11)$	$P_{45}(11)$	C(4,11)	R_4
5	s_5^0	12	$P_{51}(12)$	$P_{52}(12)$	$P_{53}(12)$	$P_{54}(12)$	$P_{55}(12)$	C(5,12)	R_5
		13	$P_{51}(13)$	$P_{52}(13)$	$P_{53}(13)$	$P_{54}(13)$	$P_{55}(13)$	C(5,13)	R_5
		14	$P_{51}(14)$	$P_{52}(14)$	$P_{53}(14)$	$P_{54}(14)$	$P_{55}(14)$	C(5,14)	R_5

CS, condition states.

A computerized elicitation program can be developed to generate a survey to address the effectiveness of possible repair actions for various major components of ship structures. Table 9.8 summarizes the outcome of implementation of the above steps. Failure probabilities can be assessed using models provided in Chapter 4.

9.4.9 Risk-Based Optimal Maintenance Policy

The data needed for determining a risk-based optimal maintenance policy for a component are summarized in Table 9.8. The objective of this particular task is to find, for a component under a particular environmental or damage category, the maintenance policy that minimizes the maintenance costs while maintaining the system below an acceptable risk level in the long run. The optimal maintenance strategy is the one that incurs the minimum total cost. An optimal maintenance policy stipulates a set of maintenance actions that must be implemented for a given component. The two main implications of an optimal policy are as follows:

- Delaying recommended actions will be more expensive in the long term.
- Performing additional maintenance actions that are considered in the model but not recommended will result in an increase in overall maintenance costs.

Four important things occur periodically with major components of a ship structure:

- Components deteriorate, resulting in transition from one CS to a worse CS.
- The existence of segments of a component in various CSs implies a risk of failure, which translates into expected failure costs.

- Maintenance actions (both minor yearly repairs and major dry dock repairs) are executed, thereby incurring costs.
- Implementation of maintenance actions yields an improvement in the CS of a component.

This information is summarized in Table 9.8. A risk-based optimal maintenance policy uses the above information to prescribe a set of maintenance actions that minimizes the maintenance costs while ensuring that the component is not subjected to an unacceptable risk of failure. This policy may be formulated again using the Markov decision model. The effects of a set of maintenance actions and the costs of those actions are propagated through a Markov chain via appropriate transition probabilities. It is assumed that a finite planning horizon can be defined and that future costs can be discounted, thereby accounting for economic inflation. The problem can be stated as follows for each component's CS: Find the set of maintenance actions that will minimize the total discounted vessel ownership costs over the long term, given that the component may deteriorate and assuming that the maintenance policy continues to be followed. The problem essentially requires minimization of the following relation (Putterman 1994; Ross 1970):

$$V(i) = C(i,a) + \alpha \sum_{j} P_{ij}(a)V(j) \tag{9.7}$$

where:

$V(i)$ is the long-term cost expected as a result of being in state i today
i is the CS observed today
$C(i,a)$ is the initial cost of action a taken in state i
α is the discount factor for a cost incurred a set number of years in the future
j is the CS predicted for a set number of years in the future
$P_{ij}(a)$ is the transition probability of CS j to CS i under action a
$V(j)$ is the long-term cost expected as of next year if transition to CS j occurs

The above formulation is a dynamic programming problem that has a variety of solution techniques, including the following:

- Method of successive iteration
- Policy iteration
- Linear programming formulation

These methods are beyond the scope of this section and are not covered here. Once the best maintenance strategy is chosen, its optimality must then be demonstrated.

9.5 Maintenance Implementation and Development of Risk-Ranking Scheme

As noted previously, selection of an optimal maintenance management policy is not only a function of potential maintenance actions but also, and perhaps more importantly, a scheduling of implementation of recommended maintenance actions. In developing an optimal policy for maintenance management, a suitable time frame for the implementation of maintenance actions must be chosen. Selection of such a time frame could be dictated by Flag Administration Officer or Classification Society requirements, elicitation of SMEs, engineering experience, and current practice, with values of 5–7 years being typical. Once a planning time frame has been selected, criteria must be chosen upon which to base maintenance implementation decisions. Implementation of maintenance actions for various system components may be based on such factors as maintenance costs or potential risk/failure costs. Alternatively, implementation may be based upon CS deterioration for each component. Using a combination of Flag Administration Officer and Classification Society requirements, SME elicitation, and experience, thresholds may be set for CS deterioration of major structural components. Alternative maintenance implementation schedules may then be compared, considering factors such as cost savings, risk reduction, and CS improvement, as well as any effects that delayed implementation may have on these factors. Combining this information with specific budgetary resources and risk tolerance levels of individual owner/operators, optimal maintenance schedules for each component may be ranked to assess both the relative urgency with which each must be implemented and the ability of each to meet the aforementioned criteria. The process is demonstrated by means of an example at the end of the section.

9.6 Optimal Maintenance Scheduling for the Overall Vessel

Upon selection of a suitable ranking criterion, the potential maintenance schedules for the various components should then be ranked using the selected criteria in conjunction with the available budget and threshold levels for risk and CS deterioration. It is important to note that the maintenance policies for individual components, developed in step 3 of ROMMSS, are optimal for only those components. When the budgetary resources are unlimited, the optimal maintenance policies for individual components can be scheduled for implementation without delay. This represents the most optimal maintenance policy for the overall vessel. However, budgetary resources are always limited; thus, an optimal maintenance strategy for the overall vessel must employ some sort of ranking scheme, focused on allocating scarce budgetary resources to those components with the most urgent needs, as defined in step 4 of ROMMSS.

Ship structural maintenance is somewhat unique in the sense that major repair actions typically require dry-docking of the vessel for extended periods of time, during which normal operational commitments of the vessel must be suspended. A maintenance implementation schedule ignorant of dry-docking could prove disastrous in terms of unnecessary ownership costs. The total maintenance and risk costs and CS deterioration for the system within the planning horizon should be closely examined. Scheduling dry-docking for only those components requiring extensive repair may help to further reduce unnecessary down time for the vessel. Other factors relating to dry-docking, such as availability and accessibility, should also be investigated thoroughly during the scheduling process.

9.7 Implementation of Maintenance Strategies and Updating System

Thus far, the ROMMSS procedures outlined in previous sections have not been physical in nature, but rather computational, employing an extensive network of modules and databases for CS transition matrices, maintenance and risk costs, risk and CS thresholds, expert opinions, Flag Administration Officer and Classification Society requirements, shipyard data, and budgetary resources. These databases have then been used to recommend an optimal maintenance management strategy, in terms of both repair action and scheduling. Upon recommendation of an optimal maintenance plan by the ROMMSS, physical implementation of its strategies is at the owner's discretion. As the strategies are implemented, the database of ship structural system should be continually updated. Updates should be made to the risk profile for the vessel and the associated maintenance and risk costs, and CS transition matrices may be revised, if necessary, to reflect the difference between assumed values and those observed during implementation. The merits in developing an advanced computational software tool for ship structural maintenance management, such as ROMMSS, lie not only in the potential cost savings for vessel owners through comprehensive maintenance optimization, but also the reduction in time and financial resources previously used to achieve a lower degree of optimization.

9.8 An Application: Optimal Maintenance Management of Ship Structures

When fully implemented as a software tool, ROMMSS consists of a database and a computational tool that ship designers, owners, managers, and operators can use to make long-term life cycle management decisions to reduce operational costs. The conceptual framework for ROMMSS can be demonstrated with an example problem. For the sake of simplicity and clarity, an existing vessel has been partitioned into its major components using the procedures outlined previously. Four major components are assumed to be afflicted by corrosion and might require major repair within the next 5 years. It is also assumed that the corroded components may be placed into one of five CS categories, as shown in Table 9.2, where CS 1 implies "as good as new" and CS 5 denotes ">40% corroded." The 14 maintenance actions (Table 9.3) are applicable to all four components. Also, it is assumed that a combination of expert elicitation, historical data, and engineering judgment has been used to define the unit failure/risk costs and unit maintenance costs for the CS degradations and maintenance actions, respectively. To keep the discussion as general as possible, the four components are hereafter referred to as simply component 1, component 2, component 3, and component 4. The assumed initial CS distributions for each of the four components are given in Table 9.9. For example, it can be seen that in year 1, 45% of component 1 is in CS 1 (CS1) and CS 2 (CS2), 5% in CS 3 (CS3) and CS 4 (CS4), and 0% in CS 5 (CS5).

The assumed unit maintenance costs and unit failure/risk costs for each component are summarized in Tables 9.10 and 9.11, respectively. The transition probability matrices for the four major components are presented in Tables 9.12 through 9.15.

TABLE 9.9

Assumed Initial Distribution of Component Condition States (CSs)

	Assumed Initial Distribution (%)				
Year 1	**CS1**	**CS2**	**CS3**	**CS4**	**CS5**
Component 1	45	45	5	5	0
Component 2	35	25	30	5	5
Component 3	5	20	45	15	15
Component 4	10	45	35	5	5

TABLE 9.10
Unit Maintenance Cost for Components

		Unit Maintenance Costs ($)			
CS	**Maintenance Action**	**Component 1**	**Component 2**	**Component 3**	**Component 4**
CS1	1	0	0	0	0
	2	1000	1100	1000	1200
CS2	3	0	0	0	0
	4	1000	1100	1100	1200
	5	2100	2200	2350	3500
CS3	6	0	0	0	0
	7	2000	2200	2300	3650
	8	2500	2750	2750	3750
CS4	9	0	0	0	0
	10	3500	3850	2750	4950
	11	2500	2750	3850	4850
CS5	12	0	0	0	0
	13	3500	3850	3850	4850
	14	4000	4400	4400	5489

CS, condition state.

TABLE 9.11
Unit Failure/Risk Cost for Components

	Unit Failure/Risk Cost ($)				
Component	**CS1**	**CS2**	**CS3**	**CS4**	**CS5**
1	500	1500	3500	4500	6500
2	550	1650	3850	4950	7100
3	550	1650	3850	4950	7100
4	550	1650	3850	6153	8178

Because it has been specified in this example that the components will require repairs within 5 years, a 5-year maintenance planning horizon is employed in ROMMSS. It is well known that, due to inflation, the costs tend to increase with time. Therefore, a 5% discounting factor is specified for the current example problem.

A ROMMSS-based maintenance management analysis of a vessel is performed with a number of objectives in mind. For the purpose of demonstration, the objectives include the following:

- Determine the optimal maintenance strategies for each of the defined components in each CS.
- Determine the CSs of each component in the event that their individual optimal maintenance policies are either implemented immediately or delayed for 1, 2, 3, 4, or 5 years within the planning period.
- Determine the risk/failure cost associated with delayed implementation of optimal maintenance policies.

TABLE 9.12

Transition Probabilities for Component 1

CS	Maintenance Action	Transition Probability (%)				
		CS1	CS2	CS3	CS4	CS5
CS1	1	90	10	0	0	0
	2	90	10	0	0	0
CS2	3	0	80	20	0	0
	4	0	80	20	0	0
	5	70	30	0	0	0
CS3	6	0	0	70	30	0
	7	70	30	0	0	0
	8	80	15	5	0	0
CS4	9	0	0	0	65	35
	10	65	20	10	5	0
	11	85	10	3	2	0
CS5	12	0	0	0	0	100
	13	65	20	10	5	0
	14	80	10	10	0	0

CS, condition state.

TABLE 9.13

Transition Probabilities for Component 2

CS	Maintenance Action	Transition Probability (%)				
		CS1	CS2	CS3	CS4	CS5
CS1	1	85	15	0	0	0
	2	95	5	0	0	0
CS2	3	0	75	25	0	0
	4	0	75	25	0	0
	5	70	30	0	0	0
CS3	6	0	0	65	35	0
	7	70	30	0	0	0
	8	80	15	5	0	0
CS4	9	0	0	0	60	40
	10	85	10	3	2	0
	11	75	25	0	0	0
CS5	12	0	0	0	0	100
	13	65	20	10	5	0
	14	95	5	0	0	0

CS, condition state.

- Determine the increase/decrease in maintenance costs associated with delayed implementation of optimal maintenance actions.
- Rank the relative importance of the components' maintenance schedule, based on failure/risk cost, maintenance cost, and CS deterioration, or a combination thereof.
- Determine the optimal time for scheduling a major dry-dock repair for the vessel.

TABLE 9.14

Transition Probabilities for Component 3

		Transition Probability (%)				
CS	**Maintenance Action**	**CS1**	**CS2**	**CS3**	**CS4**	**CS5**
CS1	1	85	15	0	0	0
	2	95	5	0	0	0
CS2	3	0	82	18	0	0
	4	0	82	18	0	0
	5	70	30	0	0	0
CS3	6	0	0	65	35	0
	7	80	20	0	0	0
	8	85	15	0	0	0
CS4	9	0	0	0	60	40
	10	85	10	3	2	0
	11	75	25	0	0	0
CS5	12	0	0	0	0	100
	13	55	0	0	45	0
	14	95	5	0	0	0

CS, condition state.

TABLE 9.15

Transition Probabilities for Component 4

		Transition Probability (%)				
CS	**Maintenance Action**	**CS1**	**CS2**	**CS3**	**CS4**	**CS5**
CS1	1	85	15	0	0	0
	2	85	15	0	0	0
CS2	3	0	82	18	0	0
	4	0	82	18	0	0
	5	80	10	10	0	0
CS3	6	0	0	65	35	0
	7	80	20	0	0	0
	8	83	11	6	0	0
CS4	9	0	0	0	60	40
	10	85	10	3	2	0
	11	84	16	0	0	0
CS5	12	0	0	0	0	100
	13	85	0	15	0	0
	14	95	5	0	0	0

CS, condition state.

These objectives are used in developing the rest of the example.

The optimal maintenance strategy for each individual component can be estimated using the dynamic programming model of ROMMSS, described previously by Equation 9.7. Generally speaking, the choice of optimal maintenance strategies differs from one component to another, and also for different CSs of a single component. The optimal policies are strongly dependent on the unit cost of maintenance, the unit failure/risk cost, and the degree of improvement in CSs of a component as a result of the implementation of a maintenance policy, which is reflected by its transition matrix. For the current system, the algorithms employed within ROMMSS (namely, successive iteration, policy iteration, and linear programming) will be developed to recommend the exact optimal maintenance policies. To proceed with the demonstration of other ROMMSS features, the optimal policies that will be assumed for each component in the current example are summarized in Table 9.16. For the sake of simplicity in demonstration, the optimal policies at each CS are assumed to be similar for all components, based on Table 9.3.

It is important to emphasize that the optimal policy suggested by ROMMSS for a given component is highly dependent on the properties of that component as specified by its maintenance cost, failure cost, and transition probabilities. Because no provision for correlation with other components is assumed, considerable effort should be expended in constructing the transition probabilities, unit maintenance costs, and unit risk or failure costs that best represent a component using a combination of SME, experience, and data obtained from previous inspection and maintenance actions. An optimal maintenance strategy for a given component implies that among all applicable maintenance actions as provided in Table 9.3, the most optimal policy represents the most efficient actions in terms of minimal CS maintenance/failure costs and CS improvement. Any other combination of maintenance actions might, in the long term, either increase the risk and/or maintenance costs or lead to less improvement in the CSs of the component.

For a planning horizon of 5 years, for example, the optimal component maintenance policies recommended by ROMMSS can be either implemented immediately or delayed for 1, 2, 3, or 4 years; moreover, the policies can be implemented for only selected components or all components. A decision regarding policy implementation must be made within the planning horizon. Constraints on available budget and resources, coupled with shipyard availability and operational commitments, greatly influence the implementation of maintenance schedules for a vessel. Immediate, delayed, and/or selective implementation of optimal policies will impact the CSs of each component, which will invariably affect the structural integrity of the vessel. Furthermore, Flag Administration Officer or Classification Society requirements for the vessel will also be affected by implementation decisions. Knowledge of the CSs of the various components should be

TABLE 9.16

Assumed Long-Term Optimal Maintenance Policies for Components

	Assumed Long-Term Optimal Maintenance Policies				
Component	**CS1**	**CS2**	**CS3**	**CS4**	**CS5**
1	1	5	7	11	13
2	1	5	7	11	14
3	1	5	7	10	14
4	1	5	7	11	13

Note: Use the Maintenance Action (MA) indicated for the corresponding Condition state (CS): CS1, no repair (MA1); CS2, spot blast/patch coating (MA5); CS3, spot blast/patch coating (MA7); CS4, cut out/weld new plate/spot blast/patch coating (MA10) or add/maintain sacrificial anode (MA11); CS5, cut out/weld new plate/spot blast/patch coating (MA13) or replace component (MA14).

considered in the decision-making process. ROMMSS facilitates the prediction of CS improvement/deterioration with or without the implementation of recommended maintenance policies. Recall that Table 9.9 gives a summary of the assumed CSs for each of the four components in year 1, prior to implementation of any maintenance policies. Tables 9.17 through 9.20 summarize the CSs of the components prior to implementation of optimal maintenance policies in the event that policy implementation is delayed for 1, 2, 3, or 4 years, respectively.

The information summarized in these tables can then be used to make risk-informed decisions. For example, Table 9.9 previously illustrated that currently (i.e., $t = 0$) 35% of component 2 is in the best CS (CS1), 25% in CS2, 30% in CS3, and 5% in both CS4 and CS5. As shown in Table 9.17, if maintenance were delayed for 1 year, then just prior to implementation of the optimal policy, 30% would be in CS1, 24% in CS2, 26% in CS3, 14% in CS4,

TABLE 9.17

Condition State (CS) Distribution If Implementation of Optimal Maintenance Policies Is Delayed 1 Year

	CS Distribution (%)				
Year 2	**CS1**	**CS2**	**CS3**	**CS4**	**CS5**
Component 1	41	41	13	5	2
Component 2	30	24	26	14	7
Component 3	4	17	33	25	21
Component 4	9	38	31	15	7

TABLE 9.18

Condition State (CS) Distribution If Implementation of Optimal Maintenance Policies Is Delayed 2 Years

	CS Distribution (%)				
Year 3	**CS1**	**CS2**	**CS3**	**CS4**	**CS5**
Component 1	36	36	16	8	5
Component 2	26	22	23	16	14
Component 3	4	15	25	24	32
Component 4	7	33	26	19	15

TABLE 9.19

Condition State (CS) Distribution If Implementation of Optimal Maintenance Policies Is Delayed 3 Years

	CS Distribution (%)				
Year 4	**CS1**	**CS2**	**CS3**	**CS4**	**CS5**
Component 1	33	32	17	10	8
Component 2	22	20	20	17	21
Component 3	3	14	19	22	42
Component 4	6	29	23	19	23

TABLE 9.20

Condition State (CS) Distribution If Implementation of Optimal Maintenance Policies Is Delayed 4 Years

	CS Distribution (%)				
Year 5	**CS1**	**CS2**	**CS3**	**CS4**	**CS5**
Component 1	29	28	18	12	12
Component 2	19	18	18	17	28
Component 3	3	12	15	19	51
Component 4	6	25	20	19	31

and 7% in CS5. If maintenance were delayed instead for 2 years according to Table 9.18, then 26% would be in CS1, 22% in CS2, 23% in CS3, 16% in CS4, and 14% in CS5. Thus, the condition of the component continues to deteriorate with increasing delay in maintenance implementation. The benefit of ROMMSS-based predictions lies in the fact that owner/operators do not need to spend a great deal of financial resources to predict an average amount of component deterioration. Furthermore, an ROMMSS forecast can serve as a guide to scheduling major inspections. If a target or threshold value is specified for the allowable percentage in the worst CS, information predicted by ROMMSS can then be used for maintenance implementation scheduling by providing the estimates of maximum allowable delay period. For example, assuming that for component 2 the maximum allowable percentage in CS5 is 15%, repair can be delayed no longer than 2 years; otherwise, CS deterioration will exceed the specified threshold for CS5. A comparative assessment of CSs of the four components with or without delayed implementation in optimal maintenance strategies can also be executed. The evaluation criterion can be, for example, the percentage of a component in CS5 without maintenance implementation.

Figure 9.9 compares all components based on the percentage of each in CS5 during each year of the assumed planning period. A closer look reveals that, without implementation of an optimal maintenance policy at any time during the planning horizon, component 3 consistently has the highest percentage of its contents in the worst CS (CS5), whereas component 1 consistently has the lowest percentage of its contents in CS5. Assuming, for example, the available maintenance budget allows for the repair of only one component

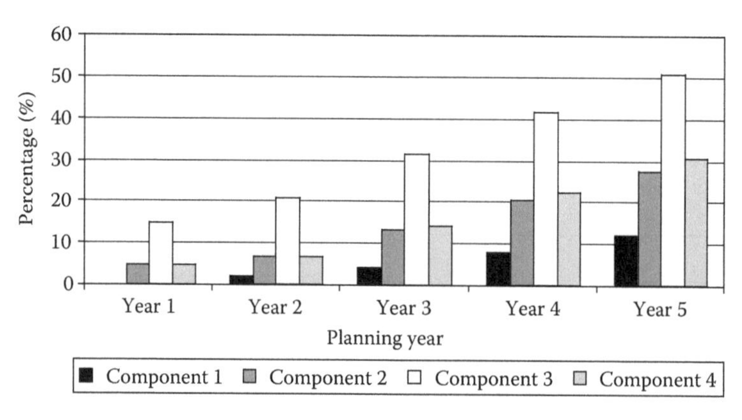

FIGURE 9.9
Variation in percentage of each component in the worst CS (CS5) with delayed implementation of optimal maintenance policies.

per year, and repair schedule prioritization is based solely on the percentage of each component in CS5, then repair of component 3 will be given top priority, followed by component 4, then component 2, and finally component 1. That is, optimal maintenance management (based on a CS5 threshold of 15%) requires that component 3 be repaired immediately, while the repair of component 1 may be delayed until the end of the assumed planning period.

The cost associated with maintenance of a ship structure is not only a function of the type of repair actions recommended for implementation, but also the manner in which such implementation is carried out. As noted in the previous section, when implementation of optimal maintenance actions is delayed, a greater fraction of a component degrades toward the worst CS, thereby implying that the costs associated with maintenance implementation will increase with delayed action. The ROMMSS strategy has been used to determine the optimal maintenance policies for each of the four components considered. Recall that the unit costs of the potential maintenance actions for each component are previously summarized in Table 9.14. The unit maintenance costs corresponding to the assumed optimal policies are given in Table 9.21.

The next question that should be answered is regarding the best time to implement the recommended policies: "Within the planning horizon, when is the most opportune time to schedule suggested repairs to each component?" The answer to this question is almost entirely dependent on the available budget. If unlimited financial resources were available, then all the components could be repaired immediately. This is rarely the case, however, as practicality requires that budgetary resources are always limited. Assuming instead that the available budget can only accommodate the repair of a single component per year within the planning horizon, then one must decide when the repair should be scheduled so as to minimize the maintenance cost. The problem then reduces to ranking the repair schedule of the component based on the associated maintenance cost. Figure 9.10 presents a summary of the maintenance cost for each component when the recommended maintenance actions are implemented within the first year or delayed for 2, 3, 4, or 5 years. It should be recalled that a 5% inflation rate (i.e., a 5% discounting factor) has been assumed during each year of the planning horizon. A careful examination of the figure shows that within each year of the planning horizon, optimal maintenance costs are highest for component 4, followed by component 3, component 2, and finally component 1, which consistently requires the least amount of money to maintain. Furthermore, when the implementation of recommended maintenance actions is delayed for any component, the increase in the maintenance cost within the first 3 years of the planning horizon is only marginal. However, beyond the third year, the maintenance costs increase dramatically, approximately doubling in each of the final 2 years of the planning horizon.

TABLE 9.21

Unit Maintenance Costs for Assumed Optimal Policies

	Unit Maintenance Costs ($)				
Component	**CS1**	**CS2**	**CS3**	**CS4**	**CS5**
1	0	2100	2000	2500	3500
2	0	2200	2200	2750	4400
3	0	2350	2300	2750	4400
4	0	3500	3650	4850	4850

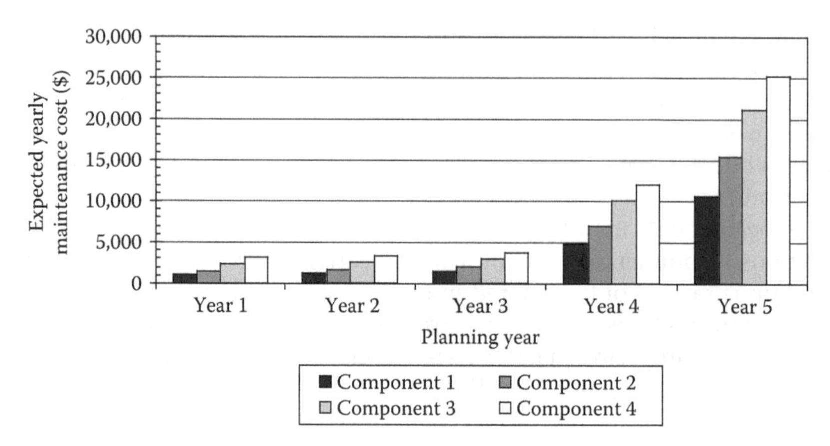

FIGURE 9.10
Variation in yearly maintenance costs during the planning horizon.

Ranking the component repairs according to dollar savings, Figure 9.10 suggests that to maximize the return on invested maintenance dollars, component 4 should be scheduled for repair implementation as soon as possible, followed by component 3, component 2, and finally component 1; moreover, implementation of maintenance actions for component 4 should not be unduly delayed. The figure also suggests that the substantial savings in maintenance costs can be realized if the optimal maintenance policies for all the components are implemented within the first 3 years of the planning horizon (starting with component 4). If implementation of maintenance actions were delayed beyond 3 or 4 years, the cost of maintenance would be more than triple or quadruple, respectively, leading to a lower return on investment and higher total ownership costs.

Scheduling the time for implementation of maintenance actions should not be based solely on the maintenance cost but should also consider the consequences of delayed implementation of optimal maintenance policies. Such consequences could be expressed in the form of an increase in anticipated risk/failure cost. Anticipated risk/failure costs such as lack of serviceability, unplanned repair and litigation, and costs associated with failure-induced environmental pollution could affect the economics of operating a vessel should they be incurred. A summary of the assumed unit failure/risk costs for each component considered in the example problem is provided in Table 9.11.

It is well known that failure/risk costs increase with delay in the implementation of optimal maintenance policies; therefore, a fundamental issue to be considered in optimal maintenance scheduling concerns the optimal time for implementation of maintenance actions for a component to ensure that the risk level, as reflected by risk/failure cost, does not exceed the allowable limits. The allowable limit of risk is a very subjective issue and is entirely dependent on the amount of risk that vessel managers, operators, and owners can tolerate. Therefore, risk tolerance thresholds must be defined (and updated, if necessary) for specific vessels. Input regarding the definition of risk thresholds can be obtained through elicitation of SMEs, historical data, and engineering experience. The major components can then be ranked according to the resulting risk levels for delayed implementation of maintenance actions. Figure 9.11 summarizes the progressive increase in failure/risk cost for each component within the planning horizon. It can be seen that this cost is a function of the timing of implementing the maintenance actions. Within the assumed planning horizon, component 3 consistently has the highest risk/failure cost when left without repair, followed by component 4, component 2, and finally component 1,

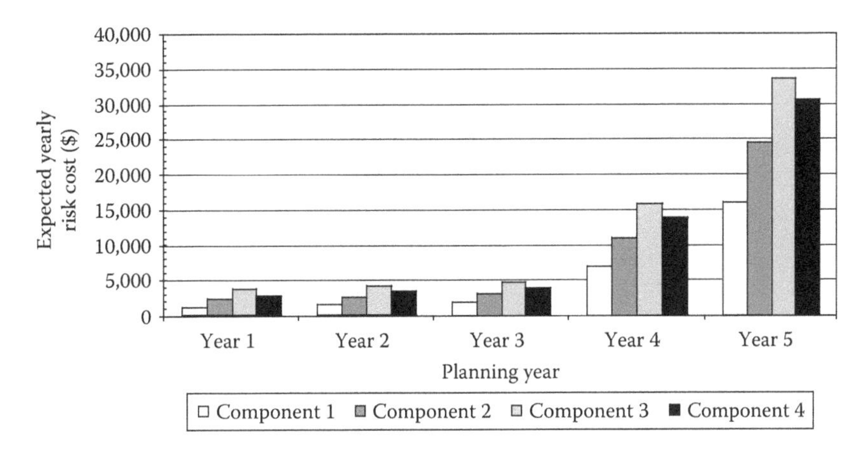

FIGURE 9.11
Variation in component risk/failure costs during the planning horizon.

which has the lowest risk/failure cost. Furthermore, similar to the trends of increasing maintenance cost with delayed maintenance implementation, as illustrated in Figure 9.11, a gradual, marginal increase in risk/failure costs occurs within the first 3 years of the planning horizon. Again, the costs for each component approximately double during each of the two remaining years of the assumed planning horizon.

To minimize the risk/failure costs of each component, Figure 9.11 suggests that repair of component 3 should be given top priority, followed by component 4 and component 2, whereas repair of component 1 may be delayed the longest. Furthermore, it is seen from the figure that if the repair operations for the components are implemented within the first 3 years, the associated risk/failure cost will generally be minimal. It is interesting to note that while an optimal maintenance repair schedule based on maintenance cost leads to the conclusion that component 4 should be repaired first; however using instead on risk/failure costs leads to giving component 3 top priority. It should also be noted that repair-scheduling conclusions based on risk/failure costs are similar to the findings based on CS deterioration. Although recommendations regarding repair scheduling based on risk/failure costs and maintenance costs appear to be conflicting, it should be noted that both recommendations have some common features. For example, both strategies suggest that repair of components 3 and 4 be given priority over repair of components 1 and 2. Furthermore, both strategies suggest that the most optimal repair time for all the components lies within the first 3 years of the assumed 5-year planning horizon, implying that implementation of repair actions should not be delayed beyond 3 years.

The decision maker, whoever it may be (manager, operator, or owner of vessel), must resolve such conflicting suggestions using his/her threshold for risk tolerance. A decision maker with a low risk tolerance will tend to follow a recommended repair schedule based on risk/failure cost, whereas one with a higher risk tolerance might prefer to execute a schedule based on minimization of maintenance costs. Alternatively, a decision maker whose risk threshold is moderate and who has the required resources available might choose to implement both recommendations simultaneously.

Ship structural systems have a unique maintenance requirement in the sense that the major implementation of maintenance repair actions generally involves dry-docking of the vessel for an extended period. During this period, normal operational commitments of the vessel must be suspended. Repair schedules based on a ranking of maintenance costs

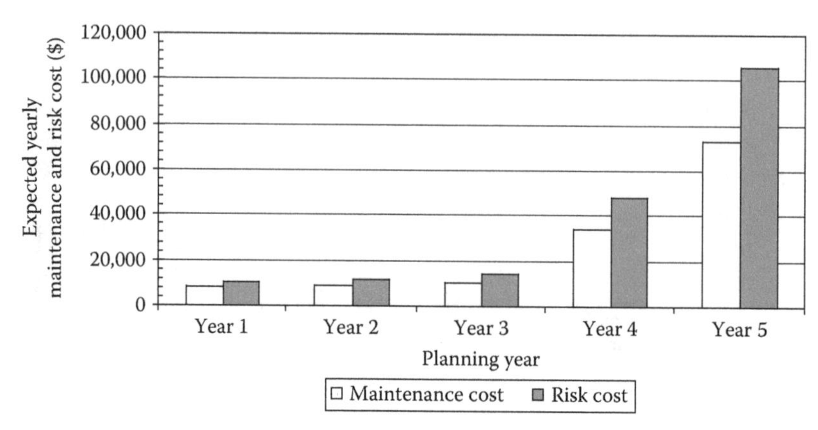

FIGURE 9.12
Expected yearly risk and maintenance costs during the planning horizon.

for the four components of the example vessel were provided previously, whereas those based instead on the ranking of risk/failure costs were recommended earlier. Considering both risk and maintenance costs for actions that are either delayed or implemented immediately and assuming that the required financial resources are available, this section poses the question: "When is the optimal time to schedule a major dry dock repair for all the components?" To facilitate optimization of a schedule for major dry-docking repairs, the total maintenance and risk costs for the system within the planning horizon, as shown in Figure 9.12, must be closely examined. The figure depicts only a marginal increase in total risk and maintenance costs for the system during the first 3 years of the assumed planning horizon, with the costs approximately doubling in each of the two remaining years. During the first 3 years, the failure/risk costs are only slightly greater than those associated with maintenance activities. During the last 2 years, however, this difference becomes rather substantial. It is therefore concluded that an optimal risk-based major dry-docking maintenance schedule for the vessel should be carried out within the first 3 years of the assumed planning period. Repair within the first year will result in the least cost, followed by repair within the second year. Any delay in repair beyond 3 years not only would lead to a significant increase in maintenance costs, but could also render the continual operation of the vessel not economical due to the significant increase in anticipated failure/risk costs.

Appendix A

Fundamentals of Probability and Statistics*

This appendix provides a summary of the fundamentals of probability and statistics for the purpose of helping readers to look up and review key equations and models for risk analysis.

CONTENTS

* This appendix is based on the book *Probability, Statistics, and Reliability for Engineers and Scientists*, 3rd ed., by B. M. Ayyub and R. H. McCuen, Chapman & Hall/CRC Press LLC, Boca Raton, FL, 2011.

A.1 Sample Spaces, Sets, and Events

Sets constitute a fundamental concept in probabilistic analysis of engineering problems. To perform probabilistic analyses of these problems, the definition of the underlying sets is essential for the establishment of a proper model and obtaining realistic results. The goal of this section is to provide the set foundation required for probabilistic analysis.

Informally, a set can be defined as a collection of elements or components. Capital letters are usually used to denote sets (e.g., A, B, X, and Y). Small letters are commonly used to denote their elements (e.g., a, b, x, and y). The following are examples of sets:

$$A = \{2,4,6,8,10\} \tag{A.1a}$$

$$B = \{b : b > 0\} \tag{A.1b}$$

where:
":" means "such that."

$$C = \{\text{Maryland, Virginia, Washington DC}\} \tag{A.1c}$$

$$D = \{\text{P, M, 2, 7, U, E}\} \tag{A.1d}$$

$$F = \{1,3,5,7,9,11,\ldots\};\ \text{the set of odd numbers} \tag{A.1e}$$

In these example sets, each set consists of a collection of elements. In set A, 2 belongs to A, and 12 does not belong to A. Using mathematical notations, this can be expressed as $2 \in A$ and $12 \notin A$.

Sets can be classified as *finite* and *infinite* sets. For example, sets A, C, and D are finite sets, whereas sets B and F are infinite sets. The elements of a set can be either *discrete* or *continuous*. For example, the elements in sets A, C, D, and F are discrete, whereas the elements in set B are continuous. A set without any elements is called a null (or empty) set and is denoted as $\varnothing$.

If every element in a set A is also a member of set B, then A is called a subset of B, which is mathematically expressed as $A \subset B$. A is contained in or equal to B, which is mathematically expressed as $A \subseteq B$ if for every a that belongs to A (i.e., $a \in A$) implies that a belongs to B (i.e., $A \in B$). Every set is considered to be a subset of itself. The null set $\emptyset$ is considered to be a subset of every set.

In engineering, the set of all possible outcomes of a system (or for an experiment) constitutes the sample space S. A sample space consists of points that correspond to all possible outcomes. Each outcome for the system should constitute a unique element in the sample space. A subset of the sample space is called an *event*. These definitions are the set basis of probabilistic analysis. An event without sample points is an empty set and is called the impossible event $\varnothing$. A set that contains all the sample points is called the certain event S. The certain event is equal to the sample space. Events and sets can be represented using spaces that are bounded by closed shapes, such as circles. These shapes are called Venn–Euler (or simply Venn) diagrams. Belonging, nonbelonging, and overlaps between events and sets can be represented by these diagrams.

In the Venn diagram shown in Figure A.1, two events (or sets) A and B that belong to a sample space S are represented. The event C is contained in B (i.e., $C \subset B$), and the event A is not equal to B (i.e., $A \neq B$). Also, the events A and B have an overlap in the sample space S.

The basic operations that can be used for sets and events are analogous to addition, subtraction, and multiplication in arithmetic calculations.

1. The *union* of events A and B, which is denoted as $A \cup B$, is the set of all elements that belong to A or B or both. Two or more events are called *collectively exhaustive* events if the union of these events results in the sample space.
2. The *intersection* of events A and B, which is denoted as $A \cap B$, is the set of all elements that belong to both A and B. Two events are termed *mutually exclusive* if the

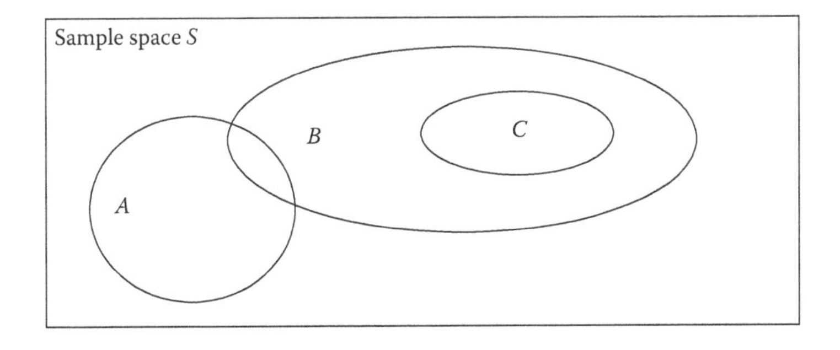

FIGURE A.1
Events.

TABLE A.1

Additional Operational Rules

Rule Type	Operations
Identity laws	$A \cup \varnothing = A, A \cap \varnothing = \varnothing, A \cup S = S, A \cap S = A$
Idempotent laws	$A \cup A = A, A \cap A = A$
Complement laws	$A \cup \bar{A} = S, A \cap \bar{A} = \varnothing, \bar{\bar{A}} = A, \bar{S} = \varnothing, \bar{\varnothing} = S$
Commutative laws	$A \cup B = B \cup A, A \cap B = B \cap A$
Associative laws	$(A \cup B) \cup C = A \cup (B \cup C), (A \cap B) \cap C = A \cap (B \cap C)$
Distributive laws	$(A \cup B) \cap C = (A \cup C) \cup (B \cap C)$
	$(A \cap B) \cup C = (A \cup C) \cap (B \cup C)$
de Morgan's law	$\overline{(A \cup B)} = \bar{A} \cap \bar{B},\ \overline{(E_1 \cup E_2 \cup \cdots \cup E_n)} = \bar{E}_1 \cap \bar{E}_2 \cap \cdots \cap \bar{E}_n$
	$\overline{(A \cap B)} = \bar{A} \cup \bar{B},\ \overline{(E_1 \cap E_2 \cap \cdots \cap E_n)} = \bar{E}_1 \cup \bar{E}_2 \cup \cdots \cup \bar{E}_n$
Combinations of laws	$\overline{(A \cup (B \cap C))} = \bar{A} \cap \overline{(B \cap C)} = (\bar{A} \cap \bar{B}) \cup (\bar{A} \cap \bar{C})$

occurrence of one event precludes the occurrence of the other event. The term can also be extended to more than two events.

3. The *difference* of events A and B, which is denoted as $A - B$, is the set of all elements that belong to A but not to B.
4. The event that contains all of the elements that do not belong to an event A is called the complement of A and is denoted as $\bar{A}$.

Table A.1 shows additional rules based on the above fundamental rules. The validity of these rules can be checked using Venn diagrams.

A.2 Mathematics of Probability

The probability of an event can be defined as the relative frequency of its occurrence or the subjective probability of its occurrence. The type of definition depends on the underlying event. For example, in an experiment that can be repeated N times with n occurrences of the underlying event, the relative frequency of occurrence can be considered as the probability of occurrence. In this case, the probability of occurrence is n/N. However, there are many problems that do not involve large numbers of repetitions, and still we are interested in estimating the probability of occurrence of some event S. For example, during the service life of an engineering product, the product either fails or does not fail in performing a set of performance criteria. The events of failure and survival are mutually exclusive and collectively exhaustive of the sample space. The probability of failure (or survival) is considered as a subjective probability. An estimate of this probability can be achieved by modeling the underlying system, its uncertainties, and performances. The resulting subjective probability is expected to reflect the status of our knowledge about the system regarding the true likelihood of occurrence of the events of interest. In this section, the mathematics of probability is applicable to both definitions; however, it is important to keep in mind both definitions so that results are not interpreted beyond the range of their validity.

In general, an axiomatic approach can be used to define probability as a function from sets to real numbers. The domain is the set of all events within the sample space of the problem, and the range consists of the numbers on the real line. For an event A, the notation $P(A)$ means the probability of occurrence of event A. The function $P()$ should satisfy the following properties:

$$0 \leq P(A) \leq 1 \quad \text{for every event } A \subseteq S \tag{A.2a}$$

$$P(S) = 1 \tag{A.2b}$$

If $A_1, A_2, \ldots, A_n$ are mutually exclusive events on S, then

$$P(A_1 \cup A_2 \cup \cdots \cup A_n) = P(A_1) + P(A_2) + \cdots + P(A_n) \tag{A.2c}$$

Computational rules can be developed based on these properties. Example rules are given as follows:

$$P(\varnothing) = 0 \tag{A.3}$$

$$P(A \cup B) = P(A) + P(B) - P(A \cap B) \tag{A.4a}$$

$$P(A \cup B \cup C) = P(A) + P(B) + P(C) - P(A \cap B) - P(A \cap C) - P(B \cap C) + P(A \cap B \cap C) \tag{A.4b}$$

$$P(\bar{A}) = 1 - P(A) \tag{A.5}$$

$$\text{If } A \subseteq B, \text{ then } P(A) \leq P(B) \tag{A.6}$$

In experiments that result in finite sample spaces, the processes of identification, enumeration, and counting are essential for the purpose of determining the probabilities of some outcomes of interest. The identification process results in defining all possible outcomes and their likelihood of occurrence. The identification of equally likely outcomes is needed to determine any probabilities of interest. The order of occurrence of the outcomes can be important in certain applications, requiring its consideration in the counting process.

The enumeration process can be performed in any systematic form that results in all possible outcomes. The multiplication principle can be used for this purpose. Let events $A_1, A_2, \ldots, A_n$ have $n_1, n_2, \ldots, n_n$ elements, respectively. Therefore, the total number of possible outcomes of selecting one element from each of $A_1, A_2, \ldots, A_n$ is the product $n_1\ n_2, \ldots n_n$, where the outcomes represent the ways to select the first element from A_1, the second element from A_2, and so on, and finally to select the nth element from A_n.

The permutation of r elements from a set of n elements is the number of arrangements that can be made by selecting r elements out of the n elements. The order of selection counts in determining these arrangements. The permutation $P_{r|n}$ of r out of n (where $r \leq n$) is as follows:

$$P_{r|n} = \frac{n!}{(n-r)!} \tag{A.7}$$

where:
$n!$ is the factorial of $n = n(n-1)(n-2) \ldots (2)(1)$

It should be noted that $0! = 1$ by convention. Equation A.7 results from the fact that there are n ways to select the first element, $(n - 1)$ ways to select the second element, $(n - 2)$ ways to select the third element, and so on to the last element (i.e., rth element).

The combination of r elements from a set of n elements is the number of arrangements that can be made by selecting r elements out of the n elements. The order of selection in this case does not count in determining these arrangements. One arrangement differs from another only if the contents of the arrangements are different. The combination $C_{r|n}$ of r out of n (where $r \leq n$) is:

$$C_{r|n} = \frac{P_{r|n}}{r!} \tag{A.8}$$

Therefore, the combination $C_{r|n}$ can be determined as follows:

$$C_{r|n} = \frac{n!}{(r!)(n-r)!} \tag{A.9}$$

It is very common to use the notation $\binom{n}{r}$ for the combination $C_{r|n}$. It can be shown that the following identity is valid:

$$\binom{n}{r} = \binom{n}{n-r} \tag{A.10}$$

The probabilities discussed earlier are based on and relate to the sample space S. However, it is common in many problems to have an interest in the probabilities of the occurrence of events that are conditioned on the occurrence of a subset of the sample space. This introduces the concept of conditional probability. For example, the probability of A given that B has occurred, denoted as $P(A|B)$, means the occurrence probability of a sample point that belongs to A given that it belongs to B. The conditional probability can be computed as follows:

$$P(A|B) = \frac{P(A \cap B)}{P(B)} \qquad \text{if } P(B) \neq 0 \tag{A.11}$$

Clearly, the underlying sample space for the conditional probability is reduced to the conditional event B. The conditional probability satisfies all the properties of probabilities. The following properties can be developed for conditional probabilities:

1. The complement of an event

$$P(\bar{A}|B) = 1 - P(A|B) \tag{A.12}$$

2. The multiplication rule for two events A and B

$$P(A \cap B) = P(A|B)P(B) \qquad \text{if } P(B) \neq 0 \tag{A.13a}$$

$$P(A \cap B) = P(B|A)P(A) \qquad \text{if } P(A) \neq 0 \tag{A.13b}$$

3. The multiplication rule for three events A, B, and C

$$P(A \cap B \cap C) = P[A|(B \cap C)]P(B|C)P(C) = P[(A \cap B)|C]P(C) \\ \text{if } P(C) \neq 0 \text{ and } P(B \cap C) \neq 0 \tag{A.14}$$

4. For mutually exclusive events A and B

$$P(A|B) = 0 \text{ and } P(B|A) = 0 \tag{A.15}$$

5. For statistically independent events A and B

$$P(A|B) = P(A),\ P(B|A) = P(B), \text{ and } P(A \cap B) = P(A)P(B) \tag{A.16a}$$

$$A \text{ and } \bar{B} \text{ are independent events} \tag{A.16b}$$

$$\bar{A} \text{ and } B \text{ are independent events} \tag{A.16c}$$

$$\bar{A} \text{ and } \bar{B} \text{ are independent events} \tag{A.16d}$$

A set of disjoint (i.e., mutually exclusive) events $A_1, A_2, \ldots, A_n$ form a partition of a sample space if $A_1 \cup A_2 \cup, \ldots, A_n = S$. An example partition is shown in Figure A.2.

If $A_1, A_2, \ldots, A_n$ represent a partition of sample space S, and E represents an arbitrary event, as shown in Figure A.3, the theorem of total probability states that

$$P(E) = P(A_1)P(E|A_1) + P(A_2)P(E|A_2) + \cdots + P(A_n)P(E|A_n) \tag{A.17}$$

This theorem is very important in computing the probability of an event E, especially in practical cases where the probability cannot be computed directly, but the probabilities of the partitioning events and the conditional probabilities can be computed.

Bayes' theorem is based on the same conditions of partitioning and events as the theorem of total probability and is very useful in computing the reverse probability of the type $P(A_i|E)$, for $i = 1, 2, \ldots, n$. The reverse probability can be computed as follows:

$$P(A_i|E) = \frac{P(A_i)P(E|A_i)}{P(A_1)P(E|A_1) + P(A_2)P(E|A_2) + \cdots + P(A_n)P(E|A_n)} \tag{A.18}$$

The denominator of this equation is $P(E)$, which is based on the theorem of total probability.

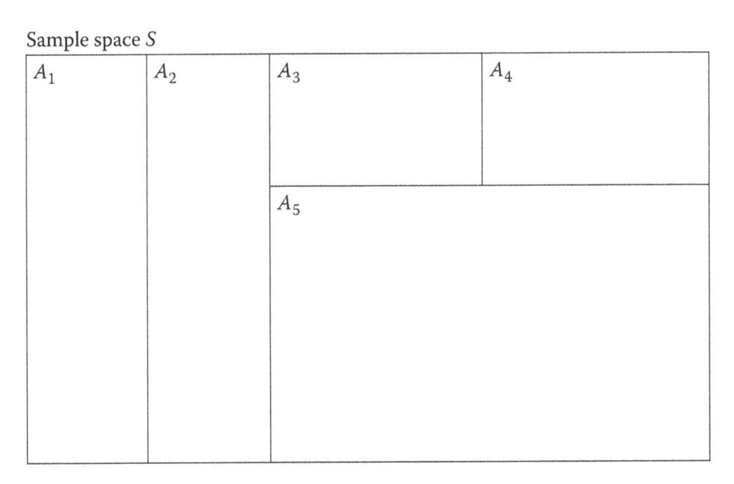

FIGURE A.2
Partitioned sample space.

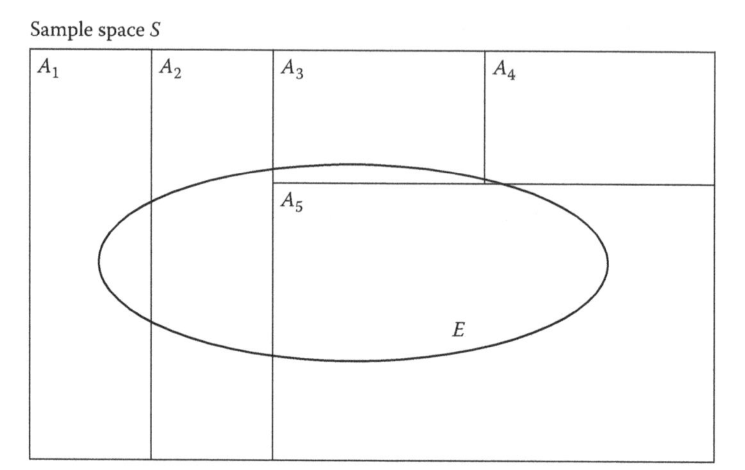

FIGURE A.3
Theorem of total probability.

A.3 Random Variables and Their Probability Distributions

A *random variable* is defined as a function that assigns a real value to every possible outcome for an engineering system. This mapping can be one to one or one to many. Based on this definition, the properties of the underlying outcomes (e.g., intersection, union, and complement) are retained in the form of, for example, overlapping ranges of real values, a combination of real ranges, and values outside these ranges. Random variables are commonly classified into two types: *discrete* or *continuous*. A discrete random variable may take on only distinct, usually integer values, for example, the outcome of a roll of a die may only take on the integer values from 1 to 6 and is, therefore, a discrete random variable. The number of floods per year at a point on a river can only take on integer values, so it is also a discrete random variable. A continuous random variable takes values within a continuum of values. For example, the average of all scores on a test having a maximum possible score of 100 may take on any value including nonintegers between 0 and 100; thus, the class average would be a continuous random variable. A distinction is made between these two types of random variables because the computations of probabilities are different for the two types.

A.3.1 Probability for Discrete Random Variables

The probability of a discrete random variable is given by the *probability mass function*, which specifies the probability that the discrete random variable X equals some value x_i and is denoted by

$$P_X(x_i) = P(X = x_i) \tag{A.19}$$

A capital X is used for the random variable, whereas an x_i is used for the ith largest value of the random variable. The probability mass function must satisfy the axioms of probability. Therefore, the probability of an event x_i must be ≤ 1 and ≥ 0, that is,

$$0 \leq P_X(x_i) \leq 1 \tag{A.20}$$

This property is valid for all possible values of the random variable X. Additionally, the sum of all possible probabilities must be equal to 1, that is,

$$\sum_{i=1}^{N} P_X(x_i) = 1 \tag{A.21}$$

where:
N is the total number of possible outcomes; for the case of the roll of a die, $N = 6$

It is often useful to present the likelihood of an outcome using the *cumulative mass function*, $F_X(x_i)$, which is given by

$$F_X(x_i) = P(X \leq x_i) = \sum_{j=1}^{i} P_X(x_j) \tag{A.22}$$

The cumulative mass function is used to indicate the probability that the random variable X is $\leq x_i$. It is inherent in the definition (Equation A.22) that the cumulative probability is defined as 0 for all the values less than the smallest x_i and 1 for all values greater than the largest value.

A.3.2 Probability for Continuous Random Variables

A *probability density function* (pdf) defines the probability of occurrence for a continuous random variable. Specifically, the probability that the random variable X lies within the interval from x_1 to x_2 is given by

$$P(x_1 \leq X \leq x_2) = \int_{x_1}^{x_2} f_X(x)\mathrm{d}x \tag{A.23}$$

where:
$f_X(x)$ is the pdf

If the interval is made infinitesimally small, x_1 approaches x_2 and $P(x_1 \leq X \leq x_2)$ approaches 0. This illustrates a property that distinguishes discrete random variables from continuous random variables. Specifically, the probability that a continuous random variable takes on a specific value equals 0; that is, probabilities for continuous random variables must be defined over an interval.

It is important to note that the integral of the pdf from $-\infty$ to $+\infty$ equals 1, that is,

$$P(-\infty < X < +\infty) = \int_{-\infty}^{+\infty} f_X(x)\mathrm{d}x = 1 \tag{A.24}$$

Also, because of Equation A.24, the following holds:

$$P(X \geq x_0) = \int_{x_0}^{+\infty} f_X(x_0)\mathrm{d}x = 1 - P(X < x_0) \tag{A.25}$$

The *cumulative distribution function* (cdf) of a continuous random variable is defined by

$$F_X(x_0) = P(X \le x_0) = \int_{-\infty}^{x_0} f_X(x)\mathrm{d}x \tag{A.26a}$$

The cdf is a nondecreasing function in that $P(X \le x_1) \le P(X \le x_2)$, where $x_1 \le x_2$. The cdf equals 0 at $-\infty$ and 1 at $+\infty$. The relationship between $f_X(x)$ and $F_X(x)$ can also be expressed as

$$f_X(x) = \frac{\mathrm{d}F_X(x)}{\mathrm{d}x} \tag{A.26b}$$

A.4 Moments

Whether summarizing a dataset or attempting to find the population, one must characterize the sample. The moments are useful descriptors of data, for example, the mean, which is a moment, is an important characteristic of a set of test scores. A moment can be referenced to any point on the measurement axis; however, the origin (i.e., zero point) and the mean are the most common reference points.

Although most data analyses use only two moments, it is important for some probabilistic and statistical studies to examine three moments:

1. *Mean*, the first moment about the origin
2. *Variance*, the second moment about the mean
3. *Skewness*, the third moment about the mean

In this section, equations and computational procedures for these moments are introduced. These moments are analogous to the area moments used to compute quantities such as the centroidal distance, the first static moment, and the moment of inertia. The respective kth moments about the origin for a continuous and a discrete random variable are

$$M'_k = \int_{-\infty}^{+\infty} x^k f_X(x)\mathrm{d}x \tag{A.27}$$

$$M'_k = \sum_{i=1}^{n} x_i^k P_X(x_i) \tag{A.28}$$

where:

X is the random variable
$f_X(x)$ is its probability density function
n is the number of elements in the underlying sample space of X
$P_X(x)$ is the probability mass function

The first moment about the origin, that is, $k = 1$ in Equations A.27 and A.28, is called the mean of X and is denoted as μ.

The respective kth moments about the mean (μ) for a continuous and a discrete random variable are as follows:

$$M_k = \int_{-\infty}^{+\infty} (x - \mu)^k f_X(x)\mathrm{d}x \tag{A.29}$$

$$M_k = \sum_{i=1}^{n} (x_i - \mu)^k P_X(x_i) \tag{A.30}$$

where:
μ is the first moment about the origin (i.e., the mean)

The above moments are considered as a special case of mathematical expectation. The mathematical expectation of an arbitrary function $g(x)$, which is a function of the random variable X, is defined, respectively, for a continuous and a discrete random variable as:

$$E[g(x)] = \int_{-\infty}^{+\infty} g(x) f_X(x)\mathrm{d}x \tag{A.31}$$

$$E[g(x)] = \sum_{i=1}^{n} g(x_i) P_X(x_i) \tag{A.32}$$

The mean value can be formally defined as the first moment measured about the origin; it is also the average of all observations on a random variable. It is important to note that the population mean is most often indicated as μ, whereas the sample mean is denoted by X. For a continuous and a discrete random variable, the mean (μ) is computed, respectively, as

$$\mu = \int_{-\infty}^{+\infty} x f_X(x)\mathrm{d}x \tag{A.33}$$

$$\mu = \sum_{i=1}^{n} x_i P_X(x_i) \tag{A.34}$$

For n observations, if all observations are given equal weights, that is, $P_X(x_i) = 1/n$, then the mean for a discrete random variable (Equation A.34) produces

$$\bar{X} = \frac{1}{n} \sum_{i=1}^{n} x_i \tag{A.35}$$

which is the average of the observed values $x_1, x_2, x_3, \ldots, x_n$.

The variance is the second moment about the mean. The variance of the population is denoted by σ^2. The variance of the sample is denoted by S^2. The units of the variance are the square of the units of the random variable; for example, if the random variable is measured in pounds per square inch (psi), the variance has units of $(\text{psi})^2$. For a continuous and a discrete random variable, respectively, the variance is computed as the second moment about the mean as follows:

$$\sigma^2 = \int_{-\infty}^{+\infty} (x - \mu)^2 f_X(x)\mathrm{d}x \tag{A.36}$$

$$\sigma^2 = \sum_{i=1}^{n} (x_i - \mu)^2 P_X(x_i) \tag{A.37}$$

when the n observations in a sample are given equal weight, that is, $P_X(x_i) = 1/n$, the variance is given by

$$S^2 = \frac{1}{n} \sum_{i=1}^{n} (x_i - \bar{X})^2 \tag{A.38}$$

The value of the variance given by Equation A.38 is biased; an unbiased estimate of the variance is given by

$$S^2 = \frac{1}{n-1} \sum_{i=1}^{n} (x_i - \bar{X})^2 \tag{A.39}$$

The variance is an important concept in probabilistic and statistical analyses because many solution methods require some measure of variance. Therefore, it is important to have a conceptual understanding of this moment. In general, it is an indicator of the closeness of the values in a sample or a population to the mean. If all values in the sample equal the mean, the sample variance would equal 0.

By definition, the standard deviation is the square root of the variance. It has the same units as the random variable and the mean; therefore, it is a better descriptor of the dispersion or spread of either a sample of data or a distribution function than the variance. The standard deviation of the population is denoted by σ, whereas the sample value is denoted by S.

The coefficient of variation (δ or Cov) is a dimensionless quantity defined as

$$\delta = \frac{\sigma}{\mu} \tag{A.40}$$

It is also used as an expression of the standard deviation in the form of a proportion of the mean. For example, consider μ and σ to be 100 and 10, respectively; therefore, $\delta = 0.1$ or 10%. In this case, the standard deviation is 10% of the mean.

The skew is the third moment measured about the mean. Unfortunately, the notation for skew is not uniform from one user to another. The sample skew can be denoted by G, whereas the skew of the population can be indicated by λ. Mathematically, it is given for a continuous and a discrete random variable, respectively, as

$$\lambda = \int_{-\infty}^{+\infty} (x - \mu)^3 f_X(x) dx \tag{A.41}$$

$$\lambda = \sum_{i=1}^{n} (x - \mu)^3 P_X(x_i) \tag{A.42}$$

It has units of the cube of the random variable; thus, if the random variable has units of pounds, the skew has units of (pounds)^3.

The skew is a measure of the lack of symmetry. A symmetric distribution has a skew of zero, whereas a nonsymmetric distribution has a positive or negative skew depending on the direction of the skewness. If the more extreme tail of the distribution is to the right, the skew is positive; if the more extreme tail is to the left of the mean, the skew is negative.

A.5 Common Discrete Probability Distributions

In this section, the Bernoulli, binomial, geometric, and Poisson distributions are discussed. The first three distributions are based on Bernoulli trials (or sequences), whereas the fourth one is not. An engineering experiment (or system) that consists of N trials is considered to result in a *Bernoulli process* (or sequence) if it satisfies the following conditions: (1) the N trials (or repetitions) are independent; (2) each trial has only two possible outcomes, say, survival (S) or failure (F); and (3) the probabilities of occurrence for the two outcomes remain constant from trial to trial. Also, the negative binomial, Pascal, and hypergeometric distributions are described. A summary of selected discrete distributions that are commonly used in reliability and risk studies is provided in Section A.7.

A.5.1 Bernoulli Distribution

For convenience, the random variable X is defined as a mapping from the sample space $\{S, F\}$ for each trial of a Bernoulli sequence to the integer values $\{1, 0\}$, with one-to-one mapping in the respective order, where, for example, S = success and F = failure. Therefore, the probability mass function is given by

$$P_X(x) = \begin{cases} p & \text{for } x = 1 \\ 1 - p & \text{for } x = 0 \\ 0 & \text{otherwise} \end{cases} \tag{A.43}$$

The probability mass function of the Bernoulli distribution is shown in Figure A.4. The mean and variance for the Bernoulli distribution are, respectively, given by

$$\mu_X = p \text{ and } \sigma_X^2 = p(1 - p) \tag{A.44}$$

A.5.2 Binomial Distribution

The underlying random variable (X) for this distribution represents the number of successes in N Bernoulli trials. The probability mass function is given by

$$P_X(x) = \begin{cases} \binom{N}{x} p^x (1 - p)^{N-x} & \text{for } x = 0, 1, 2, \ldots, N \\ 0 & \text{otherwise} \end{cases} \tag{A.45}$$

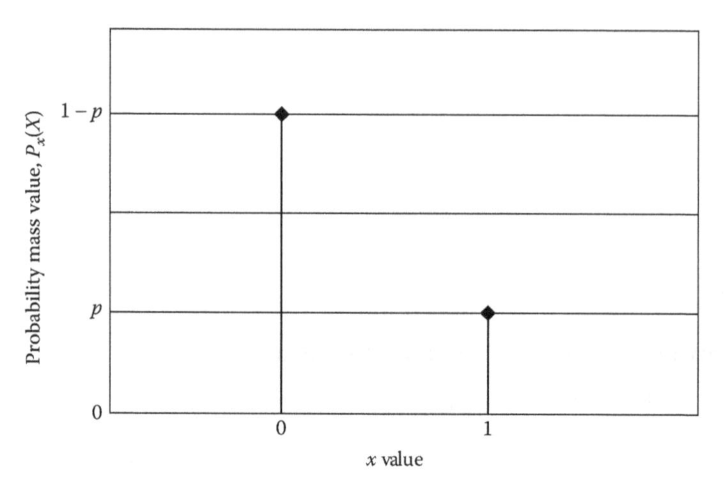

FIGURE A.4
Probability mass function of the Bernoulli distribution.

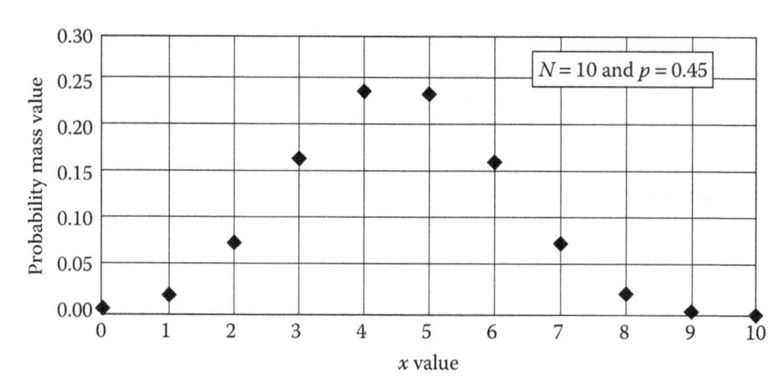

FIGURE A.5
Probability mass function of the binomial distribution.

where $\binom{N}{x}$ can be computed using Equation A.9. The probability mass and cumulative functions of an example binomial distribution are shown in Figures A.5 and A.6, respectively. The mean and variance for the binomial distribution, respectively, are given by

$$\mu_X = Np \text{ and } \sigma_X^2 = Np(1 - p) \tag{A.46}$$

A random variable can be represented by the binomial distribution, if the following three assumptions are met:

1. The distribution is based on N Bernoulli trials with only two possible outcomes.
2. The N trials are independent of each other.
3. The probabilities of the outcomes remain constant at p and $(1 - p)$ for each trial.

Therefore, the flip of a coin would meet these assumptions, but the roll of a die would not because there are six possible outcomes.

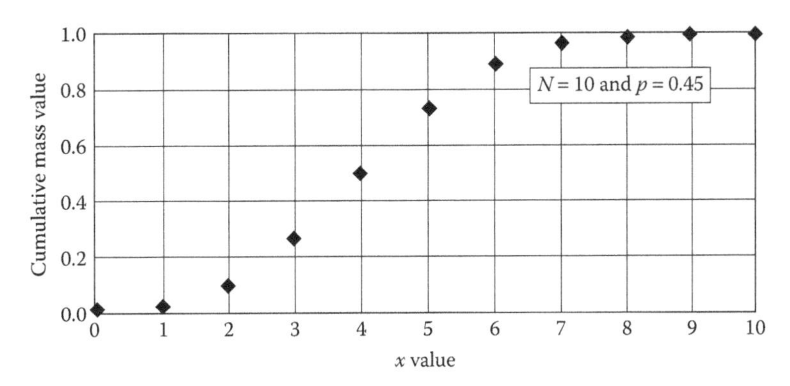

FIGURE A.6
Cumulative mass function of the binomial distribution.

A.5.3 Geometric Distribution

The underlying random variable for this distribution represents the number of Bernoulli trials that are required to achieve the first success. In this case, the number of trials needed to achieve the first success is neither fixed nor certain. The probability mass function is given by

$$P_X(x) = \begin{cases} p(1-p)^{x-1} & \text{for } x = 1, 2, 3, \ldots \\ 0 & \text{otherwise} \end{cases} \tag{A.47}$$

The probability mass function of an example geometric distribution is shown in Figure A.7. The mean and variance for the geometric distribution are, respectively, given by

$$\mu_X = \frac{1}{p} \text{ and } \sigma_X^2 = \frac{1-p}{p^2} \tag{A.48}$$

A.5.4 Poisson Distribution

The Poisson distribution is commonly used in problem solving that deals with the occurrence of some random event in the continuous dimension of time or space. For example, the number of occurrences of a natural hazard, such as earthquakes, tornadoes, or hurricanes, in some time interval, such as 1 year, can be considered as a random variable with

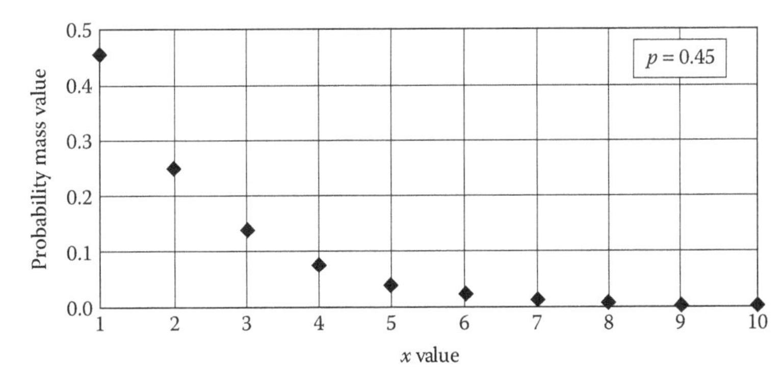

FIGURE A.7
Probability mass function of the geometric distribution.

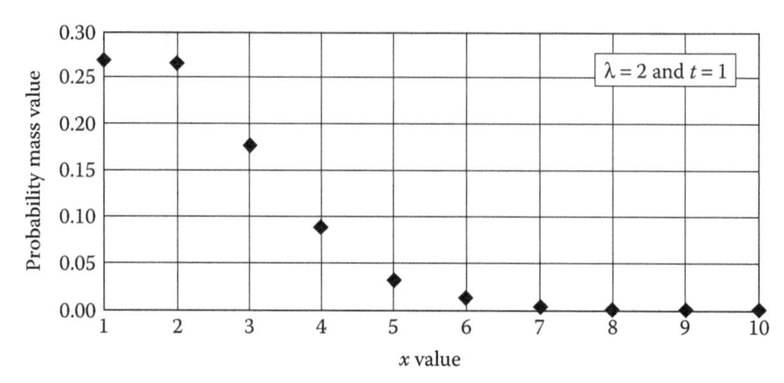

FIGURE A.8
Probability mass function of the Poisson distribution.

a Poisson distribution. In these examples, the number of occurrences in the time interval is the random variable. Therefore, the random variable is discrete, whereas its reference space (i.e., the time interval) is continuous. This distribution is considered to be the limiting case of the binomial distribution by dividing the reference space (i.e., time t) into nonoverlapping intervals of size Δt. The occurrence of the event (i.e., a natural hazard) in each interval is considered to constitute a Bernoulli sequence. The number of Bernoulli trials depends on the size of the interval Δt. By considering the limiting case where the size of the interval Δt approaches zero, the binomial distribution becomes the Poisson distribution.

The underlying random variable of this distribution is denoted by X_t, which represents the number of occurrences of an event of interest and t is the time (or space) interval. The probability mass function for the Poisson distribution is

$$P_{Xt}(x) = \begin{cases} \dfrac{(\lambda t)^x \exp(-\lambda t)}{x!} & \text{for } x = 0, 1, 2, 3,\ldots \\ 0 & \text{otherwise} \end{cases} \tag{A.49}$$

The probability mass function of an example Poisson distribution is shown in Figure A.8. The mean and variance for the Poisson distribution are, respectively, given by

$$\mu_{X_t} = \lambda t \text{ and } \sigma_{X_t}^2 = \lambda t \tag{A.50}$$

The parameter λ of the Poisson distribution represents the average rate of occurrence of the event of interest.

A.5.5 Negative Binomial and Pascal Distributions

The *negative binomial distribution* is considered a general case of the geometric distribution. Its underlying random variable is defined as the kth occurrence of an event of interest on the last trial in a sequence of X Bernoulli trials. The probability of this kth occurrence on the last trial is given by the probability mass function of the negative binomial distribution, that is,

$$P_X(x) = \begin{cases} \begin{pmatrix} x-1 \\ k-1 \end{pmatrix} p^k (1-p)^{x-k} & \text{for } x = k, k+1, k+2,\ldots \\ 0 & \text{otherwise} \end{cases} \tag{A.51}$$

The mean and variance of this distribution, respectively, are given by

$$\mu_X = \frac{k}{p} \text{ and } \sigma_X^2 = \frac{k(1-p)}{p^2} \tag{A.52}$$

The negative binomial distribution is called the *Pascal distribution* if k takes on only integer values.

A.5.6 Hypergeometric Distribution

The *hypergeometric distribution* deals with a finite population of size N, with a class of $D \leq N$ elements of the population having a property of interest (e.g., defective units or nondefective units). A random sample of size n is selected without replacement, that is, a sampled element of the population is not replaced before randomly selecting the next element of the sample. The underlying random variable, X, for this distribution is defined as the number of elements in the sample that belong to the class of interest. The probability mass function is given by

$$P_X(x) = \begin{cases} \dfrac{\dbinom{D}{x}\dbinom{N-D}{n-x}}{\dbinom{N}{n}} & \text{for } x = 0, 1, 2, \ldots, \min(n, D) \\ 0 & \text{otherwise} \end{cases} \tag{A.53}$$

The mean and variance of this distribution, respectively, are given by:

$$\mu_X = n\frac{D}{N} \text{ and } \sigma_X^2 = n\frac{D}{N}\frac{N-n}{N-1}\left(1-\frac{D}{N}\right) \tag{A.54}$$

A.6 Common Continuous Probability Distributions

In this section, several continuous distributions are discussed. The uniform distribution is very important for performing random number generation in simulation. The normal and lognormal distributions are important due to their common use and applications in engineering and economics. These two distributions also have an important and unique relationship. The importance of the exponential distribution comes from its special relation to the Poisson distribution. The triangular, gamma, Raleigh, and beta distributions are also described. Also, Student's t-distribution, the chi-squared distribution, and the F-distribution are described for their use in statistics. In addition, extreme value distributions are described. A summary of selected continuous distributions that are commonly used in reliability and risk studies is provided in Section A.7.

A.6.1 Uniform Distribution

The density function for the uniform distribution of a random variable X is given by

$$f_X(x) = \begin{cases} \dfrac{1}{b-a} & \text{for } a \le x \le b \\ 0 & \text{otherwise} \end{cases} \tag{A.55}$$

where:

a and b are real values, called *parameters*, with $a < b$

The density function for the uniform distribution takes a constant value of $1/(b - a)$ to satisfy the probability axiom that requires the area under the density function to be 1. The mean and variance for the uniform distribution, respectively, are given by

$$\mu_X = \frac{a+b}{2} \text{ and } \sigma_X^2 = \frac{(b-a)^2}{12} \tag{A.56}$$

Due to the simple geometry of the density function of the uniform distribution, it can be easily noted that its mean value and variance correspond to the centroidal distance and centroidal moment of inertia with respect to a vertical axis, respectively, of the area under the density function. This property is valid for other distributions as well. The cumulative function for the uniform distribution is a line with a constant slope and is given by

$$F_X(x) = \begin{cases} 0 & x \le a \\ \dfrac{x-a}{b-a} & a \le x \le b \\ 1 & x \ge b \end{cases} \tag{A.57}$$

A.6.2 Normal Distribution

The normal distribution (also called the *Gaussian distribution*) is widely used due to its simplicity and wide applicability. This distribution is the basis for many statistical methods. The normal density function for a random variable X is given by

$$f_X(x) = \frac{1}{\sigma\sqrt{2\pi}} \exp\left[-\frac{1}{2}\left(\frac{x-\mu}{\sigma}\right)^2\right] \qquad -\infty < x < \infty \tag{A.58}$$

It is common to use the notation $X \sim N(\mu, \sigma^2)$ to provide an abbreviated description of a normal distribution. The notation states that X is normally distributed with a mean value μ and variance σ^2. In Figure A.9, the normal distribution is used to model the concrete strength, assuming that concrete strength has a normal distribution with mean = 3.5 ksi and standard deviation = 0.2887 ksi. The density function of another normal distribution is shown in Figure A.10. The cdf of the normal distribution is given by:

$$F_X(x) = \int_{-\infty}^{+\infty} \frac{1}{\sigma\sqrt{2\pi}} \exp\left[-\frac{1}{2}\left(\frac{x-\mu}{\sigma}\right)^2\right] dx \tag{A.59}$$

The evaluation of the integral of Equation A.59 requires numerical methods for each pair (μ, σ^2). This difficulty can be reduced by performing a transformation that results in a standard normal distribution with mean $\mu = 0$ and variance $\sigma^2 = 1$, denoted as $Z \sim N(0, 1)$. Numerical

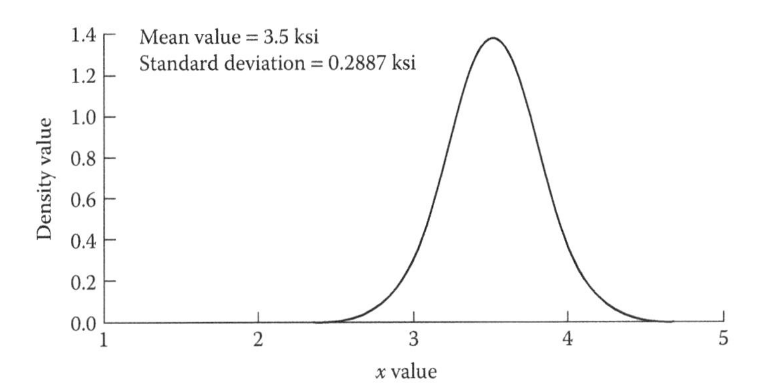

FIGURE A.9
pdf of the normal distribution.

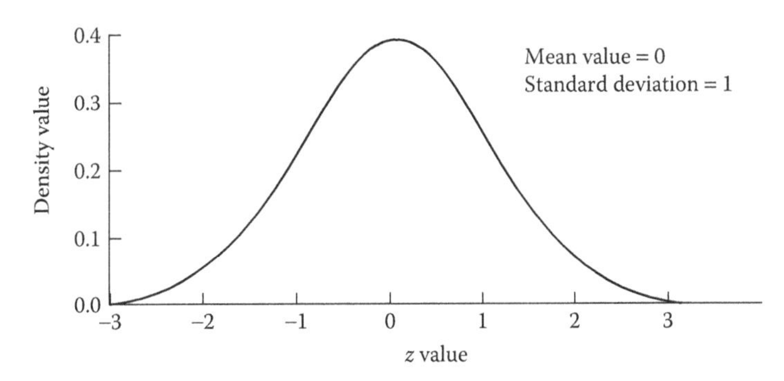

FIGURE A.10
pdf of the standard normal distribution.

integration can be used to determine the cdf of the standard normal distribution and tabulate the results as provided in probability and statistics textbooks. Using the following standard normal transformation

$$Z = \frac{X - \mu}{\sigma} \tag{A.60}$$

the density function of the standard normal is shown in Figure A.10. A special notation of $\phi(z)$ is used for the pdf of the standard normal and $\Phi(z)$ for the cdf of the standard normal. The results of the integral $\Phi(z)$ are tabulated in probability and statistics textbooks (e.g., Ayyub and McCuen 2011) or computed using functions in Microsoft Excel called NORMSDIST and NORMSINV for the cdf and its inverse, respectively. It can be shown that

$$\begin{aligned} P(a < X \le b) &= F_X(b) - F_X(a) \\ &= \Phi\left(\frac{b-\mu}{\sigma}\right) - \Phi\left(\frac{a-\mu}{\sigma}\right) \end{aligned} \tag{A.61}$$

The normal distribution has an important and useful property in the case of adding n normally distributed random variables, $X_1, X_2, \ldots, X_n$, which are not correlated, as follows:

$$Y = X_1 + X_2 + X_3 + \cdots + X_n \tag{A.62}$$

The mean and variance of Y (μ_Y and σ_Y^2, respectively) are as follows:

$$\mu_Y = \mu_{X_1} + \mu_{X_2} + \mu_{X_3} + \cdots + \mu_{X_n} \tag{A.63}$$

$$\sigma_Y^2 = \sigma_{X_1}^2 + \sigma_{X_2}^2 + \sigma_{X_3}^2 + \cdots + \sigma_{X_n}^2 \tag{A.64}$$

A.6.3 Lognormal Distribution

A random variable X is considered to have a lognormal distribution if $Y = \ln(X)$ has a normal probability distribution, where $\ln(x)$ is the natural logarithm to the base e. The density function of the lognormal distribution is given by

$$f_X(x) = \frac{1}{x\sigma_Y\sqrt{2\pi}}\exp\left[-\frac{1}{2}\left(\frac{\ln(x)-\mu_Y}{\sigma_Y}\right)^2\right] \quad \text{for } 0 < x < \infty \tag{A.65}$$

It is common to use the notation $X \sim \ln(\mu_Y, \sigma_Y^2)$ to provide an abbreviated description of a lognormal distribution. The notation states that X is lognormally distributed with the parameters μ_Y and σ_Y^2. The lognormal distribution has the following properties:

1. The values of the random variable X are positive ($x > 0$).
2. $f_X(x)$ is not a symmetric density function about the mean value μ_X.
3. The mean value μ_X and variance σ_X^2 are not equal to the parameters of the distribution (μ_Y and σ_Y^2). However, they are related to them as follows:

$$\sigma_Y^2 = \ln\left[1 + \left(\frac{\sigma_X}{\mu_X}\right)^2\right] \quad \text{and} \quad \mu_Y = \ln(\mu_X) - \frac{1}{2}\sigma_Y^2 \tag{A.66}$$

These two relations can be inverted as follows:

$$\mu_X = \exp\left(\mu_Y + \frac{1}{2}\sigma_Y^2\right) \quad \text{and} \quad \sigma_X^2 = \mu_X^2[\exp(\sigma_Y^2) - 1] \tag{A.67}$$

For a relatively small coefficient of variation δ_X [e.g., $(\sigma_X/\mu_X) \leq 0.3$], σ_Y is approximately equal to the coefficient of variation δ_X. An example density function of the lognormal distribution is shown in Figure A.11.

The cdf of the lognormal distribution can be determined based on its relationship to the normal distribution using the following transformation:

$$Z = \frac{\ln(X) - \mu_Y}{\sigma_Y} \tag{A.68}$$

Therefore, the cumulative probability is given by

$$\begin{aligned} P(a < X \leq b) &= F_X(b) - F_X(a) \\ &= \Phi\left(\frac{\ln(b) - \mu_Y}{\sigma_Y}\right) - \Phi\left(\frac{\ln(a) - \mu_Y}{\sigma_Y}\right) \end{aligned} \tag{A.69}$$

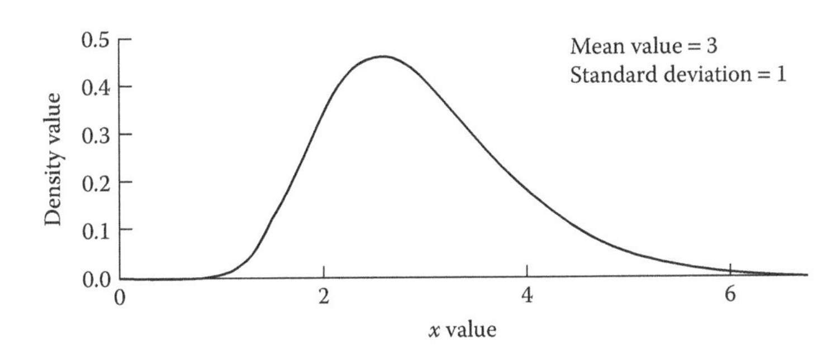

FIGURE A.11
pdf of the lognormal distribution.

A.6.4 Exponential Distribution

The importance of this distribution comes from its relationship to the Poisson distribution. For a given Poisson process, the time T between the consecutive occurrence of events has an exponential distribution with the following density function:

$$f_T(t) = \begin{cases} \lambda \exp(-\lambda t) & \text{for } t \geq 0 \\ 0 & \text{otherwise} \end{cases} \tag{A.70}$$

The cdf is given by

$$F_T(t) = 1 - \exp(-\lambda t) \tag{A.71}$$

The density and cumulative functions of the exponential distribution with $\lambda = 1$ are shown in Figures A.12 and A.13, respectively. The mean value and the variance, respectively, are given by

$$\mu_T = \frac{1}{\lambda} \text{ and } \sigma_T^2 = \frac{1}{\lambda^2} \tag{A.72}$$

Based on the means of the exponential and Poisson distributions, the *mean recurrence time* (or *return period*) is defined as $1/\lambda$.

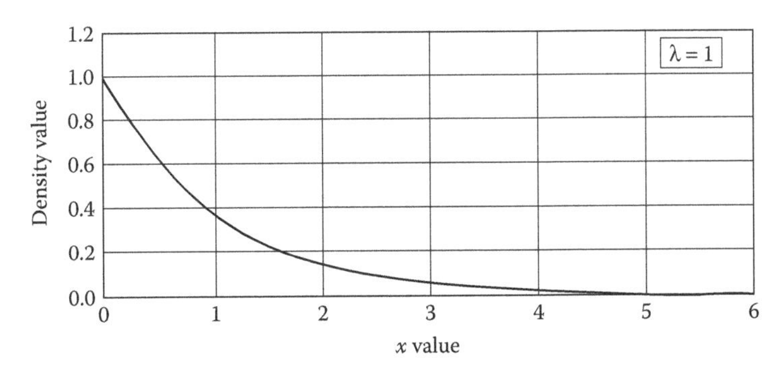

FIGURE A.12
pdf of the exponential distribution.

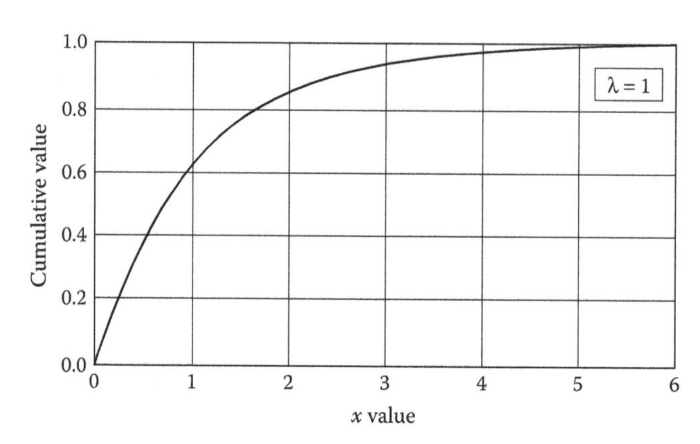

FIGURE A.13
cdf of the exponential distribution.

A.6.5 Triangular Distribution

This distribution is used to qualitatively model an uncertain variable that can be bounded between two limits, such as the duration of a construction activity. For example, the duration of a construction activity can be described by the following density function:

$$f_X(x) = \begin{cases} \dfrac{2}{b-a}\left(\dfrac{x-a}{c-a}\right) & a \le x \le c \\ \dfrac{2}{b-a}\left(\dfrac{b-x}{b-c}\right) & c \le x \le b \\ 0 & \text{otherwise} \end{cases} \tag{A.73}$$

where:
a, b, and c are lower limit, upper limit, and mode, respectively

The cdf is given by

$$F_X(x) = \begin{cases} 0 & x \le a \\ \dfrac{(x-a)^2}{(c-a)(b-a)} & a \le x \le c \\ 1 - \dfrac{(b-x)^2}{(b-c)(b-a)} & c \le x \le b \\ 1 & x \ge b \end{cases} \tag{A.74}$$

The mean (μ) and variance (σ^2) for the distribution, respectively, are given by

$$\mu_X = \frac{a+b+c}{3} \text{ and } \sigma_X^2 = \frac{a^2+b^2+c^2-ab-ac-bc}{18} \tag{A.75}$$

A.6.6 Gamma Distribution

The density function of the gamma probability distribution is given by

$$f_X(x) = \frac{v(vx)^{k-1}\exp(-vx)}{\Gamma(k)} \qquad 0 \le x \tag{A.76}$$

where:
$k > 0$ and $v > 0$ are the parameters of the distribution
The function Γ is called the *gamma function* [commonly tabulated as provided by Ayyub and McCuen (2011) or computed using a function in Microsoft Excel] and is given by

$$\Gamma(k,x) = \int_0^x \exp(-y)y^{k-1}dy \tag{A.77a}$$

$$\Gamma(k) = \int_0^\infty \exp(-y)y^{k-1}dy \tag{A.77b}$$

The cdf is given by

$$F_X(x) = \int_0^x f_X(x)dx = \frac{\Gamma(k,vx)}{\Gamma(k)} \tag{A.78}$$

The mean (μ) and variance (σ^2) for the distribution, respectively, are given by

$$\mu_X = \frac{k}{v} \text{ and } \sigma_X^2 = \frac{k}{v^2} \tag{A.79}$$

A.6.7 Rayleigh Distribution

The density function of this probability distribution is given by

$$f_X(x) = \frac{x}{\alpha^2}\exp\left[\frac{1}{2}\left(\frac{x}{\alpha}\right)^2\right] \tag{A.80}$$

where:
α is the parameter of the distribution

The cdf is given by

$$F_X(x) = 1 - \exp\left(-\frac{x^2}{2\alpha^2}\right) \tag{A.81}$$

The mean (μ) and variance (σ^2) for the distribution, respectively, are given by

$$\mu_X = \sqrt{\frac{\pi}{2}}\alpha \text{ and } \sigma_X^2 = \left(2 - \frac{\pi}{2}\right)\alpha^2 \tag{A.82}$$

For a given mean, the parameter can be computed as

$$\alpha = \sqrt{\frac{2}{\pi}}\mu_X \tag{A.83}$$

A.6.8 Beta Distribution

The beta distribution is used for modeling continuous random variables in a finite interval. The beta distribution function is also used as an auxiliary distribution in nonparametric distribution estimation and as a prior distribution in Bayesian statistical procedures.

The density function of this probability distribution is given by

$$f_X(x) = \frac{\Gamma(k+m)}{\Gamma(k)\Gamma(m)} x^{k-1}(1-x)^{m-1} \qquad \text{for } 0 \le x \le 1,\ k > 0,\ m > 0 \tag{A.84}$$

where:

k and m are the parameters of the distribution

Depending on the values of parameters k and m, the beta function takes on many different shapes. For example, if $k = m = 1$, the density function coincides with the density function of the standard uniform distribution between 0 and 1. The cdf is given by

$$F_X(x) = I_x(k,m) = \frac{\Gamma(k+m)}{\Gamma(k)\Gamma(m)} \int_0^x u^{k-1}(1-u)^{m-1}du \tag{A.85}$$

where:

I is the incomplete beta function

The mean (μ) and variance (σ^2), respectively, for the distribution, are given by

$$\mu_X = \frac{k}{k+m} \text{ and } \sigma_X^2 = \frac{km}{(k+m)^2(k+m+1)} \tag{A.86}$$

A.6.9 Statistical Probability Distributions

In statistical analysis, tables of values of Student's t-distribution, chi-squared distribution, and F-distribution are commonly used. Exceedance probability values are tabulated in textbooks on statistics, such as Ayyub and McCuen (2011).

The *Student's t-distribution* is a symmetric, bell-shaped distribution with the following density function:

$$f_T(t) = \frac{\Gamma[(k+1)/k]}{(\pi k)^{0.5}\Gamma(k/2)[1+(t^2/k)]^{0.5(k+1)}} \qquad -\infty < t < \infty \tag{A.87}$$

where:

k is a parameter of the distribution and represents the *degrees of freedom*

For $k > 2$, the mean and variance, respectively, are as follows:

$$\mu_T = 0 \text{ and } \sigma_T^2 = \frac{k}{k-2} \tag{A.88}$$

As k increases toward infinity, the variance of the distribution approaches unity, and the t distribution approaches the standard normal density function. Therefore, the t distribution has heavier tails (with more area) than the standard normal. It is of interest in statistical analysis to determine the percentage points $t_{\alpha,k}$ that correspond to the following probability:

$$\alpha = P(T > t_{\alpha,k}) \tag{A.89a}$$

or

$$\alpha = \int_{t_{\alpha,k}}^{\infty} f_T(t)\mathrm{d}t \tag{A.89b}$$

where:
 α is called the *level of significance*

These percentage points are tabulated in probability and statistics textbooks, such as Ayyub and McCuen (2011).

The *chi-squared* (χ^2) *distribution* is encountered frequently in statistical analysis, where we deal with the sum of squares of k random variables with standard normal distributions, that is,

$$\chi^2 = C = Z_1^2 + Z_2^2 + \cdots + Z_k^2 \tag{A.90}$$

where:
 C is a random variable with chi-squared distribution
 $Z_1, Z_2, \ldots Z_k$ are normally (standard normal) and independently distributed random variables

The pdf of the chi-squared distribution is

$$f_C(c) = \frac{1}{2^{0.5k}\Gamma\left(\frac{k}{2}\right)} c^{0.5k-1} \exp\left(\frac{-c}{2}\right) \quad \text{for } c > 0 \tag{A.91}$$

The distribution is defined only for positive values and has the following mean and variance, respectively:

$$\mu_C = k \text{ and } \sigma_C^2 = 2k \tag{A.92}$$

The parameter of the distribution, k, represents the degrees of freedom. This distribution is positively skewed with a shape that depends on parameter k. It is of interest in statistical analysis to determine the percentage points, $c_{\alpha,k}$, that correspond to the following probability:

$$\alpha = P(C > c_{\alpha,k}) \tag{A.93a}$$

$$\alpha = \int_{c_{\alpha,k}}^{\infty} f_C(c)\mathrm{d}c \tag{A.93b}$$

where:
 α is called the level of significance

These percentage points are tabulated in probability and statistics textbooks, such as Ayyub and McCuen (2011).

The F-*distribution* is used quite frequently in statistical analysis. It is a function of two shape parameters, $v_1 = k$ and $v_2 = u$, and has the following density function:

$$f_F(f) = \frac{\Gamma\left(\frac{u+k}{2}\right)\left(\frac{k}{u}\right)^{\frac{k}{2}}(f)^{\frac{k}{2}-1}}{\Gamma\left(\frac{k}{2}\right)\Gamma\left(\frac{u}{2}\right)\left(\frac{fk}{u}+1\right)^{\frac{U+K}{2}}} \quad \text{for } f > 0 \tag{A.94}$$

For $u > 2$, the mean and variance of this distribution, respectively, are as follows:

$$\mu_F = \frac{u}{u-2} \text{ and } \sigma_F^2 = \frac{2u^2(u+k-2)}{k(u-2)^2(u-4)} \quad \text{for } u > 4 \tag{A.95}$$

This distribution is positively skewed with a shape that depends on the parameters k and u. It is of interest in statistical analysis to determine the percentage points, $f_{\alpha,k,u}$, that correspond to the following probability:

$$\alpha = P(F > f_{\alpha,k,u}) \tag{A.96a}$$

$$= \int_{f_{\alpha,k,u}}^{\infty} f_F(x)\mathrm{d}x = \alpha \tag{A.96b}$$

where:
α is called the *level of significance*

These percentage points are tabulated in probability and statistics textbooks, such as Ayyub and McCuen (2011).

A.6.10 Extreme Value Distributions

Extreme value distributions are a class of commonly used distributions in engineering and sciences. These distributions are described in the remaining part of this section. The extreme value distributions are of three types.

Two forms of the type I extreme value (also called Gumbel) distribution can be used: the largest and smallest extreme values. The density function for the largest type I distribution of a random variable X_n is given by

$$f_{X_n}(x) = \alpha_n e^{-\alpha_n(x-u_n)} \exp[-e^{-\alpha_n(x-u_n)}] \tag{A.97}$$

where:
u_n is the location parameter of X_n
α_n is the shape parameter of X_n

The density function for the smallest type I distribution of a random variable X_1 is given by

$$f_{X_1}(x) = \alpha_1 e^{\alpha_1(x-u_1)} \exp[-e^{\alpha_1(x-u_1)}] \tag{A.98}$$

where:

u_1 is the location parameter for X_1

α_1 is the shape parameter of X_1

The cumulative function for the largest type I distribution is given by

$$F_{X_n}(x) = \exp[-e^{-\alpha_n(x-u_n)}] \tag{A.99}$$

The cumulative function for the smallest type I extreme is given by

$$F_{X_1}(x) = 1 - \exp[-e^{\alpha_1(x-u_1)}] \tag{A.100}$$

For the largest type I extreme, the mean (μ) and variance (σ^2) for the distribution, respectively, are given by

$$\mu_{X_n} = u_n + \frac{\gamma}{\alpha_n} \text{ and } \sigma^2_{X_N} = \frac{\pi}{6\alpha_n^2} \tag{A.101}$$

where:

$\pi = 3.14159$

$\gamma = 0.577216$

For the smallest type I extreme, the mean (μ) and variance (σ^2) for the distribution, respectively, are given by

$$\mu_{X_1} = u_1 - \frac{\gamma}{\alpha_1} \text{ and } \sigma^2_{X_1} = \frac{\pi^2}{6\alpha_1^2} \tag{A.102}$$

Two forms of the type II extreme value (also called Frěchet) distribution can be used: the largest and smallest extreme values. The two types are described in this section, although only the largest distribution has a common practical value. The density function for the largest type II extreme of a random variable, X_n, is given by

$$f_{X_n}(x) = \frac{k}{v_n}\left(\frac{v_n}{x}\right)^{k+1} \exp\left[-\left(\frac{v_n}{x}\right)^k\right] \tag{A.103}$$

where:

v_n is the location parameter of X_n

k is the shape parameter of X_n

The density function for the smallest type II extreme of random variable X_1 is given by

$$f_{X_1}(x) = -\frac{k}{v_1}\left(\frac{v_1}{x}\right)^{k+1} \exp\left[-\left(\frac{v_1}{x}\right)^k\right] \quad x \le 0 \tag{A.104}$$

where:

v_1 is the location parameter of X_1

k is the shape parameter of X_1

The cumulative function for the largest type II distribution is given by

$$F_{X_n}(x) = \exp\left[-\left(\frac{v_n}{x}\right)^k\right] \tag{A.105}$$

The cumulative function for the smallest type II extreme is given by

$$F_{X_1}(x) = 1 - \exp\left[-\left(\frac{v_n}{x}\right)^k\right] \qquad x \le 0 \text{ and } v_1 > 0 \tag{A.106}$$

For the largest type II extreme, the mean (μ) and variance (σ^2) for the distribution, respectively, are given by

$$\mu_{X_n} = v_n \Gamma\left(1 - \frac{1}{k}\right) \tag{A.107a}$$

$$\sigma^2_{X_n} = v_n^2\left[\Gamma\left(1 - \frac{2}{k}\right) - \Gamma^2\left(1 - \frac{1}{k}\right)\right] \qquad \text{for } k \ge 2 \tag{A.107b}$$

The coefficient of variation (δ) based on Equations A.107a and A.107b is

$$\delta^2_{X_n} = \frac{\Gamma\left(1 - \frac{2}{k}\right)}{\Gamma^2\left(1 - \frac{1}{k}\right)} - 1 \tag{A.108}$$

For the smallest type II extreme, the mean (μ) and variance (σ^2) for the distribution, respectively, are given by

$$\mu_{X_1} = v_1 \Gamma\left(1 - \frac{1}{k}\right) \tag{A.109a}$$

$$\sigma^2_{X_1} = v_1^2\left[\Gamma\left(1 - \frac{2}{k}\right) - \Gamma^2\left(1 - \frac{1}{k}\right)\right] \qquad \text{for } k \ge 2 \tag{A.109b}$$

The coefficient of variation (δ) is

$$\delta^2_{X_1} = \frac{\Gamma\left(1 - \frac{2}{k}\right)}{\Gamma^2\left(1 - \frac{1}{k}\right)} - 1 \tag{A.109c}$$

Two forms of the type III extreme value (also called Weibull) distribution can be used: the largest and smallest extreme values. These two types are described in this section. The density function for the largest type III extreme random variable, X_n, is given by

$$f_{X_n}(x) = \frac{k}{\omega - u}\left(\frac{\omega - x}{\omega - u}\right)^{k-1} \exp\left[-\left(\frac{\omega - x}{\omega - u}\right)^k\right] \qquad \text{for } x \le \omega \tag{A.110}$$

The density function for the smallest type III extreme random variable, X_1, is given by

$$f_{X_1}(x) = \frac{k}{u-\omega}\left(\frac{x-\omega}{u-\omega}\right)^{k-1} \exp\left[-\left(\frac{x-\omega}{u-\omega}\right)^k\right] \quad \text{for } x \geq \omega \tag{A.111}$$

where:
$u > 0$
$k > 0$
u is the scale parameter
k is the shape parameter
ω is the upper or lower limit on x for the largest and the smallest extreme random variable, respectively

The cdf for the largest type III extreme random variable, X_n, is given by

$$F_{X_n}(x) = \exp\left[-\left(\frac{\omega-x}{\omega-u}\right)^k\right] \quad \text{for } x \leq \omega \text{ and } k > 0 \tag{A.112}$$

The cdf for the smallest type III extreme random variable, X_1, is given by

$$F_{X_1}(x) = 1 - \exp\left[-\left(\frac{x-\omega}{u-\omega}\right)^k\right] \quad \text{for } x \geq \omega \tag{A.113}$$

For the largest type III extreme, the mean (μ) and variance (σ^2) for the distribution, respectively, are given by

$$\mu_{X_n} = \omega - (\omega - u)\Gamma\left(1 + \frac{1}{k}\right) \tag{A.114a}$$

$$\sigma_{X_n}^2 = (\omega - u)^2\left[\Gamma\left(1+\frac{2}{k}\right) - \Gamma^2\left(1+\frac{1}{k}\right)\right] \tag{A.114b}$$

For the smallest type III extreme, the mean (μ) and variance (σ^2) for the distribution, respectively, are given by

$$\mu_{X_1}(x) = \omega + (u - \omega)\left[\Gamma\left(1+\frac{1}{k}\right)\right] \tag{A.115a}$$

$$\sigma_{X_1}^2 = (u - \omega)^2\left[\Gamma\left(1+\frac{2}{k}\right) - \Gamma^2\left(1+\frac{1}{k}\right)\right] \tag{A.115b}$$

A.7 Summary of Probability Distributions

Figure A.14 provides a summary of selected discrete and continuous probability distributions that are commonly used in reliability and risk studies. The figure shows the probability function, the cumulative function, and the failure rate function for each distribution evaluated for selected parameters.

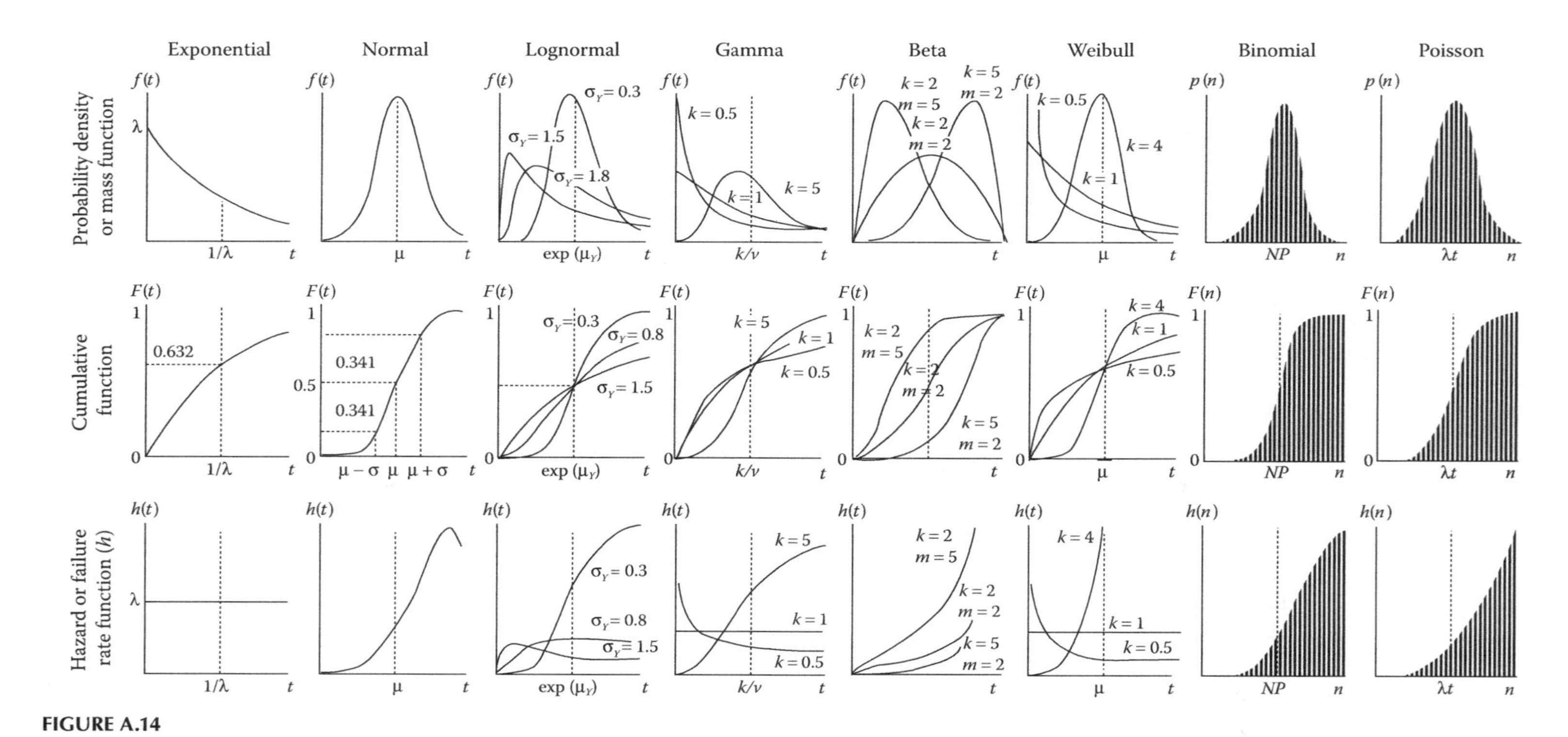

FIGURE A.14
Summary of typical probability distributions for reliability and risk studies.

A.8 Joint Random Variables and Their Probability Distributions

In some engineering applications, the outcomes, say, $E_1, E_2, \ldots, E_n$, that constitute a sample space S are mapped to an n-dimensional (n-D) space of real numbers. The functions that establish such a transformation to the n-D space are called *multiple random variables* (or *random vectors*). This mapping can be one to one or one to many.

Multiple random variables are commonly classified into two types: discrete and continuous random vectors. A discrete random vector may take on only distinct, usually integer, values, whereas a continuous random vector takes on values within a continuum of values. A distinction is made between these two types of random vectors because the computations of probabilities depend on their type.

A.8.1 Probability for Discrete Random Vectors

The probability of a discrete multiple random variable or random vector $\mathbf{X} = (X_1, X_2, \ldots, X_n)$ is given by a *joint probability mass function*. A joint mass function specifies the probability that the discrete random variable X_1 is equal to some value x_1, X_2 is equal to some value x_2, and X_n is equal to some value x_n and is denoted by

$$P_X(\mathbf{x}) = P(X_1 = x_1, X_2 = x_2, \ldots, X_n = x_n) \quad \text{(A.116)}$$

where:

$\mathbf{X}$ is a random vector that includes the random variables $(X_1, X_2, \ldots, X_n)$
$\mathbf{x}$ is a specified value for the random vectors $(x_1, x_2, \ldots, x_n)$

The probability mass function must satisfy the axioms of probability. Therefore, the probability of an event $(X_1 = x_1, X_2 = x_2, \ldots, X_n = x_n)$ must be ≤ 1, and it must be ≥ 0, that is,

$$0 \leq P = (X_1 = x_1, X_2 = x_2, \ldots, X_n = x_n) \leq 1 \quad \text{(A.117)}$$

This property is valid for all possible values of all of the random variables. Additionally, the sum of all possible probabilities must be equal to 1.

It is often useful to present the likelihood of an outcome using the *cumulative mass function*, which is given by

$$F_X(\mathbf{x}) = P(X_1 \leq x_1, X_2 \leq x_2, \ldots, X_n \leq x_n) = \sum_{\text{all } (X_1 \leq x_1,\, X_2 \leq x_2, \ldots,\, X_n \leq x_n)} P_X(x_1, x_2, x_3, \ldots, x_n) \quad \text{(A.118)}$$

The cumulative mass function is used to indicate the probability that the random variable X_1 is $\leq x_1$, X_2 is $\leq x_2$, and X_n is $\leq x_n$.

The presentation of the materials in the remaining part of this section is limited to two random variables. The presented concepts can be generalized to n random variables. Based on the definition of conditional probabilities, the conditional probability mass function $P_{X_1|X_2}(x_1|x_2)$, for two random variables X_1 and X_2, is given by

$$P_{X_1|X_2}(x_1|x_2) = \frac{P_{X_1X_2}(x_1, x_2)}{P_{X_2}(x_2)} \quad \text{(A.119)}$$

where:

$P_{X_1|X_2}(x_1|x_2)$ results in the probability of $X_1 = x_1$ given that $X_2 = x_2$
$P_{X_1X_2}(x_1,x_2)$ is the joint probability mass function of X_1 and X_2
$P_{X_2}(x_2)$ is the marginal probability mass function for X_2 that is not equal to 0

In this case, the marginal distribution is given by

$$P_{X_2}(x_2) = \sum_{\text{all } x_1} P_{X_1X_2}(x_1, x_2) \tag{A.120}$$

Similarly, the conditional probability mass function $P_{X_2|X_1}(x_2|x_1)$, for two random variables X_1 and X_2, is given by

$$P_{X_2|X_1}(x_2|x_1) = \frac{P_{X_1X_2}(x_1, x_2)}{P_{X_1}(x_1)} \tag{A.121}$$

where:

$P_{X_1}(x_1)$ is the marginal mass function and is given by

$$P_{X_1}(x_1) = \sum_{\text{all } x_2} P_{X_1X_2}(x_1, x_2) \tag{A.122}$$

The definitions provided by Equations A.119 through A.122 can be generalized for the n-D case. Based on the definition of conditional probabilities, it can be stated that if X_1 and X_2 are statistically uncorrelated random variables, then

$$P_{X_1|X_2}(x_1|x_2) = P_{X_1}(x_1) \text{ and } P_{X_2|X_1}(x_2|x_1) = P_{X_2}(x_2) \tag{A.123}$$

Therefore, using Equation A.119 or A.121, the following important relationship can be obtained:

$$P_{X_1|X_2}(x_1, x_2) = P_{X_1}(x_1)P_{X_2}(x_2) \tag{A.124}$$

A.8.2 Probability for Continuous Random Vectors

A *joint* pdf is used to define the likelihood of occurrence for a continuous random vector. Specifically, the probability that the random vector $\mathbf{X} = (X_1, X_2, \ldots, X_n)$ is within the interval from $\boldsymbol{x}^l = (x_1^u, x_2^u, x_3^u, \ldots, x_n^u)$ to $\boldsymbol{x}^u = (x_1^u, x_2^u, x_3^u, \ldots, x_n^u)$ is

$$P(\boldsymbol{x}^l \le \mathbf{X} \le \boldsymbol{x}^u) = \int_{x_1^l}^{x_1^u}\int_{x_2^l}^{x_2^u}\cdots\int_{x_n^l}^{x_n^u} f_X(\boldsymbol{x})\mathrm{d}x_1, \mathrm{d}x_2, \ldots, \mathrm{d}x_n \tag{A.125}$$

where:

$f_X(\boldsymbol{x})$ is the joint density function

It is important to note that the multiple integral of the joint pdf from $-\infty$ to $+\infty$ equals 1, that is,

$$P(-\infty < \mathbf{X} < +\infty) = \int_{-\infty}^{+\infty}\int_{-\infty}^{+\infty}\cdots\int_{-\infty}^{+\infty} f_X(\boldsymbol{x})\mathrm{d}x_1, \mathrm{d}x_2, \ldots, \mathrm{d}x_n = 1 \tag{A.126}$$

The cdf of a continuous random variable is defined by

$$F_X(x) = P(X \leq x) = \int_{-\infty}^{x_1}\int_{-\infty}^{x_2}\cdots\int_{-\infty}^{x_n} f_X(x)\mathrm{d}x_1, \mathrm{d}x_2, \ldots, \mathrm{d}x_n \tag{A.127}$$

The joint density function can be obtained from a given joint cdf by evaluating the partial derivative as follows:

$$f_{X_1X_2\ldots X_n}(x_1, x_2, \ldots x_n) = \frac{\partial^n F_{X_1X_2\ldots X_n}(x_1, x_2, \ldots, x_n)}{\partial X_1\, \partial X_2 \ldots \partial X_n} \tag{A.128}$$

The presentation of the materials in the remaining part of this section is limited to two random variables. The presented concepts can be generalized to n random variables. Based on the definition of conditional probabilities, the conditional pdf $f_{X_1|X_2}(x_1|x_2)$ for two random variables X_1 and X_2 is given by

$$f_{X_1|X_2}(x_1|x_2) = \frac{f_{X_1X_2}(x_1, x_2)}{f_{X_2}(x_2)} \tag{A.129}$$

where:

$f_{X_1X_2}(x_1,x_2)$ is the joint probability density function of X_1 and X_2

$f_{X_2}(x_2)$ is the marginal probability density function or the marginal probability function for X_2 that is not equal to 0

In this case, the marginal probability function is given by

$$f_{X_2}(x_2) = \int_{-\infty}^{+\infty} f_{X_1X_2}(x_1, x_2)\mathrm{d}x_1 \tag{A.130}$$

Similarly, the conditional pdf $f_{X_2|X_1}(x_2|x_1)$ for two random variables X_1 and X_2 is given by

$$f_{X_2|X_1}(x_2|x_1) = \frac{f_{X_1X_2}(x_1, x_2)}{f_{X_1}(x_1)} \tag{A.131}$$

where:

$f_{X_1}(x_1)$ is the marginal probability density function and is given by

$$f_{X_1}(x_1) = \int_{-\infty}^{+\infty} f_{X_1X_2}(x_1, x_2)\mathrm{d}x_2 \tag{A.132}$$

Based on the definition of conditional probabilities, it can be stated that if X_1 and X_2 are statistically uncorrelated random variables, then

$$f_{X_1|X_2}(x_1|x_2) = f_{X_1}(x_1) \text{ and } f_{X_2|X_1}(x_2|x_1) = f_{X_2}(x_2) \tag{A.133}$$

Therefore, using Equation A.129 or A.132, the following important relationship can be obtained:

$$f_{X_1|X_2}(x_1, x_2) = f_{X_1}(x_1)f_{X_2}(x_2) \tag{A.134}$$

A.8.3 Conditional Moments, Covariance, and Correlation Coefficient

In general, moments can be computed using the concept of mathematical expectation. For a continuous random vector $\mathbf{X} = \{X_1, X_2, \ldots, X_n\}$ the kth moment about the origin is given by

$$M_k' = \int_{-\infty}^{+\infty}\int_{-\infty}^{+\infty}\cdots\int_{-\infty}^{+\infty} x_1^k\, x_2^k \ldots x_n^k f_{X_1X_2\ldots X_n}(x_1, x_2, \ldots, x_n)\mathrm{d}x_1\, \mathrm{d}x_2 \ldots \mathrm{d}x_n \tag{A.135}$$

where:
$f_{X_1X_2\ldots X_n}(x_1, x_2, \ldots, x_n)$ is its joint density function

The corresponding equation for a discrete random vector $\mathbf{X}$ is

$$M_k' = \sum_{\text{all } x} x_1^k\, x_2^k \ldots x_n^k P_{X_1X_2\ldots X_n}(x_1, x_2, \ldots, x_n) \tag{A.136}$$

where:
$P_{X_1X_2\ldots X_n}(x_1, x_2, \ldots, x_n)$ is the joint probability mass function

The above moments are commonly considered special cases of mathematical expectation. The mathematical expectation of arbitrary function $g(\mathbf{X})$, a function of the random vector $\mathbf{X}$, is given by

$$E[g(\mathbf{X})] = \int_{-\infty}^{+\infty}\int_{-\infty}^{+\infty}\cdots\int_{-\infty}^{+\infty} g(\mathbf{x}) f_{X_1X_2\ldots X_n}(x_1, x_2, \ldots, x_n)\mathrm{d}x_1\, \mathrm{d}x_2 \ldots \mathrm{d}x_n \tag{A.137}$$

The corresponding equation for a discrete random vector $\mathbf{X}$ is

$$E[g(\mathbf{X})] = \sum_{\text{all } x} g(\mathbf{x}) P_{X_1X_2\ldots X_n}(x_1, x_2, \ldots, x_n) \tag{A.138}$$

For the two-dimensional case, X_1 and X_2, the conditional mean value for X_1 given that X_2 takes a value x_2, denoted by $\mu_{X_1|x_2}$, is defined in terms of the conditional mass and density functions for the discrete and continuous random variables, respectively. The conditional mean for the continuous case is

$$\mu_{X_1|x_2} = E(X_1|x_2) = \int_{-\infty}^{+\infty} x_1 f_{X_1|X_2}(x_1|x_2)\mathrm{d}x_1 \tag{A.139}$$

where:
$f_{X_1|X_2}(x_1|x_2)$ is the conditional probability density function of X_1 at a given (or specified) value of X_2

In this case, the conditional mean is the average value of the random variable X_1 given that the random variable X_2 takes the value x_2. For a discrete random variable, the conditional mean is given by

$$\mu_{X_1|x_2} = E(X_1|x_2) = \sum_{\text{all } x_1} x_1 P_{X_1|X_2}(x_1|x_2) \tag{A.140}$$

where:
$P_{X_1|X_2}(x_1|x_2)$ is the conditional probability mass function of X_1 at a given (or specified) value of X_2

For statistically uncorrelated random variables X_1 and X_2, the conditional mean of a random variable is the same as its mean, that is,

$$\mu_{X_1|x_2} = E(X_1|x_2) = E(X_1) \text{ and } \mu_{X_2|x_1} = E(X_2|x_1) = E(X_2) \quad \text{(A.141)}$$

Also, it can be shown that the expected value with respect to X_2 of the conditional mean $\mu_{X_1|X_2}$ is the mean of X_1, that is,

$$E_{X_2}(\mu_{X_1|X_2}) = E(X_1) \quad \text{(A.142)}$$

where:

E_{X_2} is the expected value with respect to X_2, that is, the variable of integration (or summation) for computing the expected value is x_2

In Equation A.142, the quantity $\mu_{X_1|X_2}$ is treated as a random variable, because conditioning is performed on the random variable X_2 (not a specified value x_2).

As previously discussed, the variance is the second moment about the mean. For two random variables, X_1 and X_2, the conditional variance $\sigma^2_{X_1|x_2}$ [or $\text{Var}(X_1|x_2)$] is computed as follows:

$$\text{Var}(X_1|x_2) = \int_{-\infty}^{+\infty} (x_1 - \mu_{X_1|x_2})^2 f_{X_1|X_2}(x_1|x_2)dx_1 \quad \text{(A.143)}$$

For a discrete variable, the conditional variance is computed by

$$\text{Var}(X_1|x_2) = \sum_{\text{all } x_1} (x_1 - \mu_{X_1|x_2})^2 P_{X_1|X_2}(x_1|x_2) \quad \text{(A.144)}$$

The variance of the random variable X_1 can also be computed using the conditional variance as follows:

$$\text{Var}(X_1) = E_{X_2}[\text{Var}(X_1|X_2)] + \text{Var}_{X_2}[E(X_1|X_2)] \quad \text{(A.145)}$$

where:

E_{X_2} is the expected value with respect to X_2

Var_{X_2} is the variance with respect to X_2, that is, the variable of integration (or summation) for computing the variance is x_2

In Equation A.145, the quantity $\text{Var}(X_1|X_2)$ is treated as a random variable because the conditioning is performed on the random variable X_2 (not value x_2).

The covariance (Cov) of two random variables, X_1 and X_2, is defined in terms of mathematical expectation as

$$\text{Cov}(X_1, X_2) = E[(X_1 - \mu_{X_1})(X_2 - \mu_{X_2})] \quad \text{(A.146)}$$

It is common to use the notation $\sigma_{X_1X_2}$, σ_{12}, or $\text{Cov}(X_1,X_2)$ for the covariance of X_1 and X_2. The covariance for two random variables can also be determined using the following equation that results from Equation A.146:

$$\text{Cov}(X_1, X_2) = E(X_1X_2) - \mu_{X_1}\mu_{X_2} \quad \text{(A.147)}$$

where:

the expected value of the product (X_1X_2) is given by

$$E(X_1X_2) = \int_{-\infty}^{+\infty}\int_{-\infty}^{+\infty} x_1x_2f_{X_1X_2}(x_1, x_2)dx_1dx_2 \quad \text{(A.148)}$$

Equation A.147 can be derived from Equation A.146 based on the definition of mathematical expectation and by separating the terms of integration. If X_1 and X_2 are statistically uncorrelated, then

$$\text{Cov}(X_1, X_2) = 0 \text{ and } E(X_1X_2) = \mu_{X_1}\mu_{X_2} \tag{A.149}$$

The correlation coefficient is defined as a normalized covariance with respect to the standard deviations of X_1 and X_2 and is given by

$$\rho_{X_1X_2} = \frac{\text{Cov}(X_1, X_2)}{\sigma_{X_1}\sigma_{X_2}} \tag{A.150}$$

The correlation coefficient ranges inclusively between −1 and +1, that is,

$$-1 \le \rho_{X_1X_2} \le +1 \tag{A.151}$$

If the correlation coefficient is 0, then the two random variables are not correlated. From the definition of correlation, in order for $\rho_{X_1X_2}$ to be 0, the Cov(X_1,X_2) must be 0. Therefore, X_1 and X_2 are statistically uncorrelated. The correlation coefficient can also be viewed as a measure of the degree of linear association between X_1 and X_2. The sign (− or +) indicates the slope for the linear association. It is important to note that the correlation coefficient does not give any indications about the presence of a nonlinear relationship between X_1 and X_2 (or the lack of it).

A.9 Functions of Random Variables

Many engineering problems deal with a dependent variable that is a function of one or more independent random variables. In this section, analytical tools for determining the probabilistic characteristics of the dependent random variable based on the given probabilistic characteristics of independent random variables and a functional relationship between them are provided. The discussion in this section is divided into the following cases: (1) probability distributions for functions of random variables and (2) approximate methods for computing the moments of functions of random variables.

A.9.1 Probability Distributions for Functions of Random Variables

A random variable X is defined as a mapping from a sample space of an engineering system or experiment to the real line of numbers. This mapping can be a one-to-one mapping or a many-to-one mapping. If Y is defined to be a dependent variable in terms of a function $Y = g(X)$, then Y is also a random variable. Assuming that both X and Y are discrete random variables and for a given probability mass function of X, $P_X(x)$, the objective here is to determine the probability mass function of Y, $P_Y(y)$. This objective can be achieved by determining the equivalent events of Y in terms of the events of X based on the given relationship between X and Y: $Y = g(X)$. For each value y_i, all of the values of x that result in y_i should be determined, say, $x_{i_1}, x_{i_2}, \ldots, x_{i_j}$. Therefore, the probability mass function of Y is given by

$$P_Y(y_i) = \sum_{k=1}^{j} P_X(x_{i_k}) \tag{A.152}$$

If X is continuous but Y is discrete, the probability mass function for Y is given by

$$P_Y(y_i) = \int_{R_e} f_X(x)\mathrm{d}x \tag{A.153}$$

where:
R_e is the region of X that defines an event equivalent to the value $Y = y_i$

If X is continuous with a given density function $f_X(x)$ and the function $g(X)$ is continuous, then $Y = g(X)$ is a continuous random variable with an unknown density function $f_Y(y)$. The density function of Y can be determined by performing the following four steps:

1. For any event defined by $Y \leq y$, an equivalent event in the space of X needs to be defined.
2. $F_Y(y) = P(Y < y)$ can then be calculated.
3. $f_Y(y)$ can be determined by differentiating $F_Y(y)$ with respect to y.
4. The range of validity of $f_Y(y)$ in the Y space should be determined.

Formally stated, if X is a continuous random variable, $Y = g(X)$ is differentiable for all x's, and $g(X)$ is either strictly (monotonically) increasing or strictly (monotonically) decreasing for all x's, then $Y = g(X)$ is a continuous random variable with the following density function:

$$f_Y(y) = \sum_{i=1}^{m} f_X[g_i^{-1}(y)]\left|\frac{\partial g_i^{-1}(y)}{\partial y}\right| \tag{A.154}$$

where:
$g_i^{-1}(y) = x_i$

The following cases are selected special functions of single and multiple random variables that are commonly used where the resulting variable (Y) can have known distribution types for some cases:

1. For multiple independent random variables $\mathbf{X} = (X_1, X_2, \ldots, X_n)$, the function $g(\mathbf{X})$ is a linear combination as given by

$$Y = g(\mathbf{X}) = a_0 + a_1X_1 + a_2X_2 + \cdots + a_nX_n \tag{A.155}$$

where:
$a_0, a_1, a_2, \ldots, a_n$ are real numbers

The mean value and variance of Y are

$$E(Y) = a_0 + a_1E(X_1) + a_2E(X_2) + \cdots + a_nE(X_n) \tag{A.156}$$

and

$$\mathrm{Var}(Y) = \sum_{i=1}^{n}\sum_{j=1}^{n} a_ia_j\mathrm{Cov}(X_i, X_j) \tag{A.157}$$

where:
$\mathrm{Cov}(X_i, X_j)$ is the covariance of X_i and X_j

It should be noted that $\text{Cov}(X_i, X_i) = \text{Var}(X_i) = \sigma^2_{X_i}$. Equation A.157 can be expressed in terms of the correlation coefficient as follows:

$$\text{Var}(Y) = \sum_{i=1}^{n}\sum_{j=1}^{n} a_i a_j \rho_{X_i X_j} \sigma_{X_i} \sigma_{X_j} \tag{A.158}$$

where:

$\rho_{X_i X_j}$ is the correlation coefficient of X_i and X_j

If the random variables of the vector $\boldsymbol{X}$ are statistically uncorrelated, then the variance of Y is

$$\text{Var}(Y) = \sum_{i=1}^{n} a_i^2 \text{Var}(X_i) \tag{A.159}$$

2. In Equations A.156 through A.159, if the random variables $X_1, X_2, X_3, \ldots, X_n$ have normal probability distributions, then Y has a normal probability distribution with a mean and variance as given by Equations A.156 through A.159. This special case was also described in Equations A.62 and A.63.
3. If X has a normal distribution, and $Y = g(X) = \exp(X)$, then Y has a lognormal distribution.
4. If $Y = X_1, X_2, X_3, \ldots, X_n$, the arithmetic multiplication of $X_1, X_2, X_3, \ldots, X_n$ with lognormal distributions, then Y has a lognormal distribution.
5. If $X_1, X_2, \ldots, X_n$ are independent random variables that have Poisson distributions with the parameters, $\lambda_1, \lambda_2, \ldots, \lambda_n$, respectively, then $Y = X_1 + X_2 + \cdots + X_n$ has a Poisson distribution with the parameter $\lambda = \lambda_1 + \lambda_2 + \cdots + \lambda_n$.

A.9.2 Approximate Methods for Computing the Moments of Functions of Random Variables

The closed-form solutions for the distribution types of dependent random variables, as well as mathematical expectation, provide solutions for the simple cases of functions of random variables. Also, they provide solutions for simple distribution types or a mixture of distribution types for the independent random variables. For cases that involve a more general function, $g(\boldsymbol{X})$, or a mixture of distribution types, these methods are not suitable for obtaining solutions due to the analytical complexity of these methods. Also, in some engineering applications, precision might not be needed. In such cases, approximate methods based on Taylor series expansion, with or without numerical solutions of needed derivatives, can be used. The use of Taylor series expansion, in this section, is divided into two types: (1) single random variable X and (2) multiple random variables (i.e., a random vector $\boldsymbol{X}$).

A.9.2.1 Single Random Variable X

The Taylor series expansion of a function $Y = g(X)$ about the mean of X, that is, $E(X)$, is given by

$$\begin{aligned} Y = g[E(X)] + [X - E(X)]\frac{dg(X)}{dX} + \frac{1}{2}[X - E(X)]^2 \frac{d^2g(X)}{dX^2} \\ + \cdots + \frac{1}{k!}[X - E(X)]^k \frac{d^kg(X)}{dX^k} + \cdots \end{aligned} \tag{A.160}$$

in which the derivatives are evaluated at the mean of X. Truncating this series at the linear terms, the *first-order mean* and *variance* of Y can be obtained by applying the mathematical expectation and variance operators, respectively. The first-order (approximate) mean is

$$E(Y) \approx g[E(X)] \tag{A.161}$$

The first-order (approximate) variance is

$$\mathrm{Var}(Y) \approx \left[\frac{\mathrm{d}g(X)}{\mathrm{d}X}\right]^2 \mathrm{Var}(X) \tag{A.162}$$

Again, the derivative in Equation A.162 is evaluated at the mean of X.

A.9.2.2 Random Vector X

The Taylor series expansion of a function $Y = g(\mathbf{X})$ about the mean values of $\mathbf{X}$, that is, $E(X_1), E(X_2), \ldots, E(X_n)$, is given by

$$\begin{aligned} Y = {} & g[E(X_1),E(X_2),\ldots,E(X_n)] + \sum_{i=1}^{n}[X_i - E(X_i)]\frac{\partial g(\mathbf{X})}{\partial X_i} \\ & + \sum_{i=1}^{n}\sum_{j=1}^{n}\frac{1}{2}[X_i - E(X_i)][X_j - E(X_j)]\frac{\partial^2 g(\mathbf{X})}{\partial X_i \partial X_j} + \cdots \end{aligned} \tag{A.163}$$

in which the derivatives are evaluated at the mean values of $\mathbf{X}$. Truncating this series at the linear terms, the *first-order mean* and *variance* of Y can be obtained by applying the mathematical expectation and variance operators, respectively. The first-order (approximate) mean is

$$E(Y) \approx g[E(X_1),E(X_2),\ldots,E(X_n)] \tag{A.164}$$

The first-order (approximate) variance is

$$\mathrm{Var}(Y) \approx \sum_{i=1}^{n}\sum_{j=1}^{n}\frac{\partial g(\mathbf{X})}{\partial X_i}\frac{\partial g(\mathbf{X})}{\partial X_j}\mathrm{Cov}(X_i, X_j) \tag{A.165}$$

in which the derivatives are evaluated at the mean values of $\mathbf{X}$, that is, $E(X_1), E(X_2), \ldots, E(X_n)$.

A.10 Samples and Populations

The data that are collected represent sample information that is not complete by itself, and predictions are not made directly from the sample. The intermediate step between sampling and prediction is identification of the underlying population. The sample is used to identify the population and then the population is used to make predictions or decisions. This sample-to-population-to-prediction sequence is true for the univariate methods of this chapter or for the bivariate and multivariate methods that follow.

A known function or model is most often used to represent the population. The normal and lognormal distributions are commonly used to model the population for a univariate problem. For bivariate and multivariate predictions, linear ($\hat{Y} = a + bX$) and power ($\hat{Y} = aX^b$) models are commonly assumed functions for representing the population, where $\hat{Y}$ is the predicted value of dependent variable Y, X is the independent random variable, and a and b are model parameters. When using a probability function to represent the population, it is necessary to estimate the parameters. For example, for the normal distribution, the location and scale parameters need to be estimated, or the mean and standard deviation, respectively. For the exponential distribution, the rate (λ) is a distribution parameter that needs to be estimated. When using the linear or power models as the population, it is necessary to estimate the coefficients a and b. In both the univariate and multivariate cases, they are called *sample estimators* of the population parameters.

A.11 Estimation of Parameters

In developing models for populations, models can be classified as univariate, bivariate, or multivariate, with parameters that provide the needed complete definition of a model. Models can have one, two, or more parameters. For example, the normal distribution as a univariate model has two parameters, the exponential distribution has one parameter, and the bivariate power model ($\hat{Y} = aX^b$) has two parameters. Samples are used to develop a model that can adequately represent the population and to estimate the parameters of the population model. The parameters can be estimated in the form of point estimates (single values) or interval estimates (ranges of values) using the samples. The equations or methods used to estimate the parameters are called *estimators*. In this section, estimators are introduced. The statistical uncertainty associated with the estimators is also discussed for statistical decision making using hypothesis testing and interval estimation.

A.11.1 Estimation of Moments

The mean or average value of n observations, if all observations are given equal weights, is given by

$$\bar{X} = \frac{1}{n}\sum_{i=1}^{n} x_i \tag{A.166}$$

where:

x_i is a sample point
$i = 1, 2, \ldots, n$

Although this moment conveys certain information about the underlying sample, it does not completely characterize the underlying variable. Two variables can have the same mean, but different histograms. For n observations in a sample that are given equal weight, the variance (S^2) is given by

$$S^2 = \frac{1}{n-1}\sum_{i=1}^{n}(x_i - \bar{X})^2 \tag{A.167}$$

The units of the variance are the square of the units of the parameter or variable x. By definition, the standard deviation (S) is the square root of the variance as follows:

$$S = \sqrt{\frac{1}{n-1}\left[\sum_{i=1}^{n} x_i^2 - \frac{1}{n}\left(\sum_{i=1}^{n} x_i\right)^2\right]} \quad \text{(A.168)}$$

The coefficient of variation (COV or δ) is a normalized quantity based on the standard deviation and mean value as

$$\text{COV} = \frac{S}{\bar{X}} \quad \text{(A.169)}$$

A.11.2 Method-of-Moments Estimation

The method of moments is one method of estimating population parameters using the moments of samples. Using the relationships between moments and parameters for various probability distributions, the parameters can be estimated based on the moments that result from sampling, such as the mean and variance. Table A.2 provides a summary of

TABLE A.2

Relationships for the Method of Moments

Distribution Type	Probability Mass or Density Function	Parameters	Relationships
(a) Discrete Distributions			
Bernoulli	$P_X(x) = \begin{cases} p & x = 1 \\ 1-p & x = 0 \\ 0 & \text{otherwise} \end{cases}$	p	$\bar{X} = p$ $S^2 = p(1-p)$
Binomial	$P_X(x) = \begin{cases} \binom{N}{x} p^x (1-p)^{N-x} & \text{for } x = 0, 1, 2, \ldots, N \\ 0 & \text{otherwise} \end{cases}$	p	$\bar{X} = Np$ $S^2 = Np(1-p)$
Geometric	$P_X(x) = \begin{cases} p(1-p)^{x-1} & x = 1, 2, 3, \ldots \\ 0 & \text{otherwise} \end{cases}$	p	$\bar{X} = 1/p$ $S^2 = (1-p)/p^2$
Poisson	$P_X(x) = \begin{cases} \frac{(\lambda t)^x \exp(-\lambda t)}{x!} & x = 0,1,2,3,\ldots \\ & \text{otherwise} \end{cases}$	λ	$\bar{X} = \lambda t$ $S^2 = \lambda t$
(b) Continuous Distributions			
Uniform	$f_X(x) = \begin{cases} \frac{1}{b-a} & \text{for } a \le x \le b \\ 0 & \text{otherwise} \end{cases}$	a, b	$\bar{X} = (a+b)/2$ $S^2 = (b-a)^2/12$
Normal	$f_X(x) = \frac{1}{\sigma\sqrt{2\pi}} \exp\left[-\frac{1}{2}\left(\frac{x-\mu}{\sigma}\right)^2\right]$ for $-\infty < x < \infty$	μ, σ	$\bar{X} = \mu$ $S^2 = \sigma^2$
Lognormal	$f_X(x) = \frac{1}{x\sigma_Y\sqrt{2\pi}} \exp\left[-\frac{1}{2}\left(\frac{\ln(x)-\mu_Y}{\sigma_Y}\right)^2\right]$ for $0<x<\infty$	μ_Y, σ_Y	$\bar{X} = \exp(\mu_Y + 0.5\sigma_Y^2)$ $S^2 = \mu_Y^2 [\exp(\sigma_Y^2) - 1]$
Exponential	$f_X(x) = \begin{cases} \lambda \exp(-\lambda X) & \text{for } x \ge 0 \\ 0 & \text{otherwise} \end{cases}$	λ	$\bar{X} = 1/\lambda$ $S^2 = 1/\lambda^2$

the relationships between the parameters of commonly used distributions, and the mean and variance. These relationships can be developed using the concepts in this appendix.

A.11.3 Maximum Likelihood Estimation

The most common statistical method of parameter estimation is the method of maximum likelihood. This method is based on the principle of calculating values of parameters that maximize the probability of obtaining the particular sample.

The likelihood of the sample is the total probability of drawing each item of the sample. The total probability is the product of all the individual item probabilities. This product is differentiated with respect to the parameters, and the resulting derivatives are set to zero to achieve the maximum.

Maximum likelihood solutions for model parameters are statistically efficient solutions, meaning that parameter values have minimum variance. This definition of a best method, however, is theoretical. Maximum likelihood solutions do not always produce solvable equations for the parameters. The following examples illustrate easy to moderately difficult solutions. For some distributions, including notably the normal distribution, the method of moments and maximum likelihood estimation produce identical solutions for the parameters.

As an example, we will find the maximum likelihood estimate of parameter λ in the density function $\lambda \exp(-\lambda x)$. Consider a sample of n items: $x_1, x_2, x_3, \ldots, x_n$. By definition the likelihood function, L, is

$$L = \prod_{i=1}^{n} \lambda \exp(-\lambda x_i) \qquad \text{(A.170)}$$

where:

$\prod$ is the product of the terms for $i = 1, 2, \ldots, n$

The product form of the function in Equation A.170 is difficult to differentiate. We make use of the fact that the logarithm of a variate must have its maximum at the same place as the maximum of the variate. Taking logarithms of Equation A.170 gives

$$\ln(L) = n\ln(\lambda) - \lambda \sum_{i=1}^{n} x_i \qquad \text{(A.171)}$$

The differential of $\ln(L)$ with respect to λ, set to 0, produces the value of the parameter that maximizes the likelihood function. The derivative is given by

$$\frac{d\ln(L)}{d\lambda} = \frac{n}{\lambda} - \sum_{i=1}^{n} x_i = 0 \qquad \text{(A.172)}$$

Equation A.172 yields the following:

$$\frac{1}{\lambda} = \frac{1}{n}\sum_{i=1}^{n} x_i = \bar{X} \qquad \text{(A.173)}$$

Thus, the maximum likelihood value of $1/\lambda$ is the mean of the sample of x's.

Consider the problem of finding the maximum likelihood value of parameter A in the density function:

$$f_X(x) = cx\exp(-Ax) \qquad \text{for } x \geq 0 \tag{A.174}$$

where:

c is a constant

To use this equation as a pdf, we must first find c from the condition for which the total probability equals 1:

$$c\int_0^{\infty} x\exp(-Ax)\mathrm{d}x = 1 \tag{A.175}$$

Solution of this equation gives $c = A^2$. Thus, the likelihood function is

$$L = \prod_{i=1}^{n} A^2 x_i \exp(-Ax_i) \tag{A.176}$$

The logarithm of this function is

$$\ln(L) = 2n\ln(A) + \sum_{i=1}^{n} \ln(x_i) - A\sum_{i=1}^{n} x_i \tag{A.177}$$

and

$$\frac{\mathrm{d}\ln(L)}{\mathrm{d}A} = \frac{2n}{A} - \sum_{i=1}^{n} x_i = 0 \tag{A.178}$$

We find that the maximum likelihood value of $1/A$ is one-half the mean of the sample.

A.12 Sampling Distributions

A.12.1 Sampling Distribution of the Mean

The sampling distribution of the mean depends on whether or not the population variance σ^2 is known. If it is known, then the mean of a random sample of size n from a population with mean μ and variance σ^2 has a normal distribution with mean μ and variance σ^2/n. The statistic Z has a standard normal distribution (i.e., mean = 0 and variance = 1) as follows:

$$Z = \frac{\bar{X} - \mu}{\sigma/\sqrt{n}} \tag{A.179}$$

If the population variance is not known, then the distribution of the mean depends on the distribution of the random variable. For a random variable with a normal distribution with mean μ, the distribution of the mean has mean μ and standard deviation $S/\sqrt{n}$. The statistic t has a t-distribution with $(n - 1)$ degrees of freedom:

$$t = \frac{\bar{X} - \mu}{S/\sqrt{n}} \tag{A.180}$$

If two independent samples of sizes n_1 and n_2 are drawn from populations with means μ_1 and μ_2 and variances σ_1^2 and σ_2^2, respectively, then the difference of the sample means $X_1 - X_2$ has a sampling distribution that is approximately normal with a mean $\mu_1 - \mu_2$ and variance $(\sigma_1^2/n_1 + \sigma_2^2/n_2)$. Thus, the statistic Z has a standard normal distribution:

$$Z = \frac{(\bar{X}_1 - \bar{X}_2) - (\mu_1 - \mu_2)}{\left(\dfrac{\sigma_1^2}{n_1} + \dfrac{\sigma_2^2}{n_2}\right)^{0.5}} \tag{A.181}$$

If the population means and variances are equal, then the Z statistic of Equation A.181 is

$$Z = \frac{(\bar{X}_1 - \bar{X}_2)}{\sigma\left(\dfrac{1}{n_1} + \dfrac{1}{n_2}\right)^{0.5}} \tag{A.182}$$

Equations A.179 through A.182 can be used to test hypotheses about the means and to form confidence intervals.

A.12.2 Sampling Distribution of the Variance

The estimated variance of a sample is a random variable, and so it has a distribution. The distribution depends on the characteristics of the underlying population from which the sample is derived. If the population is normal, then it can be shown that for the unbiased estimate of the variance, S^2, the quantity $(n - 1)S^2/\sigma^2$ is a random variable distributed as chi-square (χ^2, also C in previous sections) with $(n - 1)$ degrees of freedom. Thus, inferences about the variance of a single normally distributed population are made with

$$\chi^2 = \frac{(n-1)S^2}{\sigma^2} \tag{A.183}$$

The chi-square statistic of Equation A.183 can be used to test hypotheses about the variance of a single random variable and to form confidence intervals.

A.12.3 Sampling Distributions for Other Parameters

Any estimated quantity using a sample can be treated as a random variable, and so it has a distribution. The distribution depends on the characteristics of the underlying population from which the sample is derived. For example, the estimated correlation coefficient and the estimated parameters (or coefficients) in the regression models are treated as random variables; therefore, they are random variables and have probability distributions.

A.13 Hypothesis Testing for Means

Hypothesis testing is the formal procedure for using statistical concepts and measures in performing decision making. The following six steps can be used to make a statistical analysis of a hypothesis:

Step 1. Formulate hypotheses.

Step 2. Select the appropriate statistical model (theorem) that identifies the test statistic.

Step 3. Specify the level of significance, which is a measure of risk.

Step 4. Collect a sample of data and compute an estimate of the test statistic.

Step 5. Define the region of rejection for the test statistic.

Step 6. Select the appropriate hypothesis.

These six steps are discussed in detail in the following sections.

A.13.1 Test of the Mean with Known Population Variance

When the standard deviation of the population is known, the procedure for testing the mean is as follows:

Step 1: Formulate hypotheses. The null and alternative hypotheses must be stated in terms of the population parameter μ and the value selected for comparison, which may be denoted as μ_0. The null hypothesis should state that the mean of the population equals a preselected standard value. Acceptance of the null hypothesis implies that it is not significantly different from μ_0. Mathematically, the null hypothesis could be stated as

$$H_0 : \mu = \mu_0 \tag{A.184}$$

One of three alternative hypotheses may be selected:

$$H_{A1} : \mu < \mu_0 \quad \text{one-tailed test} \tag{A.185a}$$

$$H_{A2} : \mu > \mu_0 \quad \text{one-tailed test} \tag{A.185b}$$

$$H_{A3} : \mu \neq \mu_0 \quad \text{two-tailed test} \tag{A.185c}$$

Each of the alternative hypotheses indicates that a significant difference exists between the population mean and the standard value. The selected alternative hypothesis depends on the statement of the problem.

Step 2: Select the appropriate model. The mean, X, of a random sample is used in testing hypotheses about the population mean μ; X is itself a random variable. If the population from which the random sample is drawn has mean μ and variance σ^2, the distribution of random variable X has mean μ and variance σ^2/n for samples from infinite populations. For samples from finite populations of size N, the variance is $[\sigma^2(N - n)]/[n(N - 1)]$.

For a random sample of size n, the sample mean, X, can be used in calculating the value of test statistic z as

$$z = \frac{\bar{X} - \mu}{\sigma/\sqrt{n}} \tag{A.186}$$

where:

z is the value of a random variable whose distribution function is a standard normal

Step 3: Select the level of significance. A level of significance (α) represents the conditional probability of making an error in decision (i.e., accepting H_0 while H_0 is not true). A value of 1% can be selected for demonstration of this hypothesis test; however, in actual practice, the level selected for use should vary with the problem being studied and the impact of making an incorrect decision.

Step 4: Compute the estimate of the test statistic. A random sample consisting of 100 specimens is selected, with a computed mean of 3190 kgf. The standard deviation of the population is 160 kgf. The value of the test statistic of Equation A.186 to test for a population value of 3250 kgf is

$$z = \frac{3190 - 3250}{160/\sqrt{100}} = -3.750$$

Step 5: Define the region of rejection. For the standard normal distribution, the level of significance is the only characteristic required to determine the critical value of the test statistic. The region of rejection depends on the statement of the alternative hypothesis:

If H_A is	Then reject H_0 if
$\mu < \mu_0$	$z < -z_\alpha$
$\mu > \mu_0$	$z > z_\alpha$
$\mu \neq \mu_0$	$z < -z_{\alpha/2}$ or $z > z_{\alpha/2}$

(A.187)

Assuming a one-tailed alternative hypothesis, the critical value of z for a 1% level of significance (α) can be obtained from probability tables as

$$-z_\alpha = -\Phi^{-1}(1 - \alpha) = -2.326 \quad \text{(A.188)}$$

Thus, the region of rejection consists of all values of z less than −2.326.

Step 6: Select the appropriate hypothesis. If the computed statistic lies in the region of rejection, the null hypothesis must be rejected.

The decision criterion specified in step 3 was limited to the specification of the level of significance. If the null hypothesis was rejected for a 1% level of significance, there is a 1% chance of making a type I error; that is, there is a chance of 1 in 100 of rejection when, in fact, it is adequate. The decision criterion of step 3 did not discuss the possibility of a type II error (β). The result of a type II error would be the acceptance when in fact it is inadequate. It is common that the consequences of a type II error are probably more severe than those of a type I error. However, it is easier and more direct to specify a value for α than to specify a value for β. Error types I and II are also called manufacturer's and consumer's risks, respectively.

A.13.2 Test of the Mean with Unknown Population Variance

When the population variance is unknown, the theorem used in the preceding section is not applicable, even though the null and alternative hypotheses and the steps are the same. In such cases, a different theorem is used for testing a hypothesis about a mean. Specifically, for a random sample of size n, sample mean X and standard deviation S can be used in calculating the value of test statistic t:

$$t = \frac{\bar{X} - \mu}{S/\sqrt{n}} \quad \text{(A.189)}$$

Test statistic t is the value of a random variable having the Student's t-distribution with $\nu = n - 1$ degrees of freedom. This statistic requires that the sample be drawn from a normal population. The region of rejection depends on the level of significance, the degrees of freedom, and the statement of the alternative hypothesis:

TABLE A.3
Summary of Hypothesis Tests

H_0	Test Statistic	H_A	Region of Rejection
$\mu = \mu_0$ (σ known)	$Z = \dfrac{\bar{X} - \mu}{\sigma/n}$	$\mu < \mu_0$ $\mu > \mu_0$ $\mu \neq \mu_0$	$z < -z_\alpha$ $z > z_\alpha$ $z < -z_{\alpha/2}$ or $z > z_{\alpha/2}$
$\mu = \mu_0$ (σ unknown)	$t = \dfrac{\bar{X} - \mu}{S/n}$ $\nu = n - 1$	$\mu < \mu_0$ $\mu > \mu_0$ $\mu \neq \mu_0$	$t < -t_{\alpha,\nu}$ $t > t_{\alpha,\nu}$ $t < -t_{\alpha/2,\nu}$ or $t > t_{\alpha/2,\nu}$
$\mu_1 = \mu_2$ ($\sigma_1^2 = \sigma_2^2$, but unknown)	$t = \dfrac{\bar{X}_1 - \bar{X}_2}{S_p\left(\dfrac{1}{n_1} + \dfrac{1}{n_2}\right)^{0.5}}$ $\nu = n_1 + n_2 - 2$ $S_p^2 = \dfrac{(n_1 - 1)S_1^2 + (n_2 - 1)S_2^2}{n_1 + n_2 - 2}$	$\mu < \mu_0$ $\mu > \mu_0$ $\mu \neq \mu_0$	$t < -t_{\alpha,\nu}$ $t > t_{\alpha,\nu}$ $t < -t_{\alpha/2,\nu}$ or $t > t_{\alpha/2,\nu}$
$\sigma^2 = \sigma_0^2$	$\chi^2 = \dfrac{(n - 1)S^2}{\sigma_0^2}$ $\nu = n - 1$	$\sigma^2 < \sigma_0^2$ $\sigma^2 > \sigma_0^2$ $\sigma^2 \neq \sigma_0^2$	$\chi^2 < \chi^2_{\alpha-1,\nu}$ $\chi^2 > \chi^2_{\alpha,\nu}$ $\chi^2 < \chi^2_{1-\alpha/2,\nu}$ or $\chi^2 > \chi^2_{\alpha/2,\nu}$
$\sigma_1^2 = \sigma_2^2$ (assuming $\sigma_1^2 > \sigma_2^2$)	$F = \dfrac{S_1^2}{S_2^2}$ $\nu_1 = n_1 - 1$ $\nu_2 = n_2 - 1$	$\sigma_1^2 \neq \sigma_2^2$	$F > F_{\alpha/2,\nu_1,\nu_2}$

If H_A is	Then reject H_0 if
$\mu < \mu_0$	$t < -t_{\alpha,\nu}$
$\mu > \mu_0$	$t > t_{\alpha,\nu}$
$\mu \neq \mu_0$	$t < -t_{\alpha/2,\nu}$ or $t > t_{\alpha/2,\nu}$

(A.190)

A.13.3 Summary

Two hypothesis tests were introduced. Each test can be conducted using the six steps that are provided at the beginning of this section. In applying a hypothesis test, the important ingredients are the test statistic, the level of significance, the degrees of freedom, and the critical value of a test statistic. Table A.3 includes a convenient summary of statistical tests introduced in this section and other important tests.

A.14 Hypothesis Testing of Variances

The variance of a random sample is a measure of the dispersion of the observations about the sample mean. Although the variance is used to indicate the degree of variation about the mean, it is an important statistic in its own right. Large variation in engineering systems reflects instability or nonuniformity, both of which can be considered not to be optimal in some applications.

A.14.1 One-Sample Chi-Square Test

Consider, for example, the case of water distribution systems used for irrigation. They should be designed to distribute water uniformly over an area, such as a lawn or an agricultural field. Failure to provide a uniform application of water over the area may lead to nonoptimum grass or crop output; thus, equipment that does not apply water uniformly would probably not be purchased. A company that manufactures irrigation distribution systems wishes to determine whether a new system increases the uniformity of water application in comparison with existing models. The variance of depths of water measured at different locations in a field would serve as a measure of uniformity of water application. The following procedure is used to test for a statistical difference in the uniformity of application rates (i.e., a test of the variance of a random variable).

Step 1: Formulate hypotheses. To investigate the possibility of a significant difference existing between the variance of a population, σ^2, and the preselected standard variance value, σ_0^2, the following null hypothesis can be used:

$$H_0 : \sigma^2 = \sigma_0^2 \tag{A.191}$$

The null hypothesis can be tested against either a one-tailed or a two-tailed alternative hypothesis as follows:

$$H_{A1} : \sigma^2 < \sigma_0^2 \tag{A.192a}$$

$$H_{A2} : \sigma^2 > \sigma_0^2 \tag{A.192b}$$

$$H_{A3} : \sigma^2 \neq \sigma_0^2 \tag{A.192c}$$

Step 2: Select the appropriate model. The variance, S^2, of a random sample is a random variable itself and is used in testing the hypotheses about the variance of a population, σ^2. The sampling distribution of the estimated variance of a random sample that is drawn from a normal population has a chi-square distribution. The test statistic for testing the hypotheses is

$$\chi^2 = \frac{(n-1)S^2}{\sigma_0^2} \tag{A.193}$$

where:

χ^2 is the value of a random variable that has a chi-square distribution with $\nu = n - 1$ degrees of freedom

n is the sample size used in computing sample variance S^2

Step 3: Select the level of significance. For example, a level of significance (α) of 2.5% can be selected.

Step 4: Compute estimate of test statistic. To test the uniformity of application of water for the new irrigation system, the amount of water in each of 25 randomly placed recording devices was observed after 1 hour. The mean and standard deviation of the random sample were 0.31 and 0.063 cm/hour, respectively. The computed test statistic for a target value of 0.1^2 is

$$\chi^2 = \frac{(25-1)(0.063)^2}{(0.1)^2} = 9.526$$

Step 5: Define the region of rejection. The region of rejection for a test statistic having a chi-square distribution is a function of the level of significance, the statement of the alternative hypotheses, and the degrees of freedom. The regions of rejection for the alternative hypotheses are as follows:

If H_A is	Then reject H_0 if
$H_{A1}: \sigma^2 < \sigma_0^2$	$\chi^2 < \chi^2_{\alpha-1,\nu}$
$H_{A2}: \sigma^2 > \sigma_0^2$	$\chi^2 > \chi^2_{\alpha,\nu}$
$H_{A3}: \sigma^2 \neq \sigma_0^2$	$\chi^2 < \chi^2_{1-\alpha/2,\nu}$ or $\chi^2 > \chi^2_{\alpha/2,\nu}$

(A.194)

Step 6: Select the appropriate hypothesis. If the computed value of the test statistic is less than the critical value, the null hypothesis must be rejected.

A.14.2 Two-Sample *F* Test

For comparing the variances of two random samples, several strategies have been recommended, with each strategy being valid when the underlying assumptions hold. One of these strategies is presented here.

For a two-tailed test, an F ratio is formed as the ratio of the larger sample variance to the smaller sample variance as follows:

$$F = \frac{S_1^2}{S_2^2} \tag{A.195}$$

with $\nu_1 = n_1 - 1$ degrees of freedom for the numerator and $\nu_2 = n_2 - 1$ degrees of freedom for the denominator, where n_1 and n_2 are the sample sizes for the samples used to compute S_1^2 and S_2^2, respectively. The computed F is compared with the tabulated values for the F probability distribution tabulated in the textbooks (e.g., Ayyub and McCuen 2003), and the null hypothesis of equal variances ($H_0 : \sigma_1^2 = \sigma_2^2$) is accepted if the computed F is less than the tabulated F value for $k = \nu_1$, $u = \nu_2$, and α. If the computed F is greater than the tabulated F value, then the null hypothesis is rejected in favor of the alternative hypothesis ($H_A : \sigma_1^2 \neq \sigma_2^2$). An important note for this two-tailed test is that the level of significance is twice the value from which the tabulated F value was obtained; for example, if the 5% F table is used to obtain the critical F statistic, then the decision to accept or reject the null hypothesis is being made at a 10% level of significance. This is the price paid for using the sample knowledge that one sample has the larger variance.

For a one-tailed test, it is necessary to specify which of the two samples is expected to have the larger population variance. This must be specified prior to collecting the data. The computed F statistic is the ratio of the sample variance of the group expected to have the larger population variance to the sample variance from the second group. If it turns out that the sample variance of the group expected to have the larger variance is smaller than that of the group expected to have the smaller variance, then the computed F statistic will be <1. For a test with a level of significance equal to that shown on the table, the null hypothesis is rejected if the computed F is greater than the critical F. Because the direction is specified, the null hypothesis is accepted when the computed F is less than the critical F; the null hypothesis is rejected when the computed F is greater than the critical F.

A.14.3 Summary

Two hypothesis tests for the variance were introduced, and Table A.3 includes a summary of these tests.

A.15 Confidence Intervals

From a sample, we obtain single-valued estimates such as the mean, the variance, a correlation coefficient, or a regression coefficient. These single-valued estimates represent our best estimate of the population values, but they are the only estimates of random variables, and we know that they probably do not equal the corresponding true values. Thus, we should be interested in the accuracy of these sample estimates.

If we are only interested in whether an estimate of a random variable is significantly different from a standard of comparison, we can use a hypothesis test. However, the hypothesis test gives only a "yes" or "no" answer and not a statement of the accuracy of an estimate of a random variable, which may be the object of our attention.

Confidence intervals represent a means of providing a range of values in which the true value can be expected to lie. Confidence intervals have the additional advantage, compared with hypothesis tests, of providing a probability statement about the likelihood of correctness.

A.15.1 Confidence Interval for the Mean

The same theorems that were used for testing hypotheses on the mean are used in computing confidence intervals. In testing a hypothesis for the mean, the choice of test statistic depends on whether the standard deviation of the population, σ, is known, which is also true in computing confidence intervals. The theorem for the case where σ is known specifies a Z statistic, whereas the t statistic is used when σ is unknown; the theorems are not repeated here.

For the case where σ is known, the confidence intervals on the population mean are given by

$$\bar{X} - Z_{\alpha/2}\left(\frac{\sigma}{\sqrt{n}}\right) \le \mu \le \bar{X} + Z_{\alpha/2}\left(\frac{\sigma}{\sqrt{n}}\right) \quad \text{two-sided interval} \tag{A.196}$$

$$\bar{X} - Z_{\alpha}\left(\frac{\sigma}{\sqrt{n}}\right) \le \mu \le \infty \quad \text{lower one-sided interval} \tag{A.197}$$

$$-\infty \le \mu \le \bar{X} + Z_{\alpha}\left(\frac{\sigma}{\sqrt{n}}\right) \quad \text{upper one-sided interval} \tag{A.198}$$

where:

X is the sample mean
n is the sample size
Z_α and $Z_{\alpha/2}$ are the values of random variables having the standard normal distribution and cutting off $(1 - \alpha)$ or $(1 - \alpha/2)$ in the tail of the distribution, respectively
α is the level of significance

The confidence interval provides an interval in which we are 100(1 − α)% confident that the population value lies within the interval. The measure of dispersion is given by $\sigma/\sqrt{n}$, as $\sigma/\sqrt{n}$ is the standard error of the mean. Equation A.196 is a two-sided confidence interval, whereas Equations A.197 and A.198 are one-sided. Equation A.197 gives a lower confidence limit, with no limit on the upper side of the mean; similarly, Equation A.198 gives an upper limit, with no lower limit.

For the case where σ is unknown, the confidence intervals on the population mean are given by

$$\bar{X} - t_{\alpha/2,v}\left(\frac{S}{\sqrt{n}}\right) \le \mu \le \bar{X} + t_{\alpha/2,v}\left(\frac{S}{\sqrt{n}}\right) \quad \text{two-sided interval} \tag{A.199}$$

$$\bar{X} - t_{\alpha,v}\left(\frac{S}{\sqrt{n}}\right) \le \mu \le \infty \quad \text{lower one-sided interval} \tag{A.200}$$

$$-\infty \le \mu \le \bar{X} + t_{\alpha,v}\left(\frac{S}{\sqrt{n}}\right) \quad \text{upper one-sided interval} \tag{A.201}$$

where:

S is the sample standard deviation

$t_{\alpha,v}$ and $t_{\alpha/2,v}$ are the values of random variables having a t distribution with $v = n - 1$ degrees of freedom

The significance level (α) is used for one-sided confidence interval and α/2 is used for a two-sided confidence interval.

A.15.2 Confidence Interval for the Variance

The confidence interval on the population variance (σ^2) can be computed using the same theorem that was used in testing a hypothesis for the variance. The two-sided and one-sided confidence intervals are

$$\frac{(n-1)S^2}{\chi^2_{\alpha/2,v}} \le \sigma^2 \le \frac{(n-1)S^2}{\chi^2_{1-\alpha/2,v}} \quad \text{two-sided interval} \tag{A.202}$$

$$\frac{(n-1)S^2}{\chi^2_{\alpha,v}} \le \sigma^2 \le \infty \quad \text{lower one-sided interval} \tag{A.203}$$

$$0 \le \sigma^2 \le \frac{(n-1)S^2}{\chi^2_{1-\alpha,v}} \quad \text{upper one-sided interval} \tag{A.204}$$

where:

$\chi^2_{\alpha/2,v}$ and $\chi^2_{\alpha,v}$ are the values of a random variable having a chi-square distribution that cuts α/2 and α percent of the right tail of the distribution, respectively

$\chi^2_{1-\alpha/2,v}$ and $\chi^2_{1-\alpha,v}$ are the values of a random variable having a chi-square distribution that cuts at 1 − α/2 and 1 − α, respectively

The confidence interval provides an interval in which we are 100(1 − α)% confident that the population value lies within the interval.

Appendix B

Failure Data

This appendix provides failure data and their sources for the purpose of helping and guiding readers.

Component or Item	Failure Mode	Units	Point Estimate or Suggested Mean	Range (Low)	Range (High)	Calculated 5% Lower Limit	Calculated 95% Upper Limit	References
AC bus hardware	Failure	Hourly failure rate	1.00E−07	1.00E−08	4.00E−06	2.00E−08	5.00E−07	Modarres (1993)
Accelerometer	–	Failures per million hours	–	10	30	–	–	Smith (2001)
Accumulator	–	Hourly failure rate	5.00E−04	–	–	–	–	Anderson and Neri (1990)
Actuator	–	Hourly failure rate	–	3.00E−07	4.05E−04	–	–	Anderson and Neri (1990)
Air compressor	–	Failures per million hours	–	70	250	–	–	Smith (2001)
Air-operated valves	Failure to operate	Daily failure rate	2.00E−03	3.00E−04	2.00E−02	6.67E−04	6.00E−03	Modarres (1993)
Air-operated valves	Failure due to plugging	Daily failure rate	–	2.00E−05	1.00E−04	–	–	Modarres (1993)
Air-operated valves	Failure due to plugging	Annual failure rate	1.00E−07	–	1.00E−07	3.33E−08	3.00E−07	Modarres (1993)
Air-operated valves	Unavailability due to test and maintenance	Daily failure rate	8.00E−04	6.00E−05	6.00E−03	8.00E−05	8.00E−03	Modarres (1993)
Air-operated valves	Spurious closure	Hourly failure rate	1.00E−07	–	–	3.33E−08	3.00E−07	Modarres (1993)
Air-operated valves	Spurious open	Hourly failure rate	5.00E−07	–	–	5.00E−08	5.00E−06	Modarres (1993)
Air supply (instrument)	–	Failures per million hours	6	5	10	–	–	Smith (2001)
Alarm bell	–	Failures per million hours	–	2	10	–	–	Smith (2001)
Alarm circuit (panel)	–	Failures per million hours	–	45	–	–	–	Smith (2001)
Alarm circuit (simple)	–	Failures per million hours	–	4	–	–	–	Smith (2001)
Alarm siren	–	Failures per million hours	6	1	20	–	–	Smith (2001)
Alternator	–	Failures per million hours	–	1	9	–	–	Smith (2001)
Analyzer, Bourdon/Geiger	–	Failures per million hours	–	5	–	–	–	Smith (2001)
Analyzer, carbon dioxide	–	Failures per million hours	–	100	500	–	–	Smith (2001)
Analyzer, conductivity	–	Failures per million hours	1500	500	2000	–	–	Smith (2001)
Analyzer, dewpoint	–	Failures per million hours	–	100	200	–	–	Smith (2001)
Analyzer, Geiger	–	Failures per million hours	–	15	–	–	–	Smith (2001)

(Continued)

Component or Item	Failure Mode	Units	Point Estimate or Suggested Mean	Range (Low)	Range (High)	Calculated 5% Lower Limit	Calculated 95% Upper Limit	References
Analyzer, hydrogen sulfide	–	Failures per million hours	–	100	200	–	–	Smith (2001)
Analyzer, hydrogen	–	Failures per million hours	–	400	100	–	–	Smith (2001)
Analyzer, oxygen	–	Failures per million hours	60	50	200	–	–	Smith (2001)
Analyzer, pH	–	Failures per million hours	–	650	–	–	–	Smith (2001)
Analyzer, scintillation	–	Failures per million hours	–	20	–	–	–	Smith (2001)
Antenna	–	Failures per million hours	–	1	5	–	–	Smith (2001)
Attenuator	–	Failures per million hours	–	0.01	–	–	–	Smith (2001)
Avionics	–	Hourly failure rate	–	5.00E−04	1.00E−03	–	–	Anderson and Neri (1990)
Battery	–	Hourly failure rate	6.77E−04	–	–	–	–	Anderson and Neri (1990)
Battery	Unavailability due to test and maintenance	Daily failure rate	1.00E−03	–	–	1.00E−04	1.00E−02	Modarres (1993)
Battery charger (motor generator)	–	Failures per million hours	–	100	–	–	–	Smith (2001)
Battery charger (simple rectifier)	–	Failures per million hours	–	2	–	–	–	Smith (2001)
Battery charger (stabilized/float)	–	Failures per million hours	–	10	–	–	–	Smith (2001)
Battery, dry primary	–	Failures per million hours	–	1	30	–	–	Smith (2001)
Battery, lead	–	Failures per million hours	–	3	–	–	–	Smith (2001)
Battery, lead acid	–	Failures per million hours	1	0.5	3	–	–	Smith (2001)
Battery, lead acid (vehicle) per million miles	–	Failures per million hours	–	30	–	–	–	Smith (2001)
Battery, Ni-Cd/Ag-Zn	–	Failures per million hours	1	0.2	3	–	–	Smith (2001)
Bearing	–	Hourly failure rate	–	1.26E−05	5.32E−05	–	–	Anderson and Neri (1990)
Bearings, ball, heavy	–	Failures per million hours	–	2	20	–	–	Smith (2001)
Bearings, ball, light	–	Failures per million hours	1	0.1	10	–	–	Smith (2001)
Bearings, brush	–	Failures per million hours	–	0.5	–	–	–	Smith (2001)
Bearings, bush	–	Failures per million hours	–	0.05	0.1	–	–	Smith (2001)
Bearings, jewel	–	Failures per million hours	–	0.4	–	–	–	Smith (2001)
Bearings, roller	–	Failures per million hours	–	0.3	5	–	–	Smith (2001)
Bearings, sleeve	–	Failures per million hours	–	0.5	5	–	–	Smith (2001)
Bellows, simple expandable	–	Failures per million hours	5	2	10	–	–	Smith (2001)

(Continued)

Component or Item	Failure Mode	Units	Point Estimate or Suggested Mean	Range (Low)	Range (High)	Calculated 5% Lower Limit	Calculated 95% Upper Limit	References
Belts	–	Failures per million hours	–	4	50	–	–	Smith (2001)
Brake (magnetic)	–	Hourly failure rate	2.42E–04	–	–	–	–	Anderson and Neri (1990)
Busbars, 11 kV	–	Failures per million hours	–	0.02	0.2	–	–	Smith (2001)
Busbars, –3.3 kV	–	Failures per million hours	–	0.05	2	–	–	Smith (2001)
Busbars, –415 V	–	Failures per million hours	–	0.6	2	–	–	Smith (2001)
Capacitors, aluminum (general)	–	Failures per million hours	–	0.3	–	–	–	Smith (2001)
Capacitors, ceramic	–	Failures per million hours	0.1	0.0005	–	–	–	Smith (2001)
Capacitors, glass	–	Failures per million hours	–	0.002	–	–	–	Smith (2001)
Capacitors, mica	–	Failures per million hours	0.03	0.002	0.1	–	–	Smith (2001)
Capacitors, paper	–	Failures per million hours	0.15	0.001	–	–	–	Smith (2001)
Capacitors, plastic	–	Failures per million hours	0.01	0.001	0.05	–	–	Smith (2001)
Capacitors, Tantalum non-solderable	–	Failures per million hours	0.01	0.001	0.1	–	–	Smith (2001)
Capacitors, Tantalum Solderable	–	Failures per million hours	0.1	0.005	–	–	–	Smith (2001)
Capacitors, variable	–	Failures per million hours	0.1	0.005	2	–	–	Smith (2001)
Card reader	–	Failures per million hours	–	150	4000	–	–	Smith (2001)
Check valve	Failure to open	Daily failure rate	1.00E–04	6.00E–05	1.20E–04	3.33E–05	3.00E–04	Modarres (1993)
Check valve	Failure to close	Hourly failure rate	1.00E–03	–	–	3.33E–04	3.00E–03	Modarres (1993)
Circuit breaker	Spurious open	Hourly failure rate	1.00E–06	–	–	3.33E–07	3.00E–06	Modarres (1993)
Circuit breaker	Fail to transfer	Daily failure rate	3.00E–03	–	–	3.00E–04	3.00E–02	Modarres (1993)
Circuit breaker, >3 kV	–	Failures per million hours	–	0.5	2	–	–	Smith (2001)
Circuit breaker, <600 VA	–	Failures per million hours	–	0.5	1.5	–	–	Smith (2001)
Circuit breaker, >100 kV	–	Failures per million hours	–	3	10	–	–	Smith (2001)
Circuit protection device	–	Hourly failure rate	2.85E–05	–	–	–	–	Anderson and Neri (1990)
Clutch, friction	–	Failures per million hours	–	0.5	3	–	–	Smith (2001)
Clutch, magnetic	–	Failures per million hours	–	2.5	6	–	–	Smith (2001)
Compressor, centrifugal, turbine-driven	–	Failures per million hours	–	150	–	–	–	Smith (2001)
Compressor, electric motor-driven	–	Failures per million hours	–	100	300	–	–	Smith (2001)
Compressor, reciprocating, turbine-driven	–	Failures per million hours	–	500	–	–	–	Smith (2001)
Computer, mainframe	–	Failures per million hours	–	4000	8000	–	–	Smith (2001)

(Continued)

Component or Item	Failure Mode	Units	Point Estimate or Suggested Mean	Range (Low)	Range (High)	Calculated 5% Lower Limit	Calculated 95% Upper Limit	References
Computer, micro (CPU)	–	Failures per million hours	–	30	100	–	–	Smith (2001)
Computer, mini	–	Failures per million hours	200	100	500	–	–	Smith (2001)
Computer, programmable logic controller	–	Failures per million hours	–	20	50	–	–	Smith (2001)
Connection, flow solder	–	Failures per million hours	–	0.0003	0.001	–	–	Smith (2001)
Connections, crimped	–	Failures per million hours	–	0.0003	0.007	–	–	Smith (2001)
Connections, hand solder	–	Failures per million hours	–	0.0002	0.003	–	–	Smith (2001)
Connections, plate	–	Failures per million hours	–	0.0003	–	–	–	Smith (2001)
Connections, power cable	–	Failures per million hours	–	0.05	0.4	–	–	Smith (2001)
Connections, weld	–	Failures per million hours	–	0.002	–	–	–	Smith (2001)
Connections, wrapped	–	Failures per million hours	–	0.00003	0.001	–	–	Smith (2001)
Connectors, coaxial	–	Failures per million hours	–	0.02	0.2	–	–	Smith (2001)
Connectors, dual in-line package (DIL)	–	Failures per million hours	–	0.001	–	–	–	Smith (2001)
Connectors, Personal Computer Board (PCB)	–	Failures per million hours	–	0.0003	0.1	–	–	Smith (2001)
Connectors, pin	–	Failures per million hours	–	0.001	0.1	–	–	Smith (2001)
Connectors, pneumatic	–	Failures per million hours	–	1	–	–	–	Smith (2001)
Connectors, coaxial	–	Failures per million hours	–	0.05	–	–	–	Smith (2001)
Control/instrument (gauge)	–	Hourly failure rate	–	3.75E–05	2.70E–04	–	–	Anderson and Neri (1990)
Cooling coil	Failure to operate	Hourly failure rate	1.00E–06	–	–	3.33E–07	3.00E–06	Modarres (1993)
Cooling tower fan	Failure to start	Daily failure rate	4.00E–03	–	–	1.33E–03	1.20E–02	Modarres (1993)
Cooling tower fan	Failure to run	Hourly Failure Rate (HR)	7.00E–06	–	–	7.00E–07	7.00E–05	Modarres (1993)
Cooling tower fan	Unavailability due to test and maintenance	Daily failure rate	2.00E–03	–	–	2.00E–04	2.00E–02	Modarres (1993)
Counter (mechanical)	–	Failures per million hours	2	0.2	–	–	–	Smith (2001)
Crystal, quartz	–	Failures per million hours	0.1	0.02	0.2	–	–	Smith (2001)
Damper	Failure to open	Daily failure rate	3.00E–03	–	–	3.00E–04	3.00E–02	Modarres (1993)
DC battery	Hardware failure	Hourly failure rate	1.00E–06	–	–	3.33E–07	3.00E–06	Modarres (1993)
DC bus	Hardware failure	Hourly failure rate	1.00E–07	–	–	2.00E–08	5.00E–07	Modarres (1993)
DC bus	Unavailability due to test and maintenance	Hourly failure rate	8.00E–06	–	–	8.00E–07	8.00E–05	Modarres (1993)
DC charger	Hardware failure	Hourly failure rate	1.00E–06	–	–	3.33E–07	3.00E–06	Modarres (1993)
DC charger	Unavailability due to test and maintenance	Daily failure rate	1.00E–06	–	–	1.00E–07	1.00E–05	Modarres (1993)

(Continued)

Component or Item	Failure Mode	Units	Point Estimate or Suggested Mean	Range (Low)	Range (High)	Calculated 5% Lower Limit	Calculated 95% Upper Limit	References
DC inverter	Hardware failure	Hourly failure rate	1.00E−04	–	–	3.33E−05	3.00E−04	Modarres (1993)
DC inverter	Unavailability due to test and maintenance	Daily failure rate	1.00E−03	–	–	1.00E−04	1.00E−02	Modarres (1993)
Detectors, fire, wire/rod	–	Failures per million hours	–	10	–	–	–	Smith (2001)
Detectors, gas, pellistor	–	Failures per million hours	–	3	8	–	–	Smith (2001)
Detectors, smoke, ionization	–	Failures per million hours	–	2	6	–	–	Smith (2001)
Detectors, temperature level	–	Failures per million hours	2	0.2	8	–	–	Smith (2001)
Detectors, ultraviolet	–	Failures per million hours	–	5	15	–	–	Smith (2001)
Detectors, rate of rise (temperature)	–	Failures per million hours	–	3	9	–	–	Smith (2001)
Diesel-driven pump	Failure to start	Daily failure rate	3.00E−02	1.00E−03	1.00E−02	1.00E−02	9.00E−02	Modarres (1993)
Diesel-driven pump	Failure to run	Hourly failure rate	8.00E−04	2.00E−05	1.00E−03	8.00E−05	8.00E−03	Modarres (1993)
Diesel-driven pump	Unavailability due to test and maintenance	Daily failure rate	1.00E−02	–	–	1.00E−03	1.00E−01	Modarres (1993)
Diesel engine	–	Failures per million hours	6000	300	–	–	–	Smith (2001)
Diesel generator	Failure to start	Daily failure rate	3.00E−02	8.00E−03	1.00E−03	1.00E−02	9.00E−02	Modarres (1993)
Diesel generator	Failure to run	Hourly failure rate	2.00E−03	2.00E−04	3.00E−03	2.00E−04	2.00E−02	Modarres (1993)
Diesel generator	Unavailability due to test and maintenance	Daily failure rate	6.00E−03	−1	4.00E−02	6.00E−04	6.00E−02	Modarres (1993)
Diesel generator	–	Failures per million hours	–	125	4000	–	–	Smith (2001)
Diodes, Si-controlled rectifier (thyristor)	–	Failures per million hours	–	0.01	0.5	–	–	Smith (2001)
Diodes, Si, high power	–	Failures per million hours	0.2	0.1	–	–	–	Smith (2001)
Diodes, Si, low power	–	Failures per million hours	0.04	0.01	0.1	–	–	Smith (2001)
Diodes, varactor	–	Failures per million hours	–	0.06	0.3	–	–	Smith (2001)
Diodes, zener	–	Failures per million hours	0.03	0.005	0.1	–	–	Smith (2001)
Disk memory	–	Failures per million hours	500	100	2000	–	–	Smith (2001)
Electricity supply	–	Failures per million hours	–	100	–	–	–	Smith (2001)
Electropneumatic converter (I/P)	–	Failures per million hours	–	2	4	–	–	Smith (2001)
Explosive-operated valve	Failure to operate	Daily failure rate	3.00E−03	1.00E−03	9.00E−03	1.00E−03	9.00E−03	Modarres (1993)
Explosive-operated valve	Failure due to plugging	Daily failure rate	–	2.00E−05	1.00E−04	–	–	Modarres (1993)
Explosive-operated valve	Failure due to plugging	Annual failure rate	1.00E−07	–	1.00E−07	3.33E−08	3.00E−07	Modarres (1993)
Explosive-operated valve	Unavailability due to test and maintenance	Daily failure rate	8.00E−04	6.00E−05	6.00E−03	8.00E−05	8.00E−03	Modarres (1993)
Fan	–	Hourly failure rate	9.10E−06	–	–	–	–	Anderson and Neri (1990)

(Continued)

Component or Item	Failure Mode	Units	Point Estimate or Suggested Mean	Range (Low)	Range (High)	Calculated 5% Lower Limit	Calculated 95% Upper Limit	References
Fan	–	Failures per million hours	–	2	50	–	–	Smith (2001)
Fiber optics, cable per km	–	Failures per million hours	–	0.1	–	–	–	Smith (2001)
Fiber optics, connector	–	Failures per million hours	–	0.1	–	–	–	Smith (2001)
Fiber optics, laser	–	Failures per million hours	–	0.3	0.5	–	–	Smith (2001)
Fiber optics, LED	–	Failures per million hours	–	0.2	0.5	–	–	Smith (2001)
Fiber optics, optocoupler	–	Failures per million hours	–	0.02	0.1	–	–	Smith (2001)
Fiber optics, pin-avalanched photodiode	–	Failures per million hours	–	0.02	–	–	–	Smith (2001)
Fiber optics, Si-avalanched photodiode	–	Failures per million hours	–	0.2	–	–	–	Smith (2001)
Filter	–	Hourly failure rate	–	2.60E−05	4.96E−05	–	–	Anderson and Neri (1990)
Filter, blocked	–	Failures per million hours	1	0.5	10	–	–	Smith (2001)
Filter, leak	–	Failures per million hours	1	0.5	10	–	–	Smith (2001)
Fire sprinkler, non-operation	–	Failures per million hours	0.02	–	–	–	–	Smith (2001)
Fire sprinkler, spurious	–	Failures per million hours	0.1	0.05	0.5	–	–	Smith (2001)
Flow controller	Failure to operate	Daily failure rate	1.00E−04	–	–	3.33E−05	3.00E−04	Modarres (1993)
Flow instruments, controller	–	Failures per million hours	–	25	50	–	–	Smith (2001)
Flow instruments, DP sensor	–	Failures per million hours	–	80	200	–	–	Smith (2001)
Flow instruments, rotary meter	–	Failures per million hours	15	5	–	–	–	Smith (2001)
Flow instruments, switch	–	Failures per million hours	–	4	40	–	–	Smith (2001)
Flow instruments, transmitter	–	Failures per million hours	5	1	20	–	–	Smith (2001)
Fuse	–	Failures per million hours	–	0.02	0.5	–	–	Smith (2001)
Gasket/seal	–	Hourly failure rate	–	2.40E−06	3.16E−05	–	–	Anderson and Neri (1990)
Gaskets	–	Failures per million hours	0.4	0.05	3	–	–	Smith (2001)
Gear, assembly (proportional to size)	–	Failures per million hours	–	10	50	–	–	Smith (2001)
Gear, per mesh	–	Failures per million hours	0.5	0.05	1	–	–	Smith (2001)
Generator, AC	–	Failures per million hours	–	3	30	–	–	Smith (2001)
Generator, DC	–	Hourly failure rate	2.06E−04	–	–	–	–	Anderson and Neri (1990)
Generator, DC	–	Failures per million hours	–	1	10	–	–	Smith (2001)

(*Continued*)

Component or Item	Failure Mode	Units	Point Estimate or Suggested Mean	Range (Low)	Range (High)	Calculated 5% Lower Limit	Calculated 95% Upper Limit	References
Generator, diesel set	–	Failures per million hours	–	125	4000	–	–	Smith (2001)
Generator, motor set	–	Failures per million hours	–	30	70	–	–	Smith (2001)
Generator, turbine set	–	Failures per million hours	200	10	800	–	–	Smith (2001)
Gyroscope	–	Hourly failure rate	3.00E–04	–	–	–	–	Anderson and Neri (1990)
Heat exchanger	–	Hourly failure rate	3.84E–05	–	–	–	–	Anderson and Neri (1990)
Heat exchanger	Failure due to blockage	Hourly failure rate	5.76E–06	–	–	5.76E–07	5.76E–05	Modarres (1993)
Heat exchanger	Failure due to rupture (leakage)	Hourly failure rate	3.00E–06	–	–	3.00E–07	3.00E–05	Modarres (1993)
Heat exchanger	Unavailability due to test and maintenance	Hourly failure rate	3.00E–05	–	–	2.73E–07	3.30E–03	Modarres (1993)
Hose and fittings	–	Hourly failure rate	–	3.90E–06	3.29E–05	–	–	Anderson and Neri (1990)
Heating, Ventilation and Air conditioning (HVAC) fan	Failure to start	Daily failure rate	3.00E–04	–	–	1.00E–04	9.00E–04	Modarres (1993)
Heating, Ventilation and Air conditioning (HVAC) fan	Failure to run	Hourly failure rate	1.00E–05	–	–	3.33E–06	3.00E–05	Modarres (1993)
Heating, Ventilation and Air conditioning (HVAC) fan	Unavailability due to test and maintenance	Daily failure rate	2.00E–03	–	–	2.00E–04	2.00E–02	Modarres (1993)
Hydraulic equipment, actuator	–	Failures per million hours	–	15	–	–	–	Smith (2001)
Hydraulic equipment, actuator/damper	–	Failures per million hours	200	20	–	–	–	Smith (2001)
Hydraulic equipment, motor	–	Failures per million hours	–	5	–	–	–	Smith (2001)
Hydraulic equipment, piston	–	Failures per million hours	–	1	–	–	–	Smith (2001)
Hydraulic-operated valves	Failure to operate	Daily failure rate	2.00E–03	3.00E–04	2.00E–02	6.67E–04	6.00E–03	Modarres (1993)
Hydraulic-operated valves	Failure due to plugging	Daily failure rate	–	2.00E–05	1.00E–04	–	–	Modarres (1993)
Hydraulic-operated valves	Failure due to plugging	Annual failure rate	1.00E–07	–	1.00E–07	3.33E–08	3.00E–07	Modarres (1993)
Hydraulic-operated valves	Unavailability due to test and maintenance	Daily failure rate	8.00E–04	6.00E–05	6.00E–03	8.00E–05	8.00E–03	Modarres (1993)
Inductor	–	Failures per million hours	–	0.2	0.5	–	–	Smith (2001)
Instrument air compressor	Failure to start	Daily failure rate	8.00E–02	–	–	2.67E–02	2.40E–01	Modarres (1993)
Instrument air compressor	Failure to run	Hourly failure rate	2.00E–04	–	–	2.00E–05	2.00E–03	Modarres (1993)

(Continued)

Component or Item	Failure Mode	Units	Point Estimate or Suggested Mean	Range (Low)	Range (High)	Calculated 5% Lower Limit	Calculated 95% Upper Limit	References
Instrument air compressor	Unavailability due to test and maintenance	Daily failure rate	2.00E−03	–	–	2.00E−04	2.00E−02	Modarres (1993)
Instrumentation	Failure to operate	Hourly failure rate	3.00E−06	–	–	3.00E−07	3.00E−05	Modarres (1993)
Joints, O ring	–	Failures per million hours	–	0.2	0.5	–	–	Smith (2001)
Joints, pipe	–	Failures per million hours	–	0.5	–	–	–	Smith (2001)
Lamp, incandescent	–	Hourly failure rate	1.86E−05	–	–	–	–	Anderson and Neri (1990)
Lamps, filament	–	Failures per million hours	1	0.05	10	–	–	Smith (2001)
Lamps, neon	–	Failures per million hours	0.2	0.1	1	–	–	Smith (2001)
LCD (per character)	–	Failures per million hours	–	0.05	–	–	–	Smith (2001)
LCD (per device)	–	Failures per million hours	–	2.5	–	–	–	Smith (2001)
LED, indicator	–	Failures per million hours	–	0.06	0.3	–	–	Smith (2001)
LED, numeral (per character)	–	Failures per million hours	–	0.01	0.1	–	–	Smith (2001)
Level instruments, controller	–	Failures per million hours	–	4	20	–	–	Smith (2001)
Level instruments, indicator	–	Failures per million hours	–	1	10	–	–	Smith (2001)
Level instruments, switch	–	Failures per million hours	5	2	20	–	–	Smith (2001)
Level instruments, transmitter	–	Failures per million hours	–	10	20	–	–	Smith (2001)
Lines, communication, coaxial per km	–	Failures per million hours	–	1.5	–	–	–	Smith (2001)
Lines, communication, subsea, per kilometer	–	Failures per million hours	–	2.4	–	–	–	Smith (2001)
Lines, communication, speech channel, land	–	Failures per million hours	–	100	250	–	–	Smith (2001)
Load cell	–	Failures per million hours	–	100	400	–	–	Smith (2001)
Loudspeaker	–	Failures per million hours	–	10	–	–	–	Smith (2001)
Magnetic tape unit, including drive	–	Failures per million hours	–	200	500	–	–	Smith (2001)
Manual valve	Failure due to plugging	Daily failure rate	–	2.00E−05	1.00E−04	–	–	Modarres (1993)
Manual valve	Failure due to plugging	Annual failure rate	1.00E−07	–	1.00E−07	3.33E−08	3.00E−07	Modarres (1993)
Manual valve	Unavailability due to test and maintenance	Daily failure rate	8.00E−04	6.00E−05	6.00E−03	8.00E−05	8.00E−03	Modarres (1993)
Manual valve	Failure to open	Daily failure rate	1.00E−04	–	–	3.33E−05	3.00E−04	Modarres (1993)
Manual valve	Failure to remain closed	Daily failure rate	1.00E−04	–	–	3.33E−05	3.00E−04	Modarres (1993)

(Continued)

Component or Item	Failure Mode	Units	Point Estimate or Suggested Mean	Range (Low)	Range (High)	Calculated 5% Lower Limit	Calculated 95% Upper Limit	References
Mechanical device	–	Hourly failure rate	–	1.70E−06	9.87E−04	–	–	Anderson and Neri (1990)
Meter (moving coil)	–	Failures per million hours	–	1	5	–	–	Smith (2001)
Microwave equipment, detector/mixer	–	Failures per million hours	–	0.2	–	–	–	Smith (2001)
Microwave equipment, fixed element	–	Failures per million hours	–	0.01	–	–	–	Smith (2001)
Microwave equipment, tuned element	–	Failures per million hours	–	0.1	–	–	–	Smith (2001)
Microwave equipment, waveguide, fixed	–	Failures per million hours	–	1	–	–	–	Smith (2001)
Microwave equipment, waveguide, flexible	–	Failures per million hours	–	2.5	–	–	–	Smith (2001)
Motor-driven pump	Failure to start	Daily failure rate	3.00E−03	5.00E−04	1.00E−04	3.00E−04	3.00E−02	Modarres (1993)
Motor-driven pump	Failure to run	Hourly failure rate	3.00E−05	1.00E−06	1.00E−03	3.00E−06	3.00E−04	Modarres (1993)
Motor-driven pump	Unavailability due to test and maintenance	Daily failure rate	2.00E−03	1.00E−04	1.00E−02	2.00E−04	2.00E−02	Modarres (1993)
Motor-operated valves	Failure to operate	Daily failure rate	3.00E−03	1.00E−03	9.00E−03	3.00E−04	3.00E−02	Modarres (1993)
Motor-operated valves	Failure due to plugging	Daily failure rate	1.00E−07	2.00E−05	1.00E−04	3.33E−08	3.00E−07	Modarres (1993)
Motor-operated valves	Unavailability due to test and maintenance	Daily failure rate	8.00E−04	6.00E−05	6.00E−03	8.00E−05	8.00E−03	Modarres (1993)
Motor-operated valves	Failure to remain closed	Hourly failure rate	5.00E−07	–	–	5.00E−08	5.00E−06	Modarres (1993)
Motor-operated valves	Failure to remain open	Hourly failure rate	1.00E−07	–	–	3.33E−08	3.00E−07	Modarres (1993)
Motor, electrical, AC	–	Failures per million hours	5	1	20	–	–	Smith (2001)
Motor, electrical, DC	–	Failures per million hours	15	5	–	–	–	Smith (2001)
Motor, electrical, starter	–	Failures per million hours	–	4	10	–	–	Smith (2001)
Offsite power	Loss, other than initiator	Not listed	2.00E−04	–	–	6.67E−05	6.00E−04	Modarres (1993)
Orifice	Failure due to plugging	Daily failure rate	3.00E−04	–	–	1.00E−04	9.00E−04	Modarres (1993)
Photoelectric cell	–	Failures per million hours	–	15	–	–	–	Smith (2001)
Pneumatic equipment, connector	–	Failures per million hours	–	1.5	–	–	–	Smith (2001)
Pneumatic equipment, controller, degraded	–	Failures per million hours	–	10	20	–	–	Smith (2001)
Pneumatic equipment, controller, open or shut	–	Failures per million hours	–	1	2	–	–	Smith (2001)

(Continued)

Component or Item	Failure Mode	Units	Point Estimate or Suggested Mean	Range (Low)	Range (High)	Calculated 5% Lower Limit	Calculated 95% Upper Limit	References
Pneumatic equipment, I/P converter	–	Failures per million hours	–	2	10	–	–	Smith (2001)
Pneumatic equipment, pressure relay	–	Failures per million hours	–	20	–	–	–	Smith (2001)
Power cable per km, overhead, <600 V	–	Failures per million hours	–	0.5	–	–	–	Smith (2001)
Power cable per km, overhead, 600–15 kV	–	Failures per million hours	–	5	15	–	–	Smith (2001)
Power cable per km, overhead, >33 kV	–	Failures per million hours	–	3	7	–	–	Smith (2001)
Power cable per km, underground, <600 V	–	Failures per million hours	–	2	–	–	–	Smith (2001)
Power cable per km, underground, 600–15 kV	–	Failures per million hours	–	2	–	–	–	Smith (2001)
Power cable per km, undersea	–	Failures per million hours	–	2.5	–	–	–	Smith (2001)
Power-operated relief valve for Pressurized Water Reactor (PWR)	Failure to open on actuation	Daily failure rate	2.00E−03	–	–	6.67E−04	6.00E−03	Modarres (1993)
Power-operated relief valve for Pressurized Water Reactor (PWR)	Failure to open for pressure relief	Daily failure rate	3.00E−04	–	–	3.00E−05	3.00E−03	Modarres (1993)
Power-operated relief valve for Pressurized Water Reactor (PWR)	Failure to re-close	Daily failure rate	2.00E−03	–	–	6.67E−04	6.00E−03	Modarres (1993)
Power supply, AC/DC stabilized	–	Failures per million hours	20	5	100	–	–	Smith (2001)
Power supply, AC/DC converter	–	Failures per million hours	5	2	20	–	–	Smith (2001)
Pressure instruments, controller	–	Failures per million hours	10	1	30	–	–	Smith (2001)
Pressure instruments, indicator	–	Failures per million hours	5	1	10	–	–	Smith (2001)
Pressure instruments, sensor	–	Failures per million hours	–	2	10	–	–	Smith (2001)
Pressure instruments, switch	–	Failures per million hours	5	1	40	–	–	Smith (2001)
Pressure instruments, transmitter	–	Failures per million hours	–	5	20	–	–	Smith (2001)
Pressure regulator valve	Failure to open	Daily failure rate	2.00E−03	–	–	6.67E−04	6.00E−03	Modarres (1993)
Printed circuit board, double (plated through)	–	Failures per million hours	–	0.01	0.3	–	–	Smith (2001)
Printed circuit board, multilayer	–	Failures per million hours	–	0.07	0.1	–	–	Smith (2001)
Printed circuit board, single sided	–	Failures per million hours	–	0.02	–	–	–	Smith (2001)
Printer, line	–	Failures per million hours	–	300	1000	–	–	Smith (2001)

(Continued)

Component or Item	Failure Mode	Units	Point Estimate or Suggested Mean	Range (Low)	Range (High)	Calculated 5% Lower Limit	Calculated 95% Upper Limit	References
Pump	–	Hourly failure rate	–	1.70E–06	3.95E–04	–	–	Anderson and Neri (1990)
Pump, boiler	–	Failures per million hours	–	100	700	–	–	Smith (2001)
Pump, centrifugal	–	Failures per million hours	50	10	100	–	–	Smith (2001)
Pump, fire water, diesel	–	Failures per million hours	–	200	3000	–	–	Smith (2001)
Pump, fire water, electric	–	Failures per million hours	–	200	500	–	–	Smith (2001)
Pump, fuel	–	Failures per million hours	–	3	180	–	–	Smith (2001)
Pump, oil lubrication	–	Failures per million hours	–	6	70	–	–	Smith (2001)
Pump, vacuum	–	Failures per million hours	–	10	25	–	–	Smith (2001)
Push button	–	Failures per million hours	0.5	0.1	10	–	–	Smith (2001)
Rectifier (power)	–	Failures per million hours	–	3	5	–	–	Smith (2001)
Regulator	–	Hourly failure rate	–	3.00E–06	1.36E–04	–	–	Anderson and Neri (1990)
Relap	–	Hourly failure rate	–	1.00E–06	3.10E–05	–	–	Anderson and Neri (1990)
Relays, armature general	–	Failures per million hours	–	0.2	0.4	–	–	Smith (2001)
Relays, Safety	–	Failures per million hours	–	0.02	0.07	–	–	Smith (2001)
Relays, contractor	–	Failures per million hours	–	1	6	–	–	Smith (2001)
Relays, crystal can	–	Failures per million hours	–	0.15	–	–	–	Smith (2001)
Relays, heavy duty	–	Failures per million hours	–	2	5	–	–	Smith (2001)
Relays, latching	–	Failures per million hours	–	0.02	1.5	–	–	Smith (2001)
Relays, polarized	–	Failures per million hours	–	0.8	–	–	–	Smith (2001)
Relays, power	–	Failures per million hours	–	1	16	–	–	Smith (2001)
Relays, reed	–	Failures per million hours	0.2	0.002	2	–	–	Smith (2001)
Relays, thermal	–	Failures per million hours	–	0.5	10	–	–	Smith (2001)
Relays, time delay	–	Failures per million hours	2	0.5	10	–	–	Smith (2001)
Relief valve (not safety relief valve or PORV)	Spurious open	Hourly failure rate	3.90E–06	–	–	3.90E–07	3.90E–05	Modarres (1993)
Resistors, carbon Component	–	Failures per million hours	–	0.001	0.006	–	–	Smith (2001)
Resistors, carbon film	–	Failures per million hours	–	0.001	0.05	–	–	Smith (2001)
Resistors, metal oxide	–	Failures per million hours	0.004	0.001	0.05	–	–	Smith (2001)
Resistors, network	–	Failures per million hours	–	0.05	0.1	–	–	Smith (2001)

(*Continued*)

Component or Item	Failure Mode	Units	Point Estimate or Suggested Mean	Range (Low)	Range (High)	Calculated 5% Lower Limit	Calculated 95% Upper Limit	References
Resistors, variable comp.	–	Failures per million hours	–	0.5	1.5	–	–	Smith (2001)
Resistors, variable wire wound	–	Failures per million hours	0.05	0.02	0.5	–	–	Smith (2001)
Resistors, wire wound	–	Failures per million hours	0.005	0.001	0.5	–	–	Smith (2001)
Safety relief valve, boiling water reactor	Failure to open for pressure relief	Daily failure rate	1.00E–05	–	–	3.33E–06	3.00E–05	Modarres (1993)
Safety relief valve, boiling water reactor	Failure to open on actuation	Daily failure rate	1.00E–02	–	–	3.33E–03	3.00E–02	Modarres (1993)
Safety relief valve, boiling water reactor	Failure to re-close on pressure relief	Hourly failure rate	3.90E–06	–	–	3.90E–07	3.90E–05	Modarres (1993)
Sensor	–	Hourly failure rate	7.66E–05	–	–	–	–	Anderson and Neri (1990)
Solenoid	–	Hourly failure rate	6.56E–05	–	–	–	–	Anderson and Neri (1990)
Solenoid	–	Failures per million hours	1	0.4	4	–	–	Smith (2001)
Solenoid-operated valves	Failure to operate	Daily failure rate	2.00E–03	1.00E–03	2.00E–02	6.67E–04	6.00E–03	Modarres (1993)
Solenoid-operated valves	Failure due to plugging	Daily failure rate	–	2.00E–05	1.00E–04	–	–	Modarres (1993)
Solenoid-operated valves	Failure due to plugging	Annual failure rate	1.00E–07	–	1.00E–07	3.33E–08	3.00E–07	Modarres (1993)
Solenoid-operated valves	Unavailability due to test and maintenance	Daily failure rate	8.00E–04	6.00E–05	6.00E–03	8.00E–05	8.00E–03	Modarres (1993)
Stepper motor	–	Failures per million hours	–	0.5	5	–	–	Smith (2001)
Strainer	Failure due to plugging	Hourly failure rate	3.00E–05	–	–	3.00E–06	3.00E–04	Modarres (1993)
Structural elements	–	Hourly failure rate	–	4.00E–11	4.00E–09	–	–	Anderson and Neri (1990)
Sump	Failure due to plugging	Daily failure rate	5.00E–05	–	–	5.00E–07	5.00E–03	Modarres (1993)
Surge arrestors, >100 kV	–	Failures per million hours	–	0.5	1.5	–	–	Smith (2001)
Surge arrestors, low power	–	Failures per million hours	–	0.003	0.02	–	–	Smith (2001)
Switch	–	Hourly failure rate	–	1.86E–05	9.50E–05	–	–	Anderson and Neri (1990)
Switches per contact, Dual In Line (DIL)	–	Failures per million hours	0.5	0.03	1.8	–	–	Smith (2001)
Switches per contact, key, low power	–	Failures per million hours	–	5	10	–	–	Smith (2001)
Switches per contact, key, low power	–	Failures per million hours	–	0.003	2	–	–	Smith (2001)
Switches per contact, micro	–	Failures per million hours	–	0.1	1	–	–	Smith (2001)
Switches (per contact), pushbutton	–	Failures per million hours	1	0.2	10	–	–	Smith (2001)
Switches per contact, rotary	–	Failures per million hours	–	0.05	0.5	–	–	Smith (2001)

(Continued)

Component or Item	Failure Mode	Units	Point Estimate or Suggested Mean	Range (Low)	Range (High)	Calculated 5% Lower Limit	Calculated 95% Upper Limit	References
Switches per contact, thermal delay	–	Failures per million hours	–	0.5	3	–	–	Smith (2001)
Switches per contact, toggle	–	Failures per million hours	–	0.03	1	–	–	Smith (2001)
Synchros and resolvers	–	Failures per million hours	–	3	15	–	–	Smith (2001)
Tank	–	Hourly failure rate	–	1.09E–04	1.59E–04	–	–	Anderson and Neri (1990)
Temperature instruments, controller	–	Failures per million hours	–	20	40	–	–	Smith (2001)
Temperature instruments, pyrometer	–	Failures per million hours	–	250	1000	–	–	Smith (2001)
Temperature instruments, sensor	–	Failures per million hours	–	0.2	10	–	–	Smith (2001)
Temperature instruments, switch	–	Failures per million hours	–	3	20	–	–	Smith (2001)
Temperature instruments, transmitter	–	Failures per million hours	–	10	–	–	–	Smith (2001)
Temperature switch	Failure to operate	Daily failure rate	1.00E–04	–	–	1.00E–05	1.00E–03	Modarres (1993)
Thermionic tubes, diode	–	Failures per million hours	20	5	70	–	–	Smith (2001)
Thermionic tubes, thyratron	–	Failures per million hours	–	50	–	–	–	Smith (2001)
Thermionic tubes, triode and pentode	–	Failures per million hours	30	20	100	–	–	Smith (2001)
Thermocouple/ thermostat	–	Failures per million hours	10	1	20	–	–	Smith (2001)
Time delay relay	Fail to transfer	Hourly failure rate	3.00E–04	–	–	3.00E–05	3.00E–03	Modarres (1993)
Timer (electromechanical)	–	Failures per million hours	15	2	40	–	–	Smith (2001)
Transducer	–	Hourly failure rate	–	5.79E–05	1.00E–04	–	–	Anderson and Neri (1990)
Transfer switch	Failure to transfer	Daily failure rate	1.00E–03	–	–	3.33E–04	3.00E–03	Modarres (1993)
Transformer	Short or open	Hourly failure rate	2.00E–06	–	–	2.00E–07	2.00E–05	Modarres (1993)
Transformers, ≥415 V	–	Failures per million hours	1	0.4	7	–	–	Smith (2001)
Transformers, mains	–	Failures per million hours	0.4	0.03	0.3	–	–	Smith (2001)
Transformers, signal	–	Failures per million hours	0.2	0.005	0.3	–	–	Smith (2001)
Transistors, Si field-effect transistor high power	–	Failures per million hours	–	0.1	–	–	–	Smith (2001)
Transistors, Si field-effect transistor low power	–	Failures per million hours	–	0.05	–	–	–	Smith (2001)
Transistors, Si npn high power	–	Failures per million hours	–	0.1	0.4	–	–	Smith (2001)
Transistors, Si npn low power	–	Failures per million hours	0.05	0.01	0.2	–	–	Smith (2001)
Transmitter	Failure to operate	Hourly failure rate	1.00E–06	–	–	3.33E–07	3.00E–06	Modarres (1993)
Turbine-driven pump	Failure to start	Daily failure rate	3.00E–02	5.00E–03	9.00E–02	3.00E–03	3.00E–01	Modarres (1993)

(Continued)

Component or Item	Failure Mode	Units	Point Estimate or Suggested Mean	Range (Low)	Range (High)	Calculated 5% Lower Limit	Calculated 95% Upper Limit	References
Turbine-driven pump	Failure to run	Hourly failure rate	5.00E−03	8.00E−06	1.00E−03	5.00E−04	5.00E−02	Modarres (1993)
Turbine-driven pump	Unavailability due to test and maintenance	Daily failure rate	1.00E−02	3.00E−03	4.00E−02	1.00E−03	1.00E−01	Modarres (1993)
Turbine, steam	–	Failures per million hours	40	30	–	–	–	Smith (2001)
TV receiver (1984 figure)	–	Failures per million hours	–	2.3	–	–	–	Smith (2001)
Valve	–	Hourly failure rate	–	1.01E−05	1.34E−04	–	–	Anderson and Neri (1990)
Valve diaphragm	–	Failures per million hours	5	1	–	–	–	Smith (2001)
Valves, ball	–	Failures per million hours	3	0.2	10	–	–	Smith (2001)
Valves, butterfly	–	Failures per million hours	20	1	30	–	–	Smith (2001)
Valves, diaphragm	–	Failures per million hours	10	2.6	20	–	–	Smith (2001)
Valves, gate	–	Failures per million hours	10	1	30	–	–	Smith (2001)
Valves, globe	–	Failures per million hours	–	0.2	2	–	–	Smith (2001)
Valves, needle	–	Failures per million hours	20	1.5	–	–	–	Smith (2001)
Valves, nonreturn	–	Failures per million hours	–	1	20	–	–	Smith (2001)
Valves, plug	–	Failures per million hours	–	1	18	–	–	Smith (2001)
Valves, relief	–	Failures per million hours	–	2	8	–	–	Smith (2001)
Valves, solenoid (de-energize to trip)	–	Failures per million hours	–	1	8	–	–	Smith (2001)
Valves, solenoid (energize to trip)	–	Failures per million hours	20	8	–	–	–	Smith (2001)
Vacuum Distillation Unit (VDU)	–	Failures per million hours	200	10	500	–	–	Smith (2001)

References and Bibliography

Abraham, D.M., Bernold, L.E., and Livingston, E.E., 1989. Emulation for control system analysis in automated construction, *Journal of Computing in Civil Engineering*, 3(4), 320–332.

Al-Bahar, J. and Crandall, K., 1990. Systematic risk management approach for construction project, *Journal of Construction Engineering and Management*, 116(3), 533–547.

Albus, J.S., 1991. Outline for a theory of intelligence, *IEEE Transactions on Systems, Man and Cybernetics*, 21(3), 473–509.

Albus, J.S., 1998. *4D/RCS: A Reference Model Architecture for Demo III, Version 1.0*, NISTIR 5994, Gaithersburg, MD.

Alderson, M.R., 1981. *International Mortality Statistics*, Facts on Life, Inc., New York.

Allen, D.E., 1981. Criteria for design safety factors and quality assurance expenditure, in *Structural Safety and Reliability*, Moan, T. and Shinozuka, M., Eds., Elsevier, New York, pp. 667–678.

Al-Tabtabai, H. and Alex, A., 2000. Modeling the cost of political risk in international construction projects, *Project Management Journal*, 4(6–9), 11.

Al-Tabtabai, H., Alex, A., and Abou-alfotouh, A., 2001. Conflict resolution using cognitive analysis approach, *Project Management Journal*, 32(2), 4–5.

American Society of Civil Engineers (ASCE), 1966. Rx for risk communication: Winning strategies for project managers seeking trust and credibility within a community, by S.D. Perry, *Civil Engineering Magazine*, 66(8), 61–63, ASCE, Reston, VA.

American Society of Civil Engineers (ASCE), 2009. *Report Card for America's Infrastructure*, ASCE, Reston, VA.

American Society of Mechanical Engineers (ASME), 1993. *The Use of Decision-Analytic Reliability Methods in Codes and Standards Work*, Research Report, Vol. 23, CRTD, New York.

Ames, B., Megaw, R., and Golds, S., 1990. Ranking possible carcinogenic hazards in, *Readings in Risk*, Glickman, T.S., and Gough, M., Eds., 17–92, Resources for the Future, Washington, DC.

Anderson, R.T. and Neri, L., 1990. *Reliability-Centered Maintenance*, Elsevier, London.

Ang, A. and Tang, W., 2007. *Probability Concepts in Engineering Planning and Design*, 2nd ed., John Wiley & Sons, New York.

Ang, A.-H., Cheung, M.C., Shugar, T.A., and Fernie, J.D., 1999. Reliability-based fatigue analysis and design of floating structures, in *Proceedings of the Third International Workshop on Very Large Floating Structures*, Ertekin, R.C. and Kim, J.W., Eds., University of Hawaii, Honolulu, HI, pp. 375–380.

Ang, A.-H. and Tang, W.H., 1984. *Probability Concepts in Engineering Planning and Design: Decision, Risk, and Reliability*, Vol. II, John Wiley & Sons, New York.

Apostolakis, G. and Mosleh, A., 1979. Expert opinion and statistical evidence: An application to reactor core melt frequency, *Nuclear Science and Engineering*, 70, 135.

Arnould, R.J. and Grabowski, H., 1981. Auto safety regulation: An analysis of market failure, *Bell Journal of Economics and Management Sciences*, 12, 27–45.

Arnould, R.J. and Nichols, L.M., 1983. Wage-risk premiums and worker's compensation: A refinement of estimates of compensating wage differentials, *Journal of Political Economy*, 91, 332–340.

Arthur, W.B., 1981. The economics of risks to life, *American Economic Review*, 71(1), 54–64.

Ashley, D.D. and Bonner, J.J., 1987. Political risk in international construction, *Journal of Construction Engineering and Management*, 113(3), 447–467.

Austin, D.F., 1998. *Philosophical Analysis: A Defense by Example*, Philosophical Studies Series, Vol. 39, D. Reidel Publishing Company, Dordrecht.

Australian Safety and Compensation Council (ASCC), 2008. *The Health of Nations: The Value of a Statistical Life*, ASCC, Commonwealth of Australia.

Ayyub, B.M., 1991. Systems framework for fuzzy sets in civil engineering, *International Journal of Fuzzy Sets and Systems*, 40(3), 491–508.

Ayyub, B.M., 1992a. Generalized treatment of uncertainties in structural engineering, in *Analysis and Management of Uncertainty: Theory and Applications*, Ayyub, B.M. and Gupta, M.M., Eds., Elsevier, New York, pp. 235–246.

Ayyub, B.M., 1992b. *Fault Tree Analysis of Cargo Elevators Onboard Ships*, BMA Engineering Report, Naval Sea System Command, US Navy, Crystal City, VA.

Ayyub, B.M., 1993. *Handbook for Risk-Based Plant Integrity*, BMA Engineering Report, Chevron Research and Technology Corporation, Richmond, CA.

Ayyub, B.M., 1994. The nature of uncertainty in structural engineering, in *Uncertainty Modelling and Analysis: Theory and Applications*, Ayyub, B.M. and Gupta, M.M., Eds., North-Holland/Elsevier, Amsterdam, pp. 195–210.

Ayyub, B.M., 1997. *Guidelines for Probabilistic Risk Analysis of Marine Systems*, US Coast Guard, Washington, DC.

Ayyub, B.M., Ed., 1998. *Uncertainty Modeling and Analysis in Civil Engineering*, CRC Press, Boca Raton, FL.

Ayyub, B.M., 1999. *Guidelines on Expert-Opinion Elicitation of Probabilities and Consequences for Corps Facilities*, US Army Corps of Engineers, Institute for Water Resources, Alexandria, VA.

Ayyub, B.M., 2002. *Elicitation of Expert Opinions for Uncertainty and Risks*, CRC Press, Boca Raton, FL.

Ayyub, B.M., 2010. On uncertainty in information and ignorance in knowledge, *Journal of General Systems*, 39(4), 415–435.

Ayyub, B.M., 2012. Uncertainty analysis and risk quantification for adaptation to sea level rise, Paper CMTC-153151-PP, Carbon Management Technology Conference, ASCE, etc., Orlando, FL.

Ayyub, B.M., 2013. Systems resilience for multi-hazard environments: Definition, metrics and valuation for decision making, Working paper CTSM13-01, Center for Technology and Systems Management, University of Maryland, College Park, MD.

Ayyub, B.M. and Chao, R.-J., 1998. Uncertainty modeling in civil engineering with structural and reliability applications, in *Uncertainty Modeling and Analysis in Civil Engineering*, Ayyub, B.M., Ed., CRC Press, Boca Raton, FL, pp. 1–32.

Ayyub, B.M. and Gupta, M.M., Eds., 1997. *Uncertainty Analysis in Engineering and the Sciences: Fuzzy Logic, Statistics, and Neural Network Approach*, Kluwer Academic, Dordrecht.

Ayyub, B.M. and Haldar, A., 1984. Practical structural reliability techniques, *ASCE Journal of Structural Engineering*, 110(8), 1707–1724.

Ayyub, B.M. and Kaminskiy, M., 2001. *Assessment of Hazard Functions for Components and Systems*, Technical Report, US Army Corps of Engineers, Concord, MA.

Ayyub, B.M. and Kaminskiy, M.P., 2009. Quantitative representation of risk, in *Risk Modeling and Vulnerability Assessment Volume in the Wiley Handbook of Science and Technology for Homeland Security*, Ayyub, B.M. and Voller, G., Eds., John Wiley & Sons, Hoboken, NJ.

Ayyub, B.M. and Klir, G., 2006. *Uncertainty Modeling and Analysis in Engineering and the Sciences*, CRC Press/Chapman & Hall, Boca Raton, FL.

Ayyub, B.M. and McCuen, R.H., 2011. *Probability, Statistics, and Reliability for Engineers and Scientists*, Chapman & Hall/CRC Press, Boca Raton, FL.

Ayyub, B.M. and Moser, D.A., 2000. *Economic Consequence Assessment of Floods in the Feather River Basin of California Using Expert-Opinion Elicitation*, Technical Report, US Army Corps of Engineers, Institute for Water Resources, Alexandria, VA.

Ayyub, B.M. and Parker, L., 1494. Financing nuclear liability, *Science*, 334.

Ayyub, B.M. and White, G.J., 1990a. Life expectancy of marine structures, *Marine Structures*, 3, 301–317.

Ayyub, B.M. and White, G.J., 1990b. Structural life expectancy of marine vessels, *International Journal of Marine Structures: Design, Construction and Safety*, 3(4), 301–317.

Ayyub, B.M. and White, G.J., 1995. Probability-based life prediction, in *Probabilistic Structural Mechanics Handbook*, Sundararajan, C., Ed., Chapman & Hall, New York, pp. 416–428.

Ayyub, B.M. and Wilcox, R., 1999. *A Risk-Based Compliance Approval Process for Personal Flotation Devices*, Final Report, US Coast Guard, Office of Design and Engineering Standards, Lifesaving and Fire Safety Standards Division, Washington, DC.

Ayyub, B.M. and Wilcox, R., 2001. *A Risk-Based Compliance Approval Process for Personal Flotation Devices Using Performance Models*, BMA Engineering, Potomac, MD.

Ayyub, B.M., Assakkaf, I., Atua, K., Engle, A., Hess, P., Karaszewski, Z., Kihl, D., et al., 1998a. *Reliability-Based Design of Ship Structures: Current Practice and Emerging Technologies*, Research Report to the US Coast Guard, SNAME, T & R Report R-53.

Ayyub, B.M., Assakkaf, I., Popescu, C., and Karaszewski, Z., 1998b. *Methodology and Guidelines for Risk-Informed Analysis for the Engine Room Arrangement Model*, Technical Report No. CTSM-98–RIA-ERAM-2, Center for Technology and Systems Management, University of Maryland, College Park, MD.

Ayyub, B.M., Baecher, G., and Johnson, P., 1998c. *Guidelines for Risk-Based Assessment of Dam Safety*, US Army Corps of Engineers, Washington, DC.

Ayyub, B.M., Braileanu, H.G., and Qureshi, N., 2012. Prediction and impact of sea level rise on properties and infrastructure of Washington, DC, *Risk Analysis Journal, Society for Risk Analysis*, 32(11), 1901–1918.

Ayyub, B.M., Foster, J., and McGill, W.L., 2009a. Risk analysis of a protected hurricane-prone region I: Model development, *ASCE Natural Hazards Review*, 38–53, doi:10.1061/(ASCE)1527-6988(2009)10:2(38).

Ayyub, B.M., Gupta, M.M., and Kanal, L.N., 1992. *Analysis and Management of Uncertainty: Theory and Applications*, Elsevier, New York.

Ayyub, B.M., Guran, A., and Haldar, A., Eds., 1997. *Uncertainty Modeling in Vibration, Control, and Fuzzy Analysis of Structural Systems*, World Scientific, Singapore.

Ayyub, B.M., Koko, K.S., Akpan, U.O., and Rushton, P.A., 2002. *Risk-Based Maintenance Management System for Ship Structure*, NAVSEA SBIR Report, US Navy, Washington, DC.

Ayyub, B.M., McGill, W.L., Foster, J., and Jones, H.W., 2009b. Risk analysis of a protected hurricane-prone region II: Computations and illustrations, *ASCE Natural Hazards Review*, 54–67, doi:10.1061/(ASCE)1527-6988(2009)10:2(38).

Ayyub, B.M., McGill, W.L., and Kaminskiy, M., 2007. Critical asset and portfolio risk analysis for homeland security: An all-hazards framework, *Risk Analysis*, 27(3), 789–801.

Ayyub, B.M., Riley, B.C., and Hoge, M.T., 1996. *Expert Elicitation of Unsatisfactory-Performance Probabilities and Consequences for Civil Works Facilities*, Technical Report, US Army Corps of Engineers, Pittsburgh, PA.

Ayyub, B.M., White, G.J., Bell-Wright, T.F., and Purcell, E.S., 1990. Comparative structural life assessment of patrol boat bottom plating. *Naval Engineers Journal*, 102(3), 253–262.

Ayyub, B.M., White, G.J., and Purcell, E.S., 1989. Estimation of structural service life of ships, *Naval Engineers Journal*, 101(3), 156–166.

Baca, G.S., 1990. *Tort Liability and Risk Management*, Transportation Research Board, National Research Council, Washington, DC.

Bailey, K.D., 1994. *Methods of Social Research*, Free Press, New York.

Bajaj, D., Oluwoye, J., and Lenard, D., 1997. An analysis of contractor's approaches to risk identification in New South Wales, Australia, *Construction Management and Economics*, 15, 363–369.

Balkey, K.R., Abramson, L., Ayyub, B.M., Vic Chapman, O.J., Gore, B.F., Harris, D.O., Karydas, D., et al., 1994. *Risk-Based Inspection: Development of Guidelines, Vol. 3: Fossil-Fuel-Fired Electric Power Generating Station Applications*, CRTD, Vol. 20-0–3, ASME, Washington, DC.

Balkey, K.R., Ayyub, B.M., Vic Chapman, O.J., Gore, B.F., Harris, D.O., Karydas, D., Simonen, F.A., and Smith, H., 1991. *Risk-Based Inspection: Development of Guidelines, Vol. 1: General Document*, CRTD, Vol. 20-0–1, ASME, Washington, DC (also NUREG/GR-0005, Vol. 1, by Nuclear Regulatory Commission, Washington, DC).

Balkey, K.R., Ayyub, B.M., Vic Chapman, O.J., Gore, B.F., Harris, D.O., Phillips, J.H., Krishnan, F., et al., 1993. *Risk-Based Inspection: Development of Guidelines, Vol. 2, Part 1: Light Water Reactor Nuclear Power Plant Components*, CRTD, Vol. 20-0–2, ASME, Washington, DC (also NUREG/GR-0005, Vol. 2, Part 1, by Nuclear Regulatory Commission, Washington, DC).

Balkey, K.R., Simonen, F.A., Gold, J., Ayyub, B.M., Abramson, L., Vic Chapman, O.J., Gore, B.F., et al., 1998. *Risk-Based Inspection: Development of Guidelines, Vol. 2, Part 2: Light Water Reactor Nuclear Power Plant Components*, CRTD, Vol. 20-0-4, ASME, Washington, DC.

Baram, M.S., 1980. Cost–benefit analysis: An inadequate basis for health, safety, and environmental regulatory decision making, *Ecology Law Quarterly*, 8(3), 473–531.

Barbier, E.B., 1994. Valuing environmental functions: Tropical wetlands, *Land Economics*, 70(2), 155–173.

Barlow, R.E. and Proschan, F., 1966. Tolerance and confidence limits for classes of distributions based on failure rates, *Annals of Mathematical Statistics*, 37, 6.

Barlow, R.E. and Proschan, F., 1975. *Statistical Theory of Reliability and Life Testing, Probability Models*, Holt, Rinehart & Winston, New York.

Baum, J.W., 1994. *Value of Public Health and Safety Actions and Radiation Dose Avoided*, NUREG/CR-6212, Nuclear Regulatory Commission, Washington, DC, p. 43.

Berke, P., Kartez, J., and Wenger, D., 2008. Recovery after disaster: Achieving sustainable development, mitigation and equity, *Disasters*, 17, 93–109.

Bernstein, P., 1998. *Against the Gods: The Remarkable Story of Risk*, John Wiley & Sons, New York.

Bier, V.M., 1999. Challenges to the acceptance of probabilistic risk analysis, *Risk Analysis*, 19(4), 703–710.

Bishop, R.C., Boyle, K.J., and Welsh, M.P., 1987. Toward total economic value of Great Lakes fishery resources, *Transactions of the American Fisheries Society*, 116(3), 339–345.

Black, K., 1996. Causes of project failure: a survey of professional engineers, *PM Network*, 10(11), 21–24.

Blackman, H.S., 1997. *Human Reliability Assessment Training Course*, University of Maryland Center for Technology and Systems Management, University of Maryland, College Park, MD.

Blake, W.M., Hammond, K.R., and Meyer, G.M., 1975. An alternate approach to labour–management relations, *Administrative Science Quarterly*, 18(3), 311–327.

Blanchard, B.S., 1998. *System Engineering Management*, 2nd ed., John Wiley & Sons, New York.

Blockley, D.I., 1975. Predicting the likelihood of structural accidents, *Proceedings of the Institution of Civil Engineers*, 59(4), 659–668.

Blockley, D.I., 1979a. The calculations of uncertainty in civil engineering, *Proceedings of the Institution of Civil Engineers*, 67(2), 313–326.

Blockley, D.I., 1979b. The role of fuzzy sets in civil engineering, *Fuzzy Sets and Systems*, 2, 267–278.

Blockley, D.I., 1980. *The Nature of Structural Design and Safety*, Ellis Horwood, Chichester.

Blockley, D.I., Pilsworth, B.W., and Baldwin, J.F., 1983. Measures of uncertainty, *Civil Engineering Systems*, 1, 3–9.

Bogdanoff, J.L. and Kozin, F., 1985. *Probabilistic Models of Cumulative Damage*, John Wiley & Sons, New York.

Bostrom, N., 2002. Existential risks: Analyzing human extinction scenarios and related hazards, *Journal of Evolution and Technology*, 9(1).

Bouissac, P., 1992. The construction of ignorance and the evolution of knowledge, *University of Toronto Quarterly*, 61(4), 460–472.

Bremermann, H.J., 1962. Optimization through evolution and recombination, in *Self-Organizing Systems*, Yovits, M.C., Jacobi, G.T., and Goldstein, G.D., Eds., Spartan Books, Washington, DC, pp. 93–106.

British Standards Institute (BSI), 1996. *Guide to Project Management*, BS 6079, BSI, London, pp. 28–29.

Brown, C.B., 1979. A fuzzy safety measure, *Journal of Engineering Mechanics Division*, 105(EM5), 855–872.

Brown, C.B., 1980. The merging of fuzzy and crisp information, *Journal of Engineering Mechanics Division*, 106(EM1), 123–133.

Brown, C.B. and Yao, J.T.P., 1983. Fuzzy sets and structural engineering, *Journal of Structural Engineering*, 109(5), 1211–1225.

Burby, R., Ed., 1998. *Cooperating with Nature: Confronting Natural Hazards and Land-Use Planning for Sustainable Communities*, Joseph Henry Press, Washington, DC.

Burrage, K., 1995. Risk management in safety critical areas, *International Journal of Pressure Vessels and Piping*, 61, 229–256.

Callicott, J.B., 2004. Explicit and implicit Values in the ESA, in *The Endangered Species Act at Thirty: Retrospect and Prospects*, Davies, F., Goble, D., Heal, G., and Scott, M., Eds., Island Press, Washington, DC.

Campbell, D.T., 1974. Evolutionary epistemology, in *The Philosophy of Karl Popper*, Schilpp, P.A., Ed., Open Court Publishing, LaSalle, IL, pp. 1413–1463.

Castle, G., 1976. The 55 mph limit: A cost–benefit analysis, *Traffic Engineering*, 46, 11–14.

Chestnut, H., 1965. *Systems Engineering Tools*, John Wiley & Sons, New York.

Choudhuri, S., 1994. *Project Management*, McGraw-Hill, New York.

Clark, K., 2011. *Near Term Hurricane Models: Performance Update*, Karen Clark & Company, Boston, MA.

Clark, W.C., 1980. Witches, floods and wonder drugs, in *Societal Risk Assessment: How Safe Is Safe Enough?* Schwing, R.C. and Albers, W.A., Eds., Plenum Press, New York, 287–318.

Clotfelter, C.T. and Hahn, J.C., 1978. Assessing the national 55 mph speed limit, *Policy Sciences*, 9, 281–294.

Cohen, B.L., 1980. Society's valuation of life saving in radiation protection and other contexts, *Health Physics*, 38, 33–51.

Cohen, B.L., 1981. Long-term consequences of the linear-no-threshold-response relationship for chemical carcinogens, *Risk Analysis*, 1(4), 267–275.

Coleman, J. and Mendlow, G., 2010. Theories of tort law, in *The Stanford Encyclopedia of Philosophy*, Zalta, E.N., Ed., Fall 2010 Edition, http://plato.stanford.edu/archives/fall2010/entries/tort-theories/.

Committee on Safety Criteria for Dams, 1985. *Safety of Dams: Flood and Earthquake Criteria*, National Academy Press, Washington, DC.

Committee on the Safety of Existing Dams, 1983. *Safety of Existing Dams, Evaluation and Improvement*, National Research Council, National Academy Press, Washington, DC.

Cooke, R.M., 1986. Problems with empirical Bayes, *Risk Analysis*, 6(3), 269–272.

Cooke, R.M., 1991. *Experts in Uncertainty*, Oxford University Press, Oxford.

Coplin, W.D. and O'Leary, M.K., 1994. *The Handbook of Country and Political Risk Analysis*, Political Risk Services, New York, ISBN 1 85271 302 X.

Cournot, A., 1838. *Recherches sur les principes math ematiques de la thorie des richesses* [Researches into the Mathematical Principles of the Theory of Wealth], Hachette, Paris.

Daidola, J.C. and Basar, N.S., 1981. *Probabilistic Structural Analysis of Ship Hull Longitudinal Strength*, NTIS Publication AD-A099118, Gaithersburg, MD.

Datta, S. and Mukherjee, S.K., 2001. Developing a risk management matrix for effective project planning: An empirical study, *Project Management Journal*, 32(2), 45–55.

Datta, S. and Sridhar, A.V., 1999. *A Case Study on BSL Modernization: A Book of Selected Cases*, Vol. V, Management Training Institute, Steel Authority of India Ltd., Ranchi, India.

Datta, S., Sridhar, A.V., and Bhat, K.S., 1998. *A Case Study on DSP Modernization: A Book of Selected Cases*, Vol. IV, Management Training Institute, Steel Authority of India Ltd., Ranchi, India.

Davidson, J., Ed., 1994. *The Reliability of Mechanical Systems*, Mechanical Engineering Publications Ltd., The Institute of Mechanical Engineers, London.

De Garmo, E.P., Sullivan, W.G., Bontadelli, J.A., and Wicks, E.M., 1997. *Engineering Economy*, 3rd ed., Prentice-Hall, New York.

Deisler, P.F., 2002. A perspective: Risk analysis as a tool for reducing the risks of terrorism, *Risk Analysis*, 22(3), 405–413.

Demsetz, L.A., Cairio, R., and Schulte-Strathaus, R., 1996. *Inspection of Marine Structures*, Ship Structure Committee Report No. 389, US Coast Guard, Washington, DC.

Department of Defense (DoD), 1991. *Military Handbook 217* (MIL-HDBK-217), Department of Defense, Washington, DC.

Department of the Navy, 1998. *Top Eleven Ways to Manage Technical Risk*, NAVSO P-3686, Office of the Assistant Secretary of the Navy (RD&A), Acquisition and Business Management, Washington, DC, pp. 99–110.

Derby, S.L. and Keeney, R.L., 1990. Risk analysis: Understanding how safe is safe enough, in *Readings in Risk*, Glickman, T.S. and Gough, M., Eds., 43–49, Resources for the Future, Washington, DC.

Det Norske Veritas (DNV), 1992. *Structural Reliability Analysis of Marine Structures*, Classification Notes No. 30.6, Det Norske Veritas Classification AS, Norway.

Dexter, R.J. and Pilarski, P., 2000. *Effect of Welded Stiffeners on Fatigue Crack Growth Rate*, SSC-413, Ship Structure Committee, US Coast Guard, Washington, DC.

di Carlo, C.W., 1998. Evolutionary epistemology and the concept of ignorance, PhD dissertation, University of Waterloo, Waterloo, Ontario.

Dillon-Merrill, R.L., Parnell, G.S., and Buckshaw, D.L., 2009. Logic trees: Fault, success, attack, event, probability, and decision trees, in *Risk Modeling and Vulnerability Assessment Volume in the Wiley Handbook of Science and Technology for Homeland Security*, Ayyub, B.M. and Voller, J.C., Eds., John Wiley & Sons, Hoboken, NJ, 2009.

Douglas, J., 1985. Measuring and managing environmental risk, *EPRI Journal*, July/August, 7–13.

Drury, C.G. and Lock, M.W.B., 1996. Ergonomics in civil aircraft inspection: Human error in aircraft inspection and maintenance, paper presented to Marine Board Committee on Human Performance, Organizational Systems, and Maritime Safety, National Academy of Engineering, National Research Council, Washington, DC.

Duncan, R. and Weston-Smith, M., 1977. *The Encyclopedia of Ignorance*, Pergamon Press, New York.

Eldukair, Z.A. and Ayyub, B.M., 1991. Analysis of recent US structural and construction failures, *Journal of Performance of Constructed Facilities*, 5(1), 57–73.

Ellingwood, B.R., 1995. Engineering reliability and risk analysis for water resources investments: Role of structural degradation in time-dependent reliability analyses, Contract Report ITL-95-3, US Army Corps of Engineers, Vicksburg, MS, pp. 1–56.

Ellingwood, B.R. and Mori, Y., 1993. Probabilistic methods for condition assessment and life prediction of concrete structures in nuclear plants, *Nuclear Engineering and Design*, 142, 155–166.

Erb, C.B., Harvey, C.R., and Viskanta, T.E., 1996. Political risk, economic risk and financial risk, *Financial Analysts Journal*, http://www.duke.edu/~charvey/Country_risk/pol/pol.htm.

Etheridge, D.M., Steele, L.P., Langenfelds, R.L., Francey, R.J. Barnola, J.-M., and Morgan, V.I., 2012. Historical CO_2 records from the law dome DE08, DE08-2, and DSS ice cores, http://cdiac.ornl.gov/trends/co2/lawdome.html.

Ezell, B.C., Bennett, S.P., von Winterfeldt, D., Sokolowski, J., and Collins, A.J., 2010. Probabilistic risk analysis and terrorism risk, *Risk Analysis*, 30(4), 575–589.

Farmer, F.R., 1967a. Reactor safety and siting: A proposed risk criterion, *Nuclear Safety*, 8(6), 539–548.

Farmer, F.R., 1967b. Siting criteria—New approach, in *Proceedings of Symposium on Containment Siting of Nuclear Power Plants*, SM-89/34, International Atomic Energy Agency, Wieden, Austria.

Federal Aviation Administration, 1998. *The Economic Value for Evaluation of Federal Aviation Administration Investment and Regulatory Programs*, FAA-APO-98-8, US Department of Transportation, Washington, DC.

Feinberg, K., 2001. Final report of the special master for the September 11th victim compensation fund of 2001, Vol. I, US Department of Justice, http://www.usdoj.gov/final_report.pdf and http://www.usdoj.gov/archive/victimcompensation/index.html.

Feldman, A.D. and Owen, H.J., 1997. Communication of flood-risk information, in *Proceedings of a Hydrology and Hydraulics Workshop on Risk-Based Analysis for Flood Damage Reduction Studies*, Hydraulic Engineering Center, US Army Corps of Engineers, Davis, CA.

Fischhoff, B., 2005. The psychological perception of risk, in *Handbook of Homeland Security*, Kamien, D., Ed., McGraw-Hill, New York, 463–492.

Fragola, J.R., 1996. Reliability and risk analysis data base development: A historical perspective, *Reliability Engineering and System Safety*, 51, 125–136.

Fratzen, K.A., 2001. *Risk-Based Analysis for Environmental Managers*, Lewis Publishers, Boca Raton, FL.

Furuta, H., Fu, K.S., and Yao, J.T.P., 1985. Structural engineering applications of expert systems, *Computer Aided Design*, 17(9), 410–419.

Furuta, H., Shiraishi, N., and Yao, J.T.P., 1986. An expert system for evaluation of structural durability, *Proceedings of the Fifth OMAE Symposium*, 1, 11–15.
Garey, M.R. and Johnson, D.S., 1979. *Computers and Intractability: A Guide to the Theory of NP-Completeness*, W.H. Freeman, San Francisco, CA.
Gertamn, D.I. and Blackman, H.S., 1994. *Human Reliability and Safety Analysis Data Handbook*, John Wiley & Sons, New York.
Gilbert, S.W., 2010. *Disaster Resilience: A Guide to the Literature*, NIST Special Publication 1117, Office of Applied Economics, Engineering Laboratory, National Institute of Standards and Technology, Gaithersburg, MD.
Grose, V.L., 1987. *Managing Risk*, Obega System Group, Arlington, VA.
Gruhn, P., 1991. The pros & cons of qualitative and quantitative analysis of safety systems, *ISA Transactions*, 30(4), 79–86.
Hall, A.D., 1962. *A Method for Systems Engineering*, D. Van Nostrand Company, Princeton, NJ.
Hall, A.D., 1989. *Metasystems Methodology, A New Synthesis and Unification*, Pergamon Press, New York.
Hallden, S., 1986. *The Strategy of Ignorance: From Decision Logic to Evolutionary Epistemology*. Library of Theoria, Stockholm.
Hamley, T.C., Morgan, V.I., Thwaites, R.J., and Gao, X.Q., 1986. An ice-core drilling site at Law Dome Summit, Wilkes Land, Antarctica, Res. Note 37, Aust. Natl. Antarc. Res. Exped., Tasmania, see also http://cdiac.ornl.gov/trends/co2/lawdome.html
Hammond, K.R., McClelland, G.H., and Mumpower, J.M., 1980. *Human Judgment and Decision Making: Theories, Methods, and Procedures*, Praeger, New York.
Hanes, B.J., Chess, C., and Sandman, P.M., 1991. *Industry Risk Communication Manual: Improving Dialogue with Communities*, CRC Press, Boca Raton, FL.
Hansen, K.A., 1995. *Evaluating Technology for Marine Inspectors*, US Coast Guard Research and Development Center, Groton, CT.
Harrald, J., 1999. *The Washington State Ferry Risk Assessment*, Washington State Department of Transportation, Olympia, WA.
Hart, C.J., 1988. *A Study of the Factors Influencing the Rough Water Effectiveness of Personal Flotation Devices*, DTRC-88/026, David Taylor Research Center, US Navy, Corderock, MD.
Hausken, K., 2002. Probabilistic risk analysis and game theory, *Risk Analysis*, 22(1), 17–27.
Helmer, O., 1968. Analysis of the future: The Delphi method, and the Delphi method—An illustration, in *Technological Forecasting for Industry and Government*, Bright, J., Ed., Prentice-Hall, Englewood Cliffs, NJ.
Hofstadter, H.R., 1999. *Godel, Escher, Bach: An Eternal Golden Braid*, Basic Books, New York.
Honderich, H., 1995. *The Oxford Companion to Philosophy*, Oxford University Press, Oxford.
House of Commons Committee of Public Accounts, 2006. Channel tunnel rail link, thirty-eighth report of session 2005–06, HC 727, The Stationery Office Limited, London.
Hoyland, A. and Rausand, M., 1994. *System Reliability Theory*, John Wiley & Sons, New York.
Hubbard, D.W., 2009. *The Failure of Risk Management: Why It's Broken and How to Fix It*, John Wiley & Sons, Hoboken, NJ.
Inozu, B., 1993. Reliability, availability and maintainability databank for ships: Lesson learned, report for the School of Naval Architecture and Marine Engineering, University of New Orleans, New Orleans, LA.
Inozu, B. and Radovic, I., 1999. Practical Implementation of Shared Reliability and Maintainability Databases on Ship Machinery: Challenges and Rewards, *Transactions Institute of Marine Engineers*, 111, 121–134.
Institution of Civil Engineers and the Faculty and Institute of Actuaries, 1998. *Risk Analysis and Management for Projects*, Vol. 53, Thomas Telford Publishing, London, pp. 33–36.
Intergovernmental Panel of Climate Change (IPCC); Solomon, S., Qin, D., Manning, M., Chen, Z., Marquis, M., Averyt, K.B., Tignor, M., and Miller, H.L., Eds., 2007. Climate Change 2007: The Physical Science Basis. Contribution of Working Group I to the Fourth Assessment Report of the Intergovernmental Panel on Climate Change. Cambridge University Press, Cambridge and New York.

Intergovernmental Panel of Climate Change (IPCC), 2012. Special Report on Managing the Risks of Extreme Events and Disasters to Advance Climate Change Adaptation (SREX), Cambridge University Press, Cambridge and New York.

International Code of Safety for High-Speed Craft, 1995. *International Maritime Organization*, Annex 4, *Procedure for Failure Mode and Effects Analysis*, US Coast Guard, Washington, DC, pp. 175–185.

International Committee on Large Dams (ICOLD), 1974. *Lessons from Dam Incidents*, ICOLD, Paris.

International Committee on Large Dams (ICOLD), 1983. *Deterioration of Dams and Reservoirs*, ICOLD, Paris.

International Maritime Organization, 1995. *International Code of Safety for High-Speed Craft* (HSC Code), Resolution MSC, 36(63), adopted on May 20, 1994.

International Organization of Standardizations (ISO), 2009a. *Risk Management—Principles and Guidelines*, ISO 31000, iso.org, Geneva.

International Organization of Standardizations (ISO), 2009b, *Risk Management—Risk Assessment Techniques*, ISO Standard IEC/FDIS 31010, iso.org, Geneva.

International Organization of Standardizations (ISO), 2009c. *Risk Management—Vocabulary*, ISO Guide 73, iso.org, Geneva.

Ishizuka, M., Fu, K.S., and Yao, J.T.P., 1981. *A Rule-Inference Method for Damage Assessment*, ASCE Preprint 81-502, ASCE, St. Louis, MO.

Ishizuka, M., Fu, K.S., and Yao, J.T.P., 1983. Rule-based damage assessment system for existing structures, *Solid Mechanics Archives*, 8, 99–118.

Jannadi, M.O., 1997. Reasons for construction business failures in Saudi Arabia, *Project Management Journal*, 28(2), 32–36.

Johnson-Laird, P., 1988. *The Computer and the Mind: An Introduction to Cognitive Science*, Harvard University Press, Cambridge, MA.

Jones, R.B., 1995. *Risk-Based Management: A Reliability-Centered Approach*, Gulf Publishing, Houston, TX.

Kahn, H., 1960. *On Thermonuclear War*, Free Press, New York.

Kahn, H. and Wiener, A.J., 1967. *The Year 2000: A Framework for Speculation*, Macmillan, New York.

Kaufman, A. and Gupta, M.M., 1985. *Introduction to Fuzzy Arithmetic, Theory and Applications*, Van Nostrand Reinhold, New York.

Kaufmann, A., 1975. *Introduction to the Theory of Fuzzy Subsets*, D.L. Swanson, Trans., Academic Press, New York.

Kerzner, H., 1987. *Project Management: A Systems Approach to Planning, Scheduling, and Controlling*, Vol. 2, CBS Publishers and Distributors, New Delhi, India, pp. 71–80, 406–407.

King, L.W., Trans., 2011. Hammurabi's code of laws exploring ancient world cultures readings from the ancient near east, http://eawc.evansville.edu/anthology/hammurabi.htm.

Kirwan, B., 1992. Human error identification in human reliability assessment. Part 1: Overview of approaches, *Applied Ergonomics*, 23(5), 299–318.

Klir, G.J., 1969. *An Approach to General Systems Theory*, Van Nostrand Reinhold, New York.

Klir, G.J., 1985. *Architecture of Systems Problem Solving*, Plenum Press, New York.

Klir, G.J., 2006. *Uncertainty and Information: Generalized Information Theory*, John Wiley & Sons, Hoboken, NJ.

Klir, G.J. and Folger, T.A., 1988. *Fuzzy Sets, Uncertainty, and Information*, Prentice-Hall, Englewood Cliffs, NJ.

Klir, G.J. and Wierman, M.J., 1999. *Uncertainty-Based Information: Elements of Generalized Information Theory*, Studies in Fuzziness and Soft Computing, Physica-Verlag, New York.

Kumamoto, H. and Henley, E.J., 1996. *Probabilistic Risk Assessment and Management for Engineers and Scientists*, 2nd ed., IEEE Press, New York.

Kunreuther, H., Meyer, R., and Van den Bulte, C., 2004. Risk analysis for extreme events: Economic incentives for reducing future losses, NIST Report GCR 04-871, Gaithersburg, MD.

Landfeeld, J.S. and Seskin, E.P., 1982. The economic value of life: Linking theory to practice, *American Journal of Public Health*, 72(6), 555–566.

Langseth, H., Haugen, K., and Sandtorv, H., 1998. Analysis of OREDA data for maintenance optimization, *Reliability Engineering and System Safety*, 60, 103–110.

Leung, J., 2009. Understanding transport costs and charges: Phase 2—Value of statistical life: A meta analysis, Is the current value of safety for New Zealand too low? Technical Report, Financial and Economic Analysis Team, New Zealand Ministry of Transport.

Linstone, H.A. and Turoff, M., 1975. *The Delphi Method, Techniques and Applications*, Addison Wesley, Reading, MA.

Litai, D., 1980. A risk comparison methodology for the assessment of acceptable risk, PhD thesis, Massachusetts Institute of Technology, Cambridge, MA.

Lyu, M.R., 1996. *Handbook of Software Reliability Engineering*, McGraw Hill, New York.

Mantel, S., Meredith, J., Shafer, S., and Sutton, M., 2001. *Project Management in Practice*, John Wiley & Sons, New York, p. 69.

Marshall, K.T. and Oliver, R.M., 1995. *Decision Making and Forecasting*, McGraw-Hill, New York.

Mathews, M.K., 1990. Risk management in the transportation of dangerous goods: The influence of public perception, in *Proceedings of the National Conference on Hazardous Materials Transportation*, St. Louis, Missouri, pp. 34–37.

McGill, W.L. and Ayyub, B.M., 2009. Defeating surprise in homeland security using possibility-based risk analysis, in *Risk Modeling and Vulnerability Assessment Volume in the Wiley Handbook of Science and Technology for Homeland Security*, Ayyub, B.M. and Voller, J.G., Eds., John Wiley & Sons, Hoboken, NJ.

McGill, W.L., Ayyub, B.M., and Kaminskiy, M., 2007. A quantitative asset-level risk assessment and management framework for critical asset protection, *Risk Analysis International Journal*, 27(5), 1265–1281.

McMahon, K. and Dahdah, S., 2008. The true cost of road crashes: Valuing life and the cost of a serious injury, http://www.irap.net

Miller, J.G., 1978. *Living Systems*, McGraw-Hill, New York.

Miller, T.R., 1989. Willingness to pay comes of age: Will the system survive? *Northwestern University Law Review*, 83(4), 876–907.

Miller, T.R., 1990. The plausible range for the value-of-life-red herrings among the mackerel, *Journal of Forensic Economics*, 3(3), 17–39.

Mishan, E.J., 1971. *Cost-Benefit Analysis*, Praeger, New York.

Modarres, M., 1993. *What Every Engineer Should Know about Reliability and Analysis*, Marcel Dekker, New York.

Modarres, M., Kaminskiy, M., and Krivstov, V., 1999. *Reliability Engineering and Risk Analysis: A Practical Guide*, Marcel Decker, New York.

Modarres, M., Martz, H., and Kaminskiy, M., 1996. The accident sequence precursor analysis: Review of the methods and new insights, *Nuclear Science and Engineering*, 123, 238–258.

Moore, P.G., 1983. *Project Investment Risk: The Business of Risk*, 1st ed., Cambridge University Press, Cambridge.

Morgan, M., 1990. Choosing and managing technology induced risk, in *Readings in Risk*, Glickman, T.S. and Gough, M., Eds., 17–30, Resources for the Future, Washington, DC.

Muller, A., 1980. Evaluation of the costs and benefits of motorcycle helmet law, *Am J Public Health*, 70, 586–592.

Nagel, E. and Newman, J.R., 2001. *Godel's Proof*, New York University Press, New York.

Nash, J., 1951. Non-cooperative games, *Annals of Mathematics*, 54(2), 286–295.

National Commission on Terrorist Attacks upon the United States, 2004. The 9/11 Commission Report, Public Law 107-306, Washington, DC.

National Research Council (NRC), 1983. *Risk Assessment in the Federal Government: Managing the Process*, National Academy Press, Washington, DC.

National Research Council (NRC), 1989. *Improving Risk Communication*, National Academy Press, Washington, DC.

National Research Council (NRC), 1991a. *Ecological Risks: Perspectives from Poland to the United States*, National Academy Press, Washington, DC.

National Research Council (NRC), 1991b. *Use of Risk Analysis to Achieve Balanced Safety in Building Design and Operation*, National Academy Press, Washington, DC.

National Research Council (NRC), 1994. *Science and Judgment in Risk Assessment*, National Academy Press, Washington, DC.

National Research Council (NRC), 1995. *Flood Risk Management and the American River Basin: An Evaluation*, National Academy Press, Washington, DC.

National Research Council (NRC), 1996. *Understanding Risk: Informing Decisions in a Democratic Society*, National Academy Press, Washington, DC.

National Research Council (NRC), 2004. *Valuing Ecosystem Services: Toward Better Environmental Decision-Making*, National Academies Press, Washington, DC.

National Research Council (NRC), 2012. *Disaster Resilience: A National Imperative*, National Academies Press, Washington, DC.

Nelson, W., 1982. *Applied Life Data Analysis*, John Wiley & Sons, New York.

Nuclear Regulatory Commission (NuRC), 1975. *Reactor Safety Study*, WASH-1400, NUREG-751014, NuRC, Washington, DC.

Nuclear Regulatory Commission (NuRC), 1981. *Fault Tree Handbook*, NUREG-0492, NuRC, Washington, DC.

Nuclear Regulatory Commission (NuRC), 1983. *PRA Procedures Guide: A Guide to the Performance of Probabilistic Risk Assessments for Nuclear Power Plants*, NUREG/CR-2300, NuRC, Washington, DC.

Nuclear Regulatory Commission (NuRC), 1996. *Probabilistic Risk Analysis (PRA) Procedures Guide*, NUREG/CR 2300, NuRC, Washington, DC.

Nuclear Regulatory Commission (NuRC), 1997. *Recommendations for Probabilistic Seismic Hazard Analysis: Guidance on Uncertainty and Expert Use*, Vols. 1 and 2, NUREG/CR-6372, UCRL-ID-122160, the Senior Seismic Hazard Analysis Committee, NuRC, Washington, DC.

Olson, C.A., 1981. An analysis of wage differentials received by workers on dangerous jobs, *Journal of Human Resources*, 16, 167–185.

Omega Systems Group, 1994. *Risk Realities: Provoking a Fresh Approach to Managing Risk*, Corporate Publication, Arlington, VA.

Pal, S.K. and Skowron, A., 1999. *Rough Fuzzy Hybridization*, Springer-Verlag, New York.

Pan, Y., Birdsey, R.A., Fang, J., Houghton, R., Kauppi, P.E., Kurz, W.A., Phillips, O.L., et al., 2011. A large and persistent carbon sink in the world's forests, *Science* 133, 988–993.

Park, W., 1979. *Construction Bidding for Profit*, John Wiley & Sons, New York, pp. 168–177.

Parnell, G.S., Dillon, R.L., and Bresnick, T., 2005. Integrating risk management with homeland security and antiterrorism resource allocation decision-making, in *Handbook of Homeland Security*, Kamien, D., Ed., McGraw-Hill, New York, 431–461.

Parry, G., 1996. The characterization of uncertainty in probabilistic risk assessments of complex systems, *Reliability Engineering and System Safety*, 54, 119–126.

Pastorok, R.A., Bartell, S.M., Ferson, S., and Ginzburg, L.R., 2002. *Ecological Modeling in Risk Assessment*, Lewis Publishers, Boca Raton, FL.

Pate-Cornell, E.M., 1996a. Global risk management, *Journal of Risk and Uncertainty*, 12, 239–255.

Pate-Cornell, E.M., 1996b. Uncertainties in risk analysis: Six levels of treatment, *Reliability Engineering and System Safety*, 54(2–3), 95–111.

Pegg, N.G. and Brinkhurst, P., 1993. *Naval Ship Structures Risk Assessment Project Definition*, DREA TM/93/209, Canadian Navy, Halifax, Nova Scotia.

Persson, U., 1989. *The Value of Risk Reduction: Results of a Swedish Sample Survey*, Swedish Institute for Health Economics, University of Lund, Lund.

Pham, H., 1999. *Software Reliability*, John Wiley & Sons, New York.

Piret, N.L., 1999. The removal and safe disposal of arsenic in copper processing, *Journal of Metals*, 51(9), 16–17.

Pitz, G.F., 1992. Risk Taking, Design, and Training, in *Risk-Taking Behavior*, Yates, J.F., Ed., John Wiley & Sons, New York, pp. 283–320.

Pollalis, S.N., 2006. *Channel Tunnel Rail Link Risk Transfer and Innovation in Project Delivery*, Harvard Design School, Cambridge, MA.

Ponce, V.M., 1989. *Engineering Hydrology Principles and Practices*, Prentice-Hall, Englewood Cliffs, NJ.

Posthuma, L., Suter II, G.W., and Traas, T.P., 2002. *Species Sensitivity Distributions in Ecotoxicology*, Lewis Publishers, Boca Raton, FL.

Potter, J.M., Smith, M.L., and Panwalker, S.S., 1976. *Cost-Effectiveness of Residential Fire Detector Systems*, Texas Tech University, Lubbock, TX.

Presidential Policy Directive (PPD), 2013. Critical infrastructure security and resilience. PPD-21, http://www.whitehouse.gov/the-press-office/2013/02/12/presidential-policy-directive-critical-infrastructure-security-and-resil.

Project Management Institute, 2000. *A Guide to the Project Management Body of Knowledge* (PMBOK® Guide), Project Management Institute, Newton Square, PA, pp. 127, 129, 131–134, 137.

Puskin, J.S. and Nelson, C.B., 1989. EPA's perspective on risks from residential radon exposure, *Journal of Air Pollution Control Association*, 39(7), 915–920.

Putterman, M.L., 1994. *Markov Decision Processes*, John Wiley & Sons, New York.

Rasmussen, N.C., 1981. The application of probabilistic risk assessment techniques to energy technologies, *Annual Review of Energy*, 6, 123–138.

Reagan, R., 1985. Executive order 12498: Regulatory planning process, *Federal Register*, 50(5), January 8.

Reason, J., 1990. *Human Error*, Cambridge University Press, Cambridge.

Reason, J., 2000. Human error: Models and management, *British Medical Journal*, 320(7237), 768–770.

Rinaldi, S.M., Peerenboom, J.P., and Kelly, T.K., 2001. Identifying, understanding, and analyzing critical infrastructure interdependencies, *IEEE Control Systems Magazine*, doi:10.1109/37.969131.

Robertson, L.S., 1977. Car crashes: Perceived vulnerability and willingness to pay for crash protection, *Journal of Community Health*, 3, 136–141.

Rose, G., Hamilton, P.J.S., Colwell, L., and Shipley, M.J., 1982. A randomized control trial of anti-smoking advice: 10-year results, *Journal of Epidemiology and Community Health*, 86, 102–108.

Ross, S.M., 1970. *Introduction to Stochastic Dynamic Programming*, Academic Press, New York.

Rowe, W.D., 1977. *An Anatomy of Risk*, John Wiley & Sons, New York.

Royer, P.S., 2000. Risk management: The undiscovered dimension of project management, *Project Management Journal*, 31(1), 6–13.

Sandtorv, H.A., Hokstad, P., and Thompson, D.W., 1996. Practical experiences with a data collection project: The OREDA project, *Reliability Engineering and System Safety*, 51, 159–167.

Schall, G. and Öestergaard, C., 1991. Planning of inspection and repair of ship operation, in *Proceedings of the Marine Structural Inspection, Maintenance and Monitoring Symposium*, SNAME, New York, NY, pp. V-F-1–V-F-7.

Schelling, T.C., 1960. *Strategy of Conflict*, Harvard University Press, Cambridge, MA.

Schierow, L.-J., 1998. *Risk Analysis: Background on Environmental Protection Agency Mandates*, Environment and Natural Resources Policy Division, National Council for Science and the Environment, Washington, DC.

Shackle, G.L.S., 1970. *Decision, Order, and Time in Human Affairs*, 2nd ed., Cambridge University Press, Cambridge.

Shinozuka, P.M., 1990. Relation of inspection findings to fatigue reliability, Report No. SSC-355, Ship Structure Committee, US Coast Guard, Washington, DC.

Shinozuka, P.M. and Chen, Y.N., 1987. Reliability and durability of marine structures, *ASCE Journal of Structural Engineering*, 113(6), 1297–1314.

Shiraishi, N. and Furuta, H., 1983. Reliability analysis based on fuzzy probability, *Journal of Engineering Mechanics*, 109(6), 1445–1459.

Shiraishi, N., Furuta, H., and Sugimoto, M., 1985. Integrity assessment of structures based on extended multi-criteria analysis, in *Proceedings of the Fourth ICOSSAR*, May 27–29, 1985, Kobe, Japan.

Singh, A. and Vlatas, D.A., 1991. Using conflict management for better decision making, *Journal of Management in Engineering*, 7(1), 70–82.

Slovic, P., Fischhoff, B., and Lechtenstein, S., 1990. Rating the risks, *Readings in Risk*, Glickman, T.S. and Gough, M., Eds., 61–74, Resources for the Future, Washington, DC.

Smith, A., 1776. *The Wealth of Nations*, 1976 ed., University of Chicago Press, Chicago, IL.

Smith, D.J., 2001. *Reliability Maintainability, and Risk*, 6th ed., Butterworth-Heinemann, Oxford.
Smith, V.K., 1983. The role of site and job characteristics in hedonic wage and models, *Journal of Urban Economics*, 13(3), 296–321.
Smith, V.K. and Johnson, F.R., 1988. How do risk perceptions respond to information? The case of radon, *The Review of Economics and Statistics*, 70(1), 1–8.
Smithson, M., 1985. Towards a social theory of ignorance, *Journal of the Theory of Social Behavior*, 15, 151–172.
Smithson, M., 1988. *Ignorance and Uncertainty*, Springer-Verlag, New York.
Sober, E., 1991. *Core Questions in Philosophy*, Macmillan, New York.
Spetzler, C.S. and Stael von Holstein, C.-A.S., 1975. Probability encoding in decision analysis, *Management Science*, 22(3), 340–358.
Starr, C., 1969. Social benefit vs. technical risk. *Science*, 165, 1232–1238.
Starr, C., 1971. Benefit–cost studies in socio-technical systems, in *Perspectives on Benefit–Risk Decision Making*, Report of a Colloquium Conducted by the Committee on Public Engineering Policy, April 26–27, National Academy of Engineering, Washington, DC, pp. 17–42.
Steele, L.W., 1989. *Managing Technology: The Strategic View*, McGraw-Hill, New York, pp. 120–124.
Stillings, N., Weisler, S., Chase, C., Feinstein, M., Garfield, J., and Rissland, E., 1995. *Cognitive Science*, 2nd ed., MIT Press, Cambridge, MA.
Strange, G., 1986. *Introduction to Applied Mathematics*, Wellesley-Cambridge Press, Wellesley, MA.
Suzuki, H., 1999. Safety target of very large floating structures used as a floating airport, in *Proceedings of the Third International Workshop on Very Large Floating Structures*, September, 22–24, University of Hawaii, Honolulu, HI, pp. 607–612.
Svenson, O., 1991. The accident evolution and barrier function (AEB) model applied to incident analysis in the processing industries. *Risk Analysis*, 11(3), 499–507.
Taleb, N.N., 2004. The black swan: Why don't we learn that we don't learn. US Department of Defense Highland Forum Papers.
Thagard, P., 1996. *Mind: Introduction to Cognitive Science*, MIT Press, Cambridge, MA.
Trauth, K.M., Hora, S.C., and Guzowski, R.V., 1993. Expert judgment on markers to deter inadvertent human intrusion into the waste isolation pilot plant, Report SAND92-1382, Sandia National Laboratories, Albuquerque, NM.
Trost, W.A. and Nertney, R.J., 1985. *Barrier Analysis* (DOE 76-45/29), EG&G Idaho, Idaho Falls, ID.
Turner, J.R., 1993. *Handbook of Project Management*, McGraw-Hill, New York, pp. 7–9, 20–23, 235–267, 284–303.
T.W. Lambe Associates, 1982. Earthquake risk to patio 4 and site 400, Report to TONEN Oil Company, Ltd., Longboat Key, FL.
UK Health and Safety Executive, 1978. *An Investigation of Potential Hazards from Operations in the Canvey Island/Thurrock Area*, Her Majesty's Stationery Office, London, 192 pp.; second summary document, 39 pp.
University of New Hampshire, 2011. Global carbon cycle, http://globecarboncycle.unh.edu/CarbonCycleBackground.pdf based on http://globe.gov/science/topics/carbon-cycle#Overview.
UN Office for Disaster Risk Reduction (UNISDR), 2012. Making cities resilient: My city is getting ready! A global snapshot of how local governments reduce disaster risk, United Nations Office for Disaster Risk Reduction Report, Geneva, www.unisdr.org/campaign.
US Air Force Aeronautical Systems Division, 1973. *Aircraft Structural Integrity Program: Airplane Requirements*, MIL-STD-1530A.
US Army Corps of Engineers (USACE), 1965. *Standard Project Flood Determinations*, Civil Engineer Bulletin No. 52-8, Engineering Manual EM 1110-2-1411, USACE, Alexandria, VA.
US Army Corps of Engineers (USACE), 1975. *National Program of Inspection of Dams*, Vols. I–V, Department of the Army, Washington, DC.
US Army Corps of Engineers (USACE), 1988. *National Economic Development Procedure Manual: Urban Flood Damage*, IWR Report 88-R-2, US Army Corps of Engineers, Water Resources Support Center, Institute for Water Resources, Alexandria, VA.

US Army Corps of Engineers (USACE), 1992. *Risk Communication*, Engineering Pamphlet No. 1110-2-8, USACE, Washington, DC.

US Army Corps of Engineers (USACE), 1993. *Guidebook for Risk Perception and Communication in Water Resources*, IWR Report 93-R-13, USACE, Alexandria, VA.

US Coast Guard, International Association of Classification Societies (IACS), and Nippon Kaiji Kyokai (NKK), 1995. *Bulk Carriers Inspection Guidance for Owners/Operators, Terminal Personnel, and Crew*, Compiled for the International Maritime Organization (IMO), US Coast Guard Headquarters, Washington, DC.

US Committee on Large Dams (USCOLD), 1988. *Lessons from Dam Incidents-II*, ASCE, New York.

US Department of Transportation, 1982. *Guidance for Regulatory Evaluation: A Handbook for DOT Benefit–Cost Analysis*, Office of Industry Policy, Washington, DC.

US Interagency Advisory Committee on Water Data, Hydrology Subcommittee, 1982. *Guidelines for Determining Flood Flow Frequency*, Bulletin No. 17B, USGS, Reston, VA.

Van Leeuwen, C.J. and Hermens, J.I.M., 1995. *Risk Assessment of Chemicals: An Introduction*, Kluwer Academic, Dordrecht.

Verma, V., 1996. *Human Resources Skills for the Project Manager*, Project Management Institute, Upper Darby, PA.

Vesper, J.L., 2006. *Risk Assessment and Risk Management in the Pharmaceutical Industry Clear and Simple*, Parenteral Drug Association, Bethesda, MD.

Viscusi, W.K., 1978. Labor market valuations of life and limb: Empirical evidence and policy implications, *Public Policy*, 26(3), 339–386.

Viscusi, W.K., 1983. Alternative approaches to valuing the health impacts of accidents: Liability law and prospective evaluations, *Law and Contemporary Problems*, 47(4), 49–68.

Viscusi, W.K., 1992. *Fatal Tradeoffs: Public and Private Responsibilities for Risk*, Oxford University Press, London.

Viscusi, W.K. and Aldy, J.E., 2002. The value of a statistical life: A critical review of market estimates throughout the world, Harvard Law School John M. Olin Center for Law, Economics and Business Discussion Paper Series. Paper No. 392, http://lsr.nellco.org/harvard_olin/392.

Vo, T.V. and Balkey, K.R., 1995. Risk-based inspection and maintenance, in *Probabilistic Structural Mechanics Handbook*, Sundararajan, C., Ed., Chapman & Hall, New York, pp. 388–415.

von Eckardt, B., 1993. *What Is Cognitive Science?* MIT Press, Cambridge, MA.

von Glasersfeld, E., 1995. *Radical Constructivism: A Way of Knowing and Learning*, The Farmer Press, London.

von Neumann, J. and Morgenstern, O., 1944. *Theory of Games and Economic Behavior*, Princeton University Press, Princeton, NJ.

Wahl, T.L., 1997. Predicting embankment dam breach parameters—A needs assessment, in *XXVIIth IAHR Congress*, August 10–15, San Francisco, CA.

Waldrop, M.M., 1992. *Complexity: The Emerging Science at the Edge of Order and Chaos*, Simon & Schuster, New York.

The Washington Post, 2012. http://www.washingtonpost.com/local/trafficandcommuting/worst-case-scenario/2012/11/03/3cb1360e-260f-11e2-9313-3c7f59038d93_graphic.html, November 4.

Weaver, W., 1948. Science and complexity, *American Scientist*, 36(4), 536–544.

Weinstein, D. and Weinstein, M.A., 1978. The sociology of non-knowledge: A paradigm, in *Research in the Sociology of Knowledge, Science and Art*, Vol. 1, Jones, R.A., Ed., 151–166, JAI Press, New York.

Whipple, C., 1986. Dealing with uncertainty about risk in risk management, in *Hazards: Technology and Fairness*, National Academy of Engineering, National Research Council, Washington, DC, pp. 44–59.

White, G.J. and Ayyub, B.M., 1985. Reliability methods for ship structures. *Naval Engineers Journal*, 97(4), 86–96.

Whitehouse.gov, 2010. The ongoing administration-wide response to the deepwater BP oil spill, http://www.whitehouse.gov/blog/2010/05/05/ongoing-administration-wide-response-deepwater-bp-oil-spill.

Whitman, R.V., 1984. Evaluating calculated risk in geotechnical engineering, *ASCE Journal of Geotechnical Engineering*, 110(2), 145–186.

Wilde, G.J.S., 1988. Risk homeostasis theory and traffic accidents: Propositions, deductions and discussion of dissension in recent reactions, *Ergonomics*, 31, 441–468.
Wilson, B., 1984. *Systems: Concepts, Methodologies, and Applications*, John Wiley & Sons, New York.
Witkin, H.A. and Goodenough, D.R., 1981. *Cognitive Styles: Essence and Origins*, International Universities Press, New York.
Wohlstetter, R., 1962. *Pearl Harbor: Warning and Decision*, Stanford University Press, Stanford, CA.
Woo, G., 1999. *The Mathematics of Natural Catastrophes*, Imperial College Press, London.
Yao, J.T.P., 1979. Damage assessment and reliability evaluation of existing structures, *Engineering Structures*, 1, 245–251.
Yao, J.T.P., 1980. Damage assessment of existing structures, *Journal of Engineering Mechanics Division*, 106(EM4), 785–799.
Yao, J.T.P. and Furuta, H., 1986. Probabilistic treatment of fuzzy events in civil engineering, *Probabilistic Engineering Mechanics*, 1(1), 58–64.
Zadeh, L.A., 1965. Fuzzy sets, *Information and Control*, 8, 338–353.
Zadeh, L.A., 1968. Probability measures of fuzzy events, *Journal of Mathematical Analysis*, 23, 421–427.
Zadeh, L.A., 1973. Outline of a new approach to the analysis of complex systems and decision processes, *IEEE Transactions on Systems, Man and Cybernetics*, SMC-3(1), 28–44.
Zadeh, L.A., 1975a. The concept of linguistic variable and its application to approximate reasoning, parts I, II and III, *Information and Control*, 8, 199–249, 301–357.
Zadeh, L.A., 1975b. The concept of linguistic variable and its application to approximate reasoning, parts I, II and III, *Information and Control*, 9, 43–80.
Zadeh, L.A., 1987. Fuzzy sets as a basis for a theory of possibility, *Fuzzy Sets and Systems*, 1, 3–28.
Zadeh, L.A., Fu, K. S., Tanaka, K., and Shimara, J., 1975. *Fuzzy Sets and Their Application to Cognitive and Decision Processes*, Academic Press, New York.

Index

Note: Locators followed by '*f*' and '*t*' refer to figures and tables, respectively.

D

E

F

G

H

T

U

V

W

Z